THE NUCLEAR OVERHAUSER EFFECT IN STRUCTURAL AND CONFORMATIONAL ANALYSIS

Second Edition

Methods in Stereochemical Analysis

Series Editor

Alan P. Marchand, Denton, Texas, USA

Advisory Board

A. Greenberg, Charlotte, North Carolina, USA
I. Hargittai, Budapest, Hungary
A. R. Katritzky, Gainesville, Florida, USA
J. F. Liebman, Baltimore, Maryland, USA
E. Lippmaa, Tallinn, Estonia
L. A. Paquette, Columbus, Ohio, USA
P. von R. Schleyer, Athens, Georgia, USA
S. Sternhell, Sydney, Australia
Y. Takeuchi, Tokyo, Japan
F. Wehrli, Philadelphia, Pennsylvania, USA
D. H. Williams, Cambridge, UK
N. S. Zefirov, Moscow, Russia

Other Books in the Series:

Motohiro Nishio, Miroru Hirota, and Yoji Umezawa
The CH/π Interaction: Evidence, Nature, and Consequence

David A. Lightner and Jerome E. Gurst
Organic Conformational Analysis and Stereochemistry from Circular Dichroism Spectroscopy

Jacek Waluk
Conformational Analysis of Molecules in Excited States

Eiji Ōsawa and Osamu Yonemitsu
Carbocyclic Cage Compounds: Chemistry and Applications

Janet S. Splitter and Frantisek Turecek (editors)
Applications of Mass Spectrometry to Organic Stereochemistry

William R. Croasmun and Robert M. K. Carlson (editors)
Two-dimensional NMR Spectroscopy: Applications for Chemists and Biochemists. Second Edition

Jenny P. Glusker with Mitchell Lewis and Miriam Rossi
Crystal Structure Analysis for Chemists and Biologists

Kalevi Pihlaja and Erich Kleinpeter
Carbon-13 NMR Chemical Shifts in Structural and Stereochemical Analysis

Louis D. Quin and John G. Verkade (editors)
Phosphorus-31 NMR Spectral Properties in Compound Characterization and Structural Analysis

Eusebio Juaristi (editor)
Conformational Behavior of Six-Membered Rings: Analysis, Dynamics and Stereoelectronic Effects

THE NUCLEAR OVERHAUSER EFFECT IN STRUCTURAL AND CONFORMATIONAL ANALYSIS

Second Edition

David Neuhaus
Michael P. Williamson

A John Wiley & Sons, Inc., Publication
New York / Chichester / Weinheim / Brisbane / Singapore / Toronto

This book is printed on acid-free paper. ∞

Copyright © 2000 by Wiley-VCH, Inc. All rights reserved.

Published simultaneously in Canada.

No part of this publication may be reproduced, stored in a retrieval system or transmitted in any form or by any means, electronic, mechanical, photocopying, recording, scanning or otherwise, except as permitted under Sections 107 or 108 of the 1976 United States Copyright Act, without either the prior written permission of the Publisher, or authorization through payment of the appropriate per-copy fee to the Copyright Clearance Center, 222 Rosewood Drive, Danvers, MA 01923, (978) 750-8400, fax (978) 750-4744. Requests to the Publisher for permission should be addressed to the Permissions Department, John Wiley & Sons, Inc., 605 Third Avenue, New York, NY 10158-0012, (212) 850-6011, fax (212) 850-6008, E-Mail: PERMREQ @ WILEY.COM.

Library of Congress Cataloging-in-Publication Data:

Neuhaus, David, 1956–
    The nuclear Overhauser effect in structural and conformational analysis / David Neuhaus and Michael P. Williamson.—2nd ed.
       p. cm. — (Methods in stereochemical analysis)
    Includes bibliographical references and index.
    ISBN 0-471-24675-1 (alk. paper)
    1. Conformational analysis. 2. Overhauser effect (Nuclear physics) I. Williamson, Michael P., 1957– II. Title. III. Series.
    QD481 .N46 2000
    541.2′23—dc21   99-049630

Printed in the United States of America.

10 9 8 7 6 5 4 3 2 1

CONTENTS

PREFACE — xv

ACKNOWLEDGMENTS — xvii

PREFACE TO THE FIRST EDITION — xix

SYMBOLS, ABBREVIATIONS, AND UNITS — xxi

INTRODUCTION — xxv

PART I. THEORY — 2

CHAPTER 1. BACKGROUND — 3

1.1. Energy Levels, Populations, and Intensities / 3
1.2. Relaxation, T_1 and T_2 / 8
1.3. The Nature of Relaxation / 12
1.4. The Local Field and Dipole–Dipole Relaxation / 13
1.5. Pulses and Saturation / 15
 References / 22

CHAPTER 2. THE STEADY-STATE NOE FOR TWO SPINS   23

2.1. The Origin and Form of the NOE / 23
    2.1.1. Qualitative Considerations / 23
    2.1.2. The Solomon Equations / 27
2.2. Dependence of the NOE on Molecular Motion / 30
    2.2.1. Correlation Times, Spectral Density Functions, and Transition Probabilities / 31
    2.2.2 Anisotropic Tumbling / 38
2.3 What the Symbols Mean for Two Spins and for Many Spins / 38
    2.3.1. Relaxation Rates / 39
    2.3.2. T_1 Measurements and Cross-Relaxation / 42
2.4. Effects of Other Relaxation Sources / 46
    2.4.1. The External Relaxation Rate ρ^* / 46
    2.4.2. Intermolecular Dipole–Dipole Relaxation / 50
    2.4.3. Quadrupolar Relaxation / 52
    2.4.4. Chemical Shift Anisotropy (CSA) Relaxation / 53
    2.4.5. Scalar Relaxation / 53
    2.4.6. Spin–Rotation Relaxation / 53
2.5. The Heteronuclear NOE / 54
2.6. An Extension to the Solomon Equations / 59
References / 60

CHAPTER 3. THE STEADY-STATE NOE IN RIGID MULTISPIN SYSTEMS   62

3.1. The Equations / 63
    3.1.1. The Solomon Equations for More Than Two Spins / 63
    3.1.2. Cross-Correlation / 66
    3.1.3. Two General Solutions to the Multispin Solomon Equations / 68
        3.1.3.1. Saturation of One Spin / 68
        3.1.3.2. Saturation of All Spins Except One / 69
    3.1.4. Internuclear Distances and Steady-State NOE Enhancements / 71
3.2. What the Equations Mean / 72
    3.2.1. General: Direct Enhancements and Spin Diffusion / 72
    3.2.2. Interpretation at the Extreme Narrowing Limit ($\omega\tau_c \ll 1$) / 76

                3.2.2.1. Direct Effects / 76
                3.2.2.2. Indirect Effects / 79
                3.2.2.3. When Do Indirect Effects Matter? / 80
                3.2.2.4. Magnetic Equivalence / 82
                3.2.2.5. T_1 and the 3/2 Effect / 84
                3.2.2.6. Chemical Equivalence / 85
        3.2.3. Away from the Extreme Narrowing Limit / 86
3.3. In Practice / 91
        3.3.1. Incomplete Saturation / 92
        3.3.2. Failure to Reach Steady-State / 93
        3.3.3. Competition From Other Relaxation Sources / 94
                References / 97

CHAPTER 4. THE KINETICS OF THE NOE   98

4.1. Introduction / 98
        4.1.1. Types of Kinetic NOE Experiment / 99
        4.1.2. Overview / 100
4.2. Theory of the Kinetic NOE in a Two-Spin System / 105
4.3. Theory of the Kinetic NOE in Multispin Systems / 108
        4.3.1. Multispin Kinetics in Transient NOE Experiments / 108
        4.3.2. Multispin Kinetics in TOE Experiments / 111
4.4. Estimating Internuclear Distances / 111
        4.4.1. The Initial Rate Approximation / 112
        4.4.2. Distances From Enhancement Ratios / 113
        4.4.3. Errors in Distance Measurements Using the Initial Rate Approximation / 115
        4.4.4. Spin Diffusion, Nonlinear NOE Growth, and Interpretation / 117
4.5. More About Experiments / 122
        4.5.1. Symmetry in Kinetic NOE Experiments / 122
        4.5.2. T_1 Values as an Aid to Interpretation / 124
                References / 127

CHAPTER 5. THE EFFECTS OF EXCHANGE AND INTERNAL MOTION   129

5.1. Transfer of Saturation / 131
5.2. General Equations for the NOE in Systems of Two-Site Exchange / 136

        5.2.1. Exchange in a Two-Spin System / 136
        5.2.2. Exchange in Dimethylformamide / 143
5.3. Applications to More Complicated Cases of Exchange / 148
        5.3.1. Averaging of Rates Rather than Enhancements / 148
        5.3.2. Analyzing Conformational Equilibria / 150
                5.3.2.1. Olefinic Methoxy Conformations / 150
                5.3.2.2. Nucleotide Conformations / 151
                5.3.2.3. A Statistical Approach / 156
5.4. Estimating Flexibility Using Heteronuclear Relaxation Analysis / 158
5.5 How Internal Motions Average Internuclear Distances / 167
        5.5.1. Internal Motions Slower Than Overall Tumbling: "r^{-6} Averaging" / 171
        5.5.2. Internal Motions Faster Than Overall Tumbling / 172
5.6. Allowing for Averaging / 174
5.7. The Transferred NOE / 178
5.8. Intermolecular NOE Enhancements Involving Water / 185
        References / 187

CHAPTER 6. COMPLICATIONS FROM SPIN-SPIN COUPLING 190

6.1. Decoupling / 190
6.2. Selective Population Transfer / 191
        6.2.1. Theory / 192
        6.2.2. Consequences / 199
6.3. Effects of Cross-Correlation / 201
6.4. Strong Coupling / 204
        6.4.1. $A\{B\}$ Enhancements / 206
        6.4.2. $AB\{X\}$ Enhancements / 206
        6.4.3. Scalar Relaxation / 213
                6.4.3.1. Scalar Relaxation of the First Kind / 213
                6.4.3.2. Scalar Relaxation of the Second Kind / 216
        References / 217

PART II. EXPERIMENTAL 219

CHAPTER 7. ONE-DIMENSIONAL EXPERIMENTS 221

7.1. Sample Preparation / 221
        7.1.1. Solvent / 221
        7.1.2. Solute Concentration / 224
        7.1.3. Sample Purification / 225

7.2. Setting Up the Steady-State Difference Experiment / 227
  7.2.1. Introduction to the Difference Experiment / 227
  7.2.2. Minimizing Subtraction Artifacts / 230
  7.2.3. Automatic Multiple Experiments / 234
  7.2.4. Irradiation Power and Selectivity / 235
  7.2.5. Multiplet Irradiation and SPT Suppression / 240
  7.2.6. Timing / 243
7.3. Processing, Display, and Calculation of Results / 247
  7.3.1. General / 247
  7.3.2. Reference Deconvolution / 249
7.4. Other 1D Experiments Employing Continuous Saturation / 254
  7.4.1. CW Steady-State Integration / 254
  7.4.2. The Truncated Driven NOE (TOE) Experiment / 255
7.5. Transient Experiments / 258
  7.5.1. Selective Pulses / 259
  7.5.2. Non-Gradient Transient NOE Experiments / 260
  7.5.3. Gradient-Assisted Transient NOE Experiments / 261
    7.5.3.1. Gradient Selection / 262
    7.5.3.2. DPFGSE-NOE / 264
    7.5.3.3. GOESY / 267
    7.5.3.4. Variations / 272
    7.5.3.5. Applications and Practicalities / 274
  7.5.4. Doubly Selective Experiments / 276
  References / 279

CHAPTER 8. THE TWO-DIMENSIONAL NOESY EXPERIMENT 282

8.1. One Dimension or Two? / 282
  8.1.1. The Negative NOE Regime ($\omega\tau_c > 1.12$) / 283
  8.1.2. The Positive NOE Regime ($\omega\tau_c < 1.12$) / 283
8.2. Basic Principles / 285
8.3. Acquiring a NOESY Spectrum / 293
  8.3.1. Fixed Delays and Pulse Widths / 294
  8.3.2. Acquisition Times t_1 and t_2 and Spectral Widths SW_1 and SW_2 / 296
  8.3.3. Quadrature Detection in F_1 and F_2 / 299
  8.3.4. Phase-Sensitive NOESY / 304
8.4. Phase-Cycling, Signal Selection, and Artifact Suppression / 307
  8.4.1. Rejection of Nonlongitudinal Contributions During τ_m: J-Peak Suppression / 309

- 8.4.2. Other Forms of *J*-Peaks: Zero Quantum Coherences and Pulse Angle Effects / 310
- 8.4.3. Axial Peaks / 313
- 8.4.4. Quadrature Images / 313
- 8.4.5. t_1 Noise / 314

8.5. Data Processing / 316
- 8.5.1. Zero Filling / 317
- 8.5.2. Window Functions and Linear Prediction / 318
- 8.5.3. Symmetrization and t_1 Noise Removal / 320
- 8.5.4. Integration / 321

8.6. Variations / 322
- 8.6.1. Semi-Selective and Network-Edited Experiments / 322
- 8.6.2. Other Variants / 327

References / 328

CHAPTER 9. OTHER DEVELOPMENTS — 331

9.1. Heteronuclear NOE Enhancements / 331
- 9.1.1. Non-Specific Heteronuclear NOE Experiments / 331
- 9.1.2. Specific Heteronuclear NOE Experiments / 335

9.2. Editing and Spectral Simplification of NOE Experiments / 341
- 9.2.1. Editing Using the NOE Itself / 341
- 9.2.2. Editing Using Heteronuclear Scalar Couplings / 346
  - 9.2.2.1. X-Filtered NOE Experiments / 347
  - 9.2.2.2. X-Separated NOE Experiments / 350
- 9.2.3. Homonuclear Three-Dimensional NOE Experiments / 359
- 9.2.4. Editing Using Something Else / 362

9.3. Rotating Frame NOE Experiments / 364
- 9.3.1. Theory / 365
  - 9.3.1.1. Spin-Locking / 365
  - 9.3.1.2. Spin-Locked Transverse Dipole-Dipole Relaxation / 367
  - 9.3.1.3. Other Effects During Spin-Locking / 371
- 9.3.2. Practicalities / 379

9.4. Manipulation of $\omega\tau_c$ / 382

References / 384

PART III. APPLICATIONS 389

CHAPTER 10. APPLICATIONS OF THE NOE TO STRUCTURE ELUCIDATION 391

10.1. General Considerations / 391
   10.1.1. Why Structural and Conformational Problems Are the Same / 391
   10.1.2. Spectra and Assignments / 393
   10.1.3. Reporting Results and Interpretation / 395
   10.1.4. Miscellaneous / 397
10.2. Aromatic Substitution and Ring Fusion Patterns: Simple Cases / 398
10.3. Aromatic Substitution and Ring Fusion Patterns: More Complex Cases / 406
   10.3.1. Petroporphyrins / 407
   10.3.2. Isoquinoline and Related Alkaloids / 409
10.4. Double Bond Isomers / 415
10.5. Saturated Ring Systems: Simple Cases / 420
   10.5.1. Substituent Stereochemistry / 422
   10.5.2. Ring Fusion Stereochemistry / 432
10.6. Saturated Ring Systems: More Complex Cases / 442
   10.6.1. Pulvomycin / 444
   10.6.2. Penitrem A / 447
   10.6.3. Other Examples / 450
   References / 453

CHAPTER 11. APPLICATIONS OF THE NOE TO CONFORMATIONAL ANALYSIS 456

11.1. General Considerations / 456
   11.1.1. Why Structural and Conformational Problems are Different / 456
   11.1.2. Multiple Conformations / 458
11.2. Local Conformational Detail in Small Molecules / 459
   11.2.1. Slowly Exchanging Equilibria / 461
   11.2.2. Rapidly Exchanging Equilibria: A Hypothetical Example, $X\text{-CH}_2\text{OH}$ / 461
   11.2.3. Rapidly Exchanging Equilibria: Real Examples / 463
11.3. Conformational Analysis of Medium-Sized Molecules / 472
   References / 484

CHAPTER 12. CALCULATING STRUCTURES OF BIOPOLYMERS 485

12.1. Introduction / 485
12.2. Restraints / 486
    12.2.1. Assigning NOE Restraints / 487
        12.2.1.1. Using Only (^{1}H, ^{1}H) NOESY Data / 488
        12.2.1.2. Using Preliminary Structural Data / 489
        12.2.1.3. Using Heteronuclear Labeling / 490
    12.2.2. Measuring NOE Restraints / 491
    12.2.3. Calibrating NOE Restraints / 493
    12.2.4. Averaging in Equivalent Groups / 499
        12.2.4.1. Pseudoatom Corrections / 500
        12.2.4.2. Multiplicity Corrections / 503
        12.2.4.3. r^{-6} Summation / 506
    12.2.5. Stereoassignments and Torsion Angle Restraints / 506
    12.2.6. Other Types of Restraints / 511
12.3. Calculating Structures / 515
    12.3.1. Distance Geometry Calculations / 517
        12.3.1.1. The Exact Case / 518
        12.3.1.2. Distance Geometry Applied to NMR Structure Determination / 519
        12.3.1.3. Distance Geometry in Torsion Angle Space / 522
    12.3.2. Restrained Molecular Dynamics Calculations / 523
    12.3.3. Simulated Annealing Calculations / 527
    12.3.4. Other Methods / 529
12.4. Assessing and Describing NMR Structures / 530
    12.4.1. Global Precision: Overall Root Mean Squared Deviations / 531
    12.4.2. Local Precision: Local RMSD and Angular Order Parameters / 533
    12.4.3. Assessing the Quality of Structures / 536
12.5. Refinement / 540
    12.5.1. General / 541
    12.5.2. Specific Protocols for Refinement / 543
    References / 546

CHAPTER 13. BIOPOLYMERS 550

13.1. Peptides and Proteins / 550
    13.1.1. Assignment: Heteronuclear Methods / 554

13.1.2. Assignment: (^{1}H, ^{1}H) NOE-Based Methods / 555
13.1.3. Structure Determination of Protein Monomers / 556
    13.1.3.1. Small Rigid Unlabeled Proteins / 556
    13.1.3.2. Larger Rigid Labeled Proteins / 558
    13.1.3.3. Conformationally Mobile Proteins / 560
13.1.4. Structure Determination of Symmetric Protein Oligomers / 562
13.1.5. Through-Space Connections by Solid-State NMR Experiments with Proteins / 563

13.2. Polynucleotides / 566
  13.2.1. Structures and Conformations / 567
  13.2.2. Assignment / 574
    13.2.2.1. Duplex DNA / 574
    13.2.2.2. RNA / 576
    13.2.2.3. Other Nucleotides / 577
  13.2.3. Structure Calculation / 577
    13.2.3.1. Sequence Dependent Conformation in Duplexes / 577
    13.2.3.2. Non-Helical Conformations / 581

13.3. Oligosaccharides / 583
  13.3.1. Sequence and Linkage Determination / 584
  13.3.2. Conformation / 585

13.4. Complexes / 587
  13.4.1. Drug–Protein Complexes / 589
  13.4.2. Drug–DNA Complexes / 589
  13.4.3. Protein–Nucleic Acid Complexes / 590
  References / 592

APPENDIX I. EQUATIONS FOR ENHANCEMENTS INVOLVING GROUPS OF EQUIVALENT SPINS   596

APPENDIX II. QUANTUM MECHANICS AND TRANSITION PROBABILITIES   599

INDEX   611

PREFACE

When we completed the first edition of this book, just over 10 years ago, we did not expect to undertake rewriting it for a long time to come. It is certainly true that the theory behind the NOE has altered little in the last 10 years (as reflected by the fact that Chapters 1, 2, 3, and 6 are largely unchanged). However, there has continued to be a very rapid growth in the range of applications, particularly in the macromolecular area.

To a large extent, this has dictated the changes we have made to the book. The biggest change from the first edition is the inclusion of a completely new chapter (Chapter 12), which deals with the way NOE enhancements are used to calculate structures of biomolecules. Chapter 13, which discusses applications of the NOE to calculations of macromolecular structures, has been largely rewritten. Much the same is true for Chapter 4, which deals with the kinetics of the NOE; although the theory is largely the same, we have now placed much more emphasis on transient NOE experiments (such as 2D and gradient-selected experiments), which have seen a major increase in use over the last 10 years.

Interest in internal motions in macromolecules was the main driving force behind the changes we have made to Chapter 5 (which also includes a new section on NOE enhancements involving water, and a major update on the transferred NOE), and a large number of new approaches have been incorporated into Chapters 7 (particularly gradient enhanced experiments) and 8 (network editing). Chapter 9 ("Other Developments") remains the home for "miscellany" and contains a lot of new material, much of it concerned with the new experiments that became available in the wake of isotope labeling.

We have not significantly changed the content of the chapters dealing with applications to small molecules (Chapters 10 and 11), partly because neither of us is still active in that field, but more importantly because we feel the examples we gave in the first edition still represent a balanced selection that demonstrates coherently what is usefully possible. It is true that some types of structural problem are now more frequently solved by other means (e.g., positional isomers on aromatic rings might now be more frequently distinguished using long-range $^{13}C-^{1}H$ correlation than by using the NOE), but we felt illustration of the utility of such NOE experiments was still important.

Overall, we have thus broadened the scope of the first edition somewhat, feeling that to exclude (for example) applications of the NOE to relaxation measurements, because it does not relate directly to structure or conformation, was unnecessarily restrictive.

One of the aims we set ourselves in writing this second edition was to make the book shorter. The most casual glance will reveal that we have achieved the opposite of success in this regard. The reason, unfortunately, is simple: as George Bernard Shaw (or was it Blaise Pascal?) is supposed to have said, "I am sorry to have written you a long letter, but I did not have time to write you a short one." For this and other reasons, we feel it is safe to predict that there will be no third edition.

David Neuhaus
Michael P. Williamson

ACKNOWLEDGMENTS

We are deeply indebted to Dr. James Keeler, who has made many valuable comments on the manuscript and contributed much to the theory sections. His work on extending aspects of NOE theory remains a major influence on this book. We would also like to thank Dr. Deirdre Hickey for all her invaluable help and encouragement in preparing the manuscript, Prof. Dudley Williams, who instigated the project, Prof. Gareth Morris, for a number of useful observations and figures, and Prof. Ruth Lynden-Bell for her advice on Appendix II. Thanks also to all those who contributed ideas, comments, figures, or data, including (for the second edition) Drs. Mike Bernstein, David Case, Bob Diamond, Bob Dutnall, Philip Evans, Mark Fletcher, Charlie Hoogstraten, Richard Lewis, Beat Meier, Michael Nilges, Carol Post, and John Schwabe and (for the first edition) Andy Derome, Jeremy Everett, Duncan Farrant, Maurice Guéron, Peter Hore, Laurence Kruse, Forrest McKellar, Andrew Lane, Werner Leupin, Jeremy Sanders, Maurice Shamma, Richard Sheppard, Alan Whittle, and David Widdowson, as well as the authors and publishers who gave permission to reproduce published material. We thank all those at Wiley involved in the production of the book, particularly Barbara Goldman and Christine Punzo, both of whom showed patience well beyond the call of duty when waiting for the text, and, subsequently, in coping with all our various requests. Finally, we thank our families for putting up with so many lost evenings and weekends (again!).

PREFACE TO THE FIRST EDITION

In the past ten years a quiet revolution in the applications of NMR to organic chemistry and biochemistry has occurred. One of the main changes has been a great increase in the use of the nuclear Overhauser effect (NOE) to solve structural and conformational problems, which in turn was largely due to experimental developments such as the advent of NOE difference spectroscopy and the two-dimensional NOE experiment (NOESY). In this book we have tried to bring chemists and biochemists who are not NMR specialists up to date with these trends. The book aims to provide readers with sufficient background information to enable them to apply the NOE successfully within their own work, and also assess critically other papers in which the NOE is used to solve structural problems.

Because the NOE is transmitted directly through space, it is uniquely well suited to revealing the spatial arrangements of nuclei within molecules, but its interpretation is particularly vulnerable to the overapplication of inappropriate equations. For this reason, we have tried above all to emphasize a clear understanding of the underlying concepts, since only through an awareness of these can the most useful NOE experiments be devised and interpreted. Unavoidably, we have had to use a certain amount of mathematics to do this, but it is kept to a minimum, consistent with a clear explanation. Perhaps contrary to first appearances, most of the expressions in this book require no more than elementary mathematical skills.

As the title implies, our treatment is deliberately restricted to those aspects of the NOE that are relevant to the study of stereochemistry and conformation; topics such as dipolar interactions in solids, and the use of the NOE to study relaxation phenomena, kinetics, or exchange, have therefore been omitted.

The book is organized into three parts, dealing with theory, experimental practice, and an illustrative overview of applications of the NOE to problems in chemistry and biochemistry.

SI units have been used throughout, as far as possible. This results in occasional numerical factors in some expressions (e.g., the factor $\mu_0/4\pi$, which appears in expressions relating relaxation rates to internuclear distances).

David Neuhaus
Michael P. Williamson

SYMBOLS, ABBREVIATIONS, AND UNITS

Throughout, subscript letters are used to denote spin states and transitions; thus, N_α is the number of molecules in spin state α, and R_I is the spin–lattice relaxation rate of spin I. Superscript letters are used to denote different molecules or conformations; thus, S^a is the signal from spin S in molecules with conformation a, and R_S^b is the spin–lattice relaxation rate of spin S in molecules with conformation b. Vector quantities are written in bold.

ADC	analog-to-digital converter
AQ	acquisition time
$\gamma \mathbf{B}/2\pi$	field strength in Hz
$\gamma \mathbf{B}$	field strength in rad s^{-1}
$\mathbf{B}_0$	applied magnetic field (tesla)
$\mathbf{B}_1$	irradiating magnetic field (tesla)
CW	continuous wave
$d_{AB}(i, j)$	distance between proton A in residue i and proton B in residue j in a peptide or protein
D	dwell time in t_2 (s)
1D	one-dimensional
2D	two-dimensional
3D	three-dimensional
dB	decibels
DG	distance geometry
DPFGSE	double pulsed field gradient spin echo
DQF-COSY	double-quantum filtered 2D correlation spectroscopy
$f_I\{S\}$	fractional NOE enhancement of I on saturating S

FID	free induction decay
FT	Fourier transform
$g(\tau)$	correlation function, usually $\exp(-\tau/\tau_c)$
GOESY	gradient enhanced NOE spectroscopy
h	Planck's constant = 6.626×10^{-34} Js
$\hbar$	Planck's constant divided by 2π = 1.055×10^{-34} Js
HMQC	heteronuclear multiple quantum correlation (also coherence)
HSQC	heteronuclear single quantum correlation (also coherence)
I	nuclear spin quantum number, *or* general symbol for a nucleus; in expressions concerning NOE experiments, I generally represents the nucleus at which the enhancement is measured.
I_z	longitudinal component of I magnetization
I_z^0	equilibrium value of I_z
IN	t_1 increment in a 2D experiment (s)
ISPA	isolated spin pair approximation
J	spin–spin coupling constant (Hz)
$J(\omega)$	spectral density function, usually $2\tau_c/(1 + \omega^2\tau_c^2)$
k	Boltzmann's constant = 1.38×10^{-23} J K^{-1}
M	macroscopic magnetization vector (J T^{-1})
M_z	z component of **M**
M_{xy}	transverse component of **M**
MAS	magic angle spinning
N_i	number of molecules (population) in state i
n_i	population deviation from equilibrium of state i, namely $N_i - N_i^0$
NOE	nuclear Overhauser effect
NOESY	nuclear Overhauser effect correlation spectroscopy
R_I^{DD}	dipolar contribution to the spin–lattice relaxation rate of I, defined as $W_{0IS} + 2W_{1I} + W_{2IS} + \Sigma_x (W_{0IX} + W_{2IX})$ (cf. Eqs. 2.31 and 3.8) (s^{-1})
$R_I \equiv R_{1I}$	longitudinal relaxation rate of $I = R_I^{DD} + \rho_I^*$, roughly equal to the inverse of the selective T_1 value (cf. Section 2.3.1) (s^{-1})
R_2	transverse relaxation rate (s^{-1})
r	internuclear distance
r_{Tropp}	effective internuclear distance sensed by the NOE in the presence of rapid internal motion; most commonly used to denote the effective distance to a methyl group (cf. Section 5.5.2)
RF	radiofrequency
rmsd	root mean squared deviation (or difference); usually implies root mean squared *atomic* deviation when comparing structures
ROE	rotating frame NOE; also known as CAMELSPIN
ROESY	rotating frame 2D NOE spectroscopy
S	general symbol for a nucleus; in expressions concerning NOE experiments, S generally represents the nucleus which is saturated (S can also represent the electron spin quantum number).

	Also, generalized order parameter for describing the geometrical extent of internal motions relative to the molecular frame (cf. Section 5.4)
S^2	square of the generalised order parameter (cf. Section 5.4)
S^{ang}	angular order parameter (cf. Section 12.4.2)
S_z	longitudinal component of S magnetization
S_z^0	equilibrium value of S_z
SPT	selective population transfer
SW_1	spectral width in F_1 of a multidimensional experiment (Hz)
SW_2	spectral width in F_2 of a multidimensional experiment (Hz)
SW_3	spectral width in F_3 of a multidimensional experiment (Hz)
t_1	incremented time in 2D experiments; first incremented time in multidimensional experiments
$t_{1\,max}$	maximum time reached by t_1 in the last increment of a 2D or multidimensional experiment (s)
t_2	detection period in 2D experiments; second incremented time in multidimensional experiments
$t_{2\,max}$	maximum time reached by t_2 (s)
t_3	detection period in 3D experiments
$t_{3\,max}$	maximum time reached by t_3 (s)
T	temperature (K)
T_1	normally used to mean $T_1^{nonselective}$ (see below)
$T_1^{nonselective}$	nonselective longitudinal relaxation time, obtained by following the course of longitudinal relaxation after a nonselective pulse; approximately given by $(R_I + \sigma_{IS} + \Sigma_X \sigma_{IX})^{-1}$ (s) (cf. Section 2.3.2)
$T_1^{selective}$	selective longitudinal relaxation time, obtained by following the course of longitudinal relaxation after a selective pulse; approximately equal to R_I^{-1} (s)
$T_{1\rho}$	spin–lattice relaxation time in the rotating frame (s)
T_2	spin–spin, or transverse, relaxation time (s)
$T_{2\rho}$	spin–spin relaxation time in the rotating frame (s)
T_2^*	decay constant for free precession, which is shorter than T_2 because of inhomogeneous broadening (s)
$T_{2\rho}^*$	decay constant for free precession in the rotating frame (s)
t_D	relaxation delay (s)
TMS	tetramethylsilane
TOCSY	total correlation spectroscopy
TOE	truncated driven NOE
TPPI	time-proportional phase incrementation
TRNOE	transferred NOE
VDW	van der Waals
VT	variable temperature
W	transition probability (s^{-1})

Symbol	Description
X	general symbol for a nucleus; in expressions concerning NOE experiments in this book, X represents all nuclei other than I and S (when not italicized, X can also refer to a non-hydrogen nucleus)
X_z	longitudinal component of X magnetization
X_z^0	equilibrium value of X_z
α, β	spin states for a spin 1/2 nucleus
γ	gyromagnetic ratio (rad T^{-1} s^{-1})
η	viscosity (cP)
η_{max}	maximum enhancement attainable in a two-spin system at a given value of τ_c and ω
μ	magnetic dipole moment (J T^{-1})
μ_0	permeability constant in a vacuum = $4\pi \times 10^{-7}$ kg m s^{-2} A^{-2}
ν	precession rate (Hz)
ν_0	reference frequency of rotating frame (Hz)
ρ_{IS}	direct dipole–dipole relaxation rate between I and S, defined by $\rho_{IS} = W_{0IS} + 2W_{1I} + W_{2IS}$ [Eq. 2.30] (s^{-1})
ρ^*	external spin–lattice relaxation rate (s^{-1})
ρ_1	alternative notation for ρ_{IS}, useful when comparing the NOE with the ROE
ρ_2	direct relaxation rate in the rotating frame (s^{-1})
σ_I	chemical shift tensor for nucleus I (cf. Sections 5.4 and 6.3)
σ_{IS}	cross-relaxation rate between I and S, defined by $\sigma_{IS} = W_{2IS} - W_{0IS}$ (Eq. 2.29) (s^{-1})
σ_1	alternative notation for σ_{IS}, useful when comparing the NOE with the ROE
σ_2	cross-relaxation rate in the rotating frame (s^{-1})
τ_c	rotational correlation time (for overall molecular tumbling) (s)
τ_e	correlation time for an internal motion (cf. Section 5.4) (s)
τ_m	mixing time in a 2D or multidimensional NOE experiment (s)
τ	buildup time in kinetic NOE experiment; also general symbol for a fixed delay other than τ_m (s)
ω_0	Larmor precession frequency due to $\mathbf{B}_0$ (rad s^{-1})
ω_1	Larmor precession frequency due to $\mathbf{B}_1$ (rad s^{-1})
$\{\ \}$	denotes irradiated spin
$\langle\ \rangle$	denotes average or expectation value *or* conformational average
$\langle\vert,\vert\rangle$	Dirac notation for a wave function (ket) and its complex conjugate (bra), respectively

INTRODUCTION

The fundamentals of the NOE were described very early in the history of NMR, in a classic paper by Solomon published in 1955.[1] This paper included the first experimental demonstration of the NOE. Solomon's work followed studies on nuclear spin relaxation by Bloembergen, Purcell, and Pound,[2] and Overhauser's original prediction that saturation of electrons in a metal would produce a large polarization of the metal nuclear spins.[3,4]

Solomon's work then lay almost dormant for some years until the advent of double resonance techniques[5,6] led to the more widespread availability of spectrometers having a decoupler. Papers appeared by Kaiser on distance estimation using the intermolecular $^1H\{^1H\}$ NOE,[7] and by Lauterbur on signal enhancement in ^{13}C spectra.[8]

The first paper to demonstrate the power of the NOE in structural problems was by Anet and Bourn in 1965.[9] The next few years saw a surge in the number of papers reporting applications of the NOE, especially after publication of the book by Noggle and Schirmer.[10] On the CW spectrometers of the day, these experiments involved comparison of integrals to measure steady-state NOE enhancements.

Noggle and Schirmer's book surveyed the field as it stood in 1971, and for some years there was little that needed to be added. There were two major advances in the decade or so following 1971. The first was the application of the NOE to large molecules. In 1972, Balaram et al.[11] observed a negative NOE enhancement between a protein and a peptide binding to it, and they showed that negative enhancements are a general property of large molecules. Further applications to large molecules soon revealed the problem of spin diffusion, and the truncated driven NOE (TOE) experiment was suggested by Wagner and Wüthrich[12] as a means of getting around this difficulty.

The second major advance was the introduction of the two-dimensional NOE experiment, largely by Ernst's group. The first use of the experiment to measure NOE enhancements (as opposed to chemical exchange) was published in 1980,[13,14] and the experiment has come to play a central role in work with biological macromolecules. This area has burgeoned still further with the arrival in the late 1980s of isotopically labeled proteins and the many multidimensional NOE experiments that can be applied to them.

Both these advances could not have had the impact they did without developments in instrumental techniques. The first FT NMR spectrometers became available around 1970, and brought much improved sensitivity. To observe small NOE enhancements, one needs not only sensitivity but also instrumental stability, and this was provided by superconducting magnets, which arrived at the end of the 1970s. The third in this trio of instrumental advances was the advent of dedicated computers, which could control the cycling of decoupler frequencies and gating. A direct result of these changes was the NOE difference experiment, which emerged at about this time. A much more recent development is the use of pulsed field gradients, which have found applications in almost all parts of NMR. Gradients provide little benefit for 2D NOESY experiments, but promise to revolutionize 1D measurements.[15]

Although the theory of the NOE as applied to organic molecules has scarcely changed since 1971, its application has increased dramatically. Using the techniques available in 1971, enhancements smaller than 5% could be seen only with difficulty. Now it is not particularly difficult to detect enhancements of less than 1%, or even 0.1% using newer gradient assisted experiments. Hall and Sanders[16] point out that the maximum possible size of NOE enhancement seen between 1,3 diaxial protons in a cyclohexane chair is about 3%. This size of enhancement was just below the threshold of detection in the CW era, but it is well within modern instrumental capabilities.

REFERENCES

1. Solomon, I. *Phys. Rev.* 1955, *99*, 559.
2. Bloembergen, N.; Purcell, E. M.; Pound, R. V. *Phys. Rev.* 1948, *73*, 679.
3. Overhauser, A. W. *Phys. Rev.* 1953, *89*, 689.
4. Overhauser, A. W. *Phys. Rev.* 1953, *92*, 411.
5. Freeman, R.; Anderson, W. A. *J. Chem. Phys.* 1962, *37*, 85.
6. Hoffman, R. A.; Forsén, S. *Prog. Nucl. Magn. Reson. Spectrosc.* 1966, *1*, 15.
7. Kaiser, R. *J. Chem. Phys.* 1963, *39*, 2435.

8. Lauterbur, P. C.; quoted in Baldeschweiler, J. D.; Randall, E. W. *Chem. Rev.* 1963, *63*, 81.
9. Anet, F. A. L.; Bourn, A. J. R. *J. Am. Chem. Soc.* 1965, *87*, 5250.
10. Noggle, J. H.; Schirmer, R. E. "The Nuclear Overhauser Effect; Chemical Applications," Academic Press, New York, 1971.
11. Balaram, P.; Bothner-By, A. A.; Dadok, J. *J. Am. Chem. Soc.* 1972, *94*, 4015.
12. Wagner, G.; Wüthrich, K. *J. Magn. Reson.* 1979, *33*, 675.
13. Macura, S.; Ernst, R. R. *Mol. Phys.* 1980, *41*, 95.
14. Kumar, A.; Ernst, R. R.; Wüthrich, K. *Biochem. Biophys. Res. Commun.* 1980, *95*, 1.
15. Stott, K.; Stonehouse, J.; Keeler, J.; Hwang, T.-L.; Shaka, A. J. *J. Am. Chem. Soc.* 1995, *117*, 4199.
16. Hall, L. D.; Sanders, J. K. M. *J. Am. Chem. Soc.* 1980, *102*, 5703.

PART I

THEORY

CHAPTER 1

BACKGROUND

When one resonance in an NMR spectrum is perturbed by saturation or inversion, the net intensities of other resonances in the spectrum may change. This phenomenon is called the nuclear Overhauser effect (NOE), and its importance lies primarily in the fact that the resonances that change their intensities are due to spins *close in space* to those directly affected by the perturbation.

The NOE has its origin in the population changes brought about by a particular form of relaxation, namely dipole–dipole cross-relaxation. Before we can properly tackle the theory of the NOE, it is crucial to understand the nature of relaxation, in particular dipole–dipole cross-relaxation, and this forms the main topic of this chapter. We begin, however, with a very brief description of the NMR phenomenon itself. This is, of course, not intended to be an adequate treatment in its own right; its purpose is rather to act as a reminder of the basic concepts and to emphasize those of particular importance to the NOE.

1.1. ENERGY LEVELS, POPULATIONS, AND INTENSITIES

The NMR phenomenon is a consequence of the existence of nuclear spin. Not all nuclei have spin, but those that do also have an associated magnetic field and hence a nuclear magnetic dipole moment, μ. In much the same way as an electric dipole consists of a pair of separated, opposite electrical charges, so a magnetic dipole can be imagined as a pair of separated magnetic north and south poles, as in a bar magnet. For our relatively simple purposes, the nuclear magnetic dipole moment μ can thus be pictured as a "microscopic bar magnet," aligned with the nuclear spin axis.

When placed in a static magnetic field (B_0), such as that provided by the magnet of an NMR spectrometer, the energy of such a nucleus becomes orientation dependent; in other words, just as would a macroscopic bar magnet, the nucleus attempts (unsuccessfully, as we shall see) to align itself with its magnetic dipole parallel with the static field. To twist it into an antiparallel orientation requires an input of energy.

As the nucleus is in reality a subatomic particle rather than a macroscopic object, we must expect its behavior to be more noticeably constrained by the laws of quantum mechanics than a bar magnet would be. Specifically, the energy of the nucleus is quantized, the number of energy levels (eigenstates) being $2I + 1$ where I is the nuclear spin quantum number. We shall be directly concerned only with spin 1/2 nuclei, that is nuclei such as ^1H, ^{13}C, ^{15}N, ^{31}P, and ^{19}F for which $I = 1/2$. For such nuclei there are only two possible energy levels, called α and β. These are the low- and high-energy states, respectively, and they correspond to two particular orientations of μ relative to the static magnetic field axis.

Since the axes of the nuclear spin and the nuclear magnetic dipole are, in fact, the same, one may define a numerical (i.e., scalar) constant of proportionality between them; this ratio is called the gyromagnetic ratio, or γ. Quantum mechanics tells us that the angular momentum due to the nuclear spin has magnitude $\hbar\sqrt{I(I+1)}$, so that γ is defined by

$$|\mu| = \gamma\hbar\sqrt{I(I+1)} \tag{1.1}$$

Differences in γ between different nuclear species can be pictured as corresponding to differences in the strength of their associated magnetic dipoles. This in turn implies a difference in their NMR frequencies; the stronger a magnetic dipole is, the more energy is required to change its orientation in a static magnetic field of given strength, and, consequently, for spin 1/2 nuclei, the greater is the energy separation of the α and β states. For example, ^1H has a gyromagnetic ratio about four times that of ^{13}C. Thus, although both have $I = 1/2$ (and consequently equal angular momenta), the magnetic dipoles due to ^1H are four times as strong as those due to ^{13}C, and the resonance frequency of ^1H, proportional to the energy separation of the α and β states, is four times that of ^{13}C. The factors that determine the value of γ for a particular nuclear species, although they clearly must be related in some way to the internal nuclear structure, are first not understood, and second well outside the scope of this book!

Before leaving this discussion of the nature of the α and β states, we need one more result from quantum mechanics, namely that the projection of μ along the static magnetic field axis must be $\gamma\hbar m$, where m is the magnetic quantum number, having values $I, I-1, I-2, \ldots -I$. Although this sounds rather obscure, it is in fact nothing more than a formalization of our earlier statement that the energy of the nucleus is quantized, being restricted to $I + 1$ quantum states defined by the orientation of μ with respect to the static magnetic field axis. For spin 1/2 nuclei, these projections can, thus, only have magnitudes $(1/2)\gamma\hbar$ or $-(1/2)\gamma\hbar$, corresponding to the α and β energy states, respectively. Note that since $m < \sqrt{I(I+1)}$, these projections have magnitudes *smaller* than μ itself, implying that in either energy state the magnetic dipole μ is not aligned parallel with the field, but instead makes an angle $\cos^{-1}[m/\sqrt{I(I+1)}] = 54.73°$ with the static field axis. This leads to the most familiar visualization for the α and β states, namely the "two cone" picture shown in Figure 1.1, of which we will have more to say shortly.

We need to consider next the way in which the α and β states are populated at thermal equilibrium. Relative to kT at typical room temperatures, the energy separation of the α and β states is minute. This means, according to the Boltzmann distribution law, that the populations of the two states at thermal equilibrium will be very similar. At 400 MHz, the difference is only 1 part in 6.6×10^5. In terms of the "two cone" picture, we may represent these populations as shown in Figure 1.1. Each magnetic dipole μ is depicted as a vector starting, for visual simplicity, at a common origin. Those in the low-energy α state are evenly distributed around the surface of the upper cone, while the very slightly smaller number in the high-energy β state are distributed around the surface of the lower cone.

In terms of this representation, the net nuclear magnetization across the whole sample corresponds to the vector sum of the all the individual dipoles μ. The result of such a summation is a macroscopic magnetic moment whose length is proportional to the population difference between the α and β states at thermal equilibrium, and which is aligned parallel to the applied static field $\mathbf{B}_0$ (Fig. 1.2). The $\mathbf{B}_0$ field axis is conventionally represented as the z axis and is therefore called longitudinal, while the equilibrium magnitude of the magnetization vector is called M_z^0. This equilibrium magnetization represents the starting point for all NMR experiments.

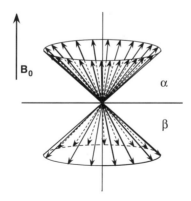

Figure 1.1. "Two-cone" representation of the α and β spin states for a single-spin system with $I = 1/2$, showing (schematically) the slightly higher population of the lower energy α state.

6 BACKGROUND

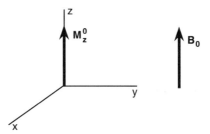

Figure 1.2. The equilibrium *macroscopic* magnetization M_z^0. This magnetization is parallel to the static field axis $\mathbf{B}_0$, and can be obtained by vector summation of the individual vectors shown in Figure 1.1.

Figure 1.2 represents the thermal equilibrium as it exists before application of a radiofrequency field. How is it that an RF field applied from coils perpendicular to the static field axis $\mathbf{B}_0$ produces an NMR signal?

The behavior of a macroscopic magnetization $\mathbf{M}$ in a magnetic field $\mathbf{B}$ is well known from classical electromagnetic theory; the field exerts a torque of $\mathbf{M} \times \mathbf{B}$ upon the magnetization $\mathbf{M}$, causing it to rotate about the field axis $\mathbf{B}$. The macroscopic vector representing $\mathbf{M}$ may be pictured as sweeping out a cone or disc about the field axis $\mathbf{B}$ in a motion referred to as precession. When the static field $\mathbf{B}_0$ is the only field present, then the equilibrium magnetization is parallel to the magnetic field direction, and precession about $\mathbf{B}_0$ corresponds only to precession of the magnetization "about its own axis." The role of the RF field (i.e., in FT spectroscopy, the pulse) is to introduce a transverse (xy) component to the total magnetic field, so that the total field axis is no longer parallel to the z axis, and the magnetization precesses about the new field axis. This topic is dealt with in greater depth in Section 1.5. Suffice it to say here that precession about $\mathbf{B}_{\text{total}}$ while the RF field is on causes the magnetization to move away from the $\mathbf{B}_0$ axis, and leaves a *transverse* component of magnetization (M_{xy}) where the RF field is switched off. Once created, the transverse magnetization M_{xy} processes about $\mathbf{B}_0$ at a frequency ν (in hertz) given by the Larmor equation:

$$\nu = \frac{\gamma}{2\pi} |\mathbf{B}_0|$$

$$= \frac{1}{h} [E_\beta - E_\alpha] \qquad (1.2)$$

This deceptively simple equation actually forms a link between the vector description of NMR and the quantum mechanical description. In the vector description, the rate of precession ν of each nucleus depends on the strengths both of the static field $\mathbf{B}_0$ and the nuclear magnetic dipole $\boldsymbol{\mu}$; the stronger either of these is, the faster the precession rate will be. The strength of $\mathbf{B}_0$ is of course a function of the spectrometer, but the strength of the nuclear magnetic dipole depends on the gyromagnetic ratio, γ. As we saw earlier, γ relates the nuclear dipole moment to the nuclear spin angular momentum, but this angular momentum is itself related to the spin *energy* of the nucleus. Thus, the precession rate can also be expressed in terms of the nuclear spin energy. This leads to the second form of Eq. 1.2, where the spin energy corresponding to resonance is set equal to the energy difference between the α and β nuclear spin states. This is an example of the Bohr frequency condition familiar from other branches of spectroscopy; an energy difference between two quantum states is related, via Planck's constant, to the frequency of radiation that can interconvert the states ($\Delta E = h\nu$). This idea of matching an energy to a frequency gap will be of central importance to the theory in Chapter 2.

It is the Larmor precession of transverse magnetization that all NMR spectrometers are designed to measure. As it rotates, the transverse magnetization induces an alternating current within the receiver coil of the spectrometer (in much the same way as a moving magnet generates current within a dynamo). This current is the NMR signal. We shall be concerned almost exclusively with Fourier transform (FT) NMR experiments, in which the RF irradiation takes the form of a very short powerful pulse. This rotates the equilibrium magnetization through an angle dependent on the length and power of the pulse, and, provided this angle is not a multiple of 180°, creates a transverse component that continues to precess about $\mathbf{B}_0$ after the pulse has ended. We may see that the intensity of the NMR signal so produced depends both on the intrinsic strength of the equilibrium magnetization (M_z^0) and also on the angle through which it is turned by the pulse.

The purpose of this rather cursory review of the NMR phenomenon was to emphasize a few points of fundamental importance to understanding the NOE. The first, and most important, is that the intensity of an NMR resonance depends on the population difference across the corresponding transitions (among other things). In the FT experiment, this population difference is, in effect, "sampled" by the pulse, which converts the longitudinal magnetization present before the pulse into a proportionately sized transverse component. (We shall see later that for coupled spin systems this view must be slightly modified, in that it is only the total intensities of entire multiplets that can be considered in this way.) The second point is that, in the absence of perturbations, these populations are given by the Boltzmann distribution law. As we shall see, the NOE produces nonequilibrium populations for certain spins, which, when sampled by a pulse, give rise to perturbed intensities for the corresponding resonances in the NMR spectrum.

Before moving on, we should pause to consider briefly the various vector models we have used to visualize the NMR phenomenon, since these underlie much of the discussion both in this and in other treatments. We must first distinguish between representation of *macroscopic* magnetizations and *molecular* magnetizations. Although it is in the nature of explanations, including this one, to start at the molecular level and work up to the macroscopic level, in reality the experimenter must always work in the opposite direction, starting from observable macroscopic phenomena. This is reflected in our models. Thus, the behavior of the *macroscopic* vector **M** corresponds to a reasonable pictorial summary of the observed behavior of the (macroscopic) nuclear magnetism during an NMR experiment (at least for spin 1/2 nuclei). The "two cone" picture, on the other hand, supposedly depicts events at the *molecular* level, but was in fact arrived at by deduction, combining a few quantum mechanical rules with the requirement that vector summation of all the individual dipoles **μ** should yield the known properties of the equilibrium macroscopic magnetization M_z^0. Thus, although it was adequate for our purpose in *visualizing* the origin of M_z^0, we should not be too surprised if this "molecular picture" becomes unwieldy, confusing, or even wrong in other situations. Almost all of the theory in subsequent sections will be presented, explicitly or implicitly, at the macroscopic level.

Turning to the macroscopic vector picture, we must expect this also to have its limitations, but fortunately these are relatively few in the area of NMR theory with which we have to deal. We should remember that the vector **M** actually represents a bulk property, evenly distributed throughout the sample but nonetheless directional. Also, our explanations have so far considered only a single resonance. In practice many resonances are usually present, resulting in many contributions to **M**. After an observe pulse each will precess at its own characteristic Larmor frequency; the sum of these contributing frequencies constitutes the free-induction decay (FID), which gives the spectrum on Fourier transformation.

1.2. RELAXATION, T_1 AND T_2

The previous section introduced the concepts of macroscopic longitudinal and transverse magnetizations. Longitudinal magnetization is aligned along the static field axis $\mathbf{B}_0$, is itself static, and has a finite equilibrium value of M_z^0. Transverse magnetization is perpendicular to $\mathbf{B}_0$, around which it precesses at the Larmor frequency ν, and it has an equilibrium value of zero.

After any disturbance brought about by the action of RF irradiation, two processes must occur for equilibrium to be restored: (i) the longitudinal magnetization must return to its original value of M_z^0, and (ii) the transverse magnetization must decay to zero. The first process is called longitudinal relaxation, and the second transverse relaxation.

As we have seen, longitudinal magnetization owes its existence entirely to the difference between the populations of the nuclear spin states α and β. It follows that longitudinal relaxation necessarily involves population changes, so that the processes that bring it about are those that induce *transitions* between the spin states. Transverse relaxation, in contrast, involves the loss and dephasing of individual contributions to the macroscopic transverse magnetization M_{xy}, so that their resultant sum becomes zero. This too will be caused by

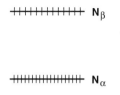

Figure 1.3. Energy level diagram for a two-level system. At thermal equilibrium, the populations N_α and N_β obey the Boltzmann distribution law.

transitions, but it will also be brought about by any process that perturbs the individual contributing Larmor frequencies. On the whole we shall have very little to say about transverse relaxation, since NOE experiments are concerned mainly with longitudinal magnetization, and longitudinal relaxation is independent of transverse relaxation.

Before going on to a more detailed picture of the causes of longitudinal relaxation, we should next deal briefly with the equations describing population changes in a two-level system, since these will reveal some useful generalizations needed later. Figure 1.3 illustrates a two-level system, in which the populations of the low- and high-energy states are N_α and N_β, respectively.

Without concerning ourselves with the details of mechanism at this stage, let there be some process that causes transitions between the two states. The number of upward and downward transitions that occur will obviously depend on the populations N_α and N_β, but it will also depend on how effective the particular mechanism involved is at causing transitions. This "efficiency" is expressed as the transition probability, W. In effect, W is a *rate constant for transitions*. The method for calculating transition probabilities is described in Appendix II, but it is worth examining here the formal definition of W, since it gives some insight into its nature:

$$W_{ij} = J(\omega) |(\langle i | \mathcal{H}_{\text{rel}} | j \rangle)^2|_{\text{average}} \tag{1.3}$$

What this slightly forbidding piece of quantum mechanics actually shows is how two spin states i and j (in this case α and β) are linked by the action of the relaxation mechanism, as characterized by its Hamiltonian, $\mathcal{H}_{\text{rel}}$. The significance of $J(\omega)$, the spectral density function for molecular tumbling, will become clear later (Section 2.2).

When the system is at thermal equilibrium, the numbers of upward and downward transitions are, by definition, equal. Away from equilibrium, however, this balance is lost, and the system returns to equilibrium at a rate governed by W, the populations, and the temperature of the surroundings (usually called the lattice in this context). The rates of change for each population are

$$\frac{dN_\alpha}{dt} = -WN_\alpha + WN_\beta + K$$

and

$$\frac{dN_\beta}{dt} = WN_\alpha - WN_\beta - K \qquad (1.4)$$

The constant K is necessary to represent the effect of the lattice in trying to impose its own temperature on the spin system (recall that W is a property of the relaxation mechanism, not the lattice). It may be evaluated using the condition that, at thermal equilibrium,

$$\frac{dN_\alpha}{dt} = 0 \quad \text{and} \quad \frac{dN_\beta}{dt} = 0 \qquad (1.5)$$

from which

$$K = WN_\alpha^0 - WN_\beta^0 \qquad (1.6)$$

where N_α^0 and N_β^0 are the equilibrium populations of the α and β states, respectively. For simplicity, we replace the absolute populations N_α and N_β with their corresponding population *differences from equilibrium*, n_α and n_β, defined by

$$n_\alpha = (N_\alpha - N_\alpha^0)$$

and

$$n_\beta = (N_\beta - N_\beta^0) \qquad (1.7)$$

Using Eqs. 1.6 and 1.7, we may now rewrite Eq. 1.4 much more conveniently as

$$\frac{dN_\alpha}{dt} = -Wn_\alpha + Wn_\beta$$

and

$$\frac{dN_\beta}{dt} = Wn_\alpha - Wn_\beta \qquad (1.8)$$

Thus, by specifying the populations as deviations from equilibrium, we automatically include the thermal effect of the lattice in our treatment, but without complicating the equations. This simplifying convention will prove very valuable in later sections.

Of course, the quantity we actually measure is the resonance intensity. As discussed in the previous section, this is proportional to the longitudinal magnetization, M_z, present immediately prior to the observation pulse, and this in turn is proportional to the population difference across the corresponding transition:

$$M_z = k(N_\alpha - N_\beta)$$

from which

$$M_z^0 = k(N_\alpha^0 - N_\beta^0) \qquad (1.9)$$

Using this centrally important definition, we may obtain from Eqs. 1.8 an expression for the rate of change of M_z:

$$\begin{aligned}\frac{dM_z}{dt} &= k\left(\frac{dN_\alpha}{dt} - \frac{dN_\beta}{dt}\right) \\ &= -2k(Wn_\alpha - Wn_\beta) \\ &= -2W(M_z - M_z^0)\end{aligned} \qquad (1.10)$$

The factors of two in Eq. 1.10 reflect the fact that each transition necessarily alters the population difference $(n_\alpha - n_\beta)$ by two.

Equation 1.10 was derived for the case of a single (unspecified) relaxation mechanism acting within a two-level system. The result, not surprisingly, is a first-order exponential decay of M_z toward its equilibrium value of M_z^0. The rate constant for this decay is $2W$, or alternatively R (also written R_1), where

$$R = 2W \qquad (1.11)$$

It may also be expressed as a characteristic time, T_1, where

$$T_1 = \frac{1}{R} \qquad (1.12)$$

Integration of Eq. 1.10 shows that the magnetization present after a time t is given by

$$\begin{aligned}(M_z^t - M_z^0) &= (M_z^{\text{initial}} - M_z^0)\exp\left(\frac{-t}{T_1}\right) \\ &= (M_z^{\text{initial}} - M_z^0)\exp(-R_1 t) \\ &= (M_z^{\text{initial}} - M_z^0)\exp(-2Wt)\end{aligned} \qquad (1.13)$$

where M_z^t is the magnetization at time t, M_z^{initial} is the magnetization present at time zero, and M_z^0 is the equilibrium magnetization.

These equations show very clearly what is meant by the longitudinal relaxation time T_1, and also how it is related to R and W *in this case*. However, as we shall see, relaxation behavior in two-spin and multispin systems is in general not strictly exponential. In such circumstances, there cannot be an exponential rate constant, so, formally at least, T_1 is no longer defined. This point will be taken up in Section 2.3.2.

1.3. THE NATURE OF RELAXATION

The equations of the previous section describe, phenomenologically, how relaxation in a single-spin system affects the populations of the spin states, and hence the resonance intensities, but they say nothing about the processes that *cause* relaxation.

As pointed out earlier, longitudinal relaxation is brought about by processes that induce transitions. Abragam shows that spontaneous emission, the process in which a molecule in an excited state simply emits a photon and thereby decays to a low-energy state, is so slow a process at NMR frequencies as to be completely irrelevant.[1] He further shows that absorption or stimulated emission involving photons from the background radiation field (i.e., the background level of "black body radiation" photons in thermal equilibrium with the temperature of the surroundings) is not a viable mechanism either.[1] In fact, the processes that cause relaxation are those of energy exchange. In these, energy is directly exchanged between one set of energy levels, here those of the nuclear spin system, and some other set of energy levels.

For the quantities of energy involved in nuclear spin relaxation in liquids, this other set of energy levels is generally provided by the various translations, rotations, and internal motions of the molecules that constitute the sample; these are what is meant by the term "lattice." Note particularly that these motions include those of the molecule in which the nuclear spin flip actually occurs. Because of the very large number of degrees of freedom that the lattice possesses, this set of energy levels is effectively a continuum. Consequently, for any given NMR transition, there will always be some possible change within the lattice that involves the same quantity of energy. A typical relaxation event might thus consist of an upward flip of a nuclear spin accompanied by a corresponding slowing of some motion of the molecule in which the flip occurred. Similarly, a downward flip would be accompanied by a corresponding acceleration of some motion of the molecule. Intermolecular relaxation is of course also possible, and cross-relaxation, as will be discussed in Chapter 2, involves the simultaneous flip of *two* spins with a corresponding loss or gain of energy by the lattice.

There is, however, another requirement for these energy exchange processes to occur. In order to mediate the energy transfer, the motion (or other change) in the lattice that is to gain or lose the energy from the nuclear spin transition must itself cause a fluctuating magnetic field at the site of the nuclear spin involved. A simple way to picture this requirement is to imagine the fluctuating field acting as a "local pulse," equivalent in its effect to that of the normal RF observe pulse from the spectrometer, but limited to affecting just the one spin in the one molecule undergoing the transition at a given moment. In acting as a "local pulse," and so causing a nuclear spin transition, the local field *transfers* the corresponding quantity of energy into (or out of) the particular motion that gave rise to the fluctuation in the local field in the first place. We shall consider the various possible origins of the local field in more detail in Sections 1.4 and 2.4.

Viewed in this way, it is easy to see what the requirements of this local fluctuating field must be. First, just as for the observe RF pulse, the local field must have a component fluctuating at the frequency corresponding to the transition to be relaxed. This requirement in particular is of crucial importance in understanding the theory in later chapters. Second, just as only x or y pulses can rotate longitudinal magnetization, so only the x and y components of the local field can cause longitudinal relaxation (the z component causes transverse relaxation).

In spite of these analogies, there is an important difference between the effects of the local fluctuating field and an RF observe pulse from the spectrometer. The RF pulse is *coherent* across the sample, whereas the local field is *incoherent*. This means that an RF pulse affects all the spins similarly and in a concerted fashion, as, for instance, when rotating the macroscopic equilibrium longitudinal magnetization into the transverse plane. The local field, in contrast, will be randomly different at the site of each spin at any instant, and so cannot create any form of order among the spins. This distinction is perhaps most clearly illustrated by the example of recovery following a population inversion (caused by a 180° pulse). Relaxation by the local field causes the inverted longitudinal magnetization to decay back through zero to its positive equilibrium value, with no transverse magnetization involved. Rotation of the inverted longitudinal magnetization back to its equilibrium value using a 180° RF pulse, in contrast, does pass through an intermediate state of macroscopic transverse magnetization, because the effect of the pulse is coherent across all the spins.

1.4. THE LOCAL FIELD AND DIPOLE–DIPOLE RELAXATION

The local fluctuating field, which progressively dominated the discussion in the previous section, is of crucial importance in relaxation theory. The field experienced by a given nucleus is considered to be the vector sum of $\mathbf{B}_0$, the applied static magnetic field, and the local field, written (not surprisingly) $\mathbf{B}_{loc}$:

$$\mathbf{B}_{total} = \mathbf{B}_0 + \mathbf{B}_{loc} \qquad (1.14)$$

$\mathbf{B}_{loc}$ consists of the varying part of the total field, and it arises due to Brownian motion and other random changes within the lattice.† Clearly, its time average must be zero:

$$\langle \mathbf{B}_{loc}(t) \rangle = 0 \qquad (1.15)$$

Other key properties of the local field, mentioned earlier, are that it is incoherent across the sample, and that it has components at many frequencies, reflecting the spectrum of frequencies present in the Brownian motion of the lattice (cf. Section 2.2).

As far as the NOE is concerned, by far the most important contribution to $\mathbf{B}_{loc}$ is the random motion of other neighboring magnetic dipoles in the same molecule. Imagine two protons H_A and H_B in the same molecule as it tumbles in solution. If the relative positions of the two protons are rigidly defined by the molecular structure, then the magnitude of the magnetic field due to H_B at the site of H_A will be constant, but its *orientation with respect to* $\mathbf{B}_0$ will vary randomly. Thus, the total field seen by H_A will include instantaneous transverse components arising from H_B, and it is these that comprise the x and y components of $\mathbf{B}_{loc}$, and that cause longitudinal dipole–dipole relaxation. Obviously, the contribution that H_A makes to $\mathbf{B}_{loc}$ at the site of H_B can be described similarly. If the two dipoles are not rigidly joined by the molecular structure then further variations in $\mathbf{B}_{loc}$ may arise through internal motions, or, for intermolecular relaxation, through relative motion of different molecules.

Such dipolar interactions are the most important way in which the motions of the molecule produce the randomly varying field $\mathbf{B}_{loc}$, and they are the basis of dipole–dipole relaxation. The manner in which this contribution to $\mathbf{B}_{loc}$ acts to transfer energy to and from the lattice was outlined in the previous section.

To put this discussion in a somewhat different context, the magnitude of $\mathbf{B}_{loc}$ from dipole–dipole interactions must be of the same order as a dipolar coupling constant, observable in spectra from solids or ordered liquids. For two protons separated by 2 Å, this coupling constant is about 8.5 kHz, that is roughly 10^{-4} times $\mathbf{B}_0$ at 100 MHz,[2] and, like dipolar cross-relaxation, the dipolar coupling constant drops off rapidly with increasing dipole–dipole separation (actually as r^{-3}). Dipolar couplings do not, of course, give rise to splittings in spectra from isotropic liquids, precisely because they are averaged to zero by molecular tumbling (cf. Eq. 1.15).

It is usual to consider $\mathbf{B}_{loc}$ as a summation of contributions from each of the various mechanisms acting to relax a particular spin. Thus, following the argument of Section 1.3, for each mechanism to work, it must couple the nuclear spin to the lattice via its own contribution to $\mathbf{B}_{loc}$. Several relaxation mechanisms other than dipole–dipole exist, and these are discussed in Section 2.4.

---

†$\mathbf{B}_{total}$ also varies because sample spinning and diffusion move the nucleus to regions of different field strength. Such variations are much slower than those considered here, and are not relevant to longitudinal relaxation.

1.5. PULSES AND SATURATION

In this final section, we consider in more detail the effects produced by the irradiating field $\mathbf{B}_1$. In order to do this, we must deal with the concept of the rotating frame, a coordinate system that rotates at the frequency (ν_0) of the $\mathbf{B}_1$ field. More detailed and rigorous treatments of this concept are available in other textbooks,[3] but the explanation that follows is intended to be sufficient to allow the principal effects of the $\mathbf{B}_1$ field to be understood. For convenience, most of the points that need to be made about the $\mathbf{B}_1$ field and its role in saturation have been collected here, although this does necessitate a considerably more involved level of discussion than elsewhere in this chapter.

As described in many textbooks on NMR, transformation into the rotating frame can be considered to introduce a "fictitious field" $2\pi\nu_0/\gamma$, where ν_0 is the frequency of rotation of the frame (in Hz). This fictitious field opposes $\mathbf{B}_0$, so as to maintain the validity of the Larmor equation (Eq. 1.2) when precession of the nuclear spins is viewed from the rotating frame. If the frame rotates at the Larmor frequency, and precession appears to have stopped, the fictitious field is equal and opposite to $\mathbf{B}_0$. In general, the effective field along the z axis *in the rotating frame* is given by

$$|\mathbf{B}_z|_{\text{eff}} = |\mathbf{B}_0| - \frac{2\pi\nu_0}{\gamma} = 2\pi\left(\frac{\nu - \nu_0}{\gamma}\right) \qquad (1.16)$$

where ν is the Larmor precession frequency as viewed in the normal laboratory frame, and γ is the gyromagnetic ratio. Note that ν and $|\mathbf{B}_0|$ will be different for the different resonances due to their chemical shifts, whereas ν_0 is a constant.

The frequency of rotation of the rotating frame is set equal to the frequency of the applied RF irradiation field ($\mathbf{B}_1$). The magnetic oscillation of this field can be resolved into two circularly counter-rotating components, one of which rotates in the same sense as the precessing nuclear spins. In the rotating frame, this correctly rotating component of the $\mathbf{B}_1$ field appears in the transverse plane as a static vector, whose magnitude depicts the *intensity* of the $\mathbf{B}_1$ field. The $\mathbf{B}_1$ field may, using simple trigonometry, be incorporated into the formula for the effective field, as seen in the rotating frame by a resonance whose Larmor frequency is ν (Eq. 1.17). It is this effective field about which the resonance precesses when the $\mathbf{B}_1$ field is present.

$$|\mathbf{B}|_{\text{eff}} = \left[\frac{4\pi^2}{\gamma^2}(\nu - \nu_0)^2 + |\mathbf{B}_1|^2\right]^{1/2} \qquad (1.17)$$

This formula tells us most of what we need to know about the $\mathbf{B}_1$ field. Suppose the $\mathbf{B}_1$ field to be applied along the x axis of the rotating frame. Spins that are exactly on resonance (i.e., precessing at ν_0) will see only the $\mathbf{B}_1$ field in the rotating frame, so that the vectors representing their (macroscopic) mag-

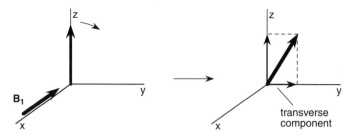

Figure 1.4. Application of an on-resonance $\mathbf{B}_1$ field causes the macroscopic magnetization vector $\mathbf{M}$ to precess about the $\mathbf{B}_1$ axis, thus creating a transverse component of magnetization.

netizations will appear to precess about the $\mathbf{B}_1$ axis, in the yz plane (Fig. 1.4). Spins that are off resonance will appear to precess around an effective field that is at an angle θ to the $\mathbf{B}_0$ axis (Fig. 1.5), where θ is given by

$$\tan \theta = \frac{(\gamma |\mathbf{B}_1|)}{2\pi(\nu - \nu_0)} \tag{1.18}$$

When their resonance offset $(\nu - \nu_0)$ is small relative to $\gamma B_1/2\pi$, this still corresponds to a substantial degree of precession about the x axis. For off-resonant spins whose resonance offset $(\nu - \nu_0)$ is large relative to $\gamma B_1/2\pi$, in contrast, the $\mathbf{B}_1$ field has little or no effect, and the spins process almost unperturbed about the z axis, even when viewed in the rotating frame.

It is worth dealing with the effect of off-resonant irradiation in slightly more detail, since this is of crucial importance when applying selective pulses or

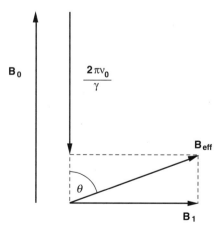

Figure 1.5. The effective field $\mathbf{B}_{\text{eff}}$. Spins that are off-resonance precess about $\mathbf{B}_{\text{eff}}$ rather than $\mathbf{B}_1$, so that they move over the surface of a cone with $\mathbf{B}_{\text{eff}}$ as its axis.

irradiation to a crowded spectrum. For a single-spin system (i.e., a spectrum comprising just one line with a resonance frequency of ν) the behavior of the corresponding magnetization vector **M** can be predicted explicitly by solving the Bloch equations for its precession around the effective field $\mathbf{B}_{\text{eff}}$. This amounts simply to finding out where **M** ends up after precessing about $\mathbf{B}_{\text{eff}}$ for a known time at a known rate (ignoring relaxation during the irradiation). If this calculation is carried out for a series of many different offset frequencies ($\nu - \nu_0$), then the results can be plotted to show the "response function" of the irradiation. For example, Figure 1.6a is a plot of M_z (calculated) against resonance offset ($\nu - \nu_0$), and it shows the extent of inversion that would be produced for a line at different resonance offsets by a pulse whose flip angle is 180° on resonance ($\nu = \nu_0$). Figure 1.6b shows the extent of the corresponding transverse magnetization produced at the same time; clearly this is zero on resonance, but not elsewhere.

We may understand the form of the curves in Figure 1.6 quite easily by taking a few specific cases. For an x pulse exactly on-resonance, the magnetization rotates in the yz plane, but for off-resonant pulses it rotates on the surface of a cone with $\mathbf{B}_{\text{eff}}$ as its axis and an internal angle (i.e., the angle between the side and axis of the cone) of θ. Here, θ is given by Eq. 1.18, and decreases with increasing resonance offset. For small offsets, the cone surface does not deviate much from the yz plane, and nearly complete inversion results. As θ decreases, the corresponding cone over which **M** moves contracts, so that the pulse creates a decreasing degree of inversion and an increasing proportion of transverse magnetization.

This explains the form of the curves in the central region of the plot, where the response function simply drops off with increasing offset. For larger offsets, the precession cone contracts sufficiently that **M** can complete more than one rotation over the cone surface during the same time that an on-resonant signal would complete half a revolution in the yz plane. This is the reason why the response function has "side lobes." Each null between lobes corresponds to a value of θ for which precession during the time of the pulse brings **M** precisely back to the z axis, following a whole number of revolutions on the cone centered on $\mathbf{B}_{\text{eff}}$. Note that, in practice, the transverse magnetization produced off resonance by a selective 180° pulse is rarely detected. This is because it does not add coherently from one scan to the next unless the selective pulse is phase coherent with the receiver, and so will average to zero during normal signal averaging. If the selective pulse is coherent with the receiver, then the transverse response can be eliminated by phase cycling or application of appropriate pulsed field gradients.

The normal RF observe pulse generally has a very high power, often in the region of 100 W. The time required for precession through 90° (i.e., the 90° pulse length) is then typically of the order of 10 μs, corresponding to a value of $\gamma B_1/2\pi$ given by $(4 \times 10\ \mu s)^{-1} = 25$ kHz. This provides enough power to excite the entire proton spectral width uniformly, even for very high field spectrometers. Broadband or composite pulse decoupling, being continuously ap-

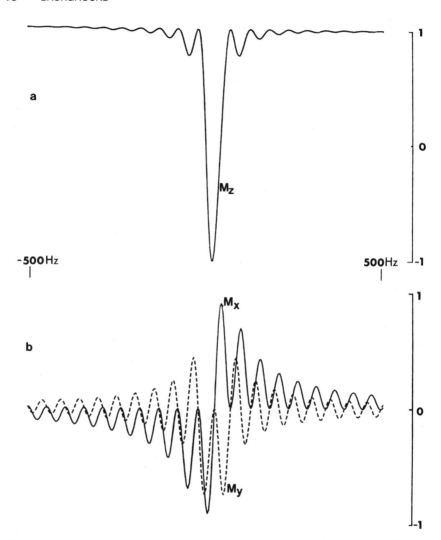

Figure 1.6. Plots showing the magnetization produced by a pulse that has a flip angle of 180° on resonance. Plot (*a*) shows the longitudinal component M_z as a function of resonance offset, while plot (*b*) shows the corresponding transverse components M_x (solid line) and M_y (dashed line). Calculated for an x pulse with $\gamma B_1/2\pi = 25$ Hz.

plied rather than pulsed, cannot use such high powers, and fields of $\gamma B_1/2\pi$ equal to a few kilohertz (typically corresponding to 1–5 W) are usual for proton decoupling during ^{13}C acquisition.

So much for high-power RF irradiation. The situation for *selective* irradiation, although governed by the same rules, is quite different. When $\gamma B_1/2\pi$ is reduced to the order of 10 Hz or so (usually corresponding to powers of the order of a milliwatt or less), Eqs. 1.16 and 1.17 predict that all resonances outside a very narrow window of a few tens of Hertz centered on ν_0 will be completely off resonance, that is, they will be unaffected by the irradiating field. The essence of selective irradiation is, thus, to use a very low power irradiation field. The on-resonance spins will also behave somewhat differently. Far from flashing around at 25 kHz, they now precess at a leisurely 10 Hz or so, and under these circumstances relaxation during the pulse becomes very important (indeed, it is the basis of the steady-state NOE experiment!). Also, the irradiation may be left on long enough that several, or even many, complete revolutions occur; in contrast, a high-power observe RF pulse is left on only long enough to cause a fraction (one quarter, for a 90° pulse) of one rotation.

The way in which magnetization behaves during selective irradiation is essentially identical to the way it behaves during free precession, except that the precession caused by the irradiation is very much slower, and is in a longitudinal plane of the rotating frame, rather than in the transverse plane as during the evolution of the FID. Just as free precession during the FID is damped by transverse relaxation and the $\mathbf{B}_0$ field inhomogeneity, so precession caused by irradiation is damped by relaxation and the $\mathbf{B}_1$ field inhomogeneity. Thus, the magnitude of the vector representing the resonance being irradiated *shrinks* as it rotates about the axis of rotation.

The characteristic time for this shrinking process is not a quantity that receives much attention in NMR textbooks (or indeed elsewhere), but it has been discussed by Torrey,[4] and by Farrar and Becker.[3] By analogy with other, better known, relaxation times, it is called $T_{2\rho}^*$. Just like T_2^*, it is composed of a relaxation part ($T_{2\rho}$) and an inhomogeneity part. The relaxation part is given by

$$\frac{1}{T_{2\rho}} = \frac{1}{2}\left(\frac{1}{T_1} + \frac{1}{T_2}\right) \tag{1.19}$$

This equation reflects the fact that the precessing magnetization spends half its time longitudinally and half its time transversely. By analogy with the definition of T_2^* (which can be found in many NMR textbooks), $T_{2\rho}^*$ can be said to be given by

$$\frac{1}{T_{2\rho}^*} = \frac{1}{T_{2\rho}} + \left(\frac{\gamma \Delta B_1}{2\pi}\right) \tag{1.20}$$

where ΔB_1 represents, somewhat loosely, the variation of the $\mathbf{B}_1$ field across the detected region of the sample. Of course, there is no reason to suppose that the decay caused by $\mathbf{B}_1$ inhomogeneity is exponential, so Eq. 1.20 represents a considerable oversimplification. It is, in fact, subject to just the same limitations of meaning as is the definition of T_2^*.

Thus, after a time that is long relative to $T_{2\rho}^*$, the net observable magnetization will have been destroyed, but it will take a longer time, long relative to $T_{2\rho}$, for true population equality across the transitions to have been reached.

Perhaps contrary to what one might at first expect, experimentally it is only the former condition, destruction of the observable magnetization, that is usually relevant, *and it is this that constitutes the definition of saturation* (at least within this book).

To understand why this should be so, it is worthwhile briefly to consider the situation as it exists after the observable magnetization has decayed to zero, but before population equality has been reached. Just as when describing the difference between T_2 and T_2^*, it is convenient to consider the sample as divided into very small regions called *isochromats*, each large enough to contribute a macroscopic magnetization vector, but small enough that the $\mathbf{B}_1$ inhomogeneity across its own volume is negligible. As shown in Figure 1.7a, after a time long with respect to $T_{2\rho}^*$, but not to $T_{2\rho}$, the individual magnetization vectors representing each isochromat are evenly distributed around a disk in the yz plane. The orientation of each vector depends on the particular local value of the $\mathbf{B}_1$ field at the corresponding point in the sample. By contrast, after a time that is long with respect to both $T_{2\rho}^*$ and $T_{2\rho}$, no net magnetization is present even within an individual isochromat (Fig. 1.7b).

What does all this mean as far as NOE enhancements are concerned? As will be shown in Section 3.3.1, enhancements observed following incomplete saturation to an extent θ (defined in Eq. 3.47) are simply scaled down by the factor θ relative to the enhancements observed for full saturation. Similarly, for partial *inversion*, a correspondingly *increased* enhancement is expected (for full inversion, the transient enhancement achieved is twice as great as that achieved by instantaneous saturation; cf. Section 4.4.1). Clearly, then, the reduced enhancements arising from those isochromats having their (S spin) magnetization z components along the $+z$ axis will be exactly counterbalanced by the increased enhancements arising from those isochromats having their (S spin) magnetization z components along the $-z$ axis (conventionally, S denotes

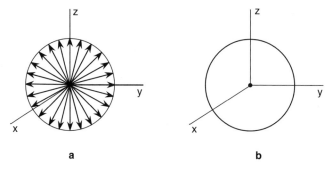

Figure 1.7. Magnetization remaining after a time that is long relative to (*a*) $T_{2\rho}^*$, and (*b*) $T_{2\rho}$. In (*a*), some magnetization remains but it has been dephased by the $\mathbf{B}_1$ inhomogeneity, so that contributions from individual isochromats cancel one another leaving no detected signal. In (*b*), irradiation has been sufficiently long that no magnetization remains even within individual isochromats.

the irradiated spin and I the observed; cf. Section 2.1). As far as we are aware, the detected result of any normal NOE experiment is unaffected by whether the sample is in a condition corresponding to Figure 1.7a or to Figure 1.7b.

If the $\mathbf{B}_1$ field is so weak that the precession rate $\gamma B_1/2\pi$ is slow with respect to $(T_{2\rho}^*)^{-1}$, then precession is, in effect, damped before it can begin, and a nonoscillatory decay of the original longitudinal magnetization takes place. In practice, the detailed behavior of the system under these conditions is likely to be a rather complicated function of the detailed nature of the $\mathbf{B}_1$ inhomogeneity. The same is true in the intermediate region, $\gamma B_1/2\pi \approx (T_{2\rho}^*)^{-1}$. These situations are of less practical interest, however, since such weak fields are usually only employed in steady-state NOE experiments, where, by definition, the path taken to reach steady state is essentially irrelevant (provided, of course, that steady state is genuinely reached).

A more important consideration for steady-state experiments is that, for still weaker $\mathbf{B}_1$ fields, a balance between the effect of irradiation and the effect of relaxation is reached, and a steady state of *partial saturation* is established. The consequences of this are discussed in Sections 2.6 and 3.3.1.

Simulations of these various forms of saturation behavior are shown in Figure 1.8.[5] The actual $\mathbf{B}_1$ field strengths at which each type of behavior occurs vary widely from case to case. The $\mathbf{B}_0$ inhomogeneity will also play a role; for one thing, irradiation at the center of an inhomogeneity-broadened resonance may not adequately saturate broad wings or humps in the lineshape. Technical details of the ways in which saturation is achieved when irradiating multiplets are to be found in Section 7.2.5.

One might suppose from all this that kinetic measurements of the NOE (during TOE experiments; cf. Section 7.4.2) are meaningful only if they are

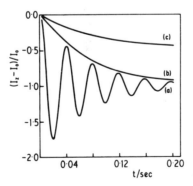

Figure 1.8. Simulations of the behavior of the magnetization due to an irradiated resonance for different irradiation power levels. The curves are calculated as $I_z = I_z^0 \theta^{-Rt} \cos \omega_1 t$, where $R = 18.9$ s^{-1}, and ω_1 is 160 rad s^{-1} (high power) in (*a*) and 8 rad s^{-1} (low power) in (*b*). In (*c*), the irradiation power is even lower (2 rad s^{-1}), and only partial saturation is achieved at steady state. See Section 7.4.2 for further discussion. Reproduced with permission from ref. 5.

made after irradiation that is long relative to T_{2p}^*. Fortunately, this is not so. When the magnetization due to the irradiated spin S is precessing under the influence of a $\mathbf{B}_1$ field, its *average z* component is zero. Thus, much as for the arguments concerning isochromats given earlier, the rapid buildup of enhancements that occurs while the S spin magnetization is close to the $-z$ axis (i.e., inverted) is balanced by the slow buildup of enhancements that occurs while the S spin magnetization is close to the $+z$ axis, and the *average* effect is equivalent to that of true saturation. Thus, one may consider saturation to be effectively instantaneous, but with the proviso that the oscillations in S_z must cause some oscillation to be superimposed on the buildup of the enhancements also. Provided the oscillations are rapid relative to the rate of NOE growth, however, the buildup of the NOE on other spins is only weakly perturbed. This topic is dealt with in greater depth in Section 7.4.2.

There is one further subtlety concerning the effect of RF irradiation during saturation that we have not considered here, or elsewhere in this book: it is possible for the RF field itself to perturb the Hamiltonian of the system, thereby changing the nature of the energy levels and potentially compromising our description of events in terms of spin-state populations. Given the very weak fields used for saturation in most NOE experiments this is not likely to cause appreciable errors, but neglect of this aspect of the theory could, for instance, lead to slight differences from the predictions for strongly coupled systems presented in Section 6.4. One way to deal with this issue would be to employ the so-called "homogeneous master-equation" formalism; interested readers are referred to references 6 and 7.

REFERENCES

1. Abragam, A. "The Principles of Nuclear Magnetism," Clarendon Press, Oxford, 1961, pp 264–268.
2. Harris, R. K. "Nuclear Magnetic Resonance Spectroscopy," Pitman, London, 1983, pp. 96ff.
3. For example: Farrar, T. C.; Becker, E.D. "Pulse and Fourier Transform NMR," Academic Press, New York, 1971.
4. Torrey, H. C. *Phys. Rev.* 1949, *76*, 1059.
5. Dobson, C. M.; Olejniczak, E. T.; Poulsen, F. M.; Ratcliffe, R. G. *J. Magn. Reson.* 1982, *48*, 97.
6. Levitt, M. H.; Di Bari, L. *Phys. Rev. Lett.* 1992, *69*, 3124.
7. Levitt, M. H.; Di Bari, L. *Bull. Magn. Reson.* 1994, *16*, 94.

CHAPTER 2

THE STEADY-STATE NOE FOR TWO SPINS

In this chapter we explain the origin of the NOE, first at a descriptive level and then in somewhat more mathematical terms, employing the classic approach described by Solomon.[1] Using the Solomon equations we examine how and why the behavior of the NOE is so intimately related to molecular motion. Finally, we examine the influence of relaxation mechanisms other than intramolecular dipole–dipole, and show how the predictions of the theory differ in heteronuclear cases.

As we shall see, in the absence of such external relaxation mechanisms, steady-state NOE enhancements in two-spin systems do not depend on molecular geometry at all. This chapter thus builds up a picture of the various *other* factors that are important in determining the sizes of enhancements. This leaves the way clear to examine in Chapter 3 the influence of molecular geometry on the sizes of steady-state enhancements in multispin systems.

2.1. THE ORIGIN AND FORM OF THE NOE

2.1.1. Qualitative Considerations

To summarize the key points of Chapter 1, we have already seen that:

1. The intensity of a resonance in an NMR spectrum is proportional to the difference in population between the two energy levels of the corresponding transition (Section 1.1).
2. The rate at which these populations return to equilibrium following a perturbation is determined by the corresponding transition probability, W (Section 1.2).

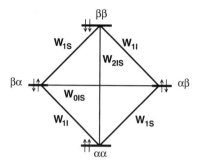

Figure 2.1. Energy level diagram for a two-spin system, showing definitions of transition probabilities and spin states. Spin states are written with the state of I first and that of S second, e.g., $\alpha\beta$ means spin I in state α, and spin S in state β.

3. For dipole–dipole relaxation, the value of W depends on (among other things) the strength of the local field component fluctuating at the frequency corresponding to the transition. This local field, $\mathbf{B}_{\text{loc}}^{\text{DD}}$, is the field at the site of one dipole (nucleus) due to the presence of the other dipole (Sections 1.3 and 1.4).
4. The frequency corresponding to a transition is, from the Bohr frequency condition ($\Delta E = h\nu$), proportional to the difference in energy between the energy levels (Section 1.1).

When we turn to a spin system comprising *two* spin 1/2 nuclei, we see that the existence of the second spin introduces two more energy levels to the energy level diagram, as shown in Figure 2.1. By convention, the two spins are usually labeled I and S, where, in an NOE experiment, I is the spin whose resonance is measured and S is the spin whose resonance is saturated.[†] We assume that these two spins are close enough together that their dipole–dipole interaction is appreciable, or in other words they are dipole–dipole coupled; however, they are not scalar coupled, that is, there is no spin–spin coupling ($J_{IS} = 0$). We further assume that both spins form part of a rigid molecule tumbling isotropically (that is, with no preferred axis of rotation), and that only dipole–dipole relaxation is appreciable.

The first thing to notice about the energy level diagram in Figure 2.1 is that there are transitions shown that involve the *simultaneous* flip of *both* spins. These are the transitions $\alpha\alpha \leftrightarrow \beta\beta$, with transition probability W_{2IS}, and

---

[†] The I and S nomenclature was first used by Solomon.[1] The reason for using I and S as symbols is not given, but it presumably derives from the conventional use of I as a symbol for the nuclear spin quantum number and S for the electron spin quantum number (recall that the original Overhauser effect[2] involved polarization of nuclei on saturation of electrons in a metal). For NOE experiments, a more evocative mnemonic is S for saturated, I for interesting.

$\alpha\beta \leftrightarrow \beta\alpha$, with transition probability W_{0IS}. These two-spin transitions, which obviously have no analogy in a single-spin system, are central to the theory of the NOE, because it is precisely these transitions that give rise to NOE enhancements, by allowing saturation of S to affect the intensity of I. Although they are "forbidden" transitions in the conventional sense that they cannot be directly excited by an RF pulse or cause directly detectable NMR signals, they are not forbidden in the context of relaxation. Another way to put this is to say that the selection rules that govern the interaction of the spin system with an external oscillating field (the RF field or the NMR signal) are different from the selection rules that apply to interactions of the spins with the lattice via energy exchange events (cf. Section 1.3).[3] The $\alpha\alpha \leftrightarrow \beta\beta$ transition is also called the double quantum, the double flip, or the W_2 transition, whereas the $\alpha\beta \leftrightarrow \beta\alpha$ transition is also called the zero quantum, flip-flop, or W_0 transition; collectively, both are known as cross-relaxation pathways.

Deferring for the moment discussion of the *values* of W_{2IS} and W_{0IS}, we now address the key question of how it is that these cross-relaxation pathways can cause NOE enhancements. The NOE enhancement $f_I\{S\}$ is defined as the fractional change in the intensity of I on saturating S:

$$f_I\{S\} = \frac{(I - I^0)}{I^0} \tag{2.1}$$

where I^0 is the equilibrium intensity of I. From Figure 2.1 we can see that the intensity of I is proportional to the sum of the population differences ($N_{\alpha\alpha} - N_{\beta\alpha}$) and ($N_{\alpha\beta} - N_{\beta\beta}$), while that of S is proportional to ($N_{\alpha\alpha} - N_{\alpha\beta}$) + ($N_{\beta\alpha} - N_{\beta\beta}$). At thermal equilibrium, for a homonuclear system, level $\alpha\alpha$ is the most populated, followed by $\alpha\beta$ and $\beta\alpha$ with essentially identical populations to one another, followed in turn by $\beta\beta$. The same population difference exists across all four single quantum transitions ($\alpha\alpha \leftrightarrow \beta\alpha$, $\alpha\alpha \leftrightarrow \alpha\beta$, $\alpha\beta \leftrightarrow \beta\beta$, and $\beta\alpha \leftrightarrow \beta\beta$), and so the equilibrium intensities I^0 and S^0 are identical (Fig. 2.2a).

When the S resonance is saturated, this *equalizes* the populations of levels $\alpha\alpha$ and $\alpha\beta$, and similarly those of levels $\beta\alpha$ and $\beta\beta$, thereby increasing the populations of levels $\alpha\beta$ and $\beta\beta$ while decreasing the populations of levels $\alpha\alpha$ and $\beta\alpha$ (Fig. 2.2b). There is no immediate change in the intensity of the I resonance, as the population differences ($N_{\alpha\alpha} - N_{\beta\alpha}$) and ($N_{\alpha\beta} - N_{\beta\beta}$) are, at this stage, unchanged.

It is clear that the W_{1I} and W_{1S} transitions can only produce *independent* spin–lattice relaxation of I and S, respectively. However, if the transition described by the probability W_{2IS} occurs, it will act to restore the populations of the $\alpha\alpha$ and $\beta\beta$ levels toward their thermal equilibrium values. Saturation of S has caused the population of level $\beta\beta$ to increase, and that of $\alpha\alpha$ to decrease; W_2 relaxation will therefore act to reverse this, leading to a decrease in the population of the $\beta\beta$ level and an increase in the population of the $\alpha\alpha$ level. This necessarily results in an increase in the population differences ($N_{\alpha\alpha} - N_{\beta\alpha}$) and ($N_{\alpha\beta} - N_{\beta\beta}$), in other words an increase in the intensity of the I resonance.

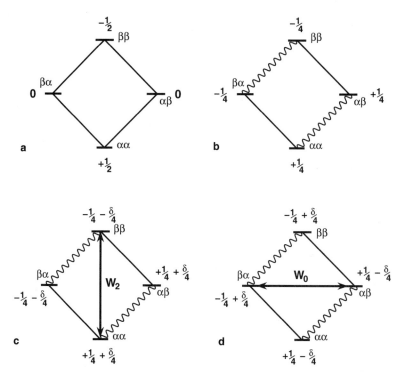

Figure 2.2. The origin of the NOE in a homonuclear two-spin system. The intensity I^0 ($=S^0$) is here represented by a total population difference of one unit, and, for convenience, the populations $N_{\alpha\beta}$ and $N_{\beta\alpha}$ are set to zero. (*a*) represents the situation at equilibrium. (*b*) represents the situation after saturation of the *S* resonance. Populations $N_{\alpha\alpha}$ and $N_{\alpha\beta}$ are equalized, as are $N_{\beta\alpha}$ and $N_{\beta\beta}$; the intensity of *I* is still one unit, but that of *S* is now zero. (*c*) shows the effect of W_2 relaxation during saturation of *S*. Population ($\delta/2$ units) has been transferred from state $\beta\beta$ to state $\alpha\alpha$, while saturation of *S* maintains the population equality across the *S* transitions. This has the effect of increasing the total intensity of *I* to $(1 + \delta)$ units, i.e., a positive NOE enhancement. (*d*) shows the effect of W_0 relaxation during saturation of *S*. Population ($\delta/2$ units) has been transferred from state $\alpha\beta$ to $\beta\alpha$, so changing their populations in the direction of equilibrium. This has the effect of decreasing the total *I* intensity to $(1 - \delta)$ units, i.e., a negative NOE enhancement.

Thus, if the W_2 transition occurs, it produces a positive NOE enhancement of the *I* signal. By an entirely analogous logic, it should be clear that "flip-flop" relaxation across the W_0 transition leads to a *negative* NOE enhancement (that is, a decrease in the intensity of the *I* signal on saturating *S*). When combined with the requirement that saturation of *S* maintains population equality across the *S* transitions, these arguments lead to the population distributions shown schematically in Figure 2.2c and d.

It is helpful to keep in mind the frequencies[†] corresponding to each of these transitions in a homonuclear system. The $\alpha\alpha \leftrightarrow \beta\alpha$ and $\alpha\beta \leftrightarrow \beta\beta$ transitions are at a frequency of ω_I; to give a concrete example, for protons in a field of 9.4 T, this is roughly 400 MHz, or $8\pi \times 10^8$ rad s^{-1}. Likewise, the $\alpha\alpha \leftrightarrow \alpha\beta$ and $\beta\alpha \leftrightarrow \beta\beta$ transitions have a frequency of ω_S, which is very close to ω_I. The $\alpha\beta \leftrightarrow \beta\alpha$ (zero quantum) transition corresponds to a frequency given by the difference $|(\omega_I - \omega_S)|$. This is simply the difference in chemical shift between the two signals, and so can be anywhere from zero up to a maximum of a few kilohertz. It is therefore a very much lower frequency than either ω_I or ω_S. The $\alpha\alpha \leftrightarrow \beta\beta$ (double quantum) transition, on the other hand, corresponds to a frequency of $(\omega_I + \omega_S)$, or roughly 800 MHz in the present example. Of course, in a *heteronuclear* system, ω_I and ω_S are very much more different from each other, and the frequency of the W_0 transition is not necessarily lower than those of the other transitions.

These frequencies are very important because, as implied earlier, one factor that determines the relative contribution that each type of transition makes to the total relaxation pattern is the strength of the local field component fluctuating at the corresponding frequency. However, before we can deal with the calculations of these contributions, and hence the actual numerical sizes of NOE enhancements, we must first put the qualitative arguments of this section on a slightly more mathematical footing.

2.1.2. The Solomon Equations

The observed intensities of I and S are proportional to I_z and S_z, the macroscopic longitudinal magnetizations due to I and S, respectively, immediately prior to application of the observe pulse. These vectors I_z and S_z are themselves proportional to the population differences between the states:

$$kI_z = N_{\alpha\alpha} - N_{\beta\alpha} + N_{\alpha\beta} - N_{\beta\beta} \tag{2.2a}$$

and

$$kS_z = N_{\alpha\alpha} + N_{\beta\alpha} - N_{\alpha\beta} - N_{\beta\beta} \tag{2.2b}$$

We need an expression for the rate of change of I_z with time. Clearly, this is given by

$$k\frac{dI_z}{dt} = \frac{dN_{\alpha\alpha}}{dt} - \frac{dN_{\beta\alpha}}{dt} + \frac{dN_{\alpha\beta}}{dt} - \frac{dN_{\beta\beta}}{dt} \tag{2.3}$$

Consideration of the ways in which population can arrive at and leave from state $\alpha\alpha$ (see Fig. 2.1) gives

---

[†] Note that we use the term "frequency of a transition" to mean the frequency separation of the energy levels, rather than the rate of spin flips across the transition.

$$\frac{dN_{\alpha\alpha}}{dt} = -(W_{1I} + W_{1S} + W_{2IS})N_{\alpha\alpha} + W_{1I}N_{\beta\alpha} + W_{1S}N_{\alpha\beta} + W_{2IS}N_{\beta\beta} + \text{constant} \quad (2.4)$$

As in Section 1.2, we may evaluate the constant by imposing the boundary condition that at thermal equilibrium $dN_{\alpha\alpha}/dt = 0$. Writing the equilibrium population of each state i as N_i^0, we have

$$(W_{1I} + W_{1S} + W_{2IS})N_{\alpha\alpha}^0 - W_{1I}N_{\beta\alpha}^0 - W_{1S}N_{\alpha\beta}^0 - W_{2IS}N_{\beta\beta}^0 = \text{constant} \quad (2.5)$$

Writing the population *differences from equilibrium* $(N_i - N_i^0)$ as n_i, we therefore have

$$\frac{dN_{\alpha\alpha}}{dt} = -(W_{1I} + W_{1S} + W_{2IS})n_{\alpha\alpha} + W_{1I}n_{\beta\alpha} + W_{1S}n_{\alpha\beta} + W_{2IS}n_{\beta\beta} \quad (2.6)$$

Similar expressions may easily be found for $dN_{\beta\alpha}/dt$, $dN_{\alpha\beta}/dt$, and $dN_{\beta\beta}/dt$, and these may be substituted into Eq. 2.3 to give

$$\begin{aligned} k\frac{dI_z}{dt} = &-(W_{1I} + W_{1S} + W_{2IS})n_{\alpha\alpha} + W_{1I}n_{\beta\alpha} + W_{1S}n_{\alpha\beta} + W_{2IS}n_{\beta\beta} \\ &+ (W_{1I} + W_{1S} + W_{0IS})n_{\beta\alpha} - W_{1I}n_{\alpha\alpha} - W_{1S}n_{\beta\beta} - W_{0IS}n_{\alpha\beta} \\ &- (W_{1I} + W_{1S} + W_{0IS})n_{\alpha\beta} + W_{1I}n_{\beta\beta} + W_{1S}n_{\alpha\alpha} + W_{0IS}n_{\beta\alpha} \\ &+ (W_{1I} + W_{1S} + W_{2IS})n_{\beta\beta} - W_{1I}n_{\alpha\beta} - W_{1S}n_{\beta\alpha} - W_{2IS}n_{\alpha\alpha} \quad (2.7) \end{aligned}$$

Eliminating the redundancies and grouping terms we get

$$k\frac{dI_z}{dt} = 2(n_{\beta\beta} - n_{\alpha\alpha})(W_{1I} + W_{2IS}) + 2(n_{\beta\alpha} - n_{\alpha\beta})(W_{1I} + W_{0IS}) \quad (2.8)$$

Combining Eqs. 2.2a and 2.2b gives

$$2(N_{\beta\beta} - N_{\alpha\alpha}) = -k(I_z + S_z)$$

and

$$2(N_{\beta\alpha} - N_{\alpha\beta}) = -k(I_z - S_z)$$

from which

$$2(n_{\beta\beta} - n_{\alpha\alpha}) = -k(I_z - I_z^0 + S_z - S_z^0)$$

and

$$2(n_{\beta\alpha} - n_{\alpha\beta}) = -k(I_z - I_z^0 - S_z + S_z^0) \quad (2.9)$$

Substituting Eqs. 2.9 into Eq. 2.8 gives

$$\frac{dI_z}{dt} = -(I_z - I_z^0 + S_z - S_z^0)(W_{1I} + W_{2IS}) - (I_z - I_z^0 - S_z + S_z^0)(W_{1I} + W_{0IS})$$

so that, finally, we get

$$\frac{dI_z}{dt} = -(I_z - I_z^0)(W_{0IS} + 2W_{1I} + W_{2IS}) - (S_z - S_z^0)(W_{2IS} - W_{0IS}) \quad (2.10)$$

This equation, often called the Solomon equation, may be considered to be the heart of NOE theory (hence our rather painstaking derivation of it). From it we may very simply derive an expression for the steady-state NOE enhancement at I on saturation of S (defined in Eq. 2.1). At steady state, $dI_z/dt = 0$ and $S_z = 0$, giving

$$0 = -(I_z - I_z^0)(W_{0IS} + 2W_{1I} + W_{2IS}) + S_z^0(W_{2IS} - W_{0IS})$$

Thus,

$$\frac{I_z - I_z^0}{S_z^0} = \frac{W_{2IS} - W_{0IS}}{W_{0IS} + 2W_{1I} + W_{2IS}} \quad (2.11)$$

Since $S_z^0 = (\gamma_S/\gamma_I)I_z^0$, we have

$$f_I\{S\} = \frac{I_z - I_z^0}{I_z^0} = \frac{\gamma_S}{\gamma_I} \frac{W_{2IS} - W_{0IS}}{W_{0IS} + 2W_{1I} + W_{2IS}} \quad (2.12)$$

The quantity $W_{2IS} - W_{0IS}$ has the same form as predicted from our qualitative arguments concerning NOE enhancements in Section 2.1.1: that is, W_{2IS} causes a positive NOE enhancement, and W_{0IS} a negative one, while W_1 transitions do not contribute. The term $(W_{2IS} - W_{0IS})$ describes the *rate of dipole–dipole transitions giving rise to an NOE enhancement*. It is often called the *cross-relaxation* rate constant, and given the symbol σ_{IS}.[1,4,5] The term $(W_{0IS} + 2W_{1I} + W_{2IS})$ is the *dipolar longitudinal relaxation* rate constant of spin I, and is given the symbol ρ_{IS}. Both these terms are discussed further in Section 2.3.1. Using these symbols, Eq. 2.12 is written as

$$f_I\{S\} = \frac{\gamma_S}{\gamma_I} \frac{\sigma_{IS}}{\rho_{IS}} \quad (2.13)$$

Before leaving the derivation of the Solomon equations, we should note that Eq. 2.8 could have been arrived at more directly. Although consideration of the population flux through every energy level is conceptually simple, it inevitably leads to rather long expressions, as is demonstrated by Eq. 2.7. When dealing with larger spin systems such an approach becomes extremely unwieldy. If we consider the definitions of transition probabilities rather more carefully, however, we may avoid the redundancies present in Eq. 2.7. Only transitions involving I can affect the value of I_z, and the populations affected by each relevant transition are known from the definitions of these transitions as summarized in the energy level diagram (Fig. 2.1). We may thus work through the energy levels and write, for each level, those terms that affect dI_z/dt:

$$k \frac{dI_z}{dt} = -2n_{\alpha\alpha}(W_{1I} + W_{2IS}) + 2n_{\beta\alpha}(W_{1I} + W_{0IS})$$
$$-2n_{\alpha\beta}(W_{1I} + W_{0IS}) + 2n_{\beta\beta}(W_{1I} + W_{2IS}) \qquad (2.14)$$

Here the factors of two arise (as before) because every transition alters the population *difference* between the affected energy levels by two, and the all-important signs are deduced from the spin state of I in the relevant population term. Those in which I is in the α state (namely $\alpha\alpha$ and $\alpha\beta$) must contribute negatively, since transitions from these states *deplete* those population differences that determine I_z; conversely those in which I is in the β state contribute positively. Equation 2.8 may now be obtained from Eq. 2.14 by a trivial rearrangement.

2.2. DEPENDENCE OF THE NOE ON MOLECULAR MOTION

In Section 1.3 we saw that a transition corresponding to a given frequency is promoted by molecular motion at the same frequency. Thus, for the example given earlier, W_0 transitions will be at their fastest when the molecule tumbles at a rate of about 1 kHz, W_1 transitions will be fastest at a tumbling rate of about 400 MHz, and W_2 transitions at a tumbling rate of about 800 MHz. On this basis, we can form a very rough idea of how NOE enhancements will change with tumbling rate. Small molecules in nonviscous solvents tumble at rates around 10^{10} Hz, while larger molecules such as proteins tumble at rates around 10^7 Hz. It therefore seems reasonable to expect that for small molecules, W_2 will be greater than W_0, and that NOE enhancements will be positive. For larger molecules, at some stage W_0 will become greater than W_2 and NOE enhancements will accordingly become negative. To determine more exactly how the NOE varies with molecular tumbling, we need to analyze how the energy available for causing relaxation transitions depends on the frequency of molecular tumbling. This dependence is described using *correlation* and *spectral density* functions.

2.2.1. Correlation Times, Spectral Density Functions, and Transition Probabilities

A correlation function simply assesses the correlation between a parameter measured at time t and the same parameter measured at time $(t + \tau)$. Mathematically the correlation function is written as

$$g(\tau) = \overline{f(t)f(t + \tau)} \tag{2.15}$$

where the bar indicates an ensemble average. Some correlation functions are shown in Figure 2.3a. The first function changes very slowly with time; in the second case there is an intermediate rate of change, and in the third case there is a rapid time dependence. By its very nature a correlation function must start at some fixed positive value (by convention unity) and decay to zero. There is no rule about the form this decay must take, but in the case of dipolar relaxation it is normally assumed to be exponential. With this assumption,

$$g(\tau) = \exp(-\tau/\tau_c) \tag{2.16}$$

where τ_c, the characteristic decay time of the correlation function, is known as the *correlation time*. Here we are concerned with isotropic molecular tumbling, and τ_c is therefore equated with the time taken for the molecule to rotate by roughly 1 radian about any axis. τ_c can be approximated to the rotational relaxation time given by Debye[6]:

$$\tau_c = 4\pi\eta a^3/3kT \tag{2.17}$$

where η is the viscosity of the solvent and a is the radius of the molecule. Using this approximation, and taking typical values of viscosity for organic solvents, one can write a *very approximate* estimate of τ_c as

$$\tau_c \simeq 10^{-12} \, W_M \tag{2.18}$$

where W_M is the molecular mass, in Daltons. For globular proteins in water, the correlation time is closer to $0.4 \times 10^{-12} \, W_M$.

τ_c can be measured by a number of methods. It is often measured using the NOE maximum itself (see below), or from other NMR relaxation parameters, for example the ^{13}C spin–lattice relaxation time, which is given by

$$\frac{1}{T_1(^{13}C)} = \left(\frac{\mu_0}{4\pi}\right)^2 \frac{N\gamma_H^2\gamma_C^2\hbar^2\tau_c}{r_{CH}^6} \tag{2.19}$$

where N is the number of attached hydrogens. Other physical methods are also available, such as fluorescence quenching.[7]

The correlation function $g(\tau)$ has a fairly simple and intelligible meaning. What we require, however, is a function not of time but of frequency. To

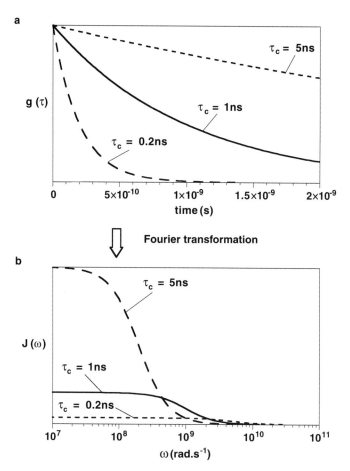

Figure 2.3. Exponentially decaying correlation functions $g(\tau)$ (panel *a*) are converted into Lorentzian spectral density functions $J(\omega)$ (panel *b*) by Fourier transformation. (Note the log scale in panel *b*.) The functions $g(\tau)$ have decay constants of $1/\tau_c$. τ_c values of 0.2 ns, 1 ns and 5 ns have been used to illustrate the behavior expected for short, intermediate, and long values of τ_c respectively. The area under the curve of $J(\omega)$ is the same in all these cases (although it appears otherwise because of the log scale for frequency).

convert to this, we must use Fourier transformation; the result for the correlation functions in Figure 2.3a is shown in Figure 2.3b. These functions are known as spectral density functions, written $J(\omega)$. There is quite a useful analogy here with the Fourier transformation of a single-line FID. For an FID that consists only of a simple exponential decay, as does $g(\tau)$, the corresponding spectrum consists of a Lorentzian line in absorption phase, centered on $\omega = 0$. The difference here, of course, is that because $g(\tau)$ decays with τ_c in a time of the order of 10^{-7}–10^{-10} s (rather than the 0.1–10 s typical for relaxation and inhomogeneity damping of an FID), the corresponding "linewidth" of the spectral density function $J(\omega)$ is of the order of hundreds of megahertz. Notice also that the (schematic) plots of spectral density functions in Figure 2.3b do not *look* much like absorption phase NMR lines, because a logarithmic scale has been used for the frequency axis. Using this analogy, it is clear that the functional form of $J(\omega)$ is a Lorentzian function[†]:

$$J(\omega) = 2\tau_c/(1 + \omega^2\tau_c^2) \qquad (2.20)$$

There is an important property of $J(\omega)$ that is most easily understood with this analogy in mind. The total integrated area of an NMR spectrum in absorption phase is given by the magnitude of the first point of the free induction decay. A similar principle holds true here: since all the functions $g(\tau)$ start at the same value (unity), all the functions $J(\omega)$ must have the same integrated area. The other significant feature of the $J(\omega)$ curves, visible in Figure 2.3b, is that they are very flat for a while, then drop off rapidly near a value of $\omega = 1/\tau_c$ (equivalent to the linewidth at half height of a Lorentzian line), and thereafter are close to zero.

What the spectral density function $J(\omega)$ tells us is the *power available from the lattice* (i.e., from molecular motions) to bring about relaxation, as a function of molecular tumbling rate. What it does not tell us is how efficiently a given relaxation mechanism can *use* this power. Clearly, this must be a property of the particular interaction responsible for the relaxation, in this case the dipole–dipole interaction. Thus, in the general expression for the transition probability W, there is both a spectral density function part and an interaction energy part, the purpose of the latter being to determine how effective the relevant interaction is at causing transitions. This was expressed in a general sense in Eq. 1.3, and is developed in greater depth in Appendix II.

From the form of the spectral density function $J(\omega)$, we can work out how the transition probability W depends on the tumbling rate τ_c for a transition with a particular frequency ω_n (see Fig. 2.4, where the frequency ω_n is chosen to be 10^9 rad s^{-1}). For long values of τ_c, the power available from the lattice is concentrated in the low-frequency motions, so $J(\omega)$ drops off before ω_n is reached (Fig. 2.4a, long-dashed line). For short τ_c, the power available from the lattice is spread more thinly across a wider range of frequencies, which again has the consequence that the value of $J(\omega)$ is low at ω_n (Fig. 2.4a, short-dashed line). $J(\omega)$ is maximal when $\tau_c \approx 1/\omega_n$ (Fig. 2.4a, solid line), so that a plot of W against τ_c has the form shown in Figure 2.4b, which is again a Lorentzian function, this time centered on ω_n. This is more commonly shown in the form of a log–log plot (Fig. 2.4c).

[†]Some authors omit the factor of 2 from the definition of $J(\omega)$. This obviously affects the appearance of equations involving $J(\omega)$, which may then require inclusion of the factor of 2 elsewhere to compensate.

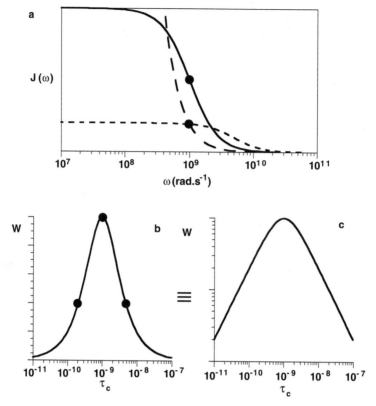

Figure 2.4. Dependence of the transition probability W on τ_c, derived using values of $J(\omega)$ at different values of τ_c. Panel a is a vertical expansion of Figure 2.3b, and shows the behavior of $J(\omega)$ for $\tau_c > 1/\omega_n$ ($\tau_c = 5$ ns; long-dashed line), $\tau_c = 1/\omega_n$ ($\tau_c = 1$ ns; solid line), and $\tau_c < 1/\omega_n$ ($\tau_c = 0.2$ ns; short-dashed line), where ω_n is the frequency of the transition for which W is to be calculated (here 10^9 rad s^{-1}). The value of W as a function of τ_c is plotted in (b), and converted to a log/log scale in (c).

Even for a first-order, isotropically tumbling, rigid system, the detailed calculations of W_{0IS}, W_{1I}, W_{1S}, and W_{2IS} from the dipolar interaction Hamiltonian are quite involved, being a function of the orientations and separation of the two dipoles (Appendix II). Nonetheless, the results are quite simple:

$$W_{0IS} = \frac{1}{20} K^2 J(\omega_I - \omega_S) = \frac{1}{10} K^2 \frac{\tau_c}{1 + (\omega_I - \omega_S)^2 \tau_c^2} \qquad (2.21a)$$

$$W_{1I} = \frac{3}{40} K^2 J(\omega_I) = \frac{3}{20} K^2 \frac{\tau_c}{1 + \omega_I^2 \tau_c^2} \qquad (2.21b)$$

DEPENDENCE OF THE NOE ON MOLECULAR MOTION

$$W_{1S} = \frac{3}{40} K^2 J(\omega_S) = \frac{3}{20} K^2 \frac{\tau_c}{1 + \omega_S^2 \tau_c^2} \qquad (2.21c)$$

$$W_{2IS} = \frac{3}{10} K^2 J(\omega_I + \omega_S) = \frac{3}{5} K^2 \frac{\tau_c}{1 + (\omega_I + \omega_S)^2 \tau_c^2} \qquad (2.21d)$$

where $K = (\mu_0/4\pi)\hbar\gamma_I\gamma_S r_{IS}^{-3}$.

These functions are plotted in Figure 2.5, at two different field strengths.

Notice that, because τ_c is defined in terms of the average time taken for the molecule to reorient through one *radian*, it is necessary to express frequencies here in rad s^{-1}, rather than in hertz. This is why the symbol ω (rather than ν) is used in expressions involving $\omega\tau_c$; a frequency in hertz may be converted to rad s^{-1} simply by multiplying by 2π (that is, $\omega = 2\pi\nu$).

When τ_c is very short, $\omega\tau_c \ll 1$, and the expression for $J(\omega)$ simplifies to become equal to just $2\tau_c$. This limit is called the *extreme narrowing limit* (because lines are narrowed by motional averaging when $\omega\tau_c \ll 1$), and small molecules in nonviscous solvents generally have correlation times in this region. Note that in the extreme narrowing limit, the ratios $W_0 : W_1 : W_2$ become 1 : (3/2) : 6.

We may now use Eqs. 2.21 to write expressions for σ_{IS} and ρ_{IS} in terms of molecular tumbling rate:

$$\sigma_{IS} = W_{2IS} - W_{0IS}$$

$$= \frac{1}{10} K^2 \tau_c \left[\frac{6}{1 + (\omega_I + \omega_S)^2 \tau_c^2} - \frac{1}{1 + (\omega_I - \omega_S)^2 \tau_c^2} \right] \qquad (2.22)$$

and

$$\rho_{IS} = W_{0IS} + 2W_{1IS} + W_{2IS}$$

$$= \frac{1}{10} K^2 \tau_c \left[\frac{1}{1 + (\omega_I - \omega_S)^2 \tau_c^2} + \frac{3}{1 + \omega_I^2 \tau_c^2} + \frac{6}{1 + (\omega_I + \omega_S)^2 \tau_c^2} \right] \qquad (2.23)$$

with K as before. (In this two-spin system, W_{1I} and W_{1IS} are synonymous, cf. Section 2.3.1.)

Finally, the fractional NOE enhancement, $f_I\{S\}$, may now also be expressed in terms of molecular tumbling rate, simply by substituting Eqs. 2.22 and 2.23 back into Eq. 2.13:

$$f_I\{S\} = \left(\frac{\gamma_S}{\gamma_I}\right) \left[\frac{6}{1 + (\omega_I + \omega_S)^2 \tau_c^2} - \frac{1}{1 + (\omega_I - \omega_S)^2 \tau_c^2} \right]$$

$$\Big/ \left[\frac{1}{1 + (\omega_I - \omega_S)^2 \tau_c^2} + \frac{3}{1 + \omega_I^2 \tau_c^2} + \frac{6}{1 + (\omega_I + \omega_S)^2 \tau_c^2} \right] \qquad (2.24)$$

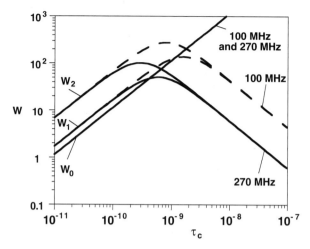

Figure 2.5. Variation of dipolar transition probabilities W_1, W_2, and W_0 with τ_c, shown for spectrometer frequencies of 100 and 270 MHz. Calculated for a pair of protons 1 Å apart. Adapted from ref. 8.

(Again, in this idealized two-spin system relaxing entirely by the dipole–dipole mechanism, $f_I\{S\}$ has the same meaning as η_{max}, the maximum theoretical two-spin enhancement, cf. Section 2.3.1.) In the homonuclear case, $\gamma_I = \gamma_S$, and there is only one frequency $\omega_I \simeq \omega_S \simeq \omega$, so that $(\omega_I - \omega_S)\tau_c$ is always much less than one. Therefore $f_I\{S\}$ (or η_{max}) simplifies to

$$f_I\{S\} = \eta_{max} = \frac{5 + \omega^2\tau_c^2 - 4\omega^4\tau_c^4}{10 + 23\omega^2\tau_c^2 + 4\omega^4\tau_c^4} \tag{2.25}$$

This function, shown in Figure 2.6, summarizes most of what we need to know concerning the variation of homonuclear NOE enhancements with molecular tumbling rate. For small molecules with short τ_c, the limiting value for η_{max} is +50%. As we shall see, competition from other relaxation mechanisms is relatively more efficient in the extreme narrowing limit, with the result that positive enhancements as large as this are rarely observed in practice (Section 2.4). For large molecules with long τ_c, the limiting value for η_{max} is −100%. Biological macromolecules or small molecules in very viscous solvents come into this category, and therefore give rise to negative enhancements. In the central region where η_{max} varies rapidly with $\omega\tau_c$, enhancements depend markedly on the spectrometer frequency as well as the tumbling rate. Many molecules of molecular weight 1000–2000 fall in this region. The variation of η_{max} with $\omega\tau_c$ is fastest around the crossover point, and there are several examples of the sign of enhancements changing with field strength, temperature, or viscosity (see also Sections 3.2.3 and 9.4).[9-11] η_{max} passes through zero when

$$5 + \omega^2\tau_c^2 - 4\omega^4\tau_c^4 = 0 \tag{2.26a}$$

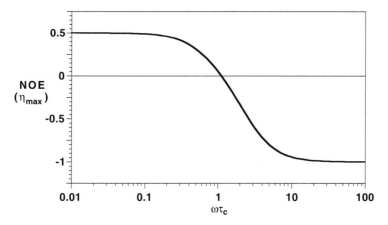

Figure 2.6. Dependence of maximum homonuclear NOE enhancement on $\omega\tau_c$. Note the log scale of $\omega\tau_c$.

that is, when

$$\omega\tau_c = (5/4)^{1/2} \simeq 1.12 \tag{2.26b}$$

At this point there arises something of a conundrum. The limiting value of η_{max} in the extreme narrowing limit is 1/2. More often than not, a factor of two in an equation has a simple physical origin that can readily be pictured. However, in the present case, η_{max} is given by a combination of other numbers:

$$\eta_{max}(\omega\tau_c \ll 1) = \frac{W_{2IS} - W_{0IS}}{W_{0IS} + 2W_{1IS} + W_{2IS}}$$

$$= \frac{6 - 1}{1 + 3 + 6} = \frac{1}{2} \tag{2.27}$$

If one pursues the origins of the numerical factors in the expressions for W_0, W_1, and W_2 (as is done in Appendix II), still more obscure numerical factors come to light, originating largely in the normalization constants for spherical harmonic functions. One is left with the uncomfortable feeling that the simple value of 1/2 for $\eta_{max}(\omega\tau_c \ll 1)$ arises either from a somewhat bizarre coincidence, or that some underlying simplicity has been thoroughly obscured by the theory. We do not know which explanation is correct; if any reader can solve this problem, we should be grateful for the answer!†

---

†Since the first edition was published, we have still not come across any simple explanation; your suggestions are once again invited.

2.2.2. Anisotropic Tumbling

If a molecule is not completely symmetrical, it will tumble at different rates about different axes of rotation. In such cases, each internuclear vector will have a different correlation time, and this complicates greatly the equations presented in the previous section. For this reason it is usually assumed that molecules do tumble isotropically. For most molecules the assumption is reasonable, and calculations of enhancements using the full equations and measured (anisotropic) correlation times give results very similar to those obtained using the isotropic equations (see e.g., ref. 12). However, for grossly asymmetric molecules, such as some polynucleotides, the anisotropy of rotation does have to be considered.[13] We shall consider here only the motion of asymmetric top molecules (that is, molecules with an axis of symmetry, e.g., $CHCl_3$, or porphyrins), since nearly all significantly asymmetric molecules can be considered as symmetric tops. For these molecules it can be shown that for homonuclear systems[14]

$$W_m(ij) = L_m K^2 \left[\left(\frac{P_{ij}\tau_P}{1 + m^2\omega^2\tau_P^2} \right) + \left(\frac{Q_{ij}\tau_Q}{1 + m^2\omega^2\tau_Q^2} \right) + \left(\frac{R_{ij}\tau_R}{1 + m^2\omega^2\tau_R^2} \right) \right] \quad (2.28)$$

with $m = 0, 1, 2$; $L_0 = 0.1$, $L_1 = 0.15$, $L_2 = 0.6$; and

$$P_{ij} = (1/4)(3\cos^2\theta_{ij} - 1)^2$$

$$Q_{ij} = 3\cos^2\theta_{ij}\sin^2\theta_{ij}$$

$$R_{ij} = (3/4)\sin^4\theta_{ij}$$

$$\tau_P = \tau_\perp$$

$$\tau_Q^{-1} = (5/6)\tau_\perp^{-1} + (1/6)\tau_\parallel^{-1}$$

$$\tau_R^{-1} = (1/3)\tau_\perp^{-1} + (2/3)\tau_\parallel^{-1}$$

where $\tau_\parallel$ is the correlation time for rotation about the symmetry axis, $\tau_\perp$ is the correlation time for rotation about an axis perpendicular to the symmetry axis, and θ is the angle between the internuclear vector and the symmetry axis. If the molecular geometry is known, $\tau_\perp$ and $\tau_\parallel$ can be measured, most simply by ^{13}C relaxation rates.[15] The effect of anisotropic rotation on the steady-state NOE is shown in Figure 2.7; similar effects are found on NOE buildup rates.

2.3. WHAT THE SYMBOLS MEAN FOR TWO SPINS AND FOR MANY SPINS

It is at about this stage in the development of the theory that the number of symbols used in the equations inevitably starts to become a little confusing. In

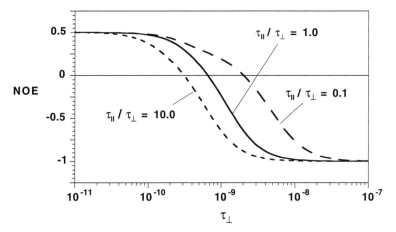

Figure 2.7. Effect of anisotropic tumbling on the homonuclear NOE enhancement for an isolated pair of spins in a symmetric top molecule. Calculations are for an angle of 90° between the internuclear vector and the symmetry axis, and a spectrometer frequency of 270 MHz ($\omega = 2\pi.270 \times 10^6$ rad s^{-1}). From ref. 8.

this section, therefore, we have assembled some points that need to be made concerning the meanings of these various symbols, as they are used in this book. Unfortunately, the symbols used in equations describing relaxation and the NOE are far from consistent throughout the literature, and so we have also pointed out a few pitfalls that may arise in comparing equations from other sources. Since this is intended primarily as a reference section, it represents a break in the discussion of the theory, which resumes in Section 2.4.

2.3.1. Relaxation Rates

The most important symbols for the NOE theory in this chapter are σ and ρ, introduced in Section 2.1.2. The rate constant σ_{IS} characterizes the *cross-relaxation rate* between two spins I and S, and it is this term that controls how *quickly* an NOE enhancement is transferred between I and S. As we use it here, σ_{IS} is positive when saturation of S causes positive NOE enhancements (that is, when $W_2 > W_0$, which is the case where $\omega\tau_c < 1.12$; cf. Section 2.2), whereas σ_{IS} is negative when saturation of S causes negative NOE enhancements (that is, when $W_2 < W_0$, which is the case where $\omega\tau_c > 1.12$). The definition of σ_{IS} is

$$\sigma_{IS} = W_{2IS} - W_{0IS}$$
$$= (1/20)K^2[6J(\omega_I + \omega_S) - J(\omega_I - \omega_S)] \qquad (2.29)$$

where, as before

$$K = \left(\frac{\mu_0}{4\pi}\right)\hbar\gamma_I\gamma_S r_{IS}^{-3}$$

and r_{IS} is the distance between the I and S spins.

The term ρ_{IS} is the *direct dipolar relaxation rate constant for relaxation of spin I by spin S*. In our idealized two-spin system relaxing entirely by the dipolar interaction between I and S, this means that ρ_{IS} is the rate constant (or, here, half the initial rate) for recovery of the I signal following *selective* inversion; this rate is faster than the NOE growth rate, except at the spin diffusion limit ($\omega\tau_c \gg 1$). Formally, ρ_{IS} is given by

$$\begin{aligned}\rho_{IS} &= W_{0IS} + 2W_{1IS} + W_{2IS}\\ &= (1/20)K^2[J(\omega_I - \omega_S) + 3J(\omega_I) + 6J(\omega_I + \omega_S)]\end{aligned} \quad (2.30)$$

with K defined as before.

Note that ρ_{IS} is not quite the same thing as ρ_{SI} (the direct relaxation rate constant for relaxation of spin S by spin I), because of the term in $J(\omega_I)$. However, for homonuclear systems, ω_I and ω_S are so similar that this distinction is irrelevant, while for heteronuclear systems it only matters if either I or S is not in the extreme narrowing limit ($\omega\tau_c \ll 1$; cf. Section 2.2). For σ_{IS}, no such complication exists, and $\sigma_{IS} = \sigma_{SI}$ under all circumstances (provided that the nuclear spin quantum numbers of I and S are equal; cf. Section 2.5).

The other rate constant that we need, although more so in Chapter 3 than here, is R_I, which characterizes the *total relaxation rate of spin I*. It is important to keep the distinction between ρ_{IS} and R_I clearly in mind. The term ρ_{IS} refers only to that *part* of the total relaxation rate of I that arises from dipole–dipole relaxation with spin S, whereas R_I represents the sum of all contributions to the relaxation rate of I. Thus in a two-spin system relaxing exclusively via the intramolecular dipole–dipole mechanism, R_I and ρ_{IS} are equal, since there are then no other contributions to R_I. However, in multispin systems R_I is given by a pairwise summation over all the two-spin terms ρ_{IS} and ρ_{IX}, where X represents all spins other than I and S within the molecule. Also, since mechanisms other than intramolecular dipole–dipole may contribute to the relaxation rate of I, it is useful to decompose R_I into an intramolecular dipole–dipole part, R_I^{DD}, and a "miscellaneous" part, called ρ_I^*:

$$\begin{aligned}R_I &= \rho_{IS} + \sum_X \rho_{IX} + \rho_I^*\\ &= R_I^{DD} + \rho_I^*\end{aligned} \quad (2.31)$$

The term ρ_I^* includes contributions from nondipolar relaxation mechanisms (such as spin–rotation, chemical shift anisotropy, and quadrupolar relaxation;

cf. Section 2.4), but, most commonly, the dominant contribution to ρ_I^* is from *intermolecular* dipole–dipole relaxation by paramagnetic species such as dissolved oxygen. Thus the superscript DD in the term R_I^{DD} is perhaps slightly misleading, and it should be remembered that it refers exclusively to *intra*molecular dipole–dipole relaxation. The same superscript is used in an entirely analogous fashion to denote the intramolecular dipolar part of other quantities in this book, e.g., $\mathbf{B}_{loc}^{DD}$, the intramolecular dipolar contribution to the local fluctuating field $\mathbf{B}_{loc}$.

Notice that the term ρ_{IS} refers only to the interaction between spins I and S. We have deliberately used a very different symbol, R_I^{DD}, for the *total* dipolar relaxation rate constant for spin I, so as to keep the distinction between "two-spin" and "many-spin" terms as clear as possible. In the literature, however, it is quite common to find the symbol ρ_I^{DD} (or just ρ_I) used to denote the total dipolar relaxation rate constant for spin I. A related point is that many equations describing NOE enhancements in the older literature are expressed in terms of ρ_{IS} rather than σ_{IS}, using the fact that at the extreme narrowing limit, $\sigma_{IS} = (1/2)\rho_{IS}$ (Section 2.2). Superficially, this can create the confusion that the NOE appears to be specified without reference to cross-relaxation; more seriously, such equations become invalid away from the extreme narrowing limit.

Just as R_I^{DD} is composed of a sum over pairwise interactions, so the dipolar single-quantum transition probability W_{1I} is also composed of a sum over pairwise contributions[†]:

$$W_{1I} = W_{1IS} + \sum_X W_{1IX} \qquad (2.32)$$

Thus, W_{1I} is related to W_{1IS} in much the same way as R_I^{DD} is related to ρ_{IS}. It is useful to retain both symbols, as each is more convenient in some circumstances. The term W_{1I} describes the total dipolar transition probability across the single-quantum transitions of I, and as such is convenient to use when deriving the multispin Solomon equations (Section 3.1). On the other hand, W_{1IS} is needed whenever the particular contribution of a given relaxation partner (here S) must be specified individually, as in the definition of ρ_{IS}. Note that in a multispin system, W_{1I} and W_{1S} are not equal, just as R_I^{DD} and R_S^{DD} are not equal. This is because the near neighbors of spin I are not the same as the near neighbors of spin S (recall that for each additional spin X, W_{1IX} and ρ_{IX} depend on r_{IX}^{-6}, whereas W_{1SX} and ρ_{SX} depend on r_{SX}^{-6}). On the other hand, W_{1IS} is the same as W_{1SI} (subject only to the same minor qualification as for the equality

---

[†]To be strictly consistent, transition probabilities such as W_{1I} should carry a superscript DD, since several mechanisms can contribute to relaxation across the single-quantum transitions. To avoid an overproliferation of subscripts and superscripts, this DD superscript will be omitted, and it should be remembered that transition probabilities refer exclusively to intramolecular dipole–dipole relaxation unless specifically indicated to the contrary (as, for instance, for W_0^{scalar} in Section 6.4.3).

of ρ_{IS} and ρ_{SI}; cf. Eq. 2.30), because for these terms only the interaction between I and S, and hence only the distance r_{IS}, is considered.

We have shown how dipolar relaxation rate constants and transition probabilities in multispin systems can usefully be decomposed into a sum of two-spin contributions. A similar idea applies to the NOE itself. In Chapter 3 it will prove a very useful concept to be able to relate enhancements in multispin systems to η_{max}, the maximum theoretical enhancement expected for a two-spin system with the same nuclear species and correlation time:

$$\eta_{max} = \frac{\gamma_S}{\gamma_I} \frac{\sigma_{IS}}{\rho_{IS}}$$

$$= \frac{\gamma_S}{\gamma_I} \frac{W_{2IS} - W_{0IS}}{W_{0IS} + 2W_{1IS} + W_{2IS}} \quad (2.33)$$

Another picture for the meaning of η_{max} is that it represents the NOE enhancement expected at I on saturation of S if all the other spins X were somehow taken away, or at least rendered magnetically inactive. It thus expresses the dependence of the NOE on τ_c, ω_I, ω_S, γ_I, and γ_S, so allowing attention to be focused on the influence of molecular geometry (that is, the arrangement of the other spins X and I and S). Unfortunately, the other main factor that determines the sizes of enhancements, namely the miscellaneous competing relaxation mechanisms collected in the term ρ_I^*, cannot be treated in quite this way, as we shall see. Thus, η_{max} can be used to refer only to the *theoretical* maximum two-spin enhancement under the assumption of exclusively dipolar relaxation. As with ρ_{IS} and R_I^{DD}, we have chosen distinctively different symbols for the two-spin term, η_{max}, and the corresponding many-spin term, $f_I\{S\}$; in the literature, however, the symbol η_{IS} is sometimes used to denote the fractional enhancement that we call $f_I\{S\}$.

2.3.2. T_1 Measurements and Cross-Relaxation

We now need to consider how the various rate constants that we have been discussing are related to experimentally measurable quantities, in particular the longitudinal relaxation time, T_1. Essentially, the link between theory and experiment is provided by the term ρ^*, the rate constant for "miscellaneous" relaxation, of which we shall have more to say in Section 2.4. Equations that lack the term ρ^* are essentially "theoretical"; they express the *expected* consequences of intramolecular dipolar relaxation and cross-relaxation, based on what we know about the dipolar interaction. On the other hand, measurement of relaxation times and steady-state NOE enhancements generally gives answers different from those predicted by the dipole-only equations. The empirical correction term ρ^* is therefore brought in to represent the "extra" relaxation, and thereby to make the equations fit the experiment. In this sense, the rates R_I^{DD} and W_{1I} may be categorized as theoretical, whereas the rate R_I may be categorized as experimental. It is thus R_I, rather than R_I^{DD}, which is related to measured T_1 values.

In a one-spin system, R_I is simply the reciprocal of the measured T_1 value for the I resonance, as summarized in Section 1.2. Usually, T_1 is derived from the results of an inversion recovery sequence, either by fitting the points to a true exponential, or, much more crudely, by measuring the time at which the inverted signal passes through zero and then using the equation $T_1 = T_{\text{null}}/\ln 2$. However, in two-spin and multispin systems where there is cross-relaxation, relaxation is necessarily nonexponential, and under these circumstances T_1 is, formally at least, no longer defined. Nonetheless, inversion-recovery experiments can always be forced to yield a result. The literature thus abounds with "approximate" T_1 values for protons in multispin systems, and this formally incorrect use of the term is probably more common than its correct usage to describe strictly exponential behavior.

In fact, this does not usually matter very much, and the deviations from exponentiality are often quite small. What is more significant is whether T_1 is measured following *selective* inversion ($T_1^{\text{selective}}$) or *nonselective* inversion ($T_1^{\text{nonselective}}$). To see why this is so, we must anticipate some results from Chapter 4. Suppose one spin (S) of a two-spin system is inverted selectively, and the system is then allowed to relax for a period τ prior to the observe pulse. As the S resonance recovers, a transient NOE enhancement appears at the I resonance (cf. Figs. 2.8b and 4.2). If this enhancement becomes appreciable, the balance of relaxation at spin I will itself be disturbed, and spin I will start to lose its enhancement via its own relaxation pathways. One such pathway is cross-relaxation with S, which leads to a "return" of intensity from I back to S; this in fact *inhibits* the later stages of the recovery of S (Fig. 2.8). Thus, the curves for both S_z and I_z are biexponential, that for S_z combining two rising exponentials, and that for I_z combining a faster rising exponential with a slower falling exponential (provided σ is positive; in the negative NOE regime the signs of these exponentials are reversed). This may be seen more clearly from the solution to the Solomon equations (Eq. 2.10) obtained under the initial conditions $S_z = -S_z^0$ and $I_z = I_z^0$:

$$\frac{S_z(t) - S_z^0}{S_z^0} = -e^{-(\rho_{IS}+\sigma_{IS})t} - e^{-(\rho_{IS}-\sigma_{IS})t}$$

$$\frac{I_z(t) - I_z^0}{I_z^0} = -e^{-(\rho_{IS}+\sigma_{IS})t} + e^{-(\rho_{IS}-\sigma_{IS})t} \quad (2.34)$$

These functions are shown in Figure 2.8, under the assumption of exclusively dipolar relaxation at the extreme narrowing limit ($\sigma_{IS} = (1/2)\rho_{IS}$). More complicated results apply if external relaxation occurs ($R_I \neq R_S$; $\rho_I^*, \rho_S^* \neq 0$), or if there are more than two spins; these cases are dealt with in Chapter 4.

The curves in Figure 2.8 show how nonexponentiality affects T_1 measurements. If a single exponential were fitted to the recovery curve for S, the value

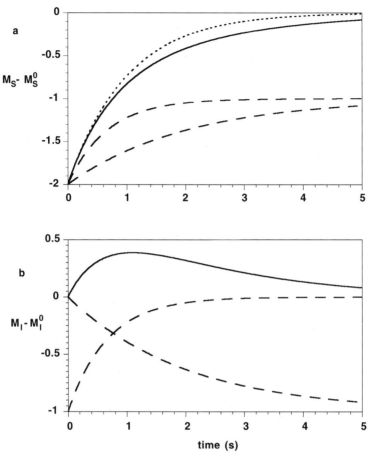

Figure 2.8. Intensity changes calculated for spins S and I following selective inversion of spin S. (a) Spin S. (b) Spin I. Intensities are shown as intensity deviations from equilibrium, and the curves are plotted from Eq. 2.34 with $\rho_{IS} = 1.0$ s^{-1}, and $\sigma_{IS} = 0.5$ s^{-1}. The solid line shows the calculated intensity time course. This is composed of the two exponentials $\exp[-(\rho_{IS} \pm \sigma_{IS})t]$, each indicated by dashed lines. Three of these have been shifted downwards for clarity (all curves approach zero for long times). The dotted line in (a) plots relaxation at a rate $2\exp[-\rho_{IS}t]$, and the difference between this line and the solid line shows how much the recovery of intensity of spin S is retarded by cross-relaxation with spin I.

of "T_1" obtained would depend on the fitting algorithm used, while the null-point method would give "T_1" as $1.443T_{\text{null}}$. Solution of the true biexponential equation (Eq. 2.34) shows that ρ_{IS}^{-1}, here equivalent to R_I^{-1} and thus the nearest concept to "T_1" in this system, is actually given by $1.308T_{\text{null}}$ (when $\sigma_{IS} = (1/2)\rho_{IS}$). This represents an error of about 10% in the determination of "T_1" by this method.

In practice it is unlikely that the deviation from exponentiality would be nearly as large as that shown in Figure 2.8. External relaxation generally increases R_I and R_S so that they become considerably larger than $2\sigma_{IS}$, with the result that the two exponential curves $\exp[-(R_S + \sigma_{IS})t]$ and $\exp[-(R_S - \sigma_{IS})t]$ become much less different from one another. In multispin systems, multiexponential curves result from cross-relaxation among all the spins, but again the total deviation from a single exponential $\exp[-R_S t]$ is usually only a few percent, and always considerably less than the NOE enhancement.

If *both* resonances of a two-spin system are inverted, then cross-relaxation from I contributes to the recovery of S from the start. Solution of the Solomon equations under the initial conditions $S_z = -S_z^0$ and $I_z = -I_z^0$ gives

$$\frac{S_z(t) - S_z^0}{S_z^0} = -2e^{-(\rho_{IS} + \sigma_{IS})t}$$

$$\frac{I_z(t) - I_z^0}{I_z^0} = -2e^{-(\rho_{IS} + \sigma_{IS})t} \qquad (2.35)$$

In contrast to the selective inversion case, recovery is truly described by a single exponential. The result found for "T_1" in a *nonselective* inversion-recovery experiment is thus $(\rho_{IS} + \sigma_{IS})^{-1}$, and the initial rate of recovery of both I and S is $2(\rho_{IS} + \sigma_{IS})$. This single-exponential behavior is only found if $R_I = R_S$; however, in practice the deviation from exponentiality is small even when $R_I \neq R_S$.

This discussion applies principally to small molecules in the positive NOE regime ($\omega\tau_c < 1.12$). For large molecules ($\omega\tau_c > 1.12$), σ_{IS} becomes large and negative, and at the spin diffusion limit σ_{IS} approaches $-R_S$. Under these circumstances, the concept of an individual "T_1" value for a particular spin breaks down.[16] For example, for our ideal two-spin system at the spin diffusion limit, inversion of S results in a two-phase recovery curve. First, very rapid initial recovery (rate constant $\rho_{SI} - \sigma_{SI} \simeq 2\rho_{SI}$) occurs while cross-relaxation with I spreads the initial population disturbance to the I resonance; this is followed by a very slow recovery of both I and S together (rate constant $\rho_{SI} + \sigma_{SI} \simeq 0$). Thus, in real cases, any selective inversion is rapidly spread and diluted around the whole spin system. Following this, all the spins recover slowly and cooperatively, while efficient cross-relaxation constantly averages their populations.

To summarize, for spin I in a multispin system:

$$T_1^{\text{(selective)}} \simeq R_I^{-1}$$

$$T_1^{\text{(nonselective)}} \simeq \left(R_I + \sigma_{IS} + \sum_X \sigma_{IX}\right)^{-1} \qquad (2.36)$$

2.4. EFFECTS OF OTHER RELAXATION SOURCES

The "miscellaneous" contribution to the relaxation of spin I, written ρ_I^*, was introduced in the previous section to represent all relaxation other than intramolecular dipole–dipole occurring for a particular spin. Later in this section we shall examine the principal mechanisms that can contribute to ρ^*, but first we consider how the existence of such competing relaxation, from whatever source, affects the NOE. In fact ρ^* is not at all a quantity amenable to theoretical prediction; rather, it is an *empirical correction term*, and can only sensibly be regarded as an unknown, determinable, for each spin, by experiment.

2.4.1. The External Relaxation Rate ρ^*

Intramolecular dipole–dipole relaxation is unique in that it connects two spins within a molecule via W_2 and W_0 cross-relaxation transitions. With the single exception of relaxation from modulated scalar couplings (which can contribute a W_0 term; Section 6.4.3), all other relaxation mechanisms contribute only to the single-quantum transition probabilities. As far as our derivation of the two-spin Solomon equations in Section 2.1.2 is concerned, this has the consequence of adding further W_1 terms (e.g., $W_1^{\text{Intermolecular DD}}$, $W_1^{\text{quadrupolar}}$, $W_1^{\text{spin-rotation}}$, etc.) of largely unknown size to the relaxation rate constant for the detected spin (I). This is represented by adding the term ρ_I^* to the intramolecular dipole–dipole direct relaxation term ρ_{IS} in the expression for the two-spin steady state NOE enhancement:

$$f_I\{S\} = \frac{\gamma_S}{\gamma_I} \frac{\sigma_{IS}}{\rho_{IS} + \rho_I^*} \qquad (2.37)$$

Quite clearly, the effect of the term ρ_I^* is to *reduce the NOE enhancement* by "diluting" the contribution of intramolecular dipole–dipole relaxation to the total relaxation of spin I. This is why experimental procedures such as degassing, which minimize the "miscellaneous" relaxation contribution, are often recommended for NOE work (Section 7.1.3); if ρ^* can be reduced, then NOE enhancements will become larger. For obvious reasons, the miscellaneous relaxation contribution ρ^* is often called the "leakage" term.

To discover by *how much* the NOE is reduced by ρ^*, we need to examine the properties of σ_{IS}, ρ_{IS}, and ρ_I^* in somewhat more detail. We saw earlier that both σ_{IS} and ρ_{IS} are proportional to the inverse sixth power of the IS internuclear distance. Clearly, ρ_I^* is independent of r_{IS}, since by definition ρ_I^* refers only to interactions of I other than that with spin S. Thus, if the IS distance is reduced, both σ_{IS} and ρ_{IS} increase; the contribution of ρ_I^* becomes relatively less significant, and the NOE increases toward its theoretical (i.e., dipole-only) limit (Fig. 2.9). Thus, we see that the two-spin steady-state NOE does depend on internuclear distance, *but only when there is external relaxation.*

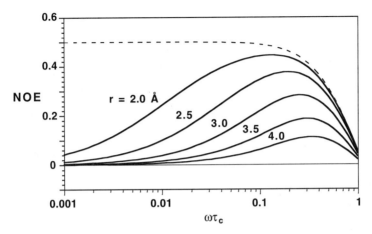

Figure 2.9. Reduction of the steady-state NOE enhancement due to leakage as a function of $\omega\tau_c$, with different internuclear distances r_{IS}. From bottom to top, r_{IS} = 4.0, 3.5, 3.0, 2.5, and 2.0 Å, with ρ_I^* held constant at 0.1 s^{-1}. The dotted line shows η_{max} (i.e., with ρ_I^* = 0). The effect of ρ^* is relatively less for shorter values of r_{IS}. Curves calculated from Eq. 2.37, for a spectrometer frequency of 300 MHz.

In addition, σ_{IS} and ρ_{IS} depend on functions of τ_c. As shown in Section 4.4.2 both $|\sigma_{IS}|$ and ρ_{IS} increase with τ_c, except for $|\sigma_{IS}|$ near the NOE zero-crossing at $\omega\tau_c \simeq 1.12$. In contrast, one does not know quite how ρ^* will vary with $\omega\tau_c$; for each spin, this will depend on the particular relaxation mechanisms that are operating. However, in many cases it may be reasonable to assume that ρ^* is largely invariant with $\omega\tau_c$. Most commonly, the dominant contribution to ρ^* is from intermolecular dipole–dipole relaxation, particularly that involving dissolved oxygen. Although τ_c for *intramolecular* interspin vectors increases with decreasing tumbling rate of the solute, this correlation time does not apply to *intermolecular* relaxation. The appropriate correlation time here is that of the vector connecting the relaxing spin on the solute molecule and the nearby oxygen molecule causing the relaxation. Since oxygen molecules always tumble rapidly, this internuclear correlation time remains very short even when the solute molecules themselves tumble slowly; thus, ρ^* may well be more or less independent of the solute tumbling rate.

Thus, if σ_{IS} and ρ_{IS} both increase with increasing τ_c but ρ_I^* does not, we may expect the influence of ρ^* to be relatively less for large, slowly tumbling molecules than for small, rapidly tumbling molecules. This is indeed the case. For large molecules in the negative NOE regime, leakage is often not significant, whereas for smaller molecules in the positive NOE regime, steady-state enhancements are generally far short of the values predicted by the dipole-only equations (Section 3.3.3). In fact, for very small molecules such as monocyclic aromatics, NOE enhancements can be very small or unmeasurable, because of the very low efficiency of cross-relaxation relative to leakage in these cases.

This is the basis of the "RABBIT" experiment (Section 9.4), in which complexation of small organic molecules to a larger inert molecule is used as a means of increasing τ_c for the organic solute, thereby increasing both σ and ρ, and hence also the steady-state NOE. These points are illustrated in Figure 2.10, which shows calculations for $f_I\{S\}$ as a function of $\omega\tau_c$, assuming that ρ_I^* is invariant with $\omega\tau_c$.

To show these effects of ρ^* on the steady-state NOE more explicitly, we may use the quantity η_{max}, the maximum theoretical two-spin steady-state enhancement, defined in Section 2.3.1. This represents the dependence of the NOE on ω_I, ω_S, γ_I, γ_S, and τ_c, so allowing attention to be concentrated on the distance-dependent terms, as we shall see in more detail in Chapter 3. We may rewrite Eq. 2.37 in terms of η_{max} and r_{IS} (using Eqs. 2.22 and 2.23) to get

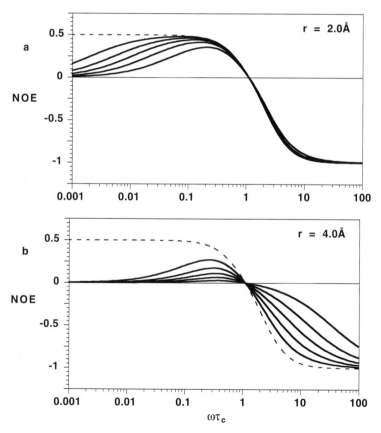

Figure 2.10. Reduction of the steady-state NOE enhancement due to leakage as a function of $\omega\tau_c$, with different values of ρ^*. The dotted curves are for $\rho_I^* = 0$, and the other curves are, from top to bottom, $\rho_I^* = 0.02, 0.05, 0.1, 0.2,$ and 0.5 s^{-1}. (a) $r_{IS} = 2.0$ Å. (b) $r_{IS} = 4.0$ Å. The spectrometer frequency for the calculation is 300 MHz.

$$f_I\{S\} = \eta_{max}\left[\frac{r_{IS}^{-6}}{r_{IS}^{-6} + \mathcal{L}_I(\omega, \tau_c)}\right] \quad (2.38)$$

Like ρ_I^*, $\mathcal{L}_I(\omega, \tau_c)$ is best regarded as an empirical correction factor. It is approximately proportional to ρ_I^*/τ_c; for small molecules ($\omega\tau_c \ll 1$) this proportionality is exact, and we may replace $\mathcal{L}_I$ by the simpler quantity a_I (following the nomenclature of Noggle and Schirmer[4]):†

$$a_I = \frac{\rho_I^*}{\gamma_I^2 \gamma_S^2 \hbar^2}\left(\frac{4\pi}{\mu_0}\right)^2 \frac{1}{\tau_c} \quad (2.39)$$

A question often asked is "over what maximum distance can an NOE enhancement be transmitted?" We can already see one reason why there is no simple answer to this—it depends on the molecular tumbling rate. For large molecules, cross-relaxation is efficient relative to leakage, and appreciable enhancements are possible over longer distances. The limiting factor here, in fact, is the confidence that can be placed on interpretation in the presence of spin diffusion (Sections 3.2 and 4.4), which in practice forces the use of kinetic experiments. For small molecules the overall geometry of the spin system plays the most important role in determining enhancements (Chapter 3), but Eq. 2.38 tells us that the smaller the molecule, and hence the larger $\mathcal{L}_I$, the less chance there is for an NOE enhancement to show up over a long internuclear distance. With these factors in mind, it is reasonable to say that an NOE enhancement is rarely useful over a distance of much more than about 5 Å.

Bell and Saunders showed that for a series of model compounds, there was a linear correlation between the steady-state NOE enhancement and the inverse sixth power of internuclear distance.[17] This result is widely quoted as demonstrating the r^{-6} dependence of the NOE, but in fact it is very far from general. Such a correlation depends fundamentally on two conditions:

1. The observed spin (I) must not have appreciable dipolar cross-relaxation involving any spin other than the saturation target (S); that is, spins I and S must form a true two-spin system.

---

†The full relationship between $\mathcal{L}_I(\omega, \tau_c)$ and ρ_I^* is given by the equation

$$\mathcal{L}_I(\omega, \tau_c) = a_I \eta_I^*$$

with a_I defined in Eq. 2.39, and η_I^* given by

$$\eta_I^* = \frac{1}{10}\left[\frac{1}{1 + (\omega_I - \omega_S)^2\tau_c^2} + \frac{3}{1 + \omega_I^2\tau_c^2} + \frac{6}{1 + (\omega_I + \omega_S)^2\tau_c^2}\right]$$

Thus η_I^* for homonuclear systems varies between 1 ($\omega\tau_c \ll 1$) and 0.1 ($\omega\tau_c \gg 1$). This rather complicated (and essentially useless) relationship between ρ_I^* and $\mathcal{L}_I$ is the price to pay for expressing the NOE directly in terms of internuclear distance.

2. The values of $\mathscr{L}_I$ must be the same for all molecules in the series (cf. Eq. 2.38).

These conditions are both extremely restrictive. There is the further problem that condition (2) cannot be tested *a priori*. Predictions based on such a correlation could thus only be trusted within a series of very similar structures where the contributions to $\mathscr{L}_I$ are always the same, and even then only when the experimental conditions are completely consistent, so as to minimize differences (e.g., in dissolved oxygen concentration) that might affect the efficiency of leakage pathways. As a means of estimating internuclear distances, this method cannot compete with kinetic NOE experiments (Chapter 4).

In the following sections, the principal mechanisms that can contribute to ρ^* are considered in turn. Our aim is to give a qualitative description of the mechanisms involved: equations for exact values of relaxation rates can be found in references 5 and 14. A more theoretical treatment for each mechanism can be found in the book by McConnell.[18]

2.4.2. Intermolecular Dipole–Dipole Relaxation

This is usually by far the most important contribution to ρ^* for all except quadrupolar nuclei (Section 2.4.3), the dominant source generally being dissolved paramagnetic species. An electron has spin $I = 1/2$, and a gyromagnetic ratio 658 times that of a proton. It is therefore very efficient as a cause of dipolar relaxation. Unpaired electrons are most commonly encountered in paramagnetic metal ions, and in dissolved molecular oxygen. Low concentrations of these paramagnetic species can severely reduce an NOE enhancement, or even destroy it altogether. Most solvents contain enough dissolved oxygen to increase the total relaxation rate by something of the order of 0.1 s^{-1}. This is enough to have an important effect for small molecules, where intramolecular dipolar relaxation is slow, but is not significant for large molecules, where intramolecular dipolar relaxation is much faster. It is often not worth removing oxygen, unless the enhancement observed is expected to be very small, or if quantitative data are required. Methods for removing paramagnetic impurities are described in Section 7.1.3.

There are experiments in which paramagnetic ions are deliberately added, and here clearly the NOE will be affected. One reason for adding paramagnetic ions is to increase the spin-lattice relaxation rate of slowly relaxing nuclei, most commonly quaternary carbons, which relax much more slowly than protonated carbons and hence appear with much lower intensity under rapid pulsing conditions. The compound normally used for this purpose is Cr(acetylacetonate)$_3$ [usually abbreviated to Cr(acac)$_3$]. The addition of Cr(acac)$_3$ often reduces the NOE enhancement for quaternary carbons to negligible levels, although it has a smaller effect on protonated carbons (see Section 2.5).

The other common use of paramagnetic ions is as shift or broadening reagents. These reagents have a lesser effect on spin–lattice relaxation rates at the concentrations used, and also tend to bind to specific regions of the molecule, which means that their effect is localized, being greatest at the binding site(s). High concentrations of ions (around 0.1M) will almost completely abolish an enhancement, but with more typical concentrations it is often still possible to observe surprisingly large enhancements, depending on the ion used, the solvent, and the distance of relevant protons from the binding site of the ion. A nice example is provided in reference 19, in which the conformation of the terpene **2.1** in $CDCl_3$ was studied. The signals from H_a and H_b could not be resolved, and so $Eu(fod)_3$ was added. The $Eu(fod)_3$ binds to the lactone carbonyl oxygen, and so relaxes H_b much more than H_a (since the relaxation rate follows an r^{-6} law). Therefore the enhancement from H_a to H_b was decreased by more than a half, whereas that from H_b to H_a was decreased by only one-sixth. The relevant theory is discussed further in references 20 and 21 (of which the latter deals mainly with the measurement of T_1 as affected by paramagnetic centers in the presence of cross-relaxation).

2.1

Some proteins contain paramagnetic centers, which have the effect of producing very large chemical shift changes (hyperfine shifts) and paramagnetic broadening. The hyperfine-shifted protons have very broad lines and short T_1 values (typically a few ms), consistent with their proximity to the paramagnetic centers, but even these protons can sometimes show small negative NOE enhancements, typically of 5% or less.[22,23]

For the general case of relaxation of a proton by dipolar interaction with nuclei having $I \geq 1/2$, the longitudinal relaxation rate is given by[8,24]

$$R_1 = \frac{2}{15}\left(\frac{\mu_0}{4\pi}\right)^2 \hbar^2 \gamma_H^2 \sum_j I_j(I_j + 1)\gamma_j^2 r_{ij}^{-6} \eta_j \tag{2.40}$$

where

$$\eta_j = \frac{\tau_c}{1 + (\omega_j - \omega_H)^2 \tau_c^2} + \frac{3\tau_c}{1 + \omega_H^2 \tau_c^2} + \frac{6\tau_c}{1 + (\omega_j + \omega_H)^2 \tau_c^2} \tag{2.41}$$

An important case of relaxation by a spin $> 1/2$ nucleus is that of relaxation by ^{14}N ($I = 1$). From Eq. 2.40, the ratio of relaxation rates due to ^{1}H and ^{14}N at the same distance to a ^{1}H nucleus is $R_1(H):R_1(N) \simeq 1:0.014$. For NH protons, the N—H distance is just over 1.0 Å, and so relaxation by a directly bonded ^{14}N has roughly the same effect on R_1 as relaxation by ^{1}H at $0.014^{-1/6}$ = 2.1 Å. Thus, amide protons are often relaxed as fast by dipole–dipole relaxation from ^{14}N as from other protons. Glickson et al.[9] observed that the maximum NOE enhancement to amide protons in valinomycin was roughly half the size of the maximum enhancement to other protons, and attributed this to ^{14}N dipolar relaxation. Others[25,26] have measured the contribution of ^{14}N dipolar relaxation in similar molecules, and verified this result.

The gyromagnetic moment of ^{2}H is only 0.15 that of ^{1}H, and therefore the relaxation rate from ^{2}H is (from Eq. 2.40) $(0.15)^2 \times 8/3 = 0.06$ that of ^{1}H at the same distance. Nevertheless, because of the high concentration of ^{2}H in deuterated solvents, relaxation from this source is sometimes significant for isolated protons in small molecules. Further consideration of this point is given in Section 7.1.1.

2.4.3. Quadrupolar Relaxation

This is normally the dominant relaxation mechanism for nuclei with $I > 1/2$, such as ^{2}H and ^{14}N. These nuclei, in addition to being dipolar, are also quadrupolar. Therefore, unlike nuclei that are only dipolar, they are affected by electric field gradients. Such gradients exist wherever the molecular electronic environment is not perfectly symmetrical, most usually because of differences in electronegativities between neighboring atoms. As a molecule tumbles, its associated electric field gradients tumble with it, and therefore the energy of each quadrupolar nucleus in the molecule fluctuates at the tumbling rate. The extent of this fluctuation depends on the size of the quadrupole moment of the nucleus, but most quadrupolar nuclei are very efficiently relaxed. This means that NOE enhancements *into* quadrupolar nuclei are usually extremely small; there are very few cases where an enhancement to a quadrupolar nucleus is useful, at least in solution. (^{6}Li is an exception, as it has a very small quadrupole moment; this is discussed in Sections 2.5 and 9.1.2.) The very rapid relaxation of quadrupolar nuclei makes their signals very hard to saturate, and hence NOE enhancements *from* quadrupoles are also rare; another reason for this is that most quadrupolar nuclei of interest have low gyromagnetic ratios, and hence the maximum enhancement expected from them, even in the absence of quadrupolar relaxation, is small.

The rapid relaxation of quadrupolar nuclei creates a modulation of the local magnetic field $\mathbf{B}_{loc}$ at other nuclei spin–spin coupled to them. It was shown in Chapter 1 that fields in the z direction affect relaxation only in the transverse plane. The field modulation caused by quadrupolar relaxation normally only affects the z component of $\mathbf{B}_{loc}$. Therefore, quadrupolar relaxation normally acts only to increase the *transverse* relaxation rates of spins coupled to a quadrupolar nucleus. Thus, although such spins may have an unusually large linewidth (e.g., NH protons coupled to ^{14}N), NOE enhancements into them are generally not affected. An exception to this is discussed in Section 6.4.3.2.

2.4.4. Chemical Shift Anisotropy (CSA) Relaxation

All bonds are anisotropic, and consequently produce a different field at a nucleus depending on the orientation of the bond relative to the applied field. The instantaneous chemical shift of the nucleus therefore depends on the orientation of the molecule. This effect is normally averaged away in solution, since the molecule tumbles at a rate of between 10^{11} and 10^6 Hz, while the chemical shift anisotropy is for many nuclei a matter of only a few kilohertz. However, nuclei still experience a rapid variation of local field from this source, and this stimulates relaxation. The effect is normally not large, since the variation of field with orientation is small. It is most significant for nuclei with very large chemical shift anisotropy, such as ^{19}F, ^{31}P, or trigonal (sp^2) ^{13}C; since the relaxation rate from this mechanism is dependent on the square of the applied field, it is likely to become more important as magnetic field strengths increase. It is also more important for more slowly tumbling molecules. It has been demonstrated to be a significant relaxation mechanism for ^{19}F in fluorinated proteins at high field strength.[27] One example concerns pyridoxal fluorophosphate bound to glycogen phosphorylase b,[28] for which the various contributions to the linewidth from chemical shift anisotropy (CSA) and dipole–dipole (DD) relaxation were found to be $\Delta\nu_{CSA}^{F} = 235$ Hz, $\Delta\nu_{CSA}^{P} = 350$ Hz, $\Delta\nu_{DD}^{FP} \simeq 20$ Hz at 254 MHz (for ^{19}F).

2.4.5. Scalar Relaxation

Scalar, or spin–spin, coupling implies that the local field at one nucleus depends on the spin state of its coupling partner, and vice versa. This is what is responsible for the splitting observed in the signals from scalar coupled nuclei. If this J coupling varies with time, then the field at both coupling partners is also modulated, and so can act as a relaxation mechanism. As in the case of quadrupolar relaxation, it is usually only the z component of the field that is modulated, so that exchange or relaxation of one spin generally produces only transverse relaxation of another spin coupled to the first. An exception to this is when the two spins are strongly coupled, in which case the scalar relaxation mechanism can actually produce a negative enhancement. This mechanism is most likely to occur in exchanging —CHOH— fragments. It is discussed more fully in Section 6.4.3.1.

2.4.6. Spin–Rotation Relaxation

Spin–rotation relaxation occurs when a molecule, or a group within a molecule, undergoes transitions between one rotational state and another. The rate of relaxation induced is related to the spin–rotation correlation time (the average time between rotational transitions), and only becomes significant for fast rotation speeds. This mechanism is therefore most effective for small symmetrical molecules, and for methyl groups. It can be important for ^{13}C, for which dipole–dipole relaxation can sometimes be relatively inefficient. An interesting example of spin–rotation relaxation is provided by a study of NOE enhancements in small aromatic systems.[29] In 3-nitro-4-methyltoluene (**2.2**), one methyl group (a) is sterically hindered, while the other is not. The carbon atom of methyl group a is relaxed largely by dipole–dipole relaxation, but the carbon atom in methyl group b receives a large contribution from spin–rotation relaxation, because of the much more rapid rotation of this group. As a result, in the noise-decoupled ^{13}C spectrum of **2.2**, the signal from methyl group b has about half the intensity of the signal from methyl group a. For similar reasons, the ^{29}Si{^{1}H} NOE enhancement in **2.3** is -241%, close to its theoretical maximum of -251%, but in the more rapidly rotating **2.4** it is only -7%.[30]

2.5. THE HETERONUCLEAR NOE

When the gyromagnetic ratios of I and S are different, Eq. 2.24 cannot be simplified to give an equation of the form of Eq 2.25, except at the extreme narrowing limit, when Eq. 2.24 simplifies to

$$\eta_{max} = \gamma_S/2\gamma_I \qquad (2.42)$$

So, for example, for ^{13}C{^{1}H} at the extreme narrowing limit, $\eta_{max} = 198.8\%$, and the ^{13}C signal is multiplied by almost 3.

The NOE enhancement from attached protons thus provides a very significant and useful increase in signal intensity; the gain is even larger for nuclei with lower gyromagnetic ratios (cf. Table 2.1). Many ^{13}C spectra are now acquired using polarization transfer sequences such as INEPT or DEPT, in which no ^{13}C{^{1}H} NOE is generated, since the carbon intensities depend on the proton populations, not on carbon. However, the simplest way to acquire ^{13}C spectra is by pulsing ^{13}C and saturating all protons during ^{13}C signal acquisition using noise decoupling,[31] which removes C—H couplings and generates an NOE enhancement for all carbon atoms. (It is occasionally useful to use single-

TABLE 2.1 Gyromagnetic Ratios and NOE Properties of Some Common Nuclei

Nucleus	$\gamma/10^7$ rad T^{-1} s^{-1} [a]	I	Max. X{^{1}H} NOE enhancement (%)	
			($\omega\tau_c \ll 1$)	($\omega\tau_c \gg 1$)
^{1}H	26.75	1/2	50	−100
^{2}H	4.11	1	[b]	[b]
^{3}H	28.54	1/2	47	−91
^{6}Li	3.94	1	339	15
^{7}Li	10.40	3/2	[b]	[b]
^{11}B	8.58	3/2	[b]	[b]
^{13}C	6.73	1/2	199	15
^{14}N	1.93	1	[b]	[b]
^{15}N	−2.71	1/2	−494	−22
^{19}F	25.18	1/2	53	−104
^{29}Si	−5.32	1/2	−251	−51
^{31}P	10.84	1/2	123	2
^{77}Se	5.12	1/2	261	16
^{113}Cd	−5.96	1/2	−224	−59
^{119}Sn	−10.02	1/2	−133	−106
^{195}Pt	5.77	1/2	232	16
^{199}Hg	4.82	1/2	277	16

[a]From ref. 15.
[b]Enhancements for quadrupolar nuclei are small, owing to rapid quadrupolar relaxation (see text).

frequency proton irradiation, giving NOE enhancements only to carbon atoms close to the irradiated proton. Use of this technique is discussed in Section 9.1.1.) Since irradiation *before* the acquisition period induces an NOE enhancement but no decoupling, and irradiation *during* acquisition causes decoupling but no NOE, it is possible to separate the two effects by "gating" the decoupler on or off for the appropriate periods (Fig. 2.11a and b). Moreover, the decoupler power required to produce an NOE is, as in homonuclear spectra, much less than that needed to decouple the protons. This means that to acquire a spectrum with both NOE and decoupling, it is possible to use lower power for the preirradiation period than for the acquisition period (Fig. 2.11c). This can be desirable, as it cuts down the considerable heating effect sometimes produced by continuous irradiation at typical decoupling powers of 2–3 W.

Since protonated carbons have at least one dipole (a proton nucleus) very close to them, their relaxation is dominated by the dipole–dipole mechanism, and the NOE enhancement they receive is very close to the maximum. For most molecules in the molecular weight range 300–3000, the dominant source of relaxation even for nonprotonated carbons is dipolar.[32] Therefore, if the intensities of nonprotonated carbon signals are found to be low, this will usually be mainly because of their long relaxation times rather than their lack of NOE.

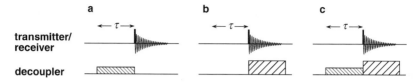

Figure 2.11. Pulse schemes for separating decoupling effects from the NOE during X nucleus acquisition. The decoupler is programmed to produce decoupling at two power levels. Suitable gating of the decoupler produces (*a*) NOE only, (*b*) proton decoupling only, or (*c*) NOE and proton decoupling.

Very small molecules (and methyl carbons) can be relaxed by spin–rotation relaxation, while in large molecules at high field, chemical shift anisotropy becomes important, especially for sp² carbons. However, in general, most carbon nuclei receive almost their full NOE enhancement, provided that long preirradiation delays are included.

A common technique in ^{13}C spectroscopy is to add a paramagnetic relaxation agent such as Cr(acac)$_3$ to reduce the spin–lattice relaxation time of quaternary carbons.† The relaxation rate varies linearly with the concentration of Cr(acac)$_3$, and can be increased a hundred-fold for quaternary carbons by addition of 0.1 M Cr(acac)$_3$.[31] The NOE enhancement therefore drops about a hundred-fold, to become less than 3%. For protonated carbons, which have much faster relaxation rates σ and ρ, the proportional effect is less, and the reduction of the enhancement is thus also less; typically a factor of about 10. Paramagnetics are discussed further in Section 2.4.2.

As $\omega\tau_c$ increases, the extreme narrowing condition no longer holds, and the magnitude of the NOE enhancement decreases, as in the homonuclear case. It can become negative, though it often does not. Three cases are shown in Figure 2.12. From Eq. 2.24, it can be calculated that if γ_S and γ_I have the same sign, the enhancement will become negative only if

$$\gamma_S/\gamma_I < 2.38 \tag{2.43}$$

The zero crossing is then given by

$$\tau_c = \left[\frac{5}{(\omega_I + \omega_S)^2 - 6(\omega_I - \omega_S)^2}\right]^{1/2} \tag{2.44}$$

For most pairs of nuclei, especially if S is ^{1}H, the ratio γ_S/γ_I is greater than 2.38, and the NOE does not change sign; in many cases the NOE just decreases to a value close to zero. This is the case, for example, for ^{1}H and ^{13}C ($\gamma_S/\gamma_I =$

---

†Tris-(2,2,6,6-tetramethyl-3,5-heptanedionate) chromium III, Cr(dpm)$_3$, has been recommended as being more soluble than Cr(acac)$_3$, more inert, and therefore more predictable in its effects. It is not readily available, but a simple synthesis is described in ref. 33.

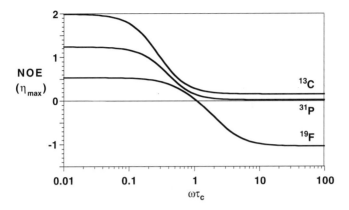

Figure 2.12. Dependence of maximum theoretical NOE enhancement on $\omega_X \tau_c$ for $X\{^1H\}$ experiments; $X = {}^{13}C$, ^{31}P, and ^{19}F.

3.98) and for ^1H and ^{31}P ($\gamma_S/\gamma_I = 2.47$), as shown in Figure 2.12.[34] Table 2.1 lists gyromagnetic ratios for common nuclei.

For *negative* ratios of γ_S/γ_I, as for example with ^1H and ^{15}N, ^{29}Si or ^{43}Ca, the NOE enhancement is negative. Generally, this is likely to be a problem if NOE enhancements of close to -100% result, in which case the signal intensity is close to zero. This will occur most often in cases in which other relaxation pathways reduce the enhancement, or where extreme narrowing conditions no longer hold. The situation can be avoided by suppressing the NOE using gated irradiation.

When γ_S/γ_I is negative, the value of ω_I is also negative. This makes the terms of Eq. 2.24 take very different values from those when ω_I is positive. Consequently, the enhancements seen as a function of $\omega\tau_c$ are not simply the inverse of those with positive ratios of gyromagnetic ratios.[35] Typical results are shown in Figure 2.13, for $S = {}^1H$, and $I = {}^{119}Sn$, ^{29}Si, and ^{15}N ($\gamma_S/\gamma_I = -2.68$, -5.03, and -9.86, respectively). For large ratios of γ_S/γ_I, the effect is roughly the mirror image of that when the signs of the γ's are the same. However, for values of γ_S/γ_I between about -3 and -1, the enhancement is roughly -100% for all values of τ_c.

$^{15}N\{^1H\}$ enhancements in amines are somewhat idiosyncratic, in that they are very dependent on pH and the state of hydration of NH groups. Thus for glycine in 12N HCl, the enhancement is -493%; at pH 7.2 it is -60%, and at pH 9.1 it is -130%.[36] The reason for this is not entirely clear, but appears in this case to be related to changes in spin–rotation relaxation rather than exchange processes.

As noted in Section 2.4.3, NOE enhancements to quadrupolar nuclei are usually very small. One exception is ^6Li, which has $I = 1$, but a very small quadrupole moment, the smallest of any quadrupolar nucleus. Quadrupolar relaxation is thus inefficient, in one case enabling an Li{H} enhancement of

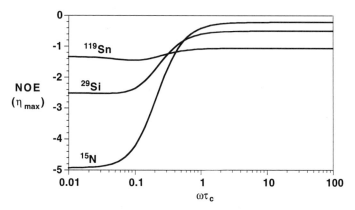

Figure 2.13. Dependence of maximum theoretical NOE enhancement on $\omega_X\tau_c$ for $X\{^1H\}$ experiments in cases where γ_X is negative; $X = {}^{119}Sn$, ${}^{29}Si$, and ${}^{15}N$.

220% to be recorded, 65% of the theoretical maximum. Measurement of the buildup rate of this enhancement permitted an estimation of the Li—H distance in a bridged boron compound (cf. Section 9.1.2).[37]

It is worth noting that many authors present the results of heteronuclear NOE experiments in the form I/I^0 (confusingly, this quantity is often called an "enhancement ratio" and given the symbol η). This quantity is related to the fractional NOE enhancement by a difference of one:

$$f_I\{S\} = \frac{I - I^0}{I^0} = \frac{I}{I^0} - 1 \tag{2.45}$$

Thus, on this scale, ${}^{13}C\{^1H\}$ values range from 2.99 at extreme narrowing to 1.15 at the slow tumbling limit, and similarly ${}^{15}N\{^1H\}$ values range from -3.94 to 0.78.

For completeness we should note that the full definitions of the relaxation rate constants for nuclei with a spin quantum number greater than 1/2 are

$$\sigma_{IS} = \tfrac{1}{15}K^2 I(I + 1)[6J(\omega_I + \omega_S) - J(\omega_I - \omega_S)]$$

$$\sigma_{SI} = \tfrac{1}{15}K^2 S(S + 1)[6J(\omega_I + \omega_S) - J(\omega_I - \omega_S)]$$

$$\rho_{IS} = \tfrac{1}{15}K^2 I(I + 1)[J(\omega_I - \omega_S) + 3J(\omega_I) + 6J(\omega_I + \omega_S)]$$

$$\rho_{SI} = \tfrac{1}{15}K^2 S(S + 1)[J(\omega_I - \omega_S) + 3J(\omega_S) + 6J(\omega_I + \omega_S)] \tag{2.46}$$

where I and S are the nuclear spin quantum numbers of nuclei I and S, and K is as before (cf. Eq. 2.29). For $I = 1/2$, these equations reduce to the equations given above (Eqs. 2.29 and 2.30).

2.6. AN EXTENSION TO THE SOLOMON EQUATIONS

If we write the Solomon equation (Eq. 2.10) for both dI_z/dt and for dS_z/dt (also using the definitions of σ and ρ to simplify the notation) we have a pair of coupled differential equations:

$$\frac{dI_z}{dt} = -\rho_{IS}(I_z - I_z^0) - \sigma_{IS}(S_z - S_z^0) \qquad (2.47)$$

$$\frac{dS_z}{dt} = -\rho_{IS}(S_z - S_z^0) - \sigma_{IS}(I_z - I_z^0) \qquad (2.48)$$

In section 2.1.2 we used Eq. 2.10 (equivalent to Eq. 2.47) to give a value for the steady-state NOE at I by setting dI_z/dt and S_z to zero and then rearranging. Bodenhausen[38] has pointed out that if one carries out the equivalent set of operations on the expression for dS_z/dt, this leads to an inconsistent and incorrect result:

Eq. 2.47 correctly yields

$$(I_z - I_z^0) = \frac{\sigma_{IS}}{\rho_{IS}}(S_z^0) \qquad (2.49)$$

Eq. 2.48 incorrectly yields

$$(I_z - I_z^0) = \frac{\rho_{IS}}{\sigma_{IS}}(S_z^0) \qquad (2.50)$$

In fact the set of just two coupled differential equations cannot correctly describe the behavior of the S spin, since the S spin precesses about the axis of the selective saturating field $\mathbf{B}_1$ producing an oscillating transverse component. For example, in the case of a $\mathbf{B}_1$ field applied along the y axis in the rotating frame, inclusion of transverse S spin magnetization in the theory leads to the following set of *three* coupled differential equations:

$$\frac{dI_z}{dt} = -\rho_{IS}(I_z - I_z^0) - \sigma_{IS}(S_z - S_z^0) \qquad (2.51)$$

$$\frac{dS_z}{dt} = -\rho_{IS}(S_z - S_z^0) - \sigma_{IS}(I_z - I_z^0) - \omega_1 S_x \qquad (2.52)$$

$$\frac{dS_x}{dt} = +\omega_1 S_z - \rho_{IS}^{tr} S_x \qquad (2.53)$$

where ω_1 is the rotation rate of the S spin magnetization about the $\mathbf{B}_1$ field axis and ρ_{IS}^{tr} is the rate constant for relaxation of the transverse component of S spin magnetization caused by dipolar relaxation from spin I.

Solution of this set of equations is a little more involved than that of the original two, and we will not work through it here. However, two points emerge: (i) *Both* Eqs. 2.51 and 2.52 now correctly yield Eq. 2.49 for the value of the steady-state NOE enhancement at the I spin; this lifts the paradox that Bodenhausen pointed out in the simpler formulation. (ii) If a subsaturating $\mathbf{B}_1$ field is used, then the steady-state S spin magnetization contributes both a longitudinal *and a transverse* component. This extended version of the theory becomes appropriate whenever the detailed behavior of the saturated spin(s) is relevant to the outcome of the experiment. Such cases include particularly the synchronous nutation experiments described in Section 7.5.4; further details may be found in reference 38.

REFERENCES

1. Solomon, I. *Phys. Rev.* 1955, *99*, 559.
2. Overhauser, A. W. *Phys. Rev.* 1953, *89*, 689; *ibid. 92*, 411.
3. Bloembergen, N.; Purcell, E. M.; Pound, R. V. *Phys. Rev.* 1948, *73*, 679 especially p. 697.
4. Noggle, J. H.; Schirmer, R. E. "The Nuclear Overhauser Effect," Academic Press, New York, 1971.
5. Jardetzky, O.; Roberts, G. C. K. "NMR in Molecular Biology," Academic Press, New York, 1981.
6. Debye, P. "Polar Molecules," Chemical Catalog Co., New York, 1929.
7. Weber, G. *Adv. Protein Chem.* 1953, *8*, 415.
8. Krishna, N. R.; Agresti, D. G.; Glickson, J. D.; Walter, R. *Biophys. J.* 1978, *24*, 791.
9. Glickson, J. D.; Gordon, S. L.; Pitner, T. P.; Agresti, D. G.; Walter, R. *Biochemistry* 1976, *15*, 5721.
10. Williamson, M. P.; Williams, D. H. *J. Chem. Soc., Chem. Commun.* 1982, 165.
11. Ley, S. V.; Neuhaus, D.; Williams, D. J. *Tetrahedron Lett.* 1982, *23*, 1207.
12. Kruse, L. I.; DeBrosse, C. W.; Kruse, C. H. *J. Am. Chem. Soc.* 1985, *107*, 5435.
13. Duben, A. J.; Hutton, W. C. *J. Am. Chem. Soc.* 1990, *112*, 5917.
14. Woessner, D. E. *J. Chem. Phys.* 1962, *37*, 647.
15. Harris, R. K. "Nuclear Magnetic Resonance Spectroscopy," Pitman, London, 1983, p. 114.
16. Kalk, A.; Berendsen, H. J. C. *J. Magn. Reson.* 1976, *24*, 343.
17. Bell, R. A.; Saunders, J. K. *Can. J. Chem.* 1970, *48*, 1114.
18. McConnell, J. "The Theory of Nuclear Magnetic Relaxation in Liquids," Cambridge University Press, Cambridge, 1987.
19. Tori, K.; Horibe, I.; Tamura, Y.; Tada, H. *J. Chem. Soc., Chem. Commun.* 1973, 620.

20. Gutowski, H. S.; Natusch, D. F. S. *J. Chem. Phys.* 1972, *57*, 1203.
21. Granot, J. *J. Magn. Reson.* 1982, *49*, 257.
22. Unger, S. W.; LeComte, J. T. J.; La Mar, G. N. *J. Magn. Reson.* 1985, *64*, 521.
23. Barbush, M.; Dixon, D. W. *Biochem. Biophys. Res. Commun.* 1985, *129*, 70.
24. Oldfield, E.; Norton, R. S.; Allerhand, A. *J. Biol. Chem.* 1975, *250*, 6368.
25. Llinás, M.; Klein, M. P.; Wüthrich, K. *Biophys. J.* 1978, *24*, 849.
26. Bleich, H. E.; Easwaran, K. R. K.; Glasel, J. A. *J. Magn. Reson.* 1978, *31*, 517.
27. Hull, W. E.; Sykes, B. D. *J. Mol. Biol.* 1975, *98*, 121.
28. Withers, S. G.; Madsen, N. B.; Sykes, B. D. *J. Magn. Reson.* 1985, *61*, 545.
29. Chazin, W. J.; Colebrook, L. D. *Magn. Reson. Chem.* 1985, *23*, 597.
30. Harris, R. K.; Kimber, B. J. *J. Chem. Soc., Chem. Commun.* 1973, 255.
31. Wehrli, F. W.; Wirthlin, T. "Interpretation of Carbon-13 NMR Spectra," Heyden, London, 1978.
32. Allerhand, A.; Doddrell, D.; Komoroski, R. *J. Chem. Phys.* 1971, *55*, 189.
33. Stille, D.; Doyle, J. R. in "Inorganic Synthesis" (J. Shreeve, Ed.) Vol. 24, Wiley, New York, 1986.
34. Doddrell, D.; Glushko, V.; Allerhand, A. *J. Chem. Phys.* 1972, *56*, 3683.
35. Werbelow, L. *J. Magn. Reson.* 1984, *57*, 136.
36. Leipert, T. K.; Noggle, J. H. *J. Am. Chem. Soc.* 1975, *97*, 269.
37. Avent, A. G.; Eaborn, C.; El-Kheli, M. N. A.; Molla, M. E.; Smith, J. D.; Sullivan, A. C. *J. Am. Chem. Soc.* 1986, *108*, 3854.
38. Boulat, B.; Bodenhausen, G. *J. Chem. Phys.* 1992, *97*, 6040.

CHAPTER 3

THE STEADY-STATE NOE IN RIGID MULTISPIN SYSTEMS

The principal motive for studying the NOE is to gain information on molecular geometry.

Our treatment of the NOE for two spins necessarily excluded any discussion of molecular geometry, since two-spin systems, by definition, have none. In effect, all *other* principal influences on the NOE were thus "factored out" and grouped together within the value of η_{max}, the maximum theoretical two-spin enhancement. Thus, we saw that η_{max} in homonuclear systems varies from $+50\%$ through 0 to -100% as the molecular tumbling rate decreases ($\omega\tau_c$ increases), and that in heteronuclear systems a further factor of γ_S/γ_I also determines the value of η_{max} (Section 2.4.1).

With the influences of these factors established, we are now in a position to concentrate on the central issue of molecular geometry. Our aim is to explain as clearly as possible, both in words and in equations, why and how steady-state NOE enhancements depend on molecular geometry. The equations will therefore be very highly idealized, since their purpose here is primarily to clarify rather than to predict; *not until the final section will equations be developed that might actually predict the numerical results of experiments.*

We begin by extending the Solomon equations to deal with a three-spin system; this treatment in fact introduces all the factors necessary to deal with a generalized *N*-spin system. As with the two-spin system, we will then explore the properties of the resulting equations that govern the NOE under differing conditions of molecular motion.

In the extreme narrowing limit we shall see that steady-state enhancements contain much explicit geometrical information, and some principles for the interpretation of data from such experiments will be developed. As the spin diffusion limit is approached, however, it will be seen that there is a progressive loss of information from the *steady-state* enhancements until, in the limit, all such enhancements in very slowly tumbling molecules become −100%. In these circumstances the steady-state NOE is essentially useless, and one must instead make kinetic measurements of NOE buildup rates, as discussed in Chapter 4.

Keeping to our policy of trying to introduce problems only one at a time, we shall assume for the moment that our spin system is first order, and that dipole–dipole relaxation predominates over all other mechanisms.

3.1. THE EQUATIONS

3.1.1. The Solomon Equations for More Than Two Spins

As with the two-spin case, the core of the theory required to understand NOE enhancements in multispin systems is contained in the relevant Solomon equations. We therefore begin this chapter by showing how these arise, using a (first-order) AMX spin system as an example. The energy level diagram and some transition probabilities for this spin system are shown in Figure 3.1. For consistency with our discussion of the two-spin case we shall refer to the three spins as I, S, and X (where, conventionally, S is saturated, and X is a third spin).

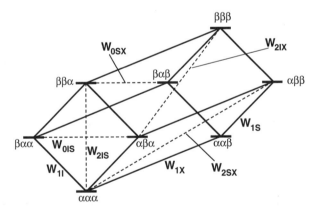

Figure 3.1. Energy level diagram for a first-order three-spin system comprising spins I, S, and X. Spin states are labeled according to the states of I (first), S (second), and X (third). For a transition between any two states that differ by the flip of only one or two spins, the corresponding transition probability can be deduced by inspection; for instance, the four transitions $\alpha\alpha\alpha \leftrightarrow \beta\alpha\alpha$, $\alpha\alpha\beta \leftrightarrow \beta\alpha\beta$, $\alpha\beta\alpha \leftrightarrow \beta\beta\alpha$, and $\alpha\beta\beta \leftrightarrow \beta\beta\beta$ all correspond to the transition probability W_{1I}, whereas the two transitions $\beta\alpha\alpha \leftrightarrow \alpha\alpha\beta$ and $\beta\beta\alpha \leftrightarrow \alpha\beta\beta$ both correspond to the transition probability W_{0IX}. Other examples are indicated on the diagram. Transitions involving the flip of more than two spins cannot be stimulated by the dipole–dipole interaction.

In order to avoid the rather long expressions that a treatment of the population flux through every energy level would generate, we shall work directly with the transition probabilities as described in Section 2.1.2. By exact analogy with the derivation of Eq. 2.14 we may write an expression for the rate of change of one population, say that of I, by considering for each energy level in turn only those transitions that affect the intensity of I:

$$k\frac{dI_z}{dt} = -2n_{\alpha\alpha\alpha}(W_{1I} + W_{2IS} + W_{2IX}) + 2n_{\beta\alpha\alpha}(W_{1I} + W_{0IS} + W_{0IX})$$

$$- 2n_{\alpha\beta\alpha}(W_{1I} + W_{0IS} + W_{2IX}) + 2n_{\beta\beta\alpha}(W_{1I} + W_{2IS} + W_{0IX})$$

$$- 2n_{\alpha\alpha\beta}(W_{1I} + W_{2IS} + W_{0IX}) + 2n_{\beta\alpha\beta}(W_{1I} + W_{0IS} + W_{2IX})$$

$$- 2n_{\alpha\beta\beta}(W_{1I} + W_{0IS} + W_{0IX}) + 2n_{\beta\beta\beta}(W_{1I} + W_{2IS} + W_{2IX}) \quad (3.1)$$

As before, the populations n_i represent population differences from equilibrium, $n_i = N_i - N_i^0$. Equation 3.1 simplifies to

$$k\frac{dI_z}{dt} = 2(n_{\beta\beta\beta} - n_{\alpha\alpha\alpha})(W_{1I} + W_{2IS} + W_{2IX})$$

$$+ 2(n_{\beta\alpha\alpha} - n_{\alpha\beta\beta})(W_{1I} + W_{0IS} + W_{0IX})$$

$$+ 2(n_{\beta\beta\alpha} - n_{\alpha\alpha\beta})(W_{1I} + W_{2IS} + W_{0IX})$$

$$+ 2(n_{\beta\alpha\beta} - n_{\alpha\beta\alpha})(W_{1I} + W_{0IS} + W_{2IX}) \quad (3.2)$$

We need next to relate these expressions to quantities that we can measure, namely the intensities of I, S, and X. Just as before, these intensities depend on the corresponding z magnetizations, and their deviations from equilibrium may be written

$$k(I_z - I_z^0) = -n_{\alpha\alpha\alpha} + n_{\beta\alpha\alpha} - n_{\alpha\beta\alpha} + n_{\beta\beta\alpha} - n_{\alpha\alpha\beta} + n_{\beta\alpha\beta} - n_{\alpha\beta\beta} + n_{\beta\beta\beta}$$

$$k(S_z - S_z^0) = -n_{\alpha\alpha\alpha} - n_{\beta\alpha\alpha} + n_{\alpha\beta\alpha} + n_{\beta\beta\alpha} - n_{\alpha\alpha\beta} - n_{\beta\alpha\beta} + n_{\alpha\beta\beta} + n_{\beta\beta\beta}$$

$$k(X_z - X_z^0) = -n_{\alpha\alpha\alpha} - n_{\beta\alpha\alpha} - n_{\alpha\beta\alpha} - n_{\beta\beta\alpha} + n_{\alpha\alpha\beta} + n_{\beta\alpha\beta} + n_{\alpha\beta\beta} + n_{\beta\beta\beta} \quad (3.3)$$

In the two-spin case, the analogous rearrangement of Eq. 2.8 into Eq. 2.10 was trivial. In the three-spin case, however, the situation is slightly more involved, and the terms of Eq. 3.2 must be permuted somewhat before the population terms can be equated with sums and differences of the intensities $(I_z - I_z^0)$, $(S_z - S_z^0)$, and $(X_z - X_z^0)$:

$$k\frac{dI_z}{dt} = 2(n_{\beta\beta\beta} - n_{\alpha\alpha\alpha} + n_{\beta\beta\alpha} - n_{\alpha\alpha\beta})(W_{1I} + W_{2IS})$$

$$+ 2(n_{\beta\alpha\alpha} - n_{\alpha\beta\beta} + n_{\beta\alpha\beta} - n_{\alpha\beta\alpha})(W_{1I} + W_{0IS})$$

$$+ 2(n_{\beta\beta\beta} - n_{\alpha\alpha\alpha} + n_{\beta\alpha\beta} - n_{\alpha\beta\alpha})W_{2IX}$$

$$+ 2(n_{\beta\alpha\alpha} - n_{\alpha\beta\beta} + n_{\beta\beta\alpha} - n_{\alpha\alpha\beta})W_{0IX} \tag{3.4}$$

from which, using Eq. 3.3, we may write:

$$\frac{dI_z}{dt} = -[(I_z - I_z^0) + (S_z - S_z^0)](W_{1I} + W_{2IS})$$

$$- [(I_z - I_z^0) - (S_z - S_z^0)](W_{1I} + W_{0IS})$$

$$- [(I_z - I_z^0) + (X_z - X_z^0)]W_{2IX}$$

$$- [(I_z - I_z^0) - (X_z - X_z^0)]W_{0IX} \tag{3.5}$$

Finally, by grouping terms in I_z, S_z, and X_z, we obtain

$$\frac{dI_z}{dt} = -(I_z - I_z^0)(2W_{1I} + W_{2IS} + W_{0IS} + W_{2IX} + W_{0IX})$$

$$- (S_z - S_z^0)(W_{2IS} - W_{0IS})$$

$$- (X_z - X_z^0)(W_{2IX} - W_{0IX}) \tag{3.6}$$

This is the Solomon equation for three spins. In fact, this treatment may be quite simply extended to deal with a generalized (first-order) spin system. The steps in such a derivation are exactly analogous to those above (the principal problem being one of notation), so we simply give the result:

$$\frac{dI_z}{dt} = -(I_z - I_z^0)\left[2W_{1I} + W_{2IS} + W_{0IS} + \sum_X (W_{2IX} + W_{0IX})\right]$$

$$- (S_z - S_z^0)(W_{2IS} - W_{0IS})$$

$$- \sum_X (X_z - X_z^0)(W_{2IX} - W_{0IX}) \tag{3.7}$$

where the summations over X include all spins other than I and S.

Note that in our derivation, only transitions involving two spins were considered. Multispin transitions, such as $n_{\alpha\alpha\alpha} \leftrightarrow n_{\beta\beta\beta}$ or $n_{\alpha\beta\alpha} \leftrightarrow n_{\beta\alpha\beta}$, are ignored since they do not correspond to dipole–dipole transitions. In this sense, the treatment of multispin NOE enhancements can be regarded as a pairwise summation over all two-spin interactions.

From Section 2.3.1 we also have that

$$R_I^{DD} = \left[2W_{1I} + W_{2IS} + W_{0IS} + \sum_X (W_{2IX} + W_{0IX}) \right]$$

$$= \rho_{IS} + \sum_X \rho_{IX} \qquad (3.8)$$

Using this expression, which in the three-spin case at least can be deduced by inspection of the energy level diagram (Fig. 3.1), and the definition of σ (Eq. 2.22), we can simplify Eq. 3.7 to

$$\frac{dI_z}{dt} = -(I_z - I_z^0)R_I^{DD} - (S_z - S_z^0)\sigma_{IS} - \sum_X (X_z - X_z^0)\sigma_{IX} \qquad (3.9)$$

3.1.2. Cross-Correlation

As we have seen, the multispin Solomon equations effectively treat relaxation behavior as a pairwise summation over all two-spin interactions. In so doing, however, one subtlety is overlooked. Although the motion of spin S relative to spin I is of itself random, and the motion of spin X relative to spin I, if considered independently, is also random, these two motions are not completely random *with respect to each other*. They cannot be, since preservation of the molecular structure necessarily limits those relative motions of S, I, and X that are possible. Such motions are said to be cross-correlated.

It is clear that cross-correlation must influence relaxation behavior to some extent. At a very qualitative level, we may take as an example the relaxation of a $^{13}CH_2$ group. We know that the relaxation of the central carbon nucleus depends on the fluctuating field it experiences due to the two protons moving around it randomly, but in a cross-correlated fashion, as the molecule tumbles. Our two-spin treatment tells us correctly what the influence of each proton is separately. However, the cross-correlation between their motion means that a simple addition of their contributions will not give quite a true picture. A simple analogy would be with the addition properties of noise, as during normal spectroscopic signal averaging. If uncorrelated samples of noise are summed, the total amplitude increases as the square root of the number of acquisitions, but if partially *correlated* samples of noise were to be summed, the total amplitude would increase differently (in this case more quickly). This is illustrated by Fig. 3.2.

In fact it proves to be quite difficult to incorporate cross-correlation effects into relaxation theory, and it can be done only by using a considerably more formal and advanced quantum-mechanical treatment than would be appropriate to describe here. Suffice it to say at this point that the modifications required to the Solomon equations can be expressed as an extra contribution to dI_z/dt, but this contribution can be formulated only in terms of variables (actually

Figure 3.2. Addition properties of (*a*) correlated and (*b*) random noise. In (*a*), the same sample of noise is added to itself, generating (top trace) noise with 2 times the original amplitude. In (*b*), independent samples of noise are added, generating (top trace) noise with $\sqrt{2}$ times the original amplitude. The difference between (*a*) and (*b*) arises because in (*a*) the noise is (completely) correlated, whereas in (*b*) it is uncorrelated.

multispin operators; cf. Section 6.2.1) that usually do not in themselves correspond to observable quantities, and are also difficult to conceptualize. The involvement of multispin operators is in a sense inevitable, since single-spin operators such as I_z and S_z cannot be expected to cope successfully with properties that depend on the simultaneous order between nuclei, which is essentially what cross-correlation is.

For these reasons, it is an almost universal practice to ignore cross-correlation completely, and work with the Solomon equations as presented in the previous section (Eq. 3.9). This practice is justified in Section 6.3, where it is shown that the net effect on ^{1}H multiplets of cross-correlation is almost always negligible. There is further discussion of cross-correlation in papers by Werbelow and Grant,[1] Tropp,[2] and Bull[3]; some indication of the theory is also given in Appendix II.

3.1.3. Two General Solutions to the Multispin Solomon Equations

Now that we have the relevant Solomon equations for multispin systems, we can use them to obtain equations defining steady-state NOE enhancements in terms of cross-relaxation and total relaxation rates. Two cases are of particular interest: saturation of only one spin and saturation of all spins except one.

3.1.3.1. Saturation of One Spin.
This is, of course, the situation encountered in the overwhelming majority of selective NOE experiments, both homonuclear and heteronuclear. If the saturated spin is S then, at steady state, $S_z = 0$ and $dI_z/dt = 0$. Substitution into the Solomon equation (Eq. 3.9) gives

$$-(I_z - I_z^0)R_I^{DD} + S_z^0 \sigma_{IS} - \sum_X (X_z - X_z^0)\sigma_{IX} = 0 \qquad (3.10)$$

We know that the equilibrium intensities I_z^0, S_z^0, and X_z^0 are related by

$$I_z^0 = \frac{\gamma_I}{\gamma_S} S_z^0 = \frac{\gamma_I}{\gamma_X} X_z^0 \qquad (3.11)$$

so that $f_I\{S\}$ and $f_X\{S\}$, the fractional enhancements of I and X on saturating S, are given by

$$f_I\{S\} = \frac{I_z - I_z^0}{I_z^0} = \frac{\gamma_S}{\gamma_I} \frac{I_z - I_z^0}{S_z^0}$$

$$f_X\{S\} = \frac{X_z - X_z^0}{X_z^0} = \frac{\gamma_S}{\gamma_X} \frac{X_z - X_z^0}{S_z^0} \qquad (3.12)$$

This allows us to rearrange Eq. 3.10 into what might be termed the "master equation" for steady-state multispin NOE enhancements:

$$f_I\{S\} = \frac{\gamma_S}{\gamma_I} \frac{1}{R_I^{DD}} \left[\sigma_{IS} - \sum_X \frac{(X_z - X_z^0)}{S_z^0} \sigma_{IX} \right]$$

(3.13)

$$= \frac{\gamma_S}{\gamma_I} \frac{1}{R_I^{DD}} \left[\sigma_{IS} - \sum_X \frac{\gamma_X}{\gamma_S} f_X\{S\} \sigma_{IX} \right]$$

This equation forms the basis for most of the remainder of this chapter, and in an ideal world would also be the basis for the interpretation of all steady-state NOE experiments. We shall return to a discussion of the *practical* utility of Eq. 3.13 at the close of this chapter, but for now we should briefly recall our initial assumptions: Eq. 3.13 relates to a first-order spin system whose relaxation occurs *entirely via the dipole–dipole mechanism*.

3.1.3.2. Saturation of All Spins Except One.

This situation might at first sight seem unlikely, but it includes a particularly common experiment in which NOE enhancements are observed; acquisition of a broad-band ^{1}H decoupled ^{13}C spectrum.

If all the spins except I are saturated then, at steady state, $S_z = 0$, $X_z = 0$ (for all X), and $dI_z/dt = 0$. Substitution into the Solomon equation (Eq. 3.9) gives

$$-(I_z - I_z^0)R_I^{DD} + S_z^0 \sigma_{IS} + \sum_X X_z^0 \sigma_{IX} = 0$$

(3.14)

from which, using Eq. 3.11, we have

$$f_I\{M\} = \frac{\gamma_S}{\gamma_I} \frac{\sigma_{IS}}{R_I^{DD}} + \sum_X \frac{\gamma_X}{\gamma_I} \frac{\sigma_{IX}}{R_I^{DD}}$$

(3.15)

where M represents all spins other than I (that is spin S and all the spins X). If all the spins M are of the same nuclear species (as in case of ^{1}H decoupled ^{13}C acquisition) then we may write

$$f_I\{M\} = \sum_M \frac{\gamma_M}{\gamma_I} \frac{\sigma_{IM}}{R_I^{DD}}$$

(3.16)

This equation, much more so than Eq. 3.13, is reminiscent of Eq. 2.33, the definition of the maximum theoretical two-spin enhancement $\eta_{\max}$. In fact, on closer inspection, we shall see that it is identical. If we substitute back the full detail of Eq. 3.16, using the definitions of R_I^{DD} (Eq. 3.8), σ_{IM} (Eq. 2.22), and the various W transition probabilities (Eqs. 2.21), then we are led to

70 THE STEADY-STATE NOE IN RIGID MULTISPIN SYSTEMS

$$f_I\{M\} = \sum_M \frac{\gamma_M}{\gamma_I} \left(\frac{W_{2IM} - W_{0IM}}{W_{0IM} + 2W_{1IM} + W_{2IM}} \right)$$

$$= \sum_M \frac{\left(\dfrac{\gamma_M}{\gamma_I}\right)\left[K_{IM}^2 \dfrac{6}{1+(\omega_I+\omega_M)^2 \tau_c^2} - K_{IM}^2 \dfrac{1}{1+(\omega_I-\omega_M)^2 \tau_c^2} \right]}{\left[K_{IM}^2 \dfrac{1}{1+(\omega_I-\omega_M)^2 \tau_c^2} + K_{IM}^2 \dfrac{3}{1+\omega_I^2 \tau_c^2} + K_{IM}^2 \dfrac{6}{1+(\omega_I+\omega_M)^2 \tau_c^2} \right]}$$

(3.17)

where $K_{IM}^2 = (\mu_0/4\pi)^2 \hbar^2 \gamma_I^2 \gamma_M^2 r_{IM}^{-6}$.

Given that variations among the frequencies ω_M are negligibly small relative to the values of $(\omega_I + \omega_M)$ and $(\omega_I - \omega_M)$, the summations over M may be carried out over the K_{IM}^2 terms only, the remaining terms being treated as multiplicative constants:

$$f_I\{M\} = \frac{\left(\dfrac{\gamma_M}{\gamma_I}\right)\left[\dfrac{6 \sum_M K_{IM}^2}{1+(\omega_I+\omega_M)^2 \tau_c^2} - \dfrac{\sum_M K_{IM}^2}{1+(\omega_I-\omega_M)^2 \tau_c^2}\right]}{\left[\dfrac{\sum_M K_{IM}^2}{1+(\omega_I-\omega_M)^2 \tau_c^2} + \dfrac{3\sum_M K_{IM}^2}{1+\omega_I^2 \tau_c^2} + \dfrac{6\sum_M K_{IM}^2}{1+(\omega_I+\omega_M)^2 \tau_c^2}\right]}$$

(3.18)

where $\sum_M K_{IM}^2 = (\mu_0/4\pi)^2 \hbar^2 \gamma_I^2 \gamma_M^2 \sum_M (r_{IM})^{-6}$.

Quite clearly, the terms K_{IM}^2 all cancel, leaving us with an expression identical to Eq. 2.24 *irrespective of the geometry of the spin system* or even of the number of spins in it. In other words, we get the maximum theoretical two-spin enhancement, η_{max}, on observing I and saturating all other spins in *any* spin system in which all the spins M are of one nuclear species and relaxation is entirely dipole-dipole:

$$f_I\{M\} = \sum_M \frac{\gamma_M}{\gamma_I} \frac{\sigma_{IM}}{R_I^{DD}} = \eta_{max} \quad (3.19)$$

This result is perhaps not of direct relevance here, but it was included to try to clarify further the relationship between η_{max} and steady-state NOE enhancements in multispin systems. We may also note that it rationalizes another familiar fact, namely that resonances due to CH, CH_2, and CH_3 groups all receive (approximately) the same NOE as one another in proton noise-decoupled ^{13}C spectra ($f_{^{13}C}\{^1H\} \approx \gamma_H/2\gamma_C \approx 1.99$ in mobile molecules). Were it not for the more effective competition of non-dipole–dipole relaxation mechanisms for quaternary carbon resonances, even these would receive the same enhancement.

3.1.4. Internuclear Distances and Steady-State NOE Enhancements

What we really want from NOE theory is an understanding of the relationship between enhancements and relative internuclear distances. The "master equation" Eq. 3.13 derived in the previous section came close to fulfilling this requirement as far as steady-state enhancements are concerned, but we still need to make explicit the distance dependencies of the various relaxation terms.

So far we have assumed only that dipole–dipole relaxation predominates and that our spin system is first order. We may write Eq. 3.13 in terms of transition probabilities:

$$f_I\{S\} = \left(\frac{\gamma_S}{\gamma_I}\right) \left[\frac{(W_{2IS} - W_{0IS}) - \sum_X (\gamma_X/\gamma_S) f_X\{S\}(W_{2IX} - W_{0IX})}{2W_{1I} + W_{2IS} + W_{0IS} + \sum_X (W_{2IX} + W_{0IX})}\right] \quad (3.20)$$

If we now assume that our spin system forms part of a rigid molecule tumbling isotropically at a rate described by a single correlation time τ_c, we may express the transition probabilities as follows:

$$W_{1I} = \frac{3}{20} K_I^2 \frac{\tau_c}{1 + \omega_I^2 \tau_c^2}$$

$$W_{2IS} = \frac{3}{5} K_{IS}^2 \frac{\tau_c}{1 + (\omega_I + \omega_S)^2 \tau_c^2}$$

$$W_{0IS} = \frac{1}{10} K_{IS}^2 \frac{\tau_c}{1 + (\omega_I - \omega_S)^2 \tau_c^2}$$

$$W_{2IX} = \frac{3}{5} K_{IX}^2 \frac{\tau_c}{1 + (\omega_I - \omega_X)^2 \tau_c^2}$$

$$W_{0IX} = \frac{1}{10} K_{IX}^2 \frac{\tau_c}{1 + (\omega_I - \omega_X)^2 \tau_c^2} \quad (3.21)$$

where

$$K_I^2 = \left(\frac{\mu_0}{4\pi}\right)^2 \left(\hbar^2 \gamma_I^2 \gamma_S^2 r_{IS}^{-6} + \sum_X \hbar^2 \gamma_I^2 \gamma_X^2 r_{IX}^{-6}\right)$$

$$K_{IS}^2 = \left(\frac{\mu_0}{4\pi}\right)^2 \hbar^2 \gamma_I^2 \gamma_S^2 r_{IS}^{-6}$$

$$K_{IX}^2 = \left(\frac{\mu_0}{4\pi}\right)^2 \hbar^2 \gamma_I^2 \gamma_X^2 r_{IX}^{-6} \quad (3.22)$$

The simplest equations arise for molecules that tumble rapidly and isotrop-

72 THE STEADY-STATE NOE IN RIGID MULTISPIN SYSTEMS

ically. As we saw in the previous chapter, for such molecules the "extreme narrowing limit," $\omega\tau_c \ll 1$, holds and all the (directly) motion-dependent terms vanish from Eqs. 3.21, leaving

$$f_I\{S\} = \frac{\gamma_S}{\gamma_I} \left[\frac{\left(\frac{3}{5} - \frac{1}{10}\right) K_{IS}^2 - \sum_X (\gamma_X/\gamma_S) f_X\{S\} \left(\frac{3}{5} - \frac{1}{10}\right) K_{IX}^2}{\frac{6}{20} K_I^2 + \left(\frac{3}{5} + \frac{1}{10}\right) K_{IS}^2 + \sum_X \left(\frac{3}{5} + \frac{1}{10}\right) K_{IX}^2} \right] \quad (3.23)$$

from which

$$f_I\{S\} = \frac{1}{2} \frac{\gamma_S}{\gamma_I} \left[\frac{r_{IS}^{-6} - \sum_X (\gamma_X/\gamma_S) f_X\{S\} r_{IX}^{-6}}{r_{IS}^{-6} + \sum_X r_{IX}^{-6}} \right] \quad (3.24)$$

This (at last) is an expression relating steady-state NOE enhancements directly with internuclear distances, albeit only for first-order spin systems in rigid, rapidly and isotropically tumbling molecules whose relaxation is entirely dipole–dipole. In the homonuclear case, Eq. 3.24 simplifies to

$$f_I\{S\} = \frac{1}{2} \left[\frac{r_{IS}^{-6} - \sum_X f_X\{S\} r_{IX}^{-6}}{r_{IS}^{-6} + \sum_X r_{IX}^{-6}} \right] \quad (3.25)$$

Another fairly simple equation arises in the case of a first-order homonuclear spin system in a rigid, isotropically tumbling molecule not necessarily at the extreme narrowing limit. Since only a single frequency ($\omega = \omega_I = \omega_S = \omega_X$) and a single correlation time are involved we may write, as in Chapter 2:

$$\eta_{max} = \left[\left(\frac{6\tau_c}{1 + 4\omega^2\tau_c^2} - \tau_c \right) \middle/ \left(\tau_c + \frac{3\tau_c}{1 + \omega^2\tau_c^2} + \frac{6\tau_c}{1 + 4\omega^2\tau_c^2} \right) \right] \quad (3.26)$$

$$= \frac{5 + \omega^2\tau_c^2 - 4\omega^4\tau_c^4}{10 + 23\omega^2\tau_c^2 + 4\omega^4\tau_c^4}$$

Without giving the full detail, we may see that a factor of η_{max} may be extracted from Eq. 3.20, leaving, by analogy with Eq. 3.25,

$$f_I\{S\} = \eta_{max} \left[\frac{r_{IS}^{-6} - \sum_X f_X\{S\} r_{IX}^{-6}}{r_{IS}^{-6} + \sum_X r_{IX}^{-6}} \right] \quad (3.27)$$

Clearly, Eq. 3.25 is one specific case of the more general Eq. 3.27. Note, however, that this separation of the motion-dependent terms into the factor η_{max} is possible only in the absence of external relaxation ($\rho^* = 0$; cf. Section 3.3.3). Other (more complicated) expressions could be deduced for more specialized situations, such as heteronuclear enhancements not at the extreme narrowing limit. We shall not deal with these cases explicitly, but it should be clear how such expressions could be derived by substituting the full detail of Eqs. 3.21 and 3.22 into Eq. 3.20.

Having now obtained our working equations, Eqs. 3.25 and 3.27, we shall next examine what they tell us about the NOE.

3.2. WHAT THE EQUATIONS MEAN

3.2.1. General: Direct Enhancements and Spin Diffusion

Perhaps the most obvious, and frequently overlooked, implication of Eqs. 3.25 and 3.27 is that the steady-state NOE is not simply an "interatomic ruler"; each enhancement depends on the relative positions of *all* the nearby spins, not merely those saturated and observed.

Furthermore, there are two distinct types of contribution to a given enhancement predicted by the equations. The first we shall call the "direct" contribution, and clearly it represents the *proportion of the total relaxation of the observed spin (I) that occurs by cross-relaxation with the saturated spin (S)*. Within Eq. 3.27 this corresponds to the term

$$f_I\{S\}_{\text{direct}} = \eta_{max} \frac{r_{IS}^{-6}}{r_{IS}^{-6} + \sum_X r_{IX}^{-6}} \qquad (3.28)$$

If this were the only contribution to NOE enhancements, matters would be much simpler, since, for one, there would be no such thing as spin diffusion (of which more shortly). As can be seen, however, Eq. 3.27 does contain another term:

$$f_I\{S\}_{\text{indirect}} = -\eta_{max} \sum_X \left[\frac{f_X\{S\} r_{IX}^{-6}}{r_{IS}^{-6} + \sum_X r_{IX}^{-6}} \right] \qquad (3.29)$$

The second term represents the "indirect" contribution to $f_I\{S\}$. Partly this corresponds to an enhancement at I that arrived via cross-relaxation of S with some third spin X *followed by* cross-relaxation of X with I. Such an indirect pathway should not surprise us; an appreciable enhancement at X means, by definition, that the populations at X are themselves not at equilibrium, and therefore cross-relaxation of X with its near neighbors, including I, is bound to perturb their populations in turn.

However, Eq. 3.29 is not necessarily limited to indirect contributions that pass through only one intermediate spin. The enhancement $f_X\{S\}$ will itself contain an indirect term, which in turn will contain still other enhancements carrying further indirect terms, and so on, so that indirect pathways over any number of intermediate spins are in fact represented. We may picture the population disturbance, initially present only at S, spreading outward through the molecule by cross-relaxation from spin to spin until, at steady state, every spin is affected to a greater or lesser extent.

This process of "propagation" of population disturbance is generally, and often somewhat loosely, referred to as *spin diffusion*. It has very markedly different properties in the positive NOE regime ($\eta_{max} > 0$) than it does in the negative NOE regime ($\eta_{max} < 0$), to the extent that many suppose it exists only in the negative NOE regime. More than anything else, it is the differing significance of spin diffusion that distinguishes the two NOE regimes, and controls the different ways in which NOE experiments must be designed in the two cases.

Before we consider these two cases in greater detail in the next sections, we shall therefore look more generally at the properties that distinguish the two NOE regimes, since they can be deduced quite simply from Eq. 3.27. In order to do this, however, we must first deal with one slight mathematical obstacle, namely that Eq. 3.27 predicts enhancements in terms not merely of distances, but also of other enhancements. If one spin in a given spin system is saturated, a set of simultaneous equations each of the form of Eq. 3.27 will exist among the various internuclear distances and enhancements. If we wish to express each enhancement in terms of internuclear distances (and η_{max}) alone, then this can be done by rearranging the simultaneous equations, but at the cost of generating some rather unwieldly expressions. Table 3.1 summarizes the results of this somewhat tedious process for a three-spin and a four-spin system.

If, for simplicity, we take the case of a three-spin system, then direct substitution of the internuclear distances into Eq. 3.30 gives

$$f_I\{S\} = \left[\frac{\eta_{max}p}{(p+q)} - \frac{\eta_{max}^2 qr}{(p+q)(q+r)}\right] \Bigg/ \left[\eta_{max} - \frac{\eta_{max}^2 q^2}{(q+1)(p+q)}\right] \quad (3.32)$$

$$= \frac{\eta_{max}(pq + pr) - \eta_{max}^2 qr}{pq + qr + pr + (1 - \eta_{max}^2)q^2}$$

where $p = r_{IS}^{-6}$, $q = r_{IX}^{-6}$, and $r = r_{SX}^{-6}$.

The geometry of the spin system is now represented within the terms p, q, and r.

It is quite satisfying to note that many important and general properties of the homonuclear steady-state NOE can be deduced simply from inspection of Eqs. 3.27 and 3.32:

1. In the positive NOE regime, that is when $\eta_{max} > 0$, most enhancements are positive, but it is possible for certain enhancements to be negative,

TABLE 3.1 Steady-State NOE Enhancements for Three- and Four-Spin Systems in the Absence of External Relaxation

For four spins I, S, X, and Y, we define $\eta_{max} = \dfrac{5 + \omega^2\tau_c^2 - 4\omega^4\tau_c^4}{10 + 23\omega^2\tau_c^2 + 4\omega^4\tau_c^4}$;

$$\sum I = r_{IX}^{-6} + r_{IS}^{-6} + r_{IY}^{-6}; \qquad \sum X = r_{IX}^{-6} + r_{XS}^{-6} + r_{XY}^{-6};$$

$$\sum Y = r_{IY}^{-6} + r_{YS}^{-6} + r_{XY}^{-6};$$

$$a = \dfrac{\eta_{max} r_{IS}^{-6}}{\sum I} \qquad d = \dfrac{\eta_{max} r_{XS}^{-6}}{\sum X} \qquad g = \dfrac{\eta_{max} r_{YS}^{-6}}{\sum Y};$$

$$b = \dfrac{\eta_{max} r_{IX}^{-6}}{\sum I} \qquad e = \dfrac{\eta_{max} r_{IX}^{-6}}{\sum X} \qquad h = \dfrac{\eta_{max} r_{IY}^{-6}}{\sum Y};$$

$$c = \dfrac{\eta_{max} r_{IY}^{-6}}{\sum I} \qquad f = \dfrac{\eta_{max} r_{XY}^{-6}}{\sum X} \qquad i = \dfrac{\eta_{max} r_{XY}^{-6}}{\sum Y};$$

For the three-spin case (i.e., ignoring spin Y):

$$f_I\{S\} = a - b(f_X\{S\}) \qquad\qquad f_I\{S\} = \dfrac{a - db}{1 - be}$$

from which (3.30)

$$f_X\{S\} = d - e(f_I\{S\}) \qquad\qquad f_X\{S\} = \dfrac{ae - d}{be - 1}$$

For the four-spin case:

$$f_I\{S\} = a - b(f_X\{S\}) - c(f_Y\{S\})$$

$$f_X\{S\} = d - e(f_I\{S\}) - f(f_Y\{S\})$$

$$f_Y\{S\} = g - h(f_I\{S\}) - i(f_X\{S\})$$

from which

$$f_I\{S\} = \dfrac{[(a - bd)(c - b/i)] - [(a - gb/i)(c - fb)]}{[(1 - eb)(c - b/i)] - [(1 - hb/i)(c - fb)]}$$

$$f_X\{S\} = \dfrac{[(a - d/e)(c - 1/h)] - [(a - g/h)(c - f/e)]}{[(b - 1/e)(c - 1/h)] - [(b - i/h)(c - f/e)]} \qquad (3.31)$$

$$f_Y\{S\} = \dfrac{[(a - d/e)(b - i/h)] - [(a - g/h)(b - 1/e)]}{[(c - f/e)(b - i/h)] - [(c - 1/h)(b - 1/e)]}$$

depending on the geometry of the spin system. The maximum positive enhancement possible is η_{max}, whereas the greatest negative enhancement possible (for $\eta_{max} > 0$) must be somewhat smaller (i.e., less in magnitude) than $-\eta_{max}^2$.

2. In the negative NOE regime, that is when $\eta_{max} < 0$, *all* enhancements must be negative. The maximum enhancement possible is again η_{max}.
3. At the extreme limit of slow tumbling, that is when $\eta_{max} = -1$, all enhancements must be -1, irrespective of the geometry of the spin system (cf. Eq. 3.32).
4. At the zero crossing of the NOE versus $\omega\tau_c$ curve, that is, when $\eta_{max} = 0$, all enhancements vanish (for a rigid molecule; see also Section 3.2.3).

3.2.2. Interpretation at the Extreme Narrowing Limit ($\omega\tau_c \ll 1$)

In this and the following sections we try to answer the question, how is the molecular geometry reflected in the observed pattern of NOE enhancements for a given spin system?

All of the theoretical examples will be calculated using Eqs. 3.30 and 3.31 (Table 3.1). Indeed, one might argue that these, and Eq. 3.27 from which they are derived, in themselves answer the question. Our approach here, however, will be more qualitative. The aim will be to show the relative importance of the various contributions to the NOE under different circumstances so that the reader may, if presented with a real molecular structure, have a fair idea of the likely relative values of enhancements without necessarily resorting to calculation.

As we have seen, a considerable simplification of the NOE equations is possible for small, rapidly tumbling molecules, since when $\omega\tau_c \ll 1$ all the motion-dependent terms vanish leaving only Eq. 3.25. This is highly desirable, since in general we will not know the value of τ_c (the correlation time for molecular tumbling), or will have at best an order of magnitude "guesstimate" for it. Provided, then, that we can reasonably assume that the extreme narrowing condition is fulfilled we may thereafter ignore τ_c. The modification required to our approach when the extreme narrowing approximation fails will be considered in Section 3.2.3.

3.2.2.1. Direct Effects.
In the majority of real cases, the enhancements that are likely to be of most interest are those between adjacent spins. For such enhancements, the major contribution predicted by Eq. 3.25 will usually be the direct one, given by Eq. 3.28 (we must say at the outset, however, that there are serious exceptions to this, as the section on indirect effects will show). The direct contribution, as we have said, reflects the *proportion* of the total relaxation of I that occurs via cross-relaxation with S. At the simplest level the shorter the internuclear distance r_{IS}, the larger the corresponding enhancement $f_I\{S\}$ is likely to be. To progress beyond this dangerously oversimplified view,

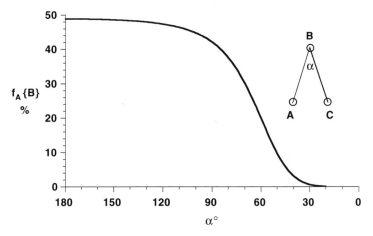

Figure 3.3. Calculated steady-state enhancement $f_A\{B\}$ at the extreme narrowing limit in a three-spin system with $r_{AB} = r_{CB}$. As the angle α decreases, so the contribution that spin C makes to the relaxation of spin A increases. For $\alpha > 90°$, spin C has little effect, but for smaller angles spin C rapidly comes to dominate the relaxation of spin A, drastically reducing the enhancement $f_A\{B\}$. External relaxation (ρ^*) was not included in these calculations.

we need next to consider the positions of other near neighbors of the observed spin, I. Clearly, the presence of spins *other* than S that cross-relax with I must diminish the enhancement $f_I\{S\}$, since they introduce distance terms that increase the denominator of Eq. 3.28 without affecting the numerator. The extent of this "dilution" increases dramatically the closer these other spins are to I, as is very clearly demonstrated by the calculated examples shown in Figures 3.3 and 3.4.

These arguments lead to some very important practical generalizations concerning steady-state NOE enhancements:

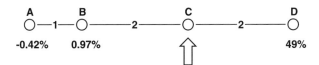

Figure 3.4. Calculated steady-state enhancements in the extreme narrowing limit on irradiating spin C in the four-spin system illustrated. Although spins B and D are equidistant from spin C, the enhancement at B is drastically reduced by the much more efficient relaxation spin B receives from spin A. Spin D, on the other hand, suffers no such loss, and receives essentially the full enhancement. This illustrates the danger of trying to equate steady-state enhancements directly with distance. External relaxation (ρ^*) was not included in these calculations; it would have the effect of reducing the difference between $f_B\{C\}$ and $f_D\{C\}$.

1. In general, enhancements are not symmetrical. In other words, an enhancement $f_A\{B\}$, measured by irradiating B and observing A, will usually be different from its reverse, $f_B\{A\}$, measured by irradiating A and observing B. This is simply because the distribution of other near neighbors around A is usually different from that around B.
2. Enhancements into methyl or methylene groups will generally be smaller than those at methine protons, other things being equal. This is because, if it is to produce an enhancement, any spin outside the methyl or methylene group must compete with very efficient cross-relaxation between the closely spaced protons within that group (cf. Section 3.2.2.4). Note, however, that an enhancement of a methyl singlet, although numerically smaller, may be *easier to measure* than a larger enhancement of a multiplet (Fig. 3.5).

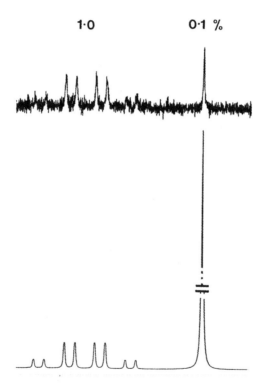

Figure 3.5. Simulated enhancements of a methyl singlet and a methylene multiplet in a noisy NOE difference spectrum. Although the methyl singlet shows the smaller fractional enhancement (0.1%, as opposed to 1% for the methylene multiplet), it is easier to measure since its intensity is concentrated into one line, giving it a much higher signal-to-noise ratio than the methylene multiplet. Reproduced with permission from ref 11.

Conversely, for a methylene group with nonequivalent protons, very large enhancements can be expected between the methylene protons.

3. Enhancements into isolated protons, such as aromatic protons between two nonprotonated *ortho* substituents, often occur over relatively long distances. This is because the other relevant interproton distances (in Eq. 3.28) are often also large in such cases. To anticipate Chapter 4, however, we should note that all long-range enhancements grow only slowly, so that a steady-state NOE is not easily obtained. Conversely, protons that *do* have one or more close neighbors generally do not show appreciable long-range enhancements.

The remainder of this book will provide many examples that bear out these extremely general principles in practice. They are in fact so general that it would be quite impossible to pick out every illustration of each point. Rather, they should be viewed as part of the basic "vocabulary" of background information that should be used whenever we interpret the results of an NOE experiment.

3.2.2.2. Indirect Effects. The simplest example of an indirectly transmitted enhancement is provided by the linear three-spin system already shown in Figure 3.3 ($\alpha = 180°$). If, instead of the central spin, one of the terminal spins (say A) is saturated, the theoretical enhancements calculated from Eq. 3.30 are as shown in Figure 3.6. The enhancement of the central spin ($f_B\{A\} = 28.29\%$) can easily be understood in terms of the direct contribution only, but that of the third spin ($f_C\{A\} = -13.16\%$) is another matter.

The mechanism whereby such negative enhancements arise has in fact already been mentioned. Just as the decoupler acts as a "source of population disturbance" in the creation of direct enhancements, so do the nonequilibrium populations of an enhanced spin themselves act to disturb the equilibrium of other nearby spins. In the example taken, the populations at spin B have been driven far from equilibrium by cross-relaxation with A, as evidenced by the enhancement $f_B\{A\}$. This change in turn disturbs the whole balance of relaxation at B, including its cross-relaxation with C, so that the population disturbance is ultimately transmitted also to C.

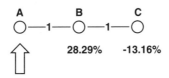

Figure 3.6. Calculated steady-state enhancements at the extreme narrowing limit on irradiating spin A in the three-spin system illustrated. The negative enhancement $f_C\{A\}$ is a three-spin effect; that is, it results from two successive cross-relaxation steps, A → B followed by B → C. External relaxation (ρ^*) was not included in this calculation.

Why is this indirect enhancement negative? To answer this we must again examine the nature of the positive NOE regime ($\eta_{max} > 0$), the name of which appears here perhaps less appropriate. The most fundamental characteristic of the positive NOE regime is not that all enhancements are positive; they are not. Rather it is that W_2, or double-quantum, cross-relaxation transitions predominate over W_0, or zero-quantum, cross-relaxation transitions (see Section 2.1.1). The key implication is that when a perturbed spin changes its populations back toward equilibrium, its cross-relaxation partners are made to change their own populations in the same *absolute* sense. For example, a saturated spin must relax by increasing the population difference between its spin states. When W_2 cross-relaxation predominates, this causes the corresponding population differences of its cross-relaxation partners to *increase* also, giving a positive enhancement for these spins. Conversely, each positively enhanced spin must itself relax by *decreasing* the population difference between its spin states. The population differences for its further cross-relaxation partners are therefore made to *decrease* also, resulting in negative indirect enhancements. The vital point is that each step along such a cross-relaxation pathway corresponds to a change in the sign of the observed enhancement. Direct enhancements are positive, indirect enhancements transmitted over one intermediate spin are negative, those transmitted over two intermediate spins are positive, and so on. This logic depends absolutely on the predominance of W_2 cross-relaxation over W_0, however. In the negative NOE regime ($\eta_{max} < 0$), W_0 cross-relaxation predominates, and there is no corresponding sign change for each step. This has drastic consequences for the conduct of NOE experiments, as will be discussed in Section 3.2.3 and Chapter 4.

3.2.2.3. When Do Indirect Effects Matter?

(We discuss here only steady-state effects in the positive NOE regime.) The indirect enhancement $f_C\{A\}$ in the earlier example owed its large size to the linear geometry of the spin system (Fig. 3.6). Not only was the direct enhancement of C thereby made negligible ($f_C\{A\}_{direct} = +0.8\%$), but also the position of B was more or less optimal for transmission of the indirect effect. It is instructive to see what becomes of the enhancement $f_C\{A\}$ as the angle α is reduced, however (Fig. 3.7). As the distance r_{AC} diminishes, the direct enhancement at C becomes more significant, and there comes a point at which it overtakes the indirect effect ($\alpha \approx 78°$). For still smaller values of α, the direct contribution rapidly comes to dominate the NOE at C, and a strong positive enhancement results.

There are two important conclusions to be drawn from this:

1. Certain geometries are associated with small or zero enhancements, not because the interacting spins are distant, but because the direct and indirect contributions *cancel each other out*.
2. Appreciable indirect effects are to be expected mainly between spins whose geometry is not too far from linear.

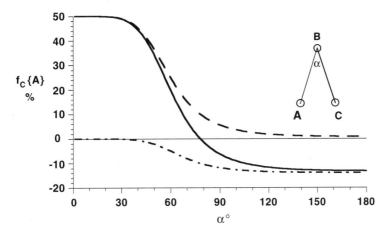

Figure 3.7. Variation in the calculated steady-state enhancement $f_C\{A\}$ at the extreme narrowing limit on changing the angle α as shown. The total enhancement is represented by the solid line, which is the sum of the direct contribution (dashed line) and the indirect contribution (dotted and dashed line). The two contributions cancel one another when the angle α is approximately 78°. External relaxation (ρ^*) was not included in these calculations.

Thus, in Figure 3.7, $f_C\{A\} = 0$ when $\alpha \simeq 78°$, in spite of the fact that r_{AC} is not particularly long. This clearly could represent a serious pitfall for the naive interpretation of NOE results! Fortunately, the curve of Figure 3.7 is steep as it passes through $f_C\{A\} = 0$, implying that a fairly small change in geometry would restore the enhancement to a measurable level. (Although Fig. 3.7 applies to one specific geometrical variation, this conclusion is more general.) Instances (reported, at least) of enhancements being *unmeasurable* owing to cancellation of direct and indirect contributions thus seem to be rare. Nonetheless, one must more generally beware of misinterpreting the *smallness* of a particular enhancement.

As regards the indirect enhancements themselves, observable indirect effects in the positive NOE regime are in practice almost invariably limited to negative enhancements transmitted via a single intermediate spin. Such enhancements are often called "three-spin effects," and they are generally much smaller than the -13.16% of our earlier theoretical example. This is mainly because the third spin usually has one or more further cross-relaxation partners, which dilute the influence of the intermediate "transmitting" spin in exactly the same way as direct enhancements are reduced by further competing cross-relaxation partners.

Another practical factor is the slow growth of indirect effects (Chapter 4), which means that experiments frequently fail to achieve steady state. In the past, this also implied that *transient* experiments, especially NOESY, usually failed to detect three-spin effects; the rapid initial population disturbance characteristic of transient experiments dissipates through direct relaxation on a timescale several times faster than the buildup rate for typical three-spin effects, so that the peak values reached for intensities of three-spin effects are very small. One would thus expect three-spin effects to be larger in steady-state experiments, where they are continuously driven throughout a long preirradiation period, than in a transient experiment, where they are not. In practice, the recent introduction of transient experiments with dramatically improved sensitivity (e.g., the DPFGSE-NOE sequence, cf. Section 7.5.3.2) implies that small three-spin effects may be most easily *detected* in these new transient experiments, despite their having a significantly smaller absolute size than in a steady-state experiment.

Three-spin effects, although rarely sought for their own sake, can be very useful. If, as is usual, a series of NOE experiments is carried out in which different resonances are separately preirradiated, then the two direct enhancements corresponding to the pathway of a particular three-spin effect will often be found as well. These observations *in combination* can give a more conclusive result than would the direct enhancements alone, since together they provide powerful evidence for the internal consistency of the data. Moreover, in cases in which one or two of the resonances on a pathway are obscured by other signals in the spectrum, a three-spin effect may be the only accessible enhancement involving the third spin, so allowing the investigation to "reach" further into a spectroscopically difficult portion of he molecule. Most frequently, three-spin effects are observed when one of the steps occurs over a very short distance (e.g., between two nonequivalent methylene protons). This is not because such geometry is especially favorable for the three spins directly involved, but rather because the "diluting" effect of other nearby spins is thereby made insignificant, at least for the short step.

It must be stressed that three-spin effects are useful only because they can be instantly recognized by their negative sign (assuming that saturation transfer is not a possibility). Sign alternation along a cross-relaxation pathway can provide still further interpretative clues, as for instance in distinguishing a very small direct enhancement from an indirect one transmitted over two intermediate spins; work on the alkaloid repanduline provides an example of this (Section 10.3.2).

Finally, heteronuclear three-spin effects of the type $^{13}C \leftarrow {}^{1}H \leftarrow \{^{1}H\}$ have been reported.[4,5] These offer the advantage of transferring interproton distance information into the often better dispersed ^{13}C spectrum, so avoiding possible overlap problems in the ^{1}H spectrum, or aiding assignments in the ^{13}C spectrum. The ^{13}C nucleus in question is generally directly bonded to the intermediate proton, so that (1) transmission of the enhancement over this step is very efficient, and (2) the carbon-proton distance is in itself of little interest. The observed ^{13}C enhancements are large and negative, but it seems unlikely that there could be any *sensitivity* advantage, given the much lower intrinsic sensitivity of carbon observation (~64-fold).

3.2.2.4. Magnetic Equivalence.
In fact, little needs to be said concerning magnetic equivalence, since the equations already developed require only slight

modification to cope with it. The difficulties that arise are generally not due to equivalence *per se*, but rather in treating the internal motions that groups of equivalent spins, particularly methyl groups, tend to indulge in. These are discussed in Chapter 5.

As a first step, the multispin Solomon equations may be very simply extended to the case in which several S spins are saturated (cf. Eq. 3.9):

$$\frac{dI_z}{dt} = -(I_z - I_z^0)R_I^{DD} - \sum_S (S_z - S_z^0)\sigma_{IS} - \sum_X (X_z - X_z^0)\sigma_{IX} \quad (3.33)$$

If the spins S are nonequivalent, this equation applies to a multiple irradiation experiment. If, as concerns us here, the spins S are equivalent, then there is only one S resonance, and the summation Σ_S runs over each identical contributing spin; these we will term individually S_i. Since these spins are indistinguishable, only their total magnetization can be observed, defined by

$$S_z = \sum_i S_{iz} \quad (3.34)$$

Rewriting Eq. 3.33 in this notation we have

$$\frac{dI_z}{dt} = -(I_z - I_z^0)R_I^{DD} - \sum_i (S_{iz} - S_{iz}^0)\sigma_{IS} - \sum_X (X_z - X_z^0)\sigma_{IX} \quad (3.35)$$

which, using Eq. 3.34, is trivially identical to Eq. 3.9. Thus, we may use Eq. 3.9 without modification to treat groups of identical S spins (or X spins for that matter), provided the magnetizations S_z and S_z^0 (or X_z and X_z^0) refer to *total z* magnetizations for those groups. Of course, under these circumstances, I_z^0, S_z^0, and X_z^0 are no longer equal to one another, and this must be taken into account when deriving expressions for the enhancements (cf. Appendix I).

In cases in which the I resonance corresponds to a group of equivalent spins, we also need to consider mutual cross-relaxation among the I spins, characterized by a rate constant σ_{II}. If there are N_I spins in group I (the observed resonance), each individual spin I_i will have $(N_I - 1)$ cross-relaxation partners within the group, and we may write for each

$$\frac{dI_{iz}}{dt} = -(I_{iz} - I_{iz}^0)R_I^{DD} - (N_I - 1)(I_{iz} - I_{iz}^0)\sigma_{II} - (S_z - S_z^0)\sigma_{IS} - \sum_X (X_z - X_z^0)\sigma_{IX}$$
(3.36)

As only the sum of all these identical contributions can be observed, all N_I identical expressions must be summed, giving

$$\frac{dI_z}{dt} = -(I_z - I_z^0)[R_I^{DD} + (N_I - 1)\sigma_{II}] - N_I(S_z - S_z^0)\sigma_{IS} - N_I \sum_X (X_z - X_z^0)\sigma_{IX}$$
(3.37)

Note that R_I^{DD}, the total dipolar relaxation rate of I, also contains a term due to mutual relaxation among the individual I_i spins:

$$R_I^{DD} = 2W_{1I} + N_S(W_{2IS} + W_{0IS}) + \sum_X N_X(W_{2IX} + W_{0IX})$$

$$+ (N_I - 1)(W_{2II} + W_{0II}) \qquad (3.38)$$

Equation 3.37 is, in effect, a version of the multispin Solomon equation, slightly revised to deal more conveniently with magnetic equivalence. We can of course use Eq. 3.37 to derive expressions for particular steady-state enhancements in terms of internuclear distances. This is a somewhat tedious process, however, and since it exactly parallels the development of Eqs. 3.25 and 3.27 from Eq. 3.9, it has been consigned to Appendix I. Here we simply quote the result for a homonuclear spin system at the extreme narrowing limit (assuming, as before, that relaxation is entirely dipole–dipole):

$$f_I\{S\} = \frac{N_S}{2} \left[\frac{\langle r_{IS}^{-6}\rangle - \sum_X (N_X/N_S) f_X\{S\}\langle r_{IX}^{-6}\rangle}{N_S\langle r_{IS}^{-6}\rangle + \sum_X N_X\langle r_{IX}^{-6}\rangle + \frac{3}{2}(N_I - 1)\langle r_{II}^{-6}\rangle} \right] \qquad (3.39)$$

where N_I, N_S, and N_X are the number of equivalent spins in each of the groups I, S, and X, respectively, and the angular brackets imply an average over internal molecular motions.

This equation calls for some comment. In particular, we have been forced to deal somewhat prematurely with internal motions, for the reason that most groups of equivalent spins owe their equivalence entirely to motional averaging. The consequences of internal motions are considered more fully in Chapter 5, but for the moment we have to anticipate a result, namely that it is the *internuclear distances*, rather than the enhancements themselves, that are averaged when internal motions are rapid on the T_1 timescale. Furthermore, the distances are not averaged directly but rather as a function that is strongly biased towards short distance. For motions slower than overall tumbling this function corresponds to averaging over the inverse sixth power of the distance ($\langle r_{AB}^{-6}\rangle$; cf. Section 5.5.1), while for motions faster than overall tumbling a more complex function is involved, based on $\langle r_{AB}^{-3}\rangle$ but dependent also on the geometry of the interaction. (We have called the relevant averaged distance in such cases r_{Tropp}; cf. Section 5.5.2.)[2]

Equation 3.39 also formalizes our earlier generalization that enhancements into methyl or methylene groups are usually small. This is because the term $(3/2)(N_I - 1)r_{II}^{-6}$ usually dominates the expression, mainly because r_{II}, the geminal interproton separation, is very short.

3.2.2.5. T_1 and the 3/2 Effect.
Equation 3.37 shows that R_I^{DD} is no longer the appropriate rate constant to describe the longitudinal relaxation behavior of

the group of equivalent I_i spins, even within the approximations on the definition of T_1 considered in Section 2.3.2. Instead we define a total dipolar rate for equivalent spins, R_{II}^{DD}:

$$R_{II}^{DD} = R_I^{DD} + (N_I - 1)\sigma_{II} \qquad (3.40)$$

For groups of equivalent spins, Eq. 3.37 shows that it is R_{II}^{DD} rather than R_I^{DD} that approximates to $(T_1)^{-1}$. This corresponds to the somewhat revised assumption that, while cross-relaxation terms with other spins (S and X) are still neglected in "defining" T_1, cross-relaxation among the identical I_i spins is accounted for (without sacrificing the assumed single-exponential behavior).

This phenomenon, namely, the generally faster relaxation of resonances due to groups of identical spins caused by their internal cross-relaxation, has somewhat confusingly become known as the "3/2 effect." The origin of this name seems to be that for the simplest case of two identical spins ($N_I = 2$):

$$(R_{II}^{DD})_{2\,\text{spin}} = R_I^{DD} + \sigma_{II} = \frac{3}{2} R_I^{DD} \qquad (3.41)$$

(at the extreme narrowing limit, for relaxation only between the I_i spins).

3.2.2.6. Chemical Equivalence.

Any spins that are indistinguishable on the NMR time scale are said to be chemically equivalent. Thus, in the previous example all three protons of a spinning methyl group are chemically equivalent, and in the symmetrically substituted cyclobutane **3.1** protons A and A' are chemically equivalent, as are protons B and B'. The requirements for *magnetic* equivalence are more restrictive, however. Magnetically equivalent spins must

3.1

not only be chemically indistinguishable, but must also couple identically to all other spins within the molecule. Thus, the spinning methyl protons are magnetically equivalent, but protons A and A' (or B and B') in compound **3.1** are not, since $J(A, B)$ is different from $J(A, B')$. More obviously, where NOE experiments are concerned, although a single average distance $\langle r_{\text{MeH}} \rangle$ is appropriate to describe cross-relaxation of a methyl group with an outside spin, six distances (r_{AB}, $r_{A'B}$, $r_{AB'}$, $r_{A'B'}$, $r_{AA'}$, and $r_{BB'}$) are required to specify cross-relaxation rates in cyclobutane **3.1** (although, from the symmetry of the molecule, we can reduce this to four since $r_{AB} = r_{A'B'}$ and $r_{A'B} = r_{AB'}$).

In this section we deal briefly with the slight extension to the theory needed for spins that are chemically but not magnetically equivalent, using cyclobutane

3.1 as an example. Only two enhancements can be measured, corresponding to

$$f_A\{B, B'\}[\equiv f_{A'}\{B, B'\} \equiv f_{A,A'}\{B, B'\}]$$

and (3.42)

$$f_B\{A, A'\}[\equiv f_{B'}\{A, A'\} \equiv f_{B,B'}\{A, A'\}]$$

Effectively, these are cases of multiple saturation, and Eq. 3.33 is appropriate. In the extreme narrowing limit, by analogy with Eq. 3.25, this leads to

$$f_I\{S\} = \frac{1}{2}\left[\frac{\sum_S r_{IS}^{-6} + \sum_X f_X\{S\}r_{IX}^{-6}}{\sum_S r_{IS}^{-6} + \sum_X r_{IX}^{-6}}\right] \quad (3.43)$$

As far as proton A is concerned, proton A' is just another near neighbor, so it joins the summation over X:

$$f_A\{B, B'\} = \frac{1}{2}\left[\frac{r_{AB}^{-6} + r_{AB'}^{-6} + f_{A'}\{B, B'\}r_{AA'}^{-6}}{r_{AB}^{-6} + r_{AB'}^{-6} + r_{AA'}^{-6}}\right] \quad (3.44)$$

Using Eq. 3.42 and rearranging:

$$f_A\{B, B'\} = \frac{1}{2}\left[\frac{r_{AB}^{-6} + r_{AB'}^{-6}}{r_{AB}^{-6} + r_{AB'}^{-6} + \frac{1}{2}r_{AA'}^{-6}}\right] \quad (3.45)$$

This example should make clear how more complex cases would be treated.

3.2.3. Away from the Extreme Narrowing Limit

Earlier we showed explicitly that (for a three-spin system) the steady-state NOE at the extreme limit of slow tumbling is always -1, *irrespective* of the geometry of the spins. In fact this conclusion is quite general for any spin system, and it renders the steady-state NOE completely useless for structure determinations of molecules large enough that $\omega\tau_c \gg 1$. In such cases, only kinetic NOE experiments can yield structural information, as discussed in Chapter 4.

This is undoubtedly the most important point that needs to be made concerning the steady-state NOE away from the extreme narrowing limit. It is still fair to ask, however, at what stage the steady-state experiment becomes useless, and what happens when deviations from extreme narrowing are only slight. It is these questions that we tackle in this section.

As shown earlier, the formal consequence of leaving the extreme narrowing limit is that the relatively simple Eq. 3.25, on which much of the discussion in the previous several sections was based, must be replaced by the more general Eq. 3.27. Using this (and Eqs. 3.30 and 3.31 derived from it) we may calculate the effect of varying $\omega\tau_c$ on the *theoretical* enhancements in simple spin systems; results for two particular geometries appear in Figures 3.8 and 3.10. (Note that external relaxation is not included in these calculations, so the enhancements predicted for rapid tumbling rates are unrealistic.)

What do these curves tell us? First, Figure 3.8 shows that deviations from the extreme narrowing condition ($\omega\tau_c \ll 1$) start to affect enhancements significantly over only about one order of magnitude from the zero crossing point of η_{max} (that is, in the range ~$0.1 < \omega\tau_c < 1.12$). This is fairly encouraging, at least for experiments in the positive NOE regime; provided that $\omega\tau_c < ~0.1$, the consequence of increasing $\omega\tau_c$ is limited mainly to the effect it has on the relative contribution from ρ^* (Section 2.4.1). Furthermore, within the range $0.1 < \omega\tau_c < 1.12$, the effect of increasing $\omega\tau_c$ is essentially like that of turning down a gross "volume control." All the enhancements are attenuated simultaneously, more or less in proportion to the decrease in η_{max}. In fact, the detailed behavior cannot be quite this simple, as is illustrated by the somewhat different curves predicted for enhancements that are negative in both regimes (e.g., $f_D\{B\}$ in Fig. 3.8). Such subtleties are unlikely to be significant in real applications, however.

Much more significant is the influence that internal motions can have when overall molecular tumbling rates are close to $\omega(\tau_c)_{overall} = 1.12$. The correlation time that is relevant for a particular dipolar interaction is actually that of the internuclear vector connecting the two interacting spins. Until now we have

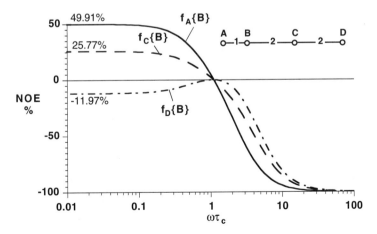

Figure 3.8. Calculated steady-state enhancements on irradiating spin B in the linear four-spin system illustrated, as a function of $\omega\tau_c$ ($\rho^* = 0$ for all spins). See text for discussion.

assumed that all such correlation times are identical to $(\tau_c)_{overall}$, but in practice there will very often be a spread of values for the various pairs of interacting nuclei, the differences between them being caused by specific internal motions or by anisotropic tumbling. One consequence of this is that the appropriate value of η_{max} will also differ for each enhancement.[†] Normally this is the least of the complications introduced by internal motions (others are considered in Chapter 5), but when $\omega(\tau_c)_{overall}$ is close to 1.12, these differences in "local η_{max}" often become the dominant factor in determining enhancements.

As an example, Figure 3.9 shows NOE spectra of a macrocyclic antibiotic, elaiophylin, in $CDCl_3$ solution.[6] Proton H_7 is at the junction of the macrocyclic ring with a more flexible side chain, and at ambient temperature the enhancement from H_7 to H_5 across the macrocyclic ring is negative. Other enhancements from H_7 are positive, however, due to the shorter correlation times appropriate in these cases. Enhancement $f_{H8}\{H_7\}$ is affected by side-chain flexibility, and enhancement $f_{Me17}\{H_7\}$ by methyl rotation. On raising the temperature by 27°C, enhancement $f_{H5}\{H_7\}$ also became positive, so proving that its original negative sign was actually due to a "local η_{max}" effect.

Similar examples have been observed in peptide spectra. Sometimes, enhancements involving only backbone protons change sign from negative to positive at higher temperatures than do enhancements involving side-chain protons; clearly, this is a result of side-chain flexibility.[7] In principle, such observations could be used as a sensitive probe for internal motions, but applications of this idea are likely to be rather limited. Of greater relevance here is the fact that, very close to the zero crossing of η_{max}, enhancements are not merely small, they are also very difficult to interpret unless one can be quite sure that internal motions or anisotropic tumbling do not complicate the issue.

So much for the positive NOE regime and the zero crossing region. Is the steady-state experiment ever useful in the negative NOE regime? As we have observed previously, a drastic loss of information is implied in the fact that all enhancements are negative in the negative NOE regime. Direct and indirect enhancements can no longer be distinguished by their signs, and this is the greatest weakness inherent in all experiments in the negative regime. Nevertheless, if we are simply asking the question "is proton *B* or proton *C* the near neighbor of proton *A*?," then an answer *might* still be available, provided $\omega\tau_c$ is not too far into the negative regime.

Figure 3.10 shows a theoretical example of such a case. The geometry was chosen such that the irradiated spin B has one close neighbor (A) and one relatively distant neighbor (C), the distant neighbor having a further spin (D) closer to it than B is. This represents about the simplest case for which the question posed above can realistically be solved. We see that the enhancement

---

[†]Strictly, η_{max} is defined only for the direct contributions to enhancements in such cases, since the indirect contributions depend on many separate dipolar interactions, each having potentially different appropriate correlation times. Equation 3.27 is not defined under these circumstances, but nonetheless the idea of a "local η_{max}" for each direct enhancement is conceptually useful.

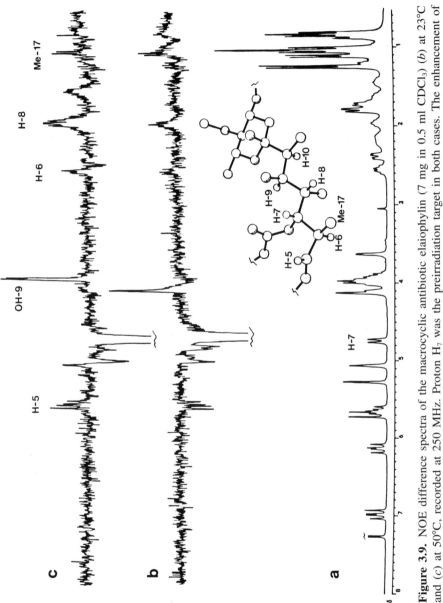

Figure 3.9. NOE difference spectra of the macrocyclic antibiotic elaiophylin (7 mg in 0.5 ml CDCl$_3$) (*b*) at 23°C and (*c*) at 50°C, recorded at 250 MHz. Proton H$_7$ was the preirradiation target in both cases. The enhancement of H$_5$ is negative at the lower temperature, because the internuclear vector connecting H$_5$ and H$_7$ has a longer correlation time than those corresponding to other enhancements from H$_7$. See text for further discussion. Reproduced with permission from ref. 6.

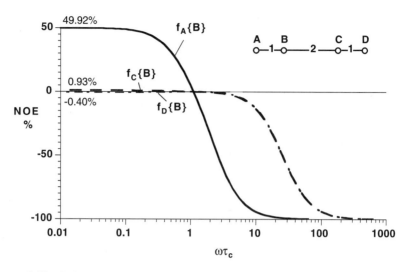

Figure 3.10. Calculated steady-state enhancements on irradiating spin B in the linear four-spin system illustrated, as a function of $\omega\tau_c$ ($\rho^* = 0$ for all spins). See text for discussion.

of the distant neighbor (C) remains very much smaller than that of the close neighbor (A) for some way into the negative regime, but that problems of interpretation become progressively more severe as $\omega\tau_c$ increases. As enhancements become more similar, so our justification for differentiating between them diminishes accordingly.

A real example is provided by some work on the alkaloid dihydrodaphnine diacetate.[8] Being a salt, this alkaloid probably aggregates in acetic acid-d_4 solution, and the measured enhancements were all negative (cf. Section 8.1.2). Sufficient differences in their magnitudes were observed that the simple question of positional isomerism (structure **3.2** and **3.3**) could be unambiguously

3.3

solved in this instance. This was perhaps more by luck than judgment, however; had the correlation time been an order of magnitude longer, the experiment might well have failed. In general, the only safe advice must be: When in the negative NOE regime, do kinetic experiments.

3.3. IN PRACTICE

Sadly, none of the equations so far developed is capable of correctly predicting the *numerical* results of NOE experiments, except at the most approximate level. In this section we must therefore tackle two questions: First, why do the equations fail, and second, what should be done about it?

Answers to the first question are relatively clear cut. At root, they all reflect the fact that the foregoing equations were very highly idealized and simplified. Each principal reason, considered below, for why measured and calculated enhancements differ thus corresponds to the failure of an assumption implicit in the equations.

The second question is necessarily much more subjective. Essentially, different approaches generally seem to reduce to one of two underlying attitudes. According to one view, experiments are designed to force the results to match the theory more closely (by, for instance, rigorous degassing, very long preirradiation periods, and so on). As we shall see, however, this approach must always force some compromise on other issues, even if only on experimental convenience.

Another approach is simply to accept the approximate nature of the equations, and to try instead to use them *only within their limitations*. According to this view, enhancements are interpreted in a semiquantitative sense only, and structural conclusions are based on a larger number of more qualitative constraints, rather than a small number of (supposedly) highly quantitated ones.

Perhaps the most compelling reason for not trying to push the equations too far is the weakness of the theory in dealing with internal motions (cf. Chapter 5). If one is to decide how rigorously an NOE experiment should be quantitated, such judgment must be based on an assessment of the weakest link in the "chain" of interpretation. Usually, this weakest link is the treatment of internal motions. As Chapter 5 will show, this is not because the relevant theory cannot be worked out, but rather that its application often requires an impossibly detailed picture of the nature of the motions themselves so as to specify their quantitative effect on enhancements.

Given that there is almost always this unavoidable limitation on the degree of quantitation that is possible, the authors' view (at least) is that overattention to quantitation in other respects can be somewhat futile. This is, of course, not intended as a licence to ignore quantitation altogether; rather, it is meant as a warning against overinterpretation. As we have said, this is a subjective issue, but we have raised it here because one's view of it is bound to affect the significance attached to the various other shortcomings about to be discussed.

3.3.1. Incomplete Saturation

It might appear trivially simple to avoid incomplete saturation, and when isolated singlets are irradiated, so it is. In cases in which the irradiation target is close to other resonances, however, a choice must be made between saturation efficiency and frequency selectivity. This is particularly true when the irradiation target is a multiplet. Experimental techniques that can help in such cases are discussed in Section 7.2.5, but even these generally lead only to fractional saturation.

Fortunately, the only real penalty of incomplete saturation is a corresponding loss in sensitivity. All the predicted enhancements are simply scaled down in exact proportion to the actual extent of saturation. More formally, if the actual degree of saturation (θ) is

$$\theta = 1 - \frac{S_z'}{S_z^0} \tag{3.46}$$

(where S_z' is the residual steady-state z magnetization of S in the presence of the subsaturating irradiation) then, by analogy with Eq. 3.10:

$$\frac{dI_z}{dt} = 0 = -(I_z - I_z^0)R_I^{DD} - (S_z' - S_z^0)\sigma_{IS} - \sum_X (X_z - X_z^0)\sigma_{IX} \tag{3.47}$$

from which (cf. Eq. 3.13)

$$f_I'\{S\} = \frac{\gamma_S}{\gamma_I} \frac{1}{R_I^{DD}} \left[\theta \sigma_{IS} - \sum_X \frac{\gamma_X}{\gamma_S} f_X'\{S\} \sigma_{IX} \right] \tag{3.48}$$

where $f_I'\{S\}$ and $f_X'\{S\}$ represent the partially excited enhancements actually measured.

Provided that the irradiated and control data are separately available, the extent of saturation can of course readily be found by comparing the intensity of the S resonance; S_z^0 is the intensity in the control spectrum and S_z' that in the irradiated spectrum. It is therefore common practice to correct measured enhancements using the experimental value of θ:

$$[f_A\{B\}]_{true} = [f_A\{B\}]_{measured}\theta^{-1} \tag{3.49}$$

Unfortunately it is often unclear whether particular enhancements in the literature have been corrected in this way or not. When the original spectra are reproduced there is no ambiguity, but when results are presented only numerically, mention should be made if correction for incomplete saturation was used.

In summary, it is almost always better simply to put up with incomplete saturation, rather than risk saturating other nearby resonances. This is particularly so in view of the serious errors of interpretation that can occur if such unwanted saturation is not recognized (cf. Section 7.2.4).

3.3.2. Failure to Reach Steady State

Almost all so-called steady-state NOE difference experiments probably fail to reach steady state. It would be too easy to say that this should be rectified by using longer preirradiation times. In practice, time, be it machine time, experimenter's time, or even compound lifetime, is invariably limited, and some compromise must be reached. Clearly, it is important to use the available time as efficiently as possible, and experimental strategies to achieve this are discussed in Section 7.2. Here we discuss only the consequences for interpretation of data obtained short of steady state.

In principle, the theory can be extended to provide a value for an enhancement at any time during its development (cf. Chapter 4). This corresponds to integrating the expression for dI_z/dt from $t = 0$ to some value t, rather than from $t = 0$ to $t = \infty$, as is implicit in the steady-state treatment. Although this is possible in principle (Section 4.3.2), it is a far more complicated procedure than simply using the steady-state equations.

Given that it is much easier to use the steady-state theory even though data may have been obtained somewhat short of steady state, how does this affect interpretation? Clearly, the enhancements most at risk in difference experiments with short preirradiation times must be those that grow slowly. As Chapter 4 will show, and as already indicated, these are the *long-range* enhancements.

We must therefore expect that three-spin effects, and direct enhancements transmitted over relatively long distances, will be partially or completely lost in experiments using short preirradiation times. Conversely, if the aim is specifically to detect long-range enhancements, which by their nature often contain the most valuable information, then experiments *must* be planned to allow long preirradiation times. Unfortunately, it is not easy to predict just how long is needed to excite a particular long-range enhancement; essentially, this information is available only from the experiment itself.

The consequences of short preirradiation are of course still more severe where quantitation is concerned. The numerical values of experimentally determined long-range enhancements are likely to fall short of the true steady-state values unless extremely long preirradiation times are used, and, again, the actual meaning of "extremely long" can really be found only by experiment. In view of this, it is perhaps wise when interpreting quantitative steady-state NOE data to adopt a "sliding scale" of skepticism, increasing in proportion to distance.

Another possible consequence of failure to reach steady state is that the magnetization due to the irradiated signal may still be oscillating at the time of the observe pulse (Section 1.5). The complications that this introduces are dealt with in Sections 4.4.3 and 7.2.6.

3.3.3. Competition from Other Relaxation Sources

Section 2.4 showed how the two-spin NOE is reduced by any form of relaxation other than intramolecular dipole–dipole. Exactly the same "diluting" effect also occurs in multispin systems, but the theory so far developed in this chapter has been limited explicitly to exclusive intramolecular dipole–dipole relaxation. In this section we must therefore look at the consequences of competing relaxation for interpretation of the NOE in multispin systems.

The principal candidates for competing relaxation mechanisms were described in Chapter 2. For homonuclear interproton NOE experiments, the main contributing sources are usually intermolecular dipole–dipole relaxation with paramagnetic impurities (dissolved oxygen or traces of metal ions), or with the bulk solvent. For enhancements into methyl groups, spin–rotation relaxation may also be important, while in heteronuclear cases, other mechanisms may be significant (Section 2.4). Furthermore, each spin in a multispin system may be affected differently by competing relaxation. Thus, spins that are not solvent exposed will in general be less severely affected by intermolecular relaxation than others, while formation of specific complexes between solute molecules and traces of paramagnetic impurities might cause fast intermolecular relaxation for particular spins.

To what extent is it possible, or even desirable, to limit these factors? As before, it is important to distinguish between *detection* and *quantitation* of enhancements. For many applications (probably for most), the numerical value of an enhancement is not in itself vital in reaching a structural conclusion; the appearance of an enhancement at one resonance rather than another is often sufficient, provided the enhancement can be shown to be genuine. When this is the case, elaborate precautions to degas and purify the sample are probably inappropriate. Newer experimental methods, particularly difference spectroscopy and gradient-assisted transient techniques (Chapter 7), have greatly improved sensitivity, and it is perfectly legitimate (in the authors' view at least) to "spend" some of this advantage on greater experimental convenience. A similar area of possible contention is in the choice of a suitable solvent; although highly deuterated solvents (such as DMSO-d_6, benzene-d_6, or acetone-d_6) do increase the intermolecular relaxation rate and thereby reduce the NOE somewhat, this disadvantage is in practice more than offset by the better spectrometer stability that results from a strong lock signal (see also Chapter 7).

Even when quantitation is required, the most determined efforts will never eliminate competing relaxation entirely,[†] and so it becomes necessary to modify the theory to cope with it. The "diluting" effect of competing relaxation can more formally be expressed in terms of the parameter ρ^* (Section 2.4). Since ρ^* represents the "additional" relaxation occurring for a particular spin, it appears as an added term in the total relaxation rate for that spin:

$$R_I = R_I^{DD} + \rho_I^* \tag{3.50}$$

We may thus modify earlier key equations in a very obvious way, as follows:
From Eq. 3.9:

$$\frac{dI_z}{dt} = -(I_z - I_z^0)(R_I^{DD} + \rho_I^*) - (S_z - S_z^0)\sigma_{IS} - \sum_X (X_z - X_z^0)\sigma_{IX} \tag{3.51}$$

from Eq. 3.13:

$$f_I\{S\} = \frac{\gamma_S}{\gamma_I} \frac{1}{R_I^{DD} + \rho_I^*} \left[\sigma_{IS} - \sum_X \frac{\gamma_X}{\gamma_S} f_X\{S\}\sigma_{IX} \right] \tag{3.52}$$

and from Eq. 3.20:

$$f_I\{S\} = \frac{\gamma_S}{\gamma_I} \left[\frac{(W_{2IS} - W_{0IS}) - \sum_X (\gamma_X/\gamma_S) f_X\{S\}(W_{2IX} - W_{0IX})}{2W_{1I} + W_{2IS} + W_{0IS} + \sum_X (W_{2IX} + W_{0IX}) + \rho_I^*} \right] \tag{3.53}$$

These equations all involve only relaxation rates or probabilities. To derive equations involving distances, we must use the functions α_I and $\mathcal{L}_I(\omega, \tau_c)$ defined in Section 2.4.1. Using these, we have from Eq. 3.25

$$f_I\{S\} = \frac{1}{2} \left[\frac{r_{IS}^{-6} - \sum_X f_X\{S\}r_{IX}^{-6}}{r_{IS}^{-6} + \sum_X r_{IX}^{-6} + a_I} \right] \tag{3.54}$$

[†]The closest approach to this ideal appears to have been an enhancement of +49% observed at H_3 on preirradiation of the 2,4-dimethoxy signal in a thoroughly de-gassed, dilute solution of 1,5-dichloro-2,4-dimethoxybenzene in $CDCl_3$.[9]

from Eq. 3.27:

$$f_I\{S\} = \eta_{max} \left[\frac{r_{IS}^{-6} - \sum_X f_X\{S\} r_{IX}^{-6}}{r_{IS}^{-6} + \sum_X r_{IX}^{-6} + \mathcal{L}_I(\omega, \tau_c)} \right] \quad (3.55)$$

and from Eq. 3.39:

$$f_I\{S\} = \frac{N_S}{2} \left[\frac{\langle r_{IS}^{-6} \rangle - \sum_X (N_X/N_S) f_X\{S\} \langle r_{IX}^{-6} \rangle}{N_S \langle r_{IS}^{-6} \rangle + \sum_X N_X \langle r_{IX}^{-6} \rangle + \frac{3}{2}(N_I - 1)\langle r_{II}^{-6} \rangle + \mathcal{L}_I(\omega, \tau_c)} \right] \quad (3.56)$$

If steady-state enhancements are to be used quantitatively to derive internuclear distances, then each value of ρ_I^* [or $\mathcal{L}_I(\omega, \tau_c)$] must be included as an unknown, to be determined by experiment. In principle, for an N-spin system, there will be up to $N(N-1)$ possible enhancements, while the equations relating them will involve $(1/2)N(N-1)$ internuclear distances and N values of ρ^* [or $\mathcal{L}(\omega, \tau_c)$]. If extreme narrowing is assumed, so that the $(1/2)N(N-1)$ potentially different values of τ_c do not need to be treated as additional unknowns, and given that only *relative* values of internuclear distances are available, this corresponds to a total of $[(1/2)N(N+1)] - 1$ unknowns specified by $N(N-1)$ equations.

The most important words in the preceding paragraph are, of course, "in principle." In practice, many of the $N(N-1)$ enhancements will be impossible to measure, either due to spectral overlap (quantitative steady-state enhancements are available only from selective irradiation experiments) or because they are too small. Generally, the problem must be limited to a few spins of interest, and a sufficient number of enhancements measured between them to determine just the relative internuclear separations of these spins.

As an aid to this, it may be convenient to use the "a" parameters to include relaxation due to all spins outside the group of interest. This corresponds to transferring some of the terms from the summations over the spins X (e.g., in the denominator of Eq. 3.54) into the terms a_I. Viewed in this context, a_I can itself be regarded as a "spurious distance" (actually the inverse sixth power of a spurious distance), corresponding to a hypothetical "extra spin" providing the "extra" relaxation represented by ρ^*. However, this interpretation is possible only for molecules within the extreme narrowing limit, and is complicated conceptually by the fact that the supposed "extra spin," unlike a real spin, only contributes to the W_1 terms. Noggle and Schirmer[10] give some discussion of the use of these equations, and some advisable precautions.

Clearly, this theoretical machinery required to cope with competing relaxation is somewhat clumsy. In practice, however, there is seldom much need to use it in earnest. In general, when internuclear distances are genuinely required, the best approach is to determine them from kinetic NOE data (Chapter 4). Applications of equations derived from Eq. 3.51 in the literature thus generally predate the kinetic experiments in use now.

In the authors' view, the most important conclusion to be drawn from this whole section is that steady-state NOE data are fundamentally ill suited to quantitative interpretation in terms of internuclear distances. This is not necessarily a loss. As we shall try to show in the Applications Section (Part III), probably the best uses of the steady-state NOE seek to avoid these limitations by using it in a more qualitative sense, to *distinguish* between various previously defined possible structures or conformations. In general, the NOE is more likely to provide a clear and reliable conclusion when a problem can be specified in terms of "yes" or "no" answers.

REFERENCES

1. Werbelow, L. G.; Grant, D. M. *Adv. Magn. Reson.* 1977, *9*, 189.
2. Tropp, J. *J. Chem. Phys.* 1980, *72*, 6035.
3. Bull, T. E. *J. Magn. Reson.* 1987, *72*, 397.
4. Cativiela, C.; Sánchez-Ferrando, F. *Magn. Reson. Chem.* 1985, *23*, 1072.
5. Kövér, K. E.; Batta, Gy. *J. Am. Chem. Soc.* 1985, *107*, 5829.
6. Ley, S. V.; Neuhaus, D.; Williams, D. J. *Tetrahedron Lett.* 1982, *23*, 1207.
7. Neuhaus, D. Unpublished results.
8. Neuhaus, D.; Rzepa, H. S.; Sheppard, R. N.; Bick, I. R. C. *Tetrahedron Lett.* 1981, *22*, 2933.
9. Bain, A. D.; Mazzola, E. P.; Page, S. W. *Magn. Reson. Chem.* 1998, *36*, 403.
10. Noggle, J. H.; Schirmer, R. E. "The Nuclear Overhauser Effect; Chemical Applications," Academic Press, New York, 1971, p. 72.
11. Sanders, J. K. M.; Mersh, J. D. *Prog. Nucl. Magn. Reson. Spectrosc.* 1982, *15*, 353.

CHAPTER 4

THE KINETICS OF THE NOE

So far, we have been concerned only with the steady-state NOE. Chapter 3 showed how the geometry of a multispin system is reflected in the pattern of steady-state enhancements observed within it, allowing deductions to be made about the relative positions of protons. The "geometry" of a true two-spin system, however, consists only of a single internuclear distance, and Chapter 2 showed that, in the absence of external relaxation, the two-spin steady-state NOE does not depend on this distance at all. Put differently, steady-state NOE enhancements are determined by *relative* values of internuclear distances, so that the single absolute distance that defines a two-spin system has no effect on the theoretical steady-state enhancement. As discussed in Sections 2.4 and 3.3.3, external relaxation does introduce some distance dependence into two-spin systems, although not usually in a particularly useful fashion.

The internuclear distance does, however, affect the *rate* at which steady state is reached. In this chapter we discuss how the NOE builds up towards steady state, or, for transient experiments, how it builds up to a maximum and then decays back to zero, and we show how rate measurements can give direct information about internuclear distances. We shall see that in the very early stages of NOE buildup, all enhancements behave as though they were in two-spin systems. When such behavior is assumed during interpretation, this is often called the "initial rate approximation," "two spin approximation," or "isolated spin pair approximation" (ISPA). In general, the initial rate of NOE buildup is unrelated to the final (steady-state) value of the enhancement, a fact that has important consequences for the interpretation of NOE spectra.

Most applications of kinetic NOE measurements are to large molecules, whose tumbling rates place them in the negative NOE regime ($\omega\tau_c > 1.12$). Steady-state enhancements are essentially useless in such cases, as previously discussed (Section 3.2.3), since the phenomenon of spin diffusion allows enhancements to dissipate through the molecule until, at the limit of slow tumbling, all steady-state enhancements are -100%. As will be shown, the great virtue of kinetic experiments is that they can allow at least partial separation of direct and indirect effects even for slowly tumbling molecules. To the extent that this approach succeeds in eliminating the influence of spin diffusion, such kinetic experiments are often simpler to interpret than are steady-state experiments with small molecules.

4.1. INTRODUCTION

Before considering the kinetic behavior of the NOE any further, we must first introduce the main experiments used to measure kinetic NOE enhancements, since the different types of experiment demand somewhat different theoretical treatments. We then present a qualitative account of some aspects of the kinetic NOE that, although it anticipates some of the results shown more formally later, may help to establish an appropriate context for the theory.

4.1.1. Types of Kinetic NOE Experiment

Kinetic NOE experiments can be divided into two types, depending on whether RF irradiation is present or absent during the period allowed for evolution of NOE enhancements.

In one type of kinetic NOE experiment, called the "truncated driven NOE" or TOE experiment, continuous irradiation of a resonance or group of resonances is applied throughout the NOE evolution period, immediately after which a pulse is applied to sample the population distributions (Fig. 4.1a). If the saturating irradiation were applied for a sufficiently long period relative to the timescale of relaxation processes, the steady-state NOE enhancements would be generated. However, for shorter irradiation times, "pre-steady-state" enhancements result, and the buildup of such enhancements as a function of irradiation time can be used to follow NOE kinetics. Of course, individual enhancements will each build up at different rates in such experiments.

The second type of kinetic NOE experiment, called a transient NOE experiment, is fundamentally different. In these experiments the system is initially perturbed, for example by selective inversion of one resonance, or (as in NOESY) by a sequence of hard pulses, and is then allowed to evolve for a period in the *absence* of a saturating field. NOE enhancements initially build up, reach a maximum, and then decay—eventually to zero if the NOE evolution period is long enough. As with TOE, individual enhancements show different kinetics, so each will have a different initial buildup rate, maximum amplitude, and time to reach maximum amplitude. At the end of the NOE evolution period, the populations are sampled by applying an observation pulse. The NOESY experiment (Fig. 4.1b) and its various multidimensional progeny

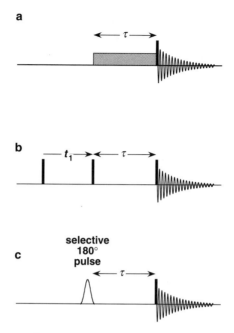

Figure 4.1. Experiments for measuring kinetic NOE enhancements. (*a*) Steady-state and truncated driven NOE (TOE) experiments, which differ only in the duration of the saturation. (*b*) Two-dimensional NOESY. (*c*) Simple 1D transient NOE experiment. In each experiment, NOE enhancements build up during the delay before the final pulse.

(Sections 9.2.2 and 9.2.3) are the most commonly used experiments of this type. Although the simple 1D transient pulse sequence (Fig. 4.1c) is not particularly widely used, inclusion of pulsed field gradients has led to various "cleaned-up versions" such as the DPFGSE-NOE experiment (cf. Section 7.5.3.2), and these have proved extremely powerful. Transient experiments can also be carried out to measure the rotating frame Overhauser effect (ROE enhancements), as discussed in Section 9.3.

In multidimensional experiments the period for NOE evolution is often called the "mixing time," denoted τ_m, since it allows mixing to occur between signals from different frequency dimensions. In this book we reserve the term "mixing time" exclusively for multidimensional experiments, and we use the symbol τ to denote the corresponding period for NOE evolution in 1D experiments (although, elsewhere, τ_m is sometimes also used in the context of 1D experiments).

4.1.2. Overview

Before embarking on a mathematical description of the kinetic NOE, it may be helpful first to obtain a qualitative picture. Figures 4.2 and 4.3 show the

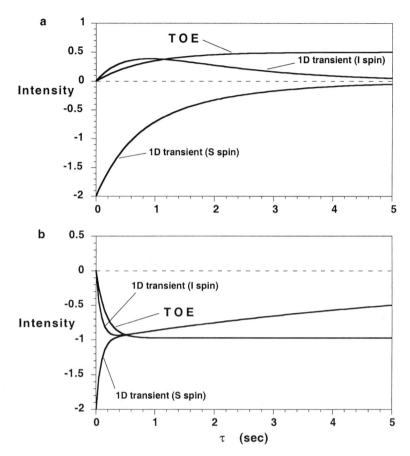

Figure 4.2. General form of NOE evolution in 1D transient NOE and TOE experiments. The recovery of the inverted signal in the transient case is also shown. Note that the vertical axis represents peak intensity in a *difference* spectrum. (*a*) $\omega\tau_c \ll 1$ (*b*) $\omega\tau_c \gg 1$.

general form expected for NOE kinetics in 1D and 2D experiments respectively. This is calculated for an ideal two-spin system with no external relaxation by using equations that will be presented in Section 4.2.

As already shown in Chapter 2, in the extreme narrowing limit ($\omega\tau_c \ll 1$), the ratio of W_0, W_1, and W_2 transition probabilities is such that direct relaxation (ρ_{IS}) competes quite effectively with cross-relaxation (σ_{IS}). The result (Figs. 4.2a and 4.3a) is that the transient NOE enhancement remains small, with a maximum value of only 38.5%. In addition, because intramolecular dipolar relaxation is slow in the extreme narrowing limit, the effect of external relaxation (ρ^*) will be relatively more important (cf. Section 2.3.1), making the values of real transient enhancements much smaller still. This is one reason

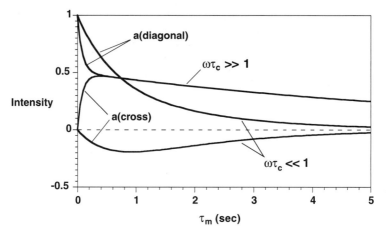

Figure 4.3. General form of NOE evolution in 2D NOESY for both $\omega\tau_c \ll 1$ and $\omega\tau_c \gg 1$. These curves are closely related to those for the transient NOE difference intensities, cf. Figure 4.2.

why transient enhancements are not widely used with small molecules, the other being that in such cases transient experiments are very time-consuming, mainly because they need very long relaxation delays to ensure re-equilibration during the time between passes through the pulse sequence.

The situation for large molecules ($\omega\tau_c \gg 1$) is quite different. Here, spin–lattice relaxation is inefficient and cross-relaxation is fast; consequently, enhancements build up much more quickly than they do for small molecules, become much larger, and remain large for longer. For instance, at $\omega\tau_c = 10$, the maximum theoretical transient NOE enhancement is -90% (cf. Section 8.1).

The initial rate of enhancement buildup in any kinetic NOE experiment can be approximated as linear; this is what is meant by the "initial rate approximation" (Section 4.4.1). At later times this linear NOE buildup falls off, either, in TOE experiments, as steady state is approached, or, in transient NOE experiments, as the enhancement approaches its maximum prior to decaying back to zero.

In multispin systems, a very important factor at later times is the appearance of indirect enhancements, which has very important consequences for interpretation. Figures 4.4 and 4.5 show calculated NOE kinetics for a four-spin system in a TOE experiment, under the conditions $\omega\tau_c < 1$ and $\omega\tau_c > 1$ respectively. In both cases, the direct enhancement $B \leftarrow \{A\}$ approaches steady state approximately exponentially, whereas there is a significant induction period in the buildup of the indirect enhancements. This is easily understood: The indirect enhancement $C \leftarrow B \leftarrow \{A\}$ can only start to build up once the direct $B \leftarrow \{A\}$ enhancement is appreciable, and similarly the enhancement $D \leftarrow C \leftarrow$

INTRODUCTION   103

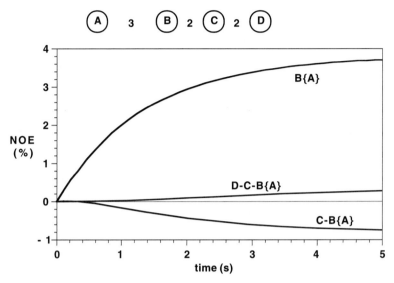

Figure 4.4. Buildup of TOE enhancements in a linear four-spin system, $r_{AB} = 3.0$ Å, $r_{BC} = r_{CD} = 2.0$ Å, $\omega\tau_c = 0.25$. The direct $B\{A\}$ enhancement is positive, the three-spin enhancement $C\text{-}B\{A\}$ is negative, the four-spin enhancement $D\text{-}C\text{-}B\{A\}$ is positive, and builds up very slowly. External relaxation (ρ^*) was set to zero in these calculations.

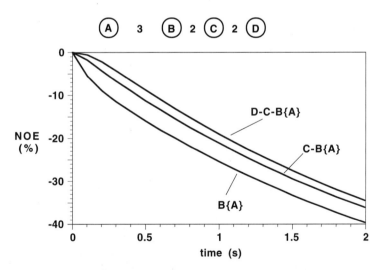

Figure 4.5. Buildup of TOE enhancements in the linear four-spin system of Figure 4.4, $\omega\tau_c = 25$. All enhancements are negative, and the longer the chain, the greater the induction period before significant buildup of NOE enhancement occurs. Note how similar in intensity the enhancements to B, C, and D are except at very short times, so that it is very difficult to deduce distance information from such experiments. External relaxation (ρ^*) was set to zero in these calculations.

$B \leftarrow \{A\}$ can only start to build up once the $C \leftarrow B \leftarrow \{A\}$ enhancement has developed significantly (similar arguments hold for enhancements in transient experiments).

Indirect enhancements in small molecules are relatively slow to build up and may be difficult to detect. For TOE experiments, simulations show it can take more than 20 s for indirect enhancements to reach 95% of their steady-state values for small molecules in organic solvents ($\omega\tau_c \ll 1$). In real experiments, external relaxation (not included in Figs. 4.4 and 4.5) will reduce enhancements still further, so, for typical irradiation times of 2–10 s, many indirect enhancements will be very much smaller than their steady-state values.

In large molecules, indirect effects build up more quickly, and can be hard to distinguish from direct effects. The most striking difference between Figures 4.4 and 4.5 is that in Figure 4.4 the three enhancements have quite different intensities, signs, and buildup rates, whereas in Figure 4.5 all three enhancements are very similar. This is a graphic illustration of the equations developed in Chapters 2 and 3. Figure 4.4 illustrates the consequences for small molecules of the smaller maximum enhancement (50% compared to 100%) and the alternation of the sign of the enhancement along a chain of spins (the "three-spin effect"; Section 3.2.2.2). In contrast, Figure 4.5 illustrates the undesirable but often unavoidable phenomenon of "spin diffusion" (Section 4.4.4), which causes enhancements to be transmitted along a chain of spins, thereby effectively removing any useful relationship between NOE intensity and distance.

The nature of spin diffusion for slowly tumbling molecules ($\omega\tau_c \gg 1$) may be further illustrated by an analogy. Under these conditions, there is an essentially exact analogy between spin diffusion and the diffusion of heat through the contents of a series of containers connected together and immersed in a bath (Fig. 4.6).[1] The irradiated spin is analogous to a container A heated to a high temperature (an infinite temperature in the case of complete saturation). Saturation of the irradiated spin is lost either by direct relaxation to the lattice at a rate characterized by R_A, or by cross-relaxation to other spins such as B or E at rates characterized by σ_{AB} and σ_{AE} (using the same symbols here for the spins as for the corresponding containers). This is analogous to heat loss from container A either to the bath or to containers B and E, at rates dependent on the thermal contact between then, and the rise in temperature of containers B and E is analogous to the corresponding NOE enhancements. The bath is

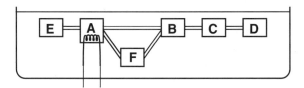

Figure 4.6. Heat diffusion analogy for spin diffusion (see text for discussion).

assumed to be sufficiently large that its temperature remains constant; for the small energies involved in NMR this represents a good approximation. Different spin–lattice relaxation rates for different spins are represented as containers having different rates of heat loss to the bath, and further spin diffusion away from spins B and E is analogous to transfer of heat to other containers. The time-course of the temperature of each container is analogous to the NOE kinetics of the corresponding signal. In a TOE experiment, the final temperature of each container (analogous to the corresponding steady-state NOE enhancement) represents the balance between heat arriving from other containers (cross-relaxation) and heat loss to the bath (spin–lattice relaxation). In a transient experiment, the temperature of container A, corresponding to the perturbed spin, is rapidly increased (in the case of inversion, formally to negative temperatures, i.e., higher than infinite), then the heating is stopped and the system left to cool. The other containers experience a transient rise in temperature as energy from A dissipates through them, before they eventually cool back down to equilibrium with the bath.

Figure 4.6 also illustrates an indirect heat transfer from A to B via F, which may transfer heat faster or slower than the direct A–B route. If we want to measure the direct A–B transfer rate (i.e., measure the direct NOE between A and B as a way of calculating their separation), we must do it before F has had a chance to "warm up," which will only be possible if at least one of the transfers A–F or F–B is significantly slower than the direct A–B transfer. Another aspect of spin diffusion is illustrated by spins B, C, and D. Clearly, on continuous heating of container A, container C cannot heat up until B is hot, and D cannot heat up until C is hot, so there is a progressive spread of heat along the chain, the approach to steady state being slower the further one gets from A.[2] In the spin-diffusion limit, contact with the lattice is weak, so the drop in spin temperature along a chain of spins is small, and eventually all the spins become saturated to a large extent. The calculated NOE enhancements in Figure 4.5 illustrate such behavior.

In the extreme narrowing limit or for ROE experiments, the heat diffusion analogy is less useful because there is no thermal effect corresponding to the alternation in sign of enhancements along a chain of spins.

4.2. THEORY OF THE KINETIC NOE IN A TWO-SPIN SYSTEM

We begin our more formal treatment of the theory of the kinetic NOE with the simplest case, namely a two-spin system undergoing a TOE experiment.

Since spin S is continuously saturated in a TOE experiment, the treatment for a two-spin system is particularly straightforward, as only spin I evolves. Calculating the time-course of the NOE therefore involves solving the single differential equation for I_z under the condition that the value of S_z is zero at all times:

$$dI_z/dt = -R_I(I_z - I_z^0) - \sigma_{IS}(S_z - S_z^0) \tag{4.1}$$

The solution for I_z at time τ under the initial condition $I_z = I_z^0$ may be found straightforwardly, and it is

$$I_z(\tau) = I_z^0 + S_z^0 \left(\frac{\sigma_{IS}}{R_I}\right)(1 - e^{-R_I\tau}) \tag{4.2}$$

Setting $S_z^0 = I_z^0$ followed by rearrangement yields the fractional NOE at time τ:

$$f_I\{S\}(\tau) = (1 - e^{-R_I\tau})\left(\frac{\sigma_{IS}}{R_I}\right) \tag{4.3}$$

We now turn to the slightly more complicated case of the transient experiment. During a 1D transient experiment with a two-spin system, both spins evolve during the NOE evolution period τ, so here calculating the NOE time-course involves solving the two simultaneous differential equations:

$$dI_z/dt = -R_I(I_z - I_z^0) - \sigma_{IS}(S_z - S_z^0)$$
$$dS_z/dt = -R_S(S_z - S_z^0) - \sigma_{IS}(I_z - I_z^0) \tag{4.4}$$

In the symmetrical case where $R_I = R_S = R$, the straightforward solution for $f_I\{S\}(\tau)$ under the initial conditions $I_z = I_z^0$ and $S_z = -S_z^0$ (that is, following inversion of spin S) gives[3]:

$$f_I\{S\}(\tau) = e^{-(R-\sigma)\tau}(1 - e^{-2\sigma\tau})$$
$$= e^{-(R-\sigma)\tau} - e^{-(R+\sigma)\tau} \tag{4.5}$$

When the two relaxation rates differ, the solution becomes somewhat more complicated[4]:

$$f_I\{S\}(\tau) = f_S\{I\}(\tau)$$
$$= (\sigma/D)(e^{-(R'-D)\tau} - e^{-(R'+D)\tau}) \tag{4.6}$$

where

$$D = \sqrt{\frac{1}{4}(R_I - R_S)^2 + \sigma_{IS}^2}$$

and

$$R' = \frac{1}{2}(R_I + R_S)$$

The solution for the NOESY experiment is essentially identical. For the symmetrical case it is

$$a_{\text{cross}}(\tau_m) = (M^0/2)(-e^{-(R-\sigma)\tau_m} + e^{-(R+\sigma)\tau_m})$$
$$a_{\text{dia}}(\tau_m) = (M^0/2)(e^{-(R-\sigma)\tau_m} + e^{-(R+\sigma)\tau_m}) \quad (4.7)$$

where $a_{\text{cross}}(\tau_m)$ and $a_{\text{dia}}(\tau_m)$ represent the intensities of the cross peaks and the diagonal peaks respectively after a mixing time of τ_m, and M^0 is the intensity of the diagonal peak at $\tau_m = 0$.[†] The factor of two relative to Eq. 4.5 arises because, on average (over all values of t_1), the degree of inversion of one spin relative to the other in a NOESY experiment is only half that in a 1D transient experiment, where one spin is fully inverted and the other is unaffected. Equations for the unsymmetrical case follow by an obvious analogy to Eq. 4.6.

Figures 4.2 and 4.3 illustrate the form of Eq. 4.5. The maximum enhancement occurs when $df/d\tau = 0$, which for the symmetrical case occurs at a time given by

$$\tau = \left(\frac{1}{2\sigma}\right) \ln\left(\frac{R+\sigma}{R-\sigma}\right) \quad (4.8)$$

At that time the enhancement is given by

$$\eta_{\text{max}}^{\text{transient}} = \left(\frac{R+\sigma}{R-\sigma}\right)^{-[(R-\sigma)/2\sigma]} - \left(\frac{R+\sigma}{R-\sigma}\right)^{-[(R+\sigma)/2\sigma]} \quad (4.9)$$

A plot of $\eta_{\text{max}}^{\text{transient}}$ against $\omega\tau_c$ is shown in Figure 4.7. The curve is very similar to that of η_{max} for the steady-state enhancement (Fig. 2.6), essentially the only difference being that the maximum transient enhancement observable in the extreme narrowing limit is 38.5%, rather than 50% as in the steady-state case. Eq. 4.9 may also be modified to calculate the maximum transient enhancement in the rotating frame, simply by replacing R and σ by their rotating frame analogs R_2 and σ_2 (see Section 9.3.1.2).

[†]Note that the overall sign of the expression for a_{cross} is opposite to those for $f_I\{S\}(\tau)$ in Eqs. 4.5 and 4.6. This simply reflects the different phasing usually applied to the two types of spectra; in NOESY spectra the diagonal is positive, whereas in a 1D transient experiment the inverted peak is conventionally phased as negative.

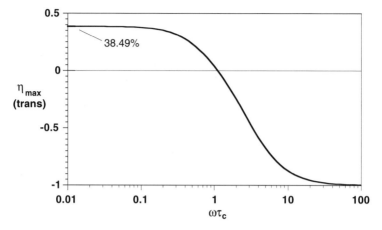

Figure 4.7. The maximum enhancement attainable in a two-spin homonuclear transient NOE experiment $\eta_{\max}^{\text{trans}}$, shown as a function of $\omega\tau_c$. Note the similarity to the curve for steady-state enhancements (Fig. 2.6), the main difference being the smaller value at the extreme narrowing limit ($\omega\tau_c \ll 1$).

4.3. THEORY OF THE KINETIC NOE IN MULTISPIN SYSTEMS

It is possible to calculate useful analytical solutions for three-spin systems,[5] but beyond this level there are only two approaches that have been widely used for the calculation of kinetic NOE enhancements; numerical integration and matrix algebra. Comparing these two approaches in detail leads to the conclusion that for calculating the result of TOE experiments, or for transient (including NOESY) experiments in which only a few spins are of interest, the numerical integration method is more efficient.[6] For calculations of transient experiments in which a large number of nuclei are of interest the relaxation matrix method is more efficient. The latter is often the case for calculations of NOE enhancements for entire macromolecules and, for these applications, relaxation matrix calculations have been overwhelmingly used. We therefore consider the relaxation matrix approach to transient experiments first.[7–9]

4.3.1. Multispin Kinetics in Transient NOE Experiments

As shown in Chapter 3, the Solomon equations for any number of spins can be written

$$dI_{nz}/dt = -(I_{nz} - I_{nz}^0)R_{I_n} - \sum_m (I_{mz} - I_{mz}^0)\sigma_{nm} \qquad (4.10)$$

Here, since no spin is saturated in a transient experiment and all spins may be observed, we have dropped the labels S and X that were used in Eq. 3.9, and relabelled all spins as I. In the expression for one given spin I_n, the summation over m therefore runs over all spins except I_n. These equations can be re-expressed in matrix form. They are usually written in a form appropriate for NOESY experiments, where they relate to cross-peak intensities as a function of the mixing time τ_m. Since cross-peak intensities directly reflect population differences from equilibrium and the terms I_{nz}^0 and I_{mz}^0 are all constants, it follows that for cross-peak intensities

$$dI_z/dt = d(I_z - I_z^0)/dt = d\mathbf{a}(\tau_m)/d\tau_m$$

and we can therefore write Eq. 4.10 in matrix form as follows:

$$\frac{d[\mathbf{a}(\tau_m)]}{d\tau_m} = -\mathbf{R}\mathbf{a}(\tau_m) \qquad (4.11)$$

where $\mathbf{a}(\tau_m)$ is the matrix of cross-peak intensities at mixing time τ_m, and $\mathbf{R}$ is the relaxation matrix given by

$$\mathbf{R} = \begin{bmatrix} R_1 & \sigma_{12} & \sigma_{13} & \cdots & \sigma_{1n} \\ \sigma_{21} & R_2 & \sigma_{23} & \cdots & \sigma_{2n} \\ \sigma_{31} & \sigma_{32} & R_3 & \cdots & \sigma_{3n} \\ \cdot & \cdot & \cdot & \cdots & \cdot \\ \cdot & \cdot & \cdot & \cdots & \cdot \\ \sigma_{n1} & \sigma_{n2} & \sigma_{n3} & \cdots & R_n \end{bmatrix} \qquad (4.12)$$

Here, the diagonal elements of the relaxation matrix are the longitudinal relaxation rate constants of each spin and the off-diagonal elements are the various cross-relaxation rate constants.

The solution of Eq. 4.11 for the cross-peak intensities at time τ_m is given by

$$\mathbf{a}(\tau_m) = \exp(-\mathbf{R}\tau_m)\mathbf{a}^0 \qquad (4.13)$$

where $\mathbf{a}^0$ is the matrix of peak intensities at the start of τ_m. Usually, $\mathbf{a}^0$ would be the diagonal matrix of equilibrium intensities, which for 2D NOESY experiments correspond to the volumes of the diagonal peaks at $\tau_m = 0$.

In order to solve this equation for the cross-peak intensities $\mathbf{a}(\tau_m)$, it is first necessary to convert the matrix $\mathbf{R}$ into a diagonal form λ, since only for diagonal matrices can an exponential be calculated directly:

$$\exp(\lambda) = \begin{bmatrix} \exp(\lambda_1) & 0 & \cdots & 0 \\ 0 & \exp(\lambda_2) & \cdots & 0 \\ \cdot & \cdot & & \cdot \\ \cdot & \cdot & & \cdot \\ 0 & 0 & \cdots & \exp(\lambda_n) \end{bmatrix} \qquad (4.14)$$

Diagonalization of $\mathbf{R}$ is possible using standard methods of matrix algebra.

The result of such a procedure is to identify the rotation matrix χ that converts $\mathbf{R}$ to its diagonalized form λ:

$$\mathbf{R} = \chi \lambda \chi^{-1} \tag{4.15}$$

The elements λ_i of the diagonal matrix λ are the eigenvalues of $\mathbf{R}$, and the columns of χ comprise the eigenvectors of $\mathbf{R}$. We may then substitute Eq. 4.15 into Eq. 4.13 as follows:

$$\begin{aligned}\mathbf{a}(\tau_m) &= \exp(-\mathbf{R}\tau_m)\mathbf{a}^0 \\ &= \exp(-\chi\lambda\chi^{-1}\tau_m)\mathbf{a}^0 \\ &= \chi \exp(-\lambda\tau_m)\chi^{-1}\mathbf{a}^0 \end{aligned} \tag{4.16}$$

Thus, once the elements of the relaxation matrix are known, the cross-peak intensities (or, for a 1D experiment, the transient NOE intensities) at any given mixing time may be calculated.

It is becoming common practice in macromolecular structural studies (particularly of nucleic acids) to use such calculations as part of structural refinement protocols. Since these calculations rely on an iterative comparison of experimental NOE intensities with those calculated from the relaxation matrix (which in turn is often calculated from the structural coordinates), this requires not only the calculation of NOE intensities from the relaxation matrix, but also the reverse process (i.e., calculation of the relaxation matrix from cross-peak intensities). Subsequent cycles of iteration, then, often replace calculated intensities by the experimental ones where possible, in order to generate an improved relaxation matrix for the next round of the cycle (cf. Section 12.5.2). Calculation of the relaxation matrix from the cross-peak intensity matrix can be achieved using essentially identical algebra. From Eq. 4.13 we can write

$$\mathbf{R} = \frac{-\ln[\mathbf{a}(\tau_m)/\mathbf{a}^0]}{\tau_m} \tag{4.17}$$

where, again, $\mathbf{a}^0$ corresponds to the equilibrium intensity matrix. We then calculate the diagonal matrix $\mathbf{L}$, which contains the eigenvalues of $\mathbf{a}(\tau_m)/\mathbf{a}^0$, and the rotation matrix $\mathbf{X}$, whose columns are the eigenvectors of $\mathbf{a}(\tau_m)/\mathbf{a}^0$, using which we can then re-express $\mathbf{R}$ as

$$\mathbf{R} = -\mathbf{X} \ln(\mathbf{L}/\tau_m)\mathbf{X}^{-1} \tag{4.18}$$

There are now a number of programs for the refinement of macromolecular structures which aim to improve the relationship between NOE intensity and distance by using algebra of this sort. Such refinement methods are considered further in Section 12.5.2.

The calculation of NOE intensities as a function of time can also be achieved using numerical integration.[6] Given the relaxation matrix and the initial intensities, the Solomon equations (Eq. 4.10) are solved numerically using a series of small time increments δt:

$$I_z(t + \delta t) = I_z(t) + (dI_z/dt)\delta t \qquad (4.19)$$

Applications of this method for the calculation of NOESY intensities in proteins are presented in reference 10, which provides a useful insight to the relationship between geometry and cross-peak intensity.

4.3.2. Multispin Kinetics in TOE Experiments

In TOE experiments, spin S is continuously irradiated throughout the NOE evolution period, and the experiment is necessarily one-dimensional. The object of calculation is to find all the enhancements $f_{Ii}\{S\}(\tau)$ after a particular evolution time τ. These enhancements may be expressed in the form of a column vector $\mathbf{f}(\tau)$ of dimension $n - 1$ (where n is the number of spins in the system including S).

The numerical integration method is easily adapted to calculating enhancements in TOE experiments, simply by resetting the intensity of the irradiated spin to zero at each time increment δt during the calculation. Adaptation of the matrix method is slightly more involved. The first step is to calculate a "reduced" relaxation matrix $\mathbf{R}'$ of dimension $(n - 1) \times (n - 1)$, which is identical to the normal relaxation matrix $\mathbf{R}$ except that it lacks the row and the column containing all the terms involving spin S. Equation 4.11 then becomes

$$\frac{d[\mathbf{f}(\tau)]}{d\tau} = \sigma - \mathbf{R}'\mathbf{f}(\tau) \qquad (4.20)$$

where σ is a column vector of dimension $n - 1$ that contains all the cross-relaxation rates σ_{IiS} involving S. Solving this equation for $\mathbf{f}(\tau)$ yields[11]

$$\mathbf{f}(\tau) = \mathbf{R}'^{-1}\sigma[1 - \chi' \exp(-\lambda'\tau)\chi'^{-1}] \qquad (4.21)$$

where χ' and λ' are the reduced analogs of their counterparts in Eq. 4.16.

4.4. ESTIMATING INTERNUCLEAR DISTANCES

Having covered the theory required to describe the full time-course of NOE enhancements in kinetic NOE experiments, we turn now to the most significant application of kinetic NOE measurements. Perhaps ironically, in the simplest approach to estimating internuclear distances, much of the foregoing theory is rendered redundant, as will be described in the following section on the initial rate approximation. However, as will be seen in Sections 4.4.2 and 4.4.3, this simplification often breaks down, at least in part, and more sophisticated approaches are then required.

4.4.1. The Initial Rate Approximation

The initial conditions for a TOE experiment (assuming instantaneous saturation) are $I_z = I_z^0$ and $S_z = 0$. Making these substitutions in Eq. 4.1, the initial rate of growth from an NOE enhancement in a two-spin system is

$$dI_z/dt|_{t\to 0} = \sigma_{IS} S_z^0 \qquad (4.22)$$

Similarly, in a 1D transient experiment with a two-spin system, the initial conditions are $I_z = I_z^0$ and $S_z = -S_z^0$, giving the initial rate of NOE growth as

$$dI_z/dt|_{t\to 0} = 2\sigma_{IS} S_z^0 \qquad (4.23)$$

For a NOESY experiment with a two-spin system, the analogous result is[12]

$$\left.\frac{d[a_{\text{cross}}(\tau_m)]}{d\tau_m}\right|_{\tau_m \to 0} = -\sigma_{IS} a^0 \qquad (4.24)$$

where a^0 is the intensity of the diagonal peak at $\tau_m = 0$.

For a multispin system (cf. Eq. 3.9) the appropriate differential equation is

$$dI_z/dt = -R_I(I_z - I_z^0) - \sigma_{IS}(S_z - S_z^0) - \sum_X \sigma_{IX}(X_z - X_z^0) \qquad (4.25)$$

Here the initial conditions for TOE are $I_z = I_z^0$, $X_z = X_z^0$ and $S_z = 0$, so that the initial rate is again

$$dI_z/dt|_{t\to 0} = \sigma_{IS} S_z^0$$

while for 1D transient and NOESY experiments, the initial rates are exactly as they were for the corresponding two-spin cases.

Thus, at the limit of short NOE evolution times in all kinetic NOE experiments, we see that *only the cross relaxation rate constant σ_{IS} controls the size of the IS enhancement*, and *all spin pairs behave as though they were isolated two-spin systems*. This is what is meant by the initial rate approximation.

These are profoundly important results. From the moment that NOE evolution starts in any kinetic NOE experiment, cross-relaxation at a rate characterized by σ_{IS} begins, but the relaxation of the enhanced signals only starts significantly to oppose this incoming population flux at later times, as the enhancements build up significant population deviations from equilibrium. Complicated dependencies on internuclear distances other than r_{IS} are all associated with the relaxation terms R_I, R_X (and, for transient experiments, R_S), which only start to influence NOE kinetics once enhancements have grown to an appreciable level. In TOE experiments, the steady state that is eventually reached represents the final balance between the rate at which the enhancement arrives at I (characterized by σ_{IS}) and the rate at which it is lost (characterized by R_I), together with the effects of indirectly transmitted enhancements. In transient experiments, eventually the enhancements decay to zero.

We may also see from this discussion what the region of validity of the initial rate approximation must be. Clearly, for TOE experiments it is valid only for a period that is short *relative to* $(R_I)^{-1}$. This means, of course, that it is valid for a shorter time for enhancements into spins that relax quickly than it is for enhancements into spins that relax slowly—a point that has important consequences for interpretation. In the case of transient experiments such as NOESY, the evolution of enhancements depends not only on R_I but also on R_S, so that the region of validity of the initial rate approximation depends on the relaxation times of both participating spins in a given enhancement. For a 1D transient NOE experiment, Eq. 4.23 shows that enhancements grow initially at a rate 2σ, rather than σ. It might be thought that this gives such experiments an advantage, in that larger enhancements are reached more quickly, so "beating" the limitations of the initial rate approximation. This is not so, however, as the corollary is that the initial rate approximation becomes invalid twice as quickly for such experiments.

4.4.2. Distances from Enhancement Ratios

The key implication of the initial rate approximation is that, while it is valid, r_{IS} is the only internuclear distance that influences the size of the NOE enhancement between I and S. Measuring the initial rate of growth of an NOE enhancement or cross peak thus gives a direct measure of σ_{IS}, and this in turn is directly related to the internuclear distance:

$$\sigma_{IS} = \zeta r_{IS}^{-6} \qquad (4.26)$$

where

$$\zeta = \left(\frac{\mu_0}{4\pi}\right)^2 \frac{\hbar^2 \gamma^4}{10} \left(\frac{6\tau_c}{1 + 4\omega^2 \tau_c^2} - \tau_c\right) \qquad (4.27)$$

This provides a very sensitive measure of internuclear distance, provided that the relationship between NOE intensity and distance can be calibrated. Since the only unknown in Eq. 4.27 is τ_c, calibration could in principle be attempted on an absolute basis; but, in practice, σ_{IS} depends very strongly on τ_c (Fig. 4.8), and τ_c is itself not at all easy to measure, making a direct calculation of ζ difficult. However, if the system under investigation is rigid and tumbles isotropically, all the internuclear vectors in the molecule will have the same correlation time. Under these circumstances, a relative calibration can be

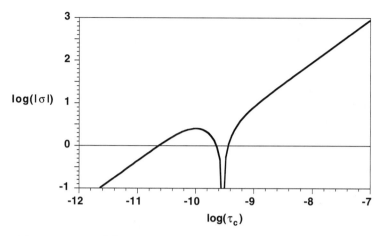

Figure 4.8. Variation of $|\sigma|$ with correlation time τ_c (cf. Eq. 4.27). The curve is calculated for two protons 2 Å apart for a 600 MHz NMR frequency ($\omega = 2\pi \times 600 \times 10^6$ rad s^{-1}). To the right of the cusp, σ is negative.

achieved by comparing the NOE intensity corresponding to the unknown distance with that for a pair of protons of known internuclear separation, r_{ref}. The unknown distance can then be calculated very simply:

$$f_I\{S\}(t) = \zeta r_{IS}^{-6} t$$

$$f_{\text{ref}}(t) = \zeta r_{\text{ref}}^{-6} t$$

where $f_{\text{ref}}(t)$ is the reference NOE intensity at time t. Taking the ratio of these expressions gives

$$\frac{f_I\{S\}(t)}{f_{\text{ref}}(t)} = \left(\frac{r_{IS}}{r_{\text{ref}}}\right)^{-6} \quad (4.28)$$

which rearranges to

$$r_{IS} = r_{\text{ref}} \left[\frac{f_I\{S\}(t)}{f_{\text{ref}}(t)}\right]^{-1/6} \quad (4.29)$$

This strikingly simple result is the basis of most distance estimation using the NOE. It is often known as the isolated spin pair approximation (ISPA), because it is only strictly valid for isolated pairs of protons (see Section 4.4.1).

The commonest application of this method is to the calibration of cross-peak intensities from 2D NOESY spectra of macromolecules, in which cases the chosen reference distances are fixed either by covalent geometry or, more often, by regular secondary structure. Such procedures are discussed in detail in Section 12.2.3. We illustrate the principle here using a surprisingly difficult problem, that of defining the geometry about a double bond where quite precise distances were required to distinguish between two isomeric structures.

Following the addition reaction of LiOCH=C(Me)SOEt with PhCH-(Me)NH$_3$Cl, a β-sulphinylenamine was produced as just one of the two possible double bond isomers, E (**4.1**) or Z (**4.2**), the problem being the definition of which isomer is actually present. Calculated steady-state NOE values for the

4.1   **4.2**

H{Me$_B$} NOE enhancement across the double bond are predicted to differ only by a factor of 3 in the alternative forms, which would be too small a difference to use safely to distinguish the two forms (particularly if only one isomer is available). However, the H$_A$—Me$_B$ internuclear distances in the two forms are significantly different (r_{AB} = 2.03 Å for the E isomer, 3.15 Å for the Z isomer, measured to the centroid of the methyl group in each case), suggesting that kinetic NOE measurements from TOE experiments could be used instead. Enhancements were measured at several time points for H$_A${Me$_B$} and also for H$_C${H$_D$}, which acted as the reference distance with a covalently fixed separation of about 1.75 Å. ^{13}C T_1 values for nearby carbons were used to demonstrate roughly similar rotational correlation times throughout this part of the molecule. Using a TOE buildup time of 0.5 s (which for this small molecule still lies within the validity range of the initial rate approximation), the enhancements found were f_A{B} = 3.4% and f_C{D} = 6.3%, giving an estimated internuclear distance r_{AB} of 1.75 × (3.4/6.3)$^{-1/6}$ = 1.94 Å. This implies an E geometry. The same conclusion was reached after estimating the distance by fitting the buildup to a simple two-spin buildup curve over time points up to 2 s (Eq. 4.3).[13]

4.4.3. Errors in Distance Measurements Using the Initial Rate Approximation

The method for estimating internuclear distances outlined in the previous section is reasonably robust, largely because it relies on the sixth root of NOE intensities. Thus, even quite large errors in intensity measurements often have a relatively small effect on the derived internuclear separation. Errors in the reference distance translate into an equivalent fractional error in the calculated distance, so that small errors have little effect (at least, relative to the other major sources of error considered below). Probably the most commonly used reference distances are those between methylene protons (r = 1.75–1.80 Å) and those between *ortho* aromatic protons (r = ca. 2.80 Å). As discussed more extensively in Section 12.2.3, these distances are not ideal for several reasons. First, they are usually significantly shorter than the unknown distance, which can introduce systematic errors into the measurement (see also the following section). Second, the protons concerned are mutually coupled, which can cause complications from strong coupling (Section 6.4.1). Third, these protons are often in parts of the molecule that are strongly affected by internal motions; for instance, mobile sidechains in proteins often have correlation times 2 to 5 times shorter than that for overall tumbling of the whole molecule. This can mean that the corresponding NOE intensities are significantly weakened or even absent, with the potential to cause major errors of interpretation. Further discussion of this issue can be found in Chapter 5 and Section 12.2.3.

Although the reference distance is generally one that is fixed by covalent geometry, the same is not usually true of the unknown distance being measured, so for these a second issue can arise, namely motional averaging of the distances themselves. In flexible molecules, distances derived for NOE measurements correspond either to $\langle r^{-6} \rangle^{-1/6}$ (where the angled brackets represent conformational averaging) for internal motions slower than overall tumbling, or r_{Tropp} for internal motions faster than overall tumbling (cf. Section 5.5.2). These averages are inherently weighted towards shorter distances, so however carefully the experimental NOE data are quantified, distances derived using Eq. 4.29 may not correspond to the true mean distance when there is internal motion present; in other words, the measured distance may be *precise* but it is not necessarily *accurate* (cf. Sections 5.6 and 12.5.1). In general, the extent of internal motions is unknown and, in principle, the weighting to short distances could mean that quite small radial displacements of a proton pair (i.e., internal motions that cause the protons to move closer together or further apart) might cause significant changes in NOE enhancements. While such distortions of the NOE could be common, calculations based on molecular dynamics trajectories for proteins have suggested that usually the effect on the NOE from this component of the change in the interproton vector is small.[14–15] These various topics are discussed more fully in Chapter 5 and Section 12.2.3.

One possible source of error specific to TOE experiments is noninstantaneous saturation. As discussed in Sections 1.5 and 7.4.2, and more fully in reference 11, when a spin is irradiated, its path to saturation follows an exponentially damped cosine wave. As long as the irradiation power is high enough, this has only a minor effect on the time-course of the enhancement, which behaves very much as if the spin had been instantaneously saturated. However, if the irradiation power is too low, the rate of the NOE buildup is significantly slower than σ, and enhancement intensities cannot easily be used to measure distances.

However, the biggest single cause of error in using the initial rate approximation for distance measurement is that, in reality, NOE buildup is nonlinear. This topic is the subject of the next section.

4.4.4. Spin Diffusion, Nonlinear NOE Growth, and Interpretation

The overwhelming majority of kinetic NOE measurements are applied to the structure determination of biomolecules using NOESY experiments. As previous sections have shown, simple-minded use of the initial rate approximation is likely to lead to significant errors due to spin diffusion, except at very short τ_m. The particular questions that then arise are: "How can spin diffusion be recognized?" "How short must τ_m be to be 'safe'?" and "How can errors from spin diffusion be minimized or avoided?"

It is often said that spin diffusion can be identified by the presence of an induction period at the start of an NOE buildup curve. Such features can indeed be seen in some of the calculated curves shown in Figures 4.4, 4.5, and 4.9, where the spin-system geometries make the induction period particularly visible, as there is essentially no direct NOE. However, it is also clear that even some of these induction periods, especially those for enhancement $C \leftarrow B \leftarrow \{A\}$ in Figure 4.5 and enhancement AC in Figure 4.9, are so short that they would probably be undetectable by experiment. For instance, the inset in Figure 4.9 shows that τ_m values of less than 20 ms would be needed to detect the induction period in this case, and measurement of a true initial rate would

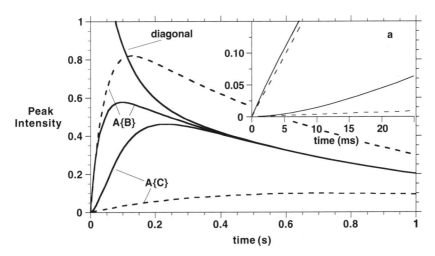

Figure 4.9. The dependence of cross-peak intensity on τ_m in a NOESY experiment. Calculated for a linear three-spin system with $r_{AB} = r_{BC} = 2.5$ Å, using $\tau_c = 5 \times 10^{-8}$ s, 500 MHz NMR frequency, $\rho^* = 10$ s^{-1}. The dashed lines show the curves for the equivalent two-spin system. The inset shows an expansion of the initial part of the curve.

require measurements at mixing times of well under 5 ms, where enhancements would be so small as to be undetectable. Thus, in practice, spin diffusion can be very difficult or impossible to spot using this approach.

Once the initial induction period is over, linear geometries of the sort just considered generally give the biggest errors due to spin diffusion, because transmission of the AC enhancement through an intermediate spin B will always make the r_{AC} distance appear shorter than its true value. In contrast, if the distance r_{AC} were comparable to or shorter than at least one of the distances r_{AB} or r_{BC} (in a nonlinear system), then the main effect of spin B would be to reduce the AC enhancement by relaxing spins A and C, thereby causing a slight *increase* in the apparent distance r_{AC}.[5] Given this link between spin diffusion and linear geometries, it follows that for an A-B-C spin diffusion pathway, cross peaks AB and AC will both be more intense than the AC cross peak. Consequently, the cross peaks most likely to arise from spin diffusion are generally the weakest ones, which is mainly why upper bounds for weak cross peaks are normally set somewhat longer than would be indicated just by simple-minded application of Eq. 4.29 (cf. Section 12.2.3).

It is also clear that such errors from spin diffusion are smaller for data collected with shorter τ_m values, as indicated by the scatter plots shown in Figure 4.10. However, even at 50 ms, the errors for the more distantly separated spin pairs (i.e., those corresponding to weak cross peaks and having more chance of arising through intervening spin diffusion) introduce a highly significant bias, and produce errors of around 10% at a true distance of 4.5 Å (equivalent to an intensity $(5/4.5)^6 = 1.9$ times bigger than that given by the initial rate approximation). Further discussion may be found in reference 16.

The take-home message from all this is that there exists no practical value of τ_m short enough that spin diffusion can be guaranteed not to cause significant deviations from idealized ISPA intensities. Indeed, when a third spin causes significant spin diffusion, there is no simple way to extract distance information from the AC enhancement, even using relaxation matrix calculations.[18] Given these limitations, one must instead find the best way of coping with spin diffusion so as to minimize its impact on interpretation. Essentially the only simple way to protect against errors from spin diffusion is to use long upper distance bounds for weak NOE cross peaks, as described above, and to attempt to use data recorded at the shortest practical mixing time to calibrate each enhancement (see Section 12.2.3). At a more sophisticated level, one may use a full relaxation matrix approach to refine iteratively the relationship between NOE intensity and distance, perhaps employing a series of NOESY experiments at different mixing times (cf. Sections 4.3.1 and 12.5.2).[8,19] Such methods can be extended to handle ROESY data (cf. Section 9.3.1.3),[20] internal motion, anisotropic tumbling, and so on.[21] Another stratagem during more conventional refinement is to use preliminary calculated structures to try to identify likely spin diffusion pathways for suspect peaks, so as to exclude the corresponding entries from the constraint list (cf. Section 12.2.3).

ESTIMATING INTERNUCLEAR DISTANCES   119

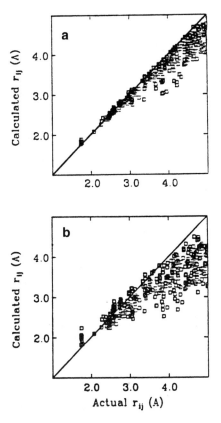

Figure 4.10. Comparison of exact distance against distance calculated using the initial rate approximation from NOESY intensities using data sets simulated for the small protein lysozyme, with mixing times (*a*) 50 ms and (*b*) 200 ms. Calculated using τ_c = 9 ns. The "reference distance" used for the initial rate approximation was 2.48 Å. Note that distances larger than this tend to be underestimated, and distances shorter than 2.48 Å tend to be overestimated. Reproduced with permission from ref. 17.

It is also worth noting here that aromatic ring motions may cause a rather insidious form of spin diffusion.[22] Suppose that there are two protons X and Y that are positioned on either side of a symmetric, fast-flipping aromatic ring, X being close to H_2 and Y close to H_6. Protons X and Y will each show a strong NOE to their corresponding aromatic proton neighbors, but the ring motion will also relay the indirect NOE interaction between X and Y, effectively bridging the 4.2 Å distance between H_2 and H_6 as though it did not exist. This still corresponds to an indirect pathway between X and Y, but it grows at a very much faster rate than would be anticipated from considering only the static positions of the four spins involved. In some examples this process has been shown to cause large errors in distance constraints, leading to large errors in calculated structures.[23]

So far we have discussed mainly deviations from linear buildup rates whereby enhancements grow faster than predicted on the basis of the two-spin approximation, due to intervening spin diffusion. However, a more common deviation from linearity is for buildup rates to become slower with time. The initially linear growth of all enhancements slows and eventually stops, in TOE experiments because they reach steady state, or in transient experiments because they reach their maximum and then decay. This is not really due to spin diffusion, in that such nonlinearity is predicted even for two-spin systems; Eqs. 4.3 and 4.5 show that it is largely governed by spin–lattice relaxation rates. Nonlinear growth can be apparent even at very early times during NOE buildup, and, as would be expected, it is most marked for spins that relax quickly, such as methylene protons.

One consequence is that the buildup of enhancements involving closely separated spins falls off more rapidly than does the buildup of those involving spins that are further apart. Thus, if an enhancement used for calibration corresponds to a significantly shorter distance than an unknown distance that is to be estimated, the enhancement used for calibration will lose more intensity due to nonlinearity than will the enhancement due to the unknown distance. When Eq. 4.29 is applied, this causes the unknown distance to be underestimated. Of course, in the less common circumstance that the reference distance is longer than the unknown, application of Eq. 4.29 would then cause the unknown distance to be overestimated. This problem and various approaches to solving it are considered further in Section 12.2.3.

Figure 4.11 illustrates how both the types of error considered in this section can interact as a function of the spin-system geometry.

Finally, it is useful to consider the effect of spin diffusion on NOE kinetics when $\omega\tau_c$ is less than 1 (i.e., for rapidly tumbling molecules); this also roughly parallels the situation for large molecules in ROESY experiments. As shown elsewhere (e.g., Sections 3.2.2 and 9.3.1) direct enhancements in such cases are positive (equivalent to saying that cross peaks in two-dimensional spectra have the opposite sign to the diagonal), whereas indirect effects are negative (i.e., cross peaks have the same sign as the diagonal). For a linear arrangement of spins A-B-C the result is that the enhancement $f_C\{A\}$ is easy to identify as being indirect because of its sign. The *direct* enhancement $f_B\{A\}$ is reduced relative to the size it would have had in the absence of spin C, due to the relaxation of B by C (this is very similar, both qualitatively and quantitatively, to what happens for direct effects in slowly tumbling molecules). When translated into distances, this has the effect that the distance r_{AB} inferred from the $f_B\{A\}$ enhancement appears to be slightly longer when there is another spin C nearby that can relax A or B. As the angle ABC is reduced, bringing C closer to A, the direct (positive) contribution to the $f_C\{A\}$ enhancement becomes larger and eventually overtakes the indirect (negative) contribution via B (cf. Section 3.2.2.3). Therefore, there is a range of geometries in which the enhancement $f_C\{A\}$ is very small and bears no obvious relationship to the distance r_{AC}.

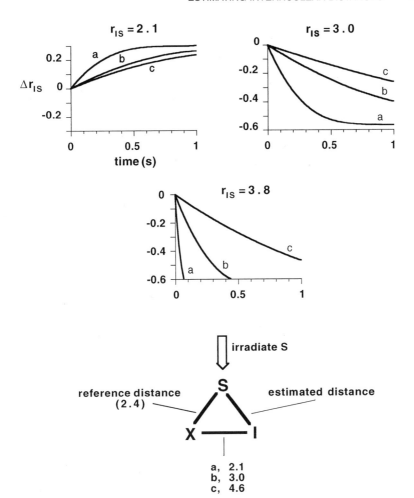

Figure 4.11. Errors in estimated distance arising from application of the initial rate approximation. Data are calculated for transient experiments on the three-spin system shown. The reference distance is $r_{SX} = 2.4$ Å, the distance being measured is r_{IS}, and r_{IX} has the values 2.1, 3.0, or 4.6 Å. The true distance r_{IS} is given above each graph, on which are plotted the errors in the estimated distance r_{IS} versus time. $\omega\tau_c = 15.7$ (i.e., $\tau_c = 5$ ns on a 500 MHz spectrometer), and external relaxation rates are set at 0.5 s^{-1} for each proton. Adapted from ref. 32.

It is thus clear that by acting to reduce the intensities of direct enhancements, spin diffusion leads to a slight increase in apparent distances between adjacent nuclei, both for small molecules ($\omega\tau_c < 1.12$) and for large ($\omega\tau_c > 1.12$). In contrast, for nuclei that are further apart, rather different effects are seen for small and large molecules. For small molecules, spin diffusion leads to a reduction in the intensity of such enhancements (apparent distances longer than the true distances), depending on the geometry. In large molecules, by contrast, spin diffusion to more distant nuclei *increases* the intensities of the corresponding enhancements, thereby giving apparent distances shorter than the true distance, as discussed above. When NOE enhancements are translated into upper bounds on distance ranges, the latter is clearly the more serious problem for structure calculations: A distance restraint that is too weak (as in the small molecule case) merely makes structure calculations less constrained, but a restraint that is too tight (as in the large molecule case) can seriously distort structures (cf. Section 12.2.3). Of course, if NOE enhancements are used to generate specific, narrow constraints on the target distance (as opposed to just the more usual restraint on upper bounds), then both kinds of spin diffusion will produce errors in structure calculations. It should, however, be evident from the rest of this book that we are not enthusiastic about using overquantified constraints.

4.5. MORE ABOUT EXPERIMENTS

In this section we collect some further points concerning experiments involving NOE kinetics.

4.5.1. Symmetry in Kinetic NOE Experiments

It is important to note that enhancements in 1D transient and NOESY experiments are symmetrical, whereas those in TOE experiments are not. In other words, in a 1D transient experiment, $f_I\{S\} = f_S\{I\}$ for all values of τ and, in NOESY, the two symmetry-related cross peaks corresponding to the *IS* interaction have the same intensity as each other for all values of τ_m, whereas in TOE experiments $f_I\{S\}$ can differ considerably from $f_S\{I\}$, depending on the geometry of the spin system.[†] In fact, the relaxation processes occurring during 1D transient and 2D NOESY experiments follow essentially identical kinetics, so these experiments produce closely related results.[25]

This difference between transient NOE experiments and TOE deserves some comment. In the TOE and steady-state experiments, the relaxation rate of the irradiated spin is not relevant to the NOE kinetics, since it is continuously saturated (Fig. 4.12a); for saturation of *S*, the direct enhancement at *I* is determined only by the ratio of σ_{IS} to R_I (cf. Eq. 4.3). If R_S is much larger than

---

[†]There are circumstances in which transient experiments are not symmetrical. The most common is when the relaxation delay is too short to allow all spins to regain thermal equilibrium between successive passes through the pulse sequence. Slowly relaxing spins then start each pass through the pulse sequence with less than their equilibrium intensity, so that enhancements from them (or, in the 2D experiment, cross peaks with an F_1 frequency corresponding to the slowly relaxing spin) are reduced in intensity.[24] In addition, although intensities may be roughly equal, the differing resolution in the two dimensions of 2D spectra may make symmetry-related peaks appear to have different peak heights (see chapter 8).

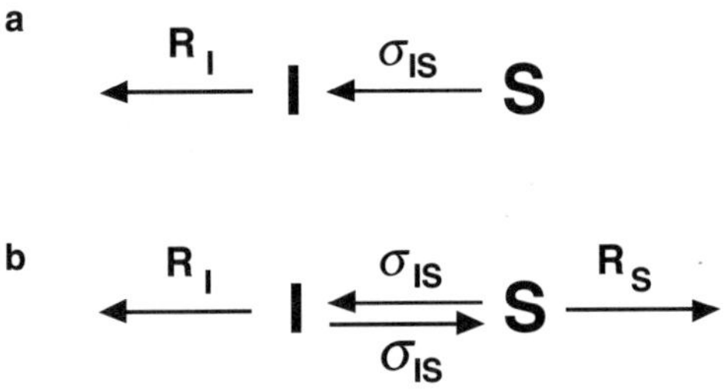

```
a           R_I        σ_IS
         ←——— I ←——— S

b           R_I        σ_IS         R_S
         ←——— I ⇌ S ———→
                    σ_IS
```

Figure 4.12. Relaxation pathways relevant during (*a*) TOE and (*b*) transient experiments. The saturated spin is S.

R_I, then clearly this will make the steady-state or TOE enhancement $f_S\{I\}$ much smaller than $f_I\{S\}$. In contrast, in the transient and NOESY experiment, neither I nor S is saturated during NOE evolution, so the enhancement at either signal is dependent on the relaxation rates of both (Fig. 4.12b). It is instructive to see how this symmetry arises in the transient experiment, even when one of the spins, say S, relaxes significantly faster than the other. On inversion of S, the enhancement at I starts to build up, but its growth soon slows because the intensity of the rapidly relaxing S signal returns quickly towards equilibrium and, therefore, ceases to drive the enhancement. Conversely, if I is inverted, the enhancement at S starts to build up at the same initial rate, but its growth is opposed by the rapid spin–lattice relaxation rate of S. In either case, the result is the same: an initial buildup at a rate σ_{IS}, followed by a decay that depends on R_I, R_S, and σ_{IS} (see Section 4.2 for details).

The symmetry of transient experiments also holds when groups of equivalent spins are present. Thus, in NOESY experiments, the two symmetry-related cross peaks between a single proton and a methyl group have the same intensity (though they may, of course, have different peak heights because of differing resolution in the two dimensions). It is readily shown[12] that if there are n_I spins I and n_S spins S, then the intensity of either IS cross peak is simply proportional to $n_I n_S$ (even though it has to be said that the first printing of the first edition of this book was in error on this point, as pointed out by Yip[26]). In the initial rate approximation (see below), a NOESY cross peak between a single proton and a methyl group would thus be three times more intense than a cross peak between two single protons for the same cross-relaxation rate, while a methyl–methyl cross peak would be nine times more intense than a cross peak between two single protons. The same applies in 1D transient experiments, in that the *absolute* NOE enhancements $(I_z - I_z^0)$ and $(S_z - S_z^0)$, analogous to the cross peak intensities in NOESY, are the same no matter how many equivalent spins n_I or n_S there may be. However, the *fractional* enhancements, which are the values normally calculated in 1D experiments, differ by the ratio n_I/n_S simply due to division by the equilibrium intensities I^0 and S^0. The same would be true in NOESY, if cross-peak intensities were measured as a fraction of one of the corresponding diagonal peaks at zero τ_m.

These differences between transient and TOE experiments are illustrated in Figure 4.13, using experimental data for 3-methylthiophene-2-carboxylic acid (**4.3**). The fractional transient enhancement $H_4\{Me\}$ is exactly three times the enhancement $Me\{H_4\}$ for all values of τ, whereas the corresponding steady-

4.3

state enhancements differ by a factor of 4.68. Figure 4.13 also illustrates some other points for which the theoretical background was developed earlier in this chapter. First, the decay phase of the transient enhancement is slower than the buildup phase (Fig. 4.13a). This is typical of all transient enhancements (indeed, by definition, it must be true), but would be much more pronounced for a molecule in the negative NOE regime. Second, the initial buildup rate in the transient experiment is twice that in the TOE experiment (cf. Section 4.4.1). Third, the maximum transient NOE enhancement in this instance is actually larger than the steady-state enhancement for $Me\{H_4\}$. Since this would not be true for an ideal two-spin system (maximum steady-state enhancement 50%, maximum transient enhancement 38.5%, Section 4.2), this behavior presumably arises because the faster initial growth in the transient experiment competes more effectively against external relaxation (ρ^*, cf. Section 2.4).

4.5.2. T_1 Values as an Aid to Interpretation

Some knowledge of T_1 values is clearly valuable in a general way when planning or interpreting NOE experiments. The time taken for enhancements to reach steady state, or for thermal equilibrium to be regained after a perturbation, is of the order of $5 \times T_1$. Similarly, the time at which the maximum NOE enhancement is reached in a transient experiment (Eq. 4.15) is of the same order as T_1, while the region of validity of the initial rate approximation is clearly also dependent upon T_1 (Section 4.4.1). Knowing T_1 values is generally useful; in addition, as we will discuss in this section, measured T_1 values, in conjunction with NOE data, can provide some structural information in their own right. In the main, this discussion applies to small and medium-sized molecules, since for large molecules ($\omega\tau_c \gg 1$), rapid spin diffusion causes the spins to relax cooperatively, largely destroying the concept of an individual T_1 for each spin (Section 2.3.2).[27]

One attraction of using T_1 values to obtain structural information in small molecules is that they can be measured very much more quickly than NOE enhancements. Against this must be set the strong disadvantage that T_1 values relate only to one signal, whereas NOE enhancements relate two signals to one another. This may sound trivial, yet it is profoundly important for interpretation.

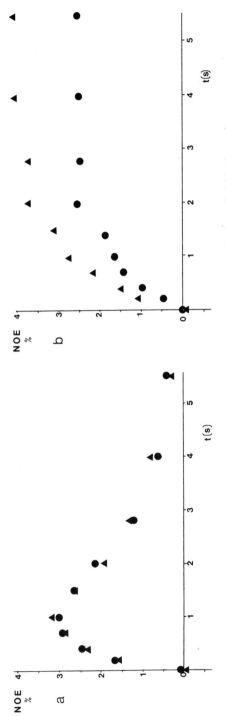

Figure 4.13. Experimental time-course of 1D NOE enhancements in 3-methylthiophene-2-carboxylic acid (**4.3**) for (*a*) transient experiments and (*b*) TOE experiments. The enhancements between H$_4$ and the methyl group are shown, with the Me{H$_4$} fractional enhancements multiplied by a factor of three to aid comparison. (▲)H$_4${Me}, (●)Me{H$_4$}. All measurements were made using a Bruker AM-300 spectrometer and selective inversions were achieved using a 22 ms decoupler pulse. Reproduced with permission from ref. 25.

It implies that interpreting T_1 values in isolation necessarily involves a much higher degree of ambiguity and supposition than does interpreting NOE enhancements, since the origin of a particular effect on T_1 must always be deduced, rather than being known directly.

Gross differences in T_1 values can be useful in several ways. Isolated protons can often be identified from their much longer T_1 values. This can be used to provide assignments if the structure is known, and also it gives prior warning that NOE enhancements into such protons will take a long time to reach steady state and may be transmitted over quite long distances. Also, partially relaxed spectra can be used to achieve limited spectral editing.[28,29]

Solution of a structural problem is generally a more demanding task than making assignments of the sort discussed above. Consequently, applications to structure elucidation where T_1 measurements have been used *in preference* to NOE experiments are rare. An interesting example is provided by the distinction of structures **4.4** and **4.5**, where the symmetry of both structures precluded any useful NOE experiments. However, measurement of T_1 for the cyclobutyl

4.4   **4.5**

methine signal showed a markedly shorter value for the *cis* compound **4.4** (T_1 = 3.4 s) than for the *trans* compound **4.5** (T_1 = 4.8 s), because of the larger number of nearby methyl protons in the former case.[30] This approach may be quite general for such symmetrical systems, but can only be expected to give a clear result when both stereoisomers are available for study.

In general, however, T_1 measurements are most useful just as a "backup" for the interpretation of NOE results. Once a structure has been proposed, it can be valuable to check that there are no obvious contradictions with the relaxation rates; in particular, signals with anomalously slow relaxation should correspond to sites relatively distant from other spins, if the structure and assignment are correct. Much as with chemical shift arguments, only relatively large differences in T_1 allow structural conclusions to be drawn with confidence.

Recently, another way of using T_1 values has been proposed, namely to specify distance constraints for structure calculations of paramagnetic proteins, so as to augment those available from the NOE.[31] Essentially, the T_1 value of a particular proton is used as a measure of the proximity of that proton to the paramagnetic center of the protein; the closer a proton is to the unpaired electron(s), the shorter its T_1 value will be. In some cases, such paramagnetic centers can be rather diffuse, as unpaired electrons may be delocalized around the ring of a heme system, or over the atoms of an iron-sulfur cluster. However, since the derived distance constraints (between the proton and some atom or atoms associated with the paramagnetic center) are generally quite loose, this probably does not matter very much. More significant is the fact, mentioned earlier, that cross-relaxation amongst spins in larger molecules will tend to destroy the concept of an individual T_1 for each spin by causing cooperative relaxation of all spins as an ensemble, undermining the basis for the proposed interpretation. The justification in this case must come down largely to a question of how different the T_1 values being interpreted actually are. The influence of spin diffusion will be to make individual T_1 values more similar, whereas the different proximities of protons to the paramagnetic center will tend to make T_1 values different. If the data show a clear and structurally plausible trend amongst the T_1 values (i.e., that those for spins close to the paramagnetic center are generally significantly shorter than others), this in itself is clear evidence that spin diffusion has not averaged away the differential effects of paramagnetic relaxation. Even though the precise extent to which spin diffusion may have lessened differences between T_1 values may be unknown, it then still seems reasonable to use the observed trends in an empirical way, and this essentially is what is done in practice.

REFERENCES

1. Abragam, A. "The Principles of Nuclear Magnetism," Clarendon Press, Oxford, 1961, p. 138.
2. Bothner-By, A. A.; Noggle, J. H. *J. Am. Chem. Soc.* 1979, *101*, 5152.
3. Solomon, I. *Phys. Rev.* 1955, *99*, 559.
4. Noggle, J. H.; Schirmer, R. E. "The Nuclear Overhauser Effect; Chemical Applications," Academic Press, New York, 1971, pp. 113–120.
5. Landy, S. B.; Rao, B. D. N. *J. Magn. Reson. B* 1993, *102*, 209.
6. Forster, M. J. *J. Comput. Chem.* 1991, *12*, 292.
7. Perrin, C. L.; Gipe, R. K. *J. Am. Chem. Soc.* 1984, *106*, 4036.
8. Olejniczak, E. T.; Gampe, R. T.; Fesik, S. W. *J. Magn. Reson.* 1986, *67*, 28.
9. Bull, T. E. *J. Magn. Reson.* 1987, *72*, 397.
10. Williamson, M. P. *Magn. Reson. Chem.* 1987, *25*, 356.
11. Dobson, C. M.; Olejniczak, E. T.; Poulsen, F. M.; Ratcliffe, R. G. *J. Magn. Reson.* 1982, *48*, 97.
12. Macura, S.; Ernst, R. R. *Mol. Phys.* 1980, *41*, 95.
13. Kozerski, L.; Kawecki, R.; Krajewski, P.; Gluzinski, P.; Pupek, K.; Hansen, P. E.; Williamson, M. P. *J. Org. Chem.* 1995, *60*, 3533.
14. Post, C. B. *J. Mol. Biol.* 1992, *224*, 1087.
15. LeMaster, D. M.; Kay, L. E.; Brünger, A. T.; Prestegard, J. H. *FEBS Lett.* 1988, *236*, 71.
16. Madrid, M.; Mace, J. E.; Jardetzky, O. *J. Magn. Reson.* 1989, *83*, 267.

17. Meadows, R. P.; Kaluarachchi, K.; Post, C. B.; Gorenstein, D. G. *Bull. Magn. Reson.* 1991, *13*, 22.
18. Clore, G. M.; Gronenborn, A. M. *J. Magn. Reson.* 1989, *84*, 398.
19. James, T. L. *Curr. Op. Struct. Biol.* 1991, *1*, 1042.
20. Allard, P.; Helgstrand, M.; Härd, T. *J. Magn. Reson.* 1997, *129*, 19.
21. Leeflang, B. R.; Kroon-Batenburg, L. M. J. *J. Biomolec. NMR* 1992, *2*, 495.
22. Koning, T. M. G.; Boelens, R.; Kaptein, R. *J. Magn. Reson.* 1990, *90*, 111.
23. Fejzo, J.; Krezel, A. M.; Westler, W. M.; Macura, S.; Markley, J. L. *Biochemistry* 1991, *30*, 3807.
24. Andersen, N. H.; Nguyen, K. T.; Hartzell, C. J.; Eaton, H. L. *J. Magn. Reson.* 1987, *74*, 195.
25. Williamson, M. P.; Neuhaus, D. *J. Magn. Reson.* 1987, *72*, 369.
26. Yip, P. F. *J. Magn. Reson.* 1990, *90*, 382.
27. Kalk, A.; Berendsen, H. J. C. *J. Magn. Reson.* 1976, *24*, 343.
28. Sanders, J. K. M.; Hunter, B. K. "Modern NMR Spectroscopy," Oxford University Press, Oxford, 1980.
29. Hall, L. D.; Sanders, J. K. M. *J. Am. Chem. Soc.* 1980, *102*, 5703.
30. Saunders, J. K.; Easton, J. W. in "Determination of Organic Structures by Physical Methods" (F. C. Nachod, J. J. Zuckerman, and E. W. Randall, eds.), Vol 6, Academic Press, New York, 1976, pp. 271–333.
31. Bertini, I.; Couture, M. M. J.; Donaire, A.; Eltis, L. D.; Felli, I. C.; Luchinat, C.; Piccioli, M.; Rosato, A. *Eur. J. Biochem.* 1996, *241*, 440.
32. Clore, G. M.; Gronenborn, A. M. *J. Magn. Reson.* 1985, *61*, 158.

CHAPTER 5

THE EFFECTS OF EXCHANGE AND INTERNAL MOTION

This chapter deals with the various effects that exchange can have on NOE enhancements. By "exchange" we mean any process (other than molecular tumbling) undergone by the solute molecules that causes their magnetic properties, particularly the rates of dipole–dipole cross-relaxation processes within them, to vary with time. This is a very broad definition. It includes not only familiar forms of exchange, such as proton exchange of labile hydrogens (OH, NH, etc.) and slow conformational equilibria (e.g., amide *cis–trans* interconversion), but also much more rapid processes. Thus, internal motions of flexible molecular fragments, and exchange of molecules between different environments (e.g., equilibrium of a molecule between its receptor-bound and free states), are also types of exchange, and are included in this chapter.

What matters as far as effects on the NOE are concerned is not so much the molecular nature of the exchange process, but rather its *rate*. The range of possible exchange rates is extremely wide, but as we shall see, the key considerations are whether the exchange is fast relative to (i) T_1 and (ii) τ_c. The consequences of this form the subject of most of this chapter, and are summarized in Figure 5.1.

The most familiar consequence of exchange in NMR is that of spectral coalescence. This occurs when separate signals from each contributing exchange form collapse to give a single, averaged set of signals, as the exchange rate becomes fast relative to the chemical shift difference between the signals of separate forms ($\delta_I^i - \delta_I^j$). It is most important to realize that coalescence has *no effect* on the equations that govern the NOE. The only relevance of coalescence is that it determines how many resonances appear in the spectrum, and thus what NOE experiments are possible.

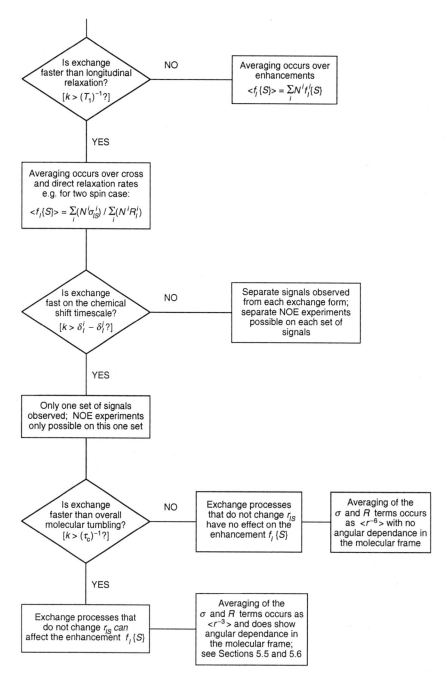

Figure 5.1. Simplified summary of the consequences of exchange for NOE enhancements. Descending the chart corresponds to increasing exchange rate, starting with very slow exchange [on a timescale of seconds; $k \ll (T_1)^{-1}$] and progressing to very fast exchange [on a timescale of picoseconds; $k \gg (\tau_c)^{-1}$]. For each "decision box" there

The sections that follow deal first with transfer of saturation in a one-spin system and then move on to the effects of simple two-site exchange in a two-spin system. This introduces all the concepts required for other cases of exchange slower than molecular tumbling. In Section 5.3 the discussion is widened to a consideration of multispin systems and multisite exchange that is fast relative to longitudinal relaxation, with particular emphasis on internal motions. In one sense, these examples are simpler than those of Section 5.2, in that the numerical value of the exchange rate constant no longer affects the result, but in another sense they are more complex, as the number and detailed nature of the contributing forms are often unknown. This places considerable limits on the interpretation of NOE enhancements under such circumstances (cf. also Section 11.1). When the exchange rate is faster than overall tumbling $[k \gg (\tau_c)^{-1}]$, as is often the case for internal motions in large molecules, further considerations apply; these are discussed in Sections 5.4 and 5.5, together with a commonly used approach for characterizing such motions using heteronuclear relaxation measurements (Section 5.4). Section 5.7 describes another important application of the NOE in exchanging systems, namely the transferred NOE (TRNOE) experiment, and finally ways in which hydration can be characterized using intermolecular NOE enhancements to water are discussed in Section 5.8.

5.1. TRANSFER OF SATURATION

Whenever a system is in the slow exchange regime on the chemical shift time scale, that is whenever there are separate signals from each contributing form, it is possible that irradiation of one resonance will cause saturation to be carried into other signals by the exchange process. This phenomenon is called transfer of saturation (or saturation transfer). Transfer of saturation is mechanistically completely distinct from the NOE, in that it is not mediated by dipole–dipole cross-relaxation (nonetheless, other authors, e.g., ref. 1, often refer to it as a form of NOE). In this section, we shall treat the simplest case for which saturation transfer can occur, that is, a one-spin system undergoing two-site exchange. No NOE enhancements need to be considered in such a system; note, however, that in larger spin systems both saturation transfer and dipole–

---

will also be a range of exchange rates where the behavior of the system passes smoothly from the slow-rate limit to the fast-rate limit. In these regions $[k \simeq (T_1)^{-1}$ and $k \simeq (\tau_c)^{-1}]$, the NOE enhancements are governed by more complex equations that involve the exchange rate explicitly; elsewhere, the equations are independent of k. Note particularly that although spectral coalescence ($k \approx \delta_I^i - \delta_I^j$) affects the number of resonances in the spectrum, and thus determines what NOE experiments are possible, it has no effect on the rate equations that govern the magnitude of NOE enhancements. All of these points are developed further in the text.

dipole cross-relaxation can occur, and the observed fractional intensity changes ($f_I\{S\}$) then become composite quantities arising from a combination of both effects (Section 5.2).

If the system is in slow exchange between two forms, "a" and "b," and the equilibrium between them is controlled by the forward and backward first-order rate constants k_1 and k_{-1}, respectively, then we have

$$a \underset{k_{-1}}{\overset{k_1}{\rightleftharpoons}} b \tag{5.1}$$

with $k_1/k_{-1} = [b]/[a]$.

Suppose a spin S gives a signal S^a when the system is in form, "a," and a signal S^b when the system is in form "b." If signal S^a is irradiated, then the populations of the two energy levels of the S^a transition are equalized; that is, there are now equal numbers of "a" molecules containing spin S in the α nuclear spin state as in the β nuclear spin state. As a consequence of the exchange equilibrium, some of these "a" molecules will turn into "b" molecules. Since the α and β populations are equal in the "a" molecules, as many S spins in the α state will undergo this change as will S spins in the β state, so the forward exchange process will tend to equalize the α and β populations of the S spins in the "b" molecules also. Opposing this trend will be spin–lattice relaxation in the "b" molecules, and the back reaction. We may show how the competition between exchange and spin–lattice relaxation controls the extent to which saturation is transferred, as follows: In exactly the same way as the Solomon equations were derived (Section 3.1), we can find the equation of motion for the S_z^b magnetization[†]:

$$\frac{dS_z^b}{dt} = -R_S^b(S_z^b - S_z^{0b}) - k_{-1}(S_z^b - S_z^{0b}) + k_1(S_z^a - S_z^{0a}) \tag{5.2}$$

On saturation of signal S^a, the steady-state intensity of S^b can be found from Eq. 5.2 by setting dS_z^b/dt and S_z^a to zero and rearranging. The fractional change at S^b is given by

$$f_S^b\{S^a\} = \frac{S_z^b - S_z^{0b}}{S_z^{0b}} = \frac{S_z^{0a}}{S_z^{0b}} \frac{-k_1}{R_S^b + k_{-1}} \tag{5.3}$$

This equation bears out exactly the qualitative argument given earlier. Notice the weighting by a factor of S_z^{0a}/S_z^{0b}, which arises because the incoming pop-

---

[†]In fact, the equilibrium magnetizations S_z^0 in the *exchange* terms are redundant, since we know that S_z^{0a} and S_z^{0b} are proportional to the concentrations [a] and [b], so that the condition for chemical equilibrium (Eq. 5.4) implies that $k_1 S_z^{0a} - k_{-1} S_z^{0b} = 0$. Nonetheless, we have chosen to retain these terms to preserve the analogy between the various contributions, and (in our view) to make the algebraic manipulations clearer.

ulation flux seen by S^b reflects not only the forward rate constant k_1, but also the relative concentrations of the two exchange forms "a" and "b." This expression may be simplified further using the condition for chemical equilibrium

$$k_1 S^{0a} = k_{-1} S^{0b} \tag{5.4}$$

to give

$$f_S^b\{S^a\} = \frac{-k_{-1}}{R_S^b + k_{-1}} \tag{5.5}$$

These expressions are valid for any ratio of concentrations of "a" and "b."

Expression 5.5 is equal to -1 when $k_{-1} \gg R_S^b$; thus, under these circumstances saturation of spin "a" produces 100% saturation of spin "b." This condition, that the exchange rate be much greater than the relaxation rate, will be met repeatedly throughout the chapter, and is referred to as being in *fast exchange on the relaxation time scale* (or T_1 time scale). Note that the two rates (k and R) defining fast exchange are always those describing the *loss* of magnetization from a particular spin. Clearly, for observation of spin a and saturation of spin b, the fast exchange condition is $k_1 \gg R_S^a$.

When k is large relative to the chemical shift difference between signals S^a and S^b, the system is in *fast exchange on the chemical shift timescale*. In this situation, the two S signals coalesce, and saturation transfer becomes meaningless. However, considerably faster rates are usually necessary to cause coalescence than are required to meet the condition of fast exchange on the relaxation timescale, so that there will often be a moderately wide range of exchange rates in which essentially complete transfer of saturation occurs between separate signals below their coalescence temperature.

$$\begin{array}{c} Me_A \\ \diagdown \\ \diagup N—C \diagdown \\ Me_B \quad\quad Me_C \end{array} \quad \diagup\!\!\!\!O$$

5.1

A few examples of saturation transfer will suffice. At low temperatures, the two methyl groups A and B of dimethylacetamide (**5.1**) give rise to two separate sharp signals, and irradiation of signal A has little effect on the intensity of signal B. As the temperature is raised, the rate of rotation about the amide bond increases. Eventually this leads to broadening and then coalescence of the two signals, but before this temperature is reached, there is a range of temperatures where irradiation of A produces a marked reduction in the intensity of B, and vice versa. The transient equivalent of this effect was demonstrated in some of the earliest NOESY experiments.[2]

In the second example, the exchange process is that of a ligand between its free and bound states. The molecule Ac-D-Ala-D-Ala binds to the antibiotic ristocetin A, and when the peptide is in two-fold excess, two pairs of alanyl signals are observed (Fig. 5.2).[3] Saturation of each signal arising from an alanyl methyl in the bound peptide causes a large decrease in intensity of the corresponding signal arising from the free peptide, thus permitting pairwise assignment of the methyl signals and, in favorable cases, allowing a determination of the rate of dissociation of the complex, using Eq. 5.2.[1]

Saturation transfer can also be seen in a wide variety of systems undergoing chemical exchange. For example, for phenols dissolved in $CDCl_3$ containing a little HOD, the phenolic proton is in exchange with the HOD molecules, and irradiation of the HOD signal reduces the intensity of the phenolic OH signal;

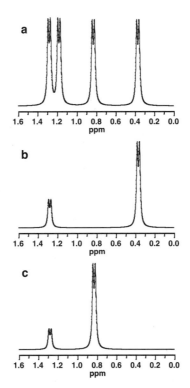

Figure 5.2. Schematic representation of a saturation transfer experiment, drawn from the data of ref. 3. Ac-D-Ala-D-Ala is in slow exchange on the chemical shift timescale between the free state and its complex with ristocetin A. When Ac-D-Ala-D-Ala is in twofold excess there are four alanyl methyl signals seen: two from the free peptide and two from the bound peptide (*a*). Saturation of the signal at 0.38 ppm, from the bound peptide, causes a 33% reduction in the intensity of the free signal at 1.27 ppm, as shown in the difference spectrum (*b*). This is due to saturation transfer and shows that these two methyl groups form an exchanging pair. Likewise, saturation of the bound signal at 0.83 ppm causes a 24% reduction in the intensity of the other free signal at 1.18 ppm (*c*).

a number of such examples have been described.[4] In aqueous solutions, exchange with the H_2O signal occurs for imino and amino signals of nucleic acids, while for proteins the same is true for OH, SH and amino, imino and guanidino NH signals of side chains, and for amide NH signals at high pH. This, of course, is very important in experiments designed to detect such resonances from H_2O solutions, since suppression of the intense solvent signal by saturation may also remove many or all of the signals of interest. Furthermore, saturation transfer from solvent to OH or NH protons can lead to NOE enhancements of near neighbors of the exchanging spins in the solute molecules,[5] an effect that has many features in common with the transferred NOE (Section 5.7) and which can complicate the interpretation of intermolecular NOE enhancements involving water (Section 5.8). On the other hand, it can also be useful. For example, saturation transfer from water to the exchangeable side-chain OH signals of tyrosines in proteins can generate an enhancement to the tyrosine ε protons, allowing the latter to be distinguished from phenylalanine aromatic signals.[6] Enhancements into OH or NH signals can also cause enhancement of the H_2O signal, although this effect is usually seen only for solutions in organic solvents where the concentration of water is low (cf. Fig. 6.12).

As we have seen, saturation transfer can, in favorable cases, be used to measure exchange rates and to make assignments. However, saturation transfer is more commonly encountered as a complication in an experiment intended to measure NOE enhancements. Under these circumstances, the main issue is to distinguish any saturation transfer signals from the NOE enhancements. For small molecules, direct NOE enhancements are positive, and although indirect (three-spin) effects do give rise to negative enhancements (Section 3.2.2.3), these are very rarely larger than a few percent. Thus the relatively intense negative signals caused by saturation transfer in the NOE difference spectrum usually stand out clearly. However, when there is exchange in the intermediate regime on the chemical shift timescale, a minor component can give a very broad signal that may be hard to spot; in such cases accidental and unrecognized saturation of this broad signal may cause effects (e.g., small negative or positive enhancements) that could easily be misinterpreted (cf. Section 7.2.4).[7] Negative enhancements due to scalar relaxation (Section 6.4.3) might appear similar to saturation transfer effects, but are extremely rare.

For larger molecules, the position is more difficult, since all enhancements are negative. Nonetheless, it is still true that saturation transfer often causes distinctly more intense signals than NOE enhancements. This distinction is particularly clear for short irradiation times (or short τ_m in NOESY), since the exchange rate is often much faster than the relaxation rate, so that the saturation transfer signal grows much faster than the NOE enhancements. In addition, the saturation transfer signals are increased by increasing the exchange rate, for example by raising the temperature, whereas the size of negative NOE enhancements often decreases on increasing the temperature. In cases of ambiguity, the rotating frame NOE experiment (Section 9.3) is often useful, since this leads to direct transverse NOE enhancements that have opposite signs to exchange signals no matter what the size of the molecule.

5.2. GENERAL EQUATIONS FOR THE NOE IN SYSTEMS OF TWO-SITE EXCHANGE

In this section, we shall show how the equations describing NOE enhancements, particularly steady-state enhancements, are modified by the presence of exchange. The nature of the exchange process in the theoretical examples will be very simple; as in the previous section, the system is in an exchange equilibrium between two forms "a" and "b," with the forward and backward rate constants k_1 and k_{-1}, respectively:

$$a \underset{k_{-1}}{\overset{k_1}{\rightleftharpoons}} b \qquad (5.1)$$

We begin with the simplest possible case, namely a two-spin system, and then move on to a slightly more complicated three-spin system, dimethylformamide. The problems of dealing with more complicated types of exchange, involving more spins and/or more contributing forms, are dealt with in Section 5.3. Cases in which the exchange rate is faster than molecular tumbling [$k \gg (\tau_c)^{-1}$] require further development of the theory, and are dealt with in Section 5.5.2.

5.2.1. Exchange in a Two-Spin System

This is the simplest spin system capable of showing both NOE enhancements and exchange, and shows all the features required for understanding more complex spin systems. The crucial equations in this section are Eqs. 5.9 and 5.12, and their analogs, Eqs. 5.14 and 5.15. From these, we can deduce that for slow exchange rates [relative to $(T_1)^{-1}$] the enhancements seen are just those expected in the absence of exchange, while for fast exchange rates [relative to $(T_1)^{-1}$] the enhancements are obtained by averaging *relaxation rates* rather than enhancements.

To take account of the exchange process, we must add new terms to the Solomon equations (Eq. 2.10) to represent the population flux between the "a" molecules and the "b" molecules. If the magnetization due to the I spins in the "a" molecules is denoted I^a, that due to I spins in the "b" molecules is denoted I^b, and similarly the two S spin contributions are denoted S^a and S^b, then the effect of the exchange is to interchange I^a and I^b, and to interchange S^a and S^b (Fig. 5.3). Provided that this exchange is slow relative to overall tumbling (i.e., $k \ll (\tau_c)^{-1}$), this leads to the following modified Solomon equations:

GENERAL EQUATIONS FOR THE NOE IN SYSTEMS OF TWO-SITE EXCHANGE   137

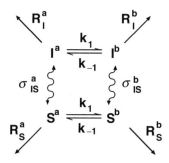

Figure 5.3. Rate constants in a two-spin system undergoing two-site exchange.

$$\frac{dI_z^a}{dt} = -R_I^a(I_z^a - I_z^{0a}) - \sigma_{IS}^a(S_z^a - S_z^{0a}) + k_{-1}(I_z^b - I_z^{0b}) - k_1(I_z^a - I_z^{0a})$$

$$\frac{dI_z^b}{dt} = -R_I^b(I_z^b - I_z^{0b}) - \sigma_{IS}^b(S_z^b - S_z^{0b}) + k_1(I_z^a - I_z^{0a}) - k_{-1}(I_z^b - I_z^{0b})$$

$$\frac{dS_z^a}{dt} = -R_S^a(S_z^a - S_z^{0a}) - \sigma_{IS}^a(I_z^a - I_z^{0a}) + k_{-1}(S_z^b - S_z^{0b}) - k_1(S_z^a - S_z^{0a})$$

$$\frac{dS_z^b}{dt} = -R_S^b(S_z^b - S_z^{0b}) - \sigma_{IS}^b(I_z^b - I_z^{0b}) + k_1(S_z^a - S_z^{0a}) - k_{-1}(S_z^b - S_z^{0b}) \quad (5.6)$$

Notice that the relaxation rate constants σ_{IS}^a, R_I^a, and R_S^a depend on r_{IS}^a, the internuclear separation between spins I and S in an "a" molecule, whereas σ_{IS}^b, R_I^b, and R_S^b depend on the distance r_{IS}^b. In general, r_{IS}^a and r_{IS}^b will be different. However, if they are equal, then $\sigma_{IS}^a = \sigma_{IS}^b$, and the formulae of this section all reduce to those for a single, rigid molecule (neglecting any possible variations in ρ^*). Similar conclusions apply to multispin systems when none of the near neighbors of I changes its relative position as a consequence of the exchange. Thus, *provided that* $k \ll (\tau_c)^{-1}$, exchange processes that do not affect the internuclear distances involved in the Solomon equations for a particular enhancement have no effect on that enhancement. This is in contrast to the situation where exchange is *faster* than overall molecular tumbling, when such exchange processes do affect enhancements (cf. Section 5.5.2).

Returning to the general case ($r_{IS}^a \ne r_{IS}^b$), the simplest experiment to consider is one in which both the S signals, S^a and S^b, are irradiated. This could correspond to a double irradiation experiment, if S^a and S^b are in slow exchange on the chemical shift timescale (i.e., $k \ll \delta_S^a - \delta_S^b$), or it could be that only a single, combined signal due to $S = S^a + S^b$ is present in the spectrum. The latter would occur if S^a and S^b are in fast exchange on the chemical shift timescale, or, in the limit, if the exchange did not affect the chemical shift of S at all ($\delta_S^a = \delta_S^b$). While these various possibilities may seem very different in

138 THE EFFECTS OF EXCHANGE AND INTERNAL MOTION

terms of the appearance of the spectrum, they are *identical* as far as the Eqs. 5.6 are concerned, and they therefore give rise to the same enhancements, $f_I^a\{S\}$ and $f_I^b\{S\}$, observed at the signals I^a and I^b, respectively.

To find the values of these enhancements at steady state, we proceed exactly as in Section 3.1.3.1. By setting of all the derivatives and both S_z^a and S_z^b to zero, and then rearranging the first two of Eqs. 5.6, we obtain

$$f_I^a\{S\} = \frac{I_z^a - I_z^{0a}}{I_z^{0a}} = \frac{\sigma_{IS}^a}{(R_I^a + k_1)} + \frac{k_1}{(R_I^a + k_1)} f_I^b\{S\}$$

and

$$f_I^b\{S\} = \frac{I_z^b - I_z^{0b}}{I_z^{0b}} = \frac{\sigma_{IS}^b}{(R_I^b + k_{-1})} + \frac{k_{-1}}{(R_I^b + k_{-1})} f_I^a\{S\} \quad (5.7)$$

where we have also used the condition for chemical equilibrium given in Eq. 5.4.

Equations 5.7 show that there are two contributions to each enhancement in this system. For example, for $f_I^a\{S\}$ the first term corresponds to the normal, direct $I \leftarrow \{S\}$ NOE enhancement for the two spins within an "a" molecule, whereas the second term corresponds to the enhancement transferred from I^a to I^b by the exchange process. This latter term is exactly analogous to transfer of saturation as dealt with in Section 5.1, except that here it is an enhancement that is transferred rather than saturation.

Equations 5.7 form a pair of simultaneous equations for $f_I^a\{S\}$ and $f_I^b\{S\}$. Solving them gives

$$f_I^a\{S\} = \frac{\sigma_{IS}^a(R_I^b + k_{-1}) + k_1 \sigma_{IS}^b}{(R_I^a + k_1)(R_I^b + k_{-1}) - k_1 k_{-1}}$$

$$f_I^b\{S\} = \frac{\sigma_{IS}^b(R_I^a + k_1) + k_{-1} \sigma_{IS}^a}{(R_I^a + k_1)(R_I^b + k_{-1}) - k_1 k_{-1}} \quad (5.8)$$

These equations demonstrate some important and general properties of the NOE in exchanging systems. As they stand, Eqs. 5.8 make no assumptions as to the size of k_1 and k_{-1} relative to R_I^a and R_I^b. However, if k_1 and k_{-1} are either much smaller or much larger than R_I^a and R_I^b, then the equations reduce to simpler forms *not involving* k_1 or k_{-1} directly. Thus, in the limit of slow exchange on the T_1 timescale ($k \ll R$), Eqs. 5.8 reduce to

$$f_I^a\{S\} = \frac{\sigma_{IS}^a}{R_I^a} \quad \text{and} \quad f_I^b\{S\} = \frac{\sigma_{IS}^b}{R_I^b} \quad (5.9)$$

while in the limit of fast exchange on the T_1 timescale ($k \gg R$), Eqs. 5.8 reduce to

$$f_I^a\{S\} = f_I^b\{S\} = \frac{k_{-1}\sigma_{IS}^a + k_1\sigma_{IS}^b}{k_{-1}R_I^a + k_1R_I^b} \quad (5.10)$$

The condition for chemical equilibrium gives us that

$$N^a = \frac{k_{-1}}{k_1 + k_{-1}} \quad \text{and} \quad N^b = \frac{k_1}{k_1 + k_{-1}} \quad (5.11)$$

where N^a and N^b are the fractional populations of forms "a" and "b," respectively, so that Eq. 5.10 becomes

$$f_I^a\{S\} = f_I^b\{S\} = \frac{N^a\sigma_{IS}^a + N^b\sigma_{IS}^b}{N^aR_I^a + N^bR_I^b} \quad (5.12)$$

Thus, in the slow exchange case, Eq. 5.9 shows that the enhancements expected are just those that would occur in the absence of exchange, while in the fast exchange case Eq. 5.12 shows that the various *relaxation rates* are averaged by the exchange process, and $f_I^a\{S\}$ and $f_I^b\{S\}$ become equal.

Notice that at no point have any assumptions been made as to whether I^a and I^b (or S^a and S^b for that matter) were in slow or fast exchange on the *chemical shift* timescale; as commented earlier, this makes no difference to the *relaxation* behavior of the system, as described by these equations. It does, however, determine how many signals are present in the spectrum, and thus what NOE measurements are possible. If I^a and I^b resonate separately ($k_1, k_{-1} \ll \delta_I^a - \delta_I^b$), then obviously the enhancements $f_I^a\{S\}$ and $f_I^b\{S\}$ can be measured separately; but if I^a and I^b are in fast exchange on the *chemical shift* timescale ($k_1, k_{-1} \gg \delta_I^a - \delta_I^b$), only a single resonance ($I = I^a + I^b$) will be present in the spectrum, and only a single, combined enhancement can be measured. This can be found from Eqs. 5.8 by adding the appropriately weighted expressions for $f_I^a\{S\}$ and $f_I^b\{S\}$:

$$f_I\{S\} = \frac{(I_z^a + I_z^b) - (I_z^{0a} + I_z^{0b})}{(I_z^{0a} + I_z^{0b})}$$

$$= \left(\frac{I_z^{0a}}{I_z^{0a} + I_z^{0b}}\right)\left(\frac{I_z^a - I_z^{0a}}{I_z^{0a}}\right) + \left(\frac{I_z^{0b}}{I_z^{0a} + I_z^{0b}}\right)\left(\frac{I_z^b - I_z^{0b}}{I_z^{0b}}\right)$$

$$= N^a f_I^a\{S\} + N^b f_I^b\{S\}$$

$$= \frac{\sigma_{IS}^a(N^aR_I^b + k_{-1}) + \sigma_{IS}^b(N^bR_I^a + k_1)}{(R_I^a + k_1)(R_I^b + k_{-1}) - k_1k_{-1}} \quad (5.13)$$

The limits of this equation for slow and fast exchange on the T_1 timescale demonstrate an important point. For $k \ll R$, Eq. 5.13 becomes

$$f_I\{S\} = N^a \left(\frac{\sigma_{IS}^a}{R_I^a}\right) + N^b \left(\frac{\sigma_{IS}^b}{R_I^b}\right) \tag{5.14}$$

while for $k \gg R$, it becomes

$$f_I\{S\} = \frac{N^a \sigma_{IS}^a + N^b \sigma_{IS}^b}{N^a R_I^a + N^b R_I^b} \tag{5.15}$$

In other words, when exchange is slow on the T_1 timescale, averaging occurs *over the enhancements themselves*, whereas when exchange is fast on the T_1 timescale, averaging occurs *over the R and σ terms directly*. In the intermediate exchange rate region, the full equation (Eq. 5.13) must be used, necessitating a knowledge of the values of k_1 and k_{-1}.

This is perhaps the most important conclusion to be drawn in this section. In order to reinforce it, we give the corresponding expressions for a multispin system undergoing two-site exchange, Eqs. 5.16 and 5.17, which may be derived identically. For the slow-exchange case, $f_I\{S\}$ is again given by an average over the individual enhancements:

$$f_I\{S\} = N^a \left(\frac{\sigma_{IS}^a - \sum_X \sigma_{IX}^a f_X\{S\}}{R_I^a}\right) + N^b \left(\frac{\sigma_{IS}^b - \sum_X \sigma_{IX}^b f_X\{S\}}{R_I^b}\right) \tag{5.16}$$

whereas in the fast exchange case, $f_I\{S\}$ is given by an average over the σ and R terms:

$$f_I\{S\} = \frac{(N^a \sigma_{IS}^a + N^b \sigma_{IS}^b) - \sum_X (N^a \sigma_{IX}^a + N^b \sigma_{IX}^b) f_X\{S\}}{N^a R_I^a + N^b R_I^b} \tag{5.17}$$

It is not difficult to see why this should be so. The relaxation rate constants R_I^a and R_I^b characterize, among other things, the rates at which enhancements $f_I^a\{S\}$ and $f_I^b\{S\}$ grow towards steady state. If exchange occurs more slowly than NOE growth, then enhancements in the separate "a" and "b" molecules will have time to approach their steady-state values uninfluenced by the exchange. On the other hand, if exchange is faster than NOE growth, relaxation will in effect take place in an "averaged molecule," since many exchange events will take place during the time required for the NOE to grow.

As implied by Figure 5.1, it is rather unlikely that a system would be in fast exchange on the chemical shift timescale and simultaneously in slow exchange on the T_1 timescale. This is partly because the chemical shift difference between the corresponding signals in "a" and "b" molecules would have to be less than $(T_1)^{-1}$, that is, typically less than about 10 Hz; and partly because the majority of conformational exchange processes that we can characterize are much faster than dipolar relaxation rates. Thus, in the majority of cases in which only one set of signals is seen, it is probable that any exchange is fast, not only on the chemical shift timescale but also on the T_1 timescale, so that averaging occurs over the relaxation rates as shown in Eqs. 5.15 and 5.17.

So far in this section, only the case in which both S signals, S^a and S^b, are saturated together has been considered. To complete the description, we must now deal with the further possibilities that arise when only one of the two S signals, say S^a, is saturated. Clearly, this case can only arise when S^a and S^b are in slow exchange on the chemical shift timescale. The relevant equations can be derived in a fashion entirely analogous to those already discussed, by setting only one of the steady-state S_z magnetizations, here S_z^a, to zero. The simultaneous equations for the three observable enhancements then become

$$f_I^a\{S^a\} = \frac{\sigma_{IS}^a}{(R_I^a + k_1)} + \frac{k_1}{(R_I^a + k_1)} f_I^b\{S^a\} \qquad (5.18a)$$

$$f_I^b\{S^a\} = \frac{-\sigma_{IS}^b}{(R_I^b + k_{-1})} f_S^b\{S^a\} + \frac{k_{-1}}{(R_I^b + k_{-1})} f_I^a\{S^a\} \qquad (5.18b)$$

$$f_S^b\{S^a\} = \frac{-k_{-1}}{(R_S^b + k_{-1})} - \frac{\sigma_{IS}^b}{(R_S^b + k_{-1})} f_I^b\{S^a\} \qquad (5.18c)$$

Each of these enhancements is composed of two contributions, and it is instructive to see where each of these comes from. For the enhancement $f_I^a\{S^a\}$, the situation is much the same as when both S signals are saturated (Eq. 5.7), in that there is a direct NOE enhancement within an "a" molecule (first term of Eq. 5.18a) to which is added a proportion of any enhancement at I^b that transfers into the I^a signal by exchange (second term of Eq. 5.18a). For the enhancement $f_I^b\{S^a\}$, by contrast, there is no direct term, since the signal S^b is not irradiated. Instead, there are two indirect pathways: An enhancement may arrive at I^b either via saturation transfer from S^a to S^b followed by an NOE step within a "b" molecule (first term of Eq. 5.18b), or via an NOE step within an "a" molecule followed by transfer of the enhancement into the I^b signal by exchange (second term of Eq. 5.18b). For the enhancement $f_S^b\{S^a\}$, the main contribution is likely to be saturation transfer directly from S^a (first term of Eq. 5.18c), while there is also a (very!) indirect contribution in which cross-relaxation passes on any enhancement at I^b into S^b (cf. Fig. 5.3).

The solutions to these simultaneous equations for the enhancements $f_I^a\{S^a\}, f_I^b\{S^a\}$, and $f_S^b\{S^a\}$ are collected in Table 5.1: similar expressions for irradiation of S^b can be obtained by permuting all "a" superscripts with "b" superscripts, and all k_1 terms with k_{-1} terms. As with the cases already discussed, these equations reduce to much simpler forms not involving k_1 or k_{-1}

TABLE 5.1 NOE Enhancements for a Two-Spin System Undergoing Two-Site Exchange

Enhancement	General Expression	Slow Exchange $[k_1, k_{-1} \ll R_I^a, R_I^b, R_S^a, R_S^b]$	Fast Exchange $[k_1, k_{-1} \gg R_I^a, R_I^b, R_S^a, R_S^b]$
$f_I^a\{S^a\}$	$\dfrac{\sigma_{IS}^a(R_I^b + k_{-1})(R_S^b + k_{-1}) - \sigma_{IS}^a \sigma_{IS}^b \sigma_{IS}^b + \sigma_{IS}^b k_1 k_{-1}}{(R_I^a + k_1)(R_I^b + k_{-1})(R_S^b + k_{-1}) - (R_I^a + k_1)\sigma_{IS}^b \sigma_{IS}^b - k_1 k_{-1}(R_S^b + k_{-1})}$	$\dfrac{\sigma_{IS}^a R_I^b R_S^b - \sigma_{IS}^a \sigma_{IS}^b \sigma_{IS}^b}{R_I^a R_I^b R_S^b - R_I^a \sigma_{IS}^b \sigma_{IS}^b} = \dfrac{\sigma_{IS}^a}{R_I^a}$	$\dfrac{\sigma_{IS}^a k_{-1}^2 + \sigma_{IS}^b k_1 k_{-1}}{R_I^a k_{-1}^2 + R_I^b k_1 k_{-1} + R_{IS}^b k_1 k_{-1} - R_S^b k_1 k_{-1}}$ $= \dfrac{N^a \sigma_{IS}^a + N^b \sigma_{IS}^b}{N^a R_I^a + N^b R_I^b}$
$f_I^b\{S^a\}$	$\dfrac{\sigma_{IS}^a k_{-1}(R_S^b + k_{-1}) + \sigma_{IS}^b k_{-1}(R_I^a + k_1)}{(R_I^a + k_1)(R_I^b + k_{-1})(R_S^b + k_{-1}) - (R_I^a + k_1)\sigma_{IS}^b \sigma_{IS}^b - k_1 k_{-1}(R_S^b + k_{-1})}$	0	$\dfrac{\sigma_{IS}^a k_{-1}^2 + \sigma_{IS}^b k_1 k_{-1}}{R_I^a k_{-1}^2 + R_I^b k_1 k_{-1} + R_{IS}^b k_1 k_{-1} - R_S^b k_1 k_{-1}}$ $= \dfrac{N^a \sigma_{IS}^a + N^b \sigma_{IS}^b}{N^a R_I^a + N^b R_I^b}$
$f_S^b\{S^a\}$	$\dfrac{k_1 k_{-1}^2 - k_{-1}(R_I^a + k_1)(R_I^b + k_{-1}) - \sigma_{IS}^a \sigma_{IS}^b k_{-1}}{(R_I^a + k_1)(R_I^b + k_{-1})(R_S^b + k_{-1}) - (R_I^a + k_1)\sigma_{IS}^b \sigma_{IS}^b - k_1 k_{-1}(R_S^b + k_{-1})}$	0	$\dfrac{-R_I^a k_{-1}^2 - R_I^b k_1 k_{-1}}{R_I^a k_{-1}^2 + R_I^b k_1 k_{-1} + R_S^b k_1 k_{-1} - R_S^b k_1 k_{-1}} = -1$

directly when the exchange is either much faster or much slower than the dipolar relaxation rates characterized by R_I^a, R_I^b, and R_S^b. These limiting expressions are also given in Table 5.1.

As before, in the slow-exchange limit, the system behaves exactly as though the exchange forms were separate components in a nonexchanging mixture. Irradiation of S^a produces the expected enhancement $f_I^a\{S^a\}$, but has no effect on either signal from the "b" molecules. Conversely, in the limit of fast exchange on the relaxation timescale [but still within slow exchange on the chemical shift timescale, i.e., $(T_1)^{-1} \ll (k_1, k_{-1}) \ll \delta_S^a - \delta_S^b$], it is immaterial which of the two S signals is irradiated. In either case, the enhancements at I^a and I^b are the same, being determined by the *averaged* cross- and direct relaxation rates. Whichever S signal is irradiated, the other is also fully saturated by saturation transfer. Not surprisingly, these enhancements in the fast-exchange regime are also identical to those observed when both S signals are irradiated (Eq. 5.12). A real example of this pattern of behavior is shown in Figure 5.4.

For cases in which exchange causes a profound change in the IS internuclear distance, it may be a reasonable approximation to say that cross-relaxation between I and S occurs only in one of the exchanging forms, say "a." Even though this may not strictly be true, or the system may comprise more than two spins, it is interesting to see what happens to the various enhancements when σ_{IS}^b is zero. Clearly, in the slow-exchange limit, the enhancement $f_I^a\{S^a\}$ is unaffected, while the enhancements $f_I^a\{S^b\}, f_I^b\{S^a\}$, and $f_I^b\{S^b\}$ are all zero. As the exchange rate increases, so the enhancement $f_I^a\{S^a\}$ diminishes and the enhancement $f_I^b\{S^a\}$ increases, because exchange starts to average the two. Although S^b and I^b do not cross-relax, the enhancements seen on irradiating S^b also become nonzero as the exchange rate rises, since saturation is transferred from S^b into S^a, where it generates an enhancement at I^a. This in turn is carried into the I^b signal via exchange. More formally, the equations in Table 5.1 can be used to show that (for the particular case where σ_{IS}^b is zero) the enhancement $f_I^a\{S^b\}$ is given, for all values of k_1 and k_{-1}, by

$$f_I^a\{S^b\} = -f_I^a\{S^a\}f_S^a\{S^b\} \tag{5.19}$$

5.2.2. Exchange in Dimethylformamide

NOE enhancements in dimethylformamide (DMF) provide a concrete example with which to illustrate the results of the previous section. Indeed, DMF has often been used as a simple prototype for describing the effect of conformational exchange on NOE enhancements.[1,9-11]

$$\text{Me}_X\diagdown\text{N}-\text{C}\diagup\overset{\displaystyle O}{\diagdown\text{H}_A}$$
$$\text{Me}_Y\diagup$$

5.2

144 THE EFFECTS OF EXCHANGE AND INTERNAL MOTION

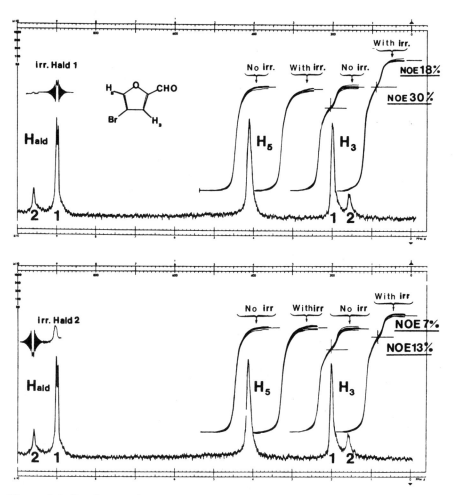

Figure 5.4. Steady-state CW NOE enhancements in an exchanging system. At −90°C in tetrahydrofuran-d_8, 2-formyl-4-bromofuran has slow enough rotation about the C_2—CHO bond that the formyl proton and H_3 each contribute separate signals from the major and minor conformers in slow exchange on the chemical shift timescale. The enhancements expected in this system can be calculated using the equations in Table 5.1 (with the formyl proton as S, and H_3 as I), from which it can be deduced that the exchange rate is of the same order as $(T_1)^{-1}$. Reproduced with permission from ref. 8.

As shown in structure **5.2**, the formyl proton is here denoted H_A, the methyl group *trans* to H_A as Me_X, and the methyl group *cis* to H_A as Me_Y. The exchange process in DMF results from rotation about the C—N amide bond, causing interchange of Me_X and Me_Y. Since the system is symmetrical, the forward and backward rate constants are identical, and will be written k. As in the previous sections, the exchange process contributes its own terms to the equations of motion for A_z, X_z, and Y_z, so that the Solomon equations modified to include exchange† become

$$dA_z/dt = -R_A(A_z - A_z^0) - \sigma_{XA}(X_z - X_z^0) - \sigma_{YA}(Y_z - Y_z^0)$$

$$dX_z/dt = -R_{XX}(X_z - X_z^0) - 3\sigma_{XA}(A_z - A_z^0) - 3\sigma_{XY}(Y_z - Y_z^0)$$
$$+ k(Y_z - Y_z^0) - k(X_z - X_z^0)$$

$$dY_z/dt = -R_{YY}(Y_z - Y_z^0) - 3\sigma_{YA}(A_z - A_z^0) - 3\sigma_{XY}(X_z - X_z^0)$$
$$+ k(X_z - X_z^0) - k(Y_z - Y_z^0) \qquad (5.20)$$

For the case of irradiating Me$_X$ ($X_z = 0$), the enhancements $f_A\{X\}$ and $f_Y\{X\}$ may be found just as in the previous section (using also the relations $X_z^0 = Y_z^0 = 3A_z^0$):

$$f_A\{X\} = \frac{3\sigma_{XA}}{R_A} - \frac{3\sigma_{YA}}{R_A} f_Y\{X\}$$

$$f_Y\{X\} = \frac{3\sigma_{XY}}{(R_{YY} + k)} - \frac{\sigma_{YA}}{(R_{YY} + k)} f_A\{X\} - \frac{k}{(R_{YY} + k)} \qquad (5.21)$$

From these simultaneous equations, and others like them for the cases of irradiating Me$_Y$ ($Y_z = 0$) and H$_A$ ($A_z = 0$), we can obtain explicit equations for all six possible enhancements in DMF; these are given in the first column of Table 5.2. Given the much longer distance from Me$_X$ to H$_A$ than from Me$_Y$ to H$_A$, it is not unreasonable to set σ_{XA} to zero, so generating the expressions in the second column of Table 5.2. The limiting forms of these in the cases of slow and fast exchange on the T_1 timescale are given in the third and fourth columns, respectively, of Table 5.2.

The qualitative message from these equations is just what one might expect. In the slow-exchange limit (column 3), the enhancements $f_A\{X\}$ and $f_X\{A\}$ are predicted to be negative, as they result entirely from three-spin effects (cf. Section 3.2.2.2). The other four enhancements are positive, and correspond to direct effects. Results of this sort would be expected for experiments with DMF at or below room temperature. However, if the temperature is raised, this increases the exchange rate and begins to modify the observed enhancements. Transfer of saturation between Me$_X$ and Me$_Y$ increases until, in the fast exchange limit (on the T_1 timescale), $f_X\{Y\} = f_Y\{X\} = -1$. Similarly, the distinction between Me$_X$ and Me$_Y$ is progressively lost from the other enhancements, so that, again in the fast exchange limit, $f_A\{X\} = f_A\{Y\}$ and $f_X\{A\} =$

---

†Magnetic equivalence within the methyl groups is included, as described in Section 3.2.2.4 and Appendix I, and σ is defined as the cross-relaxation rate between *one* proton in one group and *one* proton in another.

TABLE 5.2 NOE Enhancements in DMF

Enhancement	General Case	$\sigma_{XA} = 0$	$\sigma_{XA} = 0;\ k \ll R$	$\sigma_{XA} = 0;\ k \gg R$
$f_A\{X\}$	$\dfrac{3\sigma_{XA}(R_{YY}+k) - 9\sigma_{YA}\sigma_{XY} + 3\sigma_{YA}k}{R_A(R_{YY}+k) - 3\sigma_{YA}^2}$	$\dfrac{-3\sigma_{YA}(3\sigma_{XY}-k)}{R_A(R_{YY}+k) - 3\sigma_{YA}^2}$	$\dfrac{-9\sigma_{YA}\sigma_{XY}}{R_A R_{YY} - 3\sigma_{YA}^2}$	$\dfrac{3\sigma_{YA}}{R_A}$
$f_Y\{X\}$	$\dfrac{3\sigma_{XY}R_A - 3\sigma_{YA}\sigma_{XA} - R_A k}{R_A(R_{YY}+k) - 3\sigma_{YA}^2}$	$\dfrac{R_A(3\sigma_{XY}-k)}{R_A(R_{YY}+k) - 3\sigma_{YA}^2}$	$\dfrac{3\sigma_{XY}R_A}{R_A R_{YY} - 3\sigma_{YA}^2}$	-1
$f_A\{Y\}$	$\dfrac{3\sigma_{YA}(R_{XX}+k) - 9\sigma_{XA}\sigma_{XY} + 3\sigma_{XA}k}{R_A(R_{XX}+k) - 3\sigma_{XA}^2}$	$\dfrac{3\sigma_{YA}}{R_A}$	$\dfrac{3\sigma_{YA}}{R_A}$	$\dfrac{3\sigma_{YA}}{R_A}$
$f_X\{Y\}$	$\dfrac{3\sigma_{XY}R_A - 3\sigma_{XA}\sigma_{YA} - R_A k}{R_A(R_{XX}+k) - 3\sigma_{XA}^2}$	$\dfrac{3\sigma_{XY}-k}{R_{XX}+k}$	$\dfrac{3\sigma_{XY}}{R_{XX}}$	-1
$f_X\{A\}$	$\dfrac{\sigma_{XA}(R_{YY}+k) - 3\sigma_{XY}\sigma_{YA} + \sigma_{YA}k}{(R_{XX}+k)(R_{YY}+k) - (3\sigma_{XY}-k)^2}$	$\dfrac{-\sigma_{YA}(3\sigma_{XY}-k)}{(R_{XX}+k)(R_{YY}+k) - (3\sigma_{XY}-k)^2}$	$\dfrac{-3\sigma_{XY}\sigma_{YA}}{R_{XX}R_{YY} - 9\sigma_{XY}^2}$	$\dfrac{\sigma_{YA}}{R_{XX}+R_{YY}+6\sigma_{XY}}$
$f_Y\{A\}$	$\dfrac{\sigma_{YA}(R_{XX}+k) - 3\sigma_{XY}\sigma_{XA} + \sigma_{XA}k}{(R_{XX}+k)(R_{YY}+k) - (3\sigma_{XY}-k)^2}$	$\dfrac{\sigma_{YA}(R_{XX}+k)}{(R_{XX}+k)(R_{YY}+k) - (3\sigma_{XY}-k)^2}$	$\dfrac{\sigma_{YA}R_{XX}}{R_{XX}R_{YY} - 9\sigma_{XY}^2}$	$\dfrac{\sigma_{YA}}{R_{XX}+R_{YY}+6\sigma_{YY}}$

GENERAL EQUATIONS FOR THE NOE IN SYSTEMS OF TWO-SITE EXCHANGE   147

$f_Y\{A\}$. In practice, this limit is reached for DMF at temperatures of about 90–100°C. Note that Me$_X$ and Me$_Y$ still resonate separately at these temperatures, since the exchange rate is still slow with respect to ($\delta_X - \delta_Y$); the coalescence temperature for Me$_X$ and Me$_Y$ is considerably higher (depending on the field strength). Note also that, as one might expect, the enhancement $f_A\{Y\}$ is predicted to be independent of the exchange rate, since the exchange process does not lessen the extent of saturation of Me$_Y$.

Table 5.3 shows some experimental results obtained by Saunders and Bell[9] for the enhancements $f_A\{X\}$ and $f_A\{Y\}$ in DMF at various temperatures. As can be seen, the essential features just described for these enhancements are found in practice: The enhancement $f_A\{Y\}$ is independent of temperature, while enhancement $f_A\{X\}$ rises from a very small value at low temperature (30°C) to approach that of $f_A\{Y\}$ at high temperature (90°C). The fact that $f_A\{X\}$ is positive at 30°C probably arises from the slow but still nonzero exchange rate at this temperature.

Quite clearly, it is possible, in principle, to obtain an estimate for $\Delta E^\ddagger$, the energy barrier to amide bond rotation, by fitting data of this sort to the equations in Table 5.2. A thorough discussion of such a procedure is given in the book by Noggle and Schirmer.[1] Rearrangement of the expressions for $f_A\{X\}$ and $f_A\{Y\}$ (column 2 of Table 5.2) gives

$$\frac{f_A\{X\}}{f_A\{Y\} - f_A\{X\}} = \frac{k - 3\sigma_{XY}}{R_{YY} + 3\sigma_{XY} - (3\sigma_{YA}^2/R_A)} \quad (5.22)$$

If it is assumed that all the dipolar terms are temperature independent, then k is the only temperature-dependent quantity, and a plot of $\log[f_A\{X\}/(f_A\{Y\} - f_A\{X\})]$ against $1/T$ should give a straight line with a slope of $\Delta E^\ddagger$ [from the Arrhenius equation, $k = k_0 \exp(-\Delta E^\ddagger/RT)$].

TABLE 5.3 NOE Enhancements in DMF as a Function of Temperature[a]

$T(°C)$	$f_A\{Y\}$	$f_A\{X\}$
31	28	3
40	—	6
50	28	9
55	—	13
60	28	18
70	28	24
80	28	26
90	28	27–28

[a]Data from ref. 9. The enhancements are accurate to ±1% and the temperatures to ±3%.

While strictly such applications of the NOE are beyond our scope, a brief discussion of the difficulties of this procedure is relevant here. The main assumption contained in the fitting of the experimental temperature dependence to Eq. 5.22, which was that the relaxation terms are independent of temperature, cannot be exactly true. In particular, the ρ^* terms contained within R_A, R_{XX}, and R_{YY} are likely to vary with temperature in a way that may well differ according to field strength, and the σ terms may also vary somewhat through changes in $\omega\tau_c$. Nonetheless, the constancy of the enhancement $f_A\{Y\}$ in Table 5.3 shows that, for this data set at least, these assumptions do not appear to introduce serious error. The other important assumption is that σ_{XA} is zero. This assumption could break down if, as is quite likely, the DMF exists in solution appreciably as head-to-tail dimers. Of course, as with many such exchange processes, the numerical value of $\Delta E^\ddagger$ (estimated to be 36.4 kJ mol^{-1} by Noggle and Schirmer) would be very likely to depend on the concentration of water or traces of acid or base that might catalyze the exchange.

As far as the rest of this chapter is concerned, probably the most important aspect of the enhancements for DMF is that at high temperature (90°C), the data are inconsistent with any realistic single static structure. This is of course a general feature of the NOE in rapidly exchanging systems; when the exchange rate is sufficiently fast, the experiment gives results appropriate for an "averaged molecule," as discussed earlier (Section 5.2.1).

5.3. APPLICATIONS TO MORE COMPLICATED CASES OF EXCHANGE

In Section 5.2 we considered rather simple exchanging systems, where exchange takes place between just two clearly defined conformations in a two- or three-spin system. Even in such simple cases, the general equations turned out to be rather complex. In this section, we move on to consider more complicated cases, in which the spin systems involved are larger and may adopt a distribution over many different conformations, rather than just two.

5.3.1. Averaging of Rates Rather than Enhancements

Moving to multispin, multisite exchanging systems clearly makes analysis more difficult, but there is one factor that saves the equations from becoming impossibly complex. All the examples considered involve fast motions, particularly rotations about single bonds, which generally occur on a timescale faster than 10^6 s^{-1}; thus, for all the examples throughout Section 5.3, the exchange rate is much faster than the spin–lattice relaxation rates ($k \gg R$). This means we are in the fast-exchange limit of the general equations of Section 5.2, so that the equations are independent of the actual values of exchange rates and simply involve averaged rates of cross and direct relaxation, exactly as though NOE evolution occurred in a single "averaged molecule." As we have seen, in most cases such averaging is also fast on the chemical shift timescale, so that a single averaged spectrum results.

Under these conditions, which hold for the great majority of cases of conformational averaging in small molecules, we may simply adapt equations derived for rigid systems in previous chapters, replacing σ and R by $\langle\sigma\rangle$ and $\langle R\rangle$, the time averages of these rate constants[12]:

$$\langle\sigma_{AB}\rangle = \sum_i (N^i \sigma^i_{AB}) \quad (5.23)$$

and

$$\langle R_A\rangle = \sum_i (N^i R^i_A) \quad (5.24)$$

Thus, we may write for the two-spin steady-state homonuclear NOE enhancement

$$f_I\{S\} = \langle\sigma_{IS}\rangle/\langle R_I\rangle \quad (5.25)$$

for the multispin steady-state homonuclear NOE enhancement

$$f_I\{S\} = \langle\sigma_{IS}\rangle/\langle R_I\rangle - \sum_X (\langle\sigma_{IX}\rangle f_X\{S\}/\langle R_I\rangle) \quad (5.26)$$

and for the two-spin homonuclear NOE enhancement time-course

$$f_I\{S\}(t) = (1 - e^{-\langle R_I\rangle t})\langle\sigma_{IS}\rangle/\langle R_I\rangle \quad (5.27)$$

Similar averages may be written for other equations in Chapters 2, 3, and 4. The above equations are true even if I and S are in slow exchange on the *chemical shift* timescale,[13] since transfer of saturation (and enhancements) between conformations is complete whenever $k \gg R$, regardless of the number of signals in the spectrum. Of course, hiding within these definitions of $\langle\sigma\rangle$ and $\langle R\rangle$ is the idea of a conformationally averaged internuclear distance to the minus sixth power $\langle r_{ij}^{-6}\rangle$, which is what is "sensed" by the NOE. We shall return to this idea in Section 5.5.1.

These considerations lead to a very important point. From the averaged NMR parameters alone, one is unlikely to be able to say what conformations contribute to the ensemble. It may sometimes be possible to tell that more than one conformation must be present, for instance, if the measured NOE enhancements or coupling constants are inconsistent with any single conformation. However, knowledge of the *nature* of the individual contributing conformations must generally come from some other source external to the NMR spectra, such as molecular mechanics calculations of the low-energy conformations available to the molecule. If one is given such a set of *predefined* conformations, then one may attempt to use NOE and other NMR data to derive (approximate) relative populations for each conformation, but even then the *ability to fit NMR data using such a set of predefined conformations cannot necessarily be taken to mean that these conformations are the ones that are really present*; other choices might fit the NMR data also. This crucial point should be borne in mind for all of the examples that follow.

5.3.2. Analyzing Conformational Equilibria

In this section, we present some selected examples where the NOE has been used to assess conformational preferences. The first is a relatively simple case, where the two proposed contributing conformers probably represent the real situation quite well, but even here the analysis is not easy. The other cases are progressively more complex, because of the larger number of spins that must be considered, the greater probability that the conformers explore conformational space away from defined energy minima, and, in the last case, the larger number of contributing conformational energy minima that were considered. Taken together, these examples give a general impression of what may realistically be possible in this area. However, the difficulty of any analysis of this sort is sufficiently great that there remain rather few examples in the literature, and it is clearly the case that the detailed characterization of conformational equilibria still remains one of the principal unsolved problems in NOE interpretation.

5.3.2.1. Olefinic Methoxy Conformations.

Olefinic methoxy groups are generally thought to exist in two fairly well defined conformations (structures I and II in **5.3**), which are in fast exchange on both the chemical shift and T_1 timescales. Clearly, the ratio of the enhancements at H_A and H_B on irradiation of the methoxy signal must reflect the relative populations of the conformers I and II; but if quantitation of the populations is sought, then the other factors that influence these enhancements must be corrected for.

5.3

Since there are no other spins in the regions between either H_A or H_B and the methyl group in either conformer, indirect contributions to the enhancements will be negligible. Further, the rapid internal rotation about the $O-CH_3$ bond can largely be ignored, and the methyl protons treated to a good approximation as having a single "averaged" position in the $C=C-O$ plane. In fast exchange, averaging occurs over cross-relaxation rates, so for conformers I and II we have

$$\langle \sigma_{\text{AMe}} \rangle = N^{\text{I}} \sigma_{\text{AMe}}^{\text{I}} + N^{\text{II}} \sigma_{\text{AMe}}^{\text{II}}$$
$$\langle \sigma_{\text{BMe}} \rangle = N^{\text{I}} \sigma_{\text{BMe}}^{\text{I}} + N^{\text{II}} \sigma_{\text{BMe}}^{\text{II}} \quad (5.28)$$

But, since the distances involved are such that $\sigma_{\text{AMe}}^{\text{II}} \approx \sigma_{\text{BMe}}^{\text{I}} \approx 0$, we may write

$$\langle \sigma_{\text{AMe}} \rangle = N^{\text{I}} \sigma_{\text{AMe}}^{\text{I}}$$
$$\langle \sigma_{\text{BMe}} \rangle = N^{\text{II}} \sigma_{\text{BMe}}^{\text{II}} \quad (5.29)$$

Substituting Eq. 5.29 into Eq. 5.26 for the two steady-state NOE enhancements $f_A\{\text{Me}\}$ and $f_B\{\text{Me}\}$ (neglecting the indirect terms as mentioned above) and rearranging gives

$$\frac{N^{\text{I}}}{N^{\text{II}}} = \frac{f_A\{\text{Me}\}\langle R_A \rangle \sigma_{\text{BMe}}^{\text{II}}}{f_B\{\text{Me}\}\langle R_B \rangle \sigma_{\text{AMe}}^{\text{I}}} \quad (5.30)$$

As σ is proportional both to r^{-6} and τ_c, this can be re-expressed as

$$\frac{N^{\text{I}}}{N^{\text{II}}} = \frac{f_A\{\text{Me}\}\langle R_A \rangle \langle (r_{\text{BMe}}^{\text{II}})^{-6} \rangle (\tau_c)_{\text{BMe}}^{\text{II}}}{f_B\{\text{Me}\}\langle R_B \rangle \langle (r_{\text{AMe}}^{\text{I}})^{-6} \rangle (\tau_c)_{\text{AMe}}^{\text{I}}} \quad (5.31)$$

The relaxation rate constants $\langle R_A \rangle$ and $\langle R_B \rangle$ can be measured experimentally (e.g., by selective T_1 measurements, cf. Sections 2.3.2 and 11.2.3), and values for the correlation times (which might differ slightly because of anisotropic tumbling) can be estimated from ^{13}C T_1 values.[14] The real problem lies in the values of $r_{\text{AMe}}^{\text{I}}$ and $r_{\text{BMe}}^{\text{II}}$, which are the internuclear distances in the two *predefined* conformations, and cannot be determined from the NMR measurements independently of the population ratios. Estimates can be obtained from crystal structures, but need to be very accurate since the distances are raised to the sixth power so that small errors in r translate into large errors in the population ratio. (Also, individual distances to each methyl proton as estimated from a crystal structure would have to be combined into an appropriately averaged value before use, for instance by using an equation such as Eq. 5.53.) Thus, we see that even in this apparently simple case, where there are only two fairly well-defined conformations, population ratios cannot be derived with any great accuracy. This example is treated in greater depth in Section 11.2.3.

5.3.2.2. Nucleotide Conformations.

A nucleotide, such as uridine (**5.4**) or pseudouridine (**5.5**), consists of two more or less rigid units, the sugar and the base, linked by the glycosidic bond, about which rotation can occur. It is common to describe the conformations about the glycosidic bond in terms of two possibilities, one called *syn* (in which H6 is close to H1′) and the other called *anti* (in which the base is rotated through roughly 180°, making H6 close to H3′ and H2′; this is the conformation shown in **5.4** and **5.5**). Some experimentally determined steady-state enhancements for these compounds are shown in Table 5.4.[15]

5.4 **5.5**

This pattern of enhancements is not consistent with any one conformation for either nucleotide, so the data show that the molecules must exist in at least two conformations. These are in fast exchange on the chemical shift timescale, and almost certainly on the T_1 timescale also. However, the data show quite clearly that uridine exists *mainly* in the *anti* conformation, while pseudouridine is *mainly* in the *syn*. For most purposes, this is probably a sufficiently precise description of the situation. For any more quantitative description, we need to specify more precisely what constitutes the "*syn*" and "*anti*" conformations, that is, we need to define their respective glycosidic bond torsion angles. Armed with these definitions, we can be as quantitative as we like, but we must be aware that we are then on rather thin ice, since the exact glycosidic torsion angles chosen for the definitions of "*syn*" and "*anti*" will have a considerable influence on any derived population ratio. Thus, at the end of such a study it will not be possible to conclude that uridine necessarily exists as $x\%$ *syn* and $y\%$ *anti*, but only that such a population distribution is consistent with the NOE data *and with the chosen definitions of the contributing conformers.*

TABLE 5.4 NOE Enhancements Observed in Nucleotides 5.4 and 5.5[a]

Compound	$f_6\{1'\}$	$f_6\{2'\}$	$f_6\{3'\}$
5.4	8	23	8
5.5	26	10	9

[a]The steady-state enhancements were obtained at 100 MHz, 10°C. Data from ref. 15.

Assuming that we are prepared to accept some external definition for the exact nature of the *syn* and *anti* conformations (e.g., based on molecular mechanics calculations), then the next problem is to find out how the conformer population ratio is reflected in the NOE data. Essentially, this is the same problem as discussed in the previous section, but without the simplifications that are possible when one enhancement arises exclusively from one conformer and another arises exclusively from the other conformer. Suppose there are two steady-state enhancements $f_I\{S\}$ and $f_U\{V\}$, and that as usual X represents all the spins other than I and S, while W represents all the spins other than U and V. Denoting the contributions of the two conformations *syn* and *anti* by the superscripts a and b, respectively, each measured enhancement is given by substituting for the $\langle \sigma \rangle$ terms in Eq. 5.26:

$$f_I\{S\} = \frac{N^a \sigma_{IS}^a + N^b \sigma_{IS}^b}{\langle R_I \rangle} - \sum_X \frac{N^a \sigma_{IX}^a + N^b \sigma_{IX}^b}{\langle R_I \rangle} f_X\{S\} \qquad (5.32)$$

and

$$f_U\{V\} = \frac{N^a \sigma_{UV}^a + N^b \sigma_{UV}^b}{\langle R_U \rangle} - \sum_W \frac{N^a \sigma_{UW}^a + N^b \sigma_{UW}^b}{\langle R_U \rangle} f_W\{V\} \qquad (5.33)$$

Dividing one equation by the other and rearranging, it is easily shown that the population ratio is given by

$$\frac{N^a}{N^b} = \frac{f_I\{S\}\langle R_I \rangle A - f_U\{V\}\langle R_U \rangle B}{f_U\{V\}\langle R_U \rangle C - f_I\{S\}\langle R_I \rangle D} \qquad (5.34)$$

where[†]

$$A = \sigma_{UV}^b - \sum_W \sigma_{UW}^b f_W\{V\}$$

$$B = \sigma_{IS}^b - \sum_X \sigma_{IX}^b f_X\{S\}$$

$$C = \sigma_{IS}^a - \sum_X \sigma_{IX}^a f_X\{S\}$$

$$D = \sigma_{UV}^a - \sum_W \sigma_{UW}^a f_W\{V\} \qquad (5.35)$$

As before, the values of σ are principally dependent on the appropriate values of r^{-6}, which may be obtained by using accurate molecular models of

---

[†]Note that Eq. 5.34 becomes equivalent to Eq. 5.30 in the absence of all indirect contributions and when $\sigma_{IS}^b = \sigma_{UV}^a = 0$ (setting $S \equiv V \equiv$ Me, $I \equiv H_A$, $U \equiv H_B$, conformer a ≡ conformer I, and conformer b ≡ conformer II).

154   THE EFFECTS OF EXCHANGE AND INTERNAL MOTION

the predefined *syn* and *anti* conformations. Corrections for small variations between specific τ_c values could, if necessary, be found by analysis of ^{13}C T_1 data, and the values of $\langle R_I \rangle$ and $\langle R_U \rangle$ would have to be measured, for example, by using selective ^{1}H T_1 experiments (cf. previous section).

Clearly, some of the possible enhancements, such as $f_6\{1'\}$ and $f_6\{2'\}$, are more sensitive to the conformer distribution ratio than are others, and one would expect intuitively that these should give the most reliable values for the population ratio. Equation 5.34 shows that this is so; in cases where neither $f_I\{S\}$ nor $f_U\{V\}$ is particularly conformation dependent, the population ratio is given by a ratio of two small differences, and is consequently very sensitive to errors.

This approach is, in principle, generally applicable to any two-state conformational equilibrium in fast exchange on the chemical shift and/or T_1 timescales, but as far as we are aware it has very rarely been used, other than in the relatively favorable case of olefinic methoxy conformations as described in Sections 5.3.2.1 and 11.2.3. What has been done in the case of analyzing nucleotide conformations is to make the model rather more sophisticated by effectively including the definitions of *syn* and *anti* conformations amongst the parameters to be determined from the NOE data. Thus, Schirmer et al. analyzed the *syn*–*anti* equilibrium for 2',3'-isopropylideneuridine (**5.6**) using a five-parameter model where the variables comprised the population ratio of *syn* and *anti* conformers, the glycosidic torsion angle in each conformation, and a "width" parameter for each conformation, intended to describe the extent to

5.6

which each explores nearby conformational space through limited motion around the glycosidic bond.[12] The "width" parameters were defined in terms of assumed Gaussian distributions of glycosidic torsion angles; although this shape function was chosen for convenience, it was argued that this choice has relatively little impact on the analysis. Steady-state enhancements were calculated for each conformer distribution using Eq. 5.26, where the discrete averages such as those used in Eq. 5.32 were replaced by integrals. Thus, instead of the discrete

$$\langle \sigma_{IS} \rangle = \sum_i (N^i \sigma_{IS}^i) \quad \text{and} \quad \langle R_I \rangle = \sum_i (N^i R_I^i) \tag{5.36}$$

Schirmer et al.[12] used

$$\langle \sigma_{IS} \rangle = \frac{\int P(\chi)\sigma_{IS}(\chi)\, d\chi}{\int P(\chi)\, d\chi} \quad \text{and} \quad \langle R_I \rangle = \frac{\int P(\chi)R_I(\chi)\, d\chi}{\int P(\chi)\, d\chi} \quad (5.37)$$

where $P(\chi)$ is the probability distribution function for the glycosidic torsion angle χ. The integrals were calculated using Simpson's rule. Because the $\langle R_I \rangle$ values were calculated rather than being measured experimentally, it was necessary to add estimates for ρ^* for each proton. Because the sample was degassed, it was assumed that ρ^* would be small, and in the absence of further information it was set to the same small constant value for all protons. While this is clearly a source of error, as in reality each proton has a different ρ^*, calculations using different constant values for ρ^* showed that the final results were not greatly affected by this choice.

The best-fit parameters found in the analysis by Schirmer et al.[12] comprised an approximate population ratio of 80:20 in favor of the *syn* conformer, with χ angles of about 65° in the *syn* and 254° in the *anti* conformers,[†] and "width" parameters showing a relatively broad distribution of glycosidic angles in the *syn* conformer but a very narrow distribution in the *anti*. They also carried out calculations under the rather unrealistic assumption of slow conformational averaging on the relaxation timescale, by averaging directly the enhancements rather than the individual relaxation rates (cf. Section 5.2.1). Perhaps surprisingly, the results were very similar, with a population ratio of approximately 77:23 in favor of *syn*, and glycosidic angles only slightly shifted relative to the results under fast exchange. The biggest change was in the width of the distribution, which was significantly broader under the slow-exchange assumption than under fast exchange. This is not unexpected, given that relaxation rates in the fast-exchange limit reflect distances averaged as r^{-6}, and will be more strongly dominated by contributions from short distances than would be averages calculated directly over enhancements in the slow-exchange limit.

It is instructive to consider the interpretation of these results briefly. Clearly, there are several assumptions that may have influenced the analysis, for example, the assumptions that the ribose ring is rigid, that all protons have a uniform, small value for ρ^*, and that the population distributions have a Gaussian form. Moreover, although 14 enhancements were used for the fit, many of these are at least partially correlated with one another, and some vary very little with the glycosidic torsion angle. Thus, in reality the number of useful,

---

[†]The original paper defines the gycosidic angle in terms of a variable called Y, for which values of 355° and 166° are reported for the *syn* and *anti* conformations respectively. It would appear that χ and Y are related through the expression $\chi = \text{mod. } 360(-Y + 60°)$.

independent experimental variables is rather small to be able to define accurately such a large number of parameters. As acknowledged by Schirmer et al.,[12] the width parameters for the population distributions are meaningful only at a very qualitative level, at best. However, the population ratios and glycosidic torsion angles are more significant, and probably represent reality better than would population ratios derived from predefined, regular conformations.

5.3.2.3. A Statistical Approach. All the approaches described throughout Section 5.3.2 seek to fit NMR data from molecules undergoing a rapid conformational exchange using a plausible set of predefined conformers. A central problem in any such approach is that it is always possible to improve the fit by using more conformers, but this rapidly results in the problem becoming underdetermined. What is needed therefore is an objective criterion to determine whether expansion of the set of conformers is justified by the data. Several authors have published statistically based approaches to this issue, often aimed particularly at analyzing small flexible peptides[16–20]; we shall concentrate here on a potentially general method introduced by Kozerski et al.[21]

This method takes as its starting point conformations corresponding to energy minima from molecular mechanics calculations, and then combines these to build up an ensemble having the best fit to the NMR data (which comprise both NOE enhancements and *J*-couplings). At each stage, addition of a further conformer to the ensemble is only carried out if it makes a statistically significant improvement to the overall fit (judged by a standard *F*-test), and the process is handled in such a way that all combinations of the possible conformers are considered and their relative populations optimized. Note that the calculated values of the energies of each conformation are not used in this analysis. The underlying assumption here is that molecular mechanics calculations can probably reveal the *nature* of the contributing conformers reasonably accurately, but that such calculations are much less reliable in predicting the relative *populations* of conformers. It is therefore preferable to use the NMR data to establish the conformer populations. In a final part of the method, the NMR data may also be used to refine the structures of the individual contributing conformers to achieve a better fit with the data.

So far this particular approach has only been applied to one case, that of thedihydrobenzofuran derivative **5.7**. This molecule is the major product of the reaction of the enamineone C_2H_5—CO—$C(CH_3)$=CH—NH—$CH(CH_3)$—Ph with *p*-quinone, and it is isolated as a single diastereomer by crystallization. The first step in the process was to calculate (using the program PCMODEL[22]) all conformations of the molecule within 3 kcal mol^{-1} of the conformation with the lowest calculated energy. Since the relative stereochemistry was initially unknown, each of the four possible diastereomers (and also each invertomer at nitrogen for each diastereomer) was modelled separately. Clusters of conformations were then identified, a cluster being defined by the requirement that members of a common cluster should differ from one another by no more than 15° in any dihedral angle, and each such cluster was then taken to be represented by the lowest energy conformer within it. From among the conformers so generated, the 20 with the lowest energies were taken as the working set for the subsequent analysis; these included between 4 and 7 representative conformers for each of the possible diastereomers.

Having defined this working set of conformers, the statistical analysis was then used to determine which combination from this set best fits the NMR data. At each stage, tests were applied to determine whether addition of further conformers produced a statistically significant improvement in the overall fit to the NMR data, and for each combination of conformers examined, a set of best-fitting relative populations was derived, together with associated confidence limits. The first target was to establish the correct relative stereochemistry for the molecule. In order to achieve this, conformers corresponding to each stereochemistry were analyzed as separate self-contained sets, and the stereochemistry for which the best-fitting combination of conformers could be found was taken as correct. In the event, the statistical tests showed that one diastereomer was a clear winner. In addition, the analysis generated a composition for the best-fitting ensemble of conformers corresponding to the correct stereochemistry. This showed that only two of the seven conformers in the working set for that stereochemistry were needed to interpret the NMR data satisfactorily, both present in roughly equal amounts; both conformers had the same inversion stereochemistry at nitrogen, but differed principally in the torsion angle between the nitrogen and acyclic chiral carbon atoms.

Clearly, this method has some general limitations. It is only realistic to expect to handle a fairly small working set of possible conformers in this way, and one is necessarily reliant on the modelling program to produce energy minima that are reasonably close to reality. It may also be true that the particular case studied was quite favorable, if it is indeed true that its conformational equilibrium is dominated by only two significantly different conformers. Nonetheless, defining the relative stereochemistry at the acyclic chiral center of structure **5.7** represents quite a significant achievement, even if further details of the conformational equilibrium were to be ignored (cf. Section 10.1.1). This result also has the virtue of being testable; comparison with a crystal structure of **5.7** shows that this NMR approach did in fact generate the correct stereochemistry. Probably, methods based on trying to identify and compare *single* "best fit" conformations for each stereochemistry would not have been able to accomplish this reliably. However, only time and experience with other systems will establish how generally such statistical approaches are useful.

5.4. ESTIMATING FLEXIBILITY USING HETERONUCLEAR RELAXATION ANALYSIS

In this section, we describe a widely used approach for characterizing local flexibility in macromolecules using heteronuclear relaxation measurements, including steady-state X{^{1}H} NOE data. Although it could be argued on a narrow definition of structure determination that this topic falls outside the remit of this book, we have chosen to include it for two reasons. First, such studies of flexibility are frequently published in combination with an NOE-based structure determination, and it is important to understand the interrelationship between both sets of results. Second, the concepts required to understand such studies of flexibility are just those needed elsewhere in the book to describe the relationship between NOE intensities and flexibility, and are very conveniently developed in this context.

Throughout this book we have emphasized that NOE intensities depend upon both internuclear distances and molecular motions. Elsewhere, our attention has usually been firmly fixed on interpreting the internuclear distance dependence of NOE enhancements. However, here we concentrate on using NOE data as a source of information about molecular motion when the internuclear distance is already known, as in the case of a directly bonded proton-heteronucleus pair (e.g., r_{NH} = 1.02Å or r_{CH} = 1.08Å; but see also ref. 23). In practice, NOE data alone are not sufficient for such analysis, so additional measures of dipolar relaxation are also included, most commonly heteronuclear T_1 and T_2 values as will be described below. The essence of the approach is to measure relaxation data for a set of different sites within a molecule (e.g., all the backbone amide ^{15}N-^{1}H groups in a uniformly ^{15}N-labelled sample of a protein) and then to analyze these data in terms of some theory that dissects the relaxation into contributions from overall molecular tumbling and internal motions.

The commonest and simplest theory used to analyze local flexibility in this way is the "model-free" formalism of Lipari and Szabo, so called because it does not invoke any model to describe the form of the internal motion.[24a,24b] In this approach, the internal motion at each labelled site is characterized by just two parameters, the "generalized order parameter" S (which invariably occurs in the theory as its square, S^2) of which we shall say more in a moment, and the correlation time for the internal motion τ_e. The only other parameter needed in the simplest version of this theory is the correlation time for overall molecular tumbling. For consistency with other sections of this book, we shall symbolize this quantity as τ_c; however, it should be noted that many other publications symbolize the correlation time for overall tumbling as τ_M in order to distinguish it more clearly from τ_e, the correlation time for the internal motion.

The key concept to grasp here is that of the order parameter S, which represents the degree of angular restriction of a particular internal motion; S^2 takes values between 0 and 1. To illustrate the idea, consider an individual N—H bond vector in a protein. If the NH group is completely rigid with respect to the framework of the rest of the molecule, then there is no internal motion. This situation corresponds to an S^2 value of 1; reorientation of the N—H vector then results only from overall molecular tumbling. On the other hand, if the NH group does undergo internal motions, these will cause the orientation of the N—H vector to change relative to the molecular frame; this corresponds to an S^2 value of less than 1. The greater the angular variations caused by the internal motion, the smaller the value of S^2, and the limit of $S^2 = 0$ corresponds to movement of the N—H vector isotropically through all possible orientations relative to the rest of the molecule. It is hard to see how this latter limit could be fully attained in practice, given that the NH group must remain attached covalently to the rest of the molecule, but one can picture how, for instance, an NH group in a sizable flexible portion of a protein might sample all orientations relative to the more rigid remaining part of the protein (although not necessarily all with equal probability) through movements of a number of connecting bonds in the flexible region. Note also that, although we have used the specific example of a directly bonded NH pair to illustrate the concept of the order parameter, it applies to any internuclear vector, and can be very useful when describing the effect of internal motions on homonuclear NOE interactions that do not necessarily correspond to a fixed internuclear separation. This concept will be taken up in Section 5.6.

The meaning of the order parameter can be further clarified by looking at its effects on the correlation function $g(\tau)$ and the spectral density function $J(\omega)$. As discussed in Section 2.2.1, the correlation function describes the rate at which the orientation of some internuclear vector becomes uncorrelated relative to an arbitrary initial position as a result of motion. The definitions given earlier (Eqs. 2.16 and 2.17) corresponded only to isotropic overall motion at one rate, namely the overall tumbling rate of the whole molecule. If we now include also an internal motion, characterized by its own correlation time τ_e and value of S^2, then the definition† of $g(\tau)$ becomes[24a]

$$g(\tau) = S^2 \exp(\tau/\tau_c) + (1 - S^2)\exp(\tau/\tau_c)\exp(\tau/\tau_e)$$

$$= S^2 \exp(\tau/\tau_c) + (1 - S^2)\exp(\tau/\tau_{comb}) \tag{5.38}$$

where τ_{comb} is the combined correlation time given by

$$(\tau_{comb})^{-1} = (\tau_c)^{-1} + (\tau_e)^{-1} \tag{5.39}$$

---

†Many papers in the literature include a numerical factor of 1/5 in the definitions of $g(\tau)$ and $J(\omega)$. We have omitted this in order to maintain consistency with other parts of the book, with the consequence that the factor is incorporated instead into the relevant terms of Eqs. 5.41–5.43 for the ^{15}N T_1, T_2, and steady-state NOE. For this reason, the numerical factors in these latter equations differ from those in some of the cited publications.

The corresponding spectral density, by analogy to Eq. 2.20, is given by

$$J(\omega) = \frac{2S^2\tau_c}{1+(\omega\tau_c)^2} + \frac{2(1-S^2)\tau_{comb}}{1+(\omega\tau_{comb})^2} \qquad (5.40)$$

Figures 5.5 and 5.6 show the general form of these functions, and they clearly demonstrate a number of very important features. First, it is completely clear that *only internal motions faster than overall molecular tumbling* can contribute to the correlation function and hence affect relaxation. Overall molecular tumbling necessarily causes $g(\tau)$ to decay completely to zero, beyond which no further decay is possible; thus motions on slower time scales cannot

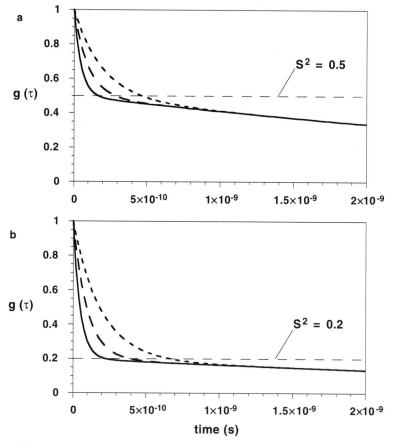

Figure 5.5. Correlation functions $g(\tau)$ for a system undergoing isotropic overall tumbling and one internal motion (Eq. 5.38), plotted for squared order parameter values of (a) $S^2 = 0.5$ and (b) $S^2 = 0.2$. The correlation times (τ_e) for the internal motion are 0.2 ns (short dashes), 0.1 ns (long dashes) and 0.05 ns (solid line), and in each case the correlation time for overall tumbling (τ_c) is 5 ns. The vertical axis is in arbitrary units.

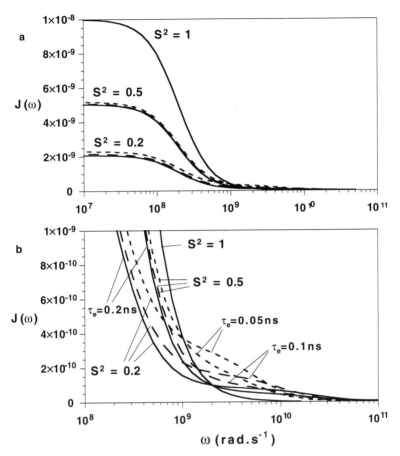

Figure 5.6. (a) Spectral density functions (Eq. 5.40; vertical axis is seconds) corresponding to the correlation functions in Figs. 5.5a ($S^2 = 0.5$) and 5.5b ($S^2 = 0.2$). As in Fig. 5.5, the correlation times (τ_e) for the internal motion are 0.2 ns (short dashes), 0.1 ns (long dashes) and 0.05 ns (solid line), and in each case the correlation time for overall tumbling (τ_c) is 5 ns. Panel (b) shows an expansion of the central region of the curves. For comparison, the spectral density function for no internal motion ($S^2 = 1$, $\tau_c = 5$ ns) is also shown.

influence $g(\tau)$ at all. This leads immediately to the principal limitation of analyzing flexibility using relaxation analysis, namely that the approach is only sensitive to internal motions faster than overall tumbling. Note also that this behavior can be demonstrated mathematically using Eqs. 5.38 and 5.39; if $\tau_e \gg \tau_c$ then $\tau_{comb} = \tau_c$, Eqs. 5.38 and 5.40 reduce directly to Eqs. 2.17 and 2.20, respectively, and all dependence on the internal motion is lost. It is also clear that if the value of S^2 were really to be zero (for a fast internal motion having $\tau_e \ll \tau_c$), then overall tumbling would make no contribution to the relaxation. Equation 5.40 would then again reduce to Eq. 2.20, but this time with τ_e as the correlation time, rather than τ_c.

Figure 5.5 also demonstrates the significance of S^2 values in a different way from the foregoing; it is the value which the correlation function approaches as a result of the internal motion. If the internal motion is much faster than overall tumbling ($\tau_e \ll \tau_c$), then the correlation function drops rapidly to a value of S^2, after which overall tumbling causes a slower subsequent decay to zero. As Figure 5.6 shows, as far as the spectral density function is concerned, this corresponds essentially to a scaling down of its amplitude by a factor of S^2. There is an additional contribution to $J(\omega)$ from the internal motion itself, but because the second term in Eq. 5.40 is effectively scaled by τ_e (since τ_{comb} is dominated by the shorter correlation time), this contribution is small and becomes still smaller for faster internal motions.

We may also see from the form of these curves that the value of S^2 has a far more profound effect on the relaxation behavior than does the value of τ_e, particularly when the internal motion is fast. For this reason, in many relaxation studies values for S^2 are the real object of the analysis, whereas values for τ_e are sometimes missing, subject to relatively large errors, or simply ignored. Indeed, in the limit of very fast internal motion, the second term of Eq. 5.40 reduces to zero, and the value of τ_e no longer influences the relaxation behavior at all. On the other hand, it is worth remembering that if significant internal motion occurs on the *same* time scale as overall tumbling, then much of the foregoing analysis breaks down. In the limit of extensive internal motions at or near this rate, even the concept of overall tumbling is not really defined any longer; if the molecule changes its structure profoundly at the same rate as it tumbles, one can no longer properly separate the two processes even conceptually, or define a molecular coordinate frame.

Armed with a suitable definition for the spectral density function, such as Eq. 5.40, it is a relatively simple matter to substitute it into equations for the heteronuclear relaxation parameters. For ^{15}N—^{1}H groups, which represent the most commonly studied case, the relevant equations are[24a]

$$\frac{1}{T_1} = R_1 = \frac{1}{20} K^2 [J(\omega_H - \omega_N) + 3J(\omega_N) + 6J(\omega_H + \omega_N)] + \frac{1}{15} c^2 J(\omega_N) \quad (5.41)$$

$$\frac{1}{T_2} = R_2 = \frac{1}{40} K^2 [4J(0) + J(\omega_H - \omega_N) + 3J(\omega_N) + 6J(\omega_H)$$

$$+ 6J(\omega_H + \omega_N)] + \frac{1}{90} c^2 [3J(\omega_N) + 4J(0)] + R_{ex} \quad (5.42)$$

$$f_N\{H\} = \frac{I - I_0}{I_0} = \frac{1}{20} \frac{\gamma_H}{\gamma_N} K^2 [6J(\omega_H + \omega_N) - J(\omega_H - \omega_N)]/R_1 \quad (5.43)$$

where, as in Chapter 2, $K = (\mu_0/4\pi)\hbar \gamma_H \gamma_N r_{HN}^{-3}$; note, however, that many authors report their NOE data as I/I^0 ratios, which correspond to $1 + f_N\{H\}$ (cf. Section 2.5). The term R_{ex} in Eq. 5.42 represents the residual contribution to the measured T_2 values from exchange, if any, (see following), and the term c represents the contribution from chemical shift anisotropy relaxation (Section 2.4.4). The term c is given by $c = \omega_N(\sigma_\parallel - \sigma_\perp)$, where the value of $(\sigma_\parallel - \sigma_\perp)$ is usually taken as 160 ppm for amide ^{15}N—^{1}H pairs.[25]

The detail of the analysis then comes down to varying the values of S^2, τ_e, and τ_c used in the spectral densities in order to achieve the best fit between data calculated using Eqs. 5.41–5.43 and the experimentally measured data. Often, this is done by minimizing an error function such as χ^2 given by

$$\chi^2 = \frac{(R_1^{exp} - R_1^{calc})^2}{(s.d.)_1^2} + \frac{(R_2^{exp} - R_2^{calc})}{(s.d.)_2^2} + \frac{(NOE^{exp} - NOE^{calc})^2}{(s.d.)_{NOE}^2} \quad (5.44)$$

where the s.d. values are standard deviations in the experimentally measured data, and their inclusion is supposed to weight the contribution of each piece of data to the fit according to the certainty with which it is known. Of course, τ_c is supposed to be the same for all sites in the molecule, whereas S^2 and τ_e are allowed to vary, so one must try to find a suitable value for τ_c before carrying out the analysis for individual sites. A commonly used approach is to measure the ratio T_1/T_2 for all sites. Several authors have shown that if $\tau_e < 0.1$ ns, $\tau_c > 1$ ns, and $R_{ex} \sim 0$, then the T_1/T_2 ratio becomes independent of S^2 and τ_e, allowing it to be used to extract a value for τ_c (see, for example, ref. 26). In practice, it is usually necessary to restrict this procedure to sites in relatively rigid parts of the molecule, since the above conditions are often not met in more flexible regions.

Generally, the pulse sequences used to measure experimental NOE, T_1 and T_2 values are modified versions of (X, ^{1}H) correlation sequences, designed in such a way that the intensities of the correlation signals depend on the relaxation parameter to be measured. We shall not describe such sequences here; a brief discussion including references and one example for measuring ^{15}N$\{^1$H$\}$ steady-state NOE values appears in Section 9.1.1.

Quite often extended versions of the formalism are used. Most commonly, the possibility of internal motions on two separate timescales is introduced.[27] If the squared order parameters for the fast and slow internal motions are, respectively, S_f^2 and S_s^2, then Eq. 5.40 becomes

$$J(\omega) = \frac{2S^2\tau_c}{1 + (\omega\tau_c)^2} + \frac{2(S_f^2 - S^2)\tau_{comb}}{1 + (\omega\tau_{comb})^2} \quad (5.45)$$

where $S^2 = S_f^2 + S_s^2$. Effectively, this modification simply relaxes the form of the second term, allowing it to become larger if this achieves a better fit overall to the data. In this variation of the formalism, τ_e now refers only to the slower of the internal motions, the value of the correlation time for the faster motion being assumed to have no effect on the relaxation. Consistent with this approach, several molecular dynamics studies have suggested that groups in the interior of proteins are likely to show an initial very rapid decay in $g(\tau)$ due to rapid highly localized atomic fluctuations, followed by a slower decay due to larger scale, more complicated internal motions.[28,29] Another common extension of the formalism is the inclusion of an empirical exchange term R_{ex} in the equation for T_2 (as already indicated in Eq. 5.42). Again, this relaxes the fit, this time effectively by allowing aberrant values of T_2 to be excluded from influencing the derived values of S^2. There are also extended forms of Eq. 5.40 that treat anisotropic tumbling, either for a symmetric top molecule (requiring two correlation times to describe overall tumbling)[24b] or for the fully asymmetric case (requiring three correlation times to describe overall tumbling).[30,31] Various criteria can be used to choose for which residues these various extended spectral density functions should be used, and for which to include an exchange term (R_{ex} in Eq. 5.42) in the fit.

Of course, increasing the number of parameters in this way will generally improve the fit, but at the cost that the problem can become rather severely under-determined. For instance, it is sometimes said that values for R_{ex} generated by the fitting procedure indicate the extent of motions on slower timescales, in the microsecond to millisecond range. While this may be true in some cases, it is hard to avoid the feeling that this is at best rather an indirect measure of such processes. To some extent, the additional parameters used in the various extended versions of the theory may be regarded as "dustbins" into which various errors due to shortcomings in the fit, the measurements or the model may accumulate; detailed interpretation of their values may well be dangerous. A possible, but rather time-consuming, way to improve the ratio of experimental data to fitting parameters is to measure relaxation data at two significantly different B_0 field strengths. This is particularly beneficial if it is intended to interpret the R_{ex} terms, since it is then possible to check whether these show the dependence on B_0^2 expected for genuine exchange contributions. A related approach is to measure the dependence of R_{ex} on the B_1 field strength in $T_{1\rho}$ experiments, or, similarly, on the 180° pulse repetition rate in T_2 experiments using a CPMG sequence.

Many studies of ^{15}N relaxation of protein backbone amide groups have appeared in the literature over recent years, mostly using the methodology outlined above. For typical well-structured small globular proteins, the results are usually rather similar. The majority of residues in regions of regular secondary structure and in the core show S^2 values in the region of 0.8–0.9, while those in flexible loops or at the chain termini (if these are not tied down in the structure) show substantially lower values. Nonetheless, studies of this sort can be very informative. Quite often, they might be undertaken to establish whether a region that is relatively poorly defined in a structure determined by NMR is genuinely mobile in solution, or whether it is disordered in the calculated structures only because few NOE restraints were identified for that region. Either answer is possible. The calcium-binding protein calbindin D_{9k} provides a good example of this.[32] NMR structure determination showed that it comprises two

"EF-hand" subdomains joined by a linker, each EF hand in turn comprising a helix-loop-helix motif with a calcium-binding site in the loop (Fig. 5.7a). However, both calcium-binding loops and the linker were relatively poorly defined in the NMR structures, as demonstrated by the local rmsd values shown in Figure 5.7b (cf. Section 12.4.2). A ^{15}N relaxation study yielded the S^2 values that are shown in Figure 5.7c. There is a clear correlation between poor local precision and reduced values of S^2 in the linker region, showing that this region is genuinely mobile in solution. However, in complete contrast, S^2 values in the calcium-binding loops are essentially identical to those for well-ordered parts of the structure. Bearing in mind our earlier caveats about the motional timescales to which ^{15}N relaxation studies are sensitive, we cannot conclusively say that this proves that only the linker region is genuinely mobile. However, it does prove that there is a clear difference between the motional properties of the linker and of the calcium-binding loops, and strongly suggests that the poor definition of the latter may be due to limitations on the number of NOE interactions that could be assigned in this region of the structure.

This study also demonstrates one great strength of analyzing a regularly repeated motif such as the backbone amide NH group. Even if individual S^2 values were to be in error, or if some systematic error should lead to all S^2 values being consistently shifted in one direction, the interpretation would be substantially unaffected because it relies mainly on the *trends* in the data along the backbone of the molecule. Figure 5.7 shows that these can be very clear indeed where there is a strong underlying cause. Another example of this comes from work on a mutant of bovine pancreatic trypsin inhibitor (BPTI), designed to mimic an intermediate on the folding pathway where only one of the three disulfide bonds in the native protein (that between Cys30 and Cys51) has formed.[34] This molecule turns out to be only partially folded (Fig. 5.8a). This is significant for its role in the folding pathway, since the next step involves formation of one of two non-native disulfide bonds. The Cys residues involved are Cys5, Cys14 and Cys38, and the S^2 values from the ^{15}N relaxation study (Fig. 5.8b) in conjunction with other data clearly show that all three are in mobile, disordered parts of the molecule. This allows the important conclusion that formation of these non-native disulfide bonds does not involve stable, non-native conformation, a point that had previously attracted some controversy.

Heteronuclear relaxation studies represent an area of continuing interest; many new developments are taking place and will doubtless continue. Other theoretical approaches have been developed and, in particular, the technique of "reduced spectral density mapping" has recently become popular. In this approach, data for the various forms of relaxation data (again usually T_1, T_2, and the NOE) are combined through a set of linear simultaneous equations to yield experimental values for $J(0)$, $J(X)$, and $J(H)$ for each residue.[35,36] Relaxation studies of other nuclei are also attracting much attention, particularly ^{13}C and ^{2}H. For molecules uniformly labelled with ^{13}C, there are severe additional problems arising from the presence of ^{13}C-labelled neighbors at most ^{13}C sites. Progress has been made for protein backbone carbonyl carbons,[37] where relax-

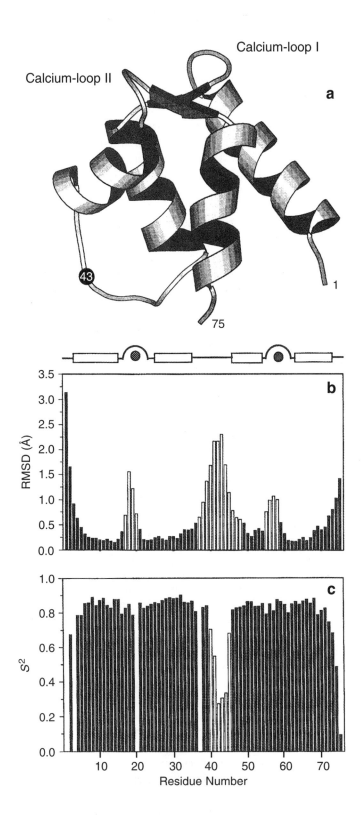

ation is dominated by the chemical shift anisotropy mechanism (Section 2.4.4), for C^α carbons,[38] and for cross-relaxation of carbonyl-C^α spin pairs.[39] Methylene and methyl groups are harder to treat, and so far this has hindered ^{13}C relaxation studies of side-chain motions. Although they are still largely in a method development phase, studies of ^{2}H relaxation, particularly of side-chain motions such as those of CH$_2$D methyl groups,[40] are currently gaining in popularity. Largely, this is because other developments in assignment and structure determination methodology have led to the more widespread availability of partially ^{2}H-labelled samples. It will be very interesting to see what developments follow in the whole area of relaxation analysis in the future.

5.5. HOW INTERNAL MOTIONS AVERAGE INTERNUCLEAR DISTANCES

In the previous section we saw how internal motions faster than overall molecular tumbling affect dipolar relaxation in cases where the internuclear distance is fixed. We now broaden the discussion to consider what happens when the length, as well as the orientation, of an internuclear vector varies as a result of an internal motion.

Unfortunately, there is no getting away from the fact that this situation demands a somewhat higher level of theory than that given in most of the rest of this book. We have already seen in the previous section that the way to treat internal motions is to incorporate their effects into a modified version of the spectral density function. If the internuclear separation varies on a timescale similar to, or faster than, that for molecular tumbling, this requires a step away from the theory presented in Chapters 2 and 3. There, it was possible simply to exclude the value of r_{IS} from the definition of $J(\omega)$, putting it instead into

←

Figure 5.7. Comparison of structural and dynamic parameters of the backbone amide nitrogens of calcium-loaded calbindin D$_{9k}$.[32] (*a*) Backbone ribbon representing the structure of calbindin D$_{9k}$, drawn using the program MOLSCRIPT.[33] The two termini (residues 1 and 75) are indicated, as are the positions of the central loop (marked using residue 43) and the two calcium binding loops. (*b*) Local rmsd (average rmsd to the mean structure, calculated per residue after a global fit; cf. Section 12.4.2) for the calculated NMR structures of calbindin D$_{9k}$, as a function of sequence. Regular secondary structure is indicated above the panel: α-helices are denoted by boxes, calcium-binding loops by semicircular arcs, and calcium by circles. (*c*) Backbone ^{15}N order parameters (S^2); the open bars denote unusual values (ignoring the termini). The linker region (approx. residues 36–45) is both mobile (it has low order parameters) and disordered in the structures (it has a high local rmsd). In contrast, the two calcium-binding loops (residues 16–24 and 55–62) have order parameters no lower than the helical regions, even though parts of them are disordered in the calculated structures. Reproduced and adapted with permission from ref. 32.

168 THE EFFECTS OF EXCHANGE AND INTERNAL MOTION

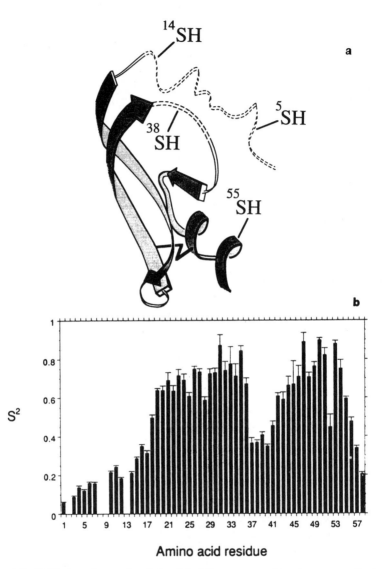

Figure 5.8. ^{15}N Relaxation study of the $(30-51)_{Ser}$ mutant of bovine pancreatic trypsin inhibitor (BPTI). (*a*) Of the three disulfide bonds in native BPTI (30–51, 5–55, and 14–38), this molecule has only that linking Cys30 and Cys51 (the other four Cys residues are mutated to Ser), so as to mimic the first key intermediate on the disulfide folding pathway. (*b*) S^2 values as a function of sequence for backbone amide NH groups in $(30-51)_{Ser}$ BPTI. Approximately the first third of the molecule is highly mobile and disordered (low values of S^2), as is the loop containing Cys38. This helps to explain how Cys5, Cys14, and Cys38 can participate in the formation of non-native disulfide bonds (5–14 and 5–38) at the next step in the folding pathway, without generating non-native folded conformation. Reproduced with permission from ref. 34.

the constant term K (as is done explicitly or implicitly, for example, in Eqs. 2.21–2.33, 2.45, and 3.17–3.27, and many other equations elsewhere in this book, including those of the previous section). In contrast, now we must develop a form of the spectral density function that incorporates the effects of changes in r_{IS}. We shall denote this form of spectral density function $J'_{ij}(\omega)$, so as to keep track of the fact that it is dimensionally distinct from our previous definitions and is specific to the ij interaction.

As we shall see, in limiting cases of slow or fast internal motion it is possible to separate out the spectral density function for overall tumbling from the expression for $J'_{ij}(\omega)$. In these circumstances, it is useful to define an effective distance as "sensed" by the NOE:

$$(r_{ij})_{\text{effective}} = \langle r_{ij}^{-6} \rangle^{-1/6} = [J'_{ij}(\omega)/J(\omega)]^{-1/6} \tag{5.46}$$

where $\langle r_{ij}^{-6} \rangle$ represents an appropriate conformational average over the different distances in different conformers. We shall see shortly what type of average is appropriate under various circumstances. (Note, however, that Eq. 5.46 is *only* valid in the limiting cases of slow or fast internal motions, since only under these circumstances is the ratio given on the right-hand side independent of ω.)

Probably the most widely used approach in this area is the "jump model" for internal motion, which treats the system as mixture of different conformers that interconvert by jumping instantaneously from one conformer to another. This approach was first extended to the case of dipolar relaxation involving varying internuclear distances by Tropp,[41] who derived expressions for the spectral density function under several different sets of conditions (arbitrary numbers, populations and interconversion rates of conformers, isotropic and anisotropic overall tumbling, etc.). Several subsequent authors have used these equations, and some have re-expressed them in various ways. We shall base our discussion on versions presented by Yip and Case,[42] whose expression for the spectral density function applicable to spins i and j in a system of N different conformations of the same molecule undergoing dynamic exchange is[†]

$$J'_{ij}(\omega) = \frac{1}{5} \sum_{n=-2}^{2} \left[\int_{-\infty}^{\infty} \sum_{\mu=1}^{N} \sum_{\nu=1}^{N} \frac{Y_{2n}(\phi^{\text{mol}}_{ij,\nu}) Y^*_{2n}(\phi^{\text{mol}}_{ij,\mu})}{r^3_{ij,\nu} r^3_{ij,\mu}} \langle P_\nu \rangle \exp(\mathbf{A}\tau)_{\mu\nu} \exp(-\tau/\tau_c) \exp(-i\omega\tau)\, d\tau \right] \tag{5.47}$$

Although this expression is of little or no practical use in itself (for reasons

---

[†]We have modified Eqs. 5.47, 5.48 and 5.51 slightly from the form in which they appear in ref. 42 to make them compatible with the form of the spectral density function used in other equations in this book. Specifically, we have changed the integral over the time-domain correlation function from the form $\int_0^\infty \cos(\omega\tau)\, d\tau$ to the form $\int_{-\infty}^\infty \exp(-i\omega\tau)\, d\tau$, which implies introduction of an additional factor of 2 into $J(\omega)$ relative to its form in ref. 42 (justified in ref. 43).

we shall see shortly), we can use it as a very powerful aid to understanding what happens to NOE interactions as a result of internal motions. While we shall make no pretense of deriving Eq. 5.47 formally, it clearly does demand some explanations and definitions.

The terms $r_{ij,\mu}$ and $r_{ij,\nu}$ are the (potentially different) internuclear distances between the same pair of spins i and j in two particular, different conformations. These conformations are labelled μ and ν (we have adopted these symbols, rather than the superscripts a and b used as labels in the two-site exchange equations of earlier sections, because here we are dealing with indices in a summation over potentially many conformations). The orientation of the ij internuclear vector in each of these two conformations is described in terms of the polar angles θ and ϕ, in a frame of reference fixed with respect to the molecule (it does not actually matter what molecular frame is chosen, as long as it is consistent throughout the calculation). The terms $Y_{2n}(\phi_{ij,\mu}^{\text{mol}})$ and $Y_{2n}^*(\phi_{ij,\nu}^{\text{mol}})$ are second-order spherical harmonic functions of these polar angles, the "mol" superscript indicates that they are defined in terms of angles in the molecular frame, and the outermost summation, $\Sigma_{n=-2}^{2}$, runs over the second index (n) of these spherical harmonic functions. These spherical harmonic functions enter the picture because they are very convenient to use when representing the angular part of the dipolar Hamiltonian. They also appear (explicitly, rather than symbolically) in Eqs. A44a–f of appendix II, and more details of this type of manipulation can be found there, as well as in reference 44. The inner pair of summations, $\Sigma_{\mu=1}^{N} \Sigma_{\nu=1}^{N}$, runs over all pairs of conformations μ and ν (and includes all cases where both indices are the same).

The lower line within the bracket of Eq. 5.47 contains all the time-dependent terms within the integral. If the term $\langle P_\nu \rangle \exp(\mathbf{A}\tau)_{\mu\nu}$ were omitted, the integral would then comprise just the Fourier transform of the exponentially decaying correlation function for isotropic overall tumbling (Eq. 2.16); in other words, it would then simply represent the spectral density function for overall tumbling (Eq. 2.20). The representation of population flux between the conformers by the term $\langle P_\nu \rangle \exp(\mathbf{A}\tau)_{\mu\nu}$ in the derivation of Eq. 5.47 is very closely analogous to that used in Chapter 4 to represent the kinetic evolution of NOE enhancements (cf. Eqs. 4.11–4.16). The term $\langle P_\nu \rangle$ represents the equilibrium population of conformer ν, and the rate matrix $\mathbf{A}$ comprises all the first-order rate constants for interconversions between conformers. Thus the combined term $\langle P_\nu \rangle \exp(\mathbf{A}\tau)_{\mu\nu}$ expresses the number of molecules initially in conformer ν that will have jumped to conformer μ during the time τ, and similarly the term $\langle P_\nu \rangle \exp(\mathbf{A}\tau)_{\nu\nu}$ expresses the number of molecules initially in conformer ν that will still remain in conformer ν at the end of the time τ.

We may now see something of what is going on inside Eq. 5.47. Dipolar relaxation is caused by changes in the orientation of the ij internuclear vector, which in turn may be caused by molecular tumbling or by jumps between conformations. The effect of such orientational changes on the dipolar interaction is expressed through the spherical harmonic functions. However, in Eq. 5.47 these functions have been transformed into the molecular frame of reference, so the contribution from overall tumbling is separated out. The effect of changes in internuclear separation appears through the terms $r_{ij,\mu}^{-3}$ and $r_{ij,\nu}^{-3}$. Perhaps the most instructive aspect of Eq. 5.47, however, is the effect that the rate matrix **A** exerts on the spectral density function. In general, it is most unlikely that one would ever know all the elements of the rate matrix **A**. However, in the limits of fast or slow internal motion relative to molecular tumbling, the rates drop out from the equations; this is rather similar to the way in which the values of the rate constants drop out from the general Eq. 5.8 to yield Eq. 5.9 for rates slow relative to T_1, and to yield Eq. 5.12 for rates fast relative to T_1 (Section 5.2.1). These two cases will be considered in turn in the following two sections.

5.5.1. Internal Motions Slower than Overall Tumbling: "r^{-6} Averaging"

For jumps between conformers on a timescale slow relative to overall tumbling, all the off-diagonal elements of **A** (i.e., the rates) are small relative to τ_c. Effectively, this removes all the mixed terms from the double summation over conformers. All the mixed terms $\exp(\mathbf{A}\tau)_{\mu\nu}(\mu \neq \nu)$ will be small, since very few molecules will jump between conformers during a time τ comparable to τ_c, while all the diagonal terms such as $\exp(\mathbf{A}\tau)_{\mu\mu}$ will be close to unity, for the same reason. [Note that, although times longer than τ_c are included in the integral, they do not contribute appreciably to the overall result because for them the value of $\exp(-\tau/\tau_c)$, also included in the integral, is very small.] More formally, the foregoing means we can approximate the term $\exp(\mathbf{A}\tau)_{\mu\nu}$ in Eq. 5.47 by the identity matrix. Note also that, provided the conformer exchange rates are *sufficiently* slow relative to overall molecular tumbling, their actual values do not matter; they only cause significant deviations from 0 (off-diagonal elements) or 1 (diagonal elements) for the terms $\exp(\mathbf{A}\tau)_{\mu\nu}$ if they become comparable to the tumbling rate. Physically, this all corresponds to saying that conformational averaging is taking place on a timescale slow relative to that for dipolar relaxation events, which effectively is defined by τ_c.

If, in addition, we impose the condition that all conformers are equally populated (as would be the case for an internal motion that interconverts members of an equivalent group of spins, such as flipping of a symmetrical aromatic ring in a protein sidechain), then we can replace all the populations $\langle P_\nu \rangle$ by $1/N$. Putting all of the above together, for the case of slow jumps between equivalent sites, Eq. 5.47 becomes

$$J'_{ij}(\omega) = \frac{1}{5N} \sum_{n=-2}^{2} \left[\sum_{\mu=1}^{N} \frac{Y_{2n}(\phi_{ij,\mu}^{\mathrm{mol}}) Y_{2n}^{*}(\phi_{ij,\mu}^{\mathrm{mol}})}{(r_{ij,\mu}^{3})^2} \int_{-\infty}^{\infty} \exp(-\tau/\tau_c)\exp(-i\omega\tau)\, d\tau \right] \quad (5.48)$$

A lot of simplifications can be made to this equation. The integral is now purely the spectral density function for overall tumbling, and because all of the spherical harmonics now relate to the same variables (i.e., the polar angles of the ij internuclear vector in just one conformation, μ) we may just add them up using the spherical harmonic addition theorem (which states that the sum of all the spherical harmonic functions of a given order is just some number, the value of which depends on how the functions were normalized). This gives

$$J'_{ij}(\omega) = \frac{1}{N}\left(\frac{2\tau_c}{1+\omega^2\tau_c^2}\right)\sum_{\mu=1}^{N} r_{ij,\mu}^{-6} \qquad (5.49)$$

This limit is very often referred to as the "r^{-6} averaging limit." The corresponding effective distance "sensed" by the NOE can be obtained following Eq. 5.46, to give[45]

$$r_{\text{effective}} = \left(\frac{1}{N}\sum_{\mu=1}^{N} r_{ij,\mu}^{-6}\right)^{-1/6} \qquad (5.50)$$

5.5.2. Internal Motions Faster than Overall Tumbling

The other limiting case of internal motion that it is essential to consider is that of fast jumps. Under these circumstances, conformational jumps bring the populations to their equilibrium values on a timescale that is short relative to τ_c. Therefore the terms $\langle P_\nu \rangle \exp(A\tau)_{\mu\nu}$ in Eq. 5.47 can be approximated as products of two equilibrium populations, $\langle P_\nu \rangle \langle P_\mu \rangle$. If, as before, we impose the condition that all conformers are equally populated (as, for instance, in the case of a methyl group), this means all the terms $\langle P_\nu \rangle \exp(A\tau)_{\mu\nu}$ become simply $1/N^2$, leading to

$$J'_{ij}(\omega) = \frac{1}{5N^2}\sum_{n=-2}^{2}\left[\sum_{\mu=1}^{N}\sum_{\nu=1}^{N}\frac{Y_{2n}(\phi_{ij,\nu}^{\text{mol}})Y_{2n}^*(\phi_{ij,\mu}^{\text{mol}})}{r_{ij,\nu}^3 r_{ij,\mu}^3}\int_{-\infty}^{\infty}\exp(-\tau/\tau_c)\exp(-i\omega\tau)\,d\tau\right] \qquad (5.51)$$

As in the case of Eq. 5.49, the part of the expression corresponding to the spectral density function for overall tumbling is now separable, so we may use Eq. 5.46 to obtain an expression for $r_{\text{effective}}$. Doing this, and also following further manipulations given by Yip and Case,[42] leads to

$$r_{\text{Tropp}} = \left[\frac{1}{5}\sum_{n=-2}^{2}\left|\frac{1}{N}\sum_{\mu=1}^{N}\frac{Y_{2n}(\phi_{ij,\mu}^{\text{mol}})}{r_{ij,\mu}^3}\right|^2\right]^{-1/6} \qquad (5.52)$$

We have here introduced the symbol r_{Tropp} to denote the value of $r_{\text{effective}}$ that results for this motional limit using Tropp's equations. This expression emphasizes the separate contributions from radial and angular terms and is thus simpler to deal with conceptually; however, again following Yip and Case, a different form can be written that is much simpler to calculate[42]

$$r_{\text{Tropp}} = \left\{ \frac{1}{2N^2} \sum_{\mu=1}^{N} \sum_{\nu=1}^{N} \left[\frac{3(\mathbf{r}_{ij,\mu} \cdot \mathbf{r}_{ij,\nu})^2 - r_{ij,\mu}^2 r_{ij,\nu}^2}{r_{ij,\mu}^5 r_{ij,\nu}^5} \right] \right\}^{-1/6} \quad (5.53)$$

By far the most important application for this equation is to methyl groups, where, of course, $N = 3$ and the three internuclear distances r_{ij} are the distances from each of the three methyl protons to some external fourth spin.

We may see from the above that there are two crucial differences between this fast motion limit and the r^{-6} averaging case considered in the previous section. First, the internuclear distances appear in Eq. 5.52 as averages over r_{ij}^{-3} rather than r_{ij}^{-6}. Second, there is a geometric dependence to the interaction, which appears in the angular (spherical harmonic) terms in Eq. 5.52 and the vector dot product in Eq. 5.53. This angular part can only reduce the size of the NOE (increase the size of r_{Tropp}) relative to that expected on the basis of r^{-3} averaging alone.

$$r_{\text{Tropp}} \geq \left(\frac{1}{N} \sum_{\mu=1}^{N} \frac{1}{r_{ij,\mu}^3} \right)^{-1/3} \quad (5.54)$$

This can be demonstrated by some straightforward manipulations from Eq. 5.53, given that the maximum value of $(\mathbf{r}_{IiS} \cdot \mathbf{r}_{IjS})^2$ is $r_{IiS}^2 r_{IjS}^2$.

In fact, the origin of the angular term in these equations is very closely related to the effects of fast internal motions considered in Section 5.4. In the case of a methyl group, fast jumps between rotamers cause the correlation function to fall very quickly to some lower value; subsequent slower decay to zero then occurs as a result of overall tumbling. The extent of the initial fall depends upon how much the orientation of the ij vector (connecting a methyl proton to a fourth spin outside the methyl group) changes as a result of jumping between rotamers (i.e., the initial fall is governed by the extent to which inequality 5.54 differs from equality). This is the information contained in the spherical harmonic terms. If the fourth spin is positioned on the axis of methyl group rotation and is quite close to the methyl protons, then the change in orientation of r_{ij} is at its greatest, and so is the loss of NOE intensity due to the angular term in Eqs. 5.52 and 5.53. This situation is equivalent to a low value of S^2 (Section 5.4). For most other sterically accessible locations the loss is much smaller, and near the plane of the methyl protons it is negligible.

There is a fairly common school of thought that summarizes the results described in this section as "when internal motion is faster than the overall tumbling, distances average as r^{-3}." This can be *nearly* true. In particular, it is nearly true in cases similar to that described at the end of the previous paragraph, where the distances to the exchanging protons change by much more than the angles; however, it should be clear from the foregoing (particularly inequality 5.54) that it is by no means generally true.

Other approaches have been taken to the theory of the averaging of NOE enhancements by fast internal motion,[46,47] but the advantage of the equations presented here, at least in the case of methyl rotations, is that they are relatively simple to calculate using a set of coordinates that are static, apart from jumps between methyl rotamers.

5.6. ALLOWING FOR AVERAGING

In the early parts of this chapter (Sections 5.1–5.3) we were largely concerned with conformational equilibria where fairly detailed structural models were available for each individual contributing conformer. The underlying question addressed was "how much can NOE studies tell us about the detailed molecular nature of complicated conformational exchange equilibria," and the answer given, broadly, was "not much." Making use of the treatment of internal motions given in Sections 5.4 and 5.5, we now consider the issue of conformational exchange from a different angle. Given that conformational exchange occurs, and that we cannot in general know much about the nature or populations of the contributing conformers, what does this imply for the interpretation of NOE data? The discussion is geared particularly toward the problems of macromolecular structure determination, where initial rates are very often used to estimate internuclear distances. Here the problem becomes "if we see an NOE enhancement between two protons in a conformationally mobile system, what can we confidently say about their separation?"

As we have seen, for a given pair of protons I and S, internal motions can change both the internuclear distance r_{IS} (a radial effect) and also the orientation of the internuclear vector relative to the molecular frame (an angular effect). As discussed in Sections 5.4 and 5.5.2, the angular component of such motions can affect NOE enhancements strongly, but only if the motions are faster than overall tumbling of the molecule. For biological macromolecules, where many internal motions are faster than overall tumbling, this can often lead to reduced values of some NOE enhancements. To give a specific example, in studying lysozyme, Olejniczak et al.[48] measured σ (i.e., the NOE enhancement after short irradiation time) for several proton pairs whose separation was fixed by covalent geometry. The data given in Table 5.5 show that measured values of σ between the $C^{\gamma 1}H_2$ protons of Ile98 and between the $C^\beta H_2$ protons of Met105

TABLE 5.5 Cross-Relaxation Rates in Lysozyme[48]

Residue	Proton Pair	Calculated σ^a	Experimental σ^b
Trp-28	$H^{\varepsilon 3}$–$H^{\varepsilon 3}$	-2.76 ± 0.55	-2.5 ± 0.3
	$H^{\varepsilon 3}$–H^η	-1.92 ± 0.4	-1.8 ± 0.2
Ile-98	$H^{\gamma 12}$–$H^{\gamma 11}$	-18.7 ± 3.4	-8.7 ± 0.9
Met-105	$H^{\beta 1}$–$H^{\beta 2}$	-18.7 ± 3.4	-6.8 ± 0.7

^{a}Using interproton distances derived from the crystal structure with standard geometry, assuming $\tau_c = 10 \pm 2$ ns, 498 MHz.
^{b}Derived from best fit to observed NOE enhancements vs. preirradiation time.

are smaller by about a factor of 2 than those calculated for a rigid model; in contrast, σ for proton pairs on the aromatic ring of Trp28 fit quite closely with calculated values for a rigid model. This presumably reflects the presence of fast local motions of the methylene groups, which would be expected to be much more extensive than internal motions (faster than overall tumbling) of the large aromatic ring of Trp. Olejniczak et al.[48] present some calculations of the possible extent of such motions required to explain the data, concluding that rapid angular fluctuations through roughly 60° may be involved.

For NOE enhancements originating in highly flexible or unfolded regions of a biomacromolecule, such effects can be so profound that they can render NOE enhancements essentially undetectable. (Of course, in such cases the problem may be compounded by the fact that few distances are consistently short enough to lead to an NOE enhancement anyway!) When the squared order parameter, S^2, for a given internuclear vector $\mathbf{r}_{ij}$ is low, and its internal motion is faster than overall molecular tumbling, then effectively it is as though the IS interaction occurred in a more rapidly tumbling (i.e, smaller) molecule. NOE interactions between protons in highly mobile regions thus build up more slowly than others because $J(\omega)$ (and therefore also σ_{IS}) is smaller (cf. Eqs. 4.26 and 4.27 and Fig. 4.8). Therefore, these interactions are less intense after a given mixing time in a transient NOE experiment than are those originating from more rigid parts of the molecule.

The clear implication from all this is that the absence or weakness of a particular NOE connectivity *cannot be taken to indicate that the two corresponding protons are necessarily far apart*. This is the origin of the almost universal practice of defining NOE restraints for structure calculations as an upper bound on distance, rather than simply defining a fixed distance restraint calculated directly from the NOE intensity (see also Sections 4.4.2 and 12.2.3).

Use of upper bounds as just outlined is by far the simplest, most widely used, and most effective measure adopted for allowing for conformational averaging during NOE studies. However, there have been some more detailed studies that cast further light on the issues involved and they deserve mention here. One of the first approaches to assessing the effect of conformational heterogeneity was the "uniform averaging" model proposed by Braun et al.[49] In this model, it is assumed that conformational averaging results in a uniform spread of values for the distance r_{ij}, between two cut-off values r_1 and r_2, and that averaging occurs over values of r_{ij}^{-6}. For a continuous distribution of distances, we can rewrite Eq. 5.37 (which dealt with a continuous distribution of angles) to give

$$\langle \sigma_{IS} \rangle = \frac{\int f(r) \sigma_{IS}(r)\, dr}{\int f(r)\, dr} \qquad (5.55)$$

where $f(r)$ is the distribution function of r, and the division by $\int f(r)\, dr$ is required for normalization. Since $\sigma = \zeta r^{-6}$ (Eq. 4.26), and $f(r)$ is here taken to be a constant A (to represent the assumed uniform distribution), this gives

$$\langle \sigma_{IS} \rangle = \frac{\zeta \int_{r_1}^{r_2} r^{-6} A\, dr}{\int_{r_1}^{r_2} A\, dr} = \frac{\zeta}{5(r_2 - r_1)} \left(\frac{1}{r_1^5} - \frac{1}{r_2^5} \right) \qquad (5.56)$$

where (Eq. 4.27)

$$\zeta = \left(\frac{\mu_0}{4\pi} \right)^2 \frac{\hbar^2 \gamma^4}{10} \left(\frac{6\tau_c}{1 + 4\omega^2 \tau_c^2} - \tau_c \right)$$

This equation expresses quantitatively what one would expect intuitively from the starting assumptions of the model: For a uniform distribution of internuclear distances averaged as r^{-6}, the dipolar relaxation would be dominated by the closest distances. Distances derived from NOE intensities under these circumstances would be shorter than the true distances and, in general, a set of such values may not be compatible with any single structure. However, in some ways this model is rather unrealistic. For one thing, any distance distribution based on a potential energy well would have lower probabilities for distances either shorter or longer than the equilibrium distance r_0, and so the shorter distances would contribute less strongly to the average than in the uniform averaging case. Also, given that the uniform averaging model does not consider the angular component of averaging and is based on r^{-6} averaging of distances, it is clear that its application is limited to internal motions slower than overall tumbling. For motions faster than overall tumbling, angular averaging can reduce enhancements significantly (see preceding and following).

Several authors have used molecular dynamics simulations of proteins to estimate the effects of fast internal motions on NOE enhancements. From a dynamics trajectory, one can calculate predicted values of NOE enhancements, explicitly taking into account the effect of internal motions during the trajectory. This approach has been adopted, for instance, by Olejniczak et al.,[29] by Palmer and Case,[50] by Post,[51] and most recently by Schneider et al.[52] Just as was discussed in Section 5.4, internal motions faster than overall tumbling cause the correlation function to decay to a "plateau value," whereafter slower decay follows as a result of overall tumbling of the molecule. If the separation of these timescales is sufficiently extreme, then one may make two simplifying assumptions: (i) the contribution to relaxation arising during the initial decay to the plateau value is short enough that it can be ignored, and (ii) overall rotation of the molecule is sufficiently slow that it is not sampled in the dynamics trajectory and thus need not be considered. Under these circumstances, the necessary equation for the "plateau value" of the internal motion correlation function (which has similarities to Eqs. 5.47 and 5.51) is

$$g'_{ij}(\tau)\big|_{\tau \to \tau_{plateau}}^{limit} = \sum_{n=-2}^{2} \left|\left\langle \frac{Y_{2n}(\Omega_{ij}, 0)}{r_{ij}^3(0)} \right\rangle\right|^2 = S_{ij}^2 \langle r_{ij}^{-3} \rangle^2 \quad (5.57)$$

where, as with $J'_{ij}(\omega)$, the prime on $g'_{ij}(\tau)$ indicates that internuclear distance is incorporated into the definition [unlike the definition of $g(\tau)$ given in Eq. 2.16], and the ij subscript indicates the value is specific to a given proton pair; the angle brackets indicate a time correlation function and Ω_{ij} represents the polar angles ϕ_{ij} and θ_{ij}. The separation into the terms S^2 (angular averaging) and $\langle r^{-3} \rangle$ (radial averaging) shown on the right of Eq. 5.57 is only possible on the assumption that the angular and radial components of averaging are uncoupled. It is also worth noting that although the discussion given here is in terms of interproton NOE interactions, the theory applies equally to heteronuclear NOE enhancements; the theory in Section 5.4 ignored radial terms only because it concerned interactions where the distance r_{ij} is fixed by covalent geometry.

Fourier transformation of $g'_{ij}(\tau)$ yields $J'_{ij}(\omega)$. Using $J'_{ij}(\omega)$, predicted values of σ_{ij} can be calculated directly using the relevant mean interproton distance from the dynamics trajectory. Post[51] also defines a radial averaging parameter R, equivalent in its effect to S^2, in the sense that it describes the scaling of the NOE resulting from averaging:

$$R = \frac{r_{model}^{-6}}{\langle r^{-3} \rangle^2} \quad (5.58)$$

This parameter depends on the choice of model structure used as a reference. In the study by Post,[51] the minimized mean structure from the dynamics trajectory coordinates was used. The combined effect of angular and radial averaging is then given by the total averaging parameter Q:

$$Q = S^2 R \quad (5.59)$$

The conclusions from such studies are not wholly consistent. As we have seen, angular averaging can only reduce NOE enhancements ($0 < S^2 < 1$). However, despite what one might expect at first sight, radial averaging can either increase or decrease enhancements (using the definition in Eq. 5.58), and the overall effect represented by Q can either be to enhance or diminish NOE intensities. LeMaster,[53] Olejniczak,[29] and Abseher[54] all suggest that there is significant cancellation between radial and angular averaging terms from internal motions, with the result that NOE estimates of distance are subject to smaller errors than one might otherwise expect. However, the study by Post,[51] employing a much larger number of interactions so as to allow a statistical evaluation (2778 single proton–single proton interactions and 1854 methyl–

single proton interactions in a dynamics simulation over a period of 102 ps for lysozyme), concluded that such cancellation was not so widespread.

Nonetheless, all authors agree that the number of cases where internal motion would lead to significant errors is fairly small. Post[51] concluded that only 11% of single proton–single proton interactions were predicted to have more than 10% distance error relative to the model structure (i.e., $Q < 0.6$ or $Q > 1.7$). On the other hand, 32% of methyl–single proton interactions had >10% error, because for these interactions the radial contribution usually reduces enhancements, reinforcing the effect of angular averaging. This does at least mean that these particular errors would mainly lead to overestimated distances, minimizing the harm done to structure determinations. Post[51] gives other more detailed breakdowns of how the different types of error are distributed over different types of interaction (interior vs. surface pairs, interresidue vs. short-range vs. long-range pairs, etc.). When a semiquantitative protocol employing division of NOE intensities into strong, medium, or weak categories is used, with each category corresponding to a particular upper bound, errors can only arise if motional averaging increases the NOE intensity to the extent that it is incorrectly placed in a higher category. Most reassuringly, the results suggest that fewer than 1% of interactions would suffer this problem, as is illustrated in Figure 5.9. Similar results were also obtained subsequently by Edmondson.[55]

These results are certainly comforting, as they suggest that the widely employed methodology for structure calculation from NOE data is unlikely to generate many significant errors. However, we should also bear in mind that these studies are necessarily limited to the rather rapid motions that can be sampled in a dynamics trajectory. While motions in the core of well-structured proteins are probably well modelled, significantly more flexible portions of molecules can give very weak NOE enhancements due to extensive angular averaging on timescales beyond those sampled in the dynamics studies. Thus, the essential motive for using upper bounds or broad distance ranges rather than precise distances when defining NOE-based restraints for structure calculations remains valid.

5.7. THE TRANSFERRED NOE

The transferred NOE or TRNOE is a particular example of the NOE in the presence of exchange, in this case the exchange of a ligand molecule between free solution and a bound state in which it is complexed to a large receptor molecule, such as a protein. To quote Clore and Gronenborn: "the aim of the TRNOE is to measure negative NOEs on the easily detectable free or [averaged] ligand resonance ... in order to obtain conformational information on the bound ligand."[13] The experiment is normally carried out in the presence of a large excess of free ligand, but still gives information on the bound state because the latter dominates the NOE kinetics; it can thus be said to provide "chemical amplification" of the NOE from the bound species.[56] It turns out to

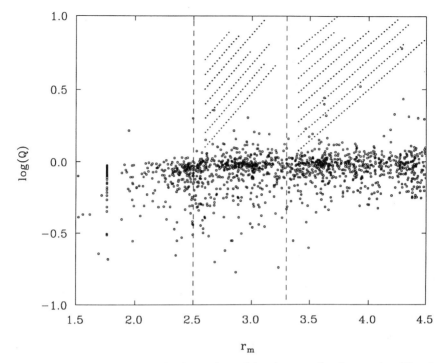

Figure 5.9. Errors in NOE-derived restraints arising from motional averaging. The plot shows values of log(Q) (the combined radial and angular averaging parameter defined in Eq. 5.59) plotted against the corresponding true distance, taking data for 1771 interior proton pairs from the molecular dynamics simulation of lysozyme referred to in the text. Broken lines at 2.5 Å and 3.3 Å separate short-, medium-, and long-distance categories, which correspond to strong, medium, and weak NOE intensities (at mixing times for which the initial rate approximation is valid). For NOE interactions corresponding to points inside the shaded areas, dynamic averaging increases cross-relaxation sufficiently that these interactions would be consigned to an incorrect category, resulting in a distance restraint that is too short. As can be seen, very few points (<1%) fall inside these shaded areas, indicating that such errors are expected to be rare. Reproduced with permission from ref. 51.

be a remarkably simple experiment to perform, and one of the few experiments capable of giving information on the conformation of bound ligands. However, it is also very hard to analyze quantitatively and, therefore, any structural information arising from the TRNOE needs to be treated very carefully.

Figure 5.10 shows the simplest system for which the theory of the TRNOE can be developed, that is, a two-spin system undergoing two-site exchange between free and bound states. Spins I and S are both protons on the ligand. The exchange equilibrium in this case is written

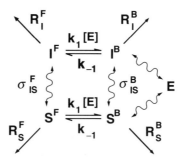

Figure 5.10. Model two-spin exchanging system involving a ligand molecule in free and bound states. Spin diffusion via the acceptor E is also indicated.

$$E + L \underset{k_{-1}}{\overset{k_1[E]}{\rightleftharpoons}} EL \tag{5.60}$$

where E represents the free acceptor (e.g., an enzyme), L the free ligand, and EL the complex between them. The equilibrium constant for this reaction (i.e., the binding constant) is

$$K = \frac{k_1}{k_{-1}} = \frac{[EL]}{[E][L]} \tag{5.61}$$

where k_1 and k_{-1} are the forward and backward rate constants, respectively, and $k_1[E]$ is therefore the pseudo-first-order forward rate constant at chemical equilibrium.

The TRNOE experiment can, in principle, involve signals that are in fast or slow exchange on the chemical shift timescale. In practice, applications are almost all to cases of fast exchange where the signals are averaged and we shall, therefore, limit our discussion to such cases. Early theoretical work on the TRNOE concentrated on 1D experiments, but more recent work has looked almost exclusively at 2D experiments, using relaxation matrix analysis (cf. Chapter 4).

The basis of the TRNOE experiment is that σ^B, the cross-relaxation rate constant in the slowly tumbling bound state, is generally much larger than σ^F, the corresponding cross-relaxation rate constant in the rapidly tumbling free state. Under conditions of fast exchange on the relaxation timescale, the averaged cross-relaxation rate $\langle\sigma\rangle$ is therefore dominated by σ^B, so that measured enhancements reflect the geometry of the ligand in its bound state. Thus, the TRNOE experiment in principle consists simply of a NOESY experiment (or, less commonly, a 1D NOE difference experiment), in which NOE enhancements measured on the ligand signals are used to provide information on the conformation of the ligand in the bound state. In practice such NOESY spectra often contain baseline distortions from the broad protein envelope, which can be attenuated by incorporating a T_2 filter.[57]

The preceding discussion makes clear the principal limitations to the applicability of the TRNOE experiment. Since we know that for fast exchange on the relaxation timescale, $\langle \sigma \rangle$ is given by the weighted average of free and bound rate constants:

$$\langle \sigma_{IS} \rangle = N^F \sigma_{IS}^F + N^B \sigma_{IS}^B \qquad (5.62)$$

the condition that $\langle \sigma \rangle$ should be dominated by σ^B simply corresponds to the inequality

$$|N^B \sigma_{IS}^B| \gg |N^F \sigma_{IS}^F| \qquad (5.63)$$

The extent to which this inequality is fulfilled depends on (i) the relative tumbling rates in the free and bound states, (ii) the distances r_{IS} in both states, and (iii) the affinity constant K for formation of the complex EL. If the complex is only *weakly* bound, than N^B will be small, and the TRNOE will disappear simply because there is insufficient ligand present in the bound state to contribute significantly to the overall relaxation. Clearly, the minimum value of K required for a viable TRNOE will vary from case to case according to the relative values of σ_{IS}^F and σ_{IS}^B. However, the more slowly tumbling the complex, the larger $|\sigma_{IS}^B|$ becomes, and consequently the more easily inequality 5.63 can be fulfilled. This is a rare example of an NMR experiment that becomes simpler as macromolecule size increases. In many real cases, the free ligand is a peptide and has a correlation time such that σ_{IS}^F is close to the crossover point, which makes inequality 5.63 particularly easy to fulfil.

At the opposite extreme, the TRNOE depends also on the *rate* of exchange between free and bound states being sufficient for an appreciable magnetization flux between them (i.e., in practice, for the off-rate to be greater than the spin–lattice relaxation rate in the bound state)

$$k_1 \gg R_S^B \qquad (5.64)$$

which is roughly equivalent to[58]

$$k_1 \gg 100 \sigma_{IS} \qquad (5.65)$$

These inequalities are likely to be broken for tightly bound complexes (very roughly they cease to be true for K stronger than about μM). Thus, the range of K for which quantitative analysis of the TRNOE is possible is rather limited and, in general, an independent estimate of the off-rate is necessary. This is illustrated by Figure 5.11, which illustrates that there is a relatively narrow region of about 2 orders of magnitude in k_{-1} (or K) within which the TRNOE is large and independent of k_{-1}. On the right of the plot, the TRNOE tends

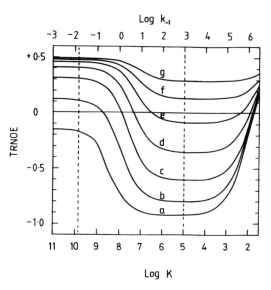

Figure 5.11. Dependence of steady-state TRNOE enhancement intensity on the ratio of total ligand to total receptor as a function of the off-rate k_{-1}. [L]/[E] = (a) 2, (b) 4, (c) 8, (d) 16, (e) 32, (f) 64, and (g) 128. For all curves $R_I^F = R_S^F = 0.5$ s^{-1}, $\sigma^F = 0.25$ s^{-1}, $R_I^B = R_S^B = 10$ s^{-1}, $\sigma^B = -10$ s^{-1}, $k_1 = 10^8$ M^{-1}s^{-1}, [E] = 5 × 10^{-4} M. The vertical dotted lines divide the regions of slow, intermediate, and fast exchange on the relaxation timescale. Reproduced with permission from ref. 13.

towards the isolated free-ligand enhancement as inequality 5.63 is broken. On the left of the plot, where inequalities 5.64 and 5.65 are broken, the TRNOE tends towards the weighted mean enhancement (cf. Eq. 5.14), where the bound form does not dominate the NOE kinetics nearly as strongly. Similar results would be obtained for 2D intensities, but with a maximum of around 38% rather than 50% (cf. Chapter 4).

Of the various parameters on which the TRNOE depends, the one that can most usefully be varied is the ratio N^F/N^B, which can be changed by changing the concentrations of ligand and receptor. The main requirement in choosing N^F/N^B is that inequality 5.63 be fulfilled. Within this limit, however, there are two reasons why a relatively large ratio of free to bound ligand concentration is desirable (ratios of around 15–20 are often used). The first is that higher concentrations of ligand give better signal-to-noise ratios; this will be especially beneficial in the common situation where the amount of receptor available (or soluble) is limited. Note that although the percentage enhancement falls as N^F/N^B increases, the *absolute* magnitude of the signal increases. The second reason is that because the bound cross-relaxation rate is effectively diluted by the fraction of free ligand, enhancements build up more slowly as the relative amount of free ligand is increased, so the experiment becomes slightly easier to carry out, and the resultant spectra may be less troubled by artifacts.

A number of theoretical accounts of the TRNOE have appeared. Since the first of these,[13] which follows an approach very similar to that used earlier in this chapter, several groups have used relaxation matrix methods, in which the standard relaxation matrix is supplemented by an exchange matrix. The simplest analysis assumes fast exchange (i.e., it assumes that inequality 5.64 holds), in which case the matrices are symmetrical.[59–62] More recent calculations relax this assumption and reveal considerable difficulties in simple-minded interpretations of TRNOE, as discussed below.[58] The TRNOE has been reviewed.[56]

The above account might imply that the TRNOE experiment is a simple and very powerful method for obtaining information on the conformation of ligands in the bound state. The reality is that although it is indeed powerful, it is by no means simple. Some problems with the method are discussed below. While these problems mean that the simple experiment should be hedged around with various control experiments, they do not preclude quantitative interpretations of TRNOE experiments and, indeed, now that a solid theoretical framework has been laid down for the experiment, we may expect further quantitative (and hopefully accurate) results.

The biggest problem with the TRNOE is the potential for spin diffusion and, in particular, spin diffusion from a bound ligand proton to the receptor and thence to a second bound ligand proton (spin diffusion within the bound ligand is less likely to be a severe problem, particularly at typical ligand/receptor ratios).[63] The result is an NOE enhancement that falsely suggests a close distance between the two ligand protons. The most commonly quoted test for this problem is to follow the build-up of the TRNOE with increase in NOESY mixing time; a TRNOE that results from spin diffusion will show a lag analogous to the lag seen in normal NOE build-up spectra (Chapter 4). In practice, however, this lag can be very hard to spot. Very similar effects can also occur when the free ligand is in large excess and tumbles rapidly, so that σ^F is positive, or when the exchange rate of the ligand is intermediate.[58,64]

One obvious way to avoid spin diffusion is to use a transferred ROESY experiment. In a transferred ROESY experiment, as for normal ROESY, "spin diffusion" or three-spin effects are of opposite sign and often very small; they are thus readily distinguished from direct effects. Such an experiment was used by Arepalli et al.[65] to show that an apparent TRNOE that had been interpreted as indicating a significant conformational change on binding was in fact due to spin diffusion. This is an elegant experiment, but removes much of the simplicity of the TRNOE approach, because in ROESY $|\sigma_{IS}^B|$ and $|\sigma_{IS}^F|$ are much more similar than they are in NOESY, so it is no longer true that $|N^B \sigma_{IS}^B| \gg |N^F \sigma_{IS}^F|$ (inequality 5.63). Therefore, careful controls are needed, generally requiring rather careful ROESY experiments on the ligand in the presence and absence of the receptor. Also, despite the fact that spin diffusion is less pronounced in ROESY than in NOESY (in that there are fewer indirect peaks), it remains true that direct cross-peak intensities can be reduced due to multispin cross-relaxation by just as much in ROESY as in NOESY (cf. Section 4.4.4). Therefore, overall, distance estimations are no less problematic for ROESY then for NOESY. It may thus be best to use NOESY for quantitative measurements, but check using ROESY for spin-diffusion problems.[66] An alternative way to avoid spin diffusion is to use perdeuterated protein, a method that may prove to be the most reliable in the long run.

A related problem with the TRNOE is that relaxation of the ligand resonances by the receptor will reduce the intensity of direct intraligand enhancements, and so increase apparent intraligand distances. Both the above problems are illustrated in Figure 5.12, where it is seen that the problem is exacerbated by slow exchange of the ligand (on the T_1 timescale).

Further problems can arise from weak secondary binding of the ligand to its receptor. This can obviously be a more severe problem at high N^F/N^B ratios. The presence of weak binding can best be detected by comparing results from

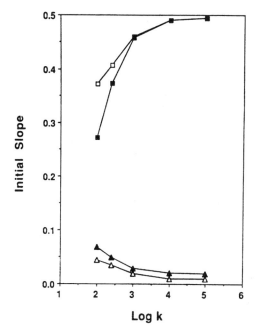

Figure 5.12. The dependence of the initial slope of TRNOESY build-up curves on the exchange rate k. NOE enhancements are calculated for a linear arrangement of three spins separated by 2.5 Å; $\nu = 500$ MHz, $\tau_c^B = 10^{-7}$ s, $\tau_c^F = 10^{-10}$ s, $N^F/N^B = 10$, $\rho^{*B} = 1.0$ s^{-1}, $\rho^{*F} = 0$. The initial slope was corrected for spin–lattice relaxation effects by dividing cross-peak intensities by the mean of the corresponding diagonal peak intensities. Results are calculated for mixing times of 10 ms (open symbols) or 100 ms (closed symbols) and are shown for the direct NOE enhancements (squares) and indirect enhancements (triangles). The slopes only approach their ideal values for $k > 10^4$. At lower exchange rates, the direct NOEs imply a much longer distance than they should, particularly for the 100 ms NOESY, while the indirect NOEs imply a much shorter distance than they should. Reproduced with permission from ref. 58.

TRNOESY experiments in the presence and absence of a strong competitive ligand. Weak secondary binding should be similar in the two cases, whereas the competitor should displace ligand from the primary binding site.[56] A further problem can arise from mobility of the ligand when bound. In particular, differential mobility of different parts of the ligand can lead to effective correlation times that are at least an order of magnitude shorter than expected, and therefore to highly misleading NOE enhancements.[61]

5.8. INTERMOLECULAR NOE ENHANCEMENTS INVOLVING WATER

Water has an important structural role in many biomolecules; for example, there are proteins that contain water molecules completely buried in internal cavities, and regular double-helical DNA has a spine of hydration down each of the grooves. It can also have an important functional role in many enzymes. It is therefore of importance to know where these hydration water molecules are and to be able to characterize their dynamics. NMR has been very useful in uncovering some of these details, in a way which is often complementary to the information provided by X-ray crystallography.

The experiments used to obtain information on hydration water are technically difficult, because exchange between hydration water and the bulk solvent is always so fast that the signals from hydration water come at the same chemical shift as bulk water. This requires careful design of solvent suppression to avoid destroying the signals of interest from protons cross-relaxing with the water, and multidimensional techniques to resolve them.[67–70]

The essential experiments used to detect and characterize water molecules are NOESY and ROESY. As shown elsewhere (Sections 4.2, 4.3, and 9.3), cross-relaxation rates in NOESY and ROESY have different dependencies on the correlation times of internuclear vectors. The cross-relaxation rate constant for ROESY is always positive (giving rise to positive NOEs, i.e., cross peaks in ROESY that have opposite phase to the diagonal), while in NOESY the cross-relaxation rate constant is positive for short correlation time but becomes negative at longer correlation time (giving rise to negative NOE signals, i.e., cross peaks in NOESY that have the same phase as the diagonal), as shown in Figure 5.13. By using a combination of NOESY and ROESY, it is therefore possible to gain information on the mobility of water molecules interacting with a biomolecule.

Cross-peaks in NOESY (or ROESY) spectra between bulk water and solute protons can arise from one of three possible mechanisms (Fig. 5.14a, b, and c respectively): direct enhancements from water; exchange-relayed enhancements (in which water protons undergo chemical exchange with solute protons, which then pass on an enhancement to other solute protons); or direct chemical exchange between water and solute.[67] The chemical exchange routes correspond to saturation transfer, as described in Section 5.1. The third route (direct chemical exchange) is easy to spot, because it gives rise to ROESY cross peaks in

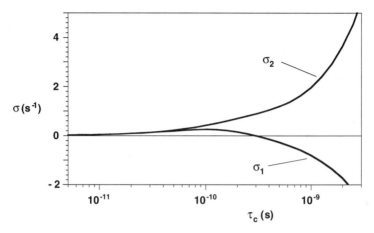

Figure 5.13. Cross-relaxation rate constants for NOE (σ_1) and ROE (σ_2) as a function of correlation time. Calculated for two protons at 600 MHz with a separation of 2.0 Å. Reproduced and modified with permission from ref. 71.

phase with the diagonal. The other two routes have so far proved indistinguishable by experiment. This means that one can only make confident assignments of direct water–solute NOE enhancements (case *a*) when the solute proton is far enough away from all exchangeable protons (in practice, Ser, Thr and Tyr hydroxyls, carboxylates, guanadinyl and amine groups) that case *b* can be ex-

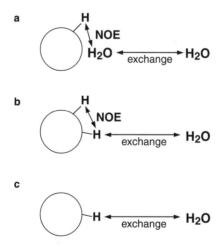

Figure 5.14. Possible origins of NOESY/ROESY cross peaks from water. (*a*) Direct NOE from hydration water molecules. (*b*) Chemical exchange from water to exchangeable solute proton, followed by intramolecular NOE ("exchange-relayed NOE"). (*c*) Chemical exchange from water to exchangeable proton. Reproduced and modified with permission from ref. 67.

cluded on purely structural grounds. This imposes a strong limitation on the number of hydration waters that can confidently be located by NMR.

However, if intermolecular NOE enhancements can be identified, the sign of the enhancement in NOESY and the relative intensity of NOESY and ROESY cross peaks allows one to make deductions about the mobility (more often described as the residence lifetime) of the water. If the NOESY cross peak is in phase with the diagonal (i.e., has a negative cross-relaxation rate constant, cf. Fig. 5.13), the effective correlation time of the solute–water vector must be long. The exact interpretation of this depends on the model used for water mobility and is an area of some debate, but essentially it implies that the residence time of the water must be longer than about 1 ns. Positive cross-relaxation rate constants in NOESY imply sub-nanosecond residence times.

These methods have been applied to a number of proteins and nucleic acids (reviewed in ref. 67). Most surface hydration water molecules exchange with lifetimes of <1 ns (but probably greater than 100 ps), while some buried water molecules can exchange more slowly, with lifetimes of between 1 ns and 1 ms.[71] Nuclear magnetic relaxation dispersion measurements suggest that the upper limit for these buried water molecules is probably in the microsecond range.[72] A number of comparisons have been made with water molecules determined by X-ray crystallography.[73] These indicate that even though water molecules may have well-defined positions in a crystal structure, individual water molecules can still exchange rapidly, and that there is often little correlation between the locations of water molecules identified by NMR and X-ray crystallography. Although water molecules are undoubtedly important for protein structure and function, they still remain rather difficult to characterize.

REFERENCES

1. Noggle, J. H.; Schirmer, R. E. "The Nuclear Overhauser Effect; Chemical Applications," Academic Press, New York, 1971.
2. Jeener, J.; Meier, B. H.; Bachmann, P.; Ernst, R. R. *J. Chem. Phys.* 1979, *71*, 4546.
3. Kalman, J. R.; Williams, D. H. *J. Am. Chem. Soc.* 1980, *102*, 906.
4. Feeney, J.; Heinrich, A. *J. Chem. Soc., Chem. Commun.* 1966, 295.
5. Stoesz, J. D.; Malinowski, D. P.; Redfield, A. G. *Biochemistry* 1979, *18*, 4669.
6. Whitehead, B.; Tessari, M.; Düx, P.; Boelens, R.; Kaptein, R.; Vuister, G. W. *J. Biomolec. NMR* 1997, *9*, 313.
7. Szántay Jr., C.; Demeter, A. *J. Magn. Reson.* 1995, *115*, 94.
8. Combrisson, S.; Roques, B.; Rigny, P.; Basselier, J. J. *Can. J. Chem.* 1971, *49*, 904.
9. Saunders, J. K.; Bell, R. A. *Can. J. Chem.* 1970, *48*, 512.
10. Bell, R. A.; Saunders, J. K. in "Topics in Stereochemistry" (N. L. Allinger and E. L. Eliel, Eds.), Vol. 7, Wiley-Interscience, New York, 1973, pp. 1–92.
11. Borzo, M.; Maciel, G. E. *J. Magn. Reson.* 1981, *43*, 175.

12. Schirmer, R. E.; Davis, J. P.; Noggle, J. H.; Hart, P. A. *J. Am. Chem. Soc.* 1972, *94*, 2561.
13. Clore, G. M.; Gronenborn, A. M. *J. Magn. Reson.* 1982, *48*, 402.
14. Kruse, L. I.; DeBrosse, C. W.; Kruse, C. H. *J. Am. Chem. Soc.* 1985, *107*, 5435.
15. Neumann, J. M.; Bernassau, J. M.; Guéron, M.; Tran-Dinh, S. *Eur. J. Biochem.* 1980, *108*, 457.
16. Landis, C.; Allured, V. S. *J. Am. Chem. Soc.* 1991, *113*, 9493.
17. Nikiforovich, G. V.; Prakash, O.; Gehrig, C. A.; Hruby, V. J. *J. Am. Chem. Soc.* 1993, *115*, 3399.
18. Wang, J.; Hodges, R. S.; Sykes, B. D. *J. Am. Chem. Soc.* 1995, *117*, 8627.
19. Cicero, D. O.; Barbato, G.; Bazzo, R. *J. Am. Chem. Soc.* 1995, *117*, 1027.
20. Bonvin, A. M. J. J.; Brünger, A. T. *J. Biomolec. NMR* 1996, *7*, 72.
21. Kozerski, L.; Krajewski, P.; Pupek, K.; Blackwell, P. G.; Williamson, M. P. *J. Chem. Soc., Perkin Trans. 2* 1997, 1811.
22. PCMODEL, Molecular Modelling Software, Serena Software, P.O. Box 3076, Bloomington, IN47402-3076, U.S.A.
23. Case, D. A. *J. Biomolec. NMR* 1999, *15*, 95.
24a. Lipari, G.; Szabo, A. *J. Am. Chem. Soc.* 1982, *104*, 4546.
24b. Lipari, G.; Szabo, A. *J. Am. Chem. Soc.* 1982, *104*, 4559.
25. Hiyama, Y.; Niu, C. H.; Silverton, J. V.; Bavoso, A.; Torchia, D. A. *J. Am. Chem. Soc.* 1988, *110*, 2378.
26. Kay, L. E.; Torchia, D. A.; Bax, A. *Biochemistry* 1989, *28*, 8972.
27. Clore, G. M.; Szabo, A.; Bax, A.; Kay, L. E.; Driscoll, P. C.; Gronenborn, A. M. *J. Am. Chem. Soc.* 1990, *112*, 4989.
28. Levy, R. M.; Karplus, M.; McCammon, J. A. *J. Am. Chem. Soc.* 1981, *103*, 994.
29. Olejniczak, E. T.; Dobson, C. M.; Karplus, M.; Levy, R. M. *J. Am. Chem. Soc.* 1984, *106*, 1923.
30. Woessner, D. E. *J. Chem. Phys.* 1962, *37*, 647.
31. Brüschweiler, R.; Liao, X.; Wright, P. E. *Science* 1995, *268*, 886.
32. Kördel, J.; Skelton, N. J.; Akke, M.; Palmer III, A. G. M.; Chazin, W. J. *Biochemistry* 1992, *31*, 4856.
33. Kraulis, P. J. *J. Appl. Crystallogr.* 1991, *24*, 946.
34. van Mierlo, C. P. M.; Darby, N. J.; Keeler, J. H.; Neuhaus, D.; Creighton, T. E. *J. Mol. Biol.* 1993, *229*, 1125.
35. Peng, J. W.; Wagner, G. *J. Magn. Reson.* 1992, *98*, 308.
36. Markus, M. A.; Dayie, K. T.; Matsudaira, P.; Wagner, G. *Biochemistry* 1996, *35*, 1722.
37. Dayie, K. T.; Wagner, G. *J. Magn. Reson. B* 1995, *109*, 105.
38. Engelke, J.; Rüterjans, H. *J. Biomolec. NMR* 1995, *5*, 173.
39. Cordier, F.; Brutscher, B.; Marion, D. *J. Biomolec. NMR* 1996, *7*, 163.
40. Muhandiram, D. R.; Yamazaki, T.; Sykes, B. D.,; Kay, L. E. *J. Am. Chem. Soc.* 1995, *117*, 11536.
41. Tropp, J. *J. Chem. Phys.* 1980, *72*, 6035.

42. Yip, P. F.; Case, D. A. in "Computational Aspects of the Study of Biological Macromolecules by Nuclear Magnetic Resonance Spectroscopy" (J. C. Hoch; F. M. Poulsen, and C. Redfield, eds.), Plenum Press, New York, 1991, pp 317–330.
43. Abragam, A. "Principles of Nuclear Magnetism," Oxford University Press, Oxford, 1961, p. 272.
44. Harris, R. K. "Nuclear Magnetic Resonance Spectroscopy," Pitman, London, 1983, Appendix 3, pp 234–238.
45. Fletcher, C. M.; Jones, D. N. M.; Diamond, R.; Neuhaus, D. *J. Biomolec. NMR* 1996, *8*, 292.
46. Pegg, D. T.; Bendall, M. R.; Doddrell, D. M. *Aust. J. Chem.* 1980, *33*, 1167.
47. Keepers, J. W.; James, T. L. *J. Magn. Reson.* 1984, *57*, 404.
48. Olejniczak, E. T.; Poulsen, F. M.; Dobson, C. M. *J. Am. Chem. Soc.* 1981, *103*, 6574.
49. Braun, W.; Bösch, C.; Brown, L. R.; Go, N.; Wüthrich, K. *Biochim. Biophys. Acta* 1981, *667*, 377.
50. Palmer III, A. G.; Case, D. A. *J. Am. Chem. Soc.* 1992, *114*, 9059.
51. Post, C. B. *J. Mol. Biol.* 1992, *224*, 1087.
52. Schneider, T. R.; Brünger, A. T.; Nilges, M. *J. Mol. Biol.* 1999, *285*, 727.
53. LeMaster, D. M.; Kay, L. E.; Brünger, A. T.; Prestegard, J. H. *FEBS Lett.* 1988, *236*, 71.
54. Abseher, R.; Lüdemann, S.; Schreiber, H.; Steinhauser, O. *J. Am. Chem. Soc.* 1994, *116*, 4006.
55. Edmondson, S. P. *J. Magn. Reson. B* 1994, *103*, 222.
56. Ni, F. *Prog. Nucl. Magn. Reson. Spectrosc.* 1994, *26*, 517.
57. Scherf, T.; Anglister, J. *Biophys. J.* 1993, *64*, 754.
58. London, R. E.; Perlman, M. E.; Davis, D. G. *J. Magn. Reson.* 1992, *97*, 79.
59. Campbell, A. P.; Sykes, B. D. *J. Magn. Reson.* 1991, *93*, 77.
60. Lippens, G. M.; Cerf, C.; Hallenga, K. *J. Magn. Reson.* 1992, *99*, 268.
61. Nirmala, N. R.; Lippens, G. M.; Hallenga, K. *J. Magn. Reson.* 1992, *100*, 25.
62. Zheng, J.; Post, C. B. *J. Magn. Reson. B* 1993, *101*, 262.
63. Jackson, P. L.; Moseley, H. N. B.; Krishna, N. R. *J. Magn. Reson. B* 1995, *107*, 289.
64. Campbell, A. P.; Sykes, B. D. *Ann. Rev. Biophys. Biomol. Struct.* 1993, *22*, 99.
65. Arepalli, S. R.; Glaudemans, C. P. J.; Daves Jr., G. D.; Kovac, P.; Bax, A. *J. Magn. Reson. B* 1995, *106*, 195.
66. Lian, L. Y.; Barsukov, I. L.; Sutcliffe, M. J.; Sze, K. H.; Roberts, G. C. K. *Methods in Enzymology* 1994, *239*, 657.
67. Otting, G. *Prog. Nucl. Magn. Reson. Spectrosc.* 1997, *31*, 259.
68. Kubinec, M. G.; Wemmer, D. E. *Curr. Opinions Struct. Biol.* 1992, *2*, 828.
69. Gerothanassis, I. P. *Prog. Nucl. Magn. Reson. Spectrosc.* 1994, *26*, 171.
70. Belton, P. S. *Prog. Biophys. Mol. Biol.* 1994, *61*, 61.
71. Otting, G.; Liepinsh, E.; Wüthrich, K. *Science* 1991, *254*, 974.
72. Venu, K.; Denisov, V. P.; Halle, B. *J. Am. Chem. Soc.* 1997, *119*, 3122.
73. Billeter, M. *Prog. Nucl. Magn. Reson. Spectrosc.* 1995, *27*, 635.

CHAPTER 6

COMPLICATIONS FROM SPIN–SPIN COUPLING

6.1. DECOUPLING

Irradiation of a multiplet during acquisition causes complete or partial decoupling of its coupling partners. In a difference decoupling experiment, the decoupled signal appears as a pattern of positive and negative peaks with a net zero intensity (Fig. 6.1a). In a pattern such as this, a small NOE enhancement is usually unrecognizable. If the digital resolution is poor, the negative wings may be partially lost, giving a signal rather hard to distinguish from an NOE enhancement, especially if the signal-to-noise ratio is low. Furthermore, the integral of a poorly digitized decoupling difference signal may not be zero, depending on where the data points fall (Fig. 6.1b). For all of these reasons, decoupling effects in NOE difference experiments are most undesirable. Fortunately, they can very easily be excluded by gating the decoupler off during acquisition, making use of the fact that decoupling disappears effectively instantaneously on switching off the decoupler, whereas the NOE is a population phenomenon and takes a time of the order of T_1 to build up or decay away.

Perhaps the most important reason for gating the decoupler off during acquisition, however, is that this eliminates Bloch–Siegert shifts. These are small movements that the decoupler causes in the shifts of all the resonances, as a result of the change that the decoupling field makes to $\mathbf{B}_{\text{eff}}$.[1] The size of a particular Bloch–Siegert shift depends on the frequency separation between the decoupler and the affected resonance, so that the Bloch–Siegert shift produced by on-resonance irradiation and by off-resonance irradiation must necessarily be different. Thus, decoupling during acquisition would result in poor subtraction, even though such differential Bloch–Siegert shifts might amount to only

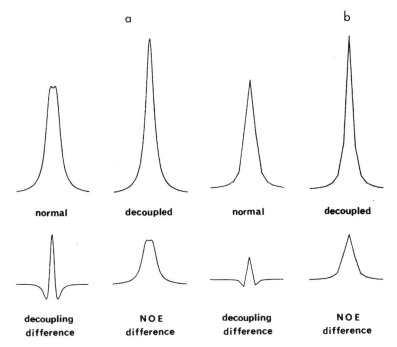

Figure 6.1. Simulated lineshapes in difference experiments at (*a*) high digital resolution, and (*b*) low digital resolution (same data). The decoupling difference pattern should have zero net integral in the absence of any NOE enhancement, but at low digital resolution the negative wings are partly lost so that a false positive integral can result.

some small fraction of a hertz at the low decoupler powers employed in NOE difference experiments.

In some older machines, switching off the decoupler may cause some transient disturbance during the very early part of the FID, resulting in a poor subtraction quality in the difference spectrum. This problem is most easily overcome by leaving a short delay between turning off the decoupler and acquiring the spectrum.

6.2. SELECTIVE POPULATION TRANSFER

When the lines of a multiplet are partially saturated to different extents, selective population transfer (SPT)[†] causes changes in the relative intensities of the multiplet components of *coupling partners* of the irradiated signal.[1,2] Thus,

---

[†]In the older literature, this effect is also known, rather confusingly, as a generalized Overhauser effect.

192 COMPLICATIONS FROM SPIN–SPIN COUPLING

when the difference spectrum is taken, each coupling partner contributes a pattern of positive and negative peaks, the overall integral of which is zero, and the lines of which (neglecting any distortions due to overlap) appear at the same frequencies as the lines in the normal multiplet. This is a quite different phenomenon from decoupling since it is purely a *population* effect. Thus once created, an SPT pattern takes a time of the order of T_1 to decay away so as to restore the equilibrium pattern.

It is important to realize that SPT effects and the NOE are *completely independent of one another* for weakly coupled systems (in the absence of cross-correlation; cf. Section 6.3). Consequently, SPT distortions at a particular multiplet do not affect the size or kinetics of any NOE enhancement that may also exist, apart from making it harder to recognize in the difference spectrum. To see why this is, we need to tackle some of the theory of SPT.

6.2.1. Theory

The origin of SPT is shown for an *AX* spin system in Figure 6.2. Saturation of *one of* the *A* transitions, for example, the 1,3 transition, causes an equalization of the populations of levels 1 and 3. This necessarily leads to a simultaneous change in the intensities of the connected *X* transitions; the population difference between levels 1 and 2 will decrease, whereas the population difference between levels 3 and 4 will increase by the same amount. When the difference spectrum is recorded, this population distribution results in a pattern of the form shown in Figure 6.3 (provided that an observe pulse with a low flip angle is used to excite the spectrum; see following). It is important to note that the intensity gained by one line is exactly matched by the intensity lost by the other, so that the *overall* integral of the *X* multiplet in the difference spectrum is zero (provided there is no NOE enhancement of *X*).

It follows that the simplest way in which to separate the NOE from SPT patterns, at least in principle, is to integrate the enhanced multiplet in the difference spectrum. However much the intensity of particular lines may be increased or decreased by SPT, these distortions balance one another out across

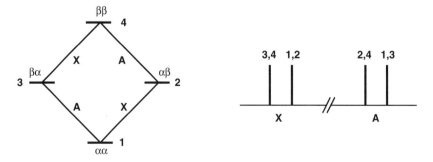

Figure 6.2. Weakly coupled two-spin system.

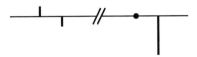

Figure 6.3. SPT effects in an *AX* system, saturating the *A* transition 1,3.

the multiplet as a whole, and the overall integral depends only on the NOE enhancement. However, this approach may be made unrealistic in practice by overlap between multiplets and the difficulty of integrating small enhancements in noisy or distorted spectra, and it is thus desirable to find a way of *suppressing* SPT, or at least of removing its effects from the difference spectrum. The practicalities of this are discussed in Section 7.2.5, but the theory that follows is intended to provide the background to how such methods work, and also to *why* the NOE is independent from SPT effects.

To progress much further in understanding SPT, we need to use the operator product formalism of Sørensen et al.[3] Although this sounds rather forbidding, it is actually quite straightforward, at least at the level at which we need to use it. We have already made extensive use of the quantities I_z and S_z in our equations, and their transverse equivalents I_x, I_y, S_x, and S_y also have fairly obvious meanings. The essence of the product operator formalism is to use these together with their various products (e.g., $2I_xS_z$, $2I_xS_y$, and $2I_zS_z$)* to describe fully the response of a (weakly coupled) spin system during an NMR experiment. The advantage of this approach is that the effects of pulses on these products are particularly easy to visualize, since each component of a particular product can be pictured separately. For example, a $90°_y$ pulse turns I_z into I_x,[†] and leaves S_y unaffected, so the effect of a $90°_y$ pulse on the product $2I_zS_y$ would be to turn it into $2I_xS_y$. Similarly, a $90°_y$ pulse would turn $2I_zS_z$ into $2I_xS_z$, and $2I_xS_z$ into $-2I_zS_x$.

We now need to consider what these products represent. Provided we only try to equate a particular operator product with a particular contribution to the spectrum, rather than trying to picture its significance at a molecular level, this can be quite simply answered using the following rules:

1. Only products having a *single transverse component* give rise to detected NMR signals (e.g., I_x, S_y, $2I_xS_z$, and $2I_zS_y$).

2. Products having more than one transverse component correspond to multiple quantum coherence, and are not detected (e.g., $2I_xS_x$, $2I_yS_x$, or for a three-spin system, $4I_xS_xX_x$ or $4I_xS_zX_x$).

---

*The normalization factors of 2 for two-spin products and 4 for three-spin products are required to maintain the internal consistency of the formalism.
†Following the sign convention of Sørensen et al., in which a positive field produces a clockwise precession as viewed up ($-\to+$) the field axis.

3. Products having one transverse term and one or several z terms represent antiphase signals. For example, $2I_xS_z$ represents transverse magnetization of I aligned along the x axis and *in antiphase with respect to the IS coupling*. In other words, the I doublet appears in the spectrum with one line up and the other line down, having an *overall* integral of zero. The equivalent pattern at S corresponds to the term $2I_zS_x$, and terms in y (e.g., $2I_yS_z$ or $2I_zS_y$) correspond to antiphase magnetization aligned along the y axis (i.e., antiphase signals in dispersion). For the term $-2I_xS_z$, the up–down pattern is inverted relative to that for $2I_xS_z$. In a three-spin system, several such antiphase products would be possible, depending on which and how many couplings appear in antiphase (Fig. 6.4).

4. Products having only z terms represent so-called "multispin order." For example, in the two-spin case $2I_zS_z$ represents the state obtained by inverting *one line* of the S doublet. Since such inversion causes simultaneous population changes across the connected I transitions, as discussed earlier, the $2I_zS_z$ state could equally well be created by inverting one of the lines of the I doublet. Note that $2I_zS_z$ is entirely a state of *longitudinal* magnetization. Just as for the antiphase single quantum terms in Figure 6.4, several permutations for such "zz" terms exist in higher spin systems, depending on which, and how many, lines are saturated or inverted.

It is these "zz" terms that are used to describe the behavior of SPT. As we have seen, NOE theory is generally expressed using single-spin longitudinal terms such as I_z and S_z. The action of the observe pulse in converting these into proportionally sized transverse components, which contribute to the signal actually detected, is usually implied rather than stated. In just the same way, "zz" terms such as $2I_zS_z$ are converted into transverse terms such as $2I_xS_z$ and $2I_zS_x$ by an observe pulse, and it is these terms that actually contribute the antiphase signals associated with SPT patterns in NOE difference spectra. However, there is one vital distinction between the effects of pulses on these different types of z magnetization. A 90° pulse creates the maximum signal from I_z or S_z, but it creates no signal whatever from zz terms. This is because they are turned into pure multiple quantum coherence, for instance $2I_zS_z$ becomes

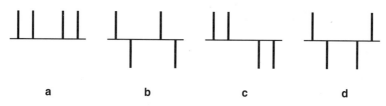

Figure 6.4. Stick spectra representing the operator products involving I_x for a three-spin system (I, S, and X, $J_{IS} < J_{IX}$). (a) I_x, (b) $2I_xS_z$, (c) $2I_xX_z$, (d) $4I_xS_zX_z$.

$2I_xS_x$ following a 90$_y^\circ$ pulse. Only for *non*-90° pulses do single quantum, detectable, antiphase terms result. Specifically, for a y pulse of flip angle α, the detectable parts of the magnetization produced from various possible z states are given by

$$I_z \to I_x \sin \alpha$$
$$2I_zS_z \to (2I_xS_z + 2I_zS_x)\sin \alpha \cos \alpha$$
$$4I_zS_zX_z \to (4I_xS_zX_z + 4I_zS_xX_z + 4I_zS_zX_x)\sin \alpha \cos^2\alpha \quad (6.1)$$

Thus, a 90° observe pulse abolishes SPT effects from the observed spectrum; this is the basis of one of the methods of SPT suppression described in Section 7.2.5.

There are a few other points worth making about these various z states before continuing. As already mentioned, it is only the *single-spin z* terms that contribute to the overall integral across a multiplet (e.g., I_z, S_z). This is why NOE theory is expressed using just these terms. Each of the possible longitudinal states has its own definition in terms of the various population differences within the spin system. For a two-spin system these definitions are

$$kI_z = N_{\beta\beta} + N_{\beta\alpha} - N_{\alpha\beta} - N_{\alpha\alpha}$$
$$kS_z = N_{\beta\beta} - N_{\beta\alpha} + N_{\alpha\beta} - N_{\alpha\alpha}$$
$$2kI_zS_z = N_{\beta\beta} - N_{\beta\alpha} - N_{\alpha\beta} + N_{\alpha\alpha} \quad (6.2)$$

while for a three-spin system they are

$$kI_z = N_{\beta\beta\beta} + N_{\beta\beta\alpha} + N_{\beta\alpha\beta} - N_{\alpha\beta\beta} + N_{\beta\alpha\alpha} - N_{\alpha\beta\alpha} - N_{\alpha\alpha\beta} - N_{\alpha\alpha\alpha}$$
$$kS_z = N_{\beta\beta\beta} + N_{\beta\beta\alpha} - N_{\beta\alpha\beta} + N_{\alpha\beta\beta} - N_{\beta\alpha\alpha} + N_{\alpha\beta\alpha} - N_{\alpha\alpha\beta} - N_{\alpha\alpha\alpha}$$
$$kX_z = N_{\beta\beta\beta} - N_{\beta\beta\alpha} + N_{\beta\alpha\beta} + N_{\alpha\beta\beta} - N_{\beta\alpha\alpha} - N_{\alpha\beta\alpha} + N_{\alpha\alpha\beta} - N_{\alpha\alpha\alpha}$$
$$2kI_zS_z = N_{\beta\beta\beta} + N_{\beta\beta\alpha} - N_{\beta\alpha\beta} - N_{\alpha\beta\beta} - N_{\beta\alpha\alpha} - N_{\alpha\beta\alpha} + N_{\alpha\alpha\beta} + N_{\alpha\alpha\alpha}$$
$$2kI_zX_z = N_{\beta\beta\beta} - N_{\beta\beta\alpha} + N_{\beta\alpha\beta} - N_{\alpha\beta\beta} - N_{\beta\alpha\alpha} + N_{\alpha\beta\alpha} - N_{\alpha\alpha\beta} + N_{\alpha\alpha\alpha}$$
$$2kS_zX_z = N_{\beta\beta\beta} - N_{\beta\beta\alpha} - N_{\beta\alpha\beta} + N_{\alpha\beta\beta} + N_{\beta\alpha\alpha} - N_{\alpha\beta\alpha} - N_{\alpha\alpha\beta} + N_{\alpha\alpha\alpha}$$
$$4kI_zS_zX_z = N_{\beta\beta\beta} - N_{\beta\beta\alpha} - N_{\beta\alpha\beta} - N_{\alpha\beta\beta} + N_{\beta\alpha\alpha} + N_{\alpha\beta\alpha} + N_{\alpha\alpha\beta} - N_{\alpha\alpha\alpha} \quad (6.3)$$

Using suitable admixtures of these terms, any population state of the system can be represented. One helpful way in which to picture each of these z states is in terms of stick spectra such as those shown in Figure 6.4, but it is vital to remember that these are purely longitudinal states. They are therefore associated equally with each of the spins included in the product, because they involve the *connected* transitions common to the interacting spins.

Suppose we saturate just one line of a multiplet. For the case of an *AMX* spin system in which each multiplet is a double doublet and the A^{13} line is saturated selectively, it is easy to show that the resulting population state is equivalent to the sum:

$$A^{13}_{\text{saturated}} \equiv A_z - 1/4(A_z + 2A_zM_z + 2A_zX_z + 4A_zM_zX_z) \qquad (6.4)$$

This illustrates a general point, namely that to describe a population state in which only one line of a multiplet is perturbed, *all* the possible longitudinal terms involving that resonance are required (except in the special case of irradiating the central line of a multiplet having an odd number of lines). Furthermore, if each line is saturated separately, and the resulting spectra are summed, only the single-spin terms will survive this addition. This is illustrated for the case of the *AMX* spin system; addition of the states corresponding to saturation of each line in turn results in a state in which A is one quarter saturated:

$$A^{13}_{\text{saturated}} \equiv A_z - 1/4(A_z + 2A_zM_z + 2A_zX_z + 4A_zM_zX_z)$$

$$A^{24}_{\text{saturated}} \equiv A_z - 1/4(A_z + 2A_zM_z - 2A_zX_z - 4A_zM_zX_z)$$

$$A^{34}_{\text{saturated}} \equiv A_z - 1/4(A_z - 2A_zM_z + 2A_zX_z - 4A_zM_zX_z)$$

$$A^{12}_{\text{saturated}} \equiv A_z - 1/4(A_z - 2A_zM_z - 2A_zX_z + 4A_zM_zX_z)$$

$$\text{Total}/4 \equiv (4A_z - A_z)/4 = 3A_z/4 \qquad (6.5)$$

This is the basis for another of the methods of SPT suppression described in Section 7.2.5. In the *summed difference spectrum*, A_z will appear undistorted with 1/4 of its normal intensity, while the coupling partners *M* and *X* will contribute no SPT patterns because all the terms involving them are cancelled out by the addition.

The discussion so far has dealt only with how *zz* states are created by uneven saturation, and converted into antiphase patterns by a non-90° pulse. In order to understand why SPT and the NOE are *independent*, we need to know how *zz* states are affected by relaxation. For the case of $2I_zS_z$ within a two-spin system, we may follow very much the same logic as in the derivation of the two-spin Solomon equation (Eq. 2.10). Following the population flux into and out of each state leads to

$$k\,d/dt(2I_zS_z) = -(W_{1I} + W_{1S} + W_2)n_{\beta\beta} + W_{1I}n_{\alpha\beta} + W_{1S}n_{\beta\alpha} + W_2 n_{\alpha\alpha}$$
$$+ W_2 n_{\beta\beta} + W_{1S}n_{\alpha\beta} + W_{1I}n_{\beta\alpha} - (W_{1I} + W_{1S} + W_2)n_{\alpha\alpha}$$
$$- W_{1S}n_{\beta\beta} - W_0 n_{\alpha\beta} + (W_0 + W_{1I} + W_{1S})n_{\beta\alpha} - W_{1I}n_{\alpha\alpha}$$
$$- W_{1I}n_{\beta\beta} + (W_0 + W_{1I} + W_{1S})n_{\alpha\beta} - W_0 n_{\beta\alpha} - W_{1S}n_{\alpha\alpha}$$
$$= 2(W_{1I} + W_{1S})(-n_{\beta\beta} + n_{\alpha\beta} + n_{\beta\alpha} - n_{\alpha\alpha}) \tag{6.6}$$

From this, given also that the equilibrium value of $2I_zS_z$ is zero, we have

$$d/dt(2I_zS_z) = -2(W_{1I} + W_{1S})(2I_zS_z - 2I_zS_z^0)$$
$$= -2(W_{1I} + W_{1S})2I_zS_z \tag{6.7}$$

By following a similar logic, and by analogy with the derivation of Eq. 3.9 from Eq. 3.1, we have for the three-spin case

$$d/dt(2I_zS_z) = -R_{IS}2I_zS_z - \sigma_{IX}2S_zX_z - \sigma_{SX}2I_zX_z \tag{6.8}$$

where

$$R_{IS} = R_I + R_S - W_{0IS} - W_{2IS}$$
$$= 2W_{1I} + 2W_{1S} + W_{0IX} + W_{2IX} + W_{0SX} + W_{2SX} \tag{6.9}$$

and similarly

$$d/dt(4I_zS_zX_z) = -R_{ISX}4I_zS_zX_z \tag{6.10}$$

where

$$R_{ISX} = 2W_{1I} + 2W_{1S} + 2W_{1X} \tag{6.11}$$

These equations illustrate several points. The most important is that terms *cross-relax only with other terms of the same order*. In other words, single-spin terms such as I_z and S_z cross-relax only with other single-spin terms, whereas two-spin terms such as $2I_zS_z$ cross-relax only with one another, and three-spin terms such as $4I_zS_zX_z$ also cross-relax only with one another. For cases in which the number of spins in the product is the same as the number of spins in the spin system (e.g., I_z in a one-spin system, $2I_zS_z$ in a two-spin system, or $4I_zS_zX_z$ in a three-spin system), no cross-relaxation involving that product is possible, and the relevant equation represents just a first-order decay with a rate constant containing only single quantum transition probabilities (W_1 terms).

In this context, we may also usefully repeat a point made in Section 3.1.2. Cross-correlation can, in principle, blur the distinctions outlined above, allowing, for instance, I_z and S_z to cross-relax with multispin zz products such as $2I_zS_z$ or $4I_zS_zX_z$. Such effects are considered further in Section 6.3.

The fact that zz terms of the same order can cross-relax with one another implies that an SPT pattern created by uneven saturation of one multiplet could, in principle, find its way via cross-relaxation into other multiplets not coupled to the first. Equation 6.8 tells us what is required for this to happen. At steady state, the derivative $d/dt(2I_zS_z)$ can be set to zero, and we have

$$2I_zS_z = (\sigma_{IX}2S_zX_z + \sigma_{SX}2I_zX_z)/R_{IS} \tag{6.12}$$

Suppose our three-spin system comprises three spins in a triangle, such that X and S are close together but with zero J coupling, while I is J-coupled to both X and S (Fig. 6.5). Uneven irradiation at X will create a nonzero steady-state contribution from the term $2I_zX_z$, but the term $2S_zX_z$ cannot be produced, since $J_{SX} = 0$. Under these circumstances, Eq. 6.12 tells us that the IS coupling can acquire some antiphase character $(2I_zS_z)$, *provided* that σ_{SX} is appreciable. In other words, the doublet due to S goes partially into antiphase as a result of cross-relaxation between $2I_zX_z$ and $2I_zS_z$, even though S is not itself coupled to the irradiated signal X.

Notice that if σ_{SX} is appreciable, there must also be an NOE enhancement between X and S. Thus, the only circumstances under which an SPT pattern can be "relayed" to a resonance not directly coupled to the saturation target arise when there is also an NOE enhancement to the same resonance. A practical example of how multiplet distortions due to such zz cross-relaxation can appear in real cases appears in Figure 6.6.

These various effects of SPT in 1D spectra have direct analogs in 2D NOESY spectra. Normally, all three pulses of the NOESY sequence are set to 90°, which means that zz terms cannot contribute. If the second and third pulses are not 90°, then extra peaks occur in the spectrum (cf. Section 8.4.2).[4] For a two-spin system, they arise as follows: I_z magnetization is turned into I_x by the first pulse (assuming a $90°_y$ pulse). During t_1, spin–spin coupling turns I_x partially into antiphase terms such as $2I_yS_z$, and a non-90° pulse then turns these partially into $2I_zS_z$. If the third pulse is also not 90°, then the zz terms present during τ_m are partially converted back into single quantum antiphase; but since $2I_zS_z$ is associated equally with both I and S, both $2I_xS_z$ and $2I_zS_x$ terms are created (Eq. 6.1). The resulting contributions to the 2D spectrum are thus es-

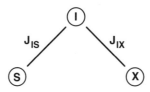

Figure 6.5. Spin system for which a "relayed SPT" effect may be appreciable. Spins X and S are close together but not J-coupled, while spin I couples to both S and X.

sentially identical to those in a corresponding double-quantum filtered COSY spectrum. Cross peaks appear linking coupling partners (i.e., at $\omega_1 = \omega_I$, $\omega_2 = \omega_S$, and at $\omega_1 = \omega_S$, $\omega_2 = \omega_I$), and both these and the corresponding diagonal contributions are in double-absorption mode, and are in antiphase with respect to the *IS* coupling in both frequency dimensions.

For higher-order spin systems, the situation is somewhat more complicated, since all orders of zz product contribute. The resulting spectrum has much in common with an E-COSY spectrum,[5] in that it resembles a superposition of all orders of multiple-quantum filtered COSY spectra. This similarity is made strongest by reducing the flip angle of the second and third pulses drastically, say to 20° or so, since then each different order of zz term makes more nearly the same contribution (cf. Eq. 6.1). Since these SPT effects do not depend on cross-relaxation, the mixing time can be made very short (but retaining the random jitter to suppress zero quantum coherence contributions, cf. Section 8.4.2). In this form, the experiment is called Z-COSY.[6] There are several experiments of this type, but since they are intended for the investigation of *J*-coupling connectivities rather than NOE enhancements, we shall not discuss them further here.

Peaks of the sort just described are the direct analogs of the SPT patterns produced at coupling partners of an unevenly saturated multiplet in a 1D experiment, and they are, therefore, present however short τ_m may be. Just as in the 1D case, however, over longer times cross-relaxation can carry intensity between these zz contributions. For our three-spin system, $2I_zS_z$ will cross-relax with $2I_zX_z$ and $2S_zX_z$ during τ_m, according to Eq. 6.8. Further discussion of the information that can be obtained from NOESY experiments having the second and third pulses different from 90° may be found in ref. 4.

6.2.2. Consequences

In summary, the important properties of SPT are as follows:

1. SPT is completely independent of the NOE, provided the NOE is measured as the *overall* change in integrated intensity across an entire multiplet. This is because patterns due to SPT have as much positive intensity as negative, and therefore have no overall integral.
2. SPT is caused in 1D experiments when a target multiplet is unevenly saturated or inverted, or in 2D experiments when non-90° pulses are used for the second and third pulses of the NOESY sequence.
3. The effects of SPT may be removed from a 1D spectrum either by using a 90° observe pulse, or else by separately irradiating each line of the target multiplet and coadding the results. These methods are considered further in Section 7.2.5.
4. Cross-relaxation among zz terms can transfer antiphase patterns created at one resonance onto its cross-relaxation partners. This can be more

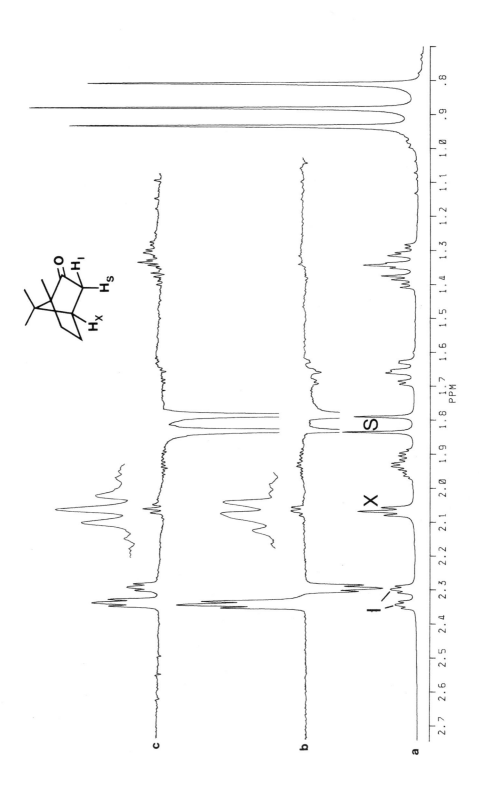

noticeable in 2D than in 1D, but such effects are suppressed if good 90° pulses are used. In 1D, "relayed" SPT distortions are possible at *enhanced* multiplets, even when these are not coupled to the irradiation target, but only under particular circumstances (cf. Figs. 6.5 and 6.6).

6.3. EFFECTS OF CROSS-CORRELATION

The theory presented so far has only considered effects arising from individual relaxation mechanisms in isolation—mainly intermolecular dipole–dipole relaxation, but also (in Chapter 2) chemical shift anisotropy (CSA) and others. We have also seen (in Chapter 3) that if spin S is dipolar coupled to I, and spin I is dipolar coupled to X, then saturation of S can give rise to an NOE enhancement of I, which in turn can then give rise to an indirect enhancement of X, via three-spin or spin diffusion effects. As already mentioned in Section 3.1.2, it is important to note that in this analysis the dipolar couplings of S to I and of I to X are treated completely independently; technically, we would say that the relaxation pathways are assumed to be *uncorrelated*, by which we mean that the rate constant for I–S cross-relaxation is not affected by the position or motion of the X spin, and conversely the cross-relaxation rate constant σ_{IX} does not depend on S. In this section we consider what are the consequences when relaxation mechanisms are *not* independent, but are correlated to each other, and we show that cross-correlated relaxation usually does not affect the measured NOE spectra appreciably, although it may have quite large effects on relaxation rates.

For two relaxation mechanisms to be cross-correlated, they must have the same time dependence. (This implies that cross-correlation cannot occur between two relaxation routes, one of which is intramolecular and the other of which is external relaxation, for example.) Furthermore, cross-correlation can only give rise to observable effects if the nuclei concerned have an observable spin–spin (or dipolar) coupling. This is because cross-correlation, which is a high-order correlation between two or more spins, gives rise to high-order magnetization modes. These are the zz antiphase terms $2I_zS_z$, $4I_zS_zX_z$, and so on, that we have seen in Section 6.2.1 in connection with the SPT. Indeed, cross-correlation effects have appearances very similar or identical to SPT effects. In particular, cross-correlation generally produces differential intensity of

←───

Figure 6.6. NOE spectra of camphor showing relayed SPT distortions. The relevant spins are labeled to be consistent with Figure 6.5. (*a*) Control spectrum. (*b*) NOE difference spectrum, obtained by saturating only the upfield line of the doublet due to spin S, and using an observe pulse with a flip angle of 20°. The SPT distortion at X is shown expanded. (*c*) As (*b*), but using a flip angle of 90°. Appreciable distortions of this sort are expected to be rare.

lines within a multiplet without changing the net intensity of the multiplet. As with the SPT, measurement of the spectrum using a 90° pulse redistributes magnetization among the magnetization modes, and so renders these antiphase cross-correlation effects unobservable. This is presumably the main reason why cross-correlation effects are not often seen, even though they can in some cases be quite large.

It can be shown[7,8] that the different magnetization modes are interconverted via relaxation as shown by the matrix equation below:

$$-\frac{d}{dt} \begin{bmatrix} E \\ A_z \\ M_z \\ X_z \\ 2A_zM_z \\ 2A_zX_z \\ 2M_zX_z \\ 4A_zM_zX_z \end{bmatrix}$$

$$= \begin{bmatrix} 1 & 0 & 0 & 0 & 0 & 0 & 0 & 0 \\ 0 & \rho_A & \sigma_{AM} & \sigma_{AX} & \Delta^A_{AM} & \Delta^A_{AX} & 0 & \delta_A \\ 0 & \sigma_{AM} & \rho_M & \sigma_{MX} & \Delta^M_{AM} & 0 & \Delta^M_{MX} & \delta_M \\ 0 & \sigma_{AX} & \sigma_{MX} & \rho_X & 0 & \Delta^X_{AX} & \Delta^X_{MX} & \delta_X \\ 0 & \Delta^A_{AM} & \Delta^M_{AM} & 0 & \rho_{AM} & \delta_A + \sigma_{MX} & \delta_M + \sigma_{AX} & \Delta^A_{AX} + \Delta^M_{MX} \\ 0 & \Delta^A_{AX} & 0 & \Delta^X_{AX} & \delta_A + \sigma_{MX} & \rho_{AX} & \delta_X + \sigma_{AM} & \Delta^A_{AM} + \Delta^X_{MX} \\ 0 & 0 & \Delta^M_{MX} & \Delta^X_{MX} & \delta_M + \sigma_{AX} & \delta_X + \sigma_{AM} & \rho_{MX} & \Delta^M_{AM} + \Delta^X_{AX} \\ 0 & \delta_A & \delta_M & \delta_X & \Delta^A_{AX} + \Delta^M_{MX} & \Delta^A_{AM} + \Delta^X_{MX} & \Delta^M_{AM} + \Delta^X_{AX} & \rho_{AMX} \end{bmatrix}$$

$$\cdot \begin{bmatrix} E \\ A_z - A_z^0 \\ M_z - M_z^0 \\ X_z - X_z^0 \\ 2A_zM_z \\ 2A_zX_z \\ 2M_zX_z \\ 4A_zM_zX_z \end{bmatrix} \quad (6.13)$$

This shows that, as we know, a nonequilibrium population of spin A relaxes towards equilibrium A magnetization with a rate $\rho_A(A_z - A_z^0)$, and cross-relaxes to M magnetization with a rate $\sigma_{AM}(A_z - A_z^0)$. Further, cross-correlation between CSA relaxation of A and dipolar A–M relaxation gives rise to the antiphase term $2A_zM_z$ with a rate $\Delta^A_{AM}(A_z - A_z^0)$, where Δ^A_{AM} is the cross-correlation rate constant between CSA and dipolar cross-relaxation, and is given (for axially symmetric CSA) by[9]:

$$\Delta_{AM}^{A} = -0.2(\mu_0/4\pi)\gamma_A^2\gamma_M\hbar\langle r_{AM}^{-3}\rangle B_0\Delta\sigma_A[\tau_c/(1+\omega_A^2\tau_c^2)](3\cos^2\phi_{AMA}-1) \quad (6.14)$$

where $\Delta\sigma_A$ is the chemical shift anisotropy of A, and ϕ_{AMA} is the angle between the unique axis of the CSA tensor σ_A and the internuclear vector $\mathbf{r}_{AM}$. A similar but more complicated equation has been presented for the more general case.[10] The chemical shift anisotropy of a nucleus is proportional to B_0, which implies that Δ_{AM}^{A} increases with B_0^2. As noted in Section 2.4.4, the chemical shift anisotropy of ^{1}H is generally small, but CSA can be much larger for heavier nuclei, particularly in anisotropic bonding arrangements, such as carbonyl carbons. Thus, cross-correlation effects associated with CSA relaxation of ^{1}H are not generally seen, but cross-correlation between CSA relaxation of X nuclei and dipole–dipole relaxation of ^{1}H can be. However, they usually lead only to intensity changes within a multiplet, and not to a net NOE enhancement. As an example of this, a large intensity distortion was observed in heteronuclear NOE spectra of a fluorinated polyaromatic compound, resulting from cross-correlation between CSA relaxation of ^{19}F and (^{19}F, ^{1}H) dipole–dipole cross relaxation.[11] The effect was large because of the unusual fixed geometry of the molecule, which resulted in a large $(3\cos^2\phi - 1)$ term; however, there was only an SPT effect, with no net NOE enhancement at ^{1}H.

Cross-correlation between AM and MX dipole–dipole relaxation gives rise to the antiphase term $4A_zM_zX_z$ at a rate $\delta_A(A_z - A_z^0)$, where δ_A is the dipolar cross-correlation rate constant. The geometric factor of δ_A has again the familiar $(3\cos^2\theta - 1)$ form:

$$\frac{\delta_A}{\sigma_{AM}} = \frac{1}{2}\left(\frac{r_{AM}}{r_{MX}}\right)^3(3\cos^2\theta - 1) \quad (6.15)$$

where θ is the angle between the AM and MX vectors. This implies that dipolar cross-correlation is strongest for a linear arrangement of spins ($\theta = 180°$).

Equations 6.14 and 6.15 both contain $(3\cos^2\theta - 1)$ terms. This implies that it should be possible to work out the angle between two internuclear vectors from the size of a dipole–dipole cross-correlation effect (i.e., from the rate of conversion of in-phase into antiphase magnetization). Similarly, the size of a CSA–dipole cross-correlation effect can reveal the angle between the internuclear vector and chemical shift anisotropy tensor (which in turn is often similar to a bond direction). Such measurements are indeed possible and have recently been exploited to obtain structural parameters,[12] but are not discussed further here as they do not fall within the scope of this book.

By contrast, effects of cross-correlation on net NOE enhancements are in general very small. It should be noted that there is no cross-correlation route that converts $(A_z - A_z^0)$ directly into M_z or X_z; in other words, cross-correlation only produces SPT effects and not net NOE enhancements. The only way in which net enhancements can be produced is by two successive cross-correlation steps, for example from A_z to $4A_zM_zX_z$, and then from $4A_zM_zX_z$ to M_z, in both cases by dipole–dipole cross-correlation. These double transfers are in general very slow, taking several seconds to develop to any significant extent. Indeed, we are not aware of any situation in which homonuclear enhancements have arisen as a result of cross-correlations, except in very contrived situations where linear arrangements of spins, very long NOE buildup times and small-angle observe pulses were used.

To summarize:

1. The observation of cross-correlation requires the two relaxation mechanisms to have the same time dependence (cf. Section 3.1.2). It also requires the spins concerned to be J-coupled. It builds up slowly (typically ≥ 0.5 s before any measurable effect can be observed).
2. Cross-correlation is for practical purposes limited to SPT-like effects, which disappear if 90° observation pulses are used.[13]
3. ^{1}H homonuclear cross-correlation is limited to dipole–dipole cross-correlations (except conceivably at very high field) because ^{1}H CSA is small, and therefore effectively to geometrical arrangements that are close to linear.
4. For heteronuclei, cross-correlation between dipolar and CSA relaxation can have significant effects on relaxation rates: for example it can markedly affect spin–lattice relaxation rates of ^{15}N and ^{1}H nuclei in amide groups of proteins.[14,15]

6.4. STRONG COUPLING

The remaining sections in this chapter are all concerned with effects that become appreciable only when the ratio of J (the spin–spin coupling constant) to δ (the difference in resonance frequency between two spins) becomes significant, or in other words, when strong coupling occurs.

The presence of strong coupling is usually clear from the additional complexity and intensity changes that it brings to the spectrum. Analysis of strongly coupled spectra generally requires computer simulation if all the shifts and couplings are required, but this aspect is not discussed here. Most of the important points *concerning the NOE* can be made using two simple examples, the *AB* case and the *ABX* case. Mathematically, calculations involving strongly coupled spectra are normally handled using density matrix theory,[16] a somewhat lengthy and involved procedure. In what follows, only the results of such theory will be discussed, together with their implications for NOE measurements.

The transitions in a two-spin *AB* system, and the corresponding NMR spectrum, are shown in Figure 6.7. In the weakly coupled limit ($J \ll \delta$), all lines have the same intensity, and the left-hand doublet is due entirely to transitions 1,2 and 3,4 involving flip of spin *B*, while the right-hand doublet is due entirely to transitions 1,3 and 2,4 involving flip of spin *A*. In quantum mechanical terms, we may say that the levels correspond simply to spin states of *A* and *B* as follows: $1 = \alpha\alpha$, $2 = \alpha\beta$, $3 = \beta\alpha$, and $4 = \beta\beta$.

STRONG COUPLING 205

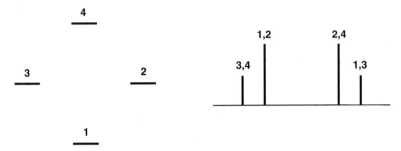

Figure 6.7. Strongly coupled two-spin system. The 1,2 and 3,4 transitions have mostly B character, and the 1,3 and 2,4 have mostly A (see text).

When strong coupling becomes appreciable, these simple interpretations break down. Levels 2 and 3 become mixed, so that their wave functions are

$$\psi_2 = (\cos\theta)\alpha\beta - (\sin\theta)\beta\alpha$$
$$\psi_3 = (\sin\theta)\alpha\beta + (\cos\theta)\beta\alpha \tag{6.16}$$

where θ is given by

$$\tan 2\theta = J/\delta. \tag{6.17}$$

The consequence of this in the spectrum is that the line frequencies change from the value predicted by first-order rules, and the intensities of the inner lines of the AB quartet grow at the expense of the outer lines, generating the familiar "roof effect" associated with strong coupling. More importantly, we may no longer assign any line in the spectrum as being *entirely* due to the flip of just one spin; the left-hand doublet now has some A spin transition character, whereas the right-hand doublet has acquired some B spin transition character. It is just this impossibility of assigning lines in the spectrum precisely to spins in the molecule that represents the heart of the problem of interpreting spectra from strongly coupled spin systems.

Nonetheless, unless the coupling becomes infinitely strong, and the spectrum collapses to an A_2 singlet, it is still possible to say that one doublet is *predominantly* associated with spin A, whereas the other is predominantly associated with spin B. This sort of reasoning is critical to the discussion that follows in Sections 6.4.1 and 6.4.2. Whenever reference is made to the "A" lines, for instance, what is meant is those lines that are *predominantly* A, but it must constantly be remembered that they carry with them a proportion of B character, which increases with the degree of strong coupling.

The mixing of wave functions has a number of consequences. First, it is clear that the enhancement seen at A on irradiation of B will be affected. Second, it is reasonable to expect that in an ABX system where A and B are strongly coupled, irradiation of *X* will produce enhancements that will in some sense be "shared" between *A* and *B*. Finally, the mixing of wavefunctions means that longitudinal scalar relaxation can become efficient so that, in addition to the effects noted above, substantial negative enhancements can be seen at *A* on saturation of *B* under certain circumstances (see also Section 2.4.5). We now deal with each of these points in turn.

6.4.1. *A{B}* Enhancements

This subject has been covered by Noggle and Schirmer,[17] and we only quote their results here. References 18 and 19 are also relevant to the discussion. The expected enhancement at *A* on saturation of *B* is given approximately by

$$f_A\{B\} = \frac{\sigma_{AB} + S^2(W_0 - W_1^* - 2W_{1A} - CW_1^*)}{R_A - S^2(W_0 - W_1^* + CW_1^*)} \quad (6.18)$$

where σ_{AB}, R_A, W_0, and W_{1A} have their normal meanings, $W_1^* = \rho^*/2$, $S = \sin 2\theta = [1 + (\delta/J)^2]^{-1/2}$, and *C* is a correlation factor, describing the degree to which both spins instantaneously see the same external fields contributing to ρ^*. When the fluctuating fields at *A* and *B* are completely correlated, *C* is equal to one; values of about 0.7 were found to fit in two cases that have been investigated.[17] Noggle and Schirmer calculated enhancements according to Eq. (6.18) for the case of $C = 1$ and $\rho^* = 4\rho$ (which probably represents something like a worst case for deviation from the weak coupling formula), and found that the reduction of the NOE from that expected in the absence of strong coupling is about 20% for $\delta/J = 6$, and 10% for $\delta/J = 10$. That is to say, two protons 0.6 ppm apart and with a 10 Hz coupling will have a mutual NOE less than normal by a factor of 20% at 100 MHz or about 7% at 200 MHz. Smaller deviations are predicted when ρ^* is smaller relative to σ_{AB}.

These deviations are fairly small, but the *A{B}* enhancement is in any case unlikely to be of much structural interest. Probably the commonest situations in which strong coupling occurs for simple organic structures at superconducting field strengths are in diastereotopic methylene groups and for adjacent aromatic ring protons. In such cases, the proximity of these protons *to one another*, which is what the *A{B}* enhancement would establish, is hardly likely to be in question. What is of more concern is the use of such enhancements to calibrate internuclear distance determinations from kinetic NOE measurements (cf. Section 4.4.2). This possible source of error is not usually considered in such calculations, but it would certainly seem prudent to avoid strongly coupled systems when choosing the reference distance for calibration.

6.4.2. *AB{X}* Enhancements

As pointed out in the previous section, enhancements *between* strongly coupled spins are in general much less likely to be of structural interest than are en-

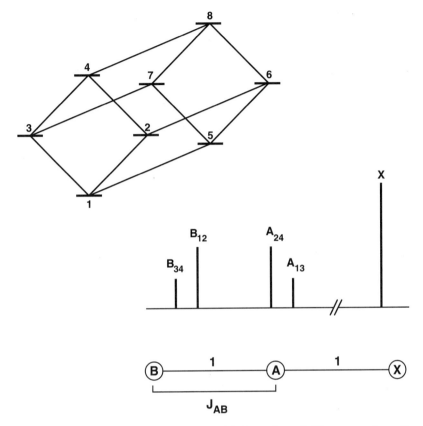

Figure 6.8. Strongly coupled ABX system, with $J_{AX} = J_{BX} = 0$. The system has a linear geometry, as indicated.

hancements *into* a strongly coupled spin system from other spins. For the particular case of enhancements involving diastereotopic methylene protons, enhancements into the methylene group as a whole may establish its gross position within a molecule, whereas any difference between the enhancements at each of the two methylene signals could be used to reveal the orientation of the group relative to other fixed points in the structure (provided stereospecific assignments are available; cf. Sections 11.2.2 and 12.2.5). Cases of this sort probably represent the commonest situations in which strong coupling is encountered in practice in NOE experiments.

As an example, we take the simplest case, which is an ABX system, where A and B are strongly coupled, but X is not spin–spin coupled to either.[20] If X is close only to A, then in the weak coupling limit, and for a rapidly tumbling molecule, saturation of X would cause a positive enhancement at A and a negative three-spin effect at B. The spin system is shown in Figure 6.8.

COMPLICATIONS FROM SPIN–SPIN COUPLING

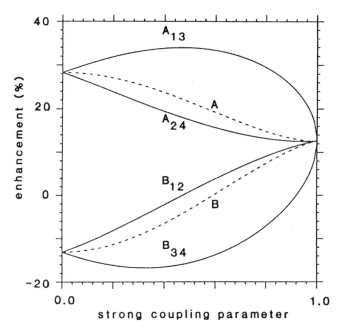

Figure 6.9. Theoretical steady-state fractional NOE enhancements calculated for the linear ABX system of Figure 6.8, with $r_{AX} = r_{AB}$, saturating X. The enhancements are plotted as a function of the strong coupling parameter $S = \sin 2\theta$. A pulse angle of 90° is used. Fractional NOE enhancements for each line (cf. Fig. 6.7) and for the total "A" and "B" multiplets (dashed lines) are shown. As S increases, the fractional enhancements on the outer lines (A_{13} and B_{34}) approach those of the inner lines, but note that the equilibrium intensities of the outer lines become smaller. Thus, in the limit of infinitely strong coupling, the outer lines are enhanced, but have zero intensity! External relaxation was not included in these calculations. Reproduced with permission from ref. 20.

There are three different aspects of the problem that must be considered when deriving the behavior of the NOE in the ABX system. The first is that as the strength of the coupling increases, so that the A resonance acquires more B character, the NOE becomes "redistributed" between the A and B signals. For the system shown in Figure 6.8, the enhancement at A becomes less positive, while the enhancement at B changes from negative to positive. In the limit of infinitely strong coupling, the system becomes A_2X, and a single positive enhancement is seen for the resulting A_2 singlet. The second aspect of the problem, which is quite independent of this and far less obvious, is that the transition probabilities W between different energy levels change as the system becomes more strongly coupled. The method for calculating values of W in strongly coupled systems is outlined in Appendix II, and is given in full in ref. 20. Third, the observe pulse causes a redistribution of magnetization between transitions (cf. Section 6.2.1),[21] and therefore the appearance of the spectrum depends on the pulse angle used.

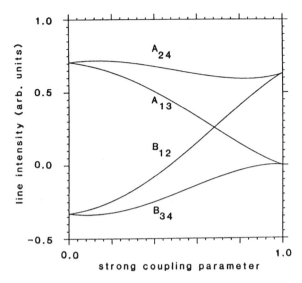

Figure 6.10. As Figure 6.9, but plotting the line intensities that would be observed in an NOE difference experiment; these differ from the fractional enhancements in that they are not divided by the control intensities. Reproduced with permission from ref. 20.

The detailed equations prove to be very complex, but their implications can be easily appreciated. Essentially, as expected from the first point described above, the NOE enhancements do get redistributed as the system becomes more strongly coupled. This is shown in Figure 6.9, where the overall steady-state enhancements at A and B become more similar at a fairly steady rate, until they eventually meet at the strong coupling limit, $\sin 2\theta = 1$. Note that the *total* enhancement ($f_A\{X\} + f_B\{X\}$) remains roughly constant throughout this process. However, the enhancements seen *at each line* behave rather differently; the fractional enhancements at the outer lines actually increase initially. This statement is rather deceptive, however, since although the *fractional* enhancements of the outer lines remain significant for large values of θ (according to Fig. 6.9), the *actual line intensities* in the difference spectrum become small fairly rapidly, because of the reduction in intensity of the outer lines in the normal spectrum as θ increases. This is shown in Figure 6.10, which now plots the actual line intensities in the difference spectrum rather than the fractional enhancements. The effect is more clearly visualized in Figure 6.11, which shows plots of calculated NOE difference spectra at various values of $\sin 2\theta$. We should note, in particular, that as $\sin 2\theta$ increases (ii) the outer line of the B multiplet in the difference spectrum shrinks quite rapidly to zero; (ii) the inner B line changes in sign and rapidly grows in intensity toward that of the inner A line, thus giving rise to a "triplet-like" pattern in the difference spectrum; and (iii) the "roof effect" (that is, the intensity asymmetry of a strongly coupled multiplet) on the A spin multiplet in the difference spectrum is reduced relative to its appearance in the normal spectrum. There is rather little pulse angle effect on this system, and other relaxation mechanisms probably do not significantly alter the general behavior, although of course the absolute size of the NOE is reduced.

Two examples suffice to give a feel for the practical consequences. They are both examples of enhancements into methylene groups. The first example, shown in Figure 6.12, is an enhancement from a methyl group to a pair of methylene protons in a modified dipeptide. In $CDCl_3$ solution the pair is strongly coupled, while in dimethyl sulfoxide solution it is much less so. Independent evidence (other enhancements and $^3J_{CH}$ couplings) shows that the conformation of this portion of the molecule is similar in the two solvents, and is predominantly fixed as shown. In the strongly coupled spectrum (Fig. 6.12a and b) a "triplet-like" enhancement appears in the difference spectrum, as described above, whereas in the less strongly coupled case (Fig. 6.12c and d) an indirect (three-spin) enhancement at H_A is probably almost exactly cancelled by a small strong coupling effect. In both cases, as expected on the basis of the discussion above, the effect of strong coupling is to "pass on" some of the strong positive enhancement seen at H_B to the H_A lines also.

The second example, shown in Figure 6.13, shows spectra recorded for 1-dehydrotestosterone. Irradiation of H_4 gives the expected strong positive enhancement to $H_{6\alpha}$, but, in addition, half of the multiplet due to $H_{6\beta}$ has a positive enhancement, while the other half has a negative enhancement. Irradiation of Me_{19} gives a similar effect in reverse, with the direct enhancement this time to $H_{6\beta}$. These enhancements can be understood once it is realized that the two protons at the 6 position are strongly coupled; in particular, the indirect enhancements can be appreciated to be strong coupling effects and therefore included in the interpretation, rather than being merely dismissed as subtraction artifacts. Although this spin system is considerably more complicated, these strong coupling effects behave just as would be expected by analogy with the simpler cases considered previously. Note in particular the marked reduction in the "roof effect" for the direct enhancement in each case.

---→

Figure 6.11. Theoretical "stick" spectra, drawn to scale, representing points on the curves of Figures 6.9 and 6.10. The upper row represent the conventional spectra, and would correspond to the control spectrum in an NOE difference experiment. The lower row are calculated NOE difference spectra, plotted with a 10-fold vertical scale expansion relative to the upper row. Note the very different behavior of the inner and outer lines of the B multiplet. The outer line rapidly shrinks to a very low intensity, whereas the inner line quickly changes from negative to positive and then grows in intensity toward the inner line of the A multiplet. In the region of intermediate strong coupling, "triplet-like" patterns result. Also, the "roof effect" for the A multiplet is substantially reduced in the NOE difference spectra when compared to the control spectra. Reproduced with permission from ref. 20.

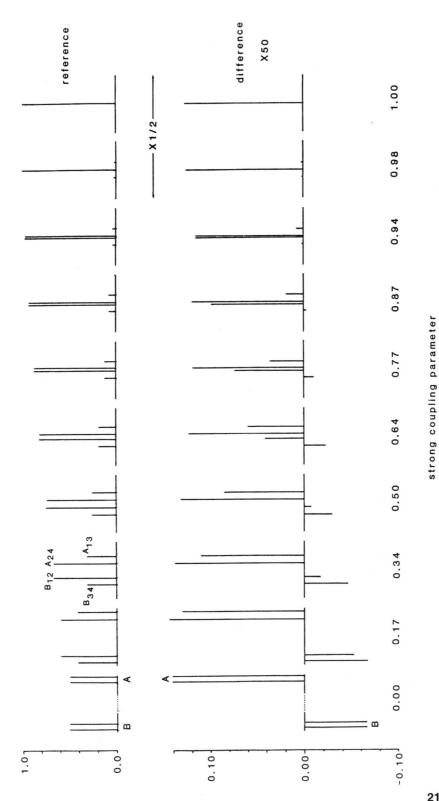

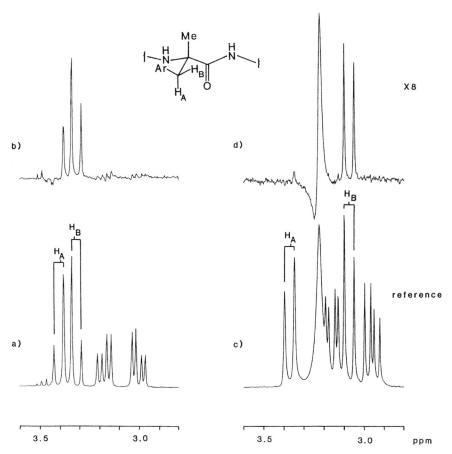

Figure 6.12. Experimental steady-state NOE difference spectra recorded for a modified dipeptide, the relevant structural fragment of which is shown. In all cases the αMe group is the irradiation target. Spectra (*a*) (the reference spectrum) and (*b*) (the NOE difference spectrum) were both recorded using CDCl$_3$ as solvent, while spectra (*c*) and (*d*) were recorded using dimethyl sulfoxide-d_6. In the latter case the degree of strong coupling between H$_A$ and H$_B$ is quite small. The observed large positive enhancement to the *B* spin multiplet is consistent with the conformation shown in the sketch, which is populated to > 90% (based on analysis of $^3J_{CH}$ couplings). It is probable that a three-spin effect on the *A* multiplet has just been overturned by the strong coupling contribution. In contrast, the spectra recorded in CDCl$_3$, in which the degree of strong coupling is very much greater, show strong positive enhancements to both lines of the *B* multiplet and to the inner line of the *A* multiplet. Note how the enhanced lines form a "triplet-like" pattern. These spectra illustrate well the effects of strong coupling discussed in the text, and also emphasize how misleading conclusions could be reached if these effects are not recognized. The large peak in the center of the spectra recorded in dimethyl sulfoxide is due to water, and it appears in the NOE difference spectrum owing to the enhancement of an amide proton followed by chemical exchange of this proton with water. Spectra recorded at 300 MHz. Reproduced with permission from ref. 20.

6.4.3. Scalar Relaxation

It was stated in Section 2.4.5 that modulation of J_{AB} by chemical exchange or rapid relaxation of A can lead to relaxation of B. Normally the only significant effect of this is transverse relaxation, but if the spins are strongly coupled, then longitudinal relaxation can also occur. Such longitudinal scalar relaxation involves only levels 2 and 3 (see Fig. 6.7), and hence the relaxation probability has the same effect as the W_0^{DD} or flip-flop term previously discussed, which is the term that gives the negative NOE. This has the interesting consequence that scalar relaxation gives rise to a *negative enhancement*, by contrast to all other nondipolar relaxation mechanisms, which merely reduce the magnitude of an existing enhancement by increasing R_l. We will normally be concerned with small molecules, and here an intensity reduction caused by scalar relaxation combined with a positive NOE enhancement will usually act just to reduce the enhancement, but some large net negative effects have been observed. The scalar relaxation rate W_0^{scalar} is given by

$$W_0^{\text{scalar}} = \frac{J_{AB}^2}{2} \frac{\tau}{1 + (\omega_A - \omega_B)^2 \tau^2} \qquad (6.19)$$

where τ is the inverse of the rate at which J is modulated (i.e., τ is the exchange rate in the case of chemical exchange or the relaxation rate in the case of rapid relaxation of A). Thus, the effects will be largest for spin systems with large coupling constants, small chemical shift differences, and an exchange rate or relaxation rate at A of the same order of magnitude as the AB chemical shift difference.

6.4.3.1. Scalar Relaxation of the First Kind.
Scalar relaxation of the first kind occurs when one partner of a strongly coupled pair is in chemical exchange at a rate comparable to $(\omega_A - \omega_B)$. The requirements for simultaneous strong coupling and fast exchange rate make this relaxation mechanism rather rare, and it is most common for —CHOH— groups. Even here, it is rather rare at superconducting field strengths.[22] As a first indicator for the possibility of scalar relaxation, the splitting pattern from the exchanging spin must be collapsed.[17] One example of a dramatic negative enhancement produced by scalar relaxation comes from reference 23, from a study of methanol at 100 MHz. The difference in chemical shifts ($\Delta\delta$) between the methyl and hydroxyl protons was 152 Hz, and the coupling constant was 5 Hz. In spite of this rather weak coupling and small coupling constant, a high scalar relaxation rate was obtained by adjusting the pH to control the ratio of exchange rate to $\Delta\delta$. At the maximum scalar relaxation rate (pH 3), an enhancement of -85% was produced at OH by saturating CH_3, and -60% at CH_3 by saturating OH (Fig. 6.14). A change of 2 pH units brings the enhancement essentially completely back to normal (i.e., a small positive NOE).

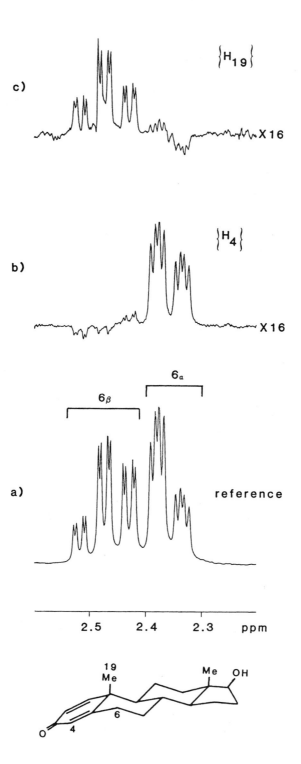

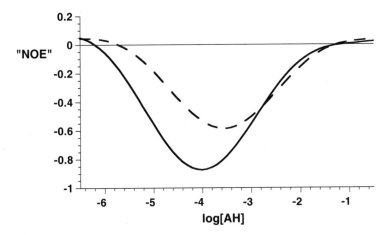

Figure 6.14. Dependence on acid molarity of enhancements observed for methanol. The experiments were carried out using methanol containing *p*-nitrobenzoic acid (data from ref. 23). Solid line: enhancement of OH signal of methanol on saturation of CH_3. Dashed line: enhancement of CH_3 on saturation of OH. The large negative enhancements are caused not by dipolar relaxation but by scalar relaxation, and are not dependent on distance (see text).

Because of the strong coupling requirement, effects due to scalar relaxation of the first kind are very rarely seen in heteronuclear systems, and only then if the gyromagnetic ratios of the two nuclei are very similar (and usually only at low field strength). However, a striking example of such a case was provided by the first experiment designed to detect the NOE, namely in liquid HF at a proton frequency of 90 MHz.[24] The very large coupling constant (estimated to be 615 Hz) was a major factor in increasing the scalar relaxation rate, which

Figure 6.13. Experimental steady-state NOE difference spectra recorded for 1-dehydrotestosterone. Spectrum (*a*) is the control, and shows that part of the spectrum where the two methylene protons at the 6-position resonate. The roof effect of the multiplets shows that strong coupling is present. In the NOE difference spectrum (*b*) H_4 is the irradiation target, and as expected a strong positive enhancement to the equatorial proton $H_{6\alpha}$ is seen. The $H_{6\beta}$ multiplet is positively enhanced on its inner lines, but negatively enhanced on its outer lines, as discussed in the text. In (*c*), Me_{19} is the irradiation target, and this time the $H_{6\beta}$ multiplet is enhanced strongly. The $H_{6\alpha}$ multiplet now has a positive enhancement on its inner lines, and negative on its outer. Note also that in both (*b*) and (*c*) the "roof effect" on the directly enhanced multiplet is substantially reduced. Spectra recorded at 300 MHz. Reproduced with permission from ref. 20.

led to a negative enhancement of up to 70%, depending on the chemical exchange rate.

6.4.3.2. Scalar Relaxation of the Second Kind.

Scalar relaxation of the second kind can occur when one partner of an *AB* system is undergoing very rapid spin–lattice relaxation, and hence the *J* coupling between them is being rapidly modulated. The usual way in which this may occur is when one partner is quadrupolar. As explained in Section 2.4.3, quadrupoles relax very rapidly, and the combination of a large one-bond coupling constant and a rapid relaxation rate can make scalar relaxation of the coupled partner very fast, but only if the two nuclei have similar gyromagnetic ratios. It is very rare that two nuclei with similar enough gyromagnetic ratios are found, and one of the few cases in which significant relaxation of this kind has been observed is in the $^{13}C-^{79}Br$ system,[25] for which $\gamma_C/\gamma_{Br} = 1.003$.

A rather different example of scalar relaxation of the second kind has been described by Johnston,[26] which may be of more generality. Here we have a strongly coupled homonuclear *AB* system, in which one partner undergoes rapid transverse relaxation due to its coupling to a quadrupolar nucleus (e.g., ^{14}N). This example (probably the case most likely to be encountered in practice) contains a CHNH fragment, where the *CH* corresponds to *A*, and the *NH* to *B*. The gyromagnetic ratios of ^{14}N and ^1H are very different, and so only transverse relaxation of *B* is promoted. However, if *A* and *B* are coupled, then the different, but correlated, transverse relaxation of *A* and *B* provides a modulation of J_{AB}, which can provide scalar relaxation of the second kind.[27] The values of *J*, τ, and $(\omega_A - \omega_B)$ are now quite different from those in the "usual" case of scalar relaxation of the second kind. *J* is much smaller, τ is much smaller (being essentially the difference in transverse relaxation rates of *A* and *B*), but $(\omega_A - \omega_B)$ is also much smaller, and so fast relaxation rates can occur. In Johnston's example, *N*-phenylformamide, J_{AB} is 12 Hz, τ is 0.0013 s, and $(\omega_A - \omega_B)$ is 67 Hz, leading to a value of W_0^{scalar} of about 1.5 s^{-1}, and a signal enhancement of about −50% for the *CH* proton of the *trans* conformer **6.1**. (As an added complication, Johnston's experiment achieved saturation of the NH of the *trans* conformer **6.1** indirectly, by irradiating the NH of the *cis* conformer **6.2**, which then undergoes saturation transfer with the *trans* form.) It should be noted that

6.1   **6.2**

a small chemical shift difference between NH and CH is needed for this phenomenon; similar effects will not occur in normal peptides, except at very low field strengths.

REFERENCES

1. Hoffman, R. A.; Forsén, S. *Prog. Nucl. Magn. Reson. Spectrosc.* 1966, *1*, 15.
2. Sanders, J. K. M.; Mersh, J. D. *Prog. Nucl. Magn. Reson. Spectrosc.* 1982, *15*, 353.
3. Sørensen, O. W.; Eich, G. W.; Levitt, M. H.; Bodenhausen, G.; Ernst, R. R. *Prog. Nucl. Magn. Reson. Spectrosc.* 1983, *16*, 163.
4. Bodenhausen, G.; Wagner, G.; Rance, M.; Sørensen, O. W.; Wüthrich, K.; Ernst, R. R. *J. Magn. Reson.* 1984, *59*, 542.
5. Griesinger, C.; Sørensen, O. W.; Ernst, R. R. *J. Am. Chem. Soc.* 1985, *107*, 6394.
6. Oschkinat, H.; Pastore, A.; Pfändler, P.; Bodenhausen, G. *J. Magn. Reson.* 1986, *69*, 559.
7. Bull, T. E. *J. Magn. Reson.* 1987, *72*, 397.
8. Kumar, A.; Madhu, P. K. *Concepts Magn. Reson.* 1996, *8*, 139.
9. Dalvit, C. *J. Magn. Reson.* 1991, *95*, 410.
10. Goldman, M. *J. Magn. Reson.* 1984, *60*, 437.
11. Keeler, J.; Sánchez-Fernando, F. *J. Magn. Reson.* 1987, *75*, 96.
12. Reif, B.; Hennig, M.; Griesinger, C. *Science* 1997, *276*, 1230.
13. Krishnan, V. V.; Kumar, A. *J. Magn. Reson.* 1991, *92*, 293.
14. Boyd, J.; Hommel, U.; Campbell, I. D. *Chem. Phys. Letts.* 1990, *175*, 477.
15. Pervushin, K.; Riek, R.; Wider, G.; Wüthrich, K. *Proc. Natl. Acad. Sci. USA* 1997, *94*, 12366.
16. Buckley, P. D.; Jolley, K. W.; Pinder, D. N. *Prog. Nucl. Magn. Reson. Spectrosc.* 1975, *10*, 1.
17. Noggle, J. H.; Schirmer, R. E. "The Nuclear Overhauser Effect; Chemical Applications," Academic Press, New York, 1971, Appendix I.
18. Noggle, J. H. *J. Chem. Phys.* 1965, *43*, 3304.
19. Freeman, R.; Wittekoek, S.; Ernst, R. R. *J. Chem. Phys.* 1970, *52*, 1529.
20. Keeler, J.; Neuhaus, D.; Williamson, M. P. *J. Magn. Reson.* 1987, *73*, 45.
21. Schäublin, S.; Höhener, A.; Ernst, R. R. *J. Magn. Reson.* 1974, *13*, 196.
22. Glickson, J. D.; Gordon, S. L.; Pitner, T. P.; Agresti, D. G.; Walter, R. *Biochemistry* 1976, *15*, 5721.
23. Fukumi, T.; Arata, Y.; Fujiwara, S. *J. Chem. Phys.* 1968, *49*, 4198.
24. Solomon, I.; Bloembergen, N, *J. Chem. Phys.* 1956, *25*, 261.
25. Levy, G. C. *J. Chem. Soc., Chem. Commun.* 1972, 352.
26. Johnston, E. R. *J. Magn. Reson.* 1984, *60*, 366.
27. Vold, R. L.; Vold, R. R. *Prog. Nucl. Magn. Reson. Spectrosc.* 1978, *12*, 79.

PART II

EXPERIMENTAL

CHAPTER 7

ONE-DIMENSIONAL EXPERIMENTS

This chapter and the two that follow are concerned with the details of how NOE experiments are carried out. The emphasis is largely on how to obtain spectra that are as free from artifacts as possible, while making the most efficient use of the available experimental time. The difficulties of *interpreting* the enhancements once they have been obtained are not discussed in these chapters, as they form the basis of Part I (Theory) and Part III (Applications) of the book.

Although the present chapter is nominally concerned with 1D experiments, many of the points in it, such as the details of sample preparation, are of a rather general nature, and apply equally to any NOE experiment. Chapters 8 and 9, therefore, concentrate just on these aspects of experimental procedure that are unique to NOESY and the other more specialized techniques discussed there. In the present chapter, we consider first requirements for sample preparation, and then how best to set up an NMR spectrometer to execute a 1D steady-state NOE difference experiment. Some other 1D experiments are covered in Section 7.4, and finally, Section 7.5 deals with transient experiments, including some new and very powerful gradient-assisted methods.

7.1. SAMPLE PREPARATION

7.1.1. Solvent

Probably the most important consideration in sample preparation is the choice of solvent, since this can affect the experiment in a number of ways. Normally, protonated solvents are avoided in all NMR experiments, largely because of the dynamic range problem and the likelihood that some signals would be obscured by the very intense solvent peak. These points are even more critical in NOE experiments, where there are still further disadvantages of protonated solvents. The very high concentration of protons in the solvent will increase the rate of *intermolecular* dipole–dipole relaxation of the solute by the solvent, so reducing the intramolecular enhancements. This is equivalent to an increase in ρ^* (cf. Sections 2.4 and 3.3.3).

If the solvent itself has exchangeable protons, then saturation transfer processes may complicate NOE experiments further. The most important case here, of course, is that of H_2O or HOD as solvent, in which saturation transfer both from solvent to solute and from solute to solvent can occur (cf. also Section 10.5.1). Problems of this sort are exacerbated by the broadness of many exchanging resonances, and it is also particularly easy to cause unwanted partial saturation of the solvent resonance, since this can have significant intensity over a broad region. Nonetheless, however great the problems of using H_2O as solvent may be, they must sometimes be preferable to the complete loss of exchangeable resonances that occurs if D_2O is used. Therefore, when enhancements involving exchangeable signals are to be studied in water, experiments have to be carried out in H_2O and in spite of the problems discussed above, remarkably good results are often obtained (cf. Chapter 13).

Another aspect of spectrometer operation affected by the choice of solvent is the lock channel. In FT spectrometers, the field-frequency lock works by continually sampling the dispersion-mode deuterium signal of the deuterated solvent. If the lock channel is exactly on resonance, then no dispersion-mode signal results. As the field drifts, however, a signal is produced, which is positive for one direction of drift and negative for the other. The intensity of this error signal increases with the extent of the drift. After suitable amplification, the error signal is returned to the field correction coils, and so acts to maintain a constant resonance frequency.† A solvent with a broad deuterium lock signal gives a smaller error signal for a given field drift than a solvent with a sharp deuterium signal. Therefore the correction of spectrometer drift is not as precise for a solvent with a broad signal, and the field-frequency lock is less effective. As we shall see in Section 7.2.1, when identical signals slightly shifted in frequency are subtracted from one another, a "dispersion-like" artifact appears, the intensity of which increases as the frequency shift between signals increases. The result is that such dispersion-like artifacts are more pronounced in solvents with broad deuterium lock signals, and conversely less pronounced in solvents with very sharp (and intense) lock signals. Therefore dimethyl sulfoxide-d_6 and acetone-d_6 are both good solvents for NOE experiments, whereas $CDCl_3$ is not as good, because it has a weaker lock signal, and D_2O is also not as good, as it has a broad lock signal. This point is well illustrated in Figure

---

†On some modern spectrometers lock channels are now provided that function in a more sophisticated way than that described here. However, our points concerning the effects of using different lock solvents remain largely valid.

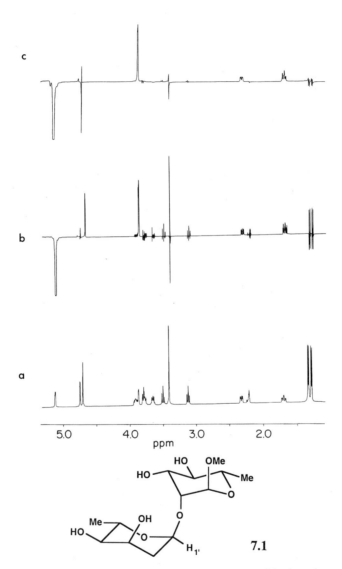

Figure 7.1. (*a*) Spectrum of methyl 2-*O*-(2,6-dideoxy-α-L-arabinohexopyranosyl)α-L-rhamnopyranoside, **7.1**, dissolved in D_2O containing 10% acetone-d_6, but locked onto D_2O; 40 mg/ml, 300 K, line broadening 0.5 Hz. (*b*) Difference NOE spectrum, saturating the H1′ signal at 5.2 ppm. Note the large subtraction artifacts. (*c*) As (*b*), except the acetone-d_6 was used as the lock signal. Using the sharper lock signal improves the subtraction quality considerably.

7.1.[1] An attempted NOE difference experiment using the saccharide **7.1** dissolved in D_2O gave very poor subtraction, and no useful information was obtained. When 10% of acetone-d_6 was added and used as the lock resonance, however, the resultant spectrum was almost free from artifacts.

D_2O is not a good solvent for NOE experiments for another reason: Its chemical shift shows a large temperature dependence. A slight shift in temperature causes a significant shift in the lock frequency, leading to very poor subtraction if temperature regulation is not close to perfect.

As shown in Section 2.2, the maximum enhancement attainable depends on τ_c, the molecular rotational correlation time, and on ω, the Larmor frequency. The enhancement is positive at low $\omega\tau_c$, passes through zero at $\omega\tau_c \cong 1.12$, and becomes negative as $\omega\tau_c$ increases further. Close to the crossover point the enhancement is very small and often not detectable. Oligopeptides and oligosaccharides, for instance, are particularly prone to "lose" their enhancements in this way. For molecules in this range of tumbling rates, the solvent viscosity will have a major effect on enhancements. Most organic solvents have low viscosity, while water is slightly more viscous, and dimethyl sulfoxide even more so. Some solvents are more viscous still, for example ethylene glycol, and, if desired, cosolvents such as vacuum oils may be added to a solution to increase its viscosity dramatically (cf. Section 9.4). A suitable choice of solvent can alter the tumbling rate sufficiently to permit detection of enhancements previously too small to be seen.

7.1.2. Solute Concentration

The only lower limit on solute concentration is that imposed by sensitivity, and this obviously depends both on available experimental time and on the strength of the enhancement being measured. In some cases, as the concentration of the solute increases, the solute begins to aggregate. The concentration at which this begins to be important depends very much on the nature of the solute and the solvent, and no hard and fast rules are possible. Many proteins start to aggregate at concentrations of less than 1 mM, particularly in low salt concentrations, while many organic molecules may not aggregate appreciably at all. Aggregation is manifested as a broadening of resonances and a shortening of T_2 (although concentration-dependent line broadening may arise in other ways, e.g., radiation damping), and may also produce chemical shift changes. It affects the NOE in two ways. First, there is a nonspecific effect, in that aggregation leads to slower tumbling and a longer τ_c. For rapidly tumbling molecules, this may result in a decrease in the intensities of steady-state enhancements because the increase in τ_c brings $\omega\tau_c$ close to 1.12, or it may result in an increase in intensities, because of the increase in ρ and σ without a proportional increase in ρ^* (cf. Section 2.4.1). For slowly tumbling molecules, it will increase spin diffusion and necessitate shorter preirradiation times. Second, aggregation implies that two or more molecules are coming together, and thus intermolecular relaxation will increase. The simplest effect of this is to increase relaxation rates of protons on the surface of the molecule, but it could also be that specific cross-relaxation pathways occur if a particular mode of aggregation is preferred, leading to potentially misleading concentration-dependent enhancements.

It is equally true that other solutes could interact with the molecule of interest, and reduce the enhancements observable by nonspecific intermolecular dipole–dipole relaxation. The effect is not likely to be large, unless these solutes are present at high concentration, but it is one reason why excessive quantities of other proton-bearing solutes (e.g., TMS) should be avoided.

7.1.3. Sample Purification

Typical precautions for any NMR experiment include removal of solid particles by filtration. Beyond this, the degree of purification needed is very dependent on the sample and on the expected difficulty of NOE measurement. The principal impurities to be considered are those that lead to rapid spin–lattice relaxation, namely paramagnetic metal ions and molecular oxygen. These species increase the overall spin–lattice relaxation rate without increasing the cross-relaxation rate, and thereby reduce the intensities of enhancements (cf. Section 2.4.2). As the rotational correlation time increases (that is, as the molecular size increases), both ρ and $|\sigma|$ increase roughly linearly with τ_c (except near $\omega\tau_c = 1.12$; see Section 2.2). This means that for large molecules the increase in relaxation rate caused by paramagnetics is relatively insignificant, and removal of paramagnetics has very little effect. Paramagnetic relaxation has a much more significant effect on enhancements in small molecules. Thus, for example, when α,2,4-trichlorotoluene (**7.2**) was dissolved in $CDCl_3$ and the CH_2 group saturated, a 3.0% enhancement was observed at H_A and no effect

7.2

was observable at H_B. After removal of oxygen and metal ions, these enhancement became 19.2% and −2.6%, respectively. The enhancements are still not maximal, largely because removal of oxygen was incomplete, but they suffice to make the point. Complete removal of oxygen, as described below, is a tedious business and is probably necessary only when looking for very small enhancements in small molecules.

Paramagnetic metal ions are more easily dealt with, usually by adding a small amount of ethylene diamine tetraacetate (EDTA), which forms strong chelating complexes with metals. Alternatively, solutions can be shaken with a chelating resin and then filtered.[2] NMR tubes should not be washed in paramagnetic solutions such as chromic acid; nitric acid is usually a good enough substitute.

Oxygen can be removed in several ways. Bubbling an inert gas through the solution is not a very effective method, as oxygen removal is rarely complete; also, the solvent tends to evaporate or even be blown out of the tube (this always seems to happen with the most precious samples!). A much more effective way of removing oxygen is the freeze-thaw cycle, in which the contents of the tube are frozen, the air space is evacuated, and then the contents are allowed to thaw. Dissolved gases then bubble off into the vacuum in the upper part of the tube, and the cycle is repeated three or four times. Apparatus for this operation is shown in Figure 7.2. It is possible to buy tubes specially designed for this purpose, some of which feature ground glass joints; alternatively, such tubes can be constructed by a glassblower. After the final evacuation, the tube is flame-sealed, taking care to make the seal symmetrical so that the tube can spin properly. A simpler apparatus is shown in Figure 7.3, which needs only a normal tube and a rubber septum cap. These caps are not airtight, however, and it is desirable to fill the tube with an inert gas rather than leaving it under vacuum, and to carry out the NOE experiment immediately after degassing; even so, oxygen exclusion is liable to be incomplete and temporary. If argon is used, this has the advantage of being denser than air, which helps it to persist for longer in the NMR tube.

We should issue a word of warning on freezing aqueous solutions—the tubes crack! This can be prevented either by tipping the tube horizontally and allowing the solution to freeze only in a layer around the surface of the tube, or by freezing the solution very slowly, for example, by suspending the tube

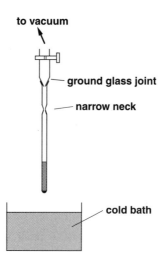

Figure 7.2. Apparatus for thorough degassing.

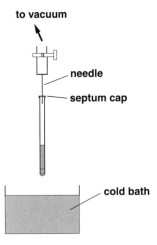

Figure 7.3. Simpler apparatus for less thorough degassing.

in the vapor space above a liquid nitrogen container, but even so it is not easy to prevent the tubes cracking.

7.2. SETTING UP THE STEADY-STATE DIFFERENCE EXPERIMENT

The pulse sequence of the steady-state difference experiment is shown in Figure 7.4. During the presaturation time τ, low-power irradiation saturates one resonance selectively. There may then be a short delay D to allow the RF circuits to settle, after which a pulse is applied and the FID collected. Acquisition may be followed by a relaxation delay RD, but as we shall see, this is quite unnecessary for the normal steady-state experiment. The sequence is repeated a number of times, and the resulting FID stored. The whole process is then repeated again, this time saturating a blank region of the spectrum, and the second FID stored separately. The frequency for this control irradiation does not matter greatly, and it may be convenient to place it well outside the usual spectral

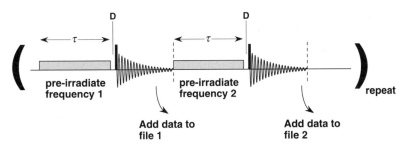

Figure 7.4. The NOE difference pulse sequence.

width so as to avoid any risk of saturating resonances. The entire cycle is then repeated many times, always coadding incoming on-resonance (i.e., preirradiated) data to the previously acquired on-resonance data, and similarly adding incoming off-resonance (control) data to previously acquired control data. The difference between the on-resonance and off-resonance data is then computed. Notice that, because the irradiation is switched off during acquisition, decoupling effects and Bloch–Siegert shifts (slight changes in resonance frequencies caused by the presence of the irradiation field) are absent. Since these are coherent effects, they disappear instantaneously as the irradiation is switched off, whereas population effects, including the NOE, persist and are sampled by the observation pulse.

Subtraction can be done equally well before Fourier transformation, using the FIDs, or afterward, using the spectra. Subtraction of spectra is often more convenient, simply because one may see more easily what is happening, and thereby avoid (some) mistakes. The only circumstance in which there is a real difference between the methods is when the spectra have a very high dynamic range, as for instance if a protonated solvent is used. Lindon and Ferrige[3] have pointed out that in such cases subtraction of FIDs is preferable, since it reduces the contribution of Fourier transform noise (errors in the computed spectrum resulting from inadequate digitization of the intensity of the FID). In the experiment as originally proposed,[4,5] subtraction took place directly in the computer memory, negated off-resonance data being added to on-resonance data on alternate scans. This method has two severe drawbacks. First, there is no control spectrum available for comparison and calibration of the enhancements (cf. Sections 3.3.1 and 7.2.4). Second, if several resonances are to be saturated, then duplicate control data must be acquired for *each* saturation site, so wasting much time. Automatic multiple experiments are considered further in Section 7.2.3.

7.2.1. Introduction to the Difference Experiment

It is often said that NOE difference spectroscopy represents a dramatic improvement over the older CW experiment (Section 7.4.1). This is true; but it is important to realize that the improvement has little to do with the fact that the results are displayed in the difference mode. The act of subtracting the spectra cannot create information (in fact the signal-to-noise ratio is degraded by a factor of $\sqrt{2}$!), so it follows that the information revealed must have been present in the original spectra. What the subtraction process in fact achieves is *clarification*. This is of much more than cosmetic value, however, in that it renders the information accessible to interpretation.

The real reason why NOE difference experiments are more sensitive than the old CW experiment is that they are *better designed*, so as to minimize systematic errors and average away random ones. This is not a criticism of earlier experimenters; the features of difference experiments that achieve this improvement became possible only with the arrival of facilities for sophisticated computer control.

The key to a good difference spectrum of any sort is to ensure that the two spectra to be subtracted differ only in the property that the experiment is designed to measure. Any differences owing to unwanted variations between the spectra are directly visible as artifacts in the difference spectrum. This is actually one of the strengths of difference spectroscopy, since the quality of subtraction (i.e., absence of artifacts) largely reveals how successful the experiment was in eliminating or averaging away errors. It is also the real reason why the results of older NOE experiments could not usefully be displayed in difference mode. Such artifacts would have completely swamped the enhancements, particularly since the irradiation was usually left on during acquisition, resulting in differential Bloch–Siegert shifts between the spectra (see preceding).

Before we consider some possible causes and cures for unwanted variation between spectra (Section 7.2.2), we should first consider how such variations will affect the appearance of a difference spectrum. Although the discussion is in terms of frequency errors between subtracted resonances, the same points apply equally to small phase errors. In fact, for very small errors, phase and frequency shifts are essentially the same thing.

Consider the effect of subtracting two peaks with different frequencies (Fig. 7.5).[6] As the frequency separation between the peaks decreases and they start to overlap, two things happen to their combined lineshape. First, the apparent separation of the peaks is distorted, becoming greater than their true separation. For two identical Lorentzian lines, the apparent separation has a limiting value of just more than half the linewidth (Fig. 7.6a); the two lines never *appear* to get closer together than this. Second, the intensity of the combined envelope decreases (Fig. 7.6b). For small relative shifts, the intensity depends roughly linearly both on frequency difference and (inversely) on linewidth.

This allows us to predict the properties of subtraction artifacts appearing in difference spectra, since they arise from subtracting identical peaks of very slightly different frequency. Subtraction artifacts appear as "dispersion-like" features, whose true lineshape is of course not dispersive at all, and whose frequency separation from positive peak to negative peak is just over half the linewidth of the relevant peak in the normal spectrum. Their overall integral is zero. The amplitude of these artifacts increases linearly with (i) the extent of the frequency error, and (ii) the sharpness of the line. Thus, the worst artifacts occur for very sharp resonances such as methyl and solvent lines, and also for temperature-sensitive resonances, whose frequencies may change slightly through the experiment. From simulations such as those in Figure 7.6, one may see that frequency errors of a few millihertz, or phase errors of fractions of a degree, will cause artifacts sufficiently intense to match genuine enhancements in the 1% range;[7] this shows just how difficult is the task of acquiring a good difference experiment.

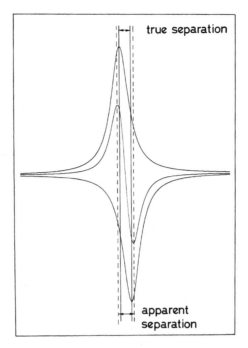

Figure 7.5. The difference between two simulated overlapping identical Lorentzian lines at slightly different frequencies is an antiphase signal of reduced intensity, with an apparent separation of the positive and negative maxima greater than the true separation. Reproduced with permission from ref. 6.

7.2.2. Minimizing Subtraction Artifacts

The generation of artifact-free difference spectra is a very severe test of a spectrometer. Almost all aspects of spectrometer instability must contribute to subtraction artifacts (just as they do to t_1 noise in 2D experiments[8]), and in view of this it is not surprising that a substantial "folklore" has arisen on the subject of minimizing them. In attempting to deal with this, we have tried to distinguish those factors that are empirically observed to be important, whether an explanation is obvious or not, from those factors whose importance is deduced from a knowledge of the workings of the spectrometer. It is also important to realize that, in both categories, these factors may vary widely and unpredictably in their relative importance from one spectrometer to another. Quite unlike the signal to thermal noise ratio, none of these factors can be improved simply by increasing the solute concentration.

Most experimental errors fall into two classes. Some are essentially random, and vary over a very short timescale, where others are associated with changes over a much longer timescale, in the range of minutes or hours. On the whole, the first category is associated with electronic instabilities, and the second with

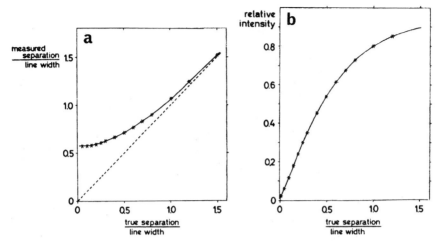

Figure 7.6. (*a*) Measured separation plotted against true separation for two overlapping identical Lorentzian lines in antiphase, normalized by linewidth. The limiting apparent separation is about 0.58 times the linewidth. (*b*) Reduction of intensity (measured as maximum vertical deviation from zero) with relative separation for two overlapping identical Lorentzian lines in antiphase. Reproduced with permission from ref. 6.

instabilities in the magnetic field, or with temperature effects (including those on the RF circuits).

Ultimately, the approach to both types of problem must be to build better spectrometers. In the meantime, there are useful ways for the practicing spectroscopist to minimize the influence of each type of error. The first category is usually tackled simply by acquiring as many transients as possible, on the basis that random error, like random noise, is reduced by coherent addition of signal. Presumably, truly random errors are reduced in proportion to the square root of the number of transients acquired. This implies that even in the rare cases where NOE enhancements can be observed after only 8 or 16 transients, much better results will be obtained with a greater number, because random changes (for example in pulse phase or power) will be averaged away.

The way to minimize the longer timescale errors is to interleave acquisition of preirradiation and control data, on the basis that each data set should then sample the *same* long-term changes in temperature or field drift. The effectiveness of this process must depend, however, on how short the intervals are between cycles of interleaved acquisition relative to the rate of drift they are supposed to sample. The shorter the cycle time, the better; but in practice a compromise must be reached, since too short a time spent at each frequency results in other problems. Typically, a cycle would consist of a few (2–4) dummy scans and perhaps 8 or 16 data acquisitions at one preirradiation frequency, followed by the same sequence at the other frequency. This cycle can be extended to include more than two frequencies in the case of automatic multiple experiments (Section 7.2.3), and the entire cycle is repeated to fill the available experimental time. For two frequencies (i.e., one preirradiation site and one control), this leads to a cycle time of the order of a minute or so, while for multiple experiments the cycle time will be proportionately longer.

The reason why this interleaving cycle time cannot usefully be made shorter is the requirement of using dummy scans. These ensure that any influences of irradiation at the previous frequency in the cycle have completely decayed away before data acquisition begins, and they are therefore vital in eliminating systematic errors. So as not to waste a disproportionate amount of time on dummy scans, it is obviously sensible to acquire a reasonable number of scans at each frequency before moving on. Given also the requirement to acquire multiples of 4 or 8 scans to accommodate the CYCLOPS phase cycle,[9] this results in the typical compromise (8 or 16 acquisitions with 2–4 dummy scans) mentioned earlier.

It is also worth considering how many of these various errors can be eliminated at source. Random variations, that is the inability of the spectrometer to do the same thing twice, have many causes. Probably the most important *symptom* is phase instability. If a supposedly identical set of spectra, each acquired using one pulse, and separated from one another by a minute or so, are acquired, processed, and phased identically, some small residual variation in phase will be observed between the spectra. Long-term drift in such phase errors may be due to temperature effects on the RF circuitry, but random changes from one spectrum to the next could be caused by any number of instabilities within the system. Improvements are steadily being introduced by the manufacturers, and the situation is markedly better now than it was at the time the first edition of this book was published; even so, the further reduction of such errors should continue to be a priority. Not merely do they contribute to poor subtraction in difference experiments, they must also contribute to t_1 noise in 2D experiments.[8]

Spinning variations can also affect the spectrum. If the spinner speed varies during the experiment, particularly if the spinner speed suddenly drops to zero or near zero, then lineshapes must be markedly affected, and subtraction will be poorer. This is an obvious problem. A less obvious difficulty is that spinning the sample induces a regular variation in the quality factor Q of the coil, due to asymmetry of the spinner and/or tube. The response to a pulse, therefore, depends on the orientation of the tube at the moment of the pulse, so that there will be a variation in intensity, and possibly phase, that changes at the spinning rate. On many spectrometers the effect will be insignificant, but it can be important if a coil with very high Q is used. It is therefore advisable to run NOE difference spectra without sample spinning.

Particularly on older spectrometers the RF circuits can take some time to settle between the end of the presaturation and the pulse.[10] It has therefore been recommended that a gap of several milliseconds be left between presaturation and the observe pulse. Quite empirically, this does not appear to be as important on more recent spectrometers.

Of the long timescale errors, the most easily identified is temperature variation of the sample. As noted in Section 7.1.1, this is most troublesome for aqueous solutions, since the resonance frequency of water is very temperature sensitive. Amide NH protons also have a large temperature dependence, and may take an hour or more to equilibrate to a given temperature. Subtraction artifacts from these signals can provide a useful and very sensitive test for the extent of temperature equilibration, and it may therefore be useful to run a "dummy" NOE experiment for half an hour or so, before starting the true experiment.

As discussed in Section 7.1.1, the field-frequency lock is derived from the dispersion-mode deuterium resonance of the solvent. It was shown in Section 7.1.1 that this means that solvents with strong, sharp lock signals give best results. It also means that the lock phase must not be altered once the experiment has started, since this would alter the "zero point" for the field-frequency lock, and so lead to a frequency shift on the observe channel. With strong lock signals, care must be taken to avoid saturating the lock channel, as this causes instability.

We saw in Section 7.2.1 that the intensity of a subtraction artifact depends on the linewidth of the corresponding resonance. It follows that subtraction artifacts can be reduced by applying a line broadening to the spectra before or after subtraction, or by changing the lineshape in other ways, for example carrying out a Lorentz-to-Gaussian transformation.

When metal objects move near the magnet, the lock signal dips markedly. This presumably disturbs the action of the lock, and therefore leads to greater "random" variation in frequency (and possibly also linewidth). For the best quality subtraction, such disturbance near the magnet should thus be avoided, for example, by acquiring the spectrum overnight. To the authors' knowledge the truth of this has never been tested, but there can certainly be no harm in acquiring spectra when no one else is around!

If the subtraction artifacts can be ascribed to a small apparent frequency shift between control and enhanced spectrum, from whatever cause, then in principle one could shift the two spectra relative to each other until the subtraction becomes optimal. The shift needed would probably be only a few millihertz, which is much less than one data point, and would therefore have to be done using computer interpolation. This approach can work[11] but is rarely worth the considerable effort involved. A similar idea is discussed further in Section 7.3.2. Alternatively, very slight phase changes deliberately introduced between the control and enhanced spectra before subtraction can sometimes be used to correct cases of poor subtraction on a trial-and-error basis.[12]

In summary, one can minimize subtraction artifacts by:

1. Using a large total number of scans
2. Recycling the frequency list, using a small number of scans at each pass (e.g., 8)

3. Regulating the temperature and allowing as long as possible for thermal equilibration before the experiment
4. Using a solvent with a sharp, intense lock signal
5. Using a moderate line broadening
6. Using a nonspinning sample

7.2.3. Automatic Multiple Experiments

This section deals with the question of how best to combine into an automatic sequence a set of NOE experiments, in each of which a different resonance is preirradiated. The advantage of such combined experiments, as mentioned in Section 7.2.1, is that (in principle) just one control spectrum will suffice to generate difference spectra from any number of preirradiation sites, saving time on unnecessary acquisition of many controls. Carried to its limit, this approach consists of an automatic sequence in which *all* suitable targets are preirradiated; Sanders[13] has called this strategy the "shotgun" approach. Although completely indiscriminate saturation undoubtedly wastes more time than it saves (particularly since even partly exposed multiplets are suitable as targets), there is much to be said for including as many preirradiation sites as practicable in an automatic sequence. If as many data as possible are available, this must increase the overall confidence in a consistent interpretation, and it also minimizes the need for further experiments, for which the experimental conditions, particularly the external relaxation rate ρ^*, may well be different.

The only important difference between an automatic multiple experiment and a single irradiation target experiment is that a multiple experiment involves a longer recycle time for interleaving the data acquisition. The consequences of this depend largely on how the repeating frequency list controlling the irradiation is organized. Entries that are close together in the list correspond to data sets that are acquired within a short time of one another on each pass around the full cycle. Those that are well separated in the list correspond to data sets that are acquired some time apart on each pass. On the basis of Section 7.2.2, we would therefore expect data corresponding to those frequencies that are close to the control entry in the irradiation list to give better subtraction than those entries that are distant from the control entry. As Figure 7.7 shows, this is indeed the case.

If many frequencies are present in the list, this problem may be quite severe, leading to severely degraded subtraction for those sets corresponding to entries most distant from the control entry in the irradiation frequency list. The problem may be eased by reducing the number of scans accumulated on each pass around the cycle (say from 16 per data set to 8), and possibly, when a single control is used, placing it in the middle of the frequency list rather than at one end may be of benefit. A more satisfactory solution is to introduce additional control entries into the irradiation frequency list. No firm guidelines can be given here, since subtraction quality is so "spectrometer specific," but in the authors' experience one control per 5–10 spectra seems to suffice and does not significantly reduce the time available for on-resonance data acquisition.

Finally, it is also useful to place a control both at the beginning and at the end of the irradiation frequency list. If for any reason the experiment should stop (or be stopped) unpredictably, this at least guarantees that any data file will contain the same number of transients as at least one of the control files.

7.2.4. Irradiation Power and Selectivity

In an ideal NOE experiment, the target resonance would be completely saturated by selective irradiation, while all other signals would be completely unaffected by the irradiating field. In reality, the lack of shift dispersion between signals means that this is often not possible. Particularly when irradiating *multiplets*, irradiation power high enough to saturate the target completely may partially saturate neighboring resonances also (Fig. 7.8). Some practical and theoretical approaches to coping with such unwanted partial saturation have been described,[14,15] but undoubtedly a better solution is to devise methods of saturation that, even if they do not fully saturate the target, at least do not saturate anything else.

This is particularly true since unwitting partial saturation of neighboring or coincident signals is probably the commonest source of serious errors in interpretation of NOE results. Enhancements that arise due to partial saturation of signals *other* than the intended target are all too easy to misinterpret. In at least two cases, incorrect structures have appeared in the literature because such enhancements were not correctly recognized,[16,17] and other instances probably remain undetected.

Before continuing, it is worth pointing out one very simple precaution that can help minimize such errors of interpretation, and that is to examine carefully the saturated target signal *in the difference spectrum*. Partial saturation of neighboring resonances should then be apparent from their presence in the difference spectrum, in addition to the desired target. This method is not foolproof, however, and cases such as accidental partial saturation of a wide, many-line multiplet coincident with a sharp methyl singlet target can be particularly difficult to spot.

The essence of all methods of selective irradiation is to use a low irradiating field strength. The behavior of the magnetization due to the target resonance, and any others near enough in frequency to be affected, during low-power irradiation was discussed in Section 1.5. The key points concerning saturation during steady-state NOE experiments are as follows:

1. Off-resonant signals are partially saturated to an extent that increases with irradiation field strength and decreases with offset from the irradiation frequency.
2. Initially, the magnetization corresponding to affected signals precesses about $\mathbf{B}_{\text{eff}}$, but these oscillations are damped with an approximate decay

236 ONE-DIMENSIONAL EXPERIMENTS

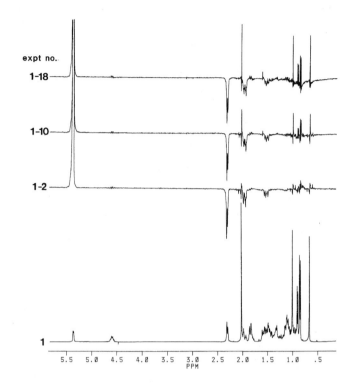

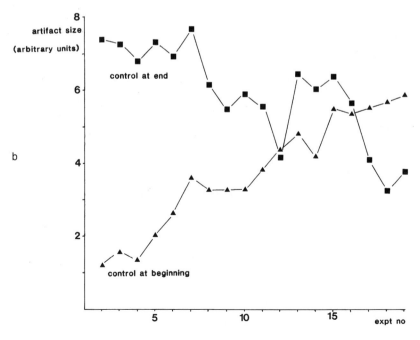

constant of $T_{2\rho}^*$ and for very weak fields or rapidly relaxing signals may not occur at all.
3. For weak fields, the target resonance itself may be only partially saturated, since a balance is reached between the effects of irradiation and relaxation.

Calibration of the irradiating field strength can be carried out in several ways, but the most direct is to measure the frequency of the oscillations that it causes. Figure 7.9 shows the on-resonance response of the $CHCl_3$ singlet to low-power pulses of different lengths. Using the positions of the various zero crossings, it is easy to determine that the 90° pulse width at this particular power setting is about 54 ms, corresponding to a field strength $\gamma B_1/2\pi$ of about 4.6 Hz. For calibrating such weak fields, resonances with long relaxation times are needed if the oscillations are to remain visible for a reasonable number of cycles. Note that the low-power pulse can also produce *transverse* components, which if not suppressed could complicate the pulse width determination (cf. Section 1.5). If the low-power pulses are produced by the same channel as the hard pulses (i.e., the observation channel), then phase cycling is needed to accomplish this (e.g., a constant phase on the low-power pulse, and the cycle x, y, −x, −y on the hard pulse and receiver). On the other hand, if the low- and high-power pulses come from different channels, their phases may not be mutually locked (i.e., they may not be phase coherent), in which case accumulation of a number of scans (16 or so) is required to average away the unwanted transverse contributions. Of course, it is also crucial for the pulse to be exactly on-resonance.

A possible problem when irradiating targets in crowded spectral regions for spinning samples is that of accidental saturation of spinning sidebands. Clearly,

←

Figure 7.7. The increase in artifact size seen when more than one frequency is saturated in an automatic multiple experiment. Cholesteryl acetate was dissolved in $CDCl_3$, and an NOE experiment was carried out, cycling repeatedly through 20 frequencies (2 dummy scans and 8 transients per site; preirradiation time 5.0 s; acquisition time 1.78 s; total time at each frequency 67.8 s at each pass around the frequency cycle). The frequency list was cycled 20 times, to give a total of 160 transients acquired at each frequency. The 20 frequencies in the frequency list were control, (5.35 ppm, 4.6 ppm)$_9$, control. Two series of 18 NOE difference spectra were produced, one series using the first file as control, and the other using the last file. Three such difference spectra are shown in (*a*), using spectrum 1 as control. There are larger artifacts in spectrum (1–18) than in (1–2), especially from the methyl groups. As a rough measure of artifact size, the average peak-to-peak height of the methyl artifacts is plotted in (*b*) (in arbitrary units) against experiment number. Two points emerge from this: (i) the intensity of the subtraction artifact increases with separation in the frequency list from the control entry; and (ii) in this particular experiment, for some reason artifacts that result when 20 is used as control are worse than when 1 is used as control.

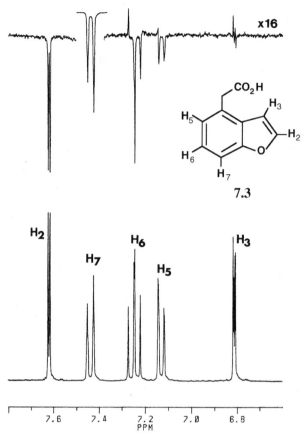

Figure 7.8. High-power saturation of a multiplet. The doublet signal from proton H_7 of benzofuran-4-acetic acid (**7.3**) has been saturated with enough irradiation power to saturate both lines almost evenly. (The saturated signal is inset at the same vertical scale as the control spectrum.) However, the selectivity at this power level is very poor, so that protons H_6, H_2, and H_5 are also partially saturated and consequently appear with negative intensity in the difference spectrum. The small positive enhancement at H_3 probably results from the partial saturation of H_2.

if two peaks are separated by one or two times the spinning rate, then saturation of one peak will necessarily result in irradiation of a spinning sideband of the other. It is fairly clear from experiments that such irradiation affects the corresponding centerband to some extent, and may thus perturb NOE enhancements also. One solution to this problem would presumably be to eliminate the spinning sidebands by improving the B_0 homogeneity (i.e., by shimming). In specific cases of difficulty, spinning sideband frequencies can also be moved around by adjusting the spinning rate. However, improvements in the design of shim systems have reached the stage over the past decade or so that spinning

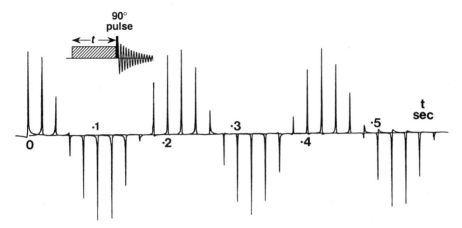

Figure 7.9. Calibration of low-power irradiation pulse length. The irradiation field is applied for a time t, and then a normal 90° transmitter pulse is used to excite the spectrum. For this experiment a sharp singlet was used (the $CHCl_3$ signal), and the transmitter was placed on resonance. The experiment was carried out using a Bruker AM-400 spectrometer, with the low-power irradiation supplied by the decoupler at an attenuation setting of 40 dB; these results show that the 90° pulse length at this setting is 54 ms. Note the very low inhomogeneity of the irradiation field (i.e., very slow damping of the oscillations).

is no longer commonly used even in high-resolution NMR, thereby eliminating this whole problem.[18]

Finally, we should deal with the issue of partial saturation and calibration of enhancements. As was shown in Section 3.3.1, the consequence of incomplete saturation is simply that NOE enhancements are scaled down in strict proportion to the actual extent of saturation (Eq. 3.49). The simplest way in which to determine the extent of saturation is to examine the irradiated signal. If its intensity in the control spectrum is S^0, in the irradiated spectrum is S', and in the difference spectrum is S^{diff}, then

$$S^{diff} = S^0 - S' \quad (7.1)$$

and the correction factor by which all enhancements should be multiplied to correct for partial saturation is*

$$[1 - (S'/S^0)]^{-1} = S^0/S^{diff} \quad (7.2)$$

---

*Strictly, for experiments in the positive NOE regime there should be a minus sign associated with S^{diff}, since it appears inverted in the difference spectrum, but we have avoided this by our definition of S^{diff} in Eq. 7.1. For experiments in the negative NOE regime the definition of S^{diff} may simply be negated.

In general, then, the percentage enhancement seen at I is given by

$$f_I\{S\} = (I^{\text{diff}}/I^0)(S^0/S^{\text{diff}}) \times 100 \qquad (7.3)$$

where I^{diff} is the intensity of I seen in the difference spectrum, and I^0 is its intensity in the control. Note that this formula is entirely unaffected by the number of protons contributing to either the S or I resonances; no further corrections are required if either I or S represents, for instance, a methyl group.

There is, however, one circumstance under which Eq. 7.3 becomes inappropriate. If the magnetization due to S is still oscillating at the time of the observe pulse, then S^{diff} as measured in the difference spectrum does not truly represent the extent of saturation throughout the irradiation, and the logic behind Eq. 7.3 breaks down. Such circumstances are normally encountered only with the relatively high-power, short-duration, irradiation used in TOE experiments. Then, as outlined in Section 1.5, the *average* extent of saturation during irradiation can generally be taken to be 100%. Genuine steady-state partial saturation, which is what Eq. 7.3 describes, is likely only when low-power, long-duration, saturation is applied in steady-state experiments. Nonetheless, if presented with a difference spectrum for which S^{diff} and S^0 are appreciably different, the only way to be completely certain as to whether Eq. 7.3 is applicable is to measure the response of S at different times, to see whether it is oscillating or not.

7.2.5. Multiplet Irradiation and SPT Suppression

As was discussed in the previous section, it is almost always preferable in steady-state NOE experiments to use low-power irradiation rather than risk unwanted saturation of nearby resonances. This has particular consequences when the target is a multiplet, since uneven partial saturation of a multiplet leads to selective population transfer (SPT; Section 6.2). This in turn causes intense antiphase patterns to appear at resonances coupled to the irradiation target, possibly obscuring NOE enhancements and so hindering interpretation (Fig. 7.10).

In fact, suppression of SPT patterns is not an easy task, since they are often many times more intense than NOE enhancements. For an enhanced multiplet to appear in an NOE difference spectrum with undistorted intensities (i.e., with the same intensity pattern as the multiplet has in the normal spectrum) the SPT patterns must therefore be suppressed by at least two orders of magnitude. In practice this is seldom achieved; the methods described below are usually only *partially* successful in removing strong SPT distortions, particularly from small enhancements.

There are two approaches to suppressing these patterns in NOE difference spectra. One is to use a single low-power selective saturation frequency that keeps the selectivity but generates SPT effects, and then to eliminate the SPT effects in the observed spectrum as far as possible by using an exact 90° pulse. The other is to cycle the saturation frequency around the multiplet during the

SETTING UP THE STEADY-STATE DIFFERENCE EXPERIMENT 241

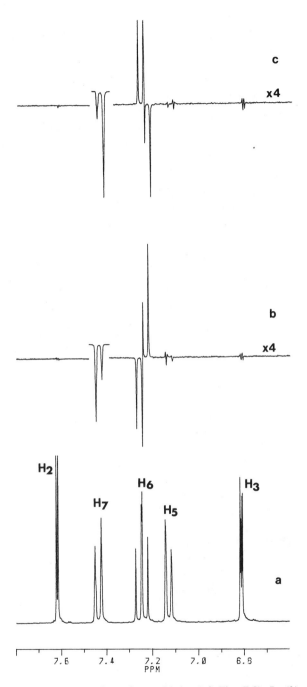

Figure 7.10. Low-power saturation of a multiplet (cf. Fig. 7.8). In (*b*), the low-field line of doublet H_7 is saturated, and in (*c*) the high-field line is saturated. There is a large SPT pattern at H_6, which has the opposite sense in the two cases, and completely obscures the NOE enhancement (cf. Fig. 7.11). Panel (*a*) shows the corresponding region of the control spectrum.

presaturation time, so as to produce an even saturation across the whole multiplet. The theoretical background to both these methods was outlined in Section 6.2.1.

Of these approaches, the first has the advantage that *even a single exposed line* of a multiplet will suffice as the preirradiation target, so that multiplets that are partially obscured by overlap present no obstacle. This method is also the simpler of the two to implement, since only one preirradiation site is required in each target multiplet. For these reasons, it is probably the technique of choice in most cases. Its disadvantage is that only a fraction of the multiplet is saturated, and therefore the saturation factor (Eq. 7.2) is often small so that all enhancements are correspondingly reduced. However, in most cases this is less of an issue than achieving selectivity of saturation.

As was discussed in Section 6.2.1, a 90° pulse appears to spread the saturation from the one line actually irradiated into the other lines of the S (irradiated) multiplet, thereby simultaneously removing the SPT distortions at coupling partners of S. This means that the technique has another use, unrelated to NOE enhancements. If the S multiplet is partly hidden by overlap in the normal spectrum, then saturation of any exposed part of it followed by excitation using a 90° pulse will reveal the whole of the S multiplet in the difference spectrum, now freed from overlap. This can be a useful trick in its own right.[19]

In practice, simply setting the pulse length to 90° may not result in particularly effective SPT suppression, due to complications from $\mathbf{B}_1$ inhomogeneity. To improve this, a composite pulse designed to compensate for $\mathbf{B}_1$ inhomogeneity can be used. Shaka et al.[20] have recommended the sequence $270_x 360_{-x} 90_y$ for this purpose, which is a composite equivalent to a 90_{-x} pulse simultaneously corrected for offset and $\mathbf{B}_1$ errors. It acts as a good 90° pulse over a larger volume of the sample than would a simple 90° pulse, and better SPT suppression results. There are two slight disadvantages of this technique: First, since a larger active sample volume is excited, good lineshape may be harder to achieve (this may be tested by recording a normal spectrum using the composite pulse when shimming); second, subtraction quality in the difference spectrum may be somewhat more susceptible to phase jitter in the RF circuits than would be the case for a simple pulse. Nonetheless, in the authors' experience, this method gives good results and is very convenient in routine use.

The second approach to SPT suppression, irradiating each line of the target multiplet in turn during the presaturation, is intrinsically more complicated and requires the whole of the target to be free from overlap (and indeed, as we shall see shortly, well separated from other signals). The irradiation is cycled repeatedly around some (or all) of the lines of a target multiplet during the preirradiation period before each scan.[21] This avoids exciting SPT distortions, since the multiplet components are (more or less) evenly saturated. The main consideration with this method is to choose an appropriate rate for cycling the irradiation around the lines of the target multiplet. There are two conflicting requirements here. Too slow a cycle rate will allow some magnetization to recover before the observe pulse, thereby reducing the saturation factor and allowing SPT distortions to reappear. On the other hand, too fast a cycle rate compromises the irradiation selectivity. As the cycle rate is increased, the selectivity of saturation goes down, so that nearby signals become partially saturated also. As an example, using a presaturation field strength of 6 Hz at a single frequency, no effect could be observed on signals 20 Hz away. However, cycling the presaturating frequency around two frequencies 6 Hz apart every millisecond allowed partial saturation to be observed 100 Hz away.[22] Empirically, we and others have observed that the best saturation is achieved using individual irradiation times corresponding to a 90° pulse at the irradiating field strength.[23] For a selective field strength of, say, 2.5 Hz, this would correspond to an irradiation time of 100 ms at each frequency during the cycling, and such an irradiation scheme would avoid saturating signals further than 25 Hz or so from the target multiplet. The practicalities of programming this experiment have been discussed.[21]

It is not necessary to irradiate every line in a many-line multiplet individually; a few strategically placed irradiation sites distributed over the multiplet may suffice, typically on the more intense lines of the multiplet so as to achieve a high saturation factor. Frequency cycling can also be used to saturate simultaneously two resonances. This is particularly useful for recording NOE spectra in H_2O, when the target resonance and solvent peak are alternately irradiated throughout the preirradiation period.[24]

A third method exists, which effectively is intermediate between the two already discussed. In this method, each line of a target multiplet is included in the irradiation frequency list for a normal NOE difference experiment, and the corresponding data sets are summed after the experiment, during data processing (Fig. 7.11).[25] This method involves rather long irradiation frequency lists, necessitating the use of several control entries in the list to maintain reasonable subtraction (Section 7.2.3), and some rather tedious manipulations of the many data files during processing. It also results in a low overall saturation factor, because each irradiation is only saturating a fraction of the multiplet. In most cases, the first method (a single irradiation frequency followed by a good 90° pulse) is therefore preferable.

7.2.6. Timing

The various times and delays in a steady-state 1D NOE difference experiment are summarized in Figure 7.4. The most important parameter as far as optimizing NOE enhancements is concerned is the preirradiation time τ. Before discussing this, however, we should first deal briefly with the others. The acquisition time AQ should be set to give a suitable compromise between sensitivity and resolution; NOE experiments are no different from any other NMR experiment in this regard. However, there is one sense in which NOE experiments differ from normal 1D acquisition, namely that AQ represents only a small fraction of the total time between transients, which is roughly $AQ + \tau$.

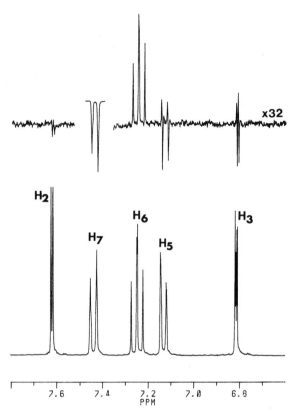

Figure 7.11. Suppression of SPT by addition of the two experiments of Figure 7.10b and c. The vertical scale of the difference spectrum has been increased 8-fold relative to Figure 7.11. The signals from H_5 and H_3, which look like subtraction artifacts, probably arise from SPT across the small four- and five-bond couplings to H_5 and H_3, respectively. These SPT patterns are not suppressed by coaddition of the two experiments, since they arise from uneven saturation across the barely resolved coupling constants within each of the two components of the H_7 doublet.

Since τ is generally several times longer than AQ (at least for small molecules, see following), the length of AQ has little effect on the overall recycle rate, and so AQ can be made as long as necessary to give the desired resolution without undue time penalty. If a higher signal-to-noise ratio is subsequently found to be desirable, this can be achieved at the expense of resolution by zeroing the later part of the FID during data processing. The reverse operation, extending the FID to improve resolution at the expense of signal-to-noise ratio, obviously necessitates rerunning the entire experiment.

For steady-state experiments, any relaxation delay between acquisition and the next pre-irradiation is, literally, a complete waste of time; if the steady state is genuinely reached by the end of τ, it makes no difference in what state the spins were at the beginning of τ. Of course, for kinetic experiments the magnetization must be allowed to relax back to equilibrium between each transient, and quite long relaxation delays are then needed. This is why kinetic experiments are so time-consuming, particularly for small molecules.

The short delay D between preirradiation and the observe pulse is present simply to allow the RF circuits to settle following the irradiation switching transient. Values of 1–5 ms are often satisfactory, but on most modern spectrometers no delay may be needed at all. This can only be tested empirically, comparing the subtraction quality obtained for different settings of D.

As we shall see, the optimum observe pulse angle is generally 90°. The reason for this will be considered in more detail below, since it is bound up closely with the choice of the period τ.

The optimum value for the preirradiation period τ depends very heavily on the molecule studied and the information required, as discussed in Section 3.3.2. If *quantitative* steady-state enhancement values are required, there is no alternative to using very long preirradiation periods. It may even be worthwhile to repeat the experiment for different τ values to demonstrate that steady state is genuinely reached. Provided that such duplicate experiments *do* give identical enhancements, they can always be pooled for better signal-to-noise during data processing.

This is a time-consuming procedure. Fortunately, as is demonstrated by the examples in Chapter 10, the majority of applications of the NOE do not require *quantitative* data of this sort. Usually, a "semi-quantitative" classification of the enhancements, such as categorizing them as strong, medium or weak, is sufficient to answer the structural problem. This means that somewhat shorter values of τ can be used, thereby increasing the total number of transients that can be collected in a given time, and improving overall sensitivity. When choosing a suitable value for τ in such cases, one must take into account the NOE kinetics. Short-range enhancements build up faster than long-range and indirect enhancements (Sections 3.3.2 and 4.1.2) so that for experiments in which τ is short there is a risk of losing the long-range and indirect enhancements. This is illustrated by the simulation shown in Figure 7.12a. The *sensitivity* gain obtained by decreasing τ and increasing the number of transients collected is illustrated by Figure 7.12b. This shows the same data as before, but now with the enhancements divided by the recycle time ($AQ + \tau$, taking a value of 2 s for AQ), to indicate the value of τ for which the maximum signal can be obtained in a given time (in this case).

It is very difficult to put specific numbers on the values of τ that should be used in practice. The exponential growth constant for a direct enhancement is R (Eq. 4.3). Thus, values of τ equal to at least five times the T_1 value for the slowest relaxing spin of interest are required for steady state to be genuinely reached. However, in cases in which precise quantitation is less crucial than sensitivity, shorter τ values suffice. For molecules in the molecular weight range roughly 100–500, preirradiation periods of about 3–5 s are often a reasonable compromise, at least when an overall impression of the enhancements is all

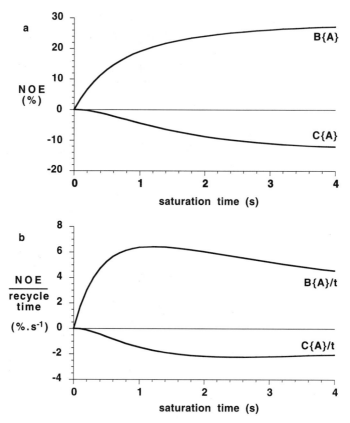

Figure 7.12. (a) Dependence of NOE enhancement on repetition rate of the pulse sequence. Simulation for a linear three-spin system C-B-{A}, with B midway between A and C: $\sigma_{AB} = \sigma_{BC} = 0.5$ s^{-1}; $\sigma_{AC} = 0.008$ s^{-1}, $R_A = R_C = 1.0$ s^{-1}, $R_B = 2.0$ s^{-1}, assuming no external relaxation. The pulse sequence simulated is [irradiate(τ) − 90° − AQ], with an acquisition time of 2.0 s, during which the irradiation is gated off. Four dummy scans are calculated to allow a pseudo–steady state to develop; larger numbers of dummy scans have no effect. In order to simulate the value for I^0 that would be used to calculate fractional enhancements in practice, the enhancements are calculated with respect to the observed peak undergoing the same pulse sequence but without saturation of A. This means that the value of the NOE enhancement at short saturation times is slightly different than it would be if a reference peak with unit intensity were used, because of the effect of rapid pulsing on all of the intensities. (b) Dependence of NOE enhancement on repetition rate. The data of (a) have been used, but now plotting enhancement divided by recycle time (= τ + AQ). The maximum of the curve therefore indicates the most "signal-efficient" saturation time for this value of AQ, other things being equal (see text).

that is required. It must be stressed, of course, that these are only very approximate guidelines, and that for molecules with long T_1 values, longer preirradiation periods will be needed.

It is sometimes wrongly supposed that, since AQ is shorter than τ, a shorter period of irradiation could be used after the first transient at a particular frequency. This is because, it is argued, the NOE would not have decayed much by the end of AQ, and would require only "topping up" by the subsequent preirradiation. This argument leads to a scheme in which, of the 8 or 16 scans acquired at each frequency on each pass around the frequency interleaving cycle (Section 7.2.3), only the first need employ the full period τ, with subsequent preirradiations being reduced in length. *This is quite wrong.* The fallacy in this argument is that it neglects the effect of the observe pulse on the populations. A 90° pulse *completely equalizes* the populations. In creating the transverse signals actually detected, it thus completely eliminates the NOE (i.e., the nonequilibrium population distribution) present at the end of τ, and the next preirradiation must create the NOE from scratch, just as did the first preirradiation. If the pulse length is reduced, then a proportion of the NOE survives, in proportion to the z component remaining after the pulse (i.e., cos α, where α is the pulse angle). To this extent it might be thought that a time-saving scheme such as that just discussed might be possible if a reduced pulse angle were used. There is some truth in this, but the gain in signal made by acquiring more transients in a given time is, in practice, more than offset by the signal loss resulting from the smaller pulse angle (the signal excited being of course proportional to sin α).

In practice, then, there is no advantage in using a non-90° pulse, and the full period τ should be used for every transient. This has the added advantage that SPT patterns are at least partially suppressed by the 90° pulse.

7.3. PROCESSING, DISPLAY, AND CALCULATION OF RESULTS

In the first part of this section are collected some general points that apply once the experiment is complete and, if appropriate, the difference spectrum has been computed. In the second part, we discuss a processing technique called reference deconvolution that can dramatically improve the quality of NOE spectra after acquisition.

7.3.1. General

The first points to check are the selectivity of saturation (Section 7.2.4) and the quality of subtraction (Section 7.2.2). It cannot be overemphasized that saturation selectivity is one of the most crucial aspects of 1D NOE experiments, and an examination of the irradiated region in the difference spectrum should always be made to check for unwanted saturation of neighboring resonances. Errors of interpretation that arise when enhancements due to such saturation are not recognized are particularly insidious, as pointed out elsewhere.

Subtraction quality can be gauged from the intensity of the "dispersion-like" artifacts resulting from residual incomplete subtraction, which can provide a very rough confidence limit on the level of enhancements that can be reliably measured in the difference spectrum. Of course, this cannot guarantee that a peak in the difference spectrum is genuinely due to an NOE enhancement; it merely shows that it arises from a systematic amplitude difference (of whatever origin) between the control and enhanced spectra, rather than a subtraction artifact. The whole issue of interpretation is deferred to Part III (Applications), and Section 10.1 in particular.

The measurement of the enhancements themselves calls for some comment. It is probably common practice to use measured peak heights in the difference and control spectra to work out numerical values for the enhancements. This practice can be dangerous. Peak heights in the difference spectrum may be considerably distorted, either by poor subtraction or by SPT patterns, even in difference spectra where the thermal noise level is negligible. In those fairly rare cases in which an enhanced signal in the difference spectrum is demonstrably free of such problems, then estimates based on peak heights may provide a "guesstimate" of the numerical values of enhancements, but it is undoubtedly better to use integration to measure the intensities in both the difference and control spectra. As previous sections have pointed out, residual SPT distortions and subtraction artifacts have no net integral, and are thus automatically discounted from such measurements, while thermal noise in the difference spectrum is at least dealt with in a consistent fashion from one signal to another. It may well be difficult to adjust the slope and bias properly when measuring integrals in noisy or distorted difference spectra, just as it is in noisy or distorted normal spectra. Unwelcome though this may be, this difficulty is no more than a genuine symptom of the fact that enhancements in such spectra are poorly determined.

Correction of the enhancements for partial saturation, using Eq. 7.3, has already been discussed in Section 7.2.4.

Finally, a few points about displaying difference spectra. By far the most intense signal in an NOE difference spectrum is the irradiation target itself, which may be several hundred times more intense than the enhanced signals of interest. There is, therefore, no point in plotting a figure to include all peaks at a constant vertical expansion, since if the irradiation target is to remain on-scale this condemns the NOE enhancements to invisibility. Nonetheless, many such figures appear in the literature. If the irradiation target is to be shown, this can be done using a small insert at reduced vertical expansion. By convention, and for clarity, difference spectra are usually plotted so that (direct) enhancements appear as positive signals, whether these correspond to positive enhancements ($\omega\tau_c < 1.12$) or negative ($\omega\tau_c > 1.12$).

7.3.2. Reference Deconvolution

Most of the undesirable features present in a 1D NMR spectrum will be present uniformly in all of the signals. The most obvious example of this is lineshape imperfection caused, for instance, by poor shimming. Although such effects may be noticeable to different extents for different signals, depending on their individual linewidths and multiplet structures, it is quite clear that the *form* of the distortion from bad shimming must essentially be the same for all lines in a given spectrum. Much the same is true for imperfections from a number of other sources, such as phase jitter in the RF, disturbance of the lock channel during acquisition, modulation by electrical mains interference, or spinning irregularities.

Reference deconvolution seeks to take advantage of this identity in the form of the distortion experienced by different signals. The technique relies on using a reference signal for which an ideal lineshape can be defined; most commonly this should be an isolated singlet with a good signal-to-noise ratio. By comparing the experimentally recorded form of this reference signal with its ideal form, one can define exactly what distortions the reference signal must have been subjected to, which in turn allows a correction function to be defined that can convert the experimental form into the ideal form. Once this correction function is known it can be applied to all the signals in the spectrum, thereby removing the experimental imperfections from them also.[26,27] Morris has named this technique FIDDLE (*F*ree *I*nduction *D*ecay *D*econvolution for *L*ineshape *E*nhancement).

In order to use this technique some additional software is required, and some small modifications should be made to the way in which the NOE difference experiment is carried out (see following). However, once these matters have been sorted out, the additional load involved in using reference deconvolution is trivial; very roughly, it involves issuing one or two extra software commands during processing and takes perhaps three or four times as long as a simple Fourier transform to carry out.[28]

In practice, this scheme is implemented as shown in Figure 7.13. Note that all data, time or frequency domain, consist throughout of *complex* points. The first stage is to separate out the contribution that the reference signal makes to the FID. First, the whole experimental FID $S_{\text{expt}}^{\text{total}}(t)$ is zero filled once to ensure that the real part of the transformed spectrum contains all the available information (cf. Section 8.5.1). After forward Fourier transformation and phasing, the reference signal is excised from (only) the absorption mode part of the spectrum; this is done by replacing the remainder of the absorption mode spectrum to left and right of the excised region with zeroes, and replacing the entire imaginary part by zeroes. This excised spectrum is then inverse Fourier transformed. Because there are only zeroes in the imaginary part of the input to this inverse transform, the resultant time domain signal is symmetric. The second half is discarded, and what remains, $S_{\text{expt}}^{\text{ref}}(t)$, corresponds to the separated contribution that the reference signal made to the original complete FID. Note that,

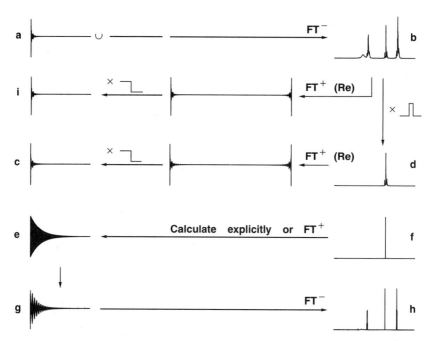

Figure 7.13. Schematic representation of the way in which reference deconvolution using FIDDLE works. Note that all the spectra and FIDs are comprised of *complex* points, even though only the real part of each is drawn here.

The experimental FID (*a*) is zero filled and forward Fourier transformed to yield the conventional spectrum (*b*), and the imaginary part (not shown) is replaced by zeroes. The reference line is then excised to yield partial spectrum (*d*), in which all regions other than that immediately around the reference line are replaced with zeroes in the real part. Inverse Fourier transformation of this part spectrum followed by deletion of the right-hand half of the resulting symmetrical FID yields (*c*), the experimental contribution to the FID arising from just the reference line (see text). Independently, one then computes the FID (*e*) that would correspond to the contribution of an idealized reference line (*f*). This can either result from inverse transformation of an ideal line, or from explicit calculation of the time-domain function. Taking the complex ratio of (c) and (e) then yields the correction function $R(t)$ (not shown), which is used to multiply the experimental FID to generate a corrected FID (g) (cf. Eq. 7.4). Forward Fourier transformation of (g) then yields the corrected spectrum (*h*). As a "safety measure," the experimental FID is actually substituted by FID (*i*), which results from passing (a) through the combination of forward Fourier transformation, selection of the real part, inverse Fourier transformation, and discarding of the second half. Conceptually, using FID (i) is identical to using FID (a), but by passing the experimental FID through the same combination of calculations as is used to generate the FID corresponding to the excised reference line, one takes care of any potential anomalies connected with subtly different implementations of the Fourier transform (e.g., different handling of first point errors, scaling, etc.).

if both the absorptive and dispersive parts of the spectrum corresponding to the excised reference signal were inverse transformed, this would yield its contribution to the normal FID directly (indeed, this was the form in which FIDDLE was originally proposed). However, this in turn would require that both the absorptive *and* dispersive parts of the reference signal were free from overlap, greatly increasing the frequency separation required between the reference and any neighboring signals. The procedure described here using only the absorptive part is equivalent to generating the imaginary part of the excised reference signal by using a Hilbert transform.[29]

Armed with the separated time-domain contribution from the reference signal $S_{\text{expt}}^{\text{ref}}(t)$, one may now use it to extract the explicit functional form of the instrumental distortions. This is done by *dividing* the corresponding ideal time domain signal $S_{\text{ideal}}^{\text{ref}}(t)$ by $S_{\text{expt}}^{\text{ref}}(t)$ (strictly, by taking the complex ratio), to yield a correction function $R(t)$ (see Eq. 7.4 to follow). The ideal time domain signal $S_{\text{ideal}}^{\text{ref}}(t)$ itself is more or less freely chosen, and is supplied by direct calculation; for instance, in the case that the desired ideal lineshape is a Lorentzian of width $(1/\pi T_2)$, $S_{\text{ideal}}^{\text{ref}}(t)$ is given by the function $\exp(-t/\pi T_2) \cdot \exp(-i\omega t)$.

The correction function $R(t)$ is the key to removing the instrumental imperfections from the remaining signals in the spectrum. One useful way to think of $R(t)$ is as the ultimate window function; multiplication by $R(t)$ has the effect of replacing the imperfect experimentally measured lineshape by the ideal lineshape defined for the reference signal, so when this is done for the complete FID $S_{\text{expt}}^{\text{ref}}(t)$ all signals are converted to the ideal lineshape.

$$S_{\text{ideal}}^{\text{total}}(t) = S_{\text{expt}}^{\text{total}} \times R(t) = S_{\text{expt}}^{\text{total}} \times \frac{S_{\text{ideal}}^{\text{ref}}(t)}{S_{\text{expt}}^{\text{ref}}(t)} \qquad (7.4)$$

As described so far, the FIDDLE technique would be just a very effective way of recovering the information content of an individual "damaged" 1D spectrum. The benefit it offers in difference spectroscopy arises when the two spectra to be subtracted are both corrected *independently* using reference deconvolution.[30] Because this enforces an identical ideal lineshape in both spectra, any unwanted distortions that affect the reference signal differently in the two spectra are removed along with all the distortions that are the same in both cases; in the former case, the form of $R(t)$ will differ slightly for the two data sets. Note particularly that such corrections include the cases of slight relative frequency or phase shift between the spectra, which are probably the predominant cause of subtraction artifacts in most NOE difference spectra (cf. Section 7.2.2).

Figure 7.14 shows an application of the FIDDLE technique to steady-state NOE difference spectroscopy. As can be seen, the usual subtraction artifacts have been eliminated to below the thermal noise level, resulting in a spectrum of remarkably high quality. In spectra of this quality, the threshold for detection of NOE enhancements can be as low as 0.02%. The region used for the reference signal (in this case the residual $CHCl_3$ signal) appears completely flat

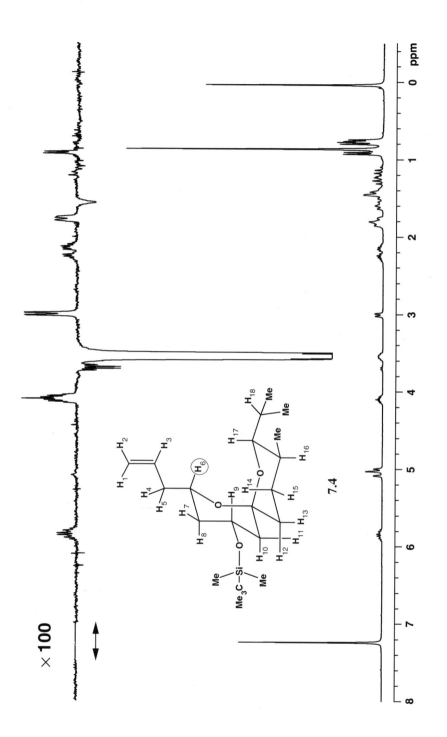

7.4

×100

252

in the difference spectrum, precisely because the reference deconvolution process has replaced it by the ideal (noise-free) lineshape in both spectra. Of course, the noise from this region has not truly disappeared entirely, since it influences the form of $R(t)$ and thereby affects the signal-to-noise ratio in the remaining parts of the spectrum.

One modification to the normal steady-state difference technique that has been recommended for experiments intended for processing using FIDDLE is to use gating rather than frequency shifting to generate the control data.[30] If off-resonance irradiation when collecting the control data generates even very small transverse magnetizations, these can persist into the acquisition time and cause unwanted systematic differences between the control and irradiated data. (Note that this effect is one of the sources of difference artifacts generally present in conventional NOE difference experiments, but it is usually swamped by other larger errors such as small frequency and phase displacements.) Since such effects vary according to offset, they cannot be wholly removed using FIDDLE; however, they can be avoided by gating the preirradiation completely off during collection of the control data (rather than relying on frequency shifting) and phase cycling the irradiation separately from the pulse during collection of the preirradiated data. Note that for this phase cycling to be effective, the preirradiation should be applied at the transmitter offset, so the acquisition window must be arranged such that the target signal is at its center. As a more convenient alternative where gradients are available, the unwanted transverse components may be dephased by a gradient applied immediately prior to the observe pulse. This approach has the advantage that only one control data set is needed, since the target is no longer constrained to be at the center of the acquisition window.

The main potential difficulty with the FIDDLE procedure is that, because it includes a division step, when $S^{\text{ref}}_{\text{expt}}(t)$ includes zeroes, $R(t)$ becomes undefined. This is one reason why all the time-domain data must be complex, since oscillation of the signal does not then introduce zeroes as it would with purely real data. It is also the reason why multiplets are difficult to use as reference signals, as the corresponding time-domain signals always contain zero crossings; however, in some cases this difficulty can be largely overcome by using careful interpolation.[31] One commonly used reference signal is the residual $CHCl_3$ line present in $CDCl_3$ solutions, but in cases where this is overlapped other singlets can be used. The TMS singlet present in many 1D ^{1}H spectra of

←

Figure 7.14. Example of a steady-state NOE difference spectrum processed using reference deconvolution. The compound studied was **7.4**, and the irradiation target was H_6 (circled). The region that was excised to provide the experimental version of the reference line (ca. 7.0–7.5 ppm) appears completely flat in the difference spectrum. Subtraction artifacts are almost completely absent from the remainder of the difference spectrum.

organic molecules can be used as a reference signal, preferably with the various long-range (^{1}H, ^{29}Si) and (^{1}H, ^{13}C) coupling satellites correctly included in the ideal reference lineshape. The only other obstacle to widespread use of the FIDDLE technique is the requirement for some nonstandard software or home-written programs; however, it is to be hoped that this will not prevent the technique gaining widespread acceptance in view of the quite dramatic improvements it can offer.

7.4. OTHER 1D EXPERIMENTS EMPLOYING CONTINUOUS SATURATION

7.4.1. CW Steady-State Integration

Although it is now seldom used in earnest, this was the method used to measure NOE enhancements in almost all applications prior to the late 1970s. It has now been all but replaced by the NOE difference technique, but since much of the earlier literature is concerned with the results of CW experiments, it is worth describing the technique briefly, and discussing the ways in which it differs from the difference experiment.

Following acquisition of a normal CW spectrum, the CW NOE experiment consists of repeated integration of one signal (the I resonance) with, alternately, on-resonance irradiation of another signal (the S resonance), and off-resonance irradiation. After several repetitions (10 or more), all the on-resonance integral values are averaged and all the off-resonance integral values are averaged. The NOE enhancement is then calculated according to Eq. 2.1. An example of such a CW NOE experiment is shown in Figure 5.4.

The CW method, in normal use, has considerably lower sensitivity than the NOE difference method; enhancements smaller than about 5% are not easy to measure reliably, whereas the difference method can, with care, reveal enhancements in the 0.1% range. This distinction arises partly because of the shortcomings of manual integration, and partly because fewer measurements are averaged to obtain the final result. This is quite understandable. In the CW method, each integral measurement requires the operator to carry it out, without the benefit of automatic control of the spectrometer. This is consistent with our earlier comment (Section 7.2) that the superiority of the NOE difference method has little to do with the fact that the results are displayed in the difference mode.

Another distinction is that, in the CW method, the irradiation is necessarily switched on during acquisition, implying that the resonances may undergo Bloch–Siegert shifts or decoupling effects. This complicates interpretation, and was the reason why the on-resonance and off-resonance frequencies were generally kept as close together as possible.

Possibly the most important, and least obvious, distinction between the CW method and the difference technique, however, is that the results of a CW NOE experiment are essentially *numerical,* whereas those of an NOE difference experiment are essentially *pictorial.* Although this sounds like a truism, it has had a considerable influence on the way in which the two sorts of experiments are executed and interpreted. In order to make the numerical results of CW experiments more reproducible, early experiments placed a very heavy emphasis on careful degassing and complete saturation of the S resonance, to produce the full enhancement. Moreover, interpretation of the results employed rather complex equations that take all relaxation mechanisms (i.e., ρ^*; Section 2.4.1) explicitly into account. In contrast, in an NOE difference experiment, the precise numerical values of the enhancements are usually of less interest. Greater emphasis is placed on *selectivity* of irradiation than on full saturation, and the reproducibility of results from one occasion to another is of less importance than the elimination of unwanted differences between the control and enhanced data in a single experiment.

Various "hybrid" FT experiments have also appeared, which combine features of the CW NOE experiment with FT detection. On the whole, these represent an intermediate stage in the evolution of the NOE experiment, and require little comment.

7.4.2. The Truncated Driven NOE (TOE) Experiment

The only difference between the truncated driven NOE (TOE) experiment and the steady-state difference experiment is that in TOE experiments the irradiation is stopped and the spectrum acquired before steady state is reached. The usefulness of TOE, as explained in Chapter 4, is that at short preirradiation times, the enhancements build up at a rate proportional to σ, which permits a direct estimate of distance.

As discussed in Section 1.5, when the irradiation is turned on, saturation is not instantaneous; the intensity of the irradiated signal approximately follows an exponentially decaying cosine wave of the form:[32]

$$I_z = I_z^0 \exp(-R_{2\rho}^* t)\cos(\omega_2 t) \qquad (7.5)$$

where $R_{2\rho}^*$ is equal to $1/T_{2\rho}^*$.

This naturally affects the build-up of the enhancements. Figure 7.15 shows some simulations, calculated by a numerical integration of the Solomon equations, as described by Chapter 4. Figure 7.15a and b show simulations for rapidly and slowly tumbling molecules, respectively, using fairly low power irradiation ($\omega_2 = 50$ rad s^{-1}, corresponding to a 180° pulse length of 62 ms, very similar to the 40 dB attenuated power level illustrated in Fig. 7.9). The oscillatory effect is very marked, and would lead to very severe errors in estimation of the true build-up rate, particularly at short times. The error is greatest when the magnetization due to the irradiated spin has rotated by roughly $(2n + 1)\pi/2$; thus, for example, in Figure 7.15b at 0.15 s, when the irradiated spin had rotated by approximately $5\pi/2$, the observed enhancement

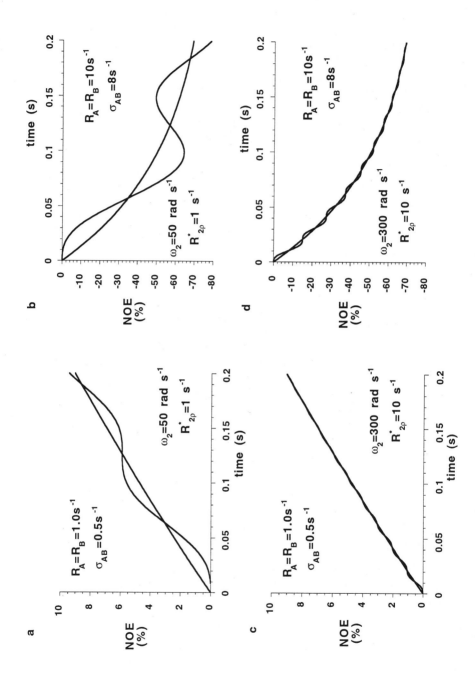

would be -49% whereas the true enhancement for instantaneous saturation would be -64%, an error of 31%. When the irradiated spin has rotated by $2n\pi/2$, the error is essentially zero. Figures 7.15c and d show simulations for higher power irradiation ($\omega_2 = 300$ rad s^{-1}, a 180° pulse length of 10.5 ms, or roughly a 35 dB attenuation on the spectrometer used for Fig. 7.9). In this case the error is much smaller and would probably be less than the experimental uncertainties of measurement.

The conclusion to be drawn from this simulation is that in TOE one should use the highest irradiation power possible consistent with selective saturation."[32] This is to be contrasted with the situation for steady-state NOE experiments, where lower powers are generally preferable. We could also add that if one is forced to use a lower power for reasons of selectivity, then the most reliable enhancements are seen for rotations of $2n\pi/2$, and the largest enhancements (which are no longer directed related to σ) are seen for rotations of $(4n-1)\pi/2$.

At extremes of high and low power, other anomalies appear. At very low power, the intensity of the irradiated signal does not oscillate, but merely decays away exponentially to zero (Section 1.5). This may occur at a rate comparable to or less than σ, which means that saturation becomes the rate-determining step in NOE buildup rate, so that NOE buildup is no longer proportional to σ. One of the results of this reduction in the buildup rate is that the distance dependence of the NOE at short times is compromised, and it becomes much harder to use enhancements to make deductions about distance.[33]

At high power levels the greatest problem is the effect of the irradiation on other signals close to the irradiation frequency. The most obvious effect is partial saturation of nearby signals, as discussed in Section 7.2.4. There are additional problems if the nearby signal also receives an NOE enhancement from the irradiated signal, since in this case the off-resonance irradiation induces oscillations in the signal intensity, as well as changing the apparent initial buildup rate (cf. Section 1.5).[15]

Oscillation in intensity of the saturated spin means that its apparent intensity in the difference spectrum is no longer a measure of the extent of saturation, and therefore should not be used as the reference intensity for calculating sizes of enhancements (Section 7.2.4). The best value to use under these circumstances is probably the intensity of S in a steady-state difference experiment

---

Figure 7.15. Simulations of the buildup of NOE enhancements with low-power irradiation. The nonoscillatory curves are the corresponding simulations for instantaneous saturation. (a) and (c): Extreme narrowing limit, $R_A = R_B = 1.0$ s^{-1}, $\sigma_{AB} = 0.5$ s^{-1}. (b) and (d): Near the spin diffusion limit, $R_A = R_B = 10$ s^{-1}, $\sigma_{AB} = -8$ s^{-1}. (a) and (b): Low-power saturating field, $\omega_2 = 50$ rad s^{-1}, $R_{2\rho}^* = 1$ s^{-1}. (c) and (d): Higher power field, $\omega_2 = 300$ rad s^{-1}, $R_{2\rho}^* = 10$ s^{-1}.

using the same power level, unless it can safely be assumed that the irradiation power is high enough to saturate the signal completely.[32]

Finally, and most importantly, we should add that for the TOE experiment (in contrast to the steady-state experiment) it *is* important for the spin to be at equilibrium before the start of the pulse sequence.[†] It is therefore necessary to wait at least $5T_1$ between each observe pulse and the start of the next irradiation. This can lead to a very slow pulsing rate for small molecules with long T_1 values. For more slowly tumbling molecules relaxation rates are faster, and the problem is less severe.

7.5. TRANSIENT EXPERIMENTS

In a 1D transient experiment, the NOE is excited not by continuous selective irradiation of the target signal S throughout the period τ, but rather by a selective perturbation of S at the start of τ. This perturbation can comprise either inversion (from an effective 180° pulse to S) or saturation (from an effective 90° pulse to S) depending on the experiment. Cross-relaxation of S during its subsequent decay back to equilibrium then causes NOE enhancements to appear during τ. Initially these build up at a rate proportional to 2σ (following inversion) or σ (following saturation); then their growth slows, the enhancements reach a maximum (each at a different time) and finally decay back to their equilibrium value of zero. The enhancements observed in a particular transient experiment thus depend on where the buildup and decay curve of the individual magnetizations has got to at the time of the observe pulse. The theory was discussed in Sections 4.2 and 4.3.1; the very close analogy between this experiment and the 2D NOESY experiment[36] is valuable to remember.

Although Solomon's first demonstration of the NOE employed a transient experiment,[37] until recently 1D transient experiments were not very widely used. Their principal area of application was similar to that for TOE experiments, namely measurement of initial rates of NOE buildup for estimating internuclear distances, but the fact that 1D transient experiments necessarily involve selective pulses used to make them somewhat more complicated to set up than TOE experiments.

However, since the mid-1990s this situation has been completely changed by the advent of some new and extremely powerful pulsed field gradient experiments to measure transient NOE enhancements. The advance offered by these experiments (for which there are no steady-state analogs) is that they largely or completely eliminate subtraction artifacts. As anyone who has read

---

[†]Simulations and experiment[34,35] have shown that *in most cases* reasonably accurate results can still be obtained for both TOE and transient experiments even with total relaxation delays of as little as T_1; however, great care is needed both to identify those cases in which larger errors may occur and to apply the necessary corrections for short relaxation delays. In quantitative applications, truncation of the relaxation delay should be used with extreme caution.

the earlier sections of this chapter (particularly Section 7.2.2) will appreciate, it is subtraction quality that sets the real limits on detectability of enhancements in conventional NOE difference experiments. Thus, even though transient NOE enhancements measured for small molecules are smaller in absolute terms than the corresponding steady-state enhancements, in practice these new gradient-assisted transient NOE experiments are significantly more sensitive than conventional steady-state NOE difference experiments because they avoid the need to batter down subtraction artifacts through prolonged signal averaging.

In the following, we shall first touch on selective pulses, which are needed for any 1D transient NOE experiment, then we will briefly discuss nongradient versions of the 1D transient experiment, before moving on to discuss the new gradient-assisted experiments.

7.5.1. Selective Pulses

In any 1D transient NOE experiment, the S signal must be perturbed using a selective pulse. This may be either a selective inversion or a selective excitation pulse, depending on the experiment. Selective pulses come in all shapes and sizes, and most modern spectrometers include facilities for generating and executing such pulses in a fairly straightforward manner. It would be inappropriate to go into detail here concerning the relative merits of the very many possible selective pulses available (these have been recently reviewed by Freeman[38]), but we will give a few generalities about using selective pulses.

Rectangular pulses used to be popular because they were so simple to implement; they only require the RF to be switched on and off at a preset power, rather than modulating its amplitude during the pulse. (A still less demanding alternative, DANTE[39] even avoided the requirement for reduced power by using a train of very short pulses alternated with delays to achieve the cumulative effect of a selective rectangular pulse.) However, hardware improvements over the past decade or so have all but eliminated these technical considerations, and shaped pulses are now generally preferred. As an illustration of the reason for this, Figure 7.16 shows the off-resonant response to a rectangular inversion pulse in comparison to that for one of the simplest shaped inversion pulses, namely a Gaussian; the rectangular pulse shows a series of undesirable sidelobes in its frequency response that are absent for the Gaussian.[20,40] When exciting or inverting a multiplet, it is particularly important that the selective pulse affect all the multiplet components as evenly as possible, so as to avoid creating unwanted phenomena such as SPT distortions or multiple quantum coherences with coupling partners. This is the essential aim behind the design of all selective pulses, even though in some experiments a certain amount of "cleaning up" can be achieved through other features of the pulse sequence (e.g., the phase cycle or gradients). Despite the plethora of publications in this area, it is probably fair to say that the Gaussian still remains the "workhorse" of selective pulses, and one seldom goes very far wrong by using it; nonetheless, many more sophisticated alternatives are available.

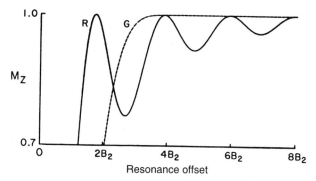

Figure 7.16. Calculated off-resonant response to spin inversion for a rectangular (R) and a Gaussian (G) pulse. The resonance offset is normalized in terms of the intensity of the radiofrequency field for the rectangular pulse. Reproduced with permission from ref. 20.

It is, of course, important to calibrate the pulse, and if possible to carry out a few simple tests to demonstrate that it has the required selectivity for its purpose (e.g., by running the NOE experiment on the sample of interest with τ set to zero in order to check that the difference spectrum contains only the S signal). Such a test can also be used to assess the extent of inversion of the S signal, which is essential information if the NOE enhancements seen in the actual experiment are to be quantified (Section 7.5.3.5).

7.5.2. Nongradient Transient NOE Experiments

The simplest form of 1D transient NOE experiment is shown in Figure 7.17a. Inversion of just the S signal is achieved using a selective 180° pulse at the

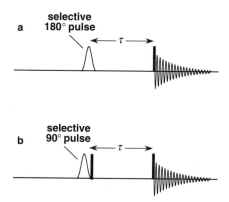

Figure 7.17. (a) Simple pulse sequence for a 1D transient NOE experiment. (b) Alternative sequence for a 1D transient NOE experiment, where inversion is achieved using a combination of a selective 90° and hard 90° pulse.[41] Hard 90° pulses are indicated by filled vertical bars.

start of the τ period; control data, for subtraction to generate the difference spectrum, is obtained either by omitting the selective pulse or, better, by moving it sufficiently off-resonance that it does not perturb any signal in the spectrum.

As for TOE, the principal application of this experiment is for the quantitation of initial buildup rates of enhancements to estimate internuclear distances. Long relaxation periods are required to re-establish thermal equilibrium between each acquisition, making the experiment very time-consuming when applied to small molecules. There is, however, a distinction from TOE that is seen when using lower power irradiation to achieve greater selectivity. In TOE, cutting back the irradiation power makes the S spin magnetization oscillate more slowly, causing larger oscillatory distortions in the NOE growth curve (Fig. 7.15). For the transient experiment, there is a different problem; the more selective (lower power) the inversion pulse, the longer it must last, with the consequence that the earliest part of the NOE buildup curve takes place during the selective pulse itself and hence cannot be observed.

The pulse sequence shown in Figure 7.17b, published by Kessler et al.,[41] removes this difficulty. The selective 180° pulse of the normal sequence (Fig. 7.17a) is here replaced by a selective 90° pulse followed by a nonselective 90° pulse. To a much greater extent than for the selective inversion sequence of Figure 7.17a, this experiment is a precise 1D analog of a 2D NOESY experiment, where frequency labelling of all signals during t_1 of the 2D experiment is replaced by selective excitation of one signal (S).

Although now largely superseded by the various gradient-assisted experiments discussed in Section 7.5.3, this experiment merits discussion because it represents the prototype for all of the later, more sophisticated, transient NOE pulse sequences. As normally implemented, it is not a difference experiment in the usual sense of subtracting a data file from a control file; rather, subtraction occurs directly as a result of phase cycling. If the first two pulses are $90°_x$ pulses, they act cumulatively to produce inversion of the S signal at the start of τ, whereas the $90°_x, 90°_{-x}$ combination leaves the S signal unaffected. The difference between the two resulting data sets accumulates directly in memory, by negating the receiver phase each time the $90°_x, 90°_{-x}$ combination of pulses is used. The phase cycling also suppresses all magnetization that was not longitudinal during τ, thereby eliminating any NOE contributions that evolved *during* the selective pulse (since the first hard 90° pulse makes these transverse during τ). This is exactly analogous to suppression of axial peaks and relay contributions by the phase cycling in a NOESY experiment (Section 8.4). On a practical level, note that this particular experiment requires that the absolute phase relationship between the low- and high-power pulses must be known and, if necessary, corrected for during implementation.

7.5.3. Gradient-Assisted Transient NOE Experiments

One of the key developments in NMR during the 1990s has been the introduction of pulsed field gradients as an element in pulse sequence design.[42] The essential idea of using pulsed field gradients is that they provide a way of distinguishing between different types of magnetization, as will be briefly outlined below. Although this concept is not new, it was only with the introduction of self-shielded gradient coils (originally developed for use in MRI) that gradient-assisted experiments became conveniently possible in high-resolution NMR spectroscopy. Triple-axis gradient systems are available, but to date most gradient-assisted spectroscopic experiments require only a single gradient direction, and given the shape of most NMR sample tubes and probes the obvious choice of direction for a single axis gradient is along the z-axis.

The following two sections describe experiments that use z-gradients very effectively to "clean up" NOE spectra. That these are transient NOE experiments is of almost entirely secondary importance; their principal *raison d'être* is to improve sensitivity by eliminating the subtraction artifacts that set the real sensitivity limit of conventional NOE difference spectra. As pointed out elsewhere (e.g., Sections 4.2, 4.3, 8.1, and 10.1), transient experiments are expected to give smaller absolute enhancements than steady-state experiments for small molecules, other things being equal. The fact that, *despite this*, these new transient experiments can give much better results with small molecules than do steady-state NOE difference experiments underlines just how successful the sequences are in achieving their goal of eliminating subtraction artifacts.

7.5.3.1. Gradient Selection.

It would be inappropriate to give a detailed account of the theory of pulsed field gradients here, but a cursory outline is needed as a basis for describing the GOESY and DPFGSE-NOE experiments; more detailed accounts can be found elsewhere.[42]

During the time that a field gradient is switched on, the precession frequency of each spin becomes a function of its position in the sample, so that by the time the gradient is switched off each precessing spin has acquired a position-dependent phase. This result is illustrated schematically in Figure 7.18. More formally, for each homonuclear coherence experiencing a z-axis gradient this position-dependent phase $\phi_g(z)$ is given by

$$\phi_g(z) = \gamma p B_g(z) \tau_g \tag{7.6}$$

where $B_g(z)$ is the strength of the gradient at position z and τ_g is its duration, γ is the gyromagnetic ratio and p is the coherence order (0 for z magnetization or zero-quantum coherence, ± 1 for normal transverse single quantum coherence, ± 2 for double quantum coherence, and so on). A similar but slightly more involved formula applies in the heteronuclear case. Note that, since z magnetization and zero quantum coherences have $p = 0$, they are unaffected by gradients.

Once dephased in this way, even a single-quantum coherence cannot give rise to an observable NMR signal unless subsequent gradients rephase it. For instance, if a gradient of equal strength and duration but opposite sign were applied immediately following the first, this would of course restore the signal.

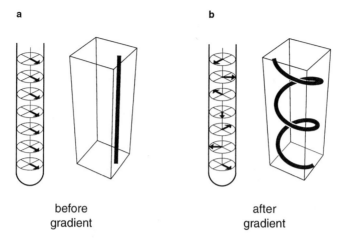

before gradient   after gradient

Figure 7.18. Position-dependent phase acquired by magnetization as a result of applying a pulsed field gradient along the z-axis, represented both using individual magnetization vectors and using a continuous "phase vector" that links the ends of all magnetization vectors. (*a*) Magnetization before the gradient. (*b*) Magnetization after the gradient. To put this diagram into a more realistic context, for a gradient of strength 10 Gauss cm^{-1} and duration 1 ms, the pitch of the magnetization spiral produced for ^{1}H single-quantum coherence would be about 0.23 mm.

However, the interest lies in manipulating the spins through the pulse sequence during the time between gradients, since later gradients can select which of the coherences then present are to be made observable. For instance, if the aim was to select magnetization that was present as a single-quantum coherence during one gradient but double-quantum during a second, the second gradient would be set to half the length (or alternatively half the strength) of the first. The first gradient is said to cause a phase-encoding of the signal, while the second decodes the signal by rephasing it. More formally, the condition for a signal to be observable (rephased) after a series of gradients is

$$\gamma p_1 B_{g1}(z)\tau_{g1} + \gamma p_2 B_{g2}(z)\tau_{g2} + \gamma p_3 B_{g3}(z)\tau_{g3} + \cdots = 0 \qquad (7.7)$$

where $B_g(z)$ and τ_g take individual values for each gradient, and for each gradient p refers to the coherence order of the particular coherence that is to be selected during that gradient.

From the standpoint of NOE experiments, the crucial difference between selection using gradients and selection using phase cycling (cf. Section 8.4) is that gradients select the desired signal *in real time*, actively destroying the unwanted signals within an individual scan. In contrast, phase cycling and difference spectroscopy rely on removing unwanted signals by taking differences between scans, which demands that the unwanted signals are identical in both scans. This in turn will only be true if the spectrometer behaves in a completely reproducible fashion from scan to scan; when it does not, the result will be subtraction artifacts.

7.5.3.2. DPFGSE-NOE.
The heart of this experiment is the DPFGSE (*D*ouble *P*ulsed *F*ield *G*radient *S*pin *E*cho) pulse sequence element, shown in Figure 7.19a. This achieves selective excitation of a chosen signal into the transverse plane, but unlike conventional selective excitation it does not leave behind any z magnetization corresponding to the unexcited signals. The authors call this process "excitation-sculpting."[43,44] The initial hard 90° pulse puts all magnetization into the transverse plane. However, only magnetization inverted by the selective 180° pulse will survive the first pair of gradients (both called G_1 as they have identical strength and length), since only for this component of magnetization will the effect of these two gradients cancel; for the all other magnetization their dephasing effect is additive. The first gradient spin echo is then immediately followed by a second, identical to the first except that, crucially, a different gradient strength (G_2) is used. The reason for using two spin echoes is that it turns out that this allows virtually any pulse to be used as the inversion element. Phase errors that might be seen with a single selective gradient spin echo are eliminated when two echoes are used, leading overall to a very clean excitation of the *S* signal with pure phase. In addition, the selectivity of the sequence is improved by repeating the spin echo. For a single pulsed gradient spin echo, the excitation profile is given by the inversion profile of the inversion pulse, whereas for a double pulsed gradient spin echo, the excitation profile is given by the square of the inversion profile of the inversion pulse.

The DPFGSE sequence can be used as the first element in a whole range of selective-excitation-based techniques. To build a 1D NOE experiment from it, one simply grafts on an NOE evolution period as shown in Figure 7.19b. The resulting "DPFGSE-NOE" sequence has much in common with the original "1D NOESY" sequence of Figure 7.17b. Essentially the only difference is that, in the former, selective excitation of *S* is achieved by the DPFGSE sequence, whereas in the latter a simple selective 90° pulse is used. However, this has the crucial consequence that immediately following the selective excitation part of the DPFGSE-NOE sequence (i.e., just prior to the 90° at the start of τ), all signals other than *S* are not merely absent from the transverse plane, their corresponding z magnetizations are also destroyed.

It is clear that, as for the 1D selective NOESY sequence of Figure 7.17b, the DPFGSE-NOE sequence is essentially a difference experiment. When the first two hard 90° pulses have the same phase, they act cumulatively to produce inversion, whereas when they have opposite phases, they cancel, restoring equilibrium z magnetization for the *S* signal. Subtraction of these two types of data occurs during the phase cycle and generates the NOE difference spectrum. However, it is important to see what is the nature of the signals that need to be cancelled by the phase cycle in the DPFGSE-NOE experiment, since in practice they can be made to be very small, and this in turn makes their as-

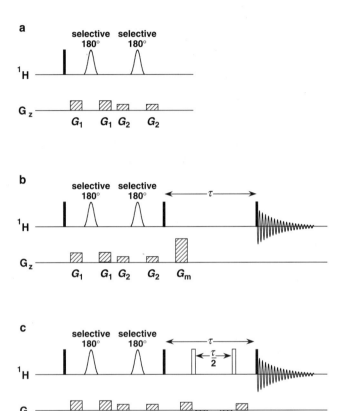

Figure 7.19. (*a*) DPFGSE (*D*ouble *P*ulsed *F*ield *G*radient *S*pin *E*cho) pulse sequence element for selective excitation. (*b*) Simple DPFGSE-NOE pulse sequence. (*c*) DPFGSE-NOE pulse sequence modified to eliminate residual subtraction artifacts. Two broadband 180° pulses are added during the NOE evolution period τ, so as to bring the recovering z magnetization back to zero amplitude exactly at the end of the NOE evolution period. This makes the signals that need to be cancelled by the phase cycle very small, and hence greatly reduces subtraction artifacts. For short NOE evolution periods, these inversion pulses are placed at 0.25τ and 0.75τ, while for longer NOE evolution periods they should be moved (as a pair, at a constant separation of 0.5τ) somewhat towards the end of the NOE evolution period. The optimal positioning can be found by trial and error, minimizing the intensity of all signals (other than S or any strongly enhanced signals) in a DPFGSE-NOE spectrum obtained with only one transient. Hard 90° pulses are represented by narrow filled vertical bars, hard 180° pulses by wider, open vertical bars.

sociated subtraction artifacts very small also. This is the key to the excellent performance offered by the DPFGSE-NOE experiment (cf. Fig. 7.20e).

Given the absence of dephasing of all magnetization other than the selectively excited S signal immediately prior to the τ period, it follows that unwanted signals can only arise after this time. In the simple DPFGSE-NOE sequence, the only source of such signals is longitudinal relaxation during τ, which acts to recreate all the z magnetizations that the DPFGSE excitation had previously destroyed. The final hard 90° pulse turns these partially recovered z magnetizations into observable signals, this being exactly analogous to the appearance of axial peaks due to longitudinal relaxation during τ_m in 2D NOESY.

The size of these unwanted "recovered" contributions will increase with the length of τ, but Stott et al.[45a] have shown that they can be largely eliminated by adding one or two 180° pulses at strategic positions during the τ period. Apart from a sign change if an odd number of 180° pulses is added, this has no effect on the detected NOE enhancements.[46] However, when added at just the right moments, the 180° pulses do greatly reduce subtraction artifacts. To show how this works, consider first the example of a single 180° pulse at the center of τ, when τ is short. Any z magnetization that recovers during the first half of τ is inverted by the 180° pulse, so that during the second half of the τ period, further longitudinal relaxation causes it to relax towards zero. If the 180° pulse is placed such that the recovered z magnetization reaches zero exactly at the end of the τ period, then effectively the phase cycle no longer has any work to do since the signals it is supposed to cancel are now absent even from individual transients. Of course, the true recovery is exponential rather than linear, so for longer NOE evolution periods the best position for a single 180° pulse (i.e., that for which the recovered z magnetizations arrive exactly *at* zero at the end of τ) will be somewhat more than halfway through the τ period. Stott et al. recommend the best compromise is to use two 180° pulses placed slightly beyond 0.25 and 0.75 of the way through the τ period (see Fig. 7.17c for further details). Using two pulses reduces the dependence of the sequence on exact values of R_1, leading to better results for samples that show a spread of R_1 values. Stott et al.[45a] also give specific practical recommendations concerning types of inversion pulses, use of split gradients around each inversion pulse to eliminate signals from pulse imperfections, suitable gradient strengths and ratios, and a number of other refinements.†

When the DPFGSE-NOE experiment is to be used for measuring initial rates of NOE growth, recovery of z magnetization will only be slight and adding broadband 180° pulses to minimize subtraction artifacts is unlikely to be necessary. For short τ periods, SPT distortions and zero-quantum coherences generated during the selective inversion pulses are more likely to be a problem.

---

†In the paper Stott et al.[45a] discuss mainly the case of a single broadband 180° during τ, but significantly better nulling of subtraction artifacts is obtained when two are used as shown in Figure 7.17c. Addition of a third such pulse brings only a small further benefit.[45b]

In a further extension of the method, Stott et al.[45a] show that these effects can largely be eliminated by placing one or two *selective* 180° pulses at the end of the τ period.

It is clear from the foregoing that setting up a DPFGSE-NOE experiment for the first time may be a slightly involved task, including perhaps trying out the effectiveness of different positions and types of 180° pulses during the NOE evolution period τ. Nonetheless, quite remarkable results can be obtained, as illustrated by the spectra shown in Figure 7.20. Enhancements below the 0.1% level, perhaps even down to the 0.01% level,[47] that previously would not have been measurable at all can now be measured in quite reasonable experiment times (dictated entirely by spectrometer sensitivity, sample strength, and the size of the NOE enhancement, rather than by the need to average away artifacts), while somewhat larger enhancements that previously would have required long experiment times to detect reliably can now often be measured in a few minutes. These experiments seem set largely to take over the field of NOE measurements in small molecules, and perhaps even to move NOE detection into a new regime of sensitivity.

7.5.3.3. GOESY. GOESY (*G*radient enhanced nuclear *O*verhauser *E*ffect *S*pectroscop*Y*) was the first gradient-assisted 1D NOE experiment to be introduced, and the simplest version of the pulse sequence is shown in Figure 7.21a.[48] The underlying idea is extremely simple: Only magnetization excited by the selective 90° pulse is phase-encoded by the gradient G_1, so only this magnetization, together with any NOE enhancements that evolved from it during τ, are rephased by gradient G_2. The purging gradient G_m provides further discrimination by eliminating any transverse coherences (except zero-quantum) present during τ. The modified version of the pulse sequence shown in Figure 7.21b replaces the selective 90° pulse for excitation by a modified DPFGSE sequence, the benefits of which were discussed in the previous section. Note, however, that here the signs of the gradients differ from those used in DPFGSE-NOE. In the present case it is the effects of the gradients on the S signal that are cumulative, while other signals are all rephased at the moment of the second hard 90° pulse. Although superficially rather different, the key property that both of these two versions of GOESY share is that only the selectively excited magnetization is phase-encoded immediately prior to the NOE evolution period.[45]

7.5

268 ONE-DIMENSIONAL EXPERIMENTS

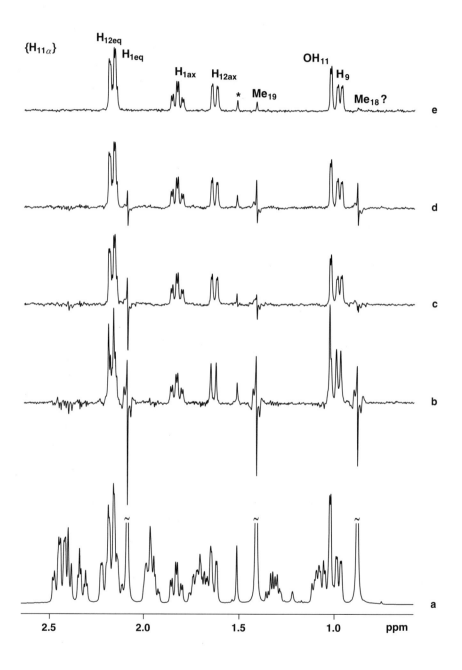

As shown in the original paper, the GOESY sequence results in very clean NOE spectra containing no measurable subtraction artifacts.[48] It turns out, however, that enhancements in GOESY spectra are very often disappointingly weak relative to those seen in conventional experiments. Partly, this is because the enhancements are transient and may have been sampled at a nonoptimum value of τ. Partly also it is because of an unavoidable loss of half the signal that is inherent in the design of the GOESY sequence. This loss arises because only one component of the dephased S spin magnetization is converted into longitudinal magnetization by the hard 90° pulse at the start of τ; the other half is left behind in the transverse plane and does not contribute to the finally detected signal.[†] In many cases, however, the predominant reason for signal loss in

---

[†] In some publications, the initial growth rate for NOE enhancements measured using GOESY is described as 0.5σ, a factor of four lower than for a conventional transient NOE experiment. The argument runs that one factor of 2 arises from the rotation of only half the dephased transverse magnetization by the 90° pulse at the start of τ (as just described in the text), while a further factor of 2 arises because the z-magnetization components that this same pulse creates include all values from $-z$ (fully inverted) to $+z$ (equilibrium). The average of this distribution is equivalent to saturation, and therefore produces only half the NOE enhancement that would be seen in a conventional transient experiment using full inversion. However, this argument is derived by consideration only of a single scan. In the conventional transient experiment (and in DPFGSE-NOE), half the scans employ full inversion of S, but the other half employ no perturbation of S, so over any even number of scans the average perturbation of S is 50% inversion, equivalent to that in GOESY. Thus, in reality, the measured initial growth rate of NOE enhancements seen in GOESY experiments corresponds to σ.

←

Figure 7.20. NOE difference experiments recorded for a 20 mM sample of 11β-hydroxyprogesterone (**7.5**) in CDCl$_3$ using a Bruker DRX500 spectrometer. (*a*) Conventional 1D spectrum. (*b*) Steady-state NOE difference spectrum. (*c*) Transient NOE difference spectrum (obtained using the sequence of Fig. 7.17a). (*d*) DPFGSE-NOE spectrum (obtained using the sequence of Fig. 7.19b). (*e*) DPFGSE-NOE spectrum with nulling of "recovered" z magnetizations during τ (obtained using the sequence of Fig. 7.19c). For each NOE spectrum, a total of 128 transients was recorded in 22 minutes and the target signal was H$_{11\alpha}$. As can be seen, spectrum (e) is essentially devoid of subtraction artifacts, and permits the observation of enhancements at Me$_{19}$ and possibly Me$_{18}$ that could not be reliably detected in the other spectra without much longer experiment durations. The signal at ~1.6 ppm marked * is an intermolecular NOE enhancement to H$_2$O dissolved in the CDCl$_3$, presumably transmitted via exchange from OH$_{11}$.

All selective pulses were Gaussians (truncated at 1%), with 10 ms duration in (c), and 5.4 ms in (d) and (e). Gradients were all of duration 1 ms, with strengths as follows: (d) $G_1 = 20\%$, $G_2 = 12\%$, $G_m = 35\%$; (e) $G_1 = 20\%$, $G_2 = 12\%$, $G_{m1} = 7\%$, $G_{m2} = -6\%$, $G_{m3} = -5\%$, $G_{m4} = 4\%$, where 100% was calibrated at 38 Gauss cm^{-1}. In (b), the saturation power was 20 Hz and the presaturation period was 5 s, while in (c)–(e) the NOE evolution period τ was 1 s, determined as giving close to the maximum enhancement. These enhancements are estimated to be in the range 0–4%. Reproduced with permission from ref. 45b.

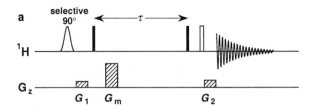

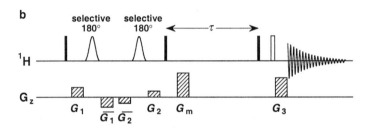

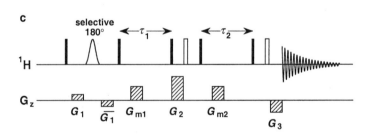

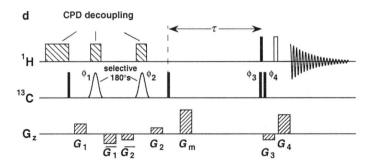

GOESY is molecular diffusion during τ. If the solute molecules move an appreciable distance during the period between gradients G_1 and G_2, this partially destroys the relationship between the phase of magnetization and its position within the sample, and thereby reduces the signal that can be rephased by gradient G_2. The problem is at its worst for small molecules in highly nonviscous solvents such as $CDCl_3$, and for long mixing times. These are not necessarily parameters that can be freely chosen to suit the requirements of GOESY, but one thing that can be done to help minimize signal loss is to use the weakest gradient strength compatible with satisfactory signal selection, since then molecular diffusion over a given distance will cause a smaller phase "error" than it would for a strong gradient. Nonetheless, the problem of signal loss due to diffusion clearly cannot be eliminated in any sequence that leaves the desired signal phase-encoded during the period of NOE evolution (τ).

Some improvement in this situation may be offered by recently reported "convection-compensated" versions of GOESY. Not all the movement of spins during τ is due to true diffusion; some of the movement can be caused by thermal convection within the solution. This movement differs fundamentally from true diffusion in that it is not random, at least over short time periods, and so in principle some or all of the signal loss that it causes can be recovered. Müller et al.[49] have described two ways of achieving this. One method depends on making a particular choice of gradient strengths, such that not only is Eq. 7.7 satisfied, but also the further condition that the velocity-dependent phases imparted by convection should sum to zero over the whole pulse sequence. This condition can be calculated straightforwardly (Müller et al.[49] give details), but it turns out that very strong gradients are needed unless τ is short, so exacerbating true diffusion losses. The second approach involves dividing the NOE evolution period of GOESY into two halves and inverting the magneti-

---

Figure 7.21. (a) Simple pulse sequence for GOESY.[48] Gradient G_2 must undo the dephasing caused by gradient G_1. (b) Pulse sequence for DPFGSE-GOESY experiment.[45a] Gradient G_3 must undo the cumulative dephasing caused by G_1, $\overline{G_1}$, $\overline{G_2}$, and G_2 (c) Pulse sequence for convention-compensated GOESY.[49] Gradient ratios are G_1: G_2:G_3 = 1:4:-2 (gradient strength G_{m1} and G_{m2} are unrelated to the others, since they are applied while the desired magnetization is along z). (d) Pulse sequence for HETGOESY. Gradients G_3 and G_4 must undo the cumulative dephasing caused by G_1, $\overline{G_1}$, $\overline{G_2}$, and G_2. The composite pulse decoupling during the relaxation delay augments sensitivity by creating a $^{13}C\{^1H\}$ NOE, while that during the selective ^{13}C 180° pulses allows for easier selective inversion by eliminating the multiplet structures of each ^{13}C resonance. Phase cycles are as follows: $\phi_1 = x, y, -x, -y$; $\phi_2 = 4(x), 4(y), 4(-x), 4(-y)$; $\phi_3 = x$; $\phi_4 = 16(x), 16(y), 16(-x), 16(-y)$; receiver $= x, -x, x, -x, -x, x, -x, x$. These phase cycles improve the performance of the 180° pulses and help eliminate SPT artifacts, but the main selection of the NOE enhanced signals is achieved by the gradients.[50] Hard 90° pulses are represented by narrow filled vertical bars, hard 180° pulses by wider, open vertical bars.

zation between them (Fig. 7.21c). The desired pathway selection is reinforced in the two halves of the experiment, but the effects of convection are approximately equal and opposite, and so largely cancel out. Unfortunately, the price for this discrimination is that another factor of two in signal strength is lost.

The benefit of using convection-compensated schemes will depend on the actual characteristics of convection in the sample, which in turn will probably depend heavily on several factors such as the VT set up (difference between sample temperature and ambient, VT gas flow rate, VT regulation loop parameters, etc.), sample depth, solvent viscosity, and probe design. At the time of writing, these experiments are too new for a clear consensus on their benefit to have emerged. Similar sequences for measuring rotating-frame enhancements (GROESY) with convection compensation were also proposed.[49]

A heteronuclear version of GOESY has also been developed for measuring $^1H\{^{13}C\}$ NOE enhancements in natural abundance materials (see Fig. 7.21d).[50] Of course, the problems of subtraction errors in difference spectroscopy are at least 100-fold more severe when measuring $^1H\{^{13}C\}$ NOE enhancements at natural abundance than they are when measuring $^1H\{^1H\}$ NOE enhancements, and in addition the enhancements are expected to be small and build up only slowly (cf. Section 9.1.2), so the benefits of GOESY are particularly significant in this case. Some results published for test experiments with gibberellic acid are shown in Figure 7.22. Possibly a convection-compensated version would offer some further improvement, particularly given the longer NOE evolution periods required for heteronuclear experiments (cf. Section 9.1.2).

Experience to date seems to be that diffusion losses in homonuclear GOESY experiments are likely to be so severe that the DPFGSE-NOE experiment is generally preferred; indeed, even in the absence of diffusion losses, DPFGSE-NOE has twice the sensitivity of GOESY (as discussed above). It is just unfortunate that the authors used up the more memorable acronym on their first offering before the preferred experiment was developed!

7.5.3.4. Variations. Although the idea of generating 1D versions of 2D experiments by replacing frequency labelling of all signals during t_1 by selective excitation of one signal is not new,[41] the ability of gradients to "clean up" such spectra has led to renewed interest. One possibility that has received attention is the 1D ROESY experiment. In principle, any of the various ROESY mixing schemes discussed in Section 9.3.2 can be combined with a suitable selective excitation scheme to yield a 1D version of the experiment. The combination of DPFGSE excitation with a simple CW spin-lock ROE evolution period has been described[51] and two papers have appeared on ROE analogs of GOESY using a phase-alternating train of hard 180° pulses for spin-locking (cf. Section 9.3.2).[52,53]

A large number of other sequences result from concatenating different mixing sequences to achieve multistage transfers (cf. Sections 8.6.2 and 9.3.2). This approach is particularly suited to applications in oligosaccharide work, since the spectra of sugars usually offer just one selective target (the anomeric

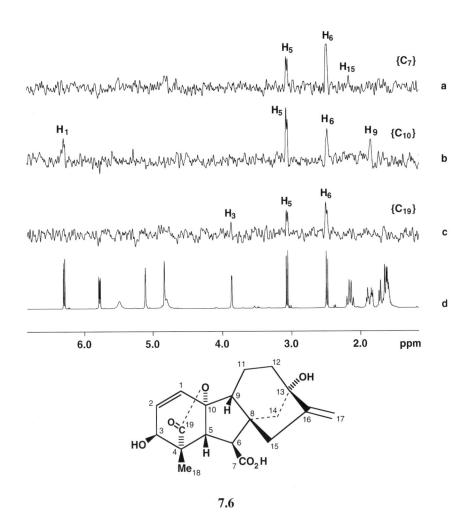

Figure 7.22. HETGOESY spectra of a ~50 mM sample of gibberellic acid (**7.6**) in DMSO-d_6, recorded using a Bruker DRX500 spectrometer. In (*d*) is shown the normal ^{1}H 1D spectrum; (*a*), (*b*) and (*c*) show HETGOESY spectra where the targets for selective excitation are, respectively, C_7, C_{10} and C_{19}. The pulse sequence was as shown in Figure 7.21d, with an NOE evolution period τ of 500 ms (determined as giving close to the maximum enhancement). A total of 8192 scans were accumulated for each ^{13}C target, requiring an overall experiment time of 8 h per spectrum.

The ^{13}C selective inversion pulses were Gaussians (truncated at 1%) with 1.2 ms duration, during which GARP decoupling (with $\gamma B/2\pi$ = 3 kHz) was applied to the protons. All gradients were of 1 ms duration. Gradient strengths: G_1 = 35%, $\overline{G_1}$ = −35%, G_2 = −25%, $\overline{G_2}$ = 25%, G_3 = 10%, G_4 = −4%, and G_5 = 26%, where 100% was calibrated at 38 Gauss cm^{-1}. Reproduced with permission from ref. 50.

signal) per spin system, with the other signals often being crowded together in an unresolved mass. One may thus use some clean selective excitation scheme to pick out a single anomeric signal and then transfer this magnetization over two or more steps to reveal its relationships with individual signals from within the unresolved part of the spectrum. Many possible combinations can be generated, and many of these have been published, including 1D NOESY-NOESY,[54] 1D NOESY-TOCSY,[54] 1D TOCSY-NOESY,[54,55] 1D COSY-NOESY,[55] 1D ROESY-TOCSY,[51,55] and 1D TOCSY-ROESY[51] (we list here only those experiments involving at least one NOE or ROE step). Generally, the most recently published sequences do not employ phase encoding during the NOE or ROE evolution periods so as to avoid the associated diffusion losses; however, several of the other sequences were published quite soon after GOESY and do use phase encoding. Combinations with heteronuclear selection steps have also been developed, for use either in labelled or natural abundance materials. Examples include 1D HSQC-ROESY, 1D HMQC-NOESY and 1D HMQC-ROESY.[56] Another example is shown in Figure 9.8.[57]

7.5.3.5. Applications and Practicalities. At the time of writing, gradient-enhanced NOE and ROE experiments such as DPFGSE-NOE and DPFGSE-ROE represent very recent developments. Probably the majority of papers concerning their use still comprise demonstrations using commercially available test samples, rather than applications to solve genuine structural problems arising during chemical research. Nonetheless, there are already enough published cases to suggest that these experiments may well largely displace conventional steady-state NOE difference spectroscopy for routine structural investigations in small and medium-sized molecules within a few years. A few representative examples, including applications to determining stereochemistry in small heterocycles[58,59] and organometallic complexes,[47] carbohydrate conformational analysis,[60] and determining tautomeric equilibria[61] can be found in the references, and an example where the technique is used to characterize a host–guest complex (the drug remacemide complexed to β-cyclodextrin) is shown in Figure 7.23.[62] It is even clearer that DPFGSE-NOE and DPFGSE-ROE are now the 1D methods of choice for measuring NOE or ROE kinetics in small molecules, for example, when determining initial growth rates.

In several of these applications, the improved sensitivity of the technique was crucial in arriving at a structural conclusion. This does, of course, lead in to the general issue of interpreting very small NOE enhancements, which will certainly be no less hazard-prone than the interpretation of larger enhancements has proven to be over the past three decades. Still greater care will be needed to make sure that conclusions are drawn from an overall pattern of NOE enhancements within a molecule, rather than doggedly using the greater sensitivity now available to hunt down one particular enhancement that is deemed to prove a particular point, regardless of its context and however small. Issues of interpretation are considered further in Part III.

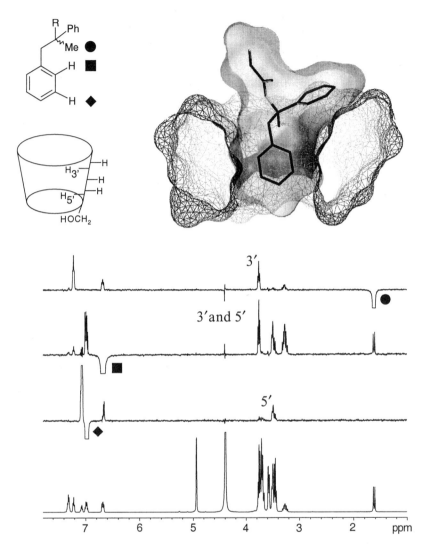

Figure 7.23. DPFGSE-NOE spectra of remacemide complexed to β-cyclodextrin, recorded at 500 MHz. The racemic drug (guest) and β-cyclodextrin (host) ratio was 1:1.3, but note that the host gives intense signals due to its degenerate structure. Intermolecular NOE enhancements are observed demonstrating that protons, ●, ■, and ♦ are located progressively deeper within the cavity. The use of the DPFGSE experiment allows clear observation of these small (ca. 0.5–1.5%) enhancements. Spectra were recorded at 47°C, using a mixing time of 1 s and a relaxation (interscan) delay of 2 s in a total experiment time for the three experiments shown of about 8.3 h. Sample concentrations were 13.78 mM (remacemide) and 18.34 mM (β-cyclodextrin) in 0.5 ml D_2O.

Another consideration arises in the reporting of results from gradient-assisted NOE experiments. Since these experiments do not give difference spectra in the conventional sense, no separate control data are available, so that it is difficult to quantify the size of enhancements or the extent of inversion of the S signal (which will of course also have relaxed during τ). Probably the best solution to this is always to record an additional data set with the NOE evolution time τ set to zero, and to use the integral of the S signal in this spectrum for normalization. (Note that this procedure, unlike comparison with a simple 1D spectrum, will automatically allow for incomplete selective excitation or inversion of S.) However, even when this is done, the fact that these are transient experiments is likely to make comparison of results between different studies more difficult than is the case with steady-state NOE experiments. The absolute sizes of transient enhancements are much more crucially dependent on $\omega\tau_c$ than are those of steady-state enhancements, so factors such as temperature, solvent viscosity, and the choice of NOE evolution period will probably have a major impact on numerical values of enhancements. Once again, the key will be to try to use patterns of data within one molecule to reach conclusions, rather than seeking to compare a few absolute numbers from different experiments.

7.5.4. Doubly Selective Experiments

All the methods described so far in this chapter are singly selective, in the sense that they involve selective perturbation of one resonance followed (after NOE evolution) by simultaneous observation of all the resulting NOE enhancements. In this section we consider methods that select not only the perturbed signal but also the enhanced signal. At first sight this might seem rather pointless, as it makes necessary a separate experiment to measure each NOE interaction. However, the underlying motive in such experiments is not to achieve spectral simplification as such, but rather to manipulate the spin system in such a way that the perturbed and enhanced spins become the only spins capable of interacting during the NOE evolution period. This isolates them from the remainder of the spin system and thereby eliminates or limits spin diffusion.

Quite a lot has been published in this area. Some methods aim to isolate just a single pair of protons in the context of a 1D experiment. These will be discussed in this section. Other methods use band-selective or heteronuclear pulses in the context of a 2D experiment to isolate particular groups of spins, either on grounds of chemical shift or of heteronuclear coupling partner, while yet other methods combine band-selective pulses with periods of spin-locking to restrict cross-relaxation between spins within the *same* region. These various 2D methods will be dealt with in Section 8.6.1.

An early approach to isolating a single spin pair (which, as ever, are referred to as I and S) during NOE evolution was to employ "synchronous nutation" during the NOE mixing period τ.[63,64] In such a scheme, the I and S spins are both made to rotate (nutate) between the $+z$ and $-z$ axes in synchrony with one another throughout τ, while other spins in the system remain unperturbed. In this way, a consistent relationship is constantly maintained between the z components of spins I and S, whereas their relationship with the z components of all other spins alternates as a result of the nutation. Therefore only the IS enhancement can build up in a consistent direction, all other interactions being suppressed because they spend alternate short periods building up in first one direction and then the opposite direction. (This is conceptually very similar to the explanation given at the end of Section 9.3.1.2 as to why transverse enhancements in ROE experiments cannot accumulate unless spin-locking is applied.) In practice, such synchronous nutation is achieved using cosine modulation of either a very low power spin-locking field[64] (Fig. 7.24a) or a train of selective 360° pulses;[63] in either case, the cosine modulation splits the RF into two sidebands that are arranged to fall on the I and S signals, and the selectivity is arranged to be good enough that all other spins are substantially unaffected. A closely related experiment called SLOESY (Selectively Locked Overhauser Effect Spectroscopy) was proposed by Bull,[65] in which two signals are selectively spin-locked allowing their mutual ROE enhancement to be observed free of spin diffusion. Note also that the theoretical treatment of these methods

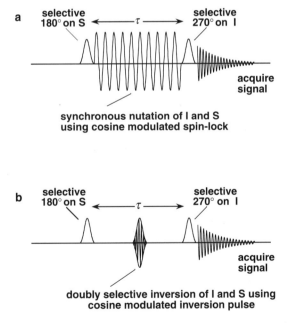

Figure 7.24. Pulse sequences for doubly selective transient NOE experiments. (*a*) Synchronous nutation, which uses a cosine-modulated spin-locking field, (*b*) QUIET NOESY, which uses a doubly selective inversion pulse at the center of τ. Both sequences use a selective inversion pulse to invert the S spin at the start of τ and a selective 270° pulse to read out the I spin magnetization at the end of τ.

requires the extended version of the Solomon equations described in Section 2.6.

The synchronous nutation method is not trivial to implement, and it has largely been overtaken by a newer variant called "QUIET NOESY" (*Q*uenching *U*ndesirable *I*ndirect *E*xternal *T*rouble in NOESY).[66-68] In this method, spin S is selectively inverted at the beginning of τ, a *doubly selective* 180° pulse is applied to both I and S at the center of τ, and then the I spin intensity is read out using a selective 270° pulse at the I spin frequency (Fig. 7.24b). Evolution of the IS interaction occurs in a consistent direction throughout τ, since both I and S are inverted by the central 180° pulse, but all interactions of I and S with other spins are largely cancelled, since they occur in opposite senses during the first and second halves of τ. Suppression of interactions outside the IS spin system may be somewhat compromised if τ is long, since then the buildup of enhancements in the two periods becomes substantially nonlinear, and cancellation between the two halves of the NOE evolution may be imperfect; however, one can improve this situation by using more than one doubly selective 180° pulse, placed symmetrically through τ.[66] In the original papers on QUIET NOESY, the I spin magnetization is always partially transferred onto a coupling partner of I using a selective TOCSY transfer step prior to observation.[66] It seems this additional step is added to improve the quality of subtraction, although it is unclear what would be the real consequences of omitting it. The additional transfer does not complicate the analysis of intensities, but of course it is only practical in cases where a convenient coupling partner of I is available. Two-dimensional variants of QUIET NOESY are considered in Section 8.6.1.

Given that only one interaction is detected per experiment using 1D doubly selective experiments, it would be completely unrealistic to expect to measure, using these techniques, all the many hundreds of NOE enhancements required during a complicated structure determination of a biomacromolecule. However, for relatively small molecules where the NOE is often employed to resolve a question of stereochemistry after most of the structure is known, elimination of spin diffusion from the measurement of a few key distances could often be a very attractive proposition. Even for macromolecules, in the later stages of a structure determination one already has a fair degree of knowledge of the structure of the system being studied, and then one can imagine several uses for doubly selective techniques. In structural regions where few NOE enhancements are measurable, accurate quantitation of these few becomes particularly important if the structure is not to become inaccurate during refinement (cf. Section 12.5). In combination with related methods that determine various relaxation times for artificially isolated spin pairs,[69] the methods of this section could also be used to study local dynamics of (^{1}H, ^{1}H) pairs in a similar fashion to the ^{15}N studies discussed in Section 5.4. Perhaps most useful of all, doubly selective methods could be applied to help in the complicated problem of accurate calibration of NOE intensities (cf. Sections 4.4 and 12.2.3). However, against these advantages is the fact that the selective inversion pulses needed in these doubly selective techniques have significant lengths, and for many macromolecules NOE evolution is too rapid for such protracted events to be usefully accommodated within typical mixing periods required in real NOE experiments. Thus, despite all of the potential uses just discussed, these methods have yet to gain widespread acceptance outside the groups that first proposed them.

REFERENCES

1. Bundle, D.; Brisson, J.-R.; Bock, K. Personal communication.
2. Irving, C. S.; Lapidot, A. *J. Am. Chem. Soc.* 1975, *97*, 5945.
3. Lindon, J. C.; Ferrige, A. G. *Prog. Nucl. Magn. Reson. Spectrosc.* 1980, *14*, 27.
4. Richarz, R.; Wüthrich, K. *J. Magn. Reson.* 1978, *30*, 147.
5. Chapman, G. E.; Abercrombie, B. D.; Cary, P. D.; Bradbury, E. M. *J. Magn. Reson.* 1978, *31*, 459.
6. Neuhaus, D.; Wagner, G.; Vašák, M.; Kägi, J. H. R.; Wüthrich, K. *Eur. J. Biochem.* 1985, *151*, 257.
7. Mersh, J. D.; Sanders, J. K. M. *J. Magn. Reson.* 1982, *50*, 289.
8. Mehlkopf, A. F.; Korbee, D.; Tiggelman, T. A.; Freeman, R. *J. Magn. Reson.* 1984, *58*, 315.
9. Hoult, D. I.; Richards, R. E. *Proc. R. Soc. Lond., Ser. A* 1975, *344*, 311.
10. Neuhaus, D.; Sheppard, R. N.; Bick, I. R. C. *J. Am. Chem. Soc.* 1983, *105*, 5996.
11. Sanders, J. K. M.; Mersh, J. D. *Prog. Nucl. Magn. Reson. Spectrosc.* 1982, *15*, 353.
12. Hull, W. E. Personal communication; Krajewski, P. Personal communication.
13. Sanders, J. K. M.; Hunter, B. K. "Modern NMR Spectroscopy," Oxford University Press, 1987, p. 189.
14. Neumann, J. M.; Bernassau, J. M.; Guéron, M.; Tran-Dinh, S. *Eur. J. Biochem.* 1980, *108*, 457.
15. Noggle, J. H. *J. Am. Chem. Soc.* 1978, *102*, 2230.
16. Guineadeau, H.; Cassels, B. K.; Shamma, M. *Heterocycles* 1982, *19*, 1009.
17. Bilton, J. N.; Broughton, H. B.; Ley, S. V.; Lidert, Z.; Morgan, E. D.; Rzepa, H. S.; Sheppard, R. N. *J. Chem. Soc., Chem. Commun.* 1985, 968.
18. Curzon, E. H.; Howarth, O. W. *J. Chem. Soc., Chem. Commun.* 1984, 1012.
19. Bauer, C.; Freeman, R. *J. Magn. Reson.* 1985, *61*, 376.
20. Shaka, A. J.; Bauer, C.; Freeman, R. *J. Magn. Reson.* 1984, *60*, 479.
21. Kinns, M.; Sanders, J. K. M. *J. Magn. Reson.* 1984, *56*, 518.
22. Krajewski, P. 1998 Personal communication.
23. Kövér, K. E. *J. Magn. Reson.* 1984, *59*, 485.
24. Williamson, M. P.; Williams, D. H. *J. Chem. Soc. Perkin Trans. 1* 1985, 949.
25. Neuhaus, D. *J. Magn. Reson.* 1983, *53*, 109.
26. Morris, G. A. *J. Magn. Reson.* 1988, *80*, 547.
27. Morris, G. A.; Barjat, H.; Horne, T. J. *Prog. Nucl. Magn. Reson. Spectrosc.* 1997, *31*, 197.

28. Morris, G. A. 1998 Personal communication.
29. Gibbs, A.; Morris, G. A. *J. Magn. Reson.* 1991, *91*, 77.
30. Morris, G. A.; Cowburn, D. *Magn. Reson. Chem.* 1989, *27*, 1085.
31. Barjat, H.; Morris, G. A.; Swanson, A. G.; Smart, S.; Williams, S. C. R. *J. Magn. Reson. A* 1995, *116*, 206.
32. Dobson, C. M.; Olejniczak, E. T.; Poulsen, F. M.; Ratcliffe, R. G. *J. Magn. Reson.* 1982, *48*, 97.
33. Bothner-By, A. A.; Noggle, J. H. *J. Am. Chem. Soc.* 1979, *101*, 5152.
34. Andersen, N. H.; Nguyen, K. T.; Hartzell, C. J.; Eaton, H. L. *J. Magn. Reson.* 1987, *74*, 195.
35. Eaton, H. L.; Andersen, N. H. *J. Magn. Reson.* 1987, *74*, 212.
36. Williamson, M. P.; Neuhaus, D. *J. Magn. Reson.* 1987, *72*, 369.
37. Solomon, I. *Phys. Rev.* 1955, *99*, 559.
38. Freeman, R. *Prog. Nucl. Magn. Reson. Spectrosc.* 1998, *32*, 59.
39. Morris, G. A.; Freeman, R. *J. Magn. Reson.* 1978, *29*, 433.
40. Bauer, C.; Freeman, R.; Frenkiel, T.; Keeler, J.; Shaka, A. J. *J. Magn. Reson.* 1984, *58*, 442.
41. Kessler, H.; Oschkinat, H.; Griesinger, C.; Bermel, W. *J. Magn. Reson.* 1986, *70*, 106.
42. Keeler, J.; Clowes, R. T.; Davis, A. L.; Laue, E. D. *Methods in Enzymology* 1994, *239*, 145.
43. Hwang, T. L.; Shaka, A. J. *J. Magn. Reson. A* 1995, *112*, 275.
44. Stott, K.; Stonehouse, J.; Keeler, J.; Hwang, T.-L.; Shaka, A. J. *J. Am. Chem. Soc.* 1995, *117*, 4199.
45a. Stott, K.; Keeler, J.; Van, Q. N.; Shaka, A. J. *J. Magn. Reson.* 1997, *125*, 302.
45b. Stott, K. PhD Thesis, University of Cambridge, 1996.
46. Goldman, M. *Quantum Description of High-Resolution NMR in Liquids.* Oxford University Press, Oxford, 1988, pp. 224–225.
47. Feher, F. J.; Hwang, T.-L.; Schwab, J. J.; Shaka, A. J.; Ziller, J. W. *Magn. Reson. Chem.* 1997, *35*, 730.
48. Stonehouse, J.; Adell, P.; Keeler, J.; Shaka, A. J. *J. Am. Chem. Soc.* 1994, *116*, 6037.
49. Jerschow, A.; Müller, N *J. Magn. Reson.* 1998, *132*, 13.
50. Stott, K.; Keeler, J. *Magn. Reson. Chem.* 1996, *34*, 554.
51. Gradwell, M. J.; Kogelberg, H.; Frenkiel, T. A. *J. Magn. Reson.* 1997, *124*, 267.
52. Dalvit, C.; Bovermann, G. *Magn. Reson. Chem.* 1995, *33*, 156.
53. Adell, P.; Parella, T.; Sánchez-Ferrando, F.; Virgili, A. *J. Magn. Reson. B* 1995, *108*, 77.
54. Uhrin, D.; Barlow, P. N. *J. Magn. Reson.* 1997, *126*, 248.
55. Adell, P.; Parella, T.; Sánchez-Ferrando, F.; Virgili, A. *J. Magn. Reson. A* 1995, *113*, 124.
56. Parella, T.; Sánchez-Ferrando, F.; Virgili, A. *J. Magn. Reson. A* 1995, *114*, 32.
57. Suzuki, M.; Neuhaus, D.; Gerstein, M.; Aimoto, S. *Protein Engineering* 1994, *7*, 461.

58. Robinson, J. K.; Lee, V.; Claridge, T. D. W.; Baldwin, J. E.; Schofield, C. J. *Tetrahedron* 1998, *54*, 981.
59. Ortuño, R. M.; Parella, T.; Planas, M.; Ventura, M. *Magn. Reson. Chem.* 1996, *34*, 983.
60. Landersjö, C.; Stenutz, R.; Widmalm, G. *J. Am. Chem. Soc.* 1997, *119*, 8695.
61. Hill, F.; Williams, D. M.; Loakes, D.; Brown, D. M. *Nucleic Acids Res.* 1998, *26*, 1144.
62. Bernstein, M. A.; Lewis, R. J. 1997 Personal communication.
63. Boulat, B.; Burghardt, I.; Bodenhausen, G. *J. Am. Chem. Soc.* 1992, *114*, 10679.
64. Burghardt, I.; Konrat, R.; Boulat, B.; Vincent, S. J. F.; Bodenhausen, G. *J. Chem. Phys.* 1993, *98*, 1721.
65. Bull, T. E. *J. Magn. Reson.* 1991, *93*, 596.
66. Zwahlen, C.; Vincent, S. J. F.; Di Bari, L.; Levitt, M. H.; Bodenhausen, G. *J. Am. Chem. Soc.* 1994, *116*, 362.
67. Schwager, M.; Bodenhausen, G. *J. Magn. Reson. B* 1996, *111*, 40.
68. Vincent, S. J. F.; Zwahlen, C.; Bodenhausen, G. *Angew. Chemie, Int. Ed. Engl.* 1994, *33*, 343.
69. Boulat, B.; Bodenhausen, G. *J. Biomolec. NMR* 1993, *3*, 335.

CHAPTER 8

THE TWO-DIMENSIONAL NOESY EXPERIMENT

8.1. ONE DIMENSION OR TWO?

Much has been written about the comparison between one-dimensional and two-dimensional experiments, both in general and specifically in relation to the NOE. The key points for NOE experiments are as follows:

1. 2D experiments do not require selective irradiation to induce NOE enhancements. In normal NOESY experiments all pulses (except for solvent presaturation, where relevant) are nonselective.
2. NOESY is a transient NOE experiment. The peak values reached by transient NOE enhancements are generally smaller than steady-state enhancements, so that NOESY is intrinsically less sensitive than a steady-state experiment. There can be no true 2D analog of the 1D steady-state experiment.
3. The rate-limiting factor in completing each type of experiment depends very much on how it is carried out, and some implementations of both experiments have become much faster in recent years. However, the need to complete the phase cycle at least once for each increment means that the minimum time required for a NOESY experiment is still usually in the range of hours, depending principally on the sensitivity and resolution (particularly F_1 resolution) required.

The consequences of these points depend entirely on the nature of the system under study, both on the complexity of the NMR spectrum and on the tumbling rate of the solute. The two NOE regimes will be considered in turn.

8.1.1. The Negative NOE Regime ($\omega\tau_c > 1.12$)

This is very much the "home territory" of the NOESY experiment, for several reasons. Steady-state experiments are of little use for large molecules, because the problem of spin diffusion renders interpretation of the enhancements difficult or (more usually) impossible (Sections 3.2.1 and 3.2.3). The lower sensitivity of transient experiments versus steady-state experiments is therefore irrelevant in these cases. As it happens, this loss is in any case minimal for slowly tumbling molecules, where σ_{IS} approaches $-\rho_{IS}$, and the maximum transient enhancement therefore approaches -100% (Sections 4.1 and 4.2).

Given that steady-state experiments are not useful, the real comparison is between 1D TOE or transient experiments and NOESY. Large molecules generally have highly complicated and overlapped spectra, and resonances are broad. These factors make selective irradiation a real problem, which gives NOESY a powerful advantage. Also, the need to complete the phase cycle on each increment of NOESY is not an obstacle here, since signal strengths are generally sufficiently weak that many transients are required in any case. Lastly, the lower resolution generally used for NOESY experiments is more likely to be acceptable in a spectrum comprised of broad resonances.

Thus, NOESY and experiments derived from it are now the methods of choice for almost all cases within the negative NOE regime. The main exceptions are molecules only just within the negative NOE regime, for which ROE experiments may be useful (Section 9.3). For molecules whose spectra afford suitable preirradiation targets, 1D experiments may sometimes be preferable if the results are to be interpreted quantitatively. This applies particularly to paramagnetic proteins, in which rapid relaxation often makes NOE enhancements involving protons close to the paramagnetic center too small to measure in 2D spectra. In such cases, 1D experiments involving irradiation of the strongly paramagnetically shifted resonances can be used to good effect to observe enhancements to neighboring protons.[1]

8.1.2. The Positive NOE Regime ($\omega\tau_c < 1.12$)

The situation for smaller molecules is less clear. Although NOESY does not very often compete favorably with 1D steady-state experiments, there are many exceptions, and to dismiss NOESY experiments entirely for molecules in the positive NOE regime is certainly too extreme a view. The problem is that it is difficult to recognize in advance which samples will give a satisfactory NOESY spectrum, and which will not. There are two points to consider here when choosing between 1D and 2D experiments: (i) the expected approximate tumbling rate of the molecules, and (ii) the number of resonances that would need to be separately preirradiated in a series of 1D experiments, and their accessibility in the spectrum.

The tumbling rate plays a crucial role. In the positive NOE regime, NOESY will work well only for molecules that tumble at a rate within a fairly narrow band, some way, but not too far, into the positive NOE regime. The reason for this was explained in Section 2.4, and is clearly shown in Figures 2.9 and 2.10. If the molecule tumbles at a rate close to the zero-crossing point ($\omega\tau_c \cong 1.12$), all enhancements will be small because η_{max} is small. On the other hand, if the tumbling rate is too fast, then dipolar relaxation becomes less efficient and ceases to compete effectively with external relaxation (ρ^*), and again enhancements are small. In Chapter 2, this argument was presented in terms of the steady-state NOE, but identical conclusions apply to transient experiments, and hence to NOESY cross-peak intensities. In fact, the range of suitable tumbling rates for NOESY is narrower than that for steady-state experiments, because the transient enhancements detected in NOESY are intrinsically smaller than steady-state enhancements. This problem is exacerbated by the fact that τ_m (the mixing period in the NOESY pulse sequence) cannot be optimal for all enhancements simultaneously, even if it were known at what time each enhancement reached its peak. Thus most cross-peak intensities will inevitably be much less than maximal, reducing sensitivity still further relative to the steady-state experiment.

It is not at all easy to predict tumbling behavior based on molecular structure. Clearly, molecular weight plays the dominant role, but other factors can be surprisingly important. Charged molecules sometimes give excellent NOESY spectra under conditions in which neutral molecules of about the same molecular weight give almost no NOESY cross-peak intensity at all; presumably the charged species impose a greater degree of local order on the surrounding solvent, and therefore cause tumbling to be slower.[2,3] Solvent viscosity and temperature are also very important (Section 9.4). Therefore, although it is clearly futile to attempt NOESY experiments with very small molecules (MW < 200), the only way to find out how "medium-sized" molecules (MW = 200–2000) will behave in NOE experiments, in the absence of prior experience with related structures, it to try one. Often the simplest test is to try a single 1D NOE difference experiment in which as least one strong enhancement is expected. The actual size (and sign) of the measured enhancement provides a rough indication of the NOE characteristics of the sample.

Although this may answer the question of whether a NOESY experiment is *feasible*, it does not show that it is *worthwhile*. This is a much more subjective question, but clearly it depends mainly on what information is sought. The greater the number of preirradiation sites required for 1D experiments, the more attractive NOESY becomes as an alternative. A related point to bear in mind is that a larger number of enhancements generally leads to a more secure interpretation, even when a small number might be thought to suffice. On the other hand, lower sensitivity in NOESY may actually result in a smaller number of enhancements being detected than in a series of 1D experiments; also the low resolution of NOESY spectra implies that multiplet structures of any signals enhanced from within regions of spectral overlap will often be resolved only in 1D experiments.

The reason often cited for turning to NOESY experiments is the avoidance of the need for selective irradiation. In fact, selectivity in 1D experiments is a real problem far less often than might be supposed. It is not necessary to saturate the whole of a multiplet to induce an NOE enhancement (Section 7.2.5), and the example of spiro-diamine **10.78** shows that even a single exposed line of a complex multiplet can suffice as a preirradiation target for a 1D experiment, albeit with reduced sensitivity. The real problem with spectra such as this is more often that of *assignment*, and although it is true that 2D experiments are the answer to this problem, the 2D experiments most useful for making assignments in crowded spectra of small molecules are not NOESY but COSY, DQF-COSY or TOCSY (cf. Section 10.1.2). Once assignments are available, preirradiation targets for 1D NOE experiments can be chosen from a position of strength, and much of the reason for running a NOESY spectrum may evaporate.

8.2. BASIC PRINCIPLES

Many explanations of 2D NMR are couched in rather mathematical terms, which give a full and precise description of the experiment but little physical insight. Rather than doing this, we aim here to provide a simpler explanation of the NOESY experiment, based on vectors representing the macroscopic magnetizations due to each spin. In fact, NOESY is one of the simplest 2D experiments to understand, largely because the processes that give rise to cross peaks can easily be pictured in terms of such vectors; in experiments such as COSY, cross peaks arise through the process of coherence transfer, which cannot easily be explained in terms of vectors.

The essential features of a NOESY experiment are illustrated in Figure 8.1, for a one-line spectrum. The pulse sequence consists of three 90° pulses separated by delays, followed by acquisition of the FID. If the FID from one such experiment is transformed to give a spectrum, the intensity of the line depends on the particular value of t_1 for that experiment, and on τ_m.[†] In the full 2D experiment, many such individual experiments are carried out. The value of t_1 is incremented in a regular manner from one experiment to the next so as to generate a *matrix* of FIDs, each individual (and separately stored) FID corresponding to a different value of t_1. Each FID is then separately Fourier transformed, resulting in a matrix of spectra. The amplitude of the line in the spectra varies regularly as a function of t_1. Figure 8.1 shows that a trace constructed along the centers of the lines in successive spectra follows a cosine wave. A second FT, this time *with respect to* t_1, extracts the frequency of this variation and results in the 2D spectrum. Note that two *independent* FTs have now been

---

[†] We have used the symbol τ_m for the mixing period in a 2D experiment, and τ for that in a 1D NOE experiment, purely for ease of discussion when comparing the two cases. No distinction in the nature of the processes occurring during τ_m and τ is intended.

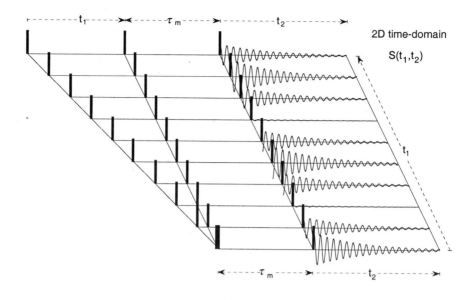

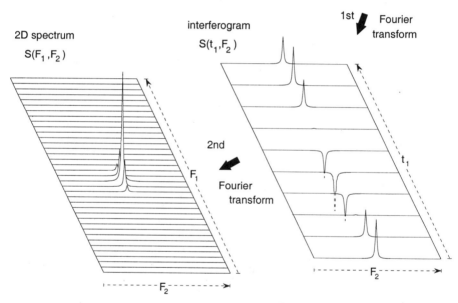

Figure 8.1. Schematic representation of the NOESY experiment for a one-line spectrum. The first 90° pulse is followed by a time t_1, which is incremented by regular steps from zero to $t_{1\mathrm{max}}$. This is followed by a second 90° pulse, a mixing time τ_m, and a third 90° pulse, after which the FID is collected during the acquisition time t_2. The intensity of each FID is a function of t_1; the experiment therefore results in a matrix of points $S(t_1, t_2)$, in which the intensity varies as cosine functions of t_1 and t_2. The first FT (with respect to t_2) turns this matrix into a matrix of points $S(t_1, F_2)$ (the interferogram), which is an array of 1D spectra in which the line intensity varies as a function of t_1. The second FT (with respect to t_1) creates the final 2D spectrum $S(F_1, F_2)$. For visual simplicity, only the first 10 increments are shown, and the incrementation of t_1 is shown greatly exaggerated.

BASIC PRINCIPLES 287

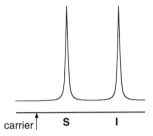

Figure 8.2. A schematic two-spin spectrum.

carried out. Starting from a data matrix that was a function of two time variables, $S(t_1, t_2)$, we first converted it to a function of one time and one frequency variable, $S(t_1, F_2)$ (often called the interferogram), and then to a function of two frequency variables, $S(F_1, F_2)$, which is the 2D spectrum. Note that, purely for visual simplicity, we have chosen to show the interferogram as starting with a positive signal. Formally, this implies that one of the three pulses of the NOESY sequence must have opposite phase to the others, and for simplicity in what follows we take this to be the second pulse. Thus the sequence that we are illustrating throughout this section is $90°_x - t_1 - 90°_{-x} - \tau_m - 90°_x - t_2$.[†]

The key feature of this experiment that we need to understand, not explained in the preceding paragraph, is the reason *why* the intensities in the interferogram (and the FIDs for that matter) vary as a regular cosine function of t_1. So as to be ready to deal with the possibility of an NOE enhancement between two spins, we shall explain this in terms of a two-spin system. As usual, the spins I and S are assumed to be spin 1/2 nuclei, close together in space but not J-coupled. For simplicity, we assume that I and S both resonate on the same side of the carrier (the transmitter frequency), so that both have the same sense of rotation when their precession is viewed in the rotating frame (Section 1.5). Also, S is closer to the carrier than is I, and the carrier position is represented as the left-hand edge of the spectrum (Fig. 8.2).

After the first $90°_x$ pulse of the sequence, I and S precess around in the xy plane (Fig. 8.3), with I precessing faster than S. The precession rate of I in the rotating frame, here denoted ω_I, is the difference in frequency between I and the carrier; similarly ω_S is the frequency difference between S and the carrier. At the end of t_1, the two spins I and S have precessed through angles $\omega_I t_1$ and $\omega_S t_1$, respectively, and a second pulse (recall this is a $90°_{-x}$ pulse) rotates the vectors representing I and S through 90° into the xz plane (Fig. 8.4). A centrally important feature of the NOESY experiment is that it is designed to record

---

[†]Elsewhere in the book (and at this point in the first edition) we base our discussion on the sequence $90°_x - t_1 - 90°_x - \tau_m - 90°_x - t_2$, which would result in all signals appearing inverted relative to the results of the $90°_x - t_1 - 90°_{-x} - \tau_m - 90°_x - t_2$ sequence. In practice, both sequences form part of the phase cycle (cf. Section 8.4), and the final algebraic sign of the whole spectrum is determined by an appropriate choice of phase correction during processing (cf. Section 8.3.4), so the distinction is not of real importance.

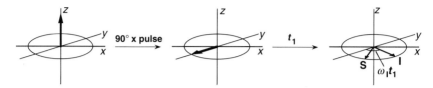

Figure 8.3. The effect of the sequence $90^\circ_x - t_1$ on equilibrium I and S magnetization. At the end of t_1 the S magnetization vector is at an angle $\omega_S t_1$ to the $-y$ axis, and the I vector is at an angle $\omega_I t_1$.

(during t_2) only that part of the magnetization that was longitudinal during the mixing period τ_m. This means that the x component of magnetization present after the second pulse must be rejected, and this is achieved using a suitable phase-cycling scheme, as described in Section 8.4.1.

The z components present after the second pulse are clearly dependent on t_1; for I, the size of this z component is given by $\cos \omega_I t_1$, and for S by $\cos \omega_S t_1$. The z components are said to be *frequency labeled*. In other words, by measuring the intensity of the z components at this point as a function of t_1, using a third 90° pulse immediately after the second to convert them into transverse signals, we could find the precession frequency that each vector had during t_1. This would be a NOESY experiment with τ_m set to zero, and the result is shown in Figure 8.5. The intensity of the signals due to I and S follow decaying cosine waves as a function of t_1, and their frequencies during t_1 are revealed by the second FT, with respect to t_1.

This is not a particularly informative experiment, since we already know the precession frequencies during t_1; they are just the same as those during t_2, since nothing has happened between t_1 and t_2 that could change them. We have thus generated a 2D spectrum that contains only peaks having the same frequency in F_1 as in F_2. These are called diagonal peaks, since they necessarily lie on the 45° diagonal of the 2D spectrum.

In a real NOESY experiment, τ_m is not zero, so that there is a fixed delay between the second pulse (which puts the frequency-labeled magnetizations onto the z axis) and the third (which reads the z magnetizations present at the end of τ_m). If nothing happens during τ_m (for example, if τ_m is very short), then we again get only diagonal peaks. Useful information only emerges if the two

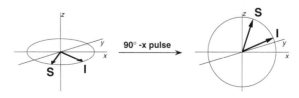

Figure 8.4. A 90°_{-x} pulse rotates I and S magnetization into the xz plane. The S vector has a component $\cos(\omega_S t_1)$ along the z axis, and the I vector has a component $\cos(\omega_I t_1)$.

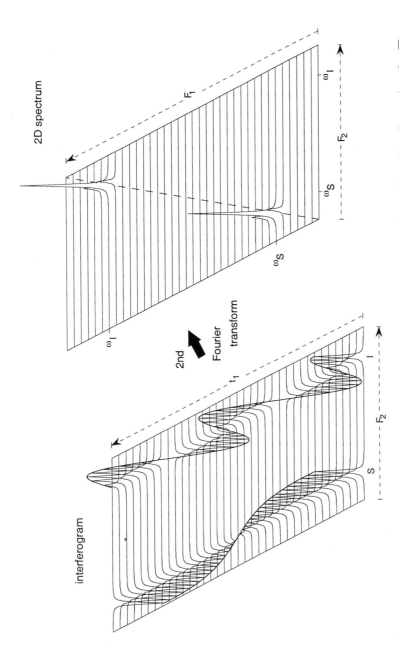

Figure 8.5. Schematic representation of a NOESY experiment with τ_m set to zero, to show the origin of the diagonal peaks. The z components of the magnetization vectors shown in Figure 8.4 are sampled as a function of t_1 by the third pulse, and after Fourier transformation with respect to t_2 give the set of spectra shown in the interferogram. Each signal is modulated by its own frequency in the t_1 dimension. Because ω_S is slower than ω_I, the intensity of the S signal varies more slowly with t_1 than does that of I. The second FT converts the modulations in t_1 into frequencies in F_1. The resultant peaks have the same frequency in F_1 and in F_2 and therefore lie on the diagonal.

spins *exchange magnetization* during τ_m, either by cross-relaxation or by chemical exchange. The way in which this happens can be pictured in terms of 1D experiments corresponding to particular t_1 values. For the case in which $t_1 = 0$, both vectors are at their equilibrium values at the start of τ_m. However, when t_1 is such that $\omega_I t_1 = \pi$, the z component of I is inverted at the start of τ_m, while the corresponding S vector, which did not precess as far during t_1, is still largely positive. This is very similar to the situation at the start of τ in a normal 1D transient NOE experiment (except that here it is I rather than S that has turned out to be inverted). Cross-relaxation during τ_m thus results in a transient NOE enhancement at S, similar to that expected in the corresponding 1D experiment.

For other values of t_1, the difference in the values of the z components of I and S will be less extreme so the enhancement will be smaller, and for later values of t_1 where S is "more inverted" than I, a transient enhancement will occur in the reverse sense, appearing at I. In general, the size and direction of the transient enhancements depend on the *difference* in the values of the z components of I and S at the start of τ_m. As we have seen, these values depend in turn on the extent of precession that I and S underwent during t_1. Thus, by the end of the mixing period τ_m, the intensity of the S vector, which was initially given just by $\cos \omega_S t_1$, has acquired an additional dependence on $\cos \omega_I t_1$. The extent (i.e., amplitude) of this new modulation at frequency ω_I depends on the size and sign of the transient enhancement. Similarly, the intensity of I at the end of τ_m has acquired an equivalent new modulation at ω_S (recall that for transient enhancements, $f_S\{I\} = f_I\{S\}$ for all values of τ; Section 4.5.1).

Figures 8.6 and 8.7 show what this implies for the 2D spectrum. The trace corresponding to the S line in the interferogram shows a major modulation at ω_S, and a minor modulation (the amplitude of which is proportional to the transient enhancement $f_S\{I\}$) at ω_I. On Fourier transformation with respect to t_1, this yields a major peak at $F_1 = \omega_S$ and a minor peak at $F_1 = \omega_I$; the major peak is the diagonal peak already encountered in Figure 8.5, while the minor peak is a *cross peak connecting I and S* (it has frequency $F_1 = \omega_I$ and $F_2 = \omega_S$). Similar conclusions apply in reverse to the trace corresponding to the I line in the interferogram, so that the full 2D spectrum appears as shown. The intensities of the two cross peaks are proportional to the intensity of the transient enhancement between I and S, and are equal provided that the spin system was genuinely at thermal equilibrium before the first 90° pulse (the consequences if this condition is not met are discussed in Section 8.3.1).

Another point to notice is that, because *all* the modulations present in the data are cosine modulations, all the peaks in the 2D spectrum have identical phase properties (although some may be negative; see following). We shall have more to say about phase properties of NOESY spectra in Section 8.3.4, but a few key points are relevant here. The first and most important is that phase-sensitive data processing is *always* strongly preferable to absolute-value processing; in absolute-value processing the phase properties of the lines are needlessly discarded with no compensating advantage in return.

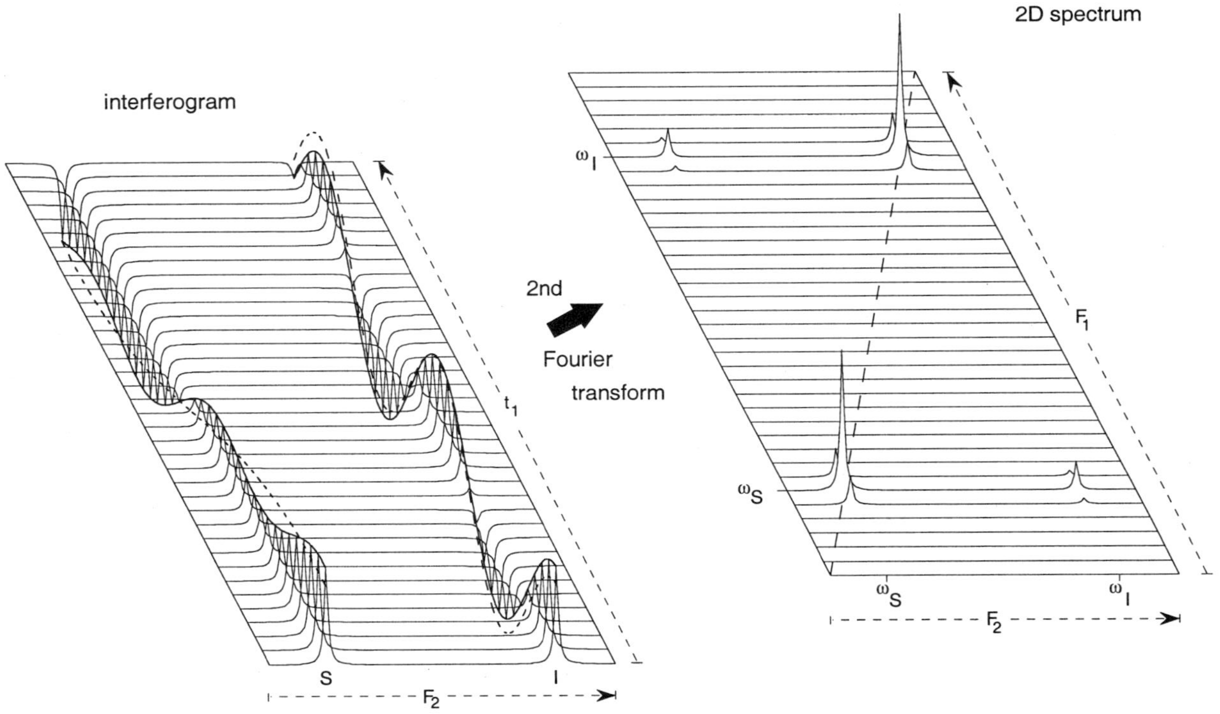

Figure 8.6. Schematic representation of a NOESY experiment with τ_m not set to zero, to show the origin of cross peaks for a molecule in the negative regime ($\omega\tau_c > 1.12$). The first part of the figure shows how cross-relaxation between I and S during the mixing time leads to an additional modulation of the interferogram. The dashed lines running back through the interferogram at the shifts of the I and S signals in F_2 represent the intensities that the signals would have had in the absence of this additional modulation (i.e., as they appeared in Fig. 8.5). After the second FT (with respect to t_1) the extra modulation produces cross peaks in the 2D spectrum.

In the negative NOE regime, cross-relaxation consists essentially of an exchange of magnetization between the two spins. The modulation therefore appears on the destination signal with the same sign as it had on the source signal; for example, when the I signal first appears inverted (a short way into t_1) the intensity of the S signal is reduced relative to the value it would have had in the absence of the additional modulation. The result is that in the 2D spectrum, the cross peaks have the same sign as the diagonal peaks (see text for further discussion).

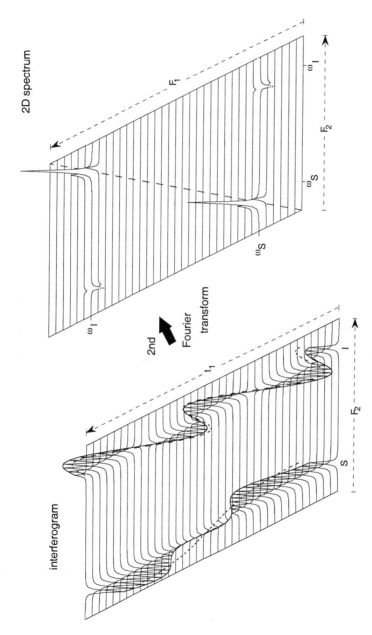

Figure 8.7. Schematic representation of a NOESY experiment with τ_m not set to zero, to show the origin of cross peaks for a molecule in the positive regime ($\omega\tau_c < 1.12$; compare also with Fig. 8.6). In the positive NOE regime, decreasing the intensity of one signal causes the intensity of its cross-relaxation partner to *increase*. The modulation therefore appears on the destination signal with a sign opposite to that it had on the source signal; for example, when the I signal first appears inverted (a short way into t_1) the intensity of the S signal is increased relative to the value it would have had in the absence of the additional modulation. The result is that in the 2D spectrum, the cross peaks have the opposite sign to the diagonal peaks (see text for further discussion).

As with 1D spectra, one may choose any desired phase for the 2D spectrum, but clearly the best is that in which all peaks appear in pure double-absorption mode. In such a presentation, the relative sign of cross peaks and diagonal peaks depends on the sign of the transient enhancement between I and S. This is explained in Figures 8.6 and 8.7. For large molecules ($\omega\tau_c > 1.12$), the transient enhancement acts to make the intensities of I and S more similar by the end of τ_m than they were at the start, meaning that the minor modulation contributes in the *same* sense (proportional to $+\cos \omega_s t_1$ for the I signal) as the major modulation ($+\cos \omega_I t_1$). In turn, this implies that *for large molecules the diagonal and cross peaks have the same sign*. Usually, both are displayed as positive (see Fig. 8.6). Conversely, by following a similar argument it is easily shown that *for small molecules the diagonal and cross peaks have opposite signs*. Thus, when the diagonal is positive, the cross peaks are negative (see Fig. 8.7). This is directly analogous to the difference in sign between directly enhanced signals and the irradiated signal in 1D NOE difference spectra of small molecules.

Chemical exchange contributes in exactly the same sense as negative NOE enhancements, that is, it gives rise to cross peaks of the same sign as the diagonal. In fact, the first publication of NOESY described its use to study chemical exchange rather than the NOE, and the experiment is still occasionally called the 2D-exchange experiment.[4] We shall not deal with applications of NOESY specifically to chemical exchange in this book.

We have seen that cross peaks are absent when the mixing period τ_m is very short. When τ_m is very long, equilibrium is restored between the second and third pulses, so that the frequency labelling is lost and *all* contributions to the 2D spectrum decay away (except for axial peaks; Section 8.4.3). The cross peaks, being by far the weakest part of the spectrum, are the first to disappear, so that the symptom of an overlong mixing period, just as for too short a mixing period, is a NOESY spectrum devoid of cross peaks. The choice of τ_m for a given case is clearly one of the most crucial decisions to be made when setting up a NOESY experiment, and is discussed further in Section 8.3.1. In general, however, it is clear that the cross-peak intensity as a function of τ_m must build up to a maximum and then decay away again, in just the same way as for transient enhancements observed in 1D experiments. Equations describing this behavior were developed in Chapter 4. These are summarized in Figure 8.8, where cross-peak intensity for a two-spin system is plotted as a function of τ_m for a small molecule and for a large molecule. Unfortunately, there can be no guarantee that even the maximum on such a curve necessarily represents an intensity strong enough to detect in a reasonable time!

8.3. ACQUIRING A NOESY SPECTRUM

General reviews of 2D NMR may be found in references 5 and 6; practical aspects of NOESY are discussed in reference 7.

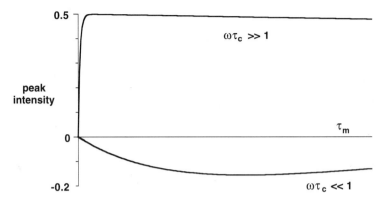

Figure 8.8. Cross-peak intensities in NOESY for a simple two-spin system with $\rho^* = 0$, as a function of mixing time. The intensities are normalized with respect to the intensity of the diagonal peak at $\tau_m = 0$. Note that with this convention the enhancement is half the size of the corresponding 1D transient enhancement, and in particular η_{max}^{trans} ($\omega\tau_c \ll 1$) = 19.2%.

There are a number of points that have to be considered when setting up a NOESY experiment, and we have classified these into two headings: "acquisition" (this section), and "selection of signals and suppression of artifacts" (Section 8.4), the latter being mostly concerned with phase cycling. In this section we deal mainly with the setting of delays, pulse widths, acquisition times, and spectral widths.

8.3.1. Fixed Delays and Pulse Widths

There are two fixed delays in the NOESY pulse sequence (Fig. 8.9), namely the mixing period τ_m and the relaxation delay t_D.

An outline of the behavior of the NOE during τ_m was given in the previous section, while more detailed equations were developed in Sections 4.2 and 4.3. The choice of a suitable mixing period depends very much on the system under study. For macromolecules, the principal issue is usually to avoid spin diffusion, by keeping τ_m sufficiently short that the initial rate approximation remains valid (Section 4.4.1). This means that NOE buildup should still be in its approximately linear phase, close to the start of the curve shown in Figure 8.8, and that cross peaks arising from multispin relaxation pathways should not

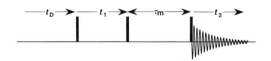

Figure 8.9. NOESY pulse sequence.

contribute. In practice, a compromise must be reached between the risk of spin diffusion and the need to have detectable intensity in the cross peaks. As usual, the weight that is placed on these points depends on the information sought; if the spectrum is required for making assignments, then a greater degree of spin diffusion may be tolerable than if quantitation to yield distance restraints is desired. In the latter case, it is desirable to use a sequence of experiments with progressively increasing τ_m values to find initial buildup rates, and so discriminate against spin diffusion. Further discussion of these issues is given in Section 4.4.

For small molecules in the positive NOE regime, sensitivity is much more likely to be the key issue. Therefore, the main consideration in setting τ_m is to be near the maximum on the NOE time course (Fig. 8.8). Although one may say that this time is of the order of T_1, or is given precisely by Eq. 4.8, this is of little help in practice, since the maximum cross-peak intensity will arise at different times for different enhancements. Short-range enhancements will reach their maximum at earlier times than long-range enhancements, and the latter may well be too weak to be detectable by NOESY even at their maxima. There is thus little option but trial and error here, with values of τ_m ranging from 0.1 s to several seconds appearing in the literature. It is, however, more economical to find out as much as possible about the NOE characteristics of such a sample using 1D experiments before turning to 2D. For instance, a 1D steady-state experiment might establish that $\omega \tau_c$ is close to 1, in which case a NOESY experiment would be futile whatever the setting of τ_m.

As in most NMR experiments, enough time should be left between passes through the pulse sequence for relaxation to restore, or nearly restore, thermal equilibrium. For NOESY, this implies that $(t_2 + t_D)$ should, ideally at least, be about 5 times T_1 for the slowest relaxing spin of interest. For macromolecules this is often not a difficult restriction, since T_1 values may be quite short, particularly at medium or low fields. However, at higher field strengths, T_1 values for macromolecules become longer, and the value of t_D may need to be checked. This may be done by recording just the first increment of the NOESY experiment separately for a series of decreasing values of t_D. Values at which the observed signal intensities are appreciably reduced are clearly too short, and a good compromise is often to set t_D to around 1.1 times the value where intensity reduction becomes noticeable. For smaller molecules in the positive NOE regime, 5 times T_1 may represent a prohibitively long relaxation delay, and some compromise may be needed. The main consequence of shortening t_D is a loss of signal intensity, although undesirable artifacts could appear in the spectrum if t_D is very short. This loss of intensity can, of course, be greater for some signals than for others, depending on the T_1 values involved, and can result in significant errors in NOE-derived distances.[8] In addition, asymmetry can appear between the intensities of symmetry-related cross-peaks on opposite sides of the diagonal wherever the relaxation rates of the spins concerned are different. This is because there is less z magnetization present from slowly relaxing spins than from rapidly relaxing spins at the time of the first 90° pulse.[9]

Setting the pulse widths is simple; all pulses should be 90°. If the first pulse is not 90° this just results in lower intensities for all peaks (except axial peaks). If the other two pulses are not 90°, then new contributions appear in the spectrum, in the form of antiphase cross-peaks linking J-coupled partners. These so-called zz peaks are discussed further in Sections 6.2 and 8.4.2. Note also that the receiver gain must be checked before starting the experiment, to make sure that the ADC is not overloaded by the signal response. For NOESY (though not for many other 2D experiments), the largest signal arises in the first t_1 increment.

It was once considered a good idea to carry out the complete 2D experiment and cycle through several times, in the same way as 1D difference NOE experiments are performed, to average out instrumental instabilities. This is now almost never done,[7] and it is better to acquire all transients for each t_1 value in one sequence. The cycling procedure is less effective than in the 1D case, since the recycle time is much longer, and it has the further disadvantage that a system failure near the end of the entire experiment may render the whole data set unusable, whereas in a consecutive experiment only the files with longer t_1 values are lost.

8.3.2. Acquisition Times t_1 and t_2 and Spectral Widths SW_1 and SW_2

Together with the length of the mixing time τ_m, the values chosen for t_{1max} and t_{2max} are probably the most critical instrumental parameters in determining the signal-to-noise ratio of a NOESY experiment.[10] Here t_{2max} is the length of the detection period or acquisition time as in 1D, while t_{1max} is the maximum time reached by the incremental period t_1. Just as in 1D spectroscopy, these times necessarily represent a compromise between the need for sensitivity and the need for resolution. However, there are two points that make the situation for 2D a little different:

1. The direct detection period t_2 can, as in 1D, be extended without necessarily increasing the overall experiment time at all. In contrast, increasing the number of increments in t_1 in order to improve resolution in F_1 *always* increases the overall experiment time proportionately. This difference arises because t_2 can be increased at the expense of the relaxation delay t_D, whereas more increments necessarily implies more experiments.
2. It used to be true that the sheer quantity of data generated in a 2D experiment imposed limitations on what was realistically possible, both in terms of data storage and data processing. This consideration has become less important over time as large capacity data storage devices and faster computers have become more widely available; however, it may still not be *desirable* to generate an inconveniently large data set.

Bearing these points in mind, some fairly clear recommendations emerge. It is almost universal practice (with good reason) to arrange for the time-domain data set to be unsymmetrical, with $t_{2\text{max}} > t_{1\text{max}}$. This means that relatively high resolution is available at least in one dimension (F_2), while the overall experiment time is not prolonged unduly. As in 1D, the best strategy is usually to record the maximum length of FID in t_2 compatible with (i) not generating an inconveniently large quantity of data, and (ii) the desired time between passes around the pulse sequence. If this is done, then sensitivity can always be regained subsequently, at the expense of resolution in F_2, by zeroing the later parts of the FIDs during data processing if necessary. However, if one adopts a similar strategy with respect to t_1 sampling, the result is that much time is wasted acquiring the later increments, which contribute relatively little (or sometimes no) signal intensity.[†] Nonetheless, once acquisition is complete, it is worth remembering that a NOESY spectrum apparently devoid of cross-peaks can sometimes be resurrected by reprocessing using only the earlier increments in t_1, thereby gaining sensitivity at the expense of resolution in F_1.

In practice, it is not normally $t_{1\text{max}}$ and $t_{2\text{max}}$ themselves that are selected *directly* when setting up an experiment, but rather the corresponding time-domain data sizes in each dimension. Common time-domain data sizes for (^{1}H, ^{1}H) NOESY experiments are 1K, or 2K complex points (see below) for t_2, with 128, 256, or 512 t_1 values. There is in fact no requirement for either of these time-domain data sizes to be powers of two; extension to the next highest power of two can always be achieved by zero filling during Fourier transformation. Thus, intermediate values are also commonly employed, particularly in t_1. If frequency-domain data with equal numbers of points in each dimension are required (e.g., if the 2D spectrum is to be symmetrized), then this too can always be achieved by zero filling (cf. Section 8.5.1).

The relationship between the time-domain data sizes and $t_{1\text{max}}$ and $t_{2\text{max}}$ is intimately tied with the values chosen for the spectral widths in F_1 and F_2, which we shall call SW_1 and SW_2, respectively. Through the Nyquist equations (Eqs. 8.2 and 8.3, following), these spectral widths determine the *dwell times* in t_1 and t_2. In t_2, the dwell time is just the interval between measurements of successive data points by the ADC, while in t_1 the equivalent parameter is the t_1 *increment*, that is, the amount by which t_1 is incremented from one experiment in the 2D matrix to the next. We shall abbreviate the dwell time in t_2 as D, and the t_1 increment as IN. Quite clearly, $t_{1\text{max}}$ and $t_{2\text{max}}$ are given by

---

[†]Since there is more signal present in early increments than in later ones, it would in principle be more efficient to acquire a larger number of transients for the earlier increments than for the later ones. Optimally, the number of transients acquired for each increment would be in proportion to $\exp(-t_1/T_2)$, where T_2 is the longest value of T_2 for any resonance of interest; in this case the procedure is equivalent to a t_1 time-domain matched filter. In practice, the need to complete whole multiples of the phase-cycle and the inconvenience of processing unevenly acquired data have so far prevented application of this idea in the context of conventional Fourier-transformation of the data. However, the closely related approach of "exponential sampling," where points in t_1 are unevenly distributed with an exponentially decreasing density, has been advocated in conjunction with maximum entropy processing.[11]

$$t_{1\max} = IN \times \text{(number of increments in } t_1\text{)}$$

and

$$t_{2\max} = D \times \text{(number of points in } t_2\text{)} \tag{8.1}$$

Just *how* the spectral width settings determine the dwell times D and IN depends on whether real data (i.e., one point for each value of t) or complex data (two points for each value of t, one real and one imaginary) are acquired. This in turn depends largely on the way in which quadrature detection is achieved, which will be touched on in the next section. Here we note that the Nyquist equation for *real* data is

$$\text{Dwell time} = 1/(2 \times \text{spectral width}) \tag{8.2}$$

while for *complex* data it is

$$\text{Dwell time} = 1/\text{spectral width} \tag{8.3}$$

These relationships apply equally in t_1, where they relate IN to SW_1, and in t_2, where they relate D to SW_2.

To summarize: In practice, one first chooses the spectral width in each dimension, and then selects the data sizes that result in the most reasonable values for $t_{1\max}$ and $t_{2\max}$, judged on the criteria given earlier. As far as t_1 is concerned, for an experiment in a given total time and using a given spectral width, this implies reaching a balance between acquiring a small number of increments in each of which many scans are coadded, or acquiring many increments each with few scans; the former choice favors sensitivity, the latter favors resolution in F_1. In t_2, as we have seen, the choice is less critical, since long t_2 acquisition times do not need to prolong the experiment.

Setting the spectral widths calls for some comment. For homonuclear experiments, it is most common to fit all of the resonances into a square frequency window ($SW_1 = SW_2$). Provided quadrature detection or its equivalent is available in both frequency dimensions (Section 8.3.3), the carrier can simply be placed at the center of the spectrum and the spectral widths both set to cover all resonances. To avoid instrumental distortions (e.g., from imperfect filters), it is advisable to leave a gap of perhaps 5–10% of the spectral width between the edge of the spectral window and the nearest resonance, although this requirement has become somewhat less critical with the advent of digital filters in some newer spectrometers. Although there is no requirement that the spectral widths SW_1 and SW_2 be equal, they are usually made so for convenience.

Occasionally, it may seem desirable to set the spectral width to exclude certain resonances, for example, if one or more uninteresting signals appear at positions shifted far away from the rest of the spectrum. This is a much more hazardous procedure in 2D than in 1D, since the filters that reject signals from outside the spectral width can operate only in F_2. Thus, even though folded resonances may be absent in a 1D test spectrum, having been removed by the filters, they will nonetheless still be excited by the pulse, and can thus still contribute cross peaks at their F_1 frequencies in a 2D spectrum. If these cross peaks lie outside the F_1 spectral window, they will fold into the spectrum with undiminished intensity, and may severely complicate interpretation.[7] As with folding in F_2, two modes of folding are possible: reflection about the nearer edge of the spectral window (Fig. 8.10a), which occurs when a real FT is used, and "wrapping around" the further edge of the spectral window (Fig. 8.10b), which occurs when a complex FT is used.[12] If we were to attempt to catalog the modes that occur on different spectrometer types (i) we would almost certainly make mistakes, and (ii) the information would rapidly become out of date as software changes are introduced. If in doubt, the only safe advice for those cases in which narrowing the spectral width is genuinely essential is to try a 2D test spectrum first, to establish how folding occurs for a known example. A possible alternative in such cases might be to record a semiselective 2D experiment, in which the signals excited in F_1 are limited by using a semiselective pulse for the first pulse of the NOESY sequence (Section 8.6.1).

8.3.3. Quadrature Detection in F_1 and F_2

The purpose of quadrature detection is to distinguish frequencies that are faster than the carrier frequency (see following) from those that are slower. Viewed in the rotating frame, this corresponds to making a distinction between *positive* and *negative* frequencies. If this is done, the carrier can be placed at the center of the spectral window, allowing the available pulse power to be used efficiently and avoiding the need to make the spectral width unnecessarily large. If quadrature is not available, one is forced to place the carrier at the edge of the spectral window, with the consequences that the pulse power is inefficiently used, and noise from the opposite side of the carrier folds into the spectrum reducing the signal-to-noise ratio by a factor of $2^{1/2}$.

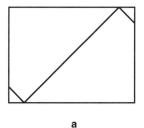
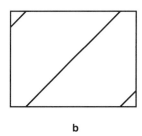

a   b

Figure 8.10. Folding of the spectrum in 2D, where the spectral width is reduced in F_1 only. This may result in (*a*) reflection about the nearer edge, or (*b*) "wrapping around" depending on the type of FT used (see text).

In order to understand how quadrature is achieved in F_1, it is first necessary to understand how it works in F_2. As detected at the probe, the NMR signal consists of a relatively narrow band of very high frequencies. Such high frequencies are impossible to digitize directly, so instead they are first compared against a reference frequency, and the *difference frequency* between the incoming signal and the reference frequency is then digitized by the ADC. This reference frequency is what is called "the carrier," and the digitized difference signal corresponds to the response of the magnetization as viewed in the rotating frame.

Quadrature detection, as implied by its name, duplicates this process of comparison using two versions of the carrier frequency that are 90° out of phase with each other. A little trigonometry shows why this works. Suppose the carrier has a frequency ϕ, and the incoming NMR signal is slightly faster; the detected signal is then $\cos(\phi + \omega)t$ where ω is the (positive) frequency shift of the NMR signal relative to the carrier. The output from the frequency comparison process consists of the product of the two input functions, but we know from standard trigonometric relations that this is the same as a mixture of the sum and difference frequencies.

$$\cos(\phi + \omega)t \cos \phi t = (1/2)\cos(2\phi + \omega)t + (1/2)\cos \omega t \qquad (8.4)$$

When we use a second version of the carrier frequency, 90° out of phase with the first, then we have

$$\cos(\phi + \omega)t \sin \phi t = (1/2)\sin(2\phi + \omega)t + (1/2)\sin \omega t \qquad (8.5)$$

In practice, the incoming signal is split and the two halves sent down separate channels, usually called x and y. Each channel is then separately compared against the carrier, one against $\cos \phi t$, and the other against $\sin \phi t$. The sum frequencies are rejected by the filters [since ω is in the kilohertz range, whereas $(2\phi + \omega)$ corresponds to hundreds of megahertz], and it is the difference frequencies that are digitized in the x and y channels, respectively (Fig. 8.11).

If we now take an NMR signal that is *slower* than the carrier by the same relative shift ω, then the signal is $\cos(\phi - \omega)t$, and the corresponding difference signals fed to the ADC are $\cos(-\omega t)$ and $\sin(-\omega t)$. Since $\cos(-\omega t) = \cos \omega t$ and $\sin(-\omega t) = -\sin \omega t$, this means that we have the same result as for the faster frequency, *except that the sign of one channel is reversed*. This is the basis for sign discrimination of frequencies with respect to the carrier. There are two methods of organizing the processing and Fourier transformation of data from the two quadrature channels so as to generate a sign-discriminated spectrum. In one method, points are sampled *simultaneously* by both channels. The time-domain data are thus organized as shown in Figure 8.12a, and they can then form the input to a *complex* Fourier transform, that is a Fourier transform that treats each time-domain point as a complex quantity $x + iy$. A com-

ACQUIRING A NOESY SPECTRUM   301

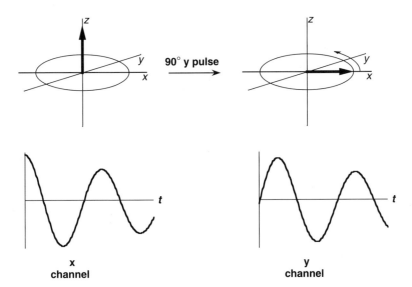

Figure 8.11. Signals recorded in the x and y quadrature channels following a 90°_y pulse. Precession of the magnetization viewed in the rotating frame results in an x component of $\cos \omega t$ and a y component of $\sin \omega t$; both are needed to achieve quadrature detection.

plex FT is intrinsically able to discriminate the direction of a rotation, since ($x + iy$) can be distinguished from ($x - iy$).

The other method of data routing and processing in common use is designed to use a *real* Fourier transform, that is a Fourier transform that treats each time-domain point as a single real number. Such a transform is intrinsically *unable* to distinguish the direction of a rotation, since the time-domain input represents only a projection of the rotation (i.e., a cosine wave). However, by manipulating the time-domain data one can avoid the need to discriminate these signs during the transform itself. This method is called time-proportional phase incrementation (TPPI), and it works as follows.[13,14]

The time-domain data are organized as shown in Figure 8.12b, that is with points from x and y channels interleaved, and with every second *pair* of points

Figure 8.12. Organization of time-domain data for quadrature detection. (*a*) Complex data (simultaneous acquisition). (*b*) Real data, as used in TPPI. For TPPI, every second pair of points is negated. The filled circles represent x channel data, and the open circles represent y channel data.

negated. This has the effect of making the receiver reference frequency *appear to rotate* from one data point to the next (relative to the rotating frame axes). The rate of this apparent rotation is one quarter of a revolution per data point, which, from the Nyquist equation for real data (Eq. 8.2), corresponds to half the spectral width. Thus, if the spectral window originally ran from $-F_S/2$ to $+F_S/2$, where F_S is the spectral width, after TPPI it runs from zero to $+F_S$. All frequencies now have the same apparent sign, and a real FT produces a spectrum in which folding about the center is avoided.

The two methods just described produce essentially identical results using essentially identical hardware; in both, the carrier can be placed centrally and the spectral width set to cover only the region of interest. Although, in the past, different spectrometer types used either method, largely accordingly to historical or commercial factors, most spectrometers are now capable of either method. Notice also that the data organization schemes in Figure 8.12 explain the difference between the Nyquist equations for real data and for complex data; although the dwell times differ by a factor of two, both schemes result in acquisition of the same quantity of data per unit time.

Turning now to 2D, the motive for using quadrature is if anything stronger, since in addition to the usual disadvantages of not doing so, three-quarters of the data are redundant when the carrier is placed at one edge of the spectral region of interest. However, if the carrier is to be placed centrally it is necessary to use *independent* quadrature in both F_1 and F_2. If an experiment were run using normal quadrature in F_2 only, the resultant 2D spectrum would appear folded about $F_1 = 0$, as shown schematically in Figure 8.13.

As in 1D, so in 2D there are two schemes of data organization for quadrature detection in F_1, one based on a complex FT in t_1, and the other on a real FT. The key, again as in 1D, is always to generate *separately* the sine and cosine

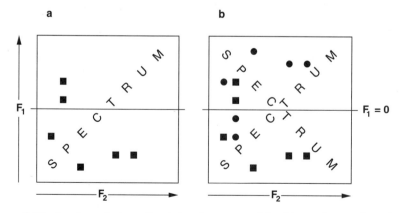

Figure 8.13. A schematic two-dimensional spectrum (*a*) with quadrature detection in F_1 and (*b*) without. The genuine cross peaks are denoted by squares, and the F_1 reflections by circles.

components of the t_1 modulations. Unlike the 1D case, however, this has nothing to do with the receiver phase, since the phases of signals in t_1 depend only on the relative phases of the pulses that flank t_1. We have seen that the NOESY experiment in its simplest form generates the t_1 signals in the form of cosine amplitude modulations. What is needed, therefore, is a second version of the NOESY experiment that generates the equivalent sine-modulated signals. This is very simply achieved by advancing the phase of the first pulse by 90° to give the sequence $90°_y - t_1 - 90°_x - \tau_m - 90°_x - t_2$. In this experiment, no magnetization is produced along the z axis by the second pulse when $t_1 = 0$, so the first increment contains no signal (provided that the phase cycling has removed all signals that were not longitudinal during τ_m). Later increments do contain a signal, since precession during t_1 creates a component of transverse magnetization perpendicular to the second pulse axis; this contributes signals proportional to $-\sin \omega t_1$.

We consider first the method for F_1 quadrature based on TPPI.[15] Here, all that is required is to advance the phase of the first pulse by 90° between each increment and the next. The t_1 modulations then appear as $+\cos \omega t_1$ (first increment), $+\sin \omega t_1$ (second increment), $-\cos \omega t_1$ (third increment), $-\sin \omega t_1$ (fourth increment), and so on throughout the whole experiment. This effectively duplicates the normal t_2 TPPI procedure.

The second method for F_1 quadrature (actually the first to be described, by States, Haberkorn, and Ruben)[16] is based on a complex FT in t_1. Here, the situation is complicated slightly by the need to provide two input functions simultaneously for the complex FT. After the first FT (and after phasing in F_2), the real part of the data consists of the absorption-mode F_2 spectra, while the imaginary part consists of the corresponding dispersion-mode F_2 spectra. Both parts are cosine modulated in t_1. However, to retain double-absorption lineshapes (which is the purpose of the whole exercise), we need to avoid incorporating the dispersion-mode spectra from the imaginary part of the data. What is done therefore is to replace the dispersion spectra in the imaginary part of the data from the first FT by the *real part of the F_2-transformed data from our second version of the NOESY experiment*. These spectra also have an absorption lineshape in F_2, but are *sine* modulated in t_1, and it is therefore these that truly correspond to the imaginary part of the t_1 time-domain data. A complex FT then correctly generates a sign-discriminated 2D spectrum with pure double-absorption phase, as for the TPPI method. Note that, exactly as in Figure 8.12, the t_1 increment *IN* is twice as long as for the TPPI method, and that sine and cosine modulated data must both be recorded for each t_1 value, rather than interleaved as for TPPI. Further discussion of the relationships between the methods, and also of their more general scope and practical implementation may be found in reference 17.

An important difference between the two procedures is that in the original States method, any axial peaks appear at the center of the spectrum in F_1 whereas in the TPPI method they appear at one edge. The latter position is by far preferable, since artifacts at the extreme edge of a spectrum are much less likely to interfere with interpretation and can if necessary be removed by time-domain manipulations such as convolution difference. A simple modification to the original States procedure was therefore introduced that moves the axial peaks to the edge of the spectrum. This hybrid method, generally called "States–TPPI," is identical to the original States procedure except that, in addition, the phases of both the first pulse and the receiver are negated each time that t_1 is incremented. The axial peaks experience only the receiver phase shifts and are therefore shifted through the Nyquist frequency, whereas other peaks experience both phase shifts and thus remain unaffected.[18]

In the first years of 2D NMR, a quite different approach to F_1 sign discrimination was used, called either echo/antiecho or N/P peak selection. As with F_1 quadrature, echo selection relies on generating the sine and cosine components of the t_1 modulations, but, unlike true quadrature methods, these components are *added* during acquisition. In this way, the amplitude modulation of the signals during t_1 is converted into a phase modulation. This provides input data suitable for a complex transform with respect to t_1, in the sense that the signs of the F_1 frequencies are determined by the direction of the phase rotation in t_1, but it makes it impossible to separate absorption and dispersion contributions in the transformed 2D spectrum. All lines then have the so-called 2D phase-twist lineshape,[19] which is in all respects undesirable for high-resolution spectroscopy.[5] Since phasing such a spectrum is pointless, normal practice was to attempt to reduce the dispersive contribution by using strong weighting functions before transformation (Section 8.5.2), and then to display the 2D spectrum in absolute value mode. Such an approach was generally only partially successful in narrowing the lines, and caused considerable degradation of the signal-to-noise ratio. The dramatic improvement in lineshape brought about by double-absorption mode NOESY spectrum can hardly be overstated.

A method of combining an echo-selected spectrum with an anti-echo-selected spectrum has been introduced that does create a pure-phase, sign-discriminated spectrum, and this is now widely used in experiments where coherence selection is brought about using pulsed field gradients during the corresponding time domain.[20] However, this approach is rarely if ever applied to NOESY (cf. Section 8.4.1).

8.3.4. Phase-Sensitive NOESY

As just mentioned, double-absorption mode phase-sensitive NOESY spectra are far superior to their absolute value counterparts, particularly in lineshape. The marked difference between the two presentations is shown in Figure 8.14, where it may be seen that the advantages of the double-absorption mode far outweigh the relatively trivial problem (for NOESY, at least) of phasing the spectra. The "mechanics" of generating a phase-sensitive NOESY spectrum having largely been covered in the previous section, we now consider some of the properties of such spectra.

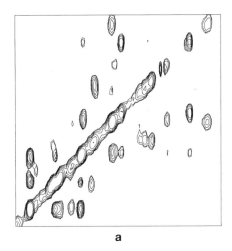 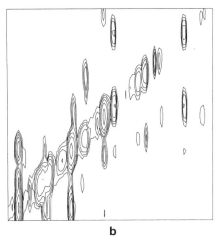

Figure 8.14. (*a*) Absorption mode and (*b*) absolute value NOESY spectra, showing the aromatic region of bovine phospholipase A_2 in D_2O. The same data set was used for both spectra. The window functions applied were approximately optimal for the two presentations, namely (in both dimensions) sine-bell shifted by $\pi/10$ for (a), and $\pi/32$ for (b). The contrast in resolution is striking. Note also how much narrower the diagonal is in the absorption mode spectrum.

The phase properties of 1D spectra are very familiar. The transformed 1D spectrum consists of two parts called real and imaginary, the two being related by a 90° phase shift. Initially, both parts contain signals that are (different) mixtures of absorption and dispersion modes, and phasing the spectrum corresponds to taking linear combinations of the real and imaginary parts until pure absorption mode has been generated in the real part. The imaginary part has then served its purpose and becomes redundant; only the real part is plotted.

In phase-sensitive 2D spectra, the same concepts apply, but now they apply equally to both dimensions. Thus, there are now four "phase quadrants" comprising the various combinations of real and imaginary parts in the two dimensions.[†] These are abbreviated RR (real F_1, real F_2), RI (real F_1, imaginary F_2), IR, and II. Phasing the 2D spectrum in F_2 involves taking linear combinations of quadrants RR and RI (and also of quadrants IR and II), while phasing in F_1 involves taking linear combinations of quadrants RR and IR (and also of RI and II). The object is to produce pure double-absorption phase in the RR quadrant. Once this has been done the other three quadrants become redundant, and only the RR quadrant is analyzed. Figure 8.15 shows the situation for a one-line spectrum after two-dimensional phasing has been completed.

[†]Note that, just as for the two parts of a 1D spectrum, each 2D phase quadrant includes the entire spectral area. Some authors use the term quadrant quite differently, to refer to the four spectral areas obtained by dividing the total 2D spectrum into four.

306 THE TWO-DIMENSIONAL NOESY EXPERIMENT

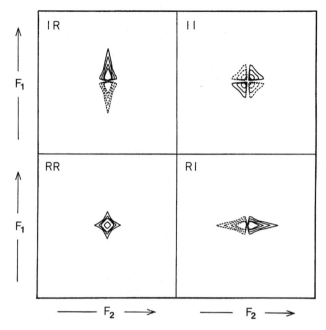

Figure 8.15. Lineshapes in the four phase quadrants of a phase-sensitive NOESY spectrum, after two-dimensional phasing has been completed, to give a double-absorption lineshape in the RR quadrant. Negative intensity is represented by dashed contours.

Phasing in either dimension is totally independent of phase in the other dimension. This may seem rather baffling, but should be clarified by the following. When examining a 1D cross-section, let us say parallel to F_2, adjusting the phase in the *same* dimension (here F_2) causes lineshape changes between absorption and dispersion in the usual way. In contrast, changing the phase along the *perpendicular* dimension (here F_1) causes only *amplitude* changes in the displayed (F_2) cross-section. Consideration of Figure 8.15 should clarify this. A helpful example is to imagine the F_2 section through the center of the double-absorption line (RR quadrant). As the phase in F_1 is changed, the intensity of the F_2 section diminishes, but its lineshape does not change. When the F_1 phase reaches pure dispersion, the F_2 section has zero intensity, and when the F_1 phase has been changed through 180°, the F_2 section is fully inverted. In fact, a 180° (zero-order) phase change in F_1 is exactly equivalent to a 180° (zero-order) phase change in F_2, this being the only case in which a phase change in one dimension has the same effect as the same change in the other dimension.

The most convenient method for finding the correct F_2 phase constants is usually to transform and phase the first data file ($t_1 = 0$, using the cosine modulated component in the case of the complex FT method) prior to full 2D transformation. If this 1D spectrum is phased for positive absorption, and the

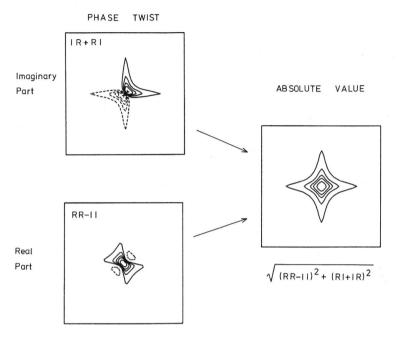

Figure 8.16. The phase-twist and absolute value 2D lineshapes, showing their relationship to the pure phase lineshapes in Figure 8.15. Negative intensity is represented by dashed contours.

resultant parameters used during 2D transformation, the 2D spectrum will appear in absorption mode with a positive diagonal, subject to later phase correction in F_1. Notice that our discussion of F_1 quadrature in the previous section made the assumption that F_2 phasing was carried out during the first transform as described earlier, rendering the RI and II quadrants unnecessary. If F_2 phasing is to be carried out *after* the 2D transform, then the RI and II quadrants must be computed also, by duplicating the processing procedures of the previous section for the imaginary parts of the F_2 spectra in the interferogram. Further discussion of phasing considerations for NOESY and other 2D experiments may be found in reference 17.

For comparison, the relationship between the four pure phase quadrants and the phase-twist lineshape is shown in Figure 8.16, together with the absolute value lineshape.

8.4. PHASE CYCLING, SIGNAL SELECTION, AND ARTIFACT SUPPRESSION

Since spectrometers are now usually supplied complete with software to acquire many 2D experiments including NOESY, and these (presumably) include correct phase-cycling programs, one might argue that there is no need to understand phase cycling at all. Apart from curiosity, there are at least two reasons for having some appreciation of its workings. The first is implied by our earlier use of the word "presumably;" when an experiment fails, some understanding of phase cycling is needed if that aspect of spectrometer operation is to be tested. Second, if one wishes to modify the experiment at all from the form in which it is supplied (by no means an uncommon or unreasonable wish), then some understanding of phase cycling is again likely to be needed. This section therefore deals with its more important aspects; it also covers some aspects of signal selection not related to phase cycling.

The first point to be made is that the complete phase cycle for an experiment performs several different tasks. These may be considered in two categories, "essential" and "nonessential." By "essential" we mean those parts of the cycle that select the types of signal that contribute to the final spectrum. In the case of NOESY, this means those parts of the phase cycle that retain z magnetizations during τ_m, rejecting other contributions. This selection process is in many ways the heart of the NOESY experiment, in that it is primarily this that distinguishes it from other three-pulse experiments such as DQF-COSY or relayed coherence transfer. Each of these other experiments uses some different phase cycling scheme to select a different contribution to the 2D spectrum; if no phase cycle were used, all these various contributions would be present. This aspect of phase cycling is also called magnetization pathway selection.

By contrast, the "nonessential" functions of phase cycling act to suppress relatively minor contributions to the spectrum arising primarily from instrumental imperfections. The most obvious example is the CYCLOPS cycle for suppressing F_2 quadrature images.[21] The key distinction is that "nonessential" parts of the phase cycle, once recognized, can if necessary be omitted to save time.

Each particular process of selection made by the phase cycling acts in much the same way as a difference experiment. Two versions of the pulse sequence, differing in their pulse and receiver phases, are alternated during signal averaging, such that the desired response reinforces on combining the signals, while the undesired response cancels. The trick is entirely in the appropriate choice of pulse and receiver phases required to bring this about, as will be illustrated below. In most schemes, each process of selection requires a pair of experiments, so that combining N selection processes involves 2^N steps.

Just as difference spectra rely on subtracting spectra having equal numbers of transients, so phase cycling depends on an equal number of transients being acquired for each phase permutation used. One must therefore set the number of scans acquired per increment to be some integer multiple of the number of steps in the phase cycle. By analogy with the fact that subtractions may (sometimes) be improved by keeping acquisition of the two components of a difference spectrum close together in time (Section 7.2.3), the phase permutations corresponding to the most important selections should be adjacent in the cycle.

When this is done, nonessential steps can be omitted simply by not executing the appropriate final portion of the phase cycle.

So much for the generalities of phase cycling. We shall now deal briefly with the various selection processes that phase cycling is required to make during a NOESY experiment, concentrating on its central role in selecting the appropriate magnetization transfer pathway. Since the same effect can often be produced by several related cycles, there is no "uniquely correct" phase cycle for many 2D experiments. The examples that follow thus represent only one way of achieving the desired selections.

8.4.1. Rejection of Nonlongitudinal Contributions During τ_m: J-Peak Suppression

A 90° pulse acting on z magnetization produces only transverse magnetization (single-quantum coherence), but a 90° pulse acting on transverse magnetization after a period of precession creates not only the desired z magnetization, but also coherence transfer among coupled single-quantum antiphase states (i.e., COSY-type transfer), together with all orders of multiple-quantum coherence.[22] All of these contributions from coupling partners are at least partially converted back into observable antiphase signals by the third pulse of the NOESY sequence, resulting in antiphase cross peaks in the 2D spectrum linking direct and relayed J-coupling partners. Although these antiphase cross peaks have a net zero integral, and could thus in principle be distinguished from NOESY cross peaks by volume integration of a phase-sensitive NOESY spectrum, in practice they are an undesirable complication, and are therefore suppressed where possible.

The key to doing this is to use the different responses that different orders of coherence show to relative phase changes in the pulses.[23] Addition of the results of the sequence $90°_x - t_1 - 90°_x - \tau_m - 90°_x - t_2$ (which we may express in shorthand as x, x, x) to the results of the sequence $(-x, -x, x)$ cancels out single quantum contributions during τ_m (so eliminating COSY and relayed coherence transfer type cross peaks). Notice, however, that the z magnetization during τ_m is *unaffected* by relative phase changes between the second and third pulses, and is therefore reinforced by adding these sequences.

Double quantum coherence, by contrast, is twice as sensitive to relative phase shifts as is single quantum coherence. It therefore also survives the addition just described, since its phase is changed through 360° by the 180° relative phase shift between the second and third pulses. In order to cancel double quantum coherence, the appropriate sequence to add is (y, y, x), where there is now a 90° relative phase shift between the second and third pulses. Combining this selection process with the previous one generates the four-step phase cycle shown in Table 8.1. Notice that the first pair of experiments cancels not only single quantum coherence contributions but also all odd orders of coherence; in fact all orders of coherence higher than triple are most unlikely to make any significant contribution.

TABLE 8.1 Phase Cycle for the Removal of Single and Double Quantum Coherence during τ_m

90	90	90	Acq	90	90	90	Acq
x	x	x	x	y	y	x	x
$-x$	$-x$	x	x	$-y$	$-y$	x	x

In all four of the sequences in Table 8.1, the longitudinal contributions during τ_m are identically reproduced, because there is always a constant phase relationship between the first and second pulses, and also between the third pulse and the receiver phase. The table also illustrates how a cycle is built up; the fourth permutation has the same relationship to the second as the third has to the first. This process of progressive permutation will be implied for later cycles, which will all be specified relative to the (x, x, x) sequence.

Another approach to selecting z magnetization during τ_m is to use pulsed field gradients, as explained in Section 7.5.3.1. In the case of NOESY, the simplest selection procedure would simply be to apply a gradient during τ_m, since z magnetization is unaffected by such a gradient, while all transverse terms (except for zero quantum coherences, see Section 8.4.2) would be dephased.[24] In practice, this is usually used *in addition* to phase cycling rather than to replace it. Partly this is because NOESY experiments are generally not so sensitive that it would be practical to acquire a spectrum using only one or two transients per increment, so there is little advantage to abandoning phase cycling. Partly also it is because such a gradient does nothing to suppress axial peaks arising during τ_m, and these can be substantial. However, this form of gradient selection is more common in multidimensional experiments, where the total number of increments collected in all dimensions is much higher, and where axial peaks could be taken care of with some residual element of phase cycling. Other forms of gradient selection can be imagined where encoding gradients before τ_m are combined with corresponding decoding gradients applied after τ_m. However, such approaches would be likely to cause substantial signal loss, both through molecular diffusion during τ_m (cf. discussion of comparable losses in GOESY, Section 7.5.3.3) and also because coherence selection during t_1 using gradients causes an unavoidable further loss of $2^{1/2}$ in signal-to-noise ratio relative to the phase-cycled methods of F_1 sign discrimination described in Section 8.3.3.

8.4.2. Other Forms of *J*-Peaks: Zero Quantum Coherences and Pulse Angle Effects

This section deals with two further types of cross peak that can occur between coupling partners, but that cannot be suppressed by phase cycling.

The first is caused by zero quantum coherences during τ_m. Unfortunately, zero quantum coherence has identical phase properties to z magnetization, as both are unaffected by relative phase changes between the second and third pulses of NOESY. However, unlike z magnetization, zero quantum coherence precesses during τ_m. Just as single quantum coherence in an AX spin system precesses at ω_A or ω_X, so double quantum coherence precesses at $(\omega_A + \omega_X)$, and zero quantum coherence at $(\omega_A - \omega_X)$. This frequency does not appear as such in the 2D spectra, because the relevant magnetization was still precessing at the single quantum frequencies ω_A and ω_X during the frequency labeling period t_1. Therefore, zero and double quantum coherence gives rise to cross peaks linking the normal (single quantum) frequencies of the participating coupling partners, just as for the DQF-COSY experiment; in fact, for very short τ_m, NOESY represents the precise zero quantum analog of DQF-COSY. However, when τ_m is not short, the precession of the zero quantum coherence during τ_m does affect the appearance of the resulting J-peaks, since their sign and intensity will depend on just where in their oscillation they happened to be caught by the end of τ_m (Fig. 8.17). This is usually the basis for their suppression.[25] If many spectra with different values of τ_m are coadded, the J-peaks will have different signs and intensities in each. They will therefore cancel in the combined spectrum, just as the average of a randomly sampled cosine wave is zero. The intensity of the NOE cross peaks, in contrast, does not oscillate, and changes only slightly with small variations in τ_m.

In practice, it is only necessary to alter τ_m through times of the order of one oscillation cycle. The slowest zero quantum precession frequency that needs to be considered is the same as the smallest frequency separation between two coupled signals that might result in a distinguishable NOE cross peak in NOESY, that is probably about 50 Hz. A random jitter in the length of τ_m covering about a 20 ms spread should therefore suppress all zero quantum J-

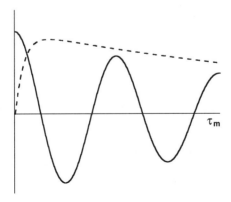

Figure 8.17. Dependence of intensity of zero-quantum coherence signals on τ_m. The dotted line shows the NOE cross-peak intensity for $\omega\tau_c \gg 1$. Zero quantum coherences in an AX spin system oscillate at $(\omega_A - \omega_X)$.

peaks further than 50 Hz from the diagonal. Ideally, the random changes should be made after each pass through the complete phase cycle (i.e., several times per increment). However, this is rarely done, the random changes in τ_m being made instead only between each t_1 increment and the next. This has the effect not of cancelling the J contributions, but rather of randomizing their F_1 frequencies. Thus the intensity that would have been present in a J-peak now contributes instead to the random t_1 noise present at the F_2 frequencies of the coupled signals, Other techniques for suppressing zero quantum J-peaks are considered in references 25–28.

Another source of J-peaks that cannot be removed by phase cycling arises when the second and third pulses of the NOESY sequence have flip angles other than 90°.[29] As discussed in Section 6.2.1, this leads to the creation of antiphase z magnetization terms of the type $I_z S_z$ during τ_m. The third pulse (if it has a flip angle other than 90°) converts these partially into transverse signals, which appear in the 2D spectrum as double-absorption mode antiphase cross peaks linking coupling partners. When there is no NOE between the coupled signals, the J-peak thus appears much as in a DQF-COSY spectrum. When superimposed on an NOE cross peak, its effect is to distort the symmetry of the multiplet structure, as shown schematically in Figure 8.18. Note that the appearance of *all possible* multiplet components in an undistorted NOESY cross peak, rather than just those connecting each component with its counterpart of corresponding spin state, is a consequence of the 90° pulses. In addition to their more familiar effects, these pulses randomize the spin states of all coupling partners each time they are applied.

Although these "zz" signals can be of interest in their own right (see Section 6.2.1), they are normally an undesirable complication. The only way to suppress them is to ensure that all pulses have exactly a 90° flip angle. Careful calibration is usually sufficient, but as an alternative the composite pulse $90°_x 90°_y$ may be substituted for the second and third pulses of NOESY.[29] This gives the sequence $90°_{\phi_1} - t_1 - 90°_{\phi_2 + \pi/2} 90°_{\phi_2} - \tau_m - 90°_{\phi_3} 90°_{\phi_3 + \pi/2} - t_2$, where ϕ_1, ϕ_2, and ϕ_3 represent the normal phase cycle for NOESY. If the 90° pulses

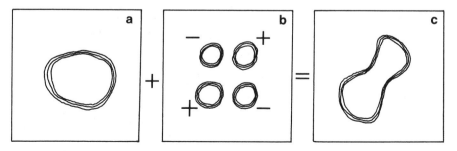

Figure 8.18. The combination of a normal NOESY cross peak (*a*), with an antiphase zz peak (*b*), gives a distorted, "tilted" peak (*c*). The lineshape in (*a*) is the result of the overlap of four in-phase multiplet components. From ref. 29.

are set incorrectly, these composite pulses compensate for the error. For instance, the J-peaks resulting from a 10% error in calibration for the simple NOESY sequence would be reduced to 3% of their intensity if the composite pulses were used with the same misset flip angle.

8.4.3. Axial Peaks

From the time of the first pulse, longitudinal relaxation acts to restore the z magnetization to its equilibrium value. Magnetization that recovers during t_1 will be made transverse by the second pulse and, after precession during τ_m, a component will survive the third pulse to contribute to the finally detected signal. Similarly, magnetization that recovers during τ_m is converted directly into detectable signals by the third pulse; in the limit of long τ_m, these are the only contributions to the spectrum. Since none of these magnetizations was transverse during t_1, they are not frequency labeled, and consequently give rise to peaks at an F_1 frequency of zero. For this reason they are called *axial peaks*. As pointed out earlier, in experiments that use either the "States" method, echo-selection, or no method at all for F_1 sign discrimination, $F_1 = 0$ corresponds to the center of the F_1 spectral width, whereas in experiments using the TPPI or States–TPPI methods, $F_1 = 0$ corresponds to one edge of the spectrum.

Axial peaks in any 2D spectrum may all be suppressed by inverting the phase of the first pulse and negating the receiver. For NOESY this corresponds to the combination $(x, x, x) - (-x, x, x)$; however, it is also possible to combine a partial suppression of axial peaks with the cycle used to suppress transverse contributions during τ_m. If the combination $(x, x, x) + (-x, -x, x)$ is used, as in Section 8.4.1, this cancels axial peaks arising during t_1, but not during τ_m. Since usually $t_{1\,max} < \tau_m$, and axial peaks arising during t_1 are attenuated by transverse relaxation during τ_m, these axial peaks are much less significant than those arising during τ_m. For this reason, the combination more commonly used is $(x, x, x) + (-x, x, -x)$, which combines rejection of transverse magnetization during τ_m with elimination of axial peaks arising during τ_m but not t_1.

8.4.4. Quadrature Images

In order to achieve sign discrimination, the methods for normal F_2 quadrature described in Section 8.3.3 depend critically on various aspects of spectrometer performance, in particular the accuracy of the 90° phase shifts both in the receiver and in the pulses, and the amplitude balance between the two quadrature channels. If any of these does not function correctly, the resulting spectrum will contain weak quadrature images corresponding to reflections of every signal about the center of the spectral window. In 1D spectroscopy, it is usual to suppress these images by cycling the pulse and the receiver in synchrony through the sequence $x, y, -x, -y$, known as CYCLOPS (*CYCL*ically *O*rdered *P*hase *S*ystem).[21] For a NOESY experiment, CYCLOPS need be applied only to the last pulse and the receiver as shown in Table 8.2, which combines all

TABLE 8.2 Phase Cycle for the Suppression of Transverse Magnetization during τ_m, Axial Peaks, and F_2 Quadrature Images

90	90	90	Acq	90	90	90	Acq
x	x	x	x	x	x	$-x$	$-x$
$-x$	x	$-x$	x	$-x$	x	x	$-x$
y	y	x	x	y	y	$-x$	$-x$
$-y$	y	$-x$	x	$-y$	y	x	$-x$
x	x	y	y	x	x	$-y$	$-y$
$-x$	x	$-y$	y	$-x$	x	y	$-y$
y	y	y	y	y	y	$-y$	$-y$
$-y$	y	$-y$	y	$-y$	y	y	$-y$

the various functions of phase cycling considered in this section. Equally, all pulses may be included to achieve the same effect; this represents a more general way to apply the CYCLOPS cycle to any 2D experiment.

The situation in F_1 is somewhat different, since F_1 quadrature depends on the separation of cosine and sine signal modulations into the results obtained from the two pulse sequences (x, x, x) and (y, x, x), respectively (Section 8.3.3). Clearly, this has nothing to do with the receiver directly, and so the cycling of receiver phases in CYCLOPS does not suppress F_1 quadrature images. This results in weak reflections of the sort shown schematically in Fig. 8.13; in a real spectrum, only the reflections of the stronger diagonal peaks are likely to be noticeable, as illustrated in Figure 8.19. If this is a problem, the only recourse is to improve the phase shifting performance and stability of the pulses, or to subtract some suitable small proportion of the reflected spectrum during data processing.[30]

8.4.5. t_1 Noise

Unlike the frequencies present during t_2, those present during t_1 are detected indirectly, being present in the data only in the form of amplitude modulations of the directly detected signals. They are therefore susceptible to interference from any factors that distort the phase, amplitude, shift, or lineshape of the signals that carry them. As a result, when a signal-bearing t_1 trace in the interferogram is transformed, these different errors for each t_1 point produce distortions and noise in the corresponding F_1 trace in the 2D spectrum. This is what is called t_1 noise. In F_1 traces, it is manifested as a significantly higher noise level for signal-bearing traces than for others, while in a contour plot it appears as streaks of noise parallel to the F_1 axis.

In general, t_1 noise includes both systematic and random contributions. The random contributions are very much like those that cause subtraction artifacts in 1D difference spectra (Section 7.2.2); presumably they are progressively

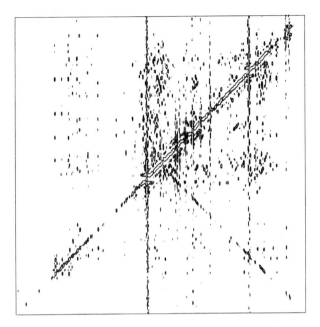

Figure 8.19. Quadrature image in F_1. The normal diagonal runs from bottom left to top right, but an image diagonal can be seen running from top left to bottom right. The sample is bovine phospholipase A_2 in D_2O.

reduced by time averaging. One source of random errors that is avoidable (and that does not decrease with time averaging) has been mentioned already; the suppression of zero quantum J-peaks by randomly varying τ_m only on each increment, rather than on each pass around the phase cycle, must add to the random t_1 noise present for coupled signals. Other random contributions are not easily removed, but are presumably minimized by improving the overall stability of the spectrometer, as discussed in Section 7.2.2. One factor that can easily be limited is temperature drift; adequate time should always be allowed for thermal equilibration before starting acquisition. Much as for subtraction artifacts, the most intense t_1 noise occurs at the F_2 frequencies corresponding to particularly sharp or intense signals.

Systematic contributions to t_1 noise can arise in various ways, but a particularly common problem is that of a "t_1 ridge." If the intensity of the first point in a 1D FID is incorrectly represented, this translates into a constant intensity offset for the whole of the corresponding frequency domain trace after Fourier transformation. In fact, it turns out that there is an anomaly concerning the way in which the first time point is handled by a digital FT.[31] As illustrated in Figure 8.20, each data point represents the integrated average of the analog signal over the sampling interval, but the *first* point represents an average over only half this interval. Consequently it should correctly be presented to the digital FT

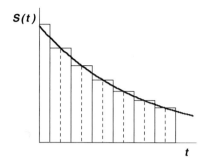

Figure 8.20. Discrete FT. The true magnetization decay is sampled at discrete time intervals, beginning at (or near) $t = 0$. The value of $S(t)$ at the first time point should have only half the weight of $S(t)$ at the other time points. From ref. 31.

with only half its measured intensity. In normal 1D acquisition, the intensity of the first point is in any case considerably diminished, primarily because of the finite response time of the analog filters, so that this refinement to the data handling is irrelevant and is omitted. However, in 2D spectra, such intensity offsets become a severe nuisance because they differ from one trace to another, so that they dominate low-level contour plots of the data. Note that for a spectrum with positive diagonal, t_1 ridges are positive, since the first t_1 data point is too intense, whereas t_2 ridges are generally negative, since the first t_2 data point is attenuated (by more than a factor of two).

Correction of t_1 ridges is simply achieved by halving the intensity of the first data file ($t_1 = 0$) prior to 2D transformation, and most manufacturers now include this correction automatically in their 2D software. Analogous correction of t_2 ridges is more difficult because the correction factor for a particular 2D data set depends on experimental variables and must be determined empirically. However, efficient algorithms for removing t_1 and t_2 ridges, together with other more generalized baseline distortions, from the fully transformed spectrum are usually available in modern processing software packages. For experiments that employ TPPI in both F_1 and F_2, an alternative approach is to change the phase cycle of the experiment such that sine modulated data are acquired rather than cosine modulated data. This avoids both t_1 and t_2 ridges, since the first time-domain points should then all have zero intensity.

A general discussion of the factors that contribute to t_1 noise may be found in reference 32.

8.5. DATA PROCESSING

The never-ending increases in the speed of computers makes this a rapidly evolving subject. Particularly for studies of macromolecules, computers are essential during interpretation, both to assist in building a structure consistent with distance restraints (cf. Chapter 12), and, to an extent, in deriving distance restraints from the spectral data (cf. Section 12.2). The most difficult part of the overall process to automate is still data reduction, that is assignment of the spectrum and location and integration of individual NOESY cross peaks. The nature of 2D and 3D NMR spectra is such that these activities necessarily involve making many somewhat subjective judgements, some of which are entrusted to a computer at some peril. Nonetheless, efforts to automate the interpretation of NOESY spectra will doubtless continue, with the ultimate aim of making the translation of raw data into a structure (almost) completely automatic.

There is little we can say about this aspect of data processing. The present section concentrates instead mainly on more technical considerations connected with the 2D Fourier transformation itself, techniques to improve the appearance of spectra, and of integrating cross peaks.

8.5.1. Zero Filling

As discussed in Section 8.3.2, the *resolution* of an NMR experiment is determined by the value of t_{max} (the length of the acquisition period), such that the minimum linewidth is given approximately by $0.6/t_{max}$.[33] However, the *representation* of this linewidth in the spectrum depends also on the number of points (N) used for the digital FT. It is important to realize that this number is not necessarily related to the number of time-domain data points actually collected (N_{TD}). When preparing the input data for the digital FT, one may either extend the FID by adding zeroes ($N > N_{TD}$), or shorten it by discarding the latter part ($N < N_{TD}$). The output from the digital FT necessarily consists of the same number of points N as the input, which (unlike N_{TD}) must be an integer power of two (if a Cooley–Tukey algorithm is used).

It turns out that when $N = N_{TD}$, information is actually discarded. The output from the digital FT consists of the real and imaginary parts of the spectrum, each of which is specified by $N/2$ points. If $N = N_{TD}$, the real and imaginary parts contain *independent* information,[34,35] so that the absorption mode spectrum contains only half of the information available from the time-domain data. However, if the value of N is set to $2N_{TD}$, by filling the FID with an equal number of zeroes, this lost information can be recovered; the real and imaginary parts of the spectrum now *each* contain all the available information, and they are no longer independent. For lines that have not decayed into the noise by the end of the FID, this means that splittings not resolved with $N = N_{TD}$ can sometimes appear in a spectrum calculated with $N = 2N_{TD}$. However, extension of the FID with zeroes beyond $2N_{TD}$ cannot reveal new information, the additional frequency-domain points being simply interpolated with a sinc function between the existing points. Although this cannot improve resolution, it does improve the representation of the spectrum. The effect on signal-to-noise ratio is somewhat less predictable, depending partly on where the data points fall across the lineshape. Zero filling can increase line-heights in cases in which

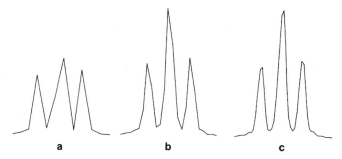

Figure 8.21. The effect of digital resolution on lineshape. (*a*) 3.13 Hz/pt. (*b*) Same data as (a), zero-filled once. Note how much better the representation of intensities is in (b). (*c*) Same data as (b), zero-filled again. The further improvement is small. Ethyl benzene in $CDCl_3$, using a cosine-bell window function to avoid sinc wiggles in (b) and (c).

the original points straddle the line (Figure 8.21).[36,37] However, zero filling beyond $4N_{TD}$ has little appreciable effect.[7] Of course, the benefits of zero filling must always be balanced against the data handling and storage penalties incurred.

8.5.2. Window Functions and Linear Prediction

Normally FIDs are weighted by multiplying them by a suitable window function prior to Fourier transformation. In NOESY, the principal reason for this is to limit the distortions that result from truncation. If the signal in an FID is still appreciable at the end of the acquisition period ($t_{1\,max}$ or $t_{2\,max}$), then lineshapes in the transformed spectrum will be convoluted with a sinc function ($\sin x/x$), as shown in Figure 8.22. The intensity of the resulting "sinc wiggles" increases with the degree of truncation. Because of the short acquisition times used in 2D, particularly in t_1, truncation is very common, and in some cases it is the actual origin of much of the t_1 noise from sharp peaks (cf. Figure 8.19). This is most clearly seen in phase-sensitive NOESY spectra, where sinc wiggles often cause the t_1 noise to oscillate regularly in sign. The effects of truncation

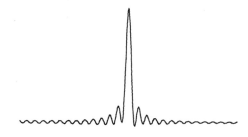

Figure 8.22. Sinc wiggles resulting from truncation.

can be much reduced by applying a window function that apodizes the data, that is, a function that decays smoothly to zero at the end of the FID.

Prior to the introduction of phase-sensitive 2D spectroscopy, the window function was also used to control the lineshape of signals in the 2D spectrum. If a bell-shaped function symmetrical about the midpoint of the acquisition period is applied, this has the effect of destroying the dispersive parts of all lines. The penalty, of course, is that it also destroys the early part of the FID, so reducing the signal-to-noise ratio if there is appreciable signal decay during the acquisition time, and distorting the relative intensities of cross peaks. Window functions that were used include pseudoecho, sine-bell, and sine-bell squared[38] (Figure 8.23). Shifted sine-bell and sine-bell squared functions represent a compromise; the greater the shift in the maximum of the bell toward the start of the FID, the better the signal-to-noise ratio, but the greater the dispersive contribution to the lineshape.

For phase-sensitive experiments, rejection of the dispersive component is not the task of the window function, so that much milder functions can be applied simply to apodize the data. A decaying exponential (i.e., line-broadening) or Lorentz-to-Gaussian transformation can be used,[39] but in many cases the simplest solution is to use a sine-bell or sine-bell squared function shifted by $\pi/2$ (or nearly $\pi/2$), since such functions have little or no rising part and necessarily decay to zero by the end of the FID. If, as is likely for unsymmetrical time-domain data sets, different window functions are used in each dimension, this makes signal intensities unsymmetrical about the diagonal, and provides a further reason why symmetrization (Section 8.5.3) may be inappropriate. In practice, the effect of various window functions on a particular data set can only be assessed by trial and error.

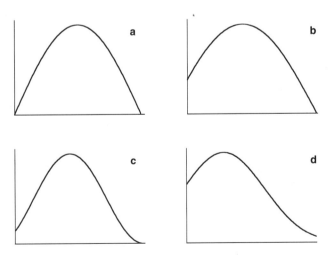

Figure 8.23. Window functions. (*a*) Sine-bell. (*b*) Sine-bell, shifted $\pi/8$. (*c*) Sine-bell squared, shifted $\pi/8$. (*d*) Lorentz-to-Gaussian transformation, with a line broadening factor of -1.5 Hz, and a maximum at $0.3\ AQ$.

When data are severely truncated, for instance in multidimensional experiments (Sections 9.2.2.2 and 9.2.3), linear prediction is widely used to improve the situation. In this technique the existing experimental time-domain data points are used to calculate extrapolated values for later time points that were not actually collected during the experiment. The theoretical basis for this method will not be described here, but suffice to say it allows an apodization function to be used that decays to zero at a time *later* than t_{max}, thereby decreasing the consequent line broadening and loss of sensitivity (relative to using the corresponding apodization function scaled to reach zero at t_{max}). Typically, linear prediction is used to double the length of the time-domain data prior to apodization, but of course the longer the time predicted, the greater the hazards of introducing artifacts; in particular, there is little point using linear prediction if the experimental data have decayed substantially prior to t_{max}. Linear prediction and zero filling may also be used in combination, thereby constructing an input to the FT that consists of experimental points followed by extrapolated points followed by zeroes. A number of parameters need to be optimized when using linear prediction. Guidance on how to do this is usually available in the relevant software manuals, but more general help on this and other aspects of data processing may be found in the book by Hoch and Stern.[40]

8.5.3. Symmetrization and t_1 Noise Removal

Symmetrization[41] is based on the fact that the desired signals in a 2D NOESY spectrum, namely the cross peaks, must always occur in pairs related by reflection about the diagonal, whereas most artifacts, in particular the t_1 noise, do not. A symmetrized spectrum is calculated by comparing the intensity of each point with its symmetry-related counterpart on the opposite side of the diagonal, taking the less intense value, and placing this value in both locations. In general, the cross peaks survive this process, but the t_1 noise, having (usually) no symmetry-related counterpart, is removed. Note that (i) for a phase-sensitive spectrum, the less intense value means that closest to the zero plane, on whichever side, and (ii) symmetrization requires that the frequency-domain data matrix has equal numbers of points in each dimension. However, the latter does *not* imply that the spectrum necessarily has symmetrical resolution before symmetrization, since this is dictated by the values of t_{1max} and t_{2max} (Section 8.3.2).

The most important point to make about symmetrization is that it is a purely *cosmetic* exercise, fundamentally incapable of ever creating new information. This must be so; in essence, the technique consists just of discarding half of the data points (the half one doesn't like). There are several specific hazards inherent in symmetrization:

1. Symmetrically disposed artifacts will survive the process. If there is a lot of t_1 noise in a spectrum, this is a real possibility. Worse still, by removing the rest of the t_1 noise, any surviving artifacts are seen out of context,

and there is little or nothing to distinguish them from genuine cross peaks. In the original spectrum, the abundance of t_1 noise would be a clear warning against false interpretation of signals at the affected F_2 frequencies.
2. Cross peaks can fail to survive symmetrization when one cross peak coincides with an intense noise peak of opposite sign. A good example of this arises for NOESY spectra of proteins recorded in H₂O. The "fingerprint" region, containing cross peaks linking NH protons to C$^\alpha$H protons, is duplicated in two spectral regions; one is reasonably free of t_1 noise, but the other can be immersed in a broad band of t_1 noise from the partially suppressed water signal, which is near the C$^\alpha$H frequencies. In an unsymmetrized spectrum, the former region (NH in F_2, C$^\alpha$H in F_1) is quite accessible to study, but on symmetrization many of the cross peaks in it disappear, either completely or partially, where single t_1 noise points happen to cancel the intensity of all or part of a cross peak.
3. Even in spectra relatively free of t_1 noise, unequal resolution in the two dimensions often causes the tails of two peaks to create an apparent cross peak in the symmetrized spectrum where none actually exists. This represents a particularly insidious hazard.

In all these cases, the problem created by symmetrization is that, although it removes many or even most artifacts from the spectrum, those that remain are now in a context that makes them extremely difficult to recognize. The cosmetic attraction of symmetrization will probably ensure that some continue to use it, but it should at least be an absolute rule never to interpret a symmetrized spectrum in isolation; the unsymmetrized spectrum should always be on hand for comparison.

An alternative approach to reducing t_1 noise is that of "t_1 ridge subtraction."[42] The systematic contribution to t_1 noise often consists largely of a constant intensity offset in signal-bearing F_1 traces, and this can be removed by subtraction. An F_2 trace that does not pass through any peaks will include a section through each t_1 ridge. If several such traces are coadded (to improve the signal-to-noise ratio) and the result subtracted from every F_2 trace in the 2D spectrum, the constant components of the t_1 ridges are eliminated. A similar approach can be used to suppress t_2 ridges, and a more sophisticated development has been proposed to take account of sinusoidal variations along the length of ridges.[30] However, improvements in baseline correction algorithms have by now rendered such techniques more or less obsolete.

8.5.4. Integration

Quantitative measurements of cross-peak intensities are difficult for several reasons. The low digital resolution means that cross peaks may be represented by only a few data points. This has two major consequences. First, the sparse representation of the data can lead to errors in peak volume estimates by as much as 100%.[43] Second, multiplet patterns, particularly in F_1, are not resolved. Unresolved multiplet structure is therefore a dominant influence on the relative heights of cross peaks, which makes peak height extremely misleading as a measure of intensity, even more so than in 1D. Window functions may also distort the relative intensities of cross peaks.

As in 1D, the correct measure of intensity should be integration, which for NOESY means volume integration of the 2D cross peaks. Routines for volume integration are usually available within manufacturers' software, and a detailed discussion here would probably be inappropriate: a comparison of several methods may be found in reference 44. The greatest problems arise in cases in which cross peaks are overlapped, and here it is clear that a fairly sophisticated approach is needed. One approach involves fitting to previously defined lineshapes and multiplet structures, themselves derived from those cross peaks that are not overlapped,[45,46] but probably the commonest approach is deconvolution using either Gaussian or Lorentzian functions (see, for example, ref. 47).

8.6. VARIATIONS

8.6.1. Semiselective and Network-Edited Experiments

The simplest forms of semiselective NOESY experiments employ a selective pulse (often a Gaussian, cf. Section 7.5.1) at the beginning of t_1 to limit the signals excited in F_1. The resultant spectrum comprises a strip of signals from the normal NOESY spectrum covering the full spectral width in F_2, but only a limited spectral width in F_1, depending on the selectivity of the first pulse.[48,49] Such experiments are formally intermediate between a normal NOESY experiment and the 1D analog of NOESY described in Section 7.5.2, and they allow high resolution to be obtained over a selected spectral window in F_1, without the usual requirement for a large number of increments and therefore a very long experiment. Although not widely used, the method can be beneficial in cases where cross peaks cannot be resolved in F_2 for some practical reason. For example, the fingerprint region of NOESY spectra of proteins in water (i.e., the region containing cross peaks between NH and $C^\alpha H$ signals) is usually accessible on only one side of the diagonal, since the symmetry-related region is obscured by t_1 noise from the partially suppressed H_2O signal. Resolving the relatively crowded $C^\alpha H$ signals therefore demands good F_1 resolution, and could benefit from a semiselective experiment. A closely related application was used in some of the first 3D experiments,[50] which employed semiselective pulses to limit the frequency range of the indirect dimensions. These are considered further in Section 9.2.3.

More recent semiselective experiments are mainly aimed at "network editing," also called MENE (*M*agnetization *E*xchange *N*etwork *E*diting), in which the object is to limit the possible pathways for spin diffusion. One-dimensional experiments that isolate the NOE evolution of a two-spin system from all other spins have already been discussed in Section 7.5.4. Here we extend that discussion to consider 2D experiments that eliminate particular *categories* of spin-diffusion pathways.

The first and simplest such experiment was MINSY (*M*ixing *I*rradiation during *NOESY*), in which continuous RF irradiation is applied during τ_m, either to a particular signal using CW irradiation, or to a group of signals using some form of composite pulse decoupling. This effectively eliminates all cross-relaxation pathways that pass through the corresponding irradiated spins.[51] The same principle applies to elimination of spin-diffusion pathways involving signals within particular frequency bands if the decoupling applied in MINSY is replaced by a train of band-selective inversion pulses; this results in the BD-NOESY (*B*lock-*D*ecoupled NOESY) experiment.[52] For example, BD-NOESY may be applied to separate a typical protein NOESY spectrum into two blocks, a "low-field block" (comprising principally the amide NH and aromatic proton resonances), and a "high-field block" (comprising principally the aliphatic CH resonances). If a band-selective inversion pulse train is applied just to the high-field block during τ_m, the resulting BD-NOESY spectrum would comprise only cross peaks between protons resonating in the low-field block, and, crucially, the intensities of these cross peaks would be undistorted by any spin-diffusion pathways involving protons resonating in the high-field block (Fig. 8.24). Effectively, the BD-NOESY sequence decomposes the relaxation matrix into two independent submatrices, and allows the NOESY spectrum corresponding to one of these submatrices to be recorded independently of the other.

Several experiments closely related to BD-NOESY have been reported. The BD-ROESY experiment is an ROE analog of BD-NOESY, and may be useful for checking for spin diffusion *within* the block of signals selected in a BD-NOESY experiment (in just the same way as comparing complete, conventional NOESY and ROESY spectra is often used to check for spin diffusion; cf. Section 9.3.1.2).[53] An experiment complementary to BD-NOESY has been devised, called CBD-NOESY (*C*omplementary *B*lock-*D*ecoupled NOESY), in which cross peaks between signals in *different* blocks can be measured free of spin-diffusion pathways involving signals within the *same* block.[54] This can be particularly helpful in reducing or eliminating spin diffusion between protons in the same methylene group in certain cases. For instance, for a pair of cross peaks from an NH signal into the two protons of a C^β methylene group, spin diffusion in a conventional NOESY experiment would usually cause the two intensities to be efficiently averaged, unless τ_m was very short. In contrast, this spin-diffusion pathway can be largely suppressed in a CBD-NOESY experiment, because the NH and $C^\beta H$ signals generally resonate in different blocks, while the two $C^\beta H$ signals resonate in the same block. CBD-NOESY works by balancing periods of normal NOE evolution with periods of ROE evolution, using the fact that, for macromolecules, spin diffusion over one step generates a sign change in ROE enhancements but not in NOE enhancements. However, empirical adjustment of the balance between ROE and NOE evolution periods may be needed.

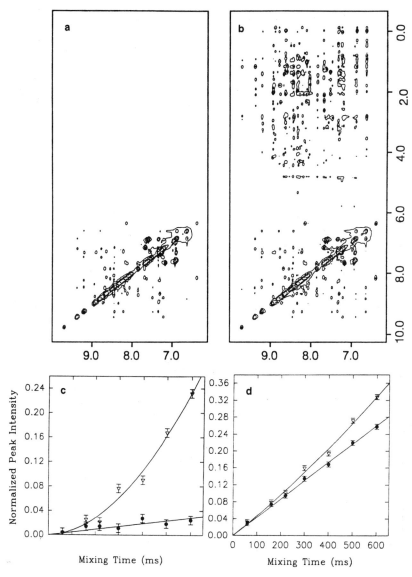

Figure 8.24. Parts of (*a*) BD-NOESY and (*b*) conventional NOESY spectra of turkey ovomucoid third domain. The absence of cross peaks in the upper part of the BD-NOESY spectrum arises because the block-decoupling scheme suppresses cross-relaxation between signals in the "low-field" (aromatic/amide) block and those in the "high-field" (aliphatic) block. There are also minor intensity differences visible between corresponding surviving cross peaks in the two spectra, which reflect the suppression of spin-diffusion pathways involving protons in the aliphatic block (see text). For example, (c) and (d) show buildup curves from conventional NOESY data (open triangles) and BD-NOESY data (filled circles) that emphasize this point. (*c*) Buildup of the enhancement between Lys29 NH and Thr30 NH is rapid and nonlinear in conventional

Another experiment that gives results somewhat similar to those of CBD-NOESY is the QUIET-BAND-NOESY experiment, developed from 1D QUIET-NOESY (Section 7.5.4).[55] A doubly band-selective 180° pulse is inserted at the center of τ_m of a NOESY experiment, so as to invert simultaneously signals within *two* chosen frequency bands, while leaving other signals untouched. The resulting NOESY spectrum contains (ideally) just cross peaks linking pairs of signals that either resonate both in the same frequency band or one in each; as before, spin-diffusion pathways involving any protons that resonate outside both of the selected frequency bands are suppressed.

A further variation on these techniques is to use heteronuclear couplings as the basis for dividing signals into different categories. If a heteronuclear filter element (in this case called a BIRD sequence[56]) is placed at the center of τ_m (Fig. 8.25), this will invert the z magnetizations of all protons coupled to the relevant X nucleus (generally ^{15}N or ^{13}C), but not those of other protons. This affects NOE evolution in very much the same way as does the doubly selective 180° pulse at the center of τ_m in a QUIET-NOESY experiment (Section 7.5.4). Thus, in the resulting NOESY spectrum two types of cross peaks will be present:

1. Cross peaks between two protons that are *both* coupled to X nuclei; these will appear inverted in the spectrum, and will be freed from the effects of spin-diffusion pathways involving protons *not* coupled to X nuclei.

2. Cross peaks between two protons *neither* of which is coupled to X nuclei; these will not appear inverted in the spectrum, and will be freed from the effects of spin-diffusion pathways involving protons that *are* coupled to X nuclei. Cross peaks between protons one of which is and one of which is not coupled to X will be absent from the spectrum.

Several experiments of this type have been proposed. Insertion of a BIRD sequence at the center of τ_m in an otherwise normal NOESY spectrum[57] probably gives the highest sensitivity (Fig. 8.25a). However, for a double-labelled protein sample, one may instead choose to combine a BIRD sequence on one nucleus, say ^{13}C, with selection of another nucleus, say ^{15}N, during one of the detection periods t_1 or t_2 (e.g., Fig. 8.25b), resulting in suppression of a different set of spin-diffusion pathways.[58] Combination of a BIRD sequence on ^{15}N with half-filters on ^{15}N in both dimensions of a 2D (^{1}H, ^{1}H) NOESY experiment has also been proposed (Fig. 8.25c),[59] but here the additional spectral simplification

---

NOESY due to spin diffusion via Lys C$^\alpha$H, but in BD-NOESY, elimination of this contribution results in a much slower buildup. (*d*) Buildup of the enhancement between Ser44 NH and Asn45 NH remains largely linear for longer in the BD-NOESY data than in the conventional NOESY case. Reproduced with permission from ref. 52.

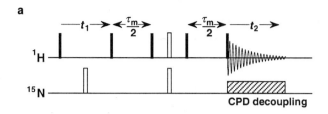

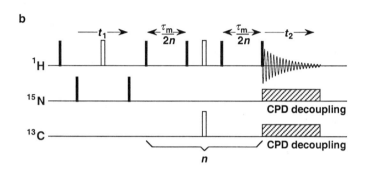

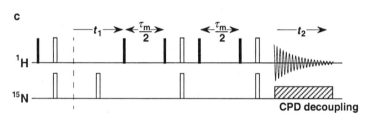

Figure 8.25. Pulse sequences for heteronuclear MENE experiments, using a BIRD sequence at the center of τ_m to achieve suppression of chosen spin-diffusion pathways (see text). (a) QUIET-BIRD-NOESY.[57] (b) (^{15}N, ^{1}H)-HMQC-(^{13}C, ^{1}H)-BD-NOESY.[58] (c) QUIET-BIRD-NOESY with two (^{15}N, ^{1}H) half-filters.[59] For simplicity, elements of these pulse sequences concerned only with solvent suppression (e.g., WATERGATE immediately prior to acquisition) are omitted in the figure. Only one BIRD sequence is shown during τ_m for (a) and (c), but in principle more can be added during τ_m for any of these experiments to achieve more efficient suppression of unwanted pathways [e.g., as illustrated in (b) for $n < 1$].

obtained by using the half-filters carries a very heavy price in terms of lost sensitivity unless the system under study relaxes unusually slowly.

Some careful work has been done to assess the effects on structure determination of using data from BD-NOESY and CBD-NOESY experiments as opposed to conventional NOESY data, both using real data[60] and simulations.[61] Structure calculations for the small protein turkey ovomucoid third domain were used as an example, and some differences were seen that were judged to be significant. In some cases these differences were local expansions of the calculated ensemble to include a wider conformational space, and the authors argue that this implies that the conventional calculations had reached an unjustifiably high (i.e., artifactual) precision. As yet, such studies have apparently not proved persuasive enough for MENE methods to have been taken up extensively outside the groups that originally proposed them. A detailed theoretical treatment of the evolution of NOE enhancements during MENE experiments is available.[58]

8.6.2. Other Variants

Two methods have been proposed for the direct determination of kinetic data from a single modified NOESY experiment. The first, called the "accordion" experiment, employs parallel incrementation of t_1 and τ_m in the pulse sequence $90° - t_1 - 90° - kt_1 - 90° - t_2$, where k is typically in the range 10–100.[62] Signals in the interferogram oscillate at their usual F_1 frequencies, but their t_1 decay envelopes depend on processes occurring both during τ_m and t_1. The F_1 lineshapes therefore contain information on the kinetics of cross-peak (and diagonal-peak) evolution and this can be extracted by fitting the lineshapes, by reverse Fourier transformation, or by normal mode analysis.[63] The method has been successfully applied to the analysis of conformational exchange in simple molecules, but is rather impractical for extension to analysis of NOE buildup curves for molecules of any complexity.

Another experiment intended to reveal information on cross-peak evolution is SKEWSY, which has the pulse sequence $180° - \tau'_m - 90° - t_1 - 90° - \tau_m - 90° - t_2$.[64] This is essentially a NOESY sequence grafted onto an inversion recovery sequence, so that each cross-peak intensity depends on the extent to which the corresponding signal in F_1 had recovered by the time of the first 90° pulse. The resulting NOESY spectrum has asymmetric cross-peak intensities that depend on the differing overall relaxation rates of the participating signals.

NOESY experiments involving delayed acquisition are common in the early literature, largely because some of the first commercially supplied 2D software included this version of NOESY instead of the normal experiment. The pulse sequence is $90° - t_1/2 - 90° - \tau_m - 90° - t_1/2 - t_2$, and the resulting spectrum bears exactly the same relationship to a normal NOESY spectrum as does a SECSY spectrum to a normal COSY spectrum. This version has no advantages over the conventional experiment, and the very considerable disadvantage that the spectrum cannot be obtained in pure double-absorption mode.

Another variation that was popular for a while was the so-called COCONOSY experiment.[65,66] This involves dual acquisition; COSY data is actually acquired *during* the τ_m period of a NOESY experiment, so that two FIDs, one of COSY data and one of NOESY data, are recorded on each pass through the pulse sequence. Although it might at first appear to offer a considerable time saving, in fact this advantage is only slight, since COSY spectra often require only a fraction of the number of transients used for NOESY. More importantly, COCONOSY can only provide COSY data, not DQF-COSY (which is usually preferable). Another drawback is that t_{2max} for the COSY data cannot be made longer than τ_m. If very accurately aligned data are required, a better option is likely to be interleaved acquisition, which allows essentially any combination of experiments. If the time is divided unevenly between them, for example by acquiring four times as many scans of NOESY data as DQF-COSY data per increment, the signal-to-noise ratio of the more insensitive experiment (in this case NOESY) can be kept nearly as high as if the whole time had been devoted to it.[67]

There are by now very many other variations and extensions of the basic NOESY experiment. Mostly, these involve various forms of editing to reduce the number of signals present in the spectrum, or else separation of signals into further dimensions. These are discussed in Section 9.2. The rotating frame 2D NOE experiment is discussed in Section 9.3.

REFERENCES

1. Banci L.; Bertini, I.; Luchinat, C. *Methods in Enzymology* 1994, *239*, 485.
2. Williams D. H. Personal communication.
3. Neuhaus, D. Unpublished results.
4. Jeener, J.; Meier, B. H.; Bachmann, P.; Ernst, R. R. *J. Chem. Phys.* 1979, *71*, 4546.
5. Bax, A. "Two-Dimensional NMR in Liquids," Delft University Press, Reidel, London, 1982.
6. Morris, G. A. in "Fourier, Hadamard and Hilbert Transforms in Chemistry" (A. G. Marshal, ed.), Plenum Press, New York, 1982, pp. 271–305.
7. Wider, G.; Macura, S.; Kumar, A.; Ernst, R. R.; Wüthrich, K. *J. Magn. Reson.* 1984, *56*, 207.
8. Dellwo, M. J.; Schneider, D. M.; Wand, A. J. *J. Magn. Reson. B* 1994, *103*, 1.
9. Andersen, N. H.; Nguyen, K. T.; Hartzell, C. J.; Eaton, H. L. *J. Magn. Reson.* 1987, *74*, 195.
10. Levitt, M. H.; Bodenhausen, G.; Ernst, R. R. *J. Magn. Reson.* 1984, *58*, 462.
11. Barna, J. C. J.; Laue, E. D.; Mayger, M. R.; Skilling, J.; Worrall, S. J. P. *J. Magn. Reson.* 1987, *73*, 69.
12. Turner, C. J.; Hill, H. D. W. *J. Magn. Reson.* 1986, *66*, 410.
13. Redfield, A. G.; Kunz, S. D. *J. Magn. Reson.* 1975, *19*, 250.
14. Bodenhausen, G.; Vold, R. L.; Vold, R. R. *J. Magn. Reson.* 1980, *37*, 93.
15. Marion, D.; Wüthrich, K. *Biochem. Biophys. Res. Commun.* 1983, *113*, 967.
16. States, D. J.; Haberkorn, R. A.; Ruben, D. J. *J. Magn. Reson.* 1982, *48*, 286.
17. Keeler, J.; Neuhaus, D. *J. Magn. Reson.* 1985, *63*, 454.
18. Marion, D.; Ikura, M.; Tschudin, R.; Bax, A. *J. Magn. Reson.* 1989, *85*, 393.

19. Bodenhausen, G.; Freeman, R.; Niedermeyer, R.; Turner, D. L. *J. Magn. Reson.* 1977, *26*, 133.
20. Ruiz-Cabello, J.; Vuister, G. W.; Moonen, C. T. W.; van Gelderen, T.; Cohen, J. S.; van Zijl, P. C. M. *J. Magn. Reson.* 1992, *100*, 282.
21. Hoult, D. I.; Richards, R. E. *Proc. R. Soc. Lond. A* 1975, *344*, 311.
22. Sørensen, O. W.; Eich, G. W.; Levitt, M. H.; Bodenhausen, G.; Ernst, R. R. *Prog. Nucl. Magn. Reson. Spectrosc.* 1983, *16*, 163.
23. Wokaun, A.; Ernst, R. R. *Chem. Phys. Lett.* 1977, *52*, 407.
24. Wagner, R.; Berger, S. *J. Magn. Reson. A* 1996, *123*, 119.
25. Macura, S.; Huang, Y.; Suter, D.; Ernst, R. R. *J. Magn. Reson.* 1981, *43*, 259.
26. Macura, S.; Wüthrich, K.; Ernst, R. R. *J. Magn. Reson.* 1982, *46*, 269.
27. Macura, S.; Wüthrich, K.; Ernst, R. R. *J. Magn. Reson.* 1982, *47*, 351.
28. Weigelt, J.; Hammarström, A.; Bermel, W.; Otting, G. *J. Magn. Reson. B* 1996, *110*, 219.
29. Bodenhausen, G.; Wagner, G.; Rance, M.; Sørensen, O. W.; Wüthrich, K.; Ernst, R. R. *J. Magn. Reson.* 1984, *59*, 542.
30. Glaser, S.; Kalbitzer, H. R. *J. Magn. Reson.* 1986, *68*, 350.
31. Otting, G.; Widmer, H.; Wagner, G.; Wüthrich, K. *J. Magn. Reson.* 1986, *66*, 187.
32. Mehlkopf, A. F.; Korbee, D.; Tiggleman, T. A.; Freeman, R. *J. Magn. Reson.* 1984, *58*, 315.
33. Aue, W. P.; Bachmann, P.; Wokaun, A.; Ernst, R. R. *J. Magn. Reson.* 1978, *29*, 523.
34. Bartholdi, E.; Ernst, R. R. *J. Magn. Reson.* 1973, *11*, 9.
35. Freeman, R. "A Handbook of NMR," Longman, London, 1988.
36. Becker, E. D.; Ferretti, J. A.; Gambhir, P. N. *Anal. Chem.* 1979, *51*, 1413.
37. Turner, D. L. *J. Magn. Reson.* 1985, *61*, 28.
38. Guéron, M. *J. Magn. Reson.* 1978, *30*, 515.
39. Lindon, J. C.; Ferrige, A. G. *J. Magn. Reson.* 1979, *36*, 277.
40. Hoch, J. C.; Stern, A. S. "NMR Data Processing," Wiley-Liss, New York, 1996.
41. Baumann, R.; Wider, G.; Ernst, R. R.; Wüthrich, K. *J. Magn. Reson.* 1981, *44*, 402.
42. Klevit, R. E. *J. Magn. Reson.* 1985, *62*, 551.
43. Weiss, G. H.; Kiefer, J. E.; Ferretti, J. A. *J. Magn. Reson.* 1992, *97*, 227.
44. Broido, M. S.; James, T. L.; Zon, G.; Keepers, J. W. *Eur. J. Biochem.* 1985, *150*, 117.
45. Denk, W.; Baumann, R.; Wagner, G. *J. Magn. Reson.* 1986, *67*, 386.
46. Holak, T. A.; Scarsdale, J. N.; Prestegard, J. H. *J. Magn. Reson.* 1987, *74*, 546.
47. Sze, K. H.; Barsukov, I. L.; Roberts, G. *J. Magn. Reson. A* 1995, *113*, 185.
48. Keeler, J. H. Personal communication.
49. Brüschweiler, R.; Madsen, J. C.; Griesinger, C.; Sørensen, O. W.; Ernst, R. R. *J. Magn. Reson.* 1987, *73*, 380.
50. Griesinger, C.; Sørensen, O. W.; Ernst, R. R. *J. Am. Chem. Soc.* 1987, *109*, 7227.
51. Massefski Jr., W.; Redfield, A. G. *J. Magn. Reson.* 1988, *78*, 150.

52. Hoogstraten, C. G.; Westler, W. M.; Macura, S.; Markley, J. L. *J. Magn. Reson. B* 1993, *102*, 232.
53. Hoogstraten, C. G.; Westler, W. M.; Mooberry, E. S.; Macura, S.; Markley, J. L. *J. Magn. Reson. B* 1995, *109*, 76.
54. Hoogstraten, C. G.; Westler, W. M.; Macura, S.; Markley, J. L. *J. Am. Chem. Soc.* 1995, *117*, 5610.
55. Vincent, S. J. F.; Zwahlen, C.; Bodenhausen, G. *J. Biomolec. NMR* 1996, *7*, 169.
56. Garbow, J. R.; Weitekamp, D. P.; Pines, A. *Chem. Phys. Lett.* 1982, *93*, 504.
57. Vincent, S. J. F.; Zwahlen, C.; Bolton, P. H.; Logan, T. M.; Bodenhausen, G. *J. Am. Chem. Soc.* 1996, *118*, 3531.
58. Zolnai, Z.; Juranic, N.; Markley, J. L.; Macura, S. *Chem. Phys.* 1995, *200*, 161.
59. Mutzenhardt, P.; Bodenhausen, G. *J. Magn. Reson.* 1998, *132*, 159.
60. Hoogstraten, C. G.; Choe, S.; Westler, W. M.; Markley, J. L. *Protein Sci.* 1995, *4*, 2289.
61. Hoogstraten, C. G.; Markley, J. L. *J. Mol. Biol.* 1996, *258*, 334.
62. Bodenhausen, G.; Ernst, R. R. *J. Am. Chem. Soc.* 1982, *104*, 1304.
63. Ernst, R. R.; Bodenhausen, G.; Wokaun, A. in "Principles of NMR in One and Two Dimensions," Oxford University Press, 1987, p. 513.
64. Bremer, J.; Mendz, G. L.; Moore, W. J. *J. Am. Chem. Soc.* 1984, *106*, 4691.
65. Haasnoot, C. A. G.; van de Ven, F. J. M.; Hilbers, C. W. *J. Magn. Reson.* 1984, *56*, 343.
66. Gurevich, A. Z.; Barsukov, I. L.; Arseniev, A. S.; Bystrov, V. F. *J. Magn. Reson.* 1984, *56*, 471.
67. Neuhaus, D.; Wagner, G.; Vašák, M.; Kägi, J. H. R.; Wüthrich, K. *Eur. J. Biochem.* 1985, *151*, 257.

CHAPTER 9

OTHER DEVELOPMENTS

9.1. HETERONUCLEAR NOE ENHANCEMENTS

This section surveys experiments concerned with measurements of heteronuclear NOE enhancements. Note that this definition excludes many heteronuclear experiments in which signals receiving (^{1}H,^{1}H) NOE enhancements are dispersed and/or filtered using heteronuclear scalar couplings; such experiments do not involve heteronuclear NOE enhancements, and are considered in Section 9.2.

9.1.1. Nonspecific Heteronuclear NOE Experiments

The simplest heteronuclear NOE experiments involve saturation of all the spins of one nuclear species, usually protons, and observation of another. This section deals with such experiments, which we shall refer to as "nonspecific" NOE experiments, since the saturation target is not a specific resonance.

Most commonly, the proton signals are saturated using either noise-modulated irradiation or composite-pulse decoupling schemes such as WALTZ16, MLEV, or GARP. Composite pulse decoupling (CPD) is generally less effective than noise-modulated irradiation in producing saturation, as judged by measuring residual proton intensities, because during CPD the magnetization remains coherent as it follows a cyclically repeated pathway designed to refocus periodically dephasing effects due to B_1 inhomogeneities and other imperfections. Nonetheless, composite pulse decoupling is highly effective in producing a heteronuclear NOE, because the trajectory that the proton magnetization follows necessarily results (over a whole number of decoupling cycles) in an average z magnetization of zero (cf. the discussion of the nature of saturation in Section 1.5). Other variations, such as using regular trains of repeated 90° or 180° proton pulses, have also sometimes been suggested, perhaps because they are easier to implement on some spectrometers.

Usually, the motive for generating nonspecific heteronuclear NOE enhancements is purely to increase sensitivity. For instance, until the advent of the DEPT and INEPT techniques for recording routine carbon spectra, almost all ^{13}C 1D spectra were recorded with direct carbon excitation and broad-band decoupling of protons, resulting in nonspecific ^{13}C{^{1}H} NOE enhancements at every signal. However, measurements of the sizes of nonspecific heteronuclear NOE enhancements can sometimes provide useful information, as the rest of this section aims to demonstrate.

The relative size of the nonspecific enhancement at a particular heteronuclear signal, X, depends on two factors: (i) the relative contribution that dipolar relaxation from (all) protons makes to the total relaxation of the X signal, and (ii) the local mobility for the corresponding X spin (strictly, the order parameter S^2 and internal correlation time τ_e for the internuclear vector linking it to each relevant proton; cf. Section 5.4). Useful applications of nonspecific heteronuclear NOE measurements are generally made in circumstances where only one of these influences can vary.

For instance, for carbon spectra of relatively small molecules in the extreme narrowing limit ($\omega_{^{13}C}\tau_c \ll 1$), the influence of local mobility is relatively unimportant. If the protons are saturated, all protonated carbons then receive essentially the full enhancement of 199%, since their relaxation is always dominated by the nearby directly attached proton(s). In contrast, quaternary carbons are relaxed significantly by other mechanisms, and therefore receive smaller nonspecific NOE enhancements (cf. Sections 2.4 and 3.1.3.2). Such competition between dipolar and other relaxation sources has sometimes been used to provide a basis for assigning quaternary carbon signals, taking the sizes of nonspecific enhancements as an approximate indicator of the proximity of the nearest protons. For example, for fluoranthene (**9.1**), the expected order of enhancements for the quaternary carbons is $C_{16} > C_{11} \simeq C_{13} > C_{15}$, and the measured enhancements can be matched with the assignments as follows: C_{16} = 152%, C_{11} and C_{13} = 135%, C_{15} = 112%.[1] Similar logic was applied to the assignment of the C_3 and C_4 signals of cephalosporins such as 7-aminodeacetoxycephalosporanic acid (**9.2**).[2] C_3 is much closer to its nearest proton than is C_4, and would therefore be expected to have a shorter T_1 and a larger enhancement on proton saturation. On this basis, the signals from the two carbons were assigned as shown in Table 9.1 for two analogs (**9.3**, **9.4**) of acid **9.2**. As a further check, single frequency irradiation of the 3-methyl protons in **9.2** gave a specific enhancement of 120% to C_3, but none to C_4, confirming the previous assignments.

A variant of this technique was proposed by Shapiro et al.,[3] based on the idea of deuteration of exchangeable sites to reduce enhancements to nearby quaternary carbons. The problem was to distinguish between the two double-

TABLE 9.1 T_1 and NOE Values for Compounds 9.3 and 9.4[a]

	9.3			9.4		
Carbon Atom	δ	$T_1(s)$	NOE(%)	δ	$T_1(s)$	NOE(%)
3	132.0	6.7	119	126.0	9.3	183
4	122.7	12.7	96	125.6	21.2	116

[a]Conditions: 22.5 MHz, 30°C, 9.3 in CD_2Cl_2 and 9.4 in a mixture of $CDCl_3$ and CD_2Cl_2. Accuracies of δ, T_1, and NOE factor about ±0.1 ppm, ±10%, and ±10%, respectively. Data from Ref. 2. T_1 measured by inversion recovery and NOE by the gated decoupler method.

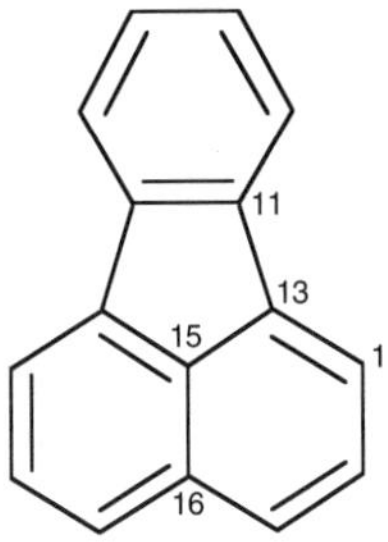

9.1

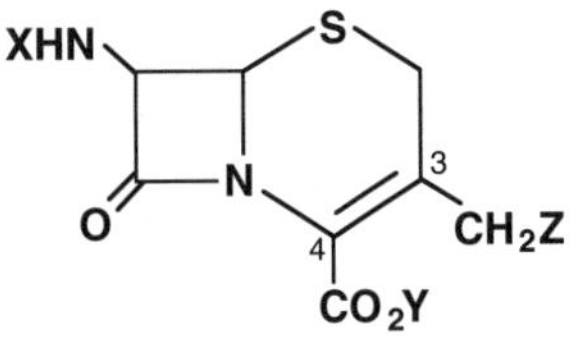

9.2   X = H, Y = H, Z = H

9.3   X = COCH(NH_2)Ph, Y = CH_3, Z = H

9.4   X = COCH$_2$-[S]-, Y = CH_3, Z = OCCH$_3$
                                        ‖
                                        O

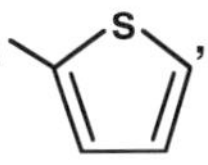

bond isomers 9.5 and 9.6. Comparison of ^{13}C spectra with noise decoupling in MeOH and in MeOD showed differences in intensity due to the loss from the

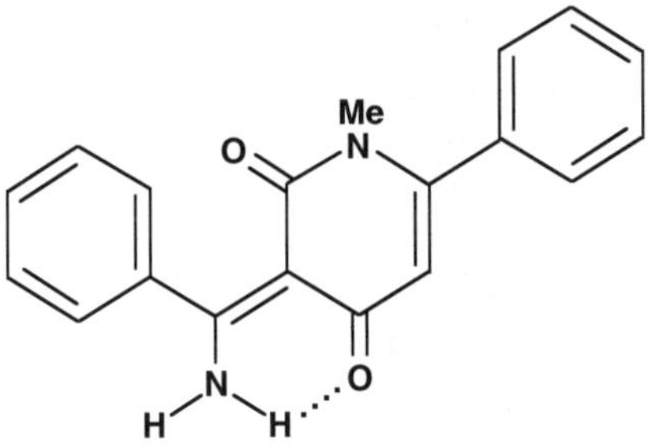

9.5 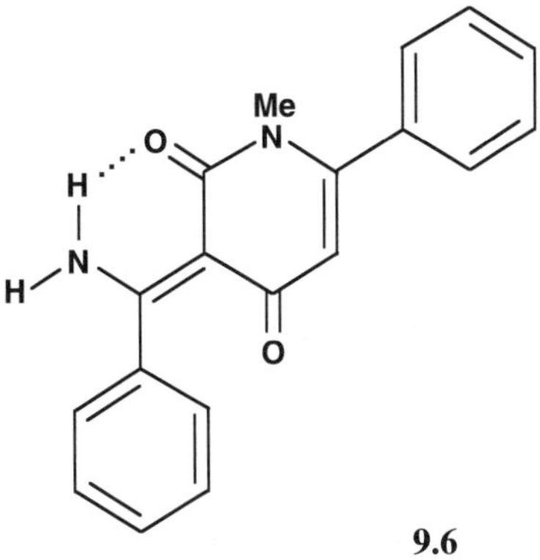   9.6

MeOD spectrum of enhancements at quaternary centers close to the NH_2 group. The correct structure was thereby shown to be 9.5. Again, this was checked by a selective experiment in which the low-field NH signal was irradiated. There are of course other methods for assignment of quaternary carbons that do not involve the NOE, of which probably the most useful are heteronuclear correlation experiments using long-range couplings.

Another, more recent, area of application for nonspecific heteronuclear NOE measurements is in studying the dynamics of isotopically labeled biomacromolecules. Here, the relaxation properties of many examples of a repeated group, most commonly the backbone amide ^{15}N—^{1}H groups of a ^{15}N labelled protein sample, are all separately measured and compared.[4] It is assumed that the relative contribution of dipolar relaxation is the same for each group, so that the variation among different ^{15}NH groups reflects only their different dynamics. The theoretical framework for such studies is outlined in Section 5.4, but it is worth briefly mentioning here the techniques used to record the relaxation parameters, particularly the ^{15}N{^{1}H} NOE. In each case, a modified 2D (^{15}N, ^{1}H) correlation spectrum is recorded in which the intensity of the cross peaks is made to depend on the relaxation parameter of interest. For instance, during the sequence for measuring ^{15}N T_1, the frequency-labeled ^{15}N magnetization is left along the z axis for a period Δ. A series of spectra is recorded, each using a different value of Δ, and the intensity decay for each cross peak calculated as a function of Δ. The ^{15}N T_1 value corresponding to each cross peak is then obtained by fitting the curve of its intensity decay. Similarly, T_2 or $T_{1\rho}$ values are extracted from a series of spectra where the ^{15}N magnetization is allowed to relax in the transverse plane during a variable delay. NOE enhancements are measured using two correlation spectra; in one of these the ^{15}N{^{1}H} NOE is fully excited by proton saturation throughout the relaxation delay and in the other it is absent (see Fig. 9.1a). The NOE enhancements are then given, for each cross peak, by the usual formula $(I-I^0)/I^0$ (although, as is common in heteronuclear studies, they are also often reported as enhancement ratios I/I^0 cf. Section 2.5). This sequence is of interest for several reasons. It represents a generalization of the NOE difference experiment in which a pulse sequence, rather than a simple pulse, is used to excite the spectrum. In this way it achieves the sensitivity advantages of proton detection, despite the fact that it is ^{15}N{^{1}H} NOE enhancements that are being measured. The sequence also serves to disperse the ^{15}N signals into a second dimension, allowing the various enhancements to be measured separately. This approach results in the restriction that only protonated ^{15}N nuclei can be studied in this experiment, although this is, of course, of no consequence for this particular application. One important subtlety when applying this sequence is to ensure that the irradiation period (and the relaxation period in the control experiment) are sufficiently long that steady state is genuinely reached, even for those NH groups that exchange relatively rapidly with the solvent (which itself will relax slowly).[5] Variations on the original pulse sequence have been published, featuring improved sensitivity and water suppression, and application of pulsed field gradients.[5,6]

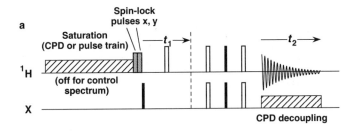

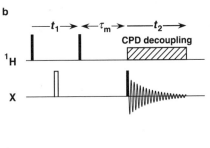

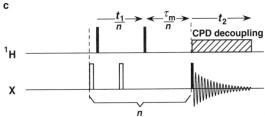

Figure 9.1. (*a*) A simple pulse sequence for measuring steady-state X{¹H} NOE values in a 2D correlation spectrum (cf. Section 5.4). In the "on-resonance" experiment, saturation is applied to protons during the whole relaxation delay, while in the "control" experiment saturation is omitted. The two spin-lock pulses are to suppress the water signal and are applied in both experiments. (*b*) A typical pulse sequence for a heteronuclear NOESY experiment, also called HOESY. Heteronuclear decoupling is applied in both dimensions. See text for further discussion. (*c*) Pulse sequence for a hybrid steady-state/transient 2D HOESY experiment. The portion of the sequence indicated by the bracket is repeated *n* times. Hard 90° pulses are represented by narrow filled vertical bars, hard 180° pulses by wider, open vertical bars.

9.1.2. Specific Heteronuclear NOE Experiments

In this section we discuss the measurement of heteronuclear NOE enhancements between specific pairs of resonances. Two factors in particular have limited the number of applications of specific heteronuclear NOE enhancements. First, for slowly tumbling molecules, η_{max} is often so small that measuring heteronuclear NOE enhancements is impractical (cf. Section 2.5). Second, for small molecules, the buildup rates for heteronuclear enhancements are generally very slow, making measurement very time-consuming and allowing leakage to compete effectively, also reducing enhancements. This slow buildup arises largely because σ_{IS} depends on γ^2 for each nucleus (Eq. 2.22); for instance, a (^{1}H—^{1}H) enhancement will build up sixteen times faster than a (^{1}H—^{13}C) enhancement over the same distance, other things being equal.

Given that sensitivity is such a crucial issue, one of the first decisions when planning any specific heteronuclear NOE experiment is which nucleus to observe and which nucleus to irradiate, or which nucleus to place in the directly detected dimension of a multidimensional experiment. Detection of the higher frequency nucleus offers a substantial sensitivity advantage in any NMR experiment, proportional to $(\gamma_I/\gamma_S)^{5/2}$, whereas the maximum observable fractional enhancement (η_{max}) is generally greater for observation of the lower-frequency nucleus (see Section 2.5 and Eq. 2.13). The former is always the stronger of these two opposing influences, making ^{1}H{X} NOE experiments intrinsically more sensitive than the corresponding X{^{1}H} experiments (assuming X to have the lower Larmor frequency). For example, the intrinsic sensitivity advantage of ^{1}H{^{13}C} NOE experiments over ^{13}C{^{1}H} NOE experiments varies between a factor of about 2.0 in the extreme narrowing limit ($\omega\tau_c \ll 1$) to about 12.6 in the slow motion limit ($\omega\tau_c \gg 1$). The corresponding advantage for ^{1}H{^{31}P} NOE experiments over ^{31}P{^{1}H} experiments is about a factor of 1.5 ($\omega\tau_c \ll 1$) or 4.8 ($\omega\tau_c \gg 1$), while for ^{1}H{^{15}N} and ^{15}N{^{1}H} experiments the factors are about 3.1 ($\omega\tau_c \ll 1$) and 83 ($\omega\tau_c \gg 1$). [These figures were derived by multiplying the relevant ratios of η_{max} values (calculated from Eq. 2.13) by the ratio $(\gamma_H/\gamma_X)^{5/2}$.]

Despite the intrinsic sensitivity advantage of ^{1}H{X} experiments, to date the majority of specific heteronuclear NOE experiments have been carried out using observation of the X nucleus and irradiation of ^{1}H. Mostly, the reasons for this are connected with the "signal-to-artifact" ratio, rather than the signal-to-thermal-noise ratio, or else with the capabilities of the spectrometer. In any ^{1}H{X} NOE experiment involving an X nucleus at low natural abundance, the enhancements must be detected in the presence of a much larger ^{1}H signal from the majority of molecules *not* containing an X nucleus at the site of interest. If suppression of this large ^{1}H signal is achieved by a difference technique, such as phase cycling, then poor subtraction or excessive t_1 noise may severely compromise the results, in practice often rendering such a ^{1}H{X} NOE experiment significantly *less* sensitive than the corresponding X{^{1}H} experiment. This is of course a general problem with all "inverse detection" (i.e., ^{1}H-detected heteronuclear) experiments using X nuclei at low natural abundance, and the results obtained depend crucially on the performance of the spectrometer and the stability of its environment. For X nuclei having a high abundance, such as ^{31}P at natural abundance or ^{13}C or ^{15}N in labeled molecules, these disadvantages of inverse detection are much reduced.

Another key consideration is to choose between a steady-state, 1D experiment, and a transient method, which may be one-, two- or multidimensional.

The generalities of this choice have already been discussed (Section 8.1), but it is important to note that the slow buildup rates of many heteronuclear enhancements shift the balance of choice significantly in favor of steady-state methods. Nonetheless, heteronuclear NOE enhancements have also been measured using 2D sequences, sometimes called HOESY experiments. A typical scheme is shown in Figure 9.1b, where heteronuclear decoupling is applied in both dimensions; note that the elimination of any antiphase terms at the end of t_1 that might lead to heteronuclear multiple quantum coherences makes any jittering of τ_m unnecessary.[7,8] Given the very low sensitivity of such experiments, it is absolutely crucial to choose a good value for the mixing time τ_m; both calculations and experiment[9] show that the optimum τ_m is shorter for one-bond enhancements than for long-range enhancements. In one case involving a compound with molecular weight about 350, the best τ_m for one-bond enhancements was found to be about 1 s, whereas for two-bond enhancements to quaternary carbons it was about 4 s. Applications of this sort seem likely to remain restricted to cases where large quantities and high concentrations of sample are available.

Hybrid steady-state/transient experiments such as that shown in Figure 9.1c have been proposed, in which the t_1 and τ_m periods are repeatedly looped to achieve higher sensitivity.[10–12] Here, the repeated 90° pulses on the ^{1}H channel have an effect not unlike that of continuous saturation, with the result that the NOE on X is driven to a larger value than it would have had in a simple HOESY experiment. Although some prior knowledge of relaxation rates is required to optimize performance, improved sensitivity can be obtained with these sequences, but at the price that quantitative relationships between cross-peak intensities and internuclear distances are (presumably) far more complicated than in a conventional HOESY experiment.

The recent development of pulsed field gradient technology in commercial spectrometers has very greatly improved the quality attainable in many inverse experiments, particularly those involving X nuclei at low natural abundance. This is largely because gradient selection of the desired signal works within each individual scan, dramatically reducing the artifact level relative to that for selection techniques based on taking the difference between successive scans or experiments. For NOE experiments, gradient techniques can only be applied to the measurement of transient NOE enhancements. Even so, it is likely that the very clean spectra, essentially devoid of subtraction artifacts, available using gradient sequences may yet make them the preferred option, even for specific heteronuclear NOE measurements, and that this in turn may lead to a shift away from X{^{1}H} steady-state experiments towards ^{1}H{X} transient experiments in the future. The recently described HETGOESY experiment (cf. Section 7.5.3.3) provides a clear pointer in this direction.[13]

There is another subtlety that needs to be considered when undertaking selective-irradiation heteronuclear steady-state NOE experiments on materials at natural abundance. To take natural abundance ^{13}C{^{1}H} experiments as an example, selective irradiation of the normal ^{12}C-bound ^{1}H resonance will not saturate the ^{1}H signals from those molecules having ^{13}C in the directly bound position. This is because the directly ^{13}C-bound proton does not resonate at that frequency, but rather at the ^{13}C satellite positions that are some 60–100 Hz [that is, $^1J(^{13}C, ^1H)/2$] away on either side of the ^{12}C-bound signal. Thus, the directly bound ^{13}C signal will not be enhanced unless very high power levels are used, or the satellites themselves are irradiated. Other ^{13}C resonances do not suffer this problem, since all (^{13}C, ^{1}H) couplings over more than one bond are so much smaller (generally <10 Hz) that the corresponding ^{1}H multiplets are saturated satisfactorily at normal power levels. As an illustration, in compound **9.7**, single-frequency irradiation of H$_4$ gave only a 29% enhancement at C$_4$, but a 90% enhancement at C$_3$.[14] When the proton frequency was

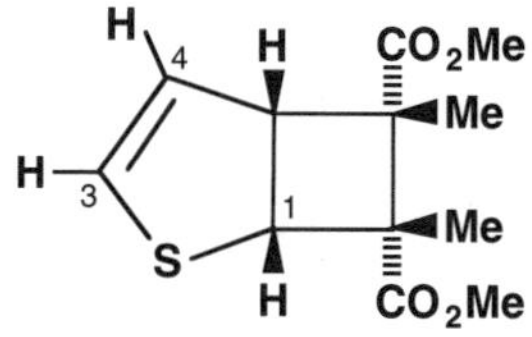

9.7

modulated with a frequency of $J/2$, then the full enhancement of 199% was seen at C$_4$ (within experimental error). However, the power levels required for the modulated irradiation were much higher, by roughly 20 dB.

In applications to problems of structure determination, probably the commonest use of heteronuclear NOE enhancements is in the identification of hydrogen bonds in peptides. Identification of amide NH groups involved in hydrogen bonds is often relatively straightforward, using measures such as temperature coefficients and exchange rates. However, it is not simple to identify the carbonyl groups to which these NH groups are hydrogen bonded. One way of achieving this is to saturate each NH proton in turn and look for enhancements into carbonyl carbons. Hydrogen bonds in valinomycin [(-L-Val-D-Hyiv-D-Val-L-Lac)$_3$, **9.8**] in CDCl$_3$ were studied in this way.[15] There are only two sets of NH protons, both of which are involved in hydrogen bonds. When

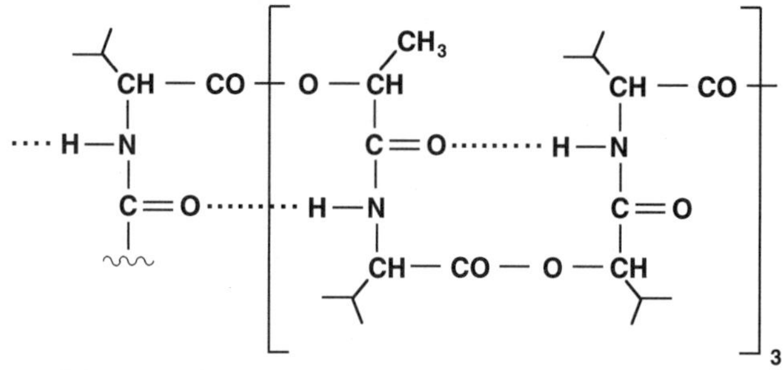

9.8

a particular NH proton signal is saturated (either with or without noise decoupling during acquisition), ^{13}C{^{1}H} enhancements occur both at the carbonyl hydrogen-bonded to that NH and at the sequentially neighboring carbonyl, the carbon of which is two bonds away from the NH proton. In this example, the enhancements to the hydrogen-bonded carbonyl were quite large, between 23% and 44%, and therefore readily observable. The backbone distance H—N—C=O is essentially constant at 2.08Å, whatever the conformation, and this is less than the distance N—$H \cdots O$=C across a hydrogen bond in solution in most cases. Therefore the largest enhancement seen is usually to the neighboring carbonyl, rather than the hydrogen-bonded one. It has been pointed out that it is not at all uncommon (especially in the neighborhood of tight turns, for example in cyclic peptides) to find several carbonyl carbons closer to an NH proton than is the hydrogen bonded one.[16] This technique, attractive though it is, must therefore be used with some caution (see also ref. 17).

Single-frequency ^{13}C{^{1}H} experiments have also been applied to the determination of the geometry of aromatic imines such as compound **9.9**; selective irradiation of a ring proton gives a positive enhancement to a CN carbon only

9.9

when the CN group is *syn* to that proton.[18] Other examples of structure determination using ^{13}C{^{1}H} enhancements may be found in references 19 and 20, and applications to the assignment of quaternary carbons are discussed in references 21 and 22.

There has previously been some interest in relayed heteronuclear NOE effects of the type ^{13}C ← ^{1}H ← {^{1}H},[23,24] which in small molecules are of course negative (three-spin) enhancements (cf. Section 3.2.2.3). The second step is most likely to involve transfer from a proton to a directly attached ^{13}C nucleus, since cross-relaxation into quaternary carbon centers would generally be too slow to allow effects of this sort to compete significantly with proton–proton spin-diffusion. Thus, the method conveys no new structural information beyond that present in the ^{1}H ← {^{1}H} enhancement, and its only advantage is in dispersing the information according to ^{13}C rather than ^{1}H chemical shifts. This is really an editing function, and may now be better carried out using specifically designed editing sequences such as those discussed in Section 9.2. Also, since the experiment involves detection of ^{13}C signals rather than ^1H signals, it suffers a significant sensitivity disadvantage.

There are several published examples of enhancements from protons to fluorine, and the $^1H\{^{19}F\}$ NOE can provide a means of editing the proton spectra of fluorine-labeled macromolecules. Proteins can be labelled fairly easily with ^{19}F (e.g., with CF_3 groups) and irradiation of the fluorine signal(s) gives enhancements to the rest of the molecule.[25] Two-dimensional NOE spectra can also be obtained.[26] The fluorine "probe" need not even be on the macromolecule, but could be on a ligand, using an intermolecular NOE to create enhancements into macromolecular proton signals.[27] However, it is important to bear in mind that, despite the similarity of $\eta_{max}(^1H, ^1H)$ to $\eta_{max}(^1H, ^{19}F)$ as a function of $\omega\tau_c$, W_0 transitions will always become more efficient for the homonuclear case when slow enough tumbling is reached. Thus, spin diffusion will become a problem for $^1H\{^{19}F\}$ NOE enhancements in large molecules, as a slow heteronuclear cross-relaxation step will be followed by more efficient, widespread interproton cross-relaxation.

A simple example of the application of fluorine–proton NOE enhancements to a structural problem in small molecules is that of **9.10**,[28] where the problem is whether the aromatic proton is at position 6 or 7. The actual substitution pattern (shown in **9.10**) is very simply demonstrated by NOE experiments in-

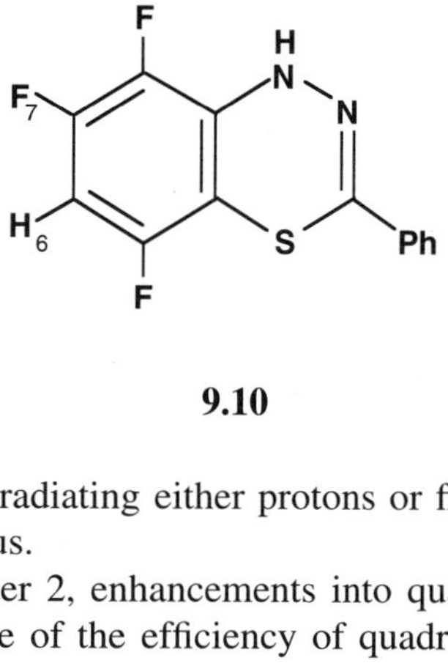

9.10

volving the NH signal, irradiating either protons or fluorine and observing the effect on the other nucleus.

As discussed in Chapter 2, enhancements into quadrupolar nuclei are generally very small, because of the efficiency of quadrupolar relaxation. An exception is ^6Li, which has the smallest quadrupole moment of any quadrupolar nucleus, and at natural abundance is 3.6 times more sensitive than ^{13}C. An

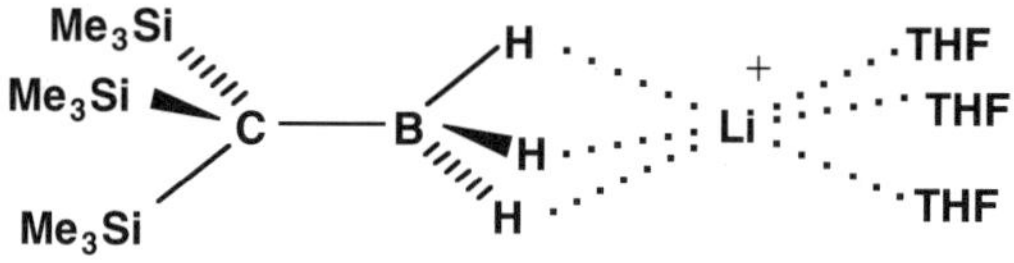

9.11

enhancement factor of 2.2 (65% of the theoretical maximum for ^{6}Li{^{1}H}) has been observed in the bridged boron compound $(Me_3Si)_3CB(\mu\text{-}H)_3Li(thf)_3$ (**9.11**), and measurement of the enhancement buildup rate permitted an estimate of the Li—H distance.[29]

9.2. EDITING AND SPECTRAL SIMPLIFICATION OF NOE EXPERIMENTS

The difference NOE experiment is in itself an editing technique, in that it selects a small subset of signals from the whole spectrum, and (usually) allows them to be viewed free from overlap. It is also possible to *combine* NOE experiments with other editing or filtering NMR techniques so as to further reduce or disperse the subset of signals recorded, or to use some other "non-NMR" approach to reduce the number of signals. In this section we consider these various topics in turn.

9.2.1. Editing Using the NOE Itself

The editing function of the difference NOE technique itself is rarely the sole object of an experiment. Nonetheless, the simplification that it implies for the interpretation of enhancements is one of the main reasons why the method is so useful and successful. Multiplet patterns revealed from within regions of spectral overlap can contain a wealth of information in addition to the magnitudes of particular enhancements, and such details often form an important part of the overall interpretation. Instances of this occur widely in the applications described elsewhere in this book, but the point is reinforced by the following example of a study of the solution conformation of (6R)-prostaglandin I$_2$ [(6R)-PGI$_2$], **9.12**.[30] Overlap in the normal 1D spectrum of (6R)-PGI$_2$ in D$_2$O at 400 MHz prevented the measurement of several key coupling constants

involving the ring protons. However, the NOE difference spectrum resulting from preirradiation of H$_{10\beta}$ (shown in Fig. 9.2) clearly reveals the multiplets due to H$_{10\alpha}$ and H$_{11}$ freed from overlap, allowing them to be analyzed. The resulting coupling constants, together with the NOE enhancements themselves,

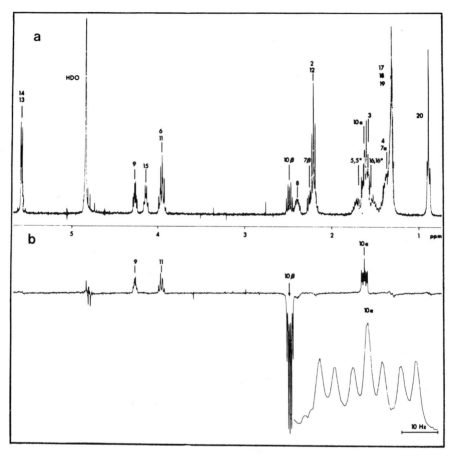

Figure 9.2. NOE difference spectrum of 20 mM (6R)-PGI$_2$ (**9.12**) in 0.2 M phosphate buffer in D$_2$O, pH 7.0, ambient temperature, 400 MHz. Preirradiation time was 12 s. (*a*) Normal spectrum and assignments. (*b*) NOE difference spectrum, preirradiating H$_{10\beta}$. The inset shows the H$_{10\alpha}$ multiplet structure revealed by the NOE experiment. Reproduced with permission from Ref. 30.

were used to establish the ring conformation. Other examples may be found in references 31 and 32, and in Figure 11.6. It is also worth remembering the closely related trick whereby the irradiation target itself, if it is only partially resolved, may be revealed free from overlap and largely undistorted in the NOE difference spectrum, provided the exposed part of it is selectively preirradiated at very low power and an exact 90° pulse is used for excitation (cf. Section 7.2.5). This method of course benefits from much higher sensitivity than would observation of the same multiplet as the recipient of an NOE enhancement in an experiment involving some other irradiation target.

The NOE can also be used to edit spectra in a rather different sense, based on differences in solution tumbling rates, using the fact that enhancements are highly dependent on $\omega\tau_c$ (Sections 2.2 and 2.5). For homonuclear systems, η_{max} varies from +50% for small $\omega\tau_c$ to −100% for large $\omega\tau_c$. Moreover, for large $\omega\tau_c$, spin diffusion leads to a rapid spread of saturation throughout the spin system (Section 3.2.3). Thus, if any signal in the spectrum of a large molecule, for instance a large protein, is saturated, spin diffusion causes the saturation to spread throughout all parts of the molecule that have a long correlation time. A spectrum acquired with presaturation applied to the envelope of protein signals will therefore only contain signals from any relatively mobile regions of the protein, the broader signals from less mobile regions being suppressed through receiving a −100% NOE enhancement.[33] The same trick can be used to suppress the "macromolecular background" signals when studying small molecules in complex biological fluids.

Similar editing can be achieved with heteronuclear spectra, using broad-band proton saturation. The $^{13}C\{^1H\}$ NOE varies from +199% for small $\omega\tau_c$ to almost zero for large $\omega\tau_c$ (cf. Section 2.5). Therefore, if a $^{13}C\{^1H\}$ *difference* spectrum is acquired, the only signals to survive will be those from relatively mobile components, since the more slowly tumbling molecules receive essentially no enhancement.[34] Similarly, the $^{15}N\{^1H\}$ NOE varies from −494% for small $\omega\tau_c$ through −100% to almost zero for large $\omega\tau_c$. This variation was put to use by Lapidot and Irving,[35] who investigated the ^{15}N spectra of ^{15}N-enriched *Escherichia coli* cells. These cells contain molecules with a wide range of τ_c values. Because rapid pulsing was used, no signals were seen from very small molecules such as cellular metabolites, since these have long ^{15}N T_1 values. Signals were also absent for very slowly tumbling molecules, such as DNA or RNA, because the ^{15}N linewidths were too large. In the absence of ^1H broadband saturation, essentially the only signals observed were from protein amide nitrogens and lysine-N^ε. In the presence of ^1H saturation, the amide backbone signals essentially disappeared, presumably because their correlation times were such that an enhancement of about −100% resulted. The observed signals were inverted, and came from protein side-chain and phosphatidylethanolamine nitrogens, which are more mobile and so receive a more negative enhancement.

This useful editing function of the NOE can also be extended and used in its own right, by combining a difference NOE experiment with other experiments. The simplest example is to combine NOE difference spectroscopy with decoupling, as shown in Figure 9.3. This technique was used during work on bovine pancreatic trypsin inhibitor (BPTI), a small globular protein of molecular weight about 6500.[36] A normal NOE difference experiment, in which the overlapping resonances at 6.75 ppm due to Tyr-21 and Tyr-35 were preirradiated, is shown in Figure 9.4b. In order to discover which, if any, of the four enhanced methyl doublets in this difference spectrum is the coupling partner of the enhanced quartet at 4.09 ppm, the *decoupled* NOE difference spectrum shown in Figure 9.4c was recorded. The collapse of the quartet on decoupling methyl 6 clearly answers this question. A very similar application of specific

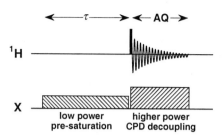

Figure 9.3. Experimental scheme for acquisition of a decoupled NOE difference spectrum. Low-power irradiation is applied during τ to generate enhancement; higher-power homogated irradiation is applied during AQ at another frequency for decoupling a particular signal during acquisition. The decoupling irradiation is applied on-resonance for the decoupling target during all transients, but the "NOE" irradiation is applied on- and off-resonance for the saturation target in different transients, as in the normal NOE difference experiment.

decoupling during acquisition, this time during t_2 of a NOESY spectrum, was made during a study of BPTI folding intermediates.[37] In this instance, it was necessary to confirm the assignment of a particularly crucial NOESY cross peak involving the methyl signal of Ala25, which was degenerate with one of the Hγ signals of Arg40. The resolution of this problem is shown in Figure 9.5.

More generally, we may note that the NOE difference experiment may, in principle, be extended to yield other hybrid sequences. In such experiments, the single observe pulse of the normal NOE difference experiment would be

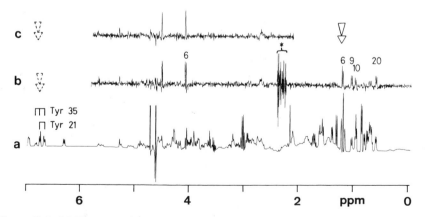

Figure 9.4. (*a*) The normal ^{1}H spectrum of BPTI in D$_2$O. (*b*) Normal NOE difference spectrum, saturating the signals at 6.75 ppm. The peaks marked with an asterisk are instrumental artifacts. (*c*) NOE difference spectrum obtained while decoupling methyl 6 (as indicated in Fig. 9.3); the resonance at 4.09 ppm is now decoupled. Reproduced with permission from Ref. 36.

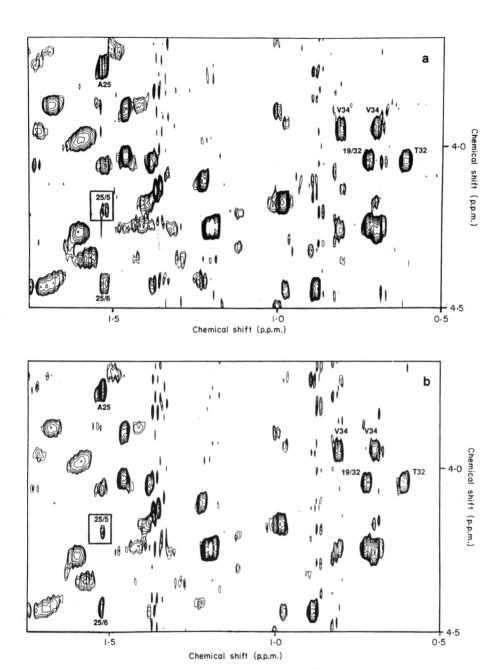

Figure 9.5. Expansions of the same region from two NOESY spectra of a mutant (C5S, C55S, M52R) of bovine pancreatic trypsin inhibitor (BPTI) designed to mimic an intermediate on the disulfide folding pathway. (*a*) Normal NOESY spectrum. The boxed cross peak has an F_2 chemical shift that could correspond either to Ala25 $C^\beta H_3$ or to one of the Arg40 H^γ protons. (*b*) NOESY spectrum obtained with decoupling of the Ala25 H^α signal during t_2. The collapse of the multiplet structure in F_2 clearly supports the assignment of this cross peak to an interaction involving Ala25 $C^\beta H_3$ (Arg40 has no signals close to Ala25 H^α). Reproduced with permission from Ref. 37.

replaced by a pulse *sequence*, perhaps intended to aid assignment of the enhanced signals. For example, if the observe pulse were replaced by a COSY or DQF-COSY pulse sequence, coupling partners of the enhanced signals could be revealed. Apart from the method for measuring nonspecific ^{15}N{^{1}H} NOE enhancements described in Section 9.1.1, few, if any, examples of such sequences appear to have been published, most probably because they are likely to suffer from poor signal-to-noise and, particularly, poor "signal-to-artifact" ratios. However, it is possible that the advent of gradient-assisted spectroscopy may change this situation, making such experiments more attractive than at present.

9.2.2. Editing Using Heteronuclear Scalar Couplings

This section deals with some of the heteronuclear experiments involving the (^{1}H, ^{1}H) NOE that have been developed recently for use with isotopically labeled biomacromolecules.

Over the past decade or so it has become feasible to incorporate stable isotopes such as ^{15}N and ^{13}C into biological macromolecules, particularly proteins, by growing genetically engineered microorganisms, often *E. coli*, in suitably enriched growth media. Overexpression of the required molecule then results in a product whose isotopic composition depends on that of the growth medium and on the biosynthetic pathways used by the microorganism. Various patterns of labeling can be obtained, but the commonest approach is to arrange for near-complete, uniform labeling throughout the molecule. To achieve this, the growth medium is enriched typically to >98% with the required isotope(s); the biosynthetic pathways in the microorganism are then essentially irrelevant, as simple mass balance ensures high incorporation throughout the product. Commonly, ^{15}N ammonium chloride is used as the nitrogen source and ^{13}C$_6$-glucose as the carbon source, although near-fully labeled rich media are also available. Proteins labeled with >98 atom% ^{15}N at all nitrogen sites have been used in structural NMR studies since about the mid 1980s, and similar levels of labeling with ^{13}C in proteins have been in use since about 1989. Labeling of RNA molecules with ^{15}N and ^{13}C was demonstrated in 1992,[38] and several groups are at present working towards viable approaches to global isotopic labeling of DNA oligomers.[39] There are reviews of labeling strategies and methodology.[40,41]

The availability of such labeled samples has led to a whole battery of new NMR experiments being developed [one of the later phases in the continuing development of which has been the assimilation of pulsed field gradients (cf. Section 7.5.3.1) into the pulse sequences].[42] Many such experiments are based purely on scalar couplings, and are intended to aid in making assignments. These experiments are often designed rather specifically around the covalent structure of the particular class of molecule to which they are applied; examples include the HNCA, HNCO, and many related sequences for making backbone assignments in labeled proteins, and the HCCH-TOCSY and HCCH-COSY sequences for making side-chain assignments.[43,44]

In contrast, heteronuclear experiments involving the NOE are often rather more general in their scope. Two classes of experiment may be distinguished. In the first, one-bond couplings to the heteronucleus provide the basis for an "NMR switch," allowing ^{1}H signals coupled to X to be discriminated from those not coupled to X, and a spectrum to be produced in which one or other class of signal has been removed. Experiments of this type are often called "filtered" experiments, but are also occasionally called "edited." In the second class of experiments, ^{1}H magnetization is frequency labeled with the chemical shift of an attached X nucleus, leading to an additional frequency dimension in some form of heteronuclear correlation spectrum (see following). The best name for the latter category is probably "X-separated" experiments, although some authors have also referred to them as "X-edited"; the latter term seems dangerously ambiguous in this context. The distinction between X-filtered and X-separated experiments is helpful to make for experiments other than just those that involve the NOE, but more general applications will not be dealt with here. X-filtered and X-separated NOE experiments will now be discussed in turn.

9.2.2.1. X-Filtered NOE Experiments.

"X-filtered" experiments are generally based on the so-called "half-filter" pulse sequence element introduced by Otting and Wüthrich,[45] shown in its simplest form in Figure 9.6. This can either be used to accept ^{1}H signals that are coupled to X while rejecting those that are not, or, by employing a different receiver phase cycle, it can be used to do the reverse, accepting ^{1}H signals that are *not* coupled to X while rejecting

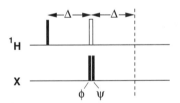

Figure 9.6. Half-filter pulse sequence element. 90° Pulses are represented by narrow filled vertical bars, 180° pulses by wider, open vertical bars. The phase cycles for the two 90° X pulses are: ϕ = +x, −x, +x, −x; ψ = +x, +x, −x, −x. (The phases of the ^{1}H 90° and ^{1}H 180° pulses can be set according to the requirements of other parts of the pulse sequence into which the half-filter is placed.) When the filter is set to retain proton signals coupled to X, the corresponding receiver phase cycle is +, −, −, +, whereas when the filter is set to reject proton signals coupled to X, the receiver phase cycle is +, +, +, +. The length of Δ is set to be approximately $1/(2J_{HX})$; typical values are around 5.5 ms for a ^{15}N half-filter [$^1J(^{15}N,^1H) \simeq 90$ Hz], or around 3.5–4 ms for a ^{13}C half-filter [$^1J(^{13}C,^1H) \simeq 130$ Hz for aliphatic CH groups, or 150–170 Hz for aromatic CH groups].

those that are. In either case, the phase cycles applied to the two 90° X pulses have the effect that these pulses either act cumulatively to give a net 180° X pulse (when ϕ and ψ reinforce), or cancel to give a net 0° X pulse (when ϕ and ψ are opposed). Proton signals that are in antiphase with respect to the (X, ^{1}H) coupling at the midpoint of the half-filter change sign on application of a 180° X pulse, so these signals will survive the phase cycle only if the receiver phase inverts whenever the two 90° X pulses reinforce. Proton signals not in antiphase with respect to the (X, ^{1}H) coupling at the midpoint of the half-filter are insensitive to phase changes of the X pulses, so such signals will cancel with this receiver phase cycle. Conversely, if it is intended to retain the non-X-coupled signals and reject the X-coupled signals, the receiver phase must be left constant through the cycling of the 90° X pulse phases. Note, however, that it is intrinsically more difficult to reject the signals coupled to X than it is to reject those that are *not* coupled to X. Signals coupled to X are only in pure antiphase at the midpoint of the half-filter if Δ is exactly equal to $[2^1J(X,^1H)]^{-1}$, and in any practical sample there may be a range of values for $^1J(X,^1H)$ for different signals, particularly if X is ^{13}C. Thus, for a half-filter sequence designed to reject ^{1}H signals coupled to X, those X-coupled signals for which the value of Δ does not correspond exactly to $[2^1J(X,^1H)]^{-1}$ will be imperfectly suppressed, since their in-phase components will survive. In contrast, in the case where the coupled signals are retained, variations in the values of $^1J(X,^1H)$ only cause intensity losses in the desired signals, rather than imperfect suppression of the undesired signals. Recently, modified half-filters have been proposed that circumvent some of these problems,[46] and improved sequences designed to purge simultaneously ^{13}C-coupled signals across a range of $^1J(^{13}C,^1H)$ couplings have also appeared.[47] In some later versions of half-filters, a 180° pulse is used in place of the two 90° pulses, and the effect of using a 0° pulse is achieved on the appropriate scans by moving the 180° pulse very far off-resonance; this approach is often necessary if the 180° pulses are selective or semiselective shaped pulses.

Half-filters may be used independently in each dimension of a multidimensional experiment, allowing a separate choice to made for each dimension as to which signals should be retained and which rejected (the name "half-filter" reflects the fact that, at the time these filters were proposed, the only multidimensional experiments in common use were two-dimensional). For instance, a possible application using a ^{15}N-labelled protein might be to insert a ^{15}N half-filter immediately prior to t_2 of a 2D NOESY experiment. If this were set to accept ^{15}N-coupled signals, the result would be a spectrum containing only signals from ^{15}N-bound protons in F_2, but with cross peaks to all of their cross-relaxation partners in F_1. Conversely, if the half-filter were set to reject ^{15}N coupled signals, then only non-^{15}N-bound protons would resonate in F_2, with all cross peaks involving these signals appearing in F_1. The sum of these two subsets of cross peaks would comprise the total set of cross peaks present in a conventional NOESY spectrum. If a further half-filter were inserted into t_1, this would allow independent selection in F_1, leading to four possible subsets of cross peaks, depending on the phase cycles used in each half-filter. "Combined" half-filters that simultaneously select ^{15}N-coupled protons *and* ^{13}C-coupled signals have been developed, and are useful when dealing with double-labeled materials.[48]

It should be noted that, because half-filters do not act during the mixing period, they do not select against certain spin-diffusion pathways that might be overlooked in a simple analysis. For instance, if ^{13}C-coupled signals are selected in both F_1 and F_2 of a 2D NOESY experiment, cross peaks may still arise through spin-diffusion pathways of the form ^{13}C—H → ^{12}C—H → ^{13}C—H. Experiments have been proposed that further select during the mixing period of NOESY to eliminate such pathways; these are examples of so-called network-editing techniques, and are discussed in Section 8.6.1.[49,50]

One of the main uses for half-filters has been in experiments involving bimolecular complexes, since it can often be arranged that one component of the complex is labeled while the other is not. Using half-filtered experiments then allows one to observe separately the intercomponent NOE interactions and each type of intracomponent NOE interaction. In the case of protein–DNA complexes, at present it would normally be the protein component that would be labeled, given the difficulty of obtaining labeled DNA; however, for protein–RNA or protein–protein complexes, either component can be labeled. Observation of signals from the labeled component offers the advantages of more rigorous signal selection (see preceding) and the added possibility of performing X-separated experiments to reduce signal overlap (see following). Therefore, when possible, it is generally worth making separate samples of the complex in which each component is labeled in turn. For trimolecular or higher complexes, more elaborate combinations of differently labeled samples would presumably be required. The approach is also useful for resolving assignment ambiguities in symmetrical dimers (cf. Section 12.2.1.3), but for symmetrical trimers or higher aggregates the half-filter approach cannot make the required distinctions between different types of symmetry-related interunit interactions to allow a completely unambiguous assignment without recourse to other information. Several structure determinations of macromolecular complexes have appeared recently in which extensive use was made of half-filtered experiments (cf. Sections 13.1.4 and 13.14).[51,52]

When using filtered 2D NOESY experiments on mixed-labeled symmetrical oligomers, it is most convenient to vary the selection employed in the F_1 and F_2 half-filters *in parallel* rather than independently. This allows both possible symmetry-related NOE pathways to contribute simultaneously, rather than having to arbitrarily select just one direction of transfer; for example, to obtain intermolecular interactions one may select simultaneously *both* the pathway (labeled → unlabeled) *and* the pathway (unlabeled → labeled) in a single experiment.[46] In such cases it can be particularly useful to adjust the ratio of the two data sets that are subtracted from one another to generate the final difference spectrum, since most artifacts do not null at an exact 1:1 ratio (and indeed different artifacts may be nulled at slightly different ratios). Of course, this approach is only possible if the individual data sets are stored separately during acquisition. An example of a spectrum obtained from such an experiment with a symmetrical dimer is shown in Figure 9.7.

An X filter can also be introduced into a 1D transient NOE experiment (the name "half-filter" seems rather inappropriate here). Such pulse sequences are useful, for instance, when a molecule has been selectively labeled at a single site and the chemical shift of the single X nucleus is not required. Figure 9.8 shows an example of such a pulse sequence, and some spectra obtained using this sequence with a small, selectively labeled peptide.[54] The action of the X filter here has the added benefit of removing many of the subtraction artifacts found in conventional 1D NOE difference spectra, making X-filtered experiments of this sort surprisingly sensitive. However, while it is one thing to detect NOE enhancements in the 0.1% range in small, conformationally flexible peptides, it is quite another to interpret them, and as yet there is very little experience available in this area.

9.2.2.2. X-Separated NOE Experiments.
"X-separated" experiments are those where the X nuclei attached to protons are frequency labeled so as to generate one or more X-nucleus chemical shift dimensions. There are a great variety of heteronuclear correlation experiments, most of which are not NOE experiments, and they may in principle have almost any number of dimensions. However, most of the important principles, as far as practical NOE experiments are concerned, can be explained using the various sequences illustrated in this section combining HSQC and NOESY.

It would be beyond the scope of our discussion to give a detailed account of the HSQC (*H*eteronuclear *S*ingle *Q*uantum *C*orrelation) pulse sequence itself, shown in Figure 9.9a, or to compare it to its subtly related counterpart HMQC (*H*eteronuclear *M*ultiple *Q*uantum *C*orrelation), to which much of what follows can equally well be applied. However, even without such a discussion it can be appreciated how the whole HSQC (or HMQC) sequence can be treated as a "building block" from which more complex sequences for NOE experiments can be constructed. In this respect, HSQC has three key features:

Figure 9.7. An example of a filtered NOESY experiment for a symmetrical dimer, in this case a fragment (residues 44–84) of the inhibitor of bovine ATP synthase.[53] This protein forms an antiparallel homodimeric coiled-coil structure in which histidine rings and hydrophobic side-chains of leucines and isoleucines (*inter alia*) interact across the inter-subunit interface. Panel (*a*) shows part of the normal NOESY spectrum while panel (*b*) shows the corresponding part of the doubly half-filtered NOESY spectrum. Ideally (b) should contain only inter-subunit interactions, and in this case this ideal is nearly reached (some artifacts appear close to the frequency of the water signal due to effects of presaturation). The filter elements used were designed to allow simultaneous selection for protons attached to ^{15}N, aliphatic ^{13}C, and histidyl ring ^{13}C. Spectra were recorded at 800 MHz and 300 K with a mixing time of 150 ms, using a 3 mM sample and a total experiment time of approximately 2 days. The protein sample comprised 50% unlabeled protein chains and 50% [^{15}N, ^{13}C] double-labeled protein chains.

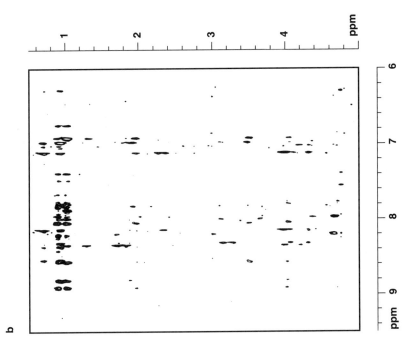
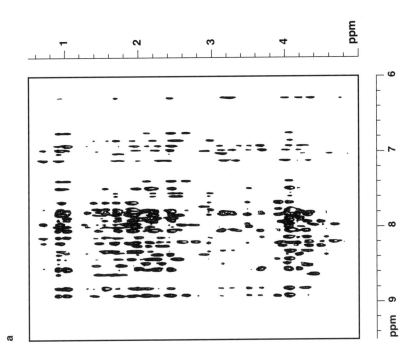

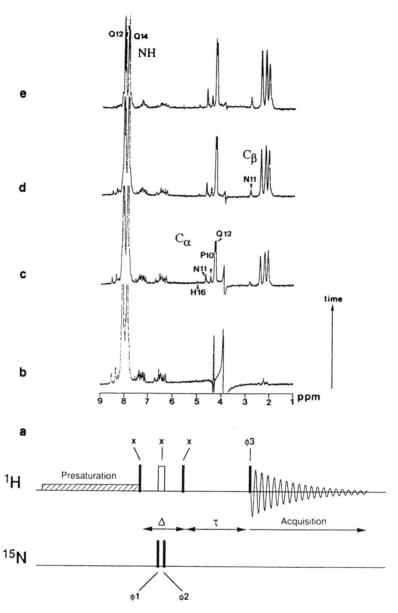

Figure 9.8. One-dimensional transient NOE experiments including an X filter. (*a*) Pulse sequence. The phase cycles used are: ϕ_1 = x, −x, x, −x; ϕ_2 = x, x, −x, −x; ϕ_3 = x, x, x, x, y, y, y, y, −x, −x, −x, −x, −y, −y, −y, −y; receiver = x, −x, −x, x, y, −y, −y, y, −x, x, x, −x, −y, y, y, −y. (*b–e*) Spectra obtained using this pulse sequence with a sample of the synthetic hexadecapeptide (Ser-Pro-Asn-Gln-Gln-Gln-His-Pro)$_2$, selectively labeled with ^{15}N at the backbone amide of Gln14 and dissolved in a mixture of 90% 1,1-[^{2}H$_2$]-2,2,2-trifluoroethanol and 10% H$_2$O at a temperature of 300K. The delay Δ was set to 5.46 ms, the mixing times (τ) were (b) 10 µs, (c) 100 ms, (d) 200 ms, and (e) 400 ms, and the spectra were recorded at 500 MHz. Reproduced with permission from Ref. 54.

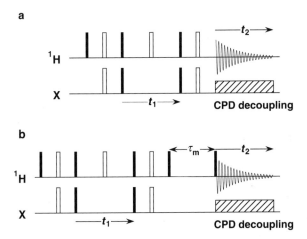

Figure 9.9. (*a*) and (*b*) Two-dimensional HSQC and HSQC-NOESY pulse sequences, respectively. 90° Pulses are represented by narrow, filled vertical bars, 180° pulses by wider, open vertical bars. See text for further discussion.

1. It correlates ^{1}H and X signals directly through one-bond (X, ^{1}H) couplings, so that X-nucleus chemical shifts appear in the indirectly detected dimension that the sequence introduces (e.g., F_1, in the case of 2D HSQC).
2. It has the full benefit of "inverse" detection, that is it starts with magnetization intensities dependent on ^{1}H populations, and it is ^{1}H transverse magnetization that is ultimately detected. This helps maximize sensitivity in any experiments derived from HSQC.
3. The magnetization present at the beginning of the detection period t_2 is transverse and is in-phase with respect to the large one-bond heteronuclear couplings; heteronuclear decoupling can therefore be applied so as to collapse the heteronuclear splittings and maximize sensitivity. Thus, the magnetization produced at the end of the sequence is qualitatively more or less equivalent to that resulting from a simple ^{1}H 90° pulse acting on z magnetization.

For X = ^{13}C or ^{15}N, the HSQC spectrum contains a single cross peak for each XH or XH$_3$ group in the molecule, while XH$_2$ groups give either a pair of cross peaks at a common X shift or else a single cross peak, depending on whether or not the two protons in the XH$_2$ group are degenerate.

The simplest combination of HSQC with NOESY is the 2D HSQC-NOESY sequence shown in Figure 9.9b. This is derived from 2D HSQC simply by adding an NOE mixing period (τ_m) immediately prior to acquisition. The last two 90° pulses are required, just as in NOESY, to rotate transverse magnetization onto the z axis at the start of τ_m and then back to the transverse plane for detection, and they are phase-cycled so as to retain only z magnetization during τ_m. The 2D HSQC-NOESY spectrum contains all the same cross peaks as does HSQC, but the intensity of each direct correlation peak is partially transferred, via the NOE, into additional cross peaks. These new cross peaks each share the same F_1 shift (i.e., X-nucleus shift) as the direct correlation peak from which they arose, but their F_2 shifts correspond to protons close in space to the proton involved in the original direct correlation peak. The information content of such a spectrum is thus rather similar to that of a (^{1}H, ^{1}H) NOESY spectrum with an X half-filter in F_1, but with the crucial difference that each X—H group contributing to the spectrum during t_1 appears at its X-nucleus chemical shift in F_1, rather than its ^{1}H chemical shift as would be the case for the half-filtered NOESY spectrum. Thus overlaps that occur between cross peaks in one spectrum will usually be absent in the other, and interpreting the two spectra *in combination* will very often resolve essentially all overlap problems amongst ^{1}H signals coupled to X nuclei. A slightly simpler approach that has essentially the same advantages is to compare an HSQC-NOESY spectrum with a normal homonuclear NOESY spectrum (acquired with X decoupling in both dimensions). Using this approach one could expect, for example, to resolve most or all problems of amide proton chemical shift degeneracy for a moderately sized ^{15}N-labeled protein by using homonuclear NOESY in conjunction with ^{15}N HSQC-NOESY (but to resolve problems of degeneracy amongst H$^\alpha$ protons or carbon-bound side-chain protons, a ^{13}C HSQC-NOESY spectrum from a ^{13}C-labelled sample would be needed).

As an alternative to collecting these spectra separately, the two experiments may be combined in a 3D experiment. Three-dimensional experiments were first introduced in 1987,[55] and represent a logical extension of the concept of 2D spectroscopy. As shown in Figure 9.10, 3D pulse sequences introduce an additional, independently incremented, indirect time dimension. There are thus three independent time variables, t_1, t_2, and t_3, of which t_3 is the directly acquired dimension, while the other two are acquired indirectly; and there are two separate mixing processes, one between t_1 and t_2 and the other between t_2 and t_3. Three-dimensional Fourier transformation then gives the 3D spectrum.

Two ways of combining HSQC and NOESY to form a 3D experiment are shown in Figure 9.10a and b. In a 3D NOESY-HSQC experiment (Fig. 9.10a), the HSQC sequence precedes the NOESY mixing period, and the result is a 3D spectrum having ^{1}H shifts in F_2 and F_3, but X shifts in F_1. The layout of signals in a 3D NOESY-HSQC spectrum is shown schematically in Figure 9.10c. It is useful to think first of the three possible projections of the 3D spectrum. The projection parallel to F_1, that is the projection in which all the F_2/F_3 planes are merged into one another, corresponds to the 2D spectrum that would be obtained if t_1 had not been incremented. Inspection of the pulse sequence shows that this would be a (^{1}H, ^{1}H) 2D NOESY spectrum with an X half-filter in the indirectly acquired dimension (i.e., the dimension corresponding to the incremented time *preceding* the NOE mixing period; Section 9.2.2.1). Starting from this viewpoint, the 3D spectrum can be pictured as a homonuclear X-filtered NOESY spectrum that has been "exploded" along F_1 according to the chemical shift of the X nucleus attached to the originating proton of each NOE interaction. This is the interpretation of the spectrum that is illustrated in Figure 9.10c. By similar reasoning, the F_1/F_2 projection (parallel to F_3) corresponds to a simple 2D HSQC spectrum, so the 3D spectrum could equally well be viewed as a 2D HSQC spectrum "exploded" along F_3 according to the NOE interactions of the proton involved in each HSQC correlation. Likewise, by considering the F_1/F_3 projection (parallel to F_2), the 3D spectrum could also be viewed as a 2D HSQC-NOESY spectrum "exploded" along F_2 according to the chemical shift of the X nucleus attached to the originating proton involved in each NOE interaction.

In contrast, if the NOESY mixing is placed before the HSQC sequence as in Figure 9.10b, the result is a 3D NOESY-HSQC experiment. As shown in Figure 9.10d, this type of 3D spectrum has ^{1}H shifts in F_1 and F_3, but X shifts in F_2, and it is now the F_1/F_3 projection that corresponds to the 2D X-filtered NOESY spectrum. Similarly, it is the F_1/F_2 projection that corresponds to a 2D HSQC-NOESY spectrum, while the F_2/F_3 projection corresponds to a simple 2D HSQC spectrum. Note that, since the HSQC sequence now *follows* the NOESY mixing period, it is the shift of the X nucleus attached to the proton *receiving* the NOE interaction that now appears in F_2 (rather than that from which the NOE interaction originates, as was the case for F_1 of 3D HSQC-NOESY). The F_1/F_3 projection thus corresponds to a homonuclear 2D NOESY spectrum with an X half-filter in the *acquired* dimension (rather than the indirectly acquired dimension, as was the case for the F_2/F_3 projection of 3D NOESY-HSQC).

Which of these two arrangements is preferable for the 3D experiment depends largely on the identity of the X nucleus and the type of sample being studied. For a ^{15}N-labelled protein, there is a clear advantage in using the 3D NOESY-HSQC experiment, because then the only ^{1}H signals actually detected during t_3 are from directly ^{15}N-bound protons. These generally resonate in the low-field part of the spectrum, so that the F_1/F_2 plane containing the intense t_1/t_2 noise from the partly suppressed water signal [i.e., the plane $F_3 = \delta(H_2O)$] falls outside the region containing cross peaks. For ^{13}C-labelled proteins the corresponding choice is less clear cut, since protons coupled to ^{13}C can cover much of the spectrum and include Hα signals very close to the water signal. It is thus quite common to carry out these experiments in D_2O rather than H_2O, although experiments in H_2O will still be needed if detection of NOE enhancements from CH to NH signals is required.

Because the incrementations of t_1 and t_2 are independent, 3D experiments can be quite lengthy; for instance, if 128 increments are acquired in both t_1 and t_2, this means a total of 16,384 increments overall. However, the advent of gradient-assisted spectroscopy has reduced this problem by allowing coherence pathway selection to be made in a much smaller number of scans (sometimes only one), thereby greatly reducing the time taken to complete a phase cycle and reducing the minimum length of many experiments.[42] A further way of addressing the implied problem of achieving adequate digitization in 3D experiments is to allow signals to fold. This is often done for heteronuclear

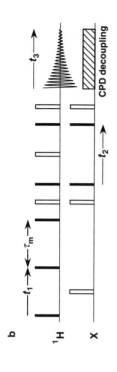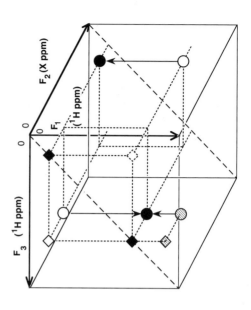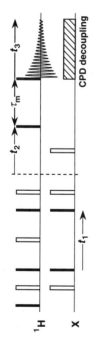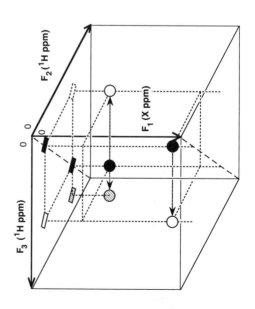

experiments where X = ^{13}C, as both ^{13}C and ^{1}H signals from aliphatic groups characteristically occupy spectral regions different from the corresponding signals from aromatic groups, so that no information need be lost if signals are folded in a carefully planned way.

Formally, the information content of a 3D NOESY-HSQC spectrum should be identical to the *combined* information content of an F_1 half-filtered (^{1}H,^{1}H) NOESY and a 2D HSQC-NOESY. In practice, however, there are significant differences between these two ways of acquiring the data. Clearly, the 3D experiment offers the considerable convenience of combining all the desired information in one spectrum. However, this convenience comes at a cost, most notably that in practice the 3D spectrum will necessarily have a significantly lower resolution than either of the 2D spectra. This loss of resolution can be a real drawback, both in resolving critical shift differences, and in that partially resolved multiplet structures can often give important additional clues to assign-

←───

Figure 9.10. (*a*) and (*b*) Three-dimensional HSQC-NOESY and NOESY-HSQC pulse sequences, respectively. 90° Pulses are represented by narrow, filled vertical bars, 180° pulses by wider, open vertical bars.

(*c*) Schematic representation of the layout of peaks in a 3D HSQC-NOESY spectrum and their relationship to the corresponding peaks in a 2D NOESY spectrum. The F_2/F_3 projection, which corresponds to a 2D (^{1}H,^{1}H) NOESY spectrum with an X nucleus half-filter in the indirectly acquired dimension (see text), is shown in the plane $F_1 = 0$ (note that the 3D spectrum itself does not contain signals in this plane, apart possibly from artifacts). This projection contains (i) a pair of diagonal peaks (black diamonds) corresponding to two protons that are close in space to one another and are each coupled to (different) X nuclei, (ii) the corresponding pair of cross peaks (white diamonds) arising through their mutual NOE interaction, and (iii) a further NOE cross peak (shaded diamond) from one of the protons to a third proton that is itself not coupled to an X nucleus. In every case, the corresponding peaks in the 3D spectrum (black, white, and shaded circles) are found in the F_1 plane at the chemical shift of the X nucleus directly attached to the proton where the NOE interaction *originates*. The arrows connecting cross-peaks in the 3D spectrum represent the directions of the NOE transfer steps that occur during the period τ_m. See text for further discussion.

(*d*) Schematic representation of the layout of peaks in a 3D NOESY-HSQC spectrum, and their relationship to the corresponding peaks in a 2D NOESY spectrum. The various diamonds, circles, and arrows, have the same meanings as in Figure 9.10c. Note, however, that now it is the F_1/F_3 projection that corresponds to an X-filtered 2D NOESY spectrum (in this case, one filtered in the directly acquired dimension), and that peaks in the 3D spectrum are now found in the F_2 plane at the chemical shift of the X nucleus directly attached to the proton that *receives* the NOE interaction. We have chosen in this diagram to maintain the same viewing orientation for the F_1, F_2, and F_3 axes as is shown in Figure 9.10c; of course, if we had instead chosen to interchange the F_1 and F_2 axes, then the layout of signals would look identical to that shown for 3D HSQC-NOESY (but note that the coordinate system would then become left-handed, rather than right-handed as it is at present). See text for further discussion.

ment in cases of ambiguity (cf. Section 12.2.1.1). Nonetheless, the relative ease of interpretation of the 3D spectrum usually makes this the preferred option. Three-dimensional NOE experiments have also been introduced in which both indirect frequency dimensions are heteronuclear.

If the resonances are sufficiently crowded, it may be worthwhile considering four-dimensional experiments such as the 4D HSQC-NOESY sequence shown in Figure 9.11a, in which each (^{1}H, ^{1}H) NOE cross peak is characterized by the shifts of *both* directly attached X nuclei, rather than just one of the X chemical shifts as in 3D HSQC-NOESY. Such experiments represent, so far, more or less the ultimate in sophistication as far as techniques to relieve overlap in NOE experiments are concerned. However, the total length of the pulse sequence can make it very insensitive for large molecules, since in such cases most of the signal created by the first pulse is destroyed by transverse relaxation before data acquisition begins. Also, the resolution available in these experiments using realistic total acquisition times is necessarily very limited. Thus, it is important to see just what the benefit of such a 4D experiment is, so as to be able to balance this against the significant disadvantages of lower sensitivity and resolution. Figure 9.11b and c illustrates the type of assignment ambiguity involved, and shows that the benefit of the 4D spectrum arises chiefly when the (^{1}H, ^{1}H) NOESY spectrum contains large numbers of superimposed cross peaks. Of course, molecules large enough to contain many overlapped NOESY cross peaks may well also be large enough that 4D experiments become insensitive due to transverse relaxation, but for several proteins in or around the 150–200 residue range, 4D techniques have proved useful.[56,57]

An elegant application of 2D X-separated techniques to the structure determination of intermediate-sized symmetric molecules at natural abundance has also been demonstrated. The first such case to appear involved the symmetrical dimer hopeaphenol (**9.13**), in which two identical subunits are connected by the central C_{8b}—$C_{8b'}$ single bond.[58] In order to establish the rotameric arrangement around this bond, the proximity of the H_{7b} and $H_{7b'}$ protons was investigated using a (^{13}C, ^{1}H) HMQC-ROESY spectrum acquired without ^{13}C decoupling in F_2. The directly bonded correlation between C_{7b} and H_{7b} gives rise, as expected, to a doublet in F_2 due to the large one-bond (^{13}C,^{1}H) coupling. However, there is also a peak observed *at the center* of this doublet, which arises via ROE enhancement of $H_{7b'}$ in the other subunit. Because ^{13}C is at natural abundance, the symmetry-related $C_{7b'}$ site will usually be occupied by a ^{12}C atom, which is why this enhanced signal lacks the splitting due to $^1J(^{13}C, ^1H)$. Other variations on this approach have appeared subsequently,[59] including a version of the experiment "cleaned up" by using pulsed field gradients and replacing the HMQC element by HSQC.

Finally, it should be pointed out that because the sequences shown in Figures 9.9, 9.10, and 9.11 are intended to be illustrative only, they each represent the simplest forms of the experiments concerned. In each case, there have been many experimental "tricks and wrinkles" suggested that improve sensitivity, water suppression, the ability to detect rapidly exchanging signals, or some other aspect of these sequences, but it would be impractical to delve any deeper into this complex and intricate area here.

9.2.3. Homonuclear Three-Dimensional NOE Experiments

Relative to heteronuclear 3D experiments, homonuclear 3D experiments suffer some severe disadvantages. Nevertheless, the first 3D experiments to be proposed were homonuclear, and such experiments, although now much less widely employed than their heteronuclear counterparts, can still be useful.

The key advantage that heteronuclear 3D experiments have over their homonuclear counterparts is that heteronuclear coherence transfer over one-bond couplings is generally very efficient. The only significant losses usually involved in HSQC or HMQC are associated with variations in the sizes of different (X, ^{1}H) one-bond coupling constants, resonance offset effects when the X nucleus has a large chemical shift range, and transverse relaxation during the fixed delays. For instance, a 3D HSQC-NOESY spectrum usually has a signal-to-noise ratio approximately comparable to a 2D NOESY spectrum acquired for the same length of time. In contrast, any homonuclear 3D experiment involves two mixing processes, *both* of which are relatively inefficient, implying that 3D cross peaks in such a spectrum will generally be weak.

Various homonuclear experiments have been proposed, differing in the nature of the two mixing processes. Mixing processes that result in in-phase multiplet structures are usually used, since the coarse digitization of most 3D experiments would lead to gross distortion and internal cancellation for any antiphase lineshapes present (cf. Figs. 7.5 and 7.6). An exception to this arises in the case of applications to samples of oligosaccharides, since then the ^{1}H chemical shift range is extremely limited, and 3D TOCSY-COSY experiments with acceptable digital resolution can be feasible.[60] More commonly, the mixing processes comprise an NOE step combined either with a TOCSY mixing step (also called HOHAHA mixing) to connect through *J*-couplings to other parts

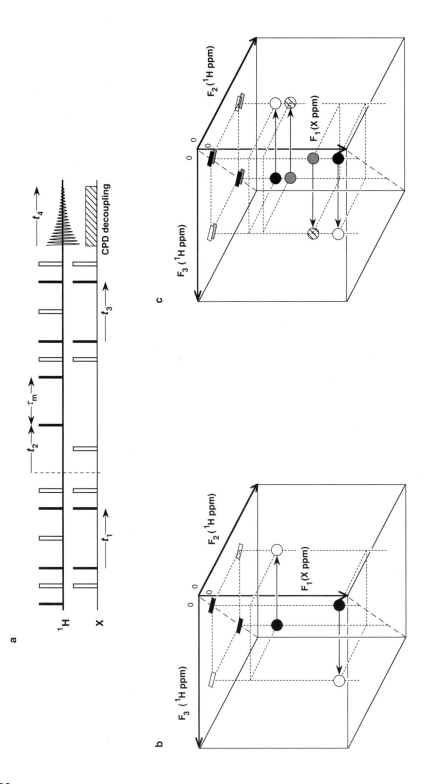

of the spin system (cf. Section 9.3.1.2),[61] or else with a second NOE step.[62] The first experiments proposed used semiselective pulses to limit the spectral width in one or more dimensions, so as to limit the difficulty of obtaining adequate digitization,[55,63] but it was soon shown that experiments including the full ^{1}H spectral width in all three dimensions were also feasible for spectra lacking antiphase lineshapes.[64] Some of these experiments are close relatives of earlier 2D experiments where two mixing steps were combined without frequency labelling the magnetization between the steps. Examples include the 2D-NOESY-COSY and 2D-COSY-NOESY relay experiments.[65]

The layout of a homonuclear 3D spectrum is somewhat more complicated to interpret than that of a heteronuclear 3D spectrum, because in the homonuclear case there is the possibility of diagonal peaks with respect to either of the mixing processes independent of the other. Thus, the various types of peak that can be present are as follows:

Figure 9.11. (a) A 4D HSQC-NOESY pulse sequence. 90° Pulses are represented by narrow, filled vertical bars, 180° pulses by wider, open vertical bars. (Although the sequence is shown using HSQC for the coherence transfer steps in order to maintain comparability with Figures 9.9 and 9.10, in practice it is probably more likely that HMQC would be used in this case.)

(b) A schematic representation of how a single ambiguous, but non-overlapped, (^{1}H,^{1}H) NOE interaction can be assigned using ^{13}C shifts from a 3D HSQC-NOESY spectrum. As in Figure 9.10, the top face of the cube shows an F_2/F_3 projection, corresponding to the 2D (^{1}H,^{1}H) NOESY spectrum with a half-filter in the indirectly acquired dimension. On this projection are shown two NOESY diagonal peaks (black diamonds) corresponding to two protons that are close in space to one another and are each coupled to (different) X nuclei, and the corresponding pair of NOESY cross peaks (white diamonds) connecting them. Although these 2D cross peaks may not be superimposed upon other cross peaks in the NOESY spectrum, there is still ambiguity as to their assignment if either of the individual protons involved has the same shift as some other proton (cf. Section 12.2.3). However, in the 3D HSQC-NOESY spectrum, the corresponding 3D cross peaks (black and white circles) appear at the shift of the X nucleus directly attached to the proton that the NOE interaction originates on. Combined analysis of all the related 3D cross peaks allows both directly attached X nucleus chemical shifts to be determined, thereby very often resolving any ambiguity in the assignment of the NOE interaction.

(c) In cases where two NOE interactions give mutually overlapped cross peaks (white diamonds and shaded diamonds) in the 2D NOESY spectrum (i.e., where *both* interacting proton shifts for one interaction are identical to their counterparts in the other) then ambiguities arise even in the analysis of the 3D HSQC-NOESY spectrum. Since two X nucleus shifts are now found for each (^{1}H,^{1}H) cross peak (and corresponding diagonal peak) in the projection, there are two possible pairings of X nucleus shifts with ^{1}H shifts for the two degenerate (^{1}H,^{1}H) NOE interactions. For pairs of proton shifts where more than two (^{1}H,^{1}H) NOESY cross peaks overlap, clearly the possible ambiguities are more numerous. It is the ambiguity between these different possible assignments that a 4D experiment would resolve.

1. Diagonal peaks (also called diagonal-diagonal peaks), corresponding to signals that did not change frequency in either mixing process: these lie along the body-diagonal of the 3D spectrum, defined by $F_1 = F_2 = F_3$.
2. First-transfer peaks, also called cross-diagonal peaks, corresponding to signals that changed frequency during the first mixing period but not the second. These lie in a diagonal plane defined by $F_2 = F_3$.
3. Second-transfer peaks, also called diagonal-cross peaks, corresponding to signals that changed frequency during the second mixing period but not the first. These lie in a diagonal plane defined by $F_1 = F_2$.
4. Back-transfer cross peaks, corresponding to signals that were transferred from one spin to another and then back again in the two mixing periods (i.e., pathways of the type $H_A \rightarrow H_B \rightarrow H_A$). These lie in the diagonal plane defined by $F_1 = F_3$.
5. 3D cross peaks, also called cross-cross peaks, corresponding to signals that changed frequency during both mixing periods. These have all three frequencies different, and their intensity is described by the product of the two transfer functions describing the two mixing processes.

Homonuclear 3D TOCSY-NOESY experiments can be useful in assigning spectra from small unlabelled DNA or RNA molecules.[66,67] Such systems generally give crowded spectra that are difficult to assign using 2D experiments alone, and are usually more difficult to obtain in an isotopically labelled form than are proteins. Furthermore, such systems also often have relatively narrow lineshapes, which improves the efficiency of the TOCSY transfer step. Applications to proteins have also been described, and it is clear that in cases where a sufficient signal-to-noise ratio can be obtained (generally from rather concentrated solutions) these techniques can provide a helpful aid to assignment and secondary structure analysis.[68] The 3D NOESY-NOESY experiment has also been shown to be of use, both in secondary structure analysis[69] and in the rather different context of assigning ambiguous NOE interactions and analyzing spin-diffusion pathways (cf. Section 12.2.1.1).[70]

9.2.4. Editing Using Something Else

One of the most fundamental ways in which to limit the set of signals enhanced in an NOE experiment is to remove undesired nuclei from the molecule. Specific replacements of protons by deuterons are widely used to simplify all types of complex proton spectra, and proton NOE spectra have received their share of attention in this context. In the past, this approach was used to make specific enhancements larger (by reducing R_I, the total relaxation rate for the observed spin, cf. Section 3.2.2.1),[71] but with the arrival of more sensitive experiments this need has essentially disappeared. Selective incorporation of deuterated amino acids into proteins from microorganisms grown in suitable media is a closely related approach (cf. Section 12.2.1.3),[72] as is the random dilution of protons by deuterons, intended to make the relaxation and coupling properties of large biomacromolecules more amenable to proton NMR in general.[73]

Another way to edit a spectrum by manipulating the sample is to compute the difference between data sets obtained from two different, but closely related, samples. In general, subtraction of spectra obtained from different samples gives poor results, since slight differences in lineshapes, shifts, temperature, and concentrations cannot be avoided. The subtraction artifacts that these cause would dominate the spectrum in most cases, but if the resonances are broad and the two samples carefully prepared to minimize unwanted differences between them, useful results can sometimes be obtained. An elegant example is provided by some work on the protein flavodoxin, the prosthetic group of which contains a quinone moiety that is diamagnetic when reduced but paramagnetic when oxidized.[74] Subtraction of a NOESY spectrum of fully reduced flavodoxin from a NOESY spectrum of partially reduced flavodoxin resulted in a spectrum containing only signals due to protons close to the quinone moiety.

Finally, we include two techniques that are essentially editing methods, namely photochemically induced dynamic nuclear polarization (photo-CIDNP), and spin-polarization induced NOE (SPI-NOE) experiments.

The CIDNP technique involves laser irradiation of a solution containing the solute of interest together with a dye, often a flavin. The laser excites electrons in the dye to a triplet state, in which form the dye can react to transfer unpaired electron density to the solute. This leads to large population changes at nuclei close to centers of unpaired electron density, often by a factor of 10 or more. The technique has been used to detect tyrosine and tryptophan residues on protein surfaces, since only these residue types react appreciably, and only surface residues can approach the dye sufficiently closely to interact. The protons whose population changes are initially changed by the dye can then undergo cross-relaxation with their near neighbors, thereby transmitting the changes to other signals in the same way as for transmission of a conventional NOE enhancement. Several examples of such photo-CIDNP induced NOE enhancements have been reported, using both 1D and 2D experiments.[75-77] Because the population changes induced by photo-CIDNP are considerably greater than those produced by conventional RF irradiation, the technique is more sensitive than normal NOE experiments—added to which the excitation has a built-in selectivity for surface groups not available from other NOE techniques. Against this, however, the method requires some expensive and rather esoteric apparatus, and unwanted additional photochemical reactions often lead to rather rapid sample deterioration.

A technique that has much in common with CIDNP at a conceptual level, although it is quite different in practice and application, is the so-called SPI-NOE experiment based on cross-relaxation of protons with laser-polarized xenon. Very substantial increases in the nuclear spin polarization of ^{129}Xe can be achieved by optical pumping: Circularly polarized laser light is used to excite rubidium atoms present in a gaseous mixture with xenon in a magnetic field, resulting in ^{129}Xe nuclear polarizations of around 10%. These massive polarizations are sufficiently long-lived that they survive transfer and freezing of the xenon into a suitably designed trap positioned above the volume of a pressure NMR tube. Upon sealing and transfer of this tube to the NMR magnet, the frozen xenon warms up and sublimes into the sample of interest in the main body of the NMR tube, where it is then available for NMR experiments. In the SPI-NOE technique, the protons are first saturated by ^{1}H 90° pulses and then allowed to recover during a period τ. Simultaneous ^{1}H and ^{129}Xe 180° pulses near the midpoint of τ isolate just the recovered ^{1}H magnetization arising through cross-relaxation with ^{129}Xe (in much the same way as the QUIET-NOESY technique isolates a particular two-spin system in a homonuclear NOE experiment; cf. Section 7.5.4). Experiments published so far involve host–guest complexes of xenon with cyclodextrin[78] and cryptophane A.[79] However, since xenon binds generally to hydrophobic sites, there is also scope for applying this technique to mapping hydrophobic surfaces or cavities in macromolecules.

9.3. ROTATING FRAME NOE EXPERIMENTS

The rotating frame NOE experiment was originally introduced by Bothner-By et al.[80] in 1984. His original name for it was CAMELSPIN (*C*ross relaxation *A*ppropriate for *M*ini-molecules *E*mu*L*ated by *SPIN*-locking), which must be one of the more thoughtful and appropriate NMR acronyms, being based on a pictorial analogy between the motions of the spins and (so it is said) a rather tricky ice-skating maneuver. However, this name has by now been displaced by the more recent acronym ROESY (*R*otating frame *O*verhauser *E*ffect *Sp*ectroscop*Y*), which was applied originally only to the 2D version of the experiment. In ROESY experiments, enhancements evolve not between elements of longitudinal magnetization, as in conventional NOE experiments, but rather between elements of transverse magnetization during a period of spin-locking. Enhancements measured in such experiments are usually called ROE (*R*otating frame *O*verhauser *E*ffect) enhancements.

The particular advantage that ROE experiments offer over their conventional NOE counterparts is that the ROE has a quite different dependence on molecular tumbling from that shown by the conventional NOE, with the result that *all* molecules behave as though they were in the positive NOE regime. In fact, for slowly tumbling molecules the maximum achievable ROE enhancement becomes *more* strongly positive than it is at the extreme narrowing limit (see following). This means that (i) there is no "difficult region" of $\omega\tau_c$ for which all enhancements are necessarily small or zero, (ii) direct and indirect enhancements can be distinguished by their signs, even for large molecules, and (iii) saturation transfer or exchange peaks have the opposite sign to direct enhancements, even for large molecules.

These advantages assume their greatest importance when dealing with molecules of intermediate size that occupy the "difficult region" of $\omega\tau_c$ close to 1.12 (Section 3.2.3), since these give small or zero enhancements in the conventional experiment, irrespective of the internuclear separations involved. Studies of macromolecules can also benefit, as spin diffusion is less significant and more recognizable in ROESY than it is in NOESY; however, for large molecules ROESY spectra usually have significantly lower sensitivity than NOESY.[81] Against these advantages there are some drawbacks, principally that several other processes can occur during the spin-locking period and give rise to cross peaks, sometimes making interpretation difficult. These points will be considered further in what follows.

In Sections 9.3.1 and 9.3.2, we have tried to summarize the main theoretical and practical considerations that are relevant to the ROESY experiment.

9.3.1. Theory

In this section we cover three topics of central importance in understanding the rotating frame experiment: (i) what is spin-locking, (ii) what is the nature of transverse cross-relaxation, and (iii) what are the other processes that occur during spin-locking.

9.3.1.1. Spin-Locking.
Spin-locking is achieved by preparing the system with a 90° pulse, and then immediately applying a strong, continuous RF field (the spin-locking field) along a transverse axis perpendicular to that of the 90° pulse. Thus a $90°_x$ pulse would be followed by a spin-locking field along the y (or $-y$) axis. Viewed in the rotating frame, the equilibrium magnetization $\mathbf{M}^0$ is first tipped over into the transverse plane by the pulse, and is then "caught" by the spin-locking field applied parallel to the new axis along which $\mathbf{M}$ now lies. It might at first appear that a $\mathbf{B}_1$ field that is colinear with $\mathbf{M}$ has no effect, but this is not so. Its role is to *suppress the relative precession* of the various magnetizations that constitute the NMR spectrum, once they are in the transverse plane.

In the absence of the $\mathbf{B}_1$ field, each signal would precess at its own Larmor frequency, just as during evolution of the FID. Viewed in the rotating frame, vectors representing individual signals at different chemical shifts would each precess at a rate determined by their particular resonance offset $(\nu - \nu_0)$. When the $\mathbf{B}_1$ field is on, however, the resonances precess around the effective field, the strength of which is given by Eq. 1.17:

$$|\mathbf{B}_{\text{eff}}| = \sqrt{4\pi^2(\nu - \nu_0)^2/\gamma^2 + |\mathbf{B}_1|^2}$$

This effective field $\mathbf{B}_{\text{eff}}$ makes an angle θ with respect to the z axis, where θ is given by Eq. 1.18:

$$\tan\theta = (\gamma|\mathbf{B}_1|)/2\pi(\nu - \nu_0)$$

If the $\mathbf{B}_1$ field is sufficiently strong for off-resonance effects to be negligible (that is, $\gamma B_1/2\pi \gg |(\nu - \nu_0)|$ and $\theta = 90°$ for the most off-resonant signal), then not only the $\mathbf{B}_1$ field but also the effective field $\mathbf{B}_{\text{eff}}$ will be colinear with the magnetization vector $\mathbf{M}$. Both fields will appear to be stationary in the rotating frame, since this rotates at ν_0, the frequency of the $\mathbf{B}_1$ field (Section 1.5).

When viewed in the rotating frame, this implies that for the vector $\mathbf{M}$, precession about $\mathbf{B}_{\text{eff}}$ consists only of precession "about its own axis," so to speak. This is very closely analogous to the situation for the equilibrium vector $\mathbf{M}^0$ when it is aligned along the axis of the static field $\mathbf{B}_0$. In fact, the analogy extends further: In some ways, the spin-locked magnetization can be considered to have been re-quantized along the spin-locking field axis, in that it behaves rather as though the spin-locking axis were a new longitudinal axis. Thus, the magnitude of the vector $\mathbf{M}$ shrinks, due to relaxation, toward a new equilibrium value determined by the $\mathbf{B}_1$ field strength. Since $\mathbf{B}_1$ is of the order of 10^5 times weaker than $\mathbf{B}_0$, there is virtually no difference between the new equilibrium populations, implying that this relaxation causes $\mathbf{M}$ to shrink essentially to zero. The characteristic time for this process is called $T_{1\rho}$, which is the rotating frame analog of T_1. This analogy between the $\mathbf{B}_0$ axis in normal NOE experiments and the $\mathbf{B}_1$ axis in ROE experiments is quite helpful when trying to understand the ROE, but we must not take it too literally. Of course, the $\mathbf{B}_0$ field is still a crucial determinant of the behavior of the spins, and it is not true to say that the original eigenstates of the system (α and β) are simply replaced by new ones determined by the $\mathbf{B}_1$ field alone. As in other experiments where spins evolve in the presence of a strong RF field, a higher level of theory is required to describe the true situation completely.

Of course, it is only when viewed in the rotating frame that the spin-locked magnetization appears to be static. In reality, it is still precessing about the z-axis at a rate ν_0. Nonetheless, even when viewed in the laboratory frame, the various contributions to the magnetization are still held colinear with the $\mathbf{B}_1$ field provided it is strong enough; the *relative* precession of the different signals is abolished.

In reality, it is impractical to apply $\mathbf{B}_1$ fields so strong that off-resonance effects are genuinely negligible, since this would involve continuous application of hundreds of watts of RF power. For the weaker fields of $\gamma B_1/2\pi = 1 - 10$ kHz commonly practicable on high-resolution spectrometers, each spin experiences an effective field that is tilted through an angle θ depending on its resonance offset (Eq. 1.18). Only the component of magnetization along this tilted field becomes spin-locked; the other component, perpendicular to the tilted field, rotates about $\mathbf{B}_{\text{eff}}$ and may be neglected, since it dephases rapidly as a result of the inhomogeneity of the $\mathbf{B}_1$ field. The spin-locked magnetization, being locked along a tilted axis, has both a transverse component $M \sin \theta$ and a longitudinal component $M \cos \theta$. The discussion of this section applies only to the transverse component; the longitudinal component undergoes normal longitudinal relaxation.[82] Of course, the relative sizes of these components vary across the spectrum; this complicates matters when dealing with cross-relaxation (cf. Eqs. 9.7 and 9.9).

9.3.1.2. Spin-Locked Transverse Dipole–Dipole Relaxation.

In this section we deal with the important properties of the ROE and its relationship to the conventional NOE. Unfortunately, the level of theory required to derive expressions for the rates of spin-locked transverse dipole–dipole relaxation is considerably more involved than that for normal longitudinal relaxation. The presence of the spin-locking field changes the nature of the energy levels, making arguments based on the usual energy levels and their populations inadequate. Instead, one can use a more sophisticated approach based on an operator description of relaxation, where individual operators are used to represent not only the parts of the dipolar Hamiltonian (as in Appendix II), but also the various components of the density matrix that fully describes the state of the spin system.[83] A detailed description would be well beyond the scope of this book, but it is important to note that this method is completely independent of population arguments, and can also be used to derive expressions given elsewhere in this book for longitudinal relaxation rates. Using this approach, one may derive expressions for σ_2 and ρ_2, the rotating frame analogs of the normal longitudinal dipole–dipole cross-relaxation rate σ, and the total dipole–dipole relaxation rate ρ, for a two-spin homonuclear spin system:

$$\sigma_2 = \frac{1}{40} K^2 \left[\frac{9}{2} J(2\omega_1) - \frac{1}{2} J(0) + 6J(\omega_0) \right] \tag{9.1}$$

$$\rho_2 = \frac{1}{40} K^2 \left[\frac{9}{2} J(2\omega_1) + \frac{1}{2} J(0) + 9J(\omega_0) + 6J(2\omega_0) \right] \tag{9.2}$$

where, as before,

$$K = (\mu_0/4\pi)\hbar\gamma^2 r_{IS}^{-3}$$

and

$$J(\omega_i) = 2\tau_c/(1 + \omega_i^2 \tau_c^2)$$

Note that ω_1 is the frequency at which the spins *precess about* the $\mathbf{B}_1$ field, and is usually of the order of 1–10 kHz (although in the context of spectral density functions it would normally be expressed in radians per second; cf. Section 2.2). The frequency of the $\mathbf{B}_1$ field *itself* is equal to the rate at which the spins precess about the static field $\mathbf{B}_0$, that is the frequency ν_0, which is of the order of hundreds of megahertz. From now on, to avoid confusion when comparing longitudinal with transverse terms within this section, we shall refer to the longitudinal relaxation terms σ and ρ as σ_1 and ρ_1 respectively.

For comparison, we also give the corresponding expressions for the longitudinal rates σ_1 and ρ_1 (Eqs. 2.22 and 2.23), recast into a similar form:

$$\sigma_1 = \frac{1}{20} K^2[-J(\omega_I - \omega_S) + 6J(2\omega_0)] \tag{9.3}$$

$$\rho_1 = \frac{1}{20} K^2[J(\omega_I - \omega_S) + 3J(\omega_0) + 6J(2\omega_0)] \tag{9.4}$$

Quite clearly, the frequencies $2\omega_1$ and $|(\omega_I - \omega_S)|$ will both be in the range of a few kilohertz or less, so that for any molecule likely to be studied in solution

$$2\omega_1 \tau_c \ll 1 \quad \text{and} \quad |(\omega_I - \omega_S)|\tau_c \ll 1 \tag{9.5}$$

from which

$$J(2\omega_1) = J(\omega_I - \omega_S) = J(0) = 2\tau_c \tag{9.6}$$

In other words, for such samples the extreme narrowing condition is always valid *with respect to* ω_1.

A number of important conclusions can be drawn from these equations that allow the key differences between NOESY and ROESY experiments to be deduced:

1. Since both of the first two spectral densities in Eq. 9.1 are within extreme narrowing with respect to $\omega_1\tau_c$, they can be combined to give a single, positive, term $[4J(0) = 8\tau_c]$, with the result that σ_2 *must be positive for all practical values of* $\omega_0\tau_c$. Only in the limit of tumbling so slow that $\omega_1\tau_c > 2^{1/2}$ could σ_2 become negative, and for B_1 fields in the region of 10 kHz this implies correlation times in the range of tens of microseconds.

2. In the extreme narrowing limit with respect to $\omega_0\tau_c$ [that is $\omega_0\tau_c \ll 1$; $J(\omega_0) = J(2\omega_0) = 2\tau_c$], normal and rotating frame relaxation rates (and hence NOE and ROE enhancements) become the same. (The same is true for T_1, T_2, and $T_{1\rho}$.)

3. In the slow motion limit with respect to ω_0 [that is $\omega_0\tau_c \gg 1$; $J(\omega_0)$, $J(2\omega_0) \ll J(0)$, $J(\omega_I - \omega_S)$], σ_2 has twice the magnitude of σ_1. Thus for large molecules, ROE enhancements grow initially at twice the rate of the corresponding NOE enhancements. This means that if ROESY and NOESY spectra of a large molecule are to be compared quantitatively, the ROESY spectrum should have half as long a mixing time as the corresponding NOESY experiment.

4. Away from these two limits, the maximum transient enhancement predicted by Eqs. 9.1 and 9.2 follows the form shown in Figure 9.12a. To facilitate comparisons between ROE and NOE experiments, Figure 9.12b also shows the ratio σ_2/σ_1 as a function of $\omega_0\tau_c$.

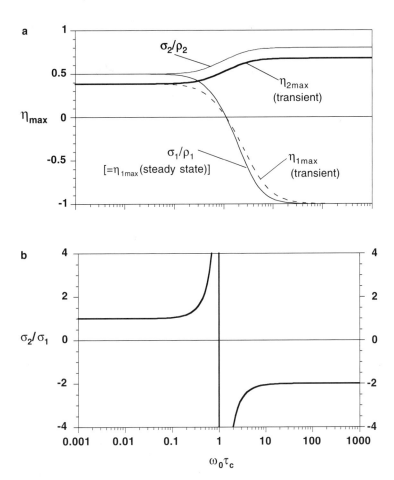

Figure 9.12. (a) The bold line shows the maximum transient ROE enhancement [η_{2max}(transient)] for a homonuclear two-spin system relaxing exclusively via the intramolecular dipole–dipole mechanism, plotted as a function of $\omega_0\tau_c$. This curve was calculated using Eq. 4.9, substituting σ_2 for σ_1, and ρ_2 for ρ_1, using Eqs. 9.1 and 9.2 to calculate σ_2 and ρ_2 and assuming that $\omega_1\tau_c \ll 1$. For comparison, the maximum longitudinal transient NOE enhancement [η_{1max}(transient)] is also shown (dashed line), as are the ratios σ_2/ρ_2 and σ_1/ρ_1 (the latter represents the maximum longitudinal steady-state NOE enhancement; Section 2.2.1). Although there is no rotating frame analog of the steady-state NOE experiment, the ratio σ_2/ρ_2 is still of interest since it gives an indication of how much leakage can be expected at each step along a spin-diffusion pathway. See text for further discussion. (b) Ratio σ_2/σ_1 as a function of $\omega_0\tau_c$. For a given internuclear interaction in a small molecule the longitudinal and transverse cross-relaxation rates are equivalent, whereas for the same interaction in a large molecule σ_2 exceeds σ_1 by a factor of two.

5. For the ROE in the slow motion limit $|\rho_2| > |\sigma_2|$, whereas for the NOE in the slow motion limit, $|\rho_1| = |\sigma_1|$. This implies that spin diffusion can never become as dominant in ROE experiments as it does in NOE experiments with large molecules, because, unlike the NOE, even in the slow motion limit there is always some loss through leakage for every cross-relaxation step along an ROE pathway. Furthermore, what little detectable spin diffusion does occur in ROE experiments is therefore largely limited to a single diffusion step (i.e., to three-spin effects; cf. Section 3.2.2.2), and these interactions produce signals opposite in sign to direct ROE interactions, allowing them to be easily recognized.

6. For large molecules, NOE interactions generally build up much faster than they decay, whereas for ROE interactions the rates of build up and decay are more similar [this is not immediately clear from these equations alone, but may be seen from Eqs. 4.5, 4.6, and 4.7 and their analogs (obviously derived) for the rotating frame experiment].[84,85] Thus ROESY experiments with large molecules are generally significantly less sensitive than corresponding NOESY experiments and, in general, have more in common with NOESY experiments on small molecules. Although ROESY experiments are more sensitive for large molecules than for small, the results of ROESY experiments are generally much less dependent on molecular tumbling rate than are those of NOESY experiments, even away from the region $\omega_0 \tau_c = 1.12$.

Before leaving this section, there is one other central question to consider about the ROESY experiment, namely, why is the spin-locking field required at all? To answer this, consider an experiment in which a given resonance S is first selectively inverted, then all the resonances are excited by a 90° pulse and allowed to evolve for a period τ before the FID is acquired (Fig. 9.13a). If

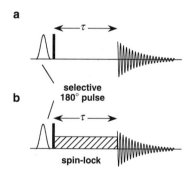

Figure 9.13. (*a*) Hypothetical 1D transverse NOE experiment with no spin-locking. A resonance is selectively inverted, then a nonselective 90° pulse is applied and evolution continues for a period τ before acquisition. This does not result in any transverse enhancement (see text). (*b*) Simple 1D ROESY experiment. This is similar to experiment (a), except that a spin-locking field is applied during the period τ.

another data set is then acquired, identical except that no selective inversion is applied, a difference spectrum could be computed from the two data sets. Aside from the very considerable phase distortions that the free precession prior to acquisition would cause, the resulting difference spectrum would be analogous to one obtained from a normal transient NOE difference experiment (Section 7.5.2), where the period τ corresponds to the time during which enhancements would evolve in the normal experiment. However, for our hypothetical transverse experiment, no enhancements would be seen, and it is important to understand why this is so. For any cross-relaxation partner, I, of the inverted resonance, the difference in the precession frequencies of I and S implies that there must necessarily be a constantly changing phase relationship between the vectors representing their magnetizations during the time τ. On average, they will therefore spend as much time with the colinear components of their magnetizations *opposed* as with them aligned. Thus, although transverse dipole–dipole cross-relaxation will occur between I and S, *no net enhancement can accrue*, because for half the time (whenever the colinear components of the I and S magnetizations are opposed) cross-relaxation increases the intensity of I, while for the other half (whenever the colinear components of the I and S magnetizations are aligned) it decreases the intensity of I.

We may now see the necessity for the spin-locking field. By abolishing the *relative* precession of the resonances, even though they still precess at ν_0, a constant phase relationship is imposed between them. This allows enhancements to build up consistently in one direction, as in the normal NOE experiment. In fact, the 1D ROESY experiment is nothing more than our hypothetical experiment of the previous paragraph, except that the period of free precession is replaced by a period of spin-locking (Fig. 9.13b).

9.3.1.3. Other Effects During Spin-Locking.
Several effects other than the ROE itself can occur during the spin-locking period of a ROESY experiment and it is these that make the interpretation of ROESY spectra potentially more complicated than that of NOESY experiments. The additional effects that we need to consider are (i) longitudinal cross-relaxation (i.e., the normal NOE), (ii) "TOCSY" type transfer through J-couplings, and (iii) chemical exchange. In addition, cross peaks can arise through multistep pathways involving various combinations of these mechanisms.

Longitudinal cross-relaxation occurs in ROESY experiments as a consequence of resonance offset effects. As pointed out earlier, practicable RF field strengths for spin-locking fields in ROESY experiments are generally set in the range approximately 1–10 kHz, and at these field strengths the tilt angle of the effective field seen by off-resonant signals is appreciable. For instance, for a signal 5 ppm away from the transmitter frequency on a 500 MHz spectrometer, the tilt angle in a 10 kHz field would be 76°, while for a 2 kHz field it would be 39° (cf. Eq. 1.18); recall that the tilt angle varies from 90° for an on-resonance signal to a value of 0° for a signal that is completely off-resonance). The consequence is that when off-resonant signals are spin-locked in a ROESY experiment, the corresponding magnetizations have both transverse and longitudinal components. It turns out that these components relax completely independently of one another, so that the overall relaxation behavior comprises simply the addition of a longitudinal part and a transverse part. One may thus define an effective cross-relaxation rate σ_{eff}, which is a composite of transverse and longitudinal terms, depending on the tilt angles θ_I and θ_S of the effective fields seen by I and S, respectively:[82]

$$(\sigma_{\text{eff}})_{IS} = \sin\theta_I \sin\theta_S \sigma_2 + \cos\theta_I \cos\theta_S \sigma_1 \qquad (9.7)$$

Clearly, σ_{eff} will take different values for different pairs of signals, depending on their chemical shifts in the spectrum. Allard et al.[86] have presented a useful formalism for calculating the time-course of magnetization in off-resonance ROESY experiments based on the "homogeneous master equation" approach (cf. Section 1.5). Their method is necessarily more computer intensive (by a factor of 27) than the relaxation matrix calculations used for purely longitudinal relaxation (cf. Section 4.3), because it must treat all three magnetization components for every spin (it therefore also requires all chemical shifts to be provided in the input data). However, as it scales linearly with the number of spins in the system, it is nonetheless possible to apply this approach to large molecules. In its simplest form the method neglects effects due to scalar couplings (i.e., TOCSY transfer; see following). A complete version of the theory is also presented that includes scalar couplings, but this does not scale linerly and thus is not easily applied to large molecules.[86]

The other important effect during spin-locking is so-called "TOCSY" transfer, also called "HOHAHA" transfer (for *HO*monuclear *HA*rtmann-*HA*hn), which occurs through one or several homonuclear *J*-couplings. (Note however that Hartmann–Hahn transfer in solids, from which the name HOHAHA was derived, is mediated by heteronuclear dipole–dipole coupling, rather than homonuclear scalar coupling as in this case). The full theory of TOCSY transfer is beyond the scope of this book, but essentially it arises because, during the spin-locking period, the spins are forced to precess at the same frequency as their various coupling partners, and therefore behave as though they were all strongly coupled together. This results in mixing of the wavefunctions and allows a net transfer of magnetization between the coupled spins that would have been forbidden had the spins precessed at their normal (different) frequencies. In a 2D TOCSY experiment, this results in cross peaks connecting coupling partners over either one coupling or a chain of couplings.

Another aspect of TOCSY transfer that we should consider is that it is oscillatory in nature and passes through intermediate antiphase states. Figure 9.14 shows how, following a selective inversion of one signal in an AX spectrum, TOCSY transfer causes the inversion to oscillate between the coupling partners as a function of the spin-locking period τ. At the midpoint of this oscillation, both signals appear in the spectrum in antiphase dispersion, and it turns out that this corresponds to magnetization that existed as spin-locked zero quantum coherence at the moment that the spin-locking field ceased. In multispin systems the situation is far more complex (and can only be analyzed numerically), and the transfer of magnetization amongst the various spins depends on the spin-locking field strength, the individual offsets of all the signals, the magnitudes of all the J-couplings, the spin-system topology, and the spin-locked relaxation behavior of all the spins.[87] However, although the transfer of magnetization around the spin system is complicated, similar principles apply. Thus TOCSY transfer in ROESY spectra may cause double-absorption cross peaks with an inphase multiplet structure (if the magnetization happened to be at the end of a period of oscillation at the end of the spin-lock period), or double-dispersion cross peaks with an antiphase multiplet structure (if the magnetization was at the midpoint of an oscillation at the end of the spin-lock period), or some mixture of the two. Note that spin-locked zero-quantum contributions are usually more intense in ROESY spectra than in TOCSY, since, in TOCSY, measures are usually taken to suppress zero-quantum coherences (e.g., using z-filters flanking the spin-lock)[88] whereas in ROESY they are not.

A key requirement for TOCSY transfer to occur is that the effective fields experienced by the coupling partners should be the same. For a weak CW (i.e., constant phase and continuous) spin-locking field, this condition is met for pairs of signals whose cross peaks lie close to the diagonal or the anti-diagonal of the spectrum, since in these regions the tilt angles θ_I and θ_S will be similar (cf. Fig. 9.14a). Away from these regions, TOCSY transfer becomes less efficient (cf. Fig. 9.14b and c), but the stronger the spin-locking field becomes, the more the regions of efficient TOCSY transfer along the diagonal and anti-diagonal will "expand" into the rest of the spectral area. In general, the condition for effective TOCSY transfer with a weak CW spin-locking field is given by[89]:

$$\frac{2\pi|(\nu_A - \nu_0)^2 - (\nu_X - \nu_0)^2|}{\gamma|\mathbf{B}_1|} < |J_{AX}| \qquad (9.8)$$

In TOCSY experiments, the whole intention is to produce TOCSY transfer between spins throughout the spectrum, regardless of offset, so weak CW fields are not very suitable. Instead, TOCSY experiments employ stronger spin-locking fields (typically 10 kHz or so, with a transmitter frequency in the centre of the spectrum) and composite-pulse decoupling schemes such as MLEV, WALTZ, or DIPSI in order to match the effective fields seen by a much greater proportion of the signals. However, in ROESY experiments, TOCSY transfer is an unwanted complication. Therefore its effects are often minimized by deliberately choosing a weak CW spin-locking field (typically 2 kHz or so) and placing the transmitter frequency somewhat to one side of the spectral area of interest. Note that when this is done, no cross peaks can lie close to the anti-diagonal (since by definition the anti-diagonal crosses the diagonal at the transmitter frequency, which is outside the region containing peaks). The only pairs of signals that can undergo mutual TOCSY transfer are then limited to those giving cross peaks close to the diagonal.

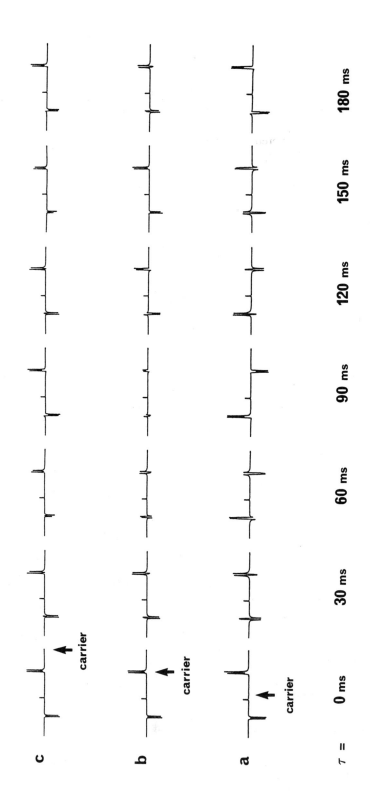

Putting all this together, we may see that elimination of TOCSY transfer and elimination of NOE cross-relaxation are fundamentally contradictory goals.[90] When the spin-locking field is made strong, NOESY cross-relaxation is disfavored so that cross-relaxation is predominantly due to the ROE, but TOCSY transfer is favored. Conversely, when weaker spin-locking fields are used, TOCSY transfer is not a problem, but cross-relaxation may include a substantial longitudinal component and there will be substantial resonance-offset effects on ROE intensities.

The consequences of this depend largely on the tumbling regime of the solute. Probably the majority of applications of ROESY are to molecules of "intermediate" size (i.e., $\omega_0 \tau_c \approx 1$) where conventional NOE experiments give very little signal. In these circumstances, contributions from longitudinal cross-relaxation are by definition not significant, and the best solution is to run ROESY with a weak off-resonance field (or some variant of this; see Section 9.3.2) so as to suppress TOCSY transfer. For small molecules in the positive NOE regime, it makes little sense to run a ROESY spectrum in any case; as we saw in Section 9.3.1.2, the behavior of longitudinal and transverse relaxation become identical in the extreme narrowing limit, so the NOESY and ROESY spectra would be identical except for the presence of additional artifacts (such as TOCSY transfer) in the case of ROESY. The most difficult case arises for large molecules. As we saw in Section 9.3.1.2, ROESY experiments are likely to be less sensitive than NOESY when $\omega_0 \tau_c \gg 1$, and the motive for running ROESY is likely to be to obtain cross-relaxation rates free of spin-diffusion. These are just the circumstances under which one would least wish

←
_____

Figure 9.14. One dimensional experiment to demonstrate TOCSY transfer in an AX system. The downfield doublet is selectively inverted, then all magnetization is made transverse using a $90°_x$ pulse, and a continuous y phase spin-locking field is applied for a time τ prior to recording the FID. In (a), the transmitter is placed centrally between the two signals, resulting in equal effective fields for the two signals. The inversion oscillates between A and X at full amplitude as a function of τ, achieving complete transfer at a time $\tau \simeq 1/2J$ (= 100 ms). In (b), the transmitter is placed on-resonance with the upfield doublet, resulting in a difference of effective field strength ($|\gamma \mathbf{B}_{\text{eff}}^A/2\pi| - |\gamma \mathbf{B}_{\text{eff}}^X/2\pi|$) of about 4.5 Hz between the two signals. The oscillation is reduced to about half its original amplitude, and is slightly faster. In (c), the transmitter was placed upfield of the X resonance by an equivalent amount again, resulting in a difference of effective field strengths of about 9 Hz. The oscillation is much reduced in amplitude (it is no longer sufficient to cause the resonances to change sign), and is noticeably faster than in (a) or (b). Spectra were recorded at 300 MHz for 3-methylthiophene-2-carboxylic acid in CDCl$_3$ solution, plotting the region 7.7–6.7 ppm in all cases ($J_{4,5}$ = 5.0 Hz). A transmitter field strength of $\gamma B_1/2\pi$ = 2.5 kHz was used for the hard pulses and spin-locking, and a 50 ms decoupler pulse was used for the selective inversion. The small signal between the doublets arises from residual CHCl$_3$.

376 OTHER DEVELOPMENTS

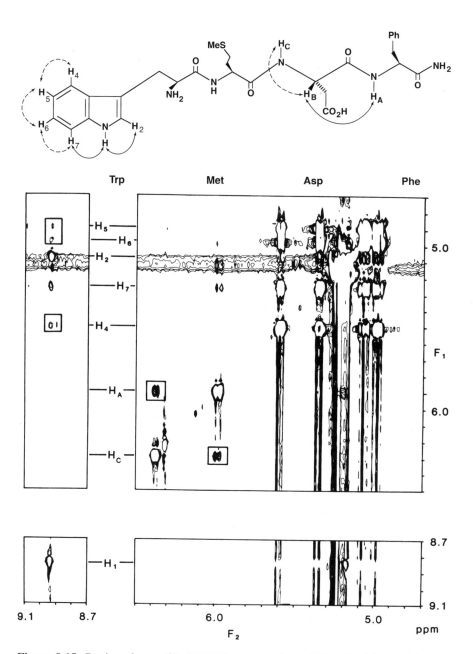

Figure 9.15. Regions from a 2D ROESY spectrum (τ_m = 200 ms) of the peptide Trp-Met-Asp-Phe-NH$_2$ in dimethylsulfoxide-d_6 solution. Positive and negative levels are here plotted without distinction, but were of course distinguishable in the original (colored) plot. All genuine ROE and TOCSY-relayed ROE cross peaks are negative, whereas diagonal and TOCSY transfer peaks are positive. Several TOCSY-relayed ROE peaks are visible (boxed in the figure). Those connecting the Asp3 NH and Phe4 NH

to mix in a proportion of longitudinal relaxation. Possible approaches to this problem in terms of practical pulse sequences are considered in Section 9.3.2.

Exchange contributions behave identically in ROESY and NOESY, so that, rather like TOCSY transfer, they give rise to cross peaks with the same phase and sign as the diagonal. For large molecules, this means that cross peaks due to exchange are easier to identify in ROESY (where they have opposite sign to the ROE cross peaks) than they are in NOESY (where they have the same sign as the NOE cross peaks). Further, by switching between ROE and NOE (i.e., transverse and longitudinal) cross-relaxation during the mixing time in a ratio of 2:1 (see Section 9.3.1.2), cross-relaxation is strongly suppressed, giving rise to 2D spectra in which cross peaks are almost purely due to exchange.[91] A similar idea lies behind the CBD-NOESY experiment mentioned in Section 8.6.1.

Finally, we must mention multistep pathways. In NOESY, multistep pathways are limited to spin-diffusion peaks (which in this context we could call NOE–NOE transfer) and combinations of saturation transfer with NOE transfer (cf. Section 5.2). Similar contributions can also occur in ROESY, but in ROESY there is the additional possibility of multistep pathways involving both ROE and TOCSY transfer, which give rise to so-called "false" or "TOCSY-relayed" ROE cross peaks.[92] Such pathways involve both a through-space step and a through-coupling step, so they have the potential to complicate interpretation severely if any cross peaks originating from TOCSY-relayed ROE transfer are not correctly identified. To make matters worse, TOCSY-relayed ROE cross peaks have an essentially identical appearance to genuine ROE cross peaks, having the same sign, phase, and multiplet structure as the latter. Figure 9.15 shows a case where several TOCSY-relayed ROE peaks were observed in the ROESY spectrum of a tetrapeptide, and it is clear that a naïve interpretation of these cross peaks (particularly those linking Asp3 NH to Phe4 NH) could easily have led to completely erroneous conclusions about the conformation of this molecule.

The clue to the recognition and suppression of TOCSY-relayed ROE cross peaks lies in the offset dependence of the TOCSY transfer step. As discussed

---

signals result from a TOCSY step linking Asp3 NH to Asp3 H$^\alpha$ (dashed arrow on structure) and an ROE step linking Asp3 H$^\alpha$ to Phe4 NH (solid arrow on structure). This particular TOCSY transfer pathway is efficient because the carrier was placed almost exactly midway between the Asp3 NH and H$^\alpha$ signals. If the carrier is placed elsewhere, the TOCSY transfer between Asp3 NH to Asp3 H$^\alpha$ becomes inefficient and the TOCSY-relayed ROE cross peak linking Asp3 NH to Phe4 NH disappears. The TOCSY-relayed ROE peaks connecting Trp H$_1$ to H$_4$, H$_5$, and H$_6$ on the Trp indole ring result from an ROE step linking H$_1$ and H$_7$ (solid arrow on structure) and a TOCSY transfer amongst the H$_4$, H$_5$, H$_6$ and H$_7$ signals (dashed arrows on structure), all of which have similar offsets. Spectra were recorded at 300 MHz. Reproduced with permission from Ref. 92.

previously, TOCSY transfer with a weak CW field is only efficient near the diagonal and anti-diagonal of the 2D spectrum, so that the spins linked by a TOCSY step must all lie within these regions if TOCSY-relayed ROE is to occur. Thus, by using a weak off-resonant CW field to minimize TOCSY transfer, one simultaneously minimizes the extent of TOCSY-relayed ROE transfer and, in cases of difficulty, one may check further by running experiments using different spin-lock frequencies. One particular reason (with hindsight!) why TOCSY-relayed cross peaks were observed in the spectrum of Figure 9.15 is that the experiment employed a transmitter frequency placed at the center of the spectrum.

To sum up, the various types of cross peak that may occur in a ROESY spectrum where the diagonal is phased for positive double-absorption are as follows:

1. *Transverse NOE transfer* (negative, double-absorption cross peaks with an in-phase multiplet structure). These occur between protons close together in space, and their detection is the object of the ROESY experiment.
2. *TOCSY transfer* (these can vary between positive, double-absorption cross peaks with an in-phase multiplet structure to double-dispersion cross peaks with an antiphase multiplet structure). These occur between protons within a common scalar coupling network and can correspond either to multiple relay steps or to transfers between direct coupling partners. For ROESY experiments using a weak CW spin-locking field, these are generally confined to cross peaks near the diagonal or anti-diagonal of the spectrum. If the weak spin-locking field is applied outside the region containing signals, TOCSY transfer is only relevant for cross peaks near the diagonal.
3. *Saturation transfer* (positive, double-absorption cross peaks with an in-phase multiplet structure). These occur via chemical or conformational exchange processes in just the same way as in conventional NOE experiments.
4. *Two-step transverse NOE transfer* (positive, double-absorption cross peaks with an in-phase multiplet structure). These are directly analogous to the three-spin effects seen in normal NOE experiments with small molecules (Section 3.2.2.2). They are expected to be appreciable only for large molecules having short $T_{1\rho}$ values, in other words the sort of molecules for which spin diffusion is appreciable in conventional experiments. Detectable ROE pathways over more than one intermediate spin are rare.
5. *Other multistep transfers* (properties depend on pathway; e.g., a combination of ROE and TOCSY transfer would give negative, double-absorption cross peaks with an in-phase multiplet structure). Although they are not very common, these are potentially rather insidious, since

they can have very similar properties to genuine, single-step ROE cross-peaks.

9.3.2. Practicalities

In principle, any experiment that involves an NOE mixing period (i.e., any transient NOE experiment) can be modified to include instead an ROE mixing period. There are thus potentially a very large number of experiments based on ROESY, (e.g., combining a ROESY step with a through coupling relay step,[93] or with a heteronuclear editing step[54]) and it would be impractical to try to discuss all of them here. Instead we shall examine the simple 1D and 2D ROESY experiments, concentrating on the hardware requirements and the different techniques used during spin-locking to try to achieve an optimal balance between the various mixing processes discussed in Section 9.3.1.3.

Before going on, however, three important points should be made about ROESY experiments. First, one should note that even 1D ROESY is necessarily always a *transient* experiment. There can never be a rotating frame analog of the steady-state NOE experiment, because the spin-locking abolishes the relative precession of the resonances, so that there is no possibility of selective irradiation during evolution of the transverse enhancements. Second, 2D ROESY (or experiments derived from it) should *always* be carried out in the phase-sensitive mode. As we have seen, interpretation of ROESY spectra relies heavily on a knowledge of the relative signs (and sometimes phases) of the various contributing signals, and to throw this information away needlessly by using absolute value data processing would be foolish in the extreme. Third, there is no heteronuclear analog of the ROESY experiment. This is because the requirement of the spin-locking field is that it be strong enough to suppress the relative precession of all the signals. In a heteronuclear case, that would require a $\mathbf{B}_1$ field strength in the order of tens or hundreds of megahertz, which is some three orders of magnitude more powerful than it is practical to apply.

When the ROESY experiment was first introduced, it placed heavy demands on the hardware of the commercial spectrometers that were then available. More recent spectrometers often have very fast power switching and continuously rated linear amplifiers available as standard, so that implementation of the ROESY sequence becomes essentially trivial. However, for the benefit of owners of older spectrometers, we shall briefly review some of the issues that can arise and how they may be tackled.

The first question is where to get the spin-locking field itself. The essential requirements of the spin-locking field are (i) it must be phase-coherent with the receiver,[94] (ii) it must have the same frequency as the excitation pulse and be phase coherent with it, and (iii) the RF circuits, especially the amplifiers and the probe, must be able to cope with the power level. The simplest solution, when possible, is to use the same RF channel for the spin lock as for the excitation pulses, since this automatically guarantees that points (i) and (ii) are taken care of. However, on some older spectrometers the transmitter channel may have very limited facilities for power control. If a reduced, medium-level power output is available, for instance from bypassing the final stage of amplification, it may be possible to use this in a continuous mode to provide the spin-locking field; however, one is then often compelled to use the same medium-level power for the excitation pulses. Nonetheless, this can often represent an acceptable compromise. Alternatively, the normal full transmitter power may be used to provide an intermittent spin-locking field with an acceptably low duty cycle (ca. 10% in practice).[95] This may require some modification of the circuits that are designed to protect the hardware from abuse, and clearly any such approach requires caution.

If the transmitter channel cannot be used to supply the spin-locking field, another channel must be used, and on older spectrometers there is usually a channel specifically designed to supply continuous RF at a wide variety of power settings for decoupling. It may be possible to use this decoupler channel to provide both spin-locking field and excitation pulses, either at the same power level as one another or, if fast switching is available, using a higher power for the excitation pulses. However, with this approach it is essential that the decoupler be made phase-coherent with the receiver, which on some spectrometers may require a modification. Another alternative is to use the decoupler for the spin-locking field and the transmitter for the excitation pulses. This approach avoids any need for fast switching, and gives freedom to set the spin-locking power as desired, but it requires that both channels be phase-coherent with one another and the receiver (again, this may require hardware modification), and also that the two channels can be combined before entering the probe (e.g., using a power combiner or a low-loss high-power directional coupler).

Having considered hardware requirements, the next question is what pulse sequence to use. Figure 9.16 shows various pulse sequences for 2D ROESY experiments. In the original sequence introduced by Bothner-By (Fig. 9.16a), mixing is achieved using a continuous period of constant-phase spin-locking.[80] One of the first improvements was the introduction of hard 90° pulses flanking the spin-lock (with the same phase as the spin-lock) in order to reduce resonance offset effects (Fig. 9.16b).[96] As explained earlier, off-resonant signals in ROESY experiments experience a tilted spin-locking field, and only the component parallel to that tilted field survives. Similarly, when the spin-locking field ceases, the surviving magnetization now lies along a tilted spin-lock axis, so only the projection of this magnetization onto the transverse plane gives rise to detectable signal. Overall, for a cross peak between signals I and S, these two projection processes cause an intensity loss given by

$$a'_{IS} = (\sin \theta_I \sin \theta_S) a_{IS} \qquad (9.9)$$

where a'_{IS} and a_{IS} are the cross-peak intensities with and without the projection loss, respectively, and θ_I and θ_S are the tilt angles of the two signals. The sequence of Figure 9.16b, proposed by Griesinger and Ernst,[96] recovers this

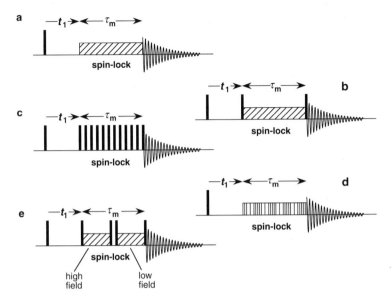

Figure 9.16. 2D ROESY pulse sequences. Sequence (*a*) is the original CAMELSPIN experiment of Bothner-By. Sequence (*b*) is an offset-compensated experiment, while sequences (*c*), (*d*) (T-ROESY), and (*e*) (JS-ROESY) are designed to reduce or eliminate the extent of TOCSY transfer. In general, the phase of the first pulse relative to the spin-locking field or pulses is not crucial, since it only controls the zero-order phase in F_1. In sequences (b) and (e), the phase of the spin-lock and the hard pulses around it are the same. In sequence (c), all hard pulses have identical phase and have flip angle $\ll \pi$. In sequence (d), the mixing sequence is either $180°_x 180°_{-x}$ or $180°_x 180°_{-x} 360°_x 360°_{-x} 180°_x 180°_{-x}$. In sequence (e), the spin-locking field is applied at different frequencies during the two halves of the mixing time, as indicated. Also, there are several variants of the pulse sequence that differ in the detail of how the offset compensation pulses at the center of the mixing time are applied. Details of these are given in Ref. 90.

loss completely by rotating the plane containing the signals through 90° immediately prior to (and following) the spin–locking period. Note, however, that the offset dependence of the cross-relaxation itself, as defined by Eq. 9.7, is not changed at all by this pulse sequence (or indeed by any pulse sequence).

The way this sequence works may be visualized by considering an off-resonant signal I with a tilt angle θ_I. Although the additional 90° pulses have no effect on intensities during the first t_1 increment, for some particular later increment the I magnetization will have precessed during t_1 through exactly the same angle as the tilt angle θ_I. In that particular t_1 increment, the first 90° pulse has the effect of turning the I magnetization exactly onto the spin-lock axis, so there is then no loss due to the first projection term for that t_1 increment. Comparing this situation with that for an on-resonance signal, both signals show an oscillatory variation of amplitude as a function of t_1. The only difference is that for the on-resonance signal, the first maximum in this oscillation is at $t_1 = 0$, whereas for an off-resonance signal, the first maximum is at some later time dependent on its offset. Thus the two cases differ only by an offset-dependent phase shift in F_1. Similarly, the second 90° pulse will rotate all the signals present in the xz plane at the end of the spin-locking period into the xy plane. This avoids amplitude losses due to the second projection term, replacing it with an offset-dependent phase gradient in F_2. Two practical considerations concerning this experiment are worth mentioning. First, it is of course completely pointless to attempt it if one is using 90° pulses at the same power level as the spin-locking field, perhaps due to hardware limitations on an older spectrometer. Second, one needs to be aware that there may be an absolute phase difference between the hard pulses and the spin-locking field, arising due to the different amplification and/or wiring paths followed by the RF at the different power levels. Any such phase difference would need to be determined experimentally and accounted for when setting up the ROESY experiment.

All of the remaining sequences in Figure 9.16 were proposed with the aim of suppressing or reducing the extent of TOCSY transfer in ROESY experiments. The sequence shown in Figure 9.16c was proposed by Kessler et al.[95] but soon afterwards it was shown that such an interrupted spin-lock is fully equivalent to a continuous spin-lock at lower power,[97] and therefore offers no advantage in suppressing TOCSY transfer [specifically, an interrupted spin-lock with pulses of length τ_p and flip angle θ separated by gaps of duration τ is fully equivalent to a continuous spin-locking field of strength $\gamma B_1/2\pi = \theta/(\tau_p + \tau)$]. The sequence called T-ROESY (Fig. 9.16d) was proposed by Hwang and Shaka, and this is effective at suppressing TOCSY transfer. However, cross-relaxation rates in T-ROESY are an equal mixture of ROE and NOE components, and also, the cross-relaxation rates measured are four times lower than those in conventional ROESY.[99] The sequence called JS-ROESY (for "Jump-Symmetrized ROESY"; Fig. 9.16e) is a variation on an earlier sequence (Fig. 9.16b). For JS-ROESY, a weak, off-resonance spin-locking field is used so as to minimize TOCSY transfer, but for the first half of the mixing period spin-locking is applied off-resonance to one side of the spectrum, while for the second half it is applied off-resonance to the other side.[90] This leads to a much reduced offset dependence of cross-peak total intensity (relative to the sequence of Fig. 9.16b, using the same spin-locking field strength and offset magnitude), but it does not alter the fact that cross peaks arise due to an offset-dependent mixture of ROE and NOE transfer. The hard pulses shown in Figure 9.16e, like those in Figure 9.16b are used to avoid the intensity loss associated with the projection terms in Eq. 9.9.

9.4. MANIPULATION OF $\omega\tau_c$

Both the steady-state NOE and the rate of buildup and decay of the NOE are very dependent on τ_c, the rotational correlation time, and on ω, the Larmor frequency. In Section 2.2 it was shown how the NOE goes through zero at $\omega\tau_c \approx 1.12$, and in Chapter 4 it was shown how the cross-relaxation rate σ does the same, but that ρ increases fairly steadily with $\omega\tau_c$. Thus, it is useful to be able to vary $\omega\tau_c$, so as to avoid the condition $\omega\tau_c \approx 1.12$, and also to make the NOE growth rate faster, if $\omega\tau_c$ is small. For kinetic NOE experiments designed to measure buildup rates, it is clear that the NOE time-course varies markedly with $\omega\tau_c$, and that the most convenient values to have are probably in the range $0.1 < \omega\tau_c < 10$, excepting of course values too close to $\omega\tau_c \approx 1.12$. This section discusses ways of altering $\omega\tau_c$.

Larger variations of $\omega\tau_c$ can normally be achieved by varying τ_c than by varying ω. Nevertheless, changes in ω can be very effective in avoiding the condition $\omega\tau_c \approx 1.12$. Obviously, ω can be altered by changing the field strength: thus, for example, an η_{max} of zero at 400 MHz can be converted to a respectable η_{max} of -12% at 500 MHz. In a sense, the rotating frame experiment (Section 9.3) represents another way in which ω can be changed, since it is largely ω_1 rather than ω_0 that determines the behavior of the NOE in this experiment.

There can, however, be more scope for variation by changing τ_c. This depends on the effective microscopic viscosity of the solvent and on the size of the molecule, approximately following the Debye equation

$$\tau_c = 4\pi\eta a^3/3kT \tag{9.10}$$

where η is the viscosity of the solvent and a is the radius of the molecule. There are therefore three variables that can be altered, namely, η, a, and T. Of these, η and T are interdependent, and the simplest way of changing the viscosity is to change the temperature (Section 3.2.3). The viscosity of water decreases by a factor of 6 between 0°C and 100°C. Less polar solvents tend to change rather less, and some highly associated solvents change more. More substantial changes in viscosity may be achieved by varying the solvent (Section 7.1.1), as shown in Table 9.2. Use of the very viscous solvents, often as

TABLE 9.2   Viscosity of Solvents at 25°C (cP)

Acetone	0.32
Chloroform	0.54
Methanol	0.55
Water	0.89
Dimethyl sulfoxide	1.98
Dimethyl sulfoxide/water 2:3	3.80
Tetramethylene sulfone	9.03[a]
Ethylene glycol	17.0
Voltalef 10S[b]	1550

[a] At 35°C.
[b] This is a vacuum oil, formula $(CFCl.CCl_2)_n$, unfortunately no longer available.[100]

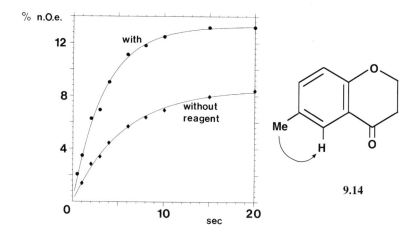

Figure 9.17. Buildup of the enhancement shown on structure **9.14** in the presence and absence of the diamagnetic reagent La(FOD)$_3$. Reproduced with permission from Ref. 107.

cosolvents rather than on their own, has proved very useful for kinetic experiments on small molecules, which would otherwise have been very slow.[100–102] The slightly less viscous solvents are useful for molecules that would be close to $\omega\tau_c \approx 1.12$ in normal solvents.[103–105] One should always be aware, however, that a change in solvent may lead to a change in conformation. It has been pointed out that solvents with decreased hydrogen bonding ability, such as tetramethylene sulfone, encourage the formation of intramolecular hydrogen bonding, and therefore induce more structure in the solute than would be present in more polar solvents.[104,106]

A more subtle, but less generally applicable, way to increase τ_c for a small molecule is to bind it, preferably noncovalently, to a larger and therefore more slowly tumbling molecule. The nature of the slowly tumbling molecule is not critical, provided that it binds to and does not alter the conformation of the molecule of interest. Molecules that have been suggested include a metalloporphyrin (which can coordinate to nucleophilic groups), a cyclodextrin or crown ether, which can form inclusion complexes, and a diamagnetic reagent.[107] An example is shown in Figure 9.17. In the presence of La(FOD)$_3$, not only do enhancements in compound **9.14** grow faster (and decay faster), but they also become larger, because other relaxation mechanisms involved in leakage become relatively less important. This experiment is called RABBIT (*R*elaxation *A*ided *By* *BI*nding *T*ightly).

REFERENCES

1. Jones, A. J.; Grant, D. M.; Kuhlmann, K. F. *J. Am. Chem. Soc.* 1969, *91*, 5013.
2. Tori, K.; Nishikawa, J.; Takeuchi, Y. *Tetrahedron Lett.* 1981, *22*, 2793.

3. Shapiro, M. J.; Kolpak, M. X.; Lemke, T. L. *J. Org. Chem.* 1984, *49*, 187.
4. Kay, L. E.; Torchia, D. A.; Bax, A. *Biochemistry* 1989, *28*, 8972.
5. Neuhaus, D.; van Mierlo, C. P. M. *J. Magn. Reson.* 1992, 100, 221.
6. Dayie, K. T.; Wagner, G. *J. Magn. Reson. A* 1994, *111*, 121.
7. Rinaldi, P. L. *J. Am. Chem. Soc.* 1983, *105*, 5167.
8. Yu, C.; Levy, G. C. *J. Am. Chem. Soc.* 1984, *106*, 6533.
9. Kövér, K. E.; Batta, G. *J. Magn. Reson.* 1986, *69*, 344.
10. Bigler, P.; Müller, C. *J. Magn. Reson.* 1988, *79*, 45.
11. Müller, C.; Bigler, P. *J. Magn. Reson.* 1989, *84*, 585.
12. Kövér, K. E.; Batta, G. *J. Magn. Reson.* 1988, *79*, 206.
13. Stott, K.; Keeler, J. *Magn. Reson. Chem.* 1996, *34*, 554.
14. Leon, V.; Bolivar, R. A.; Tasayco, M. L.; Gonzales, R.; Rivas, C. *Org. Magn. Reson.* 1983, *21*, 470.
15. Khaled, M. A.; Watkins, C. L. *J. Am. Chem. Soc.* 1983, *105*, 3363.
16. Loosli, H.-R.; Kessler, H.; Oschkinat, H.; Weber, H.-P.; Petcher, T. J.; Widmer, A. *Helv. Chim. Acta* 1985, *68*, 682.
17. Gibbons, W.; Mascagni, P.; Niccolai, N.; Rossi, C. in "Peptides 1984," Proc. 18th Eur. Pept. Symp., Djurönäset, Sweden (U. Ragnarsson, ed.), Almqvist and Wiksell, Stockholm, 1984, p. 555.
18. Neidlein, R.; Kramer, W.; Ullrich, V. *Helv. Chim. Acta* 1986, *69*, 898.
19. Aldersley, F. M.; Dean, F. M.; Mann, B. E. *J. Chem. Soc., Chem. Commun.* 1983, 107.
20. Kakinuma, K.; Imamura, N.; Ikekawa, N.; Tanaka, H.; Minami, S.; Omura, S. *J. Am. Chem. Soc.* 1980, *102*, 7493.
21. Sánchez-Ferrando, F. *Magn. Reson. Chem.* 1985, *23*, 185.
22. Cativiela, C.; Sánchez-Ferrando, F. *Magn. Reson. Chem.* 1985, *23*, 1072.
23. Kövér, K. E.; Batta, G. *J. Am. Chem. Soc.* 1985, *107*, 5829.
24. Kövér, K. E.; Batta, G. *J. Chem. Soc., Chem. Commun.* 1986, 647.
25. Ando, M. E.; Gerig, J. T. *Biochemistry* 1982, *21*, 2299.
26. Hammond, S. J. *J. Chem. Soc., Chem. Commun.* 1984, 712.
27. Cairi, M.; Gerig, J. T. *J. Magn. Reson.* 1985, *62*, 131.
28. Bell, R. A.; Saunders, J. K. in "Topics in Sterochemistry" (N. L. Alinger and E. L. Eliel, eds.), Vol. 7, Wiley-Interscience, New York, 1973, p. 86.
29. Avent, A. G.; Eaborn, C.; El-Kheli, M. N. A.; Molla, M. E.; Smith, J. D.; Sullivan, A. C. *J. Am. Chem. Soc.* 1986, *108*, 3854.
30. Kotovych, G.; Aarts, G. H. M. *Can. J. Chem.* 1980, *58*, 2649.
31. Hunter, B. K.; Hall, L. D.; Sanders, J. K. M. *J. Chem. Soc., Perkin Trans. 1* 1983, 657.
32. Grode, S. H.; Mowery, M. R. *J. Magn. Reson.* 1997, *126*, 142.
33. Akasaka, K.; Konrad, M.; Goody, R. S. *FEBS Lett.* 1978, *96*, 287.
34. Barnard, G. N.; Sanders, J. K. M. *J. Magn. Reson.* 1987, *74*, 372.
35. Lapidot, A.; Irving, C. S. *Proc. Natl. Acad. Sci. U.S.A.* 1977, *74*, 1988.
36. Richarz, R.; Wüthrich, K. *J. Magn. Reson.* 1978, *30*, 147.

37. van Mierlo, C. P. M.; Darby, N. J.; Neuhaus, D.; Creighton, T. E. *J. Mol. Biol.* 1991, *222*, 353.
38. Nikonowicz, E. P.; Pardi, A. *Nature* 1992, *355*, 184.
39. Zimmer, D. P.; Crothers, D. M. *Proc. Natl. Acad. Sci. U.S.A.* 1995, *92*, 3091.
40. McIntosh, L. P.; Dahlquist, F. W. *Quart. Rev. Biophys.* 1990, *23*, 1.
41. LeMaster, D. M. *Prog. Nucl. Magn. Reson. Spectrosc.* 1994, *26*, 371.
42. Keeler, J.; Clowes, R. T.; Davis, A. L.; Laue, E. D. *Methods in Enzymology* 1994, *239*, 145.
43. Gronenborn, A. M.; Clore, G. M. *Crit. Rev. Biochem. Mol. Biol.* 1995, *30*, 351.
44. Pelton, J. G.; Wemmer, D. E. *Ann. Rev. Phys. Chem.* 1995, *46*, 139.
45. Otting, G.; Wüthrich, K. *Quart. Rev. Biophys.* 1990, *23*, 39.
46. Folmer, R. H. A.; Hilbers, C. W.; Konings, R. N. H.; Hallenga, K. *J. Biomolec. NMR* 1995, *5*, 427.
47. Zwahlen, C.; Legault, P.; Vincent, S. J. F.; Greenblatt, J.; Konrat, R.; Kay, L. E. *J. Am. Chem. Soc.* 1997, *119*, 6711.
48. Burgering, M.; Boelens, R.; Kaptein, R. *J. Biomolec. NMR* 1993, *3*, 709.
49. Mutzenhardt, P.; Bodenhausen, G. *J. Magn. Reson.* 1998, *132*, 159.
50. Zolnai, Z.; Juranic, N.; Markley, J. L.; Macura, S. *Chem. Phys.* 1995, *200*, 161.
51. Allain, F. H.-T.; Gubser, C. C.; Howe, P. W. A.; Nagai, K.; Neuhaus, D.; Varani, G. *Nature* 1996, *380*, 646.
52. Werner, M. H.; Huth, J. R.; Gronenborn, A. M.; Clore, G. M. *Cell* 1995, *81*, 705.
53. Gordon-Smith, D. Personal communication.
54. Suzuki, M.; Neuhaus, D.; Gerstein, M.; Aimoto, S. *Protein Engineering* 1994, *7*, 461.
55. Griesinger, C.; Sørensen, O.W.; Ernst, R. R. *J. Am. Chem. Soc.* 1987, *109*, 7227.
56. Clore, G. M.; Kay, L. E.; Bax, A.; Gronenborn, A. M. *Biochemistry* 1991, *30*, 12.
57. Vuister, G. W.; Clore, G. M.; Gronenborn, A. M.; Powers, R.; Garrett, D. S.; Tschudin, R.; Bax, A. *J. Magn. Reson. B* 1993, *101*, 210.
58. Kawabata, J.; Fukushi, E.; Mizutani, J. *J. Am. Chem. Soc.* 1992, *114*, 1115.
59. Wagner, R.; Berger, S. *Magn. Reson. Chem.* 1997, *35*, 199.
60. Homans, S. W. *J. Magn. Reson.* 1990, *90*, 557.
61. Oschkinat, H.; Cieslar, C.; Gronenborn, A. M.; Clore, G. M. *J. Magn. Reson.* 1989, *81*, 212.
62. Boelens, R.; Vuister, G. W.; Koning, T. M. G.; Kaptein, R. *J. Am. Chem. Soc.* 1989, *111*, 8525.
63. Oschkinat, H.; Cieslar, C.; Holak, T. A.; Clore, G. M.; Gronenborn, A. M. *J. Magn. Reson.* 1989, *83*, 450.
64. Vuister, G. M.; Boelens, R.; Kaptein, R. *J. Magn. Reson.* 1988, *80*, 176.
65. Wagner, G. *J. Magn. Reson.* 1984, *57*, 497.
66. Mooren, M. M. W.; Hilbers, C. W.; van der Marel, G. A.; van Boom, J. H.; Wijmenga, S. S. *J. Magn. Reson.* 1991, *94*, 101.
67. Wijmenga, S. S.; Heus, H. A.; Werten, B.; van der Marel, G. A.; van Boom, J. H.; Hilbers, C. W. *J. Magn. Reson. B* 1994, *103*, 134.
68. Oschkinat, H.; Cieslar, C.; Griesinger, C. *J. Magn. Reson.* 1990, *86*, 453.

69. Breg, J. N.; Boelens, R.; Vuister, G. W.; Kaptein, R. *J. Magn. Reson.* 1990, *87*, 646.
70. Habazettl, J.; Ross, A.; Oschkinat, H.; Holak, T. A. *J. Magn. Reson.* 1992, *97*, 511.
71. Noggle, J. H.; Schirmer, R. E. "The Nuclear Overhauser Effect; Chemical Applications," Academic Press, New York, 1971.
72. Arrowsmith, C. H.; Pachter, R.; Altman, R. B.; Iyer, S.B.; Jardetzky, O. *Biochemistry* 1990, *29*, 6332.
73. Nietlispach, D.; Clowes, R. T.; Broadhurst, R. W.; Ito, Y.; Keeler, J.; Kelly, M.; Ashurst, J.; Oschkinat, H.; Domaille, P. J.; Laue, E. *J. Am. Chem. Soc.* 1996, *118*, 407.
74. Moonen, C. T. W.; Scheek, R. M.; Boelens, R.; Müller, F. *Eur. J. Biochem.* 1984, *141*, 323.
75. Bargon, J.; Gardini, G. P. *J. Am. Chem. Soc.* 1979, *101*, 7732.
76. Hore, P. J.; Egmond, M. R.; Edzes, H. T.; Kaptein, R. *J. Magn. Reson.* 1982, *49*, 122.
77. Scheek, R. M.; Stob, S.; Boelens, R.; Dijkstra, K.; Kaptein, R. *J. Am. Chem. Soc.* 1985, *107*, 705.
78. Song, Y.-Q.; Goodson, B. M.; Taylor, R. E.; Laws, D. D.; Navon, G.; Pines, A. *Angew. Chemie, Int. Ed. Engl.* 1997, *36*, 2368.
79. Luhmer, M.; Goodson, B. M.; Song, Y.-Q.; Laws, D. D.; Kaiser, L.; Cyrier, M. C.; Pines, A. *J. Am. Chem. Soc.* 1999, *121*, 3502.
80. Bothner-By, A. A.; Stephens, R. L.; Lee, J.; Warren, C. D.; Jeanloz, R. W. *J. Am. Chem. Soc.* 1984, *106*, 811.
81. Bauer, C.J.; Frenkiel, T. A.; Lane, A. N. *J. Magn. Reson.* 1990, *87*, 144.
82. Davis, D. G. *J. Am. Chem. Soc.* 1987, *109*, 3471.
83. Abragam, A. "Principles of Nuclear Magnetism," Clarendon Press, Oxford, 1961, pp. 278ff; Keeler, J. H. personal communication.
84. Marion, D. *FEBS Lett.* 1985, *192*, 99.
85. Macura, S.; Ernst, R. R. *Mol. Phys.* 1980, *41*, 95.
86. Allard, P.; Helgstrand, M.; Härd, T. *J. Magn. Reson.* 1997, *129*, 19.
87. Chung, C.-W.; Keeler, J.; Wimperis, S. *J. Magn. Reson. A* 1995, *114*, 188.
88. Rance, M. *J. Magn. Reson.* 1987, *74*, 557.
89. Davis, D. G.; Bax, A. *J. Am. Chem. Soc.* 1985, *107*, 2820.
90. Schleucher, J.; Quant, J.; Glaser, S. J.; Griesinger, C. *J. Magn. Reson. A* 1995, *112*, 144.
91. Fejzo, J.; Westler, W. M.; Macura, S.; Markley, J. L. *J. Am. Chem. Soc.* 1990, *112*, 2574.
92. Neuhaus, D.; Keeler, J. H. *J. Magn. Reson.* 1986, *68*, 568.
93. Kessler, H.; Gemmecker, G.; Haase, B. *J. Magn. Reson.* 1988, *77*, 401.
94. Rance, M.; Cavanagh, J. *J. Magn. Reson.* 1990, *87*, 363.
95. Kessler, H.; Griesinger, C.; Kerssebaum, R.; Wagner, K.; Ernst, R. R. *J. Am. Chem. Soc.* 1987, *109*, 607.
96. Griesinger, C.; Ernst, R. R. *J. Magn. Reson.* 1987, *75*, 261.

97. Bax, A. *J. Magn. Reson.* 1988, *77*, 134.
98. Hwang, T.-L.; Shaka, A. J. *J. Am. Chem. Soc.* 1992, *114*, 3157.
99. Ravikumar, M.; Bothner-By, A. A. *J. Am. Chem. Soc.* 1993, *115*, 7537.
100. Williamson, M. P.; Williams, D. H. *J. Chem. Soc., Chem. Commun.* 1981, 165.
101. Williamson, M. P.; Neuhaus, D. *J. Magn. Reson.* 1987, *72*, 369.
102. Luck, L. A.; Landis, C. R. *Organometallics* 1992, *11*, 1003.
103. Bothner-By, A. A.; Johner, P. E. *Biophys. J.* 1978, *24*, 779.
104. Fesik, S. W.; Olejniczak, E. T. *Magn. Reson. Chem.* 1987, *25*, 1046.
105. Kartha, G.; Bhandary, K. K.; Kopple, K. D.; Go, A.; Zhu, P.-P. *J. Am. Chem. Soc.* 1984, *106*, 3844.
106. Gierasch, L. M. Personal communication.
107. Mersh, J. D.; Sanders, J. K. M. *J. Magn. Reson.* 1983, *51*, 345.

PART III

APPLICATIONS

CHAPTER 10

APPLICATIONS OF THE NOE TO STRUCTURE ELUCIDATION

10.1. GENERAL CONSIDERATIONS

Although the discovery of the NOE dates from the mid-1950s, the first chemical applications were those reported by Anet and Bourn in 1965.[1] Strictly, these involved only spectroscopic assignments, but the potential of the NOE for structural and conformational studies was clearly shown, and many other examples soon followed. The situation as it existed 6 years later was well summarized in Noggle and Schirmer's book of 1971,[2] at which time it was still possible to collect together all reported applications of the NOE.

In the period since then, applications have burgeoned. In many ways, the situation at present is analogous to that which existed in the late 1960s with respect to NMR spectroscopy itself. Applications to large-scale structure determinations of complex natural products, or to conformational analysis of biopolymers, retain a "high profile" in the literature. At least as important, however, is the far larger number of applications to simpler problems. These may be found scattered widely throughout the chemical literature, and frequently appear only as a short section in a paper primarily devoted to some other aspect of research. Almost any issue of a major chemical journal now contains many such "lower profile" applications of the NOE.

This great diversity of applications shows the very considerable success of the technique, but it also poses problems when one attempts to review the field. It would now be impossible to attempt to cite all papers in which the NOE has been used, and even to cover all fields of application would be uncomfortably close to attempting to review all of chemistry. Fortunately, from the standpoint of devising and interpreting the NOE experiments themselves, many of these applications are much more similar than they are different. If, for example, one wishes to establish the relative orientation of two substituents attached to a rigid, saturated six-membered ring, it make little difference whether the ring forms part of a steroid, a terpene, a sugar, an organometallic complex, or simply a cyclohexane; the principles underlying interpretation will be much the same.

With this in mind, the organizational scheme we have adopted here, although superficially in terms of structural features, is actually intended to be in increasing order of *interpretative complexity*, introducing the more complicated aspects of NOE experiments only as they are required for the various categories. Even within this framework, our approach must be illustrative only. A relatively small number of examples has been selected in the hope that by discussing these in greater detail than would be possible for a larger number, a clearer picture may emerge. For this reason, a major criterion in choosing some examples has had to be familiarity; any bias toward work with which the authors were associated reflects this alone.

We have deliberately retained the same set of examples for Chapters 10 and 11 in the second edition as we used in the first edition. To some extent this reflects the fact that neither of the authors still works in the field of small-molecule structure determination. More significantly, however, it reflects our belief that the underlying principles of interpretation remain essentially unaltered by the many technical developments that have taken place in the past decade or so, and that the set of examples we compiled for the first edition still illustrates these underlying principles clearly. Perhaps the main shift in emphasis in current applications is that regiochemical problems in essentially "flat" molecules (such as the examples described in Sections 10.2, 10.3, and 10.4) are now rather more likely to be addressed by using long-range (^{13}C, ^{1}H) correlation experiments than by using NOE studies. Nonetheless, we have chosen to retain these sections since (i) we wished to make readers aware of the possibilities for using the NOE in such cases, and (ii) they illustrate an important area of the older literature.

10.1.1. Why Structural and Conformational Problems are the Same

In this book, applications of the NOE to small and medium-sized molecules have been divided into chapters on structure elucidation and conformational analysis. However, a far more important distinction where the NOE is concerned is that between rigid and flexible molecules. To a crude first approximation, to be explored in more detail below, NOE studies of rigid molecules or molecular regions often yield *both* structural and conformational results simultaneously, whereas NOE studies of highly flexible molecules or fragments are more likely to fail altogether, yielding *neither* structure *nor* conformation.

Consider the molecular fragment —CH(OH)—CH(Me)—. If this fragment forms part of a rigid ring system, then NOE experiments are likely to show which protons around the ring are close together and which are not. From this information, one could possibly deduce not only the relative stereochemistry of the OH and Me groups (a structural conclusion), but also the conformation of the ring. In contrast, if the fragment forms part of an acyclic chain, it is likely that internal motions would average the interproton enhancements over a wide range of contributing conformers. No single conformation would account for such data, and these internal motions would probably also destroy the *structural* specificity of the enhancements. Either of the two possible diastereomeric structures for the —CH(OH)—CH(Me)— fragment would then yield a similar pattern of enhancements.

There are, however, cases in which such an acyclic segment does adopt a single preferred conformation. Generally, this occurs only in molecules large enough to be strongly stabilized by noncovalent interactions, such as a highly specific pattern of internal hydrogen bonds. Examples of such behavior are provided by the macrolides elaiophylin (Section 3.2.3) and pulvomycin (Section 10.6.1). In a sense, even these cases can be regarded as cyclic, since each stabilizing hydrogen bond itself completes a ring within the structure, and it is these "extra" rings that constrain the conformation.

In view of this, it is no coincidence that the various section headings within this chapter, although superficially representing structural types, actually represent the principal ways in which rigidity may be imparted to a molecule. Furthermore, we may see that the close similarity between structural and conformational studies using the NOE arises largely because of the difficulty in treating flexible molecules. Almost all applications of the NOE to conformational analysis assume, explicitly or implicitly, that there is a single conformation to be found, and then treat the problem almost exactly as if it were one of structure. Not surprisingly, the success of this approach depends heavily on the validity of the initial assumption. This is discussed in greater depth in Chapter 11 (and in Section 5.3.2).

10.1.2. Spectra and Assignments

Although the discussion in later sections necessarily concentrates on structural features, the factors that dictate how an experiment is planned, and whether it succeeds, depend primarily on the *spectroscopic* characteristics of the molecule under study.

The most important aspect of an NMR spectrum, from the standpoint of designing NOE experiments, is the extent of overlap between resonances. However, it is important to distinguish between overlap that results simply from the overcrowding of first-order multiplets, and overlap that occurs in highly second-order spectra. For example, the high-field spectrum of a steroid consists of a large number of overlapping resonances, but when techniques are applied that reveal individual multiplet structures (e.g., 2D-*J* spectroscopy), almost all are seen to be essentially first order. A conventional interpretation of the spectrum then becomes possible.[3] In contrast, the spectrum of a long-chain paraffin, although it also consists of many overlapping resonances, contains very little information, since all of the resonances are combined into one relatively featureless, strongly coupled envelope. Only in the former type of overlap can

NOE experiments (or indeed any conventional high-resolution experiment) be expected to be useful.

Obviously, to be irradiated selectively, a multiplet must be at least partially free from overlap. Selectivity is probably the most important consideration when setting up an NOE experiment, in that misinterpretation of enhancements resulting from saturation of signals other than the intended target leads to the most insidious errors. However, as the example of spiro-diamine **10.78** shows, *even one exposed line* of a multiplet can be sufficient to use as a target. Experimental techniques to reduce the SPT effects that result from such line-selective saturation are discussed in Section 7.2.5. If selective irradiation really is impossible, one may decide to use the 2D experiment. A discussion of the rationale for choosing between 1D and 2D NOE experiments is given in Section 8.1.

The choice of preirradiation targets for a 1D experiment must depend very much on what is known about the problem, so that generalizations are perhaps of little value. However, unless one is quite certain that a particular set of irradiation targets will yield the required information, it is best to err on the side of including extra targets. Much unsuspected information comes to light in this way, and one (sometimes) avoids the need to compare data recorded on different occasions, as inevitably happens when interpretation of an incomplete set of experiments suggests further irradiation targets. More importantly, the confidence that can be placed in an interlocking, mutually consistent set of many enhancements is generally much higher than if only a few are available. However, this "shotgun" approach[4] must be balanced against the available experiment time!

Frequently, complex proton spectra consist of a jumble of many resonances within one spectral region, with a smaller number of exposed resonances elsewhere. For example, the high-field region of the spectra of many terpenoid natural products and steroids includes a mass of overlapping methine and methylene multiplets, while the low-field region contains a smaller number of separately resolved olefinic, aromatic, or other resonances that are shifted downfield. In such circumstances, these separately resolved peaks are usually good candidates for preirradiation, even if their relevance to the problem at hand may not at first be apparent. Results from these experiments can then "lead in" to the more complex regions of the spectrum.

This leads naturally to the topic of assignment. The importance of correct assignments can never be overstated. As in all aspects of high-resolution NMR, resonance assignments are the key that unlocks all of the information present in the spectrum; without them there is nothing. For simple spectra with only a few widely dispersed resonances, assignment is more often than not a trivial exercise, but for complex natural products, and still more for biopolymers, it can be a very severe problem. Methods for assigning complex spectra form a vast topic in their own right, and it would be impossible to cover them in detail here. Techniques applicable to biopolymers are very closely tied to the overall process of conformational analysis using the NOE and are therefore dealt with separately in the appropriate context (Sections 13.1.1, 13.1.2, and 13.2.2). More generally, however, a few important points may be made here.

Ideally, one would prefer all assignments to be available *before* the NOE data are brought to bear, so as to minimize the possibilities for later ambiguities. Furthermore, whenever possible, the basis for assignment should be *coupling connectivities*. Matching the topology of a coupling network with that expected for the structure (in as much as it is known) must always be a more fail-safe procedure than interpreting the chemical shifts of individual resonances, if only because shift arguments become more tenuous the smaller the shift difference that is supposedly explained. Methods for deciphering coupling connectivities in proton spectra are legion, but one of the most powerful techniques (in the authors' view) is still the phase-sensitive double-quantum filtered (DQF) COSY experiment.[5,6] Analysis of cross-peak multiplet structures in such spectra yields, at best, the full information content that would have been generated by all possible decoupling experiments in the absence of overlap.[7] Another very powerful approach is to use long-range (^{13}C, ^{1}H) correlation experiments of the HMBC type,[8] which can "reach" across sparsely protonated fragments of a molecule to connect otherwise isolated homonuclear proton spin-systems. Gradient-assisted versions of these experiments have become popular, since these are very much more convenient and sensitive for use with samples containing ^{13}C at natural abundance.[9] The combination of experiments such as HMBC and phase-sensitive DQF-COSY with steady-state 1D difference NOE experiments represents an exceptionally powerful strategy for the structure elucidation of small and medium-sized molecules.

In practice, of course, the NOE data must often be used to establish both structure and the last of the assignments simultaneously. In these cases, a clear head is required to keep track of the various consequences of each step in the interpretation!

10.1.3. Reporting Results and Interpretation

The clearest way to report results of either a 1D difference or a 2D NOESY experiment is to reproduce the spectra. Either type of experiment yields data that are, initially at least, pictorial rather than numerical in nature, and reducing these to a set of numbers inevitably involves a certain amount of interpretation, if only in the selection of which signals to report. Furthermore, the appearance of those regions of the spectrum in which there are no enhancements is a powerful indicator of the quality of the experiment.

Sadly, this approach to reporting is quite impractical. Routine applications of the NOE would soon fill up the organic chemical literature, and by now 1D difference experiments are sufficiently familiar that spectra are generally shown only when there is some unexpected or unusual feature present.

The most obvious way in which to condense data reporting is to report the magnitudes of the enhancements. This immediately raises the issue of quantitation. Measurement of percentage enhancements in 1D spectra (including the method of correcting for partial saturation) was dealt with in Sections 3.3.1 and 7.3.1, and the volume integration of 2D cross peaks in Section 8.5.4. However, as was pointed out, the numerical values of enhancements extracted from any such experiment may vary greatly according to the precise details of sample preparation and pulse sequence timings (particularly where some of the newer gradient-assisted NOE experiments are concerned; see Section 7.5.3.5). Therefore, unless the most stringent experimental precautions are taken, the old view that measured NOE enhancements represent characteristic and reproducible molecular parameters is no longer usefully true in a quantitative sense.

Given this, to what extent is it worthwhile to report such numbers? Essentially, the only value in quoting enhancements numerically is to facilitate comparison *within the same data set*. This places a heavy responsibility on authors to ensure that data exhibited together are genuinely comparable in this way, or at the very least to point out when they are not. The examples of chiloenamine (**10.33**) and santiagonamine (**10.34**) (Section 10.3.2) emphasize this point well.

An attractive alternative to numerical reporting is to specify the enhancements more qualitatively, for instance, as strong, medium, or weak. This involves a greater degree of prior interpretation by the experimenter, who must decide where to define the "boundaries" of each category, but it has the advantage that these terms then acquire a *relative* significance within a particular data set. Another advantage of such qualitative reporting is that it helps discourage later overinterpretation. Two related points are also important here: (i) when numerical enhancements are quoted, it should be stated whether or not these are corrected for partial saturation (cf. Sections 3.3.1 and 7.3.1), and (ii) when qualitative categories are used it is helpful if at least a rough numerical definition of each category is given, since different numerical ranges may be appropriate for different data sets.

Probably the most effective way to present NOE data for small and medium-sized molecules is by using an annotated structure diagram, on which each enhancement is depicted as an arrow starting at the irradiated spin and ending at the observed spin. This is particularly efficient when enhancements are reported using qualitative categories, since each category can simply be specified using arrows of a different linestyle. For larger molecules with many enhancements, such diagrams become too crowded for clarity. The data must then either be split between several diagrams or consigned to a table, but important steps in the interpretation can always be reinforced by showing partial structures annotated with key enhancements.

One problem with the use of annotated structural diagrams is that they can lend an air of unjustified respectability to an incorrect conclusion. This is because, by their nature, they emphasize how the data are compatible with the proposed structure, but show little or nothing of how *alternative* structures were discounted. In practice, discussion of interpretative logic is very often excluded from manuscripts by the pressure to include other material. This is unfortunate, since interpretation of NOE experiments is all too subjective and fallible a process; some discussion of the reasoning is desirable in all but the simplest cases.

Finally, a few words about using negative results (i.e., absences of enhancements) are needed. It is often said that the absence of an enhancement should never be used to prove anything. This is certainly too extreme a view, since probably the majority of applications of the NOE depend simply on detecting an enhancement at one proton rather than another. The underlying truth, however, is that the absence of a particular enhancement may be due to a variety of causes *other than* a long distance between the spins involved. Only when these other possible causes have been ruled out may the absence of an enhancement justifiably be taken as evidence for a long distance. For convenience, the principal alternative causes for missing enhancements are listed here (explanations for each may be found in the sections indicated):

1. All enhancements are quenched by other relaxation mechanisms (dissolved paramagnetic metal ions, oxygen, unpaired electrons, in the solute itself, etc.; see Section 2.4).
2. All enhancements are close to zero because the tumbling rate of the molecule is near $\omega\tau_c = 1.12$ (Section 2.2). Note that if internal motions are present, particular enhancements may be precisely zero while others are weak but still present (Section 3.2.3).
3. For large molecules such as proteins, local differences in internal mobility may be more extreme, leading to weak or absent enhancements from mobile regions, but strong (negative) enhancements from rigid regions (Section 5.6). Losses of this sort affect all the enhancements into the particular signals involved.
4. Some or all enhancements are absent due to an inappropriate choice of experimental parameters (Chapters 7 and 8).
5. A particular enhancement is close to zero because appreciable direct and indirect effects cancel (this applies only in the positive NOE regime; cf. Section 3.2.2.3).
6. A particular enhancement is close to zero because the relevant observed spin has some other efficient relaxation pathway (Sections 2.4 and 3.3.3). This can usually be checked by testing whether any other enhancements occur into this spin (as in entry 3).
7. Particular enhancements may be reduced or completely obscured by saturation transfer if the observed spin undergoes exchange (Sections 5.1 and 5.2). This exchange need not necessarily be directly with the irradiated spin.

10.1.4. Miscellaneous

Applications employing different types of experiment are not segregated in Chapters 10, 11, and 12. The relative fields of application of 1D steady-state, kinetic, and 2D NOESY experiments are discussed and compared elsewhere, in Sections 3.3, 4.1, and 8.1. Almost all of the examples are of homonuclear interproton experiments, reflecting the overwhelming preponderance of such applications in the literature. Further discussion of heteronuclear experiments may be found in Sections 9.1 and 3.2.2.3.

All the compound names and numbering schemes have been taken verbatim from the original papers.

10.2. AROMATIC SUBSTITUTION AND RING FUSION PATTERNS: SIMPLE CASES

In this section we shall see the NOE at work on some apparently very simple problems. They appear simple because, as discussed earlier, the *level* of information required to solve them is very low. Essentially, each problem is reduced to a series of questions to which we require an answer of only "yes" or "no." For instance, is group X adjacent to proton Y, or not? Answers to this sort of question are particularly easy to obtain for aromatic ring systems because they are, to all intents and purposes, flat, leading to a minimum of interpretation for the NOE data. Groups either are adjacent or they are not, and distances between nonadjacent sites are usually large enough that the corresponding (direct) enhancements are very small. Thus, a very clear pattern of strong enhancements between adjacent groups and protons around the periphery of the ring system generally emerges, leading to an equally clear, and often considerably overdetermined, structural conclusion.

All of the molecules in this section are small enough to be within the positive NOE regime ($\omega\tau_c < 1.12$), and for the extreme narrowing approximation ($\omega\tau_c \ll 1$) to be valid (Section 2.2). Also their spectra are sufficiently simple that 2D methods are not needed; even though there may be spectral crowding in some cases, there are sufficient resonances clear of overlap for 1D NOE difference experiments to be the method of choice. Nonetheless, in a few of the examples shown, the older method of comparing integrals was used, either with FT or CW detection.

Of course, not all problems of this type are soluble using the NOE. It would, for instance, be futile to try to use the NOE to distinguish isomers in which Cl, Br, and NO_2 substituents were simply interchanged. Cases in which the NOE is useful are generally those in which there is what might be called an "intermediate" level of protonation. By this we mean molecules having too few adjacent aromatic protons for the structural problem to be solved using only their mutual J-coupling network, but enough protons, particularly on the substituents, for there to be useful NOE enhancements between substituents and aromatic protons.

For all the apparent simplicity of the examples that follow, each would be very much more difficult to solve by other means. The key role of the NOE is to bridge "gaps" in the structure over which there are no (^{1}H, ^{1}H) spin–spin couplings, so it follows that other methods that achieve this might also be helpful. In particular, two- or three-bond (^{13}C, ^{1}H) couplings, or one-bond (^{13}C, ^{13}C) couplings, very often offer information parallel to that from NOE measurements. Apart from these methods, it should be clear that the only alternative to NOE measurements in most of the cases below would be an X-ray structure determination or a chemical degradation.

This point is very well illustrated by the first example, which involves the differentiation of π and τ substituted histidine residues (Fig. 10.1).[10] This had previously been an area fraught with difficulties; chemical degradations are successful only in special cases, and predictions based on empirical rules are unreliable. In contrast, 1D difference NOE experiments gave very clear and unambiguous results for those cases in which the N substituent was of the type RCH$_2$—. Exactly as expected, irradiation of either H$_2$ or H$_5$ produced an enhancement of the N—CH$_2$ signal for the τ substituted isomer **10.1**, whereas only irradiation of H$_2$ produced a similar enhancement for the π substituted isomer **10.2** (Fig. 10.1). A similar problem was identifying which of the two possible products, **10.3** or **10.4**, actually results from reaction of styrene with 4-methoxyphenol.[11] Enhancements observed from the OMe resonance to H$_3$ and H$_5$, and from the OH resonance to H$_6$, quite conclusively established structure **10.3** as correct.

The application of essentially identical logic allowed the site of trimethylsilylation in the protected indole chromium tricarbonyl complex **10.5** (Scheme

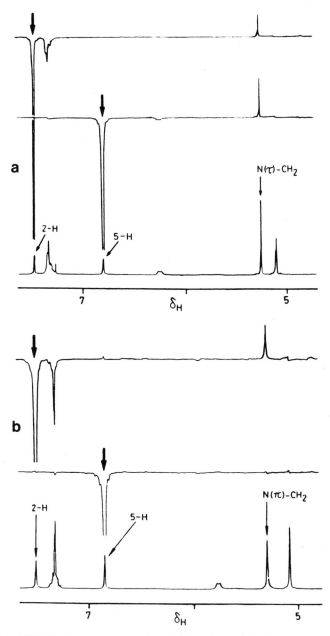

Figure 10.1. NOE difference spectra (low-field region only) of (*a*) τ-substituted histidine **10.1**, and (*b*) π-substituted histidine **10.2** (both samples about 50 mM in CDCl$_3$), recorded at 300 MHz (preirradiation 5 s). In (a), preirradiation of either H$_2$ or H$_5$ causes enhancement of the NCH$_2$ signal, whereas in (b) only preirradiation of H$_2$ does so. Reproduced with permission from ref. 10.

10.1) to be identified.[12] For the (decomplexed) product, irradiation of the trimethylsilyl singlet resulted in direct enhancements of two aromatic signals (H_3 and H_5); similarly, irradiation of the triisopropyl *CH* signal caused enhancements of H_2 and H_7 (Fig. 10.2). This clearly established the 4-substituted structure **10.6** as correct; the alternative 7-substituted product **10.7** would have exhibited only one enhanced aromatic signal in either experiment.

All of these examples clearly illustrate how a structural and an assignment problem can be solved simultaneously, but in each case an unambiguous starting point for the interpretation was required to "lead in" to the rest of the structure. This role was fulfilled by, respectively, the N—CH_2 signal in the histidine case, the OMe signal in the methoxyphenol case, and the (trivially assignable) trialkylsilyl signals in the indole case. This is, in fact, a very general feature in the structural interpretation of NOE experiments, but as will be seen, the required starting point is not always so obvious.

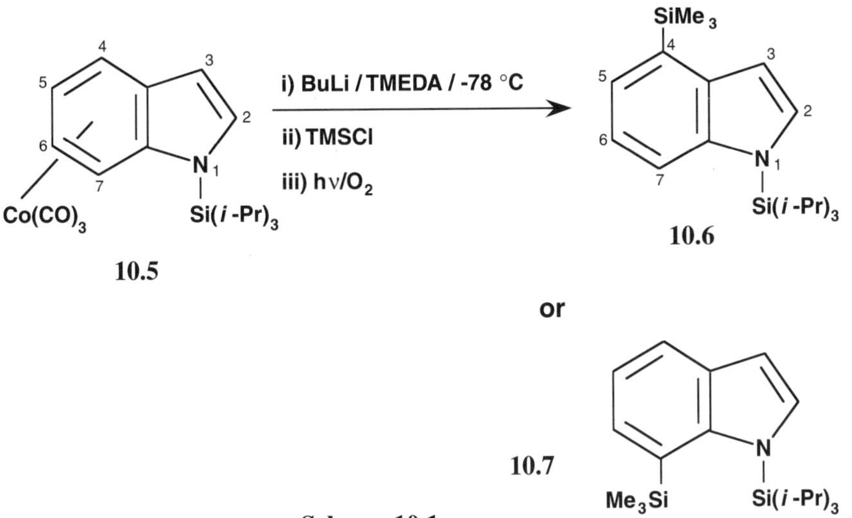

Scheme 10.1

The indole example also demonstrates a particular strength of the NOE in this context, namely its ability to identify *peri* related groups on adjacent rings in a fused aromatic ring system. Ambiguities that require this ability are extremely common; in fact all the remaining examples in this section and the next make use of it, either to establish the relative positions of substituents in different rings, or to establish the ring fusion pattern itself.

As a further example of this, enhancements between *peri* related protons were crucial in establishing the position of the amino group in the aminoquinoline **10.8**.[13] Analysis of shifts and couplings (particularly *meta* couplings) showed the presence of the two fragments **10.9** and **10.10**. The role of the NOE was to establish which of the two possible ways of interconnecting these fragments was correct, the key observation being the strong mutual enhancements observed between H_C and H_D. The method was also extended to deal with the

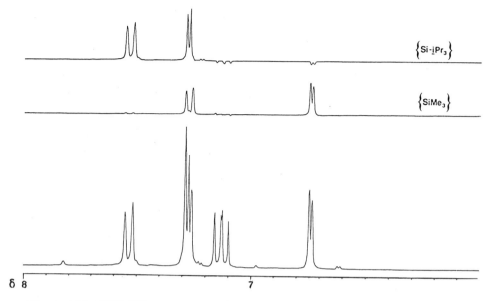

Figure 10.2. NOE difference spectra (aromatic region only) of *N*-triisopropylsilyl-4-trimethylsilylindole **10.6** in CDCl$_3$, recorded at 250 MHz (preirradiation 5 s). These enhancements clearly demonstrate that the TMS group is present at the 4 position of the indole rather than the 7 position. Reproduced with permission from ref. 61.

2-oxo and 4-oxodihydronitroquinolines **10.11** and **10.12**. Here the N substituent was deliberately added in order to replace the original NH group by a protonated "handle" for the NOE experiments. Thus, irradiation of the N—CH$_2$ singlet caused enhancement of H$_8$, allowing the relative connectivity between the two rings to be deduced as before. These experiments used the integral comparison method.

Although it was used in this case, chemical modification of exchangeable protons is by no means a general requirement. As discussed in Chapter 5, experiments in which exchangeable OH or NH protons are the preirradiation targets may be complicated by saturation transfer effects, but unless two or more NH or OH protons within the solute undergo appreciable saturation transfer *with one another*, this is unlikely to confuse the qualitative interpretation much. Saturation transfer to the residual water resonance in an organic solvent, although common, is usually irrelevant to the outcome of the experiment (unless the water concentration and saturation transfer rate are sufficiently high that intermolecular cross-relaxation of the water with other solute spins becomes a problem). A more significant problem is likely to be the accidental irradiation of the residual water line itself; an example of this is given in Section 10.5.1.

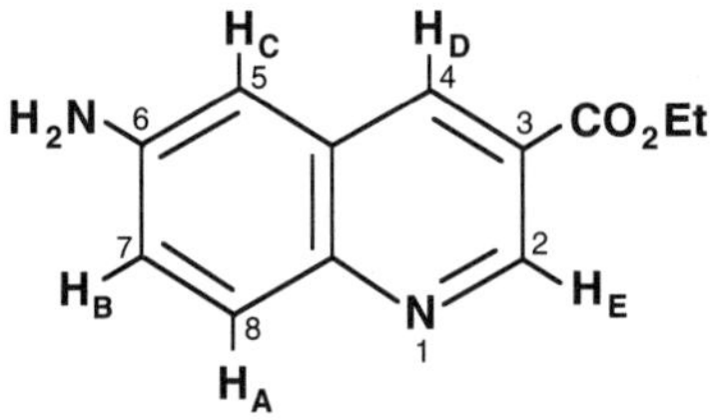

10.8 (structure with H_2N-6, H_C-5, H_D-4, CO_2Et-3, H_E-2, N-1, H_A-8, H_B-7)

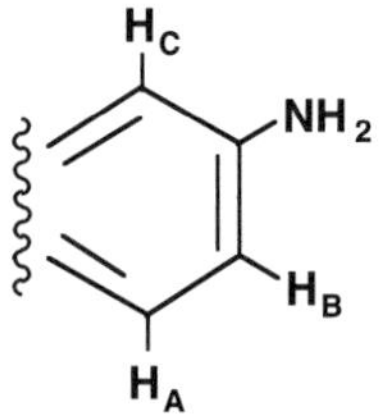

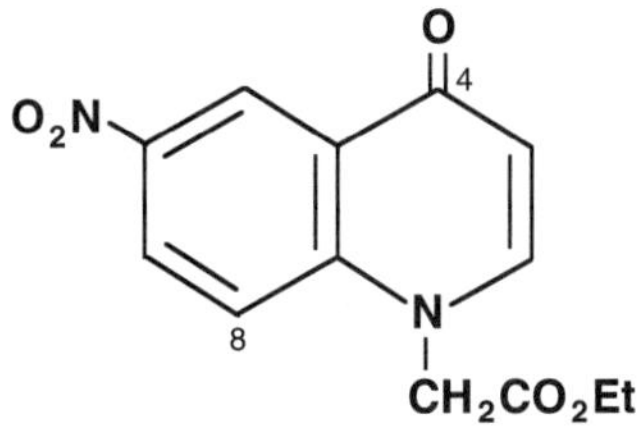

10.9

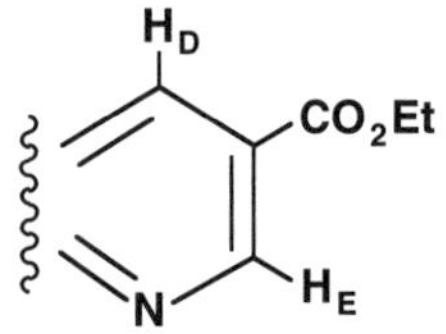

10.10

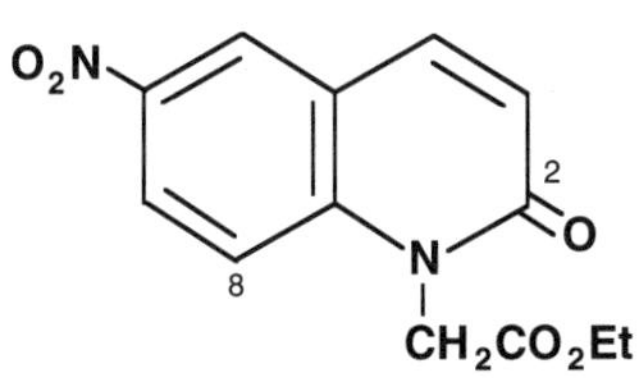

10.11 (6-nitro-quinolin-2(1H)-one with N-CH_2CO_2Et)

10.12 (6-nitro-quinolin-4(1H)-one with N-CH_2CO_2Et)

The methoxyphenol case (**10.3**) provides one example of a successful NOE experiment employing an exchangeable proton as the preirradiation target, and another is provided by work on the thienopyrroles **10.13** and **10.14**, derived

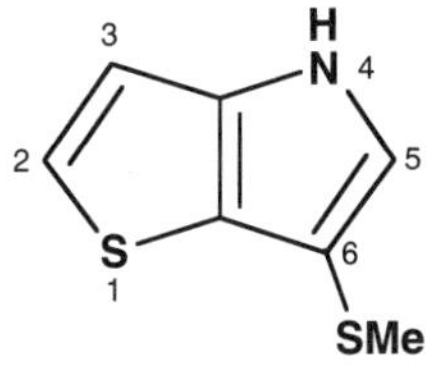

10.13 (thienopyrrole with SMe at position 6)

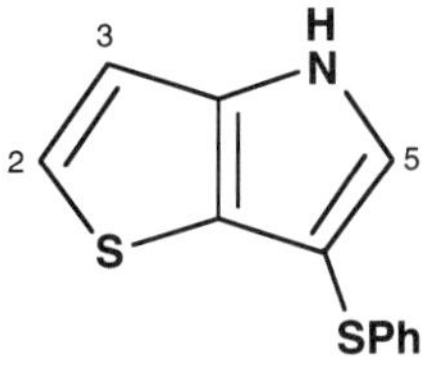

10.14 (thienopyrrole with SPh)

from photolysis of thienothiazines **10.15**.[14] Here, the position of the thioether substituent was established by preirradiating the NH resonance, resulting in enhancements of H_5 and H_3 that ruled out the alternative 5-substituted isomers. A smaller enhancement at H_3 than at H_5 reflects the fact that H_3 has another near neighbor, H_2, whereas H_5 does not. These results establish that SR must have migrated in preference to CO_2Et in the presumed intermediates **10.16** (Scheme 10.2). Note that, in order to avoid having to rely on the *absence* of

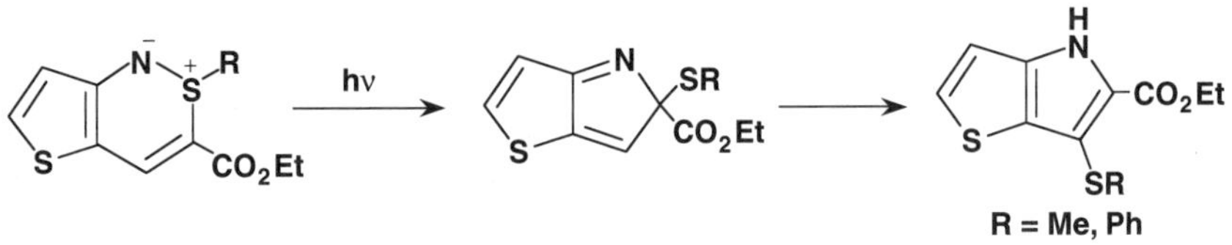

10.15 → hv → **10.16** → **10.17**

R = Me, Ph

Scheme 10.2

404 APPLICATIONS OF THE NOE TO STRUCTURE ELUCIDATION

an enhancement between the NH and SR groups, the first formed thienopyrrole esters (**10.17**) had to be hydrolyzed and decarboxylated to yield compounds **10.13** and **10.14**.

Another example that takes advantage of the editing ability of the difference experiment is the differentiation of the isomeric cyclopentanaphthalenes **10.19** and **10.20**.[15] Compound **10.19** was isolated following photolysis of 4a-methyl-4a*H*-fluorene (**10.18**), and was then converted into a 2:1 mixture of compounds **10.19** and **10.20** (by treatment with butyl lithium) (Scheme 10.3). This mixture was used for blank experiments, to confirm both the structure determination of compound **10.19** and the photostability of compound **10.20**. This problem is exactly analogous to the previous two examples in its interpretative logic. Enhancements observed on preirradiating the methyl and methylene signals of each compound establish which groups are *peri* related, and hence determine the structure (Fig. 10.3). Note the large SPT effects between the methylene and olefinic protons, which show that symmetrical but incomplete saturation of a multiplet (even if incompletely resolved) does not abolish SPT, but instead produces symmetrical SPT patterns for the coupling partners in the difference spectrum. No attempt was made to suppress SPT in these cases (for further discussion of SPT effects and their suppression, see Section 6.2).

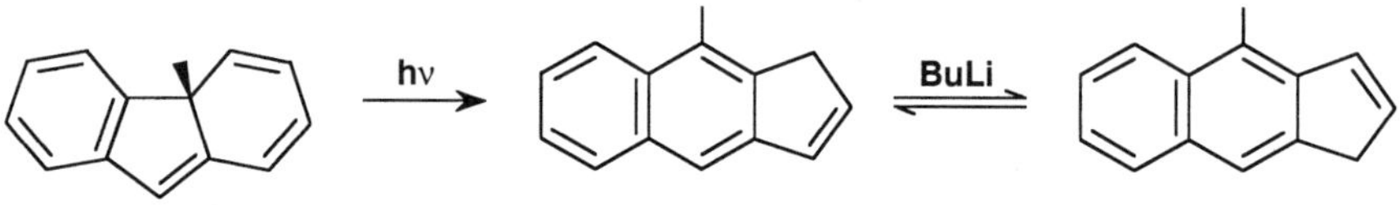

Scheme 10.3

An important point that this example highlights is the ability of NOE difference spectroscopy to yield structural information for individual components in mixtures. Provided the preirradiation target itself is free from overlap, the degree of overlap in the rest of the spectrum is, in principle at least, irrelevant. In practice, results from within regions of overlap depend very much on the quality of subtraction, so, as always, effects that are small relative to the resonances being subtracted will require long experiment times to detect. Thus, for example, a 5% enhancement for the minor component in a 10:1 mixture

Figure 10.3. NOE difference spectra of a mixture of 9-methylcyclopenta[*b*]naphthalene **10.19** and 4-methylcyclopenta[*b*]naphthalene **10.20** in CDCl$_3$, recorded at 250 MHz (preirradiation 5 s). Enhancements of the CH$_2$ signal in **10.19** and of H$_3$ in **10.20** on preirradiation of each corresponding methyl signal allowed the resonances of each compound to be assigned, showing that **10.19** is the major component of the mixture. Reproduced with permission from ref. 44.

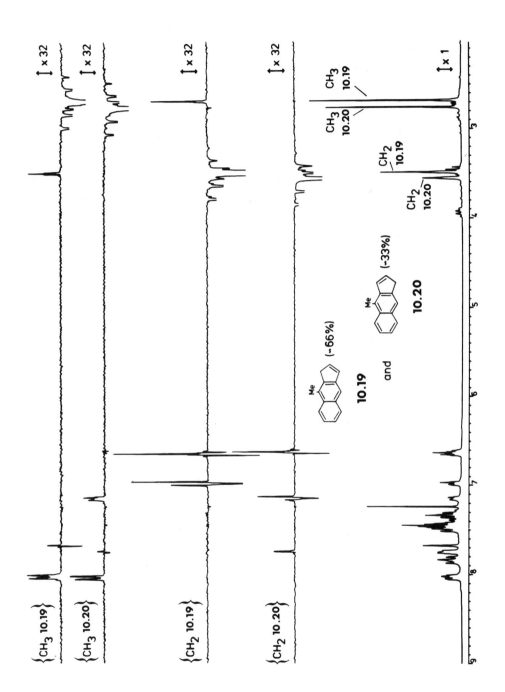

would be as hard to detect (when overlapped with signals of the major component) as would a 0.5% enhancement for the major component. Nonetheless, experiments of this sort can be extremely valuable, since they avoid the need to carry out possibly very difficult separations. A similar example appears at the end of Section 10.3.

A further case in which *peri* effects are important is that of distinguishing the "linear" and "angular" ring fusion isomers **10.22** and **10.23**.[16] Condensation of the 1′-methylthiaminium ion **10.21** with 4-methyl-2-amino-pyridine gave a tricyclic product, for which either structure was plausible (Scheme 10.4). However, a strong enhancement (24%) from the H_5 methylene signal to H_7 effectively ruled out the angular isomer, leaving structure **10.22** as the correct choice. These experiments used the integral comparison method, and the original paper includes a detailed description of the NOE measurements.

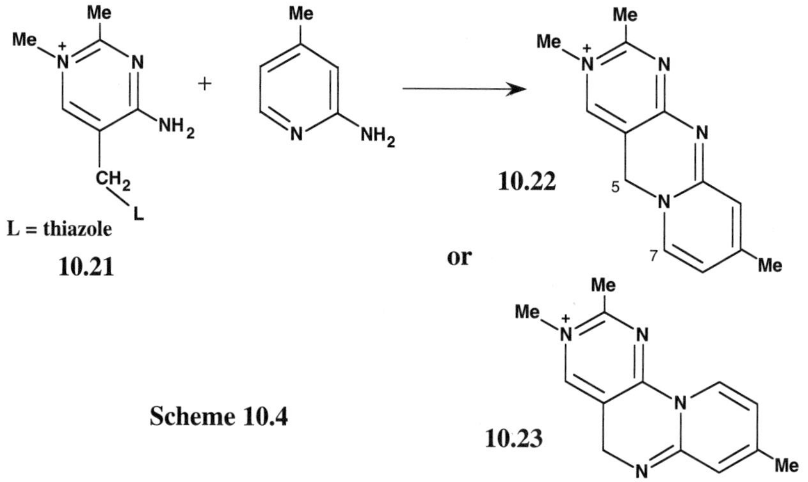

Scheme 10.4

10.3. AROMATIC SUBSTITUTION AND RING FUSION PATTERNS: MORE COMPLEX CASES

The examples in this section are just as simple to interpret as those in the previous section. The only sense in which they are more complex is that, here, a whole sequence of structural conclusions is required to piece together a molecule, rather like fitting together a jig-saw puzzle, whereas the previous examples each required the resolution of only a single ambiguity.

Again, 1D methods are generally used, although it is arguable that 2D NOESY or ROESY experiments might be of some benefit for some of the petroporphyrin examples, where crucial shift differences are often very small. As discussed in Section 8.1.2, it is very hard to say in advance whether or not such 2D methods will yield results for molecules of this size. In almost all cases the molecules are in the positive NOE regime ($\omega\tau_c < 1.12$), although the

AROMATIC SUBSTITUTION AND RING FUSION PATTERNS: MORE COMPLEX CASES

extreme narrowing approximation ($\omega\tau_c \ll 1$) may be beginning to break down (Section 3.2.3). However, since numerical interpretation of the enhancement data is scarcely necessary, this is of little concern.

Two principal areas of application stand out from the literature, namely petroporphyrins and isoquinoline alkaloids. These will be discussed in turn.

10.3.1. Petroporphyrins

Petroporphyrins are highly defunctionalized porphyrins found in geological samples, and are thought to be derived from natural chlorophylls during formation of oil from ancient plant material. Unlike porphyrins and chlorophylls derived from present-day plant sources, they are almost all devoid of stereochemical ambiguities, and consist of a fully unsaturated ring system with only very simple substituents. Structure determination of petroporphyrins has generated considerable interest, since they offer a means of characterizing the geological samples from which they were isolated.

Generally, structure determination of petroporphyrins by NOE reduces to a "two-dimensional" problem of organizing correctly the substituents around the ring system periphery. Information as to what substituents are present will normally have been found by other means, particularly mass spectroscopy and normal 1D NMR spectroscopy. The logic by which the NOE experiments are devised and interpreted is essentially identical to that used in the examples of the previous section. Enhancements between the *meso* protons and β protons of the substituents, and also between substituents on a common pyrrole ring, are used to discover which groups are adjacent, and hence to assemble the whole structure.

A few examples of structure determination of petroporphyrins by NOE are summarized in Table 10.1. C_{32} etioporphyrin (**10.24**) and C_{32} deoxophylloerythroetioporphyrin (DPEP porphyrin) (**10.25**) are the two major petroporphyrins from Yorkshire Marl slate (enhancements between substituents on common pyrrole rings were also observed for these porphyrins, and were presumably used in the interpretation, but were not reported in detail.)[17] Porphyrin **10.26** is the monoacetylation product of C_{30} etioporphyrin from Swiss Serpiano oil shale,[18] and C_{35} DPEP porphyrin (**10.27**) originates from immature Messel oil shale.[19]

Of course, the structure determination of more complex porphyrins and related materials is in many ways similar to these examples. Differences arise mainly because of the wider possibilities for *stereochemical* ambiguity that such molecules may possess. The more the porphyrin ring system becomes reduced, substituted, and functionalized, the more complex the NOE (and other) experiments necessarily become. An example of such a very highly modified porphyrin (Factor F-430) appears in Section 10.6.3, but for now we include one example of a "flat" porphyrin from a nongeological source. Compound **10.28** was isolated from rat hepatic microsomal cytochrome P-450 after treat-

TABLE 10.1 NOE Enhancement Data for Petroporphyrins 10.24–10.27

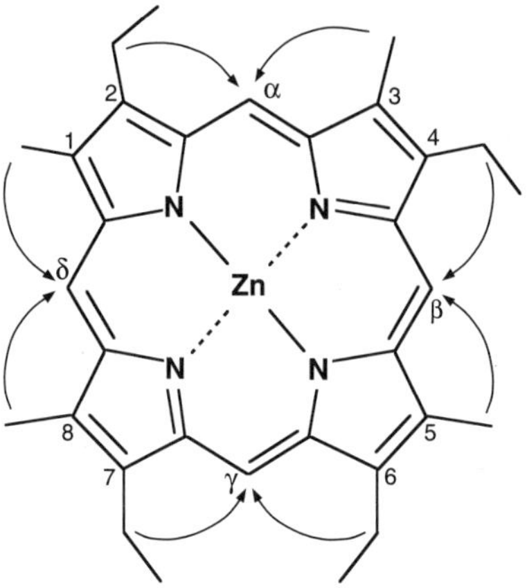

10.24
(ref. 17)

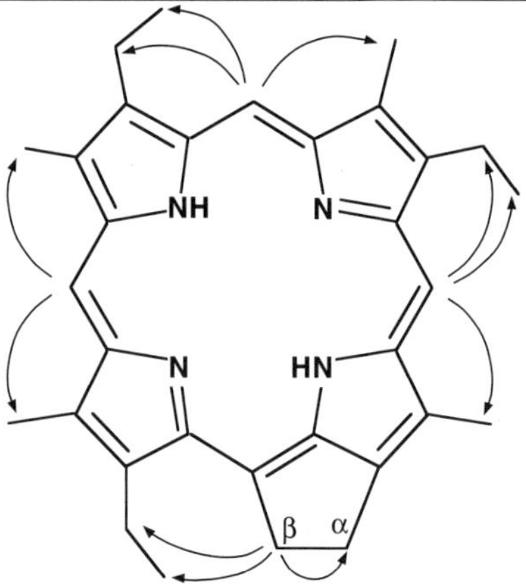

10.25
(ref. 17)

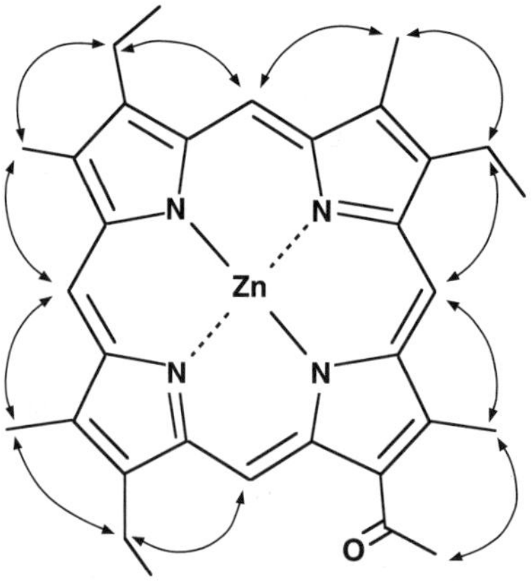

10.26
(ref. 18)

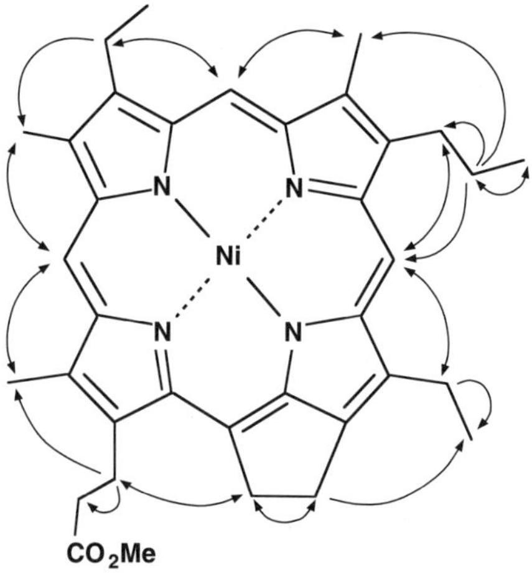

10.27
(ref. 19)

ment with propyne gas, and its structure determination followed a course very similar to the previous examples. In this case, long-range (^{1}H, ^{1}H) couplings (indicated J on **10.28**) were also used to relate methyl and vinyl substituents on common pyrrole rings, and the N-(2-oxopropyl) substituent was placed using a chemical shift argument.[20]

10.28   CO₂Me   CO₂Me

10.3.2. Isoquinoline and Related Alkaloids

Isoquinoline alkaloids pose a much greater variety of structural problems than do the petroporphyrin examples of the previous section. Few of them are completely aromatic, and many possess one or more chiral centers, leading to the possibility of stereochemical ambiguities. Nonetheless, a large part of their structure determination often comes down to resolving ambiguities of positional isomerism on aromatic rings, and it is these aspects in particular that we consider below.

The complexity of the NOE interpretation in these examples depends very much on how much information has already been gathered from other sources. In some cases, only a single ambiguity remains to be solved, while in others the relative placings of several groups may be required. For some of the related examples, the structure of the ring system itself may be unknown, and here interpretation of the NOE becomes inevitably somewhat more involved.

The first two examples show cases in which the solution to a single ambiguity was sought. For the bisbenzylisoquinoline alkaloid temuconine, mass spectroscopy and normal 1D NMR spectroscopy limit the possible structures to two, **10.29** and **10.30**.[21] In particular, mass spectroscopy revealed the masses of the two isoquinoline units ($m/z = 206$ and $m/z = 192$) resulting from cleavage at the C_1—C_α and $C_{1'}$—$C_{\alpha'}$ bonds, while NMR allowed the oxygenation pattern of each aromatic ring to be deduced. However, the relative attachment of these two isoquinoline units onto the remaining central diaryl ether fragment was unclear. This information was provided by the NOE results summarized around the correct structure, **10.29**. These results establish an almost unbroken pattern of connectivities around the peripheries of both isoquinoline units and the "outer ends" of the diaryl ether fragment, simultaneously yielding assignments for many of the aliphatic protons and confirming the sites of methoxyl substitution. The relative attachment of the isoquinoline units is revealed by a large number of mutually consistent enhancements involving the benzyl methylene links, and also by enhancements directly between the aromatic protons on the diaryl ether and isoquinoline fragments. This mass of interlocking NOE data might at first appear somewhat redundant, since several enhancements actually resolve the original ambiguity, but in fact such a complete study is always the wisest course. In this context, we may note that the alternative structure **10.30** had originally been assigned on the basis of a less complete set of NOE data[22]; this was largely because an enhancement of H_{10} originally thought to be due to saturation of a supposed methoxy signal at C_7 was in fact caused by partial saturation of the coincident multiplet due to H_1.[23]

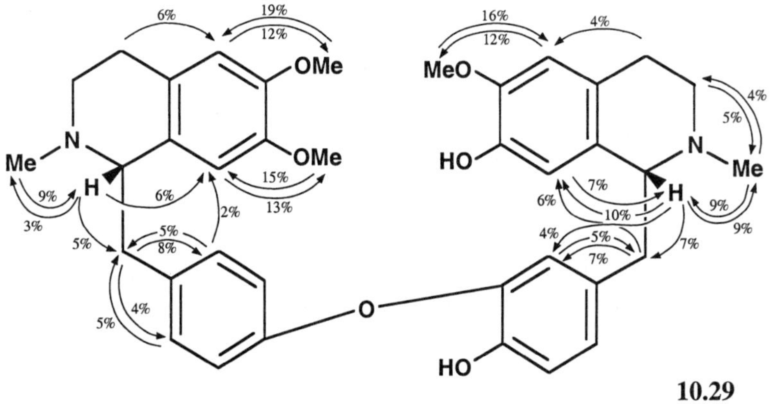

10.29

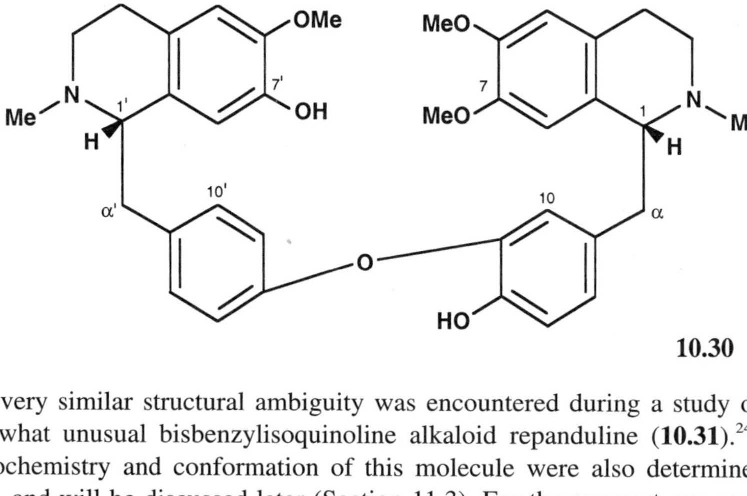

10.30

A very similar structural ambiguity was encountered during a study of the somewhat unusual bisbenzylisoquinoline alkaloid repanduline (**10.31**).[24] The stereochemistry and conformation of this molecule were also determined by NOE, and will be discussed later (Section 11.3). For the moment, we consider just the distinction of two possible structures **10.31** and **10.32**, which differ

AROMATIC SUBSTITUTION AND RING FUSION PATTERNS: MORE COMPLEX CASES   411

only in the relative positions of the spiro link and the methylene group in ring G. This problem had for many years eluded solution by chemical degradation or spectroscopic means, although the bulk of the structure was established.

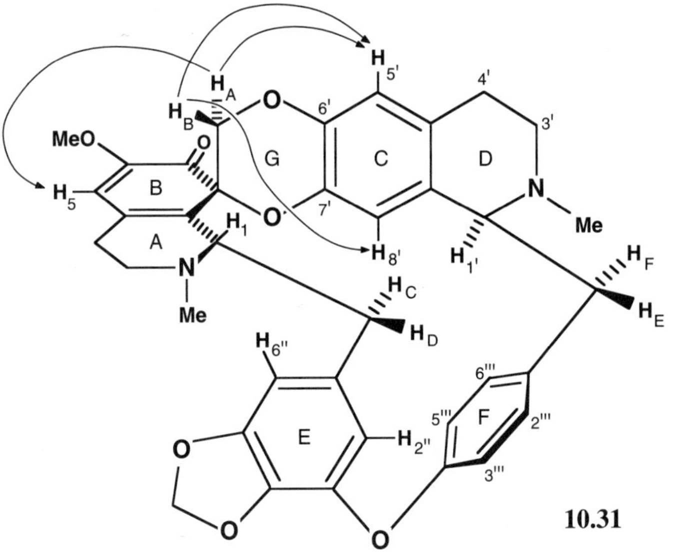

10.31

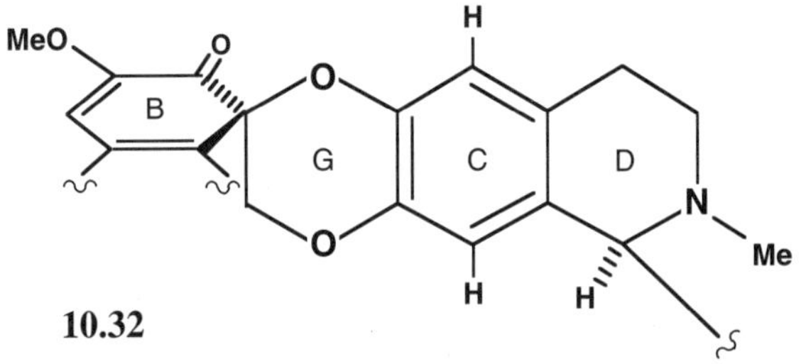

10.32

The interpretation here was very similar to that of the previous example. First, rigorous assignments had to be found for protons $H_{5'}$ and $H_{8'}$. These were easily established using enhancements from $H_{5'}$ to $H_{4'}$ protons, and from $H_{8'}$ to $H_{1'}$; the $H_{1'}$ resonance was clearly distinguishable from the $H_{4'}$ resonances by its shift and multiplet pattern. Next, protons H_A and H_B were separately preirradiated, to discover which of the aromatic protons $H_{5'}$ or $H_{8'}$ was the closer to the OCH_2 group, and hence to distinguish which of the structures **10.31** and **10.32** was correct (Fig. 10.4).

This example requires some further comment, for two reasons. First, the interpretation must necessarily take account of the conformation of ring G. Of the two methylene protons, models show that the pseudoaxial one (H_B) is

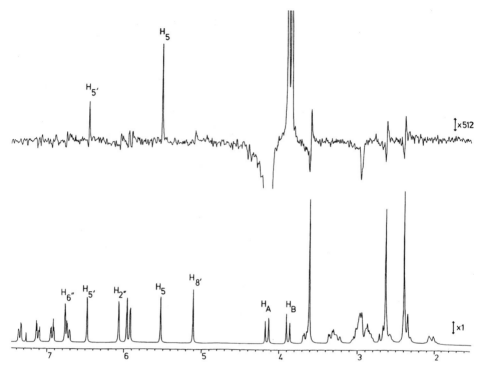

Figure 10.4. NOE difference spectra of repanduline **10.31** (0.26 M in CDCl$_3$) preirradiating H$_A$; recorded at 250 MHz (preirradiation 5 s). The spectrum results from the combination of a total of 32,000 transients (8,000 with preirradiation of each line of the H$_A$ doublet, and 16,000 controls) acquired over a weekend. The crucial H$_{5'}${H$_A$} enhancement of 0.18% shows the correct substitution pattern in ring C to be as in **10.31** rather than **10.32**. See text for further discussion. Reproduced with permission from ref. 24.

roughly equidistant from H$_{5'}$ and H$_{8'}$, whereas the pseudoequatorial one (H$_A$) is significantly closer to H$_{5'}$ (~4.5 Å) than to H$_{8'}$ (~6 Å) in all reasonable conformations of structure **10.31** (these distances are reversed for structure **10.32**). These NOE experiments thus served a dual role. They established the relative assignments of H$_A$ and H$_B$, since preirradiation of H$_B$ (not shown in Fig. 10.4) enhanced both H$_{5'}$ and H$_{8'}$, whereas preirradiation of H$_A$ enhanced only one (this relative assignment for H$_A$ and H$_B$ was reinforced by several other enhancements). More significantly, the experiments established structure **10.31** as correct, since the proton enhanced on preirradiation of H$_B$ was H$_{5'}$, rather than H$_{8'}$ as would have been required by structure **10.32**.

Perhaps the most significant feature of these experiments, however, and the reason for their inclusion here, is the very small size of the enhancements. Given the distances involved, the smallness of these effects is entirely as expected, but one may reasonably ask: to what extent is the interpretation of such very small enhancements justified?

During the work on repanduline, this issue was attacked at several levels. First, the enhancements were reproducible (an important point). That the enhancements were genuine relaxation phenomena was supported by the observation that they increased when the sample was degassed prior to measurement (e.g., $H_{5'}\{H_A\}$ increased from 0.18 to 0.26% following a triple freeze-thaw degassing cycle under nitrogen). The plausibility of the crucial $H_{5'}\{H_A\}$ enhancement can also be assessed by inspection of the remainder of the difference spectrum in which it was measured. Residual artifacts from poor subtraction of the sharp methyl lines are about ±0.04%, which sets an approximate confidence limit for the detection of enhancements. All other effects in the difference spectrum are fully accounted for by the proposed structure [direct enhancements at H_B (14%) and H_5 (0.42%), three-spin effect to H_1 transmitted via H_B]. Finally, and perhaps most importantly, the long-range enhancements fitted well into the overall picture of many enhancements, large and small, recorded for the repanduline molecule as a whole. The structural conclusion of this work was (later!) confirmed by X-ray diffraction, and by analysis of $^3J(^{13}C, ^1H)$ coupling constants.

This example well illustrates how, with added care, 1D difference experiments can considerably extend the useful range of the NOE, allowing an investigation to "see" across longer distances than would be feasible with other NOE experiments. However, experimental design and, particularly, careful interpretation become absolutely crucial in such cases.

The final examples in this section are of highly modified alkaloids that, although related to isoquinoline alkaloids, themselves lack the isoquinoline ring system. In these cases, NOE experiments play a more central role in the overall structure determination process, since they are used to establish the ring fusion pattern itself.

For chiloenamine (**10.33**), the structures of the side-chains were easily deduced from mass spectroscopy, IR spectroscopy, and normal 1D NMR spectroscopy. A dimethylaminoethyl group was clearly identified from the mass spectrum (base peak at m/z = 58; $[CH_2N(Me)_2]^+$), while the CH—CH_2—CO_2Me and OMe groupings were identified by NMR, and the presence of an OH group was confirmed by an acetylation reaction. The problem of organizing these side chains around a suitable aromatic skeleton, within the constraints of the molecular formula $C_{20}H_{23}O_6N$, was fully solved using the NOE results summarized around the structure. No other reasonable structure fits all these data simultaneously.[25]

A very similar approach was used to solve the structure of santiagonamine (**10.34**)[26] The various side-chains were identified as before, and the presence

414 APPLICATIONS OF THE NOE TO STRUCTURE ELUCIDATION

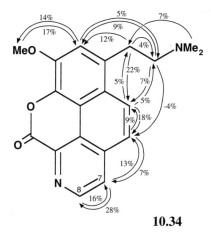

10.33

of a pyridine ring was inferred from the characteristic shifts ($\delta 9.15$ and $\delta 8.06$) and mutual coupling ($J = 5.2$ Hz) of H_7 and H_8. The lactone carbonyl was identified by IR spectroscopy [ν_{max} (CHCl$_3$) = 1755 cm^{-1}], and the molecular formula ($C_{19}H_{18}O_3N_2$) was found by mass spectroscopy. Starting from these data, the NOE results shown lead clearly to the proposed structure; the only plausible alternative, in which the sense of attachment of the lactone unit is reversed, was (reasonably enough, in this particular case) ruled out on biogenetic grounds.[†] The high sensitivity of the NOE difference experiment was also put to use in this study; these data were obtained from just 1 mg of solute.[23]

10.34

These two examples also illustrate an important point made earlier, namely that the *absolute* magnitudes of enhancements have little meaning in themselves. All the reported enhancements for chiloenamine are considerably smaller than the corresponding enhancements for santiagonamine. Given the

---

[†]Assignments for the C$^\alpha$H$_2$ and C$^\beta$H$_2$ resonances are reversed relative to those in ref. 26, following personal communication from the authors.[23]

similarity of their structures, this must almost certainly be due to some difference in preparation or experimental conditions on the two occasions. It does not matter much what this difference was, nor is it necessarily a problem; the point is that in each case a structural conclusion was reached entirely by comparing the *relative* values of various enhancements within one data set. This also emphasizes the importance of experimental design in minimizing the extent of such systematic errors within a single investigation. If different enhancements for the same substance were to show similar variations unrelated to structure, perhaps because they were recorded from different samples, then any comparative interpretation would be totally misleading.

10.4. DOUBLE BOND ISOMERS

Problems involving double bond isomerism are nearly as common as those involving aromatic positional isomers. They are also nearly as simple to solve using the NOE.

Once again, interpretation is greatly simplified by the fact that the double bond unit is flat. As with the examples of aromatic substitution patterns, the NOE is at its most useful where there is an "intermediate" level of protonation. Here we may be rather more specific about what is meant by this. Clearly, most applications will be to tri- or tetra-substituted double bonds having a high proportion of proton-bearing substituents. *Cis-* and *trans-*1,2-disubstituted olefins can usually be distinguished by their characteristically different coupling constants ($J_{cis} \sim$ 7–11 Hz, $J_{trans} \sim$ 12–18 Hz), while for 1,1-disubstituted or monosubstituted olefins no ambiguities of positional isomerism can arise. However, because the double bond unit is smaller than the aromatic ring systems considered earlier, the conformation of any flexible side chains present may become a more significant factor in the interpretation of enhancements, particularly if the protons involved are more than one atom removed from the double bond.

One further complication makes interpretation of the NOE slightly more difficult here. A short through-space contact between two groups, as determined by the NOE, does not distinguish whether the groups are *cis* related on adjacent carbon atoms, or both bonded to the same carbon atom. Although some slight difference in internuclear distances might be expected between the two cases, everything from previous chapters should make it clear that such a difference would be far too slender to be distinguished by NOE measurements, particularly if there is any flexibility at all in the rest of the molecule. Consequently, NOE results in isolation leave an ambiguity between compounds related by a *trans*-1,2 exchange of substituents; Figure 10.5 illustrates this.

Generally, the information required to solve this ambiguity has to be supplied from some other source. Very often, the chemical nature of the problem is such that only one of these possible structures needs to be considered, par-

416 APPLICATIONS OF THE NOE TO STRUCTURE ELUCIDATION

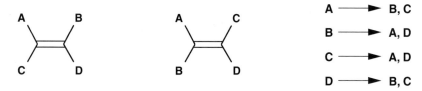

Figure 10.5. NOE enhancement data cannot distinguish between isomeric olefins related by a 1,2-*trans* interchange of substituents.

ticularly if the double bond participates in a ring system. In some simple cases, however, other experimental evidence may be needed.

The first example illustrates these points well. Addition of methanesulfenyl chloride to 3,3-dimethylbutyne is a complicated reaction that yields up to three olefinic products, depending on the conditions.[27] The first step in identifying these is to make an assumption based on mechanistic knowledge; since the reaction is expected to be a 1,2 addition, the SMe and Cl groups may be expected to be at opposite ends of the double bond in the product. This reduces the number of possible olefinic structures from six (**10.35–10.40**) to four (**10.35–10.38**). NOE experiments go some way toward identifying the three products from within these four possibilities, as summarized in Table 10.2.

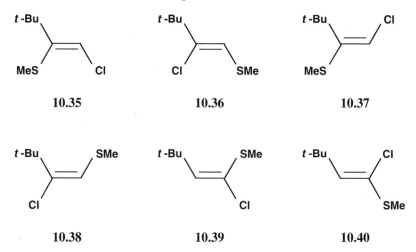

Structures **10.35** and **10.36** are uniquely identified by these enhancements, but structures **10.37** and **10.38** cannot be distinguished, since they are related by a *trans*-1,2 exchange of substituents, as discussed earlier. To complete the structure determination, therefore, further evidence is required, and this is supplied by chemical degradation. Oxidation followed by catalytic hydrogenation converts compounds **10.35** and **10.37** into sulfoxide **10.41**, whereas compounds **10.36** and **10.38–10.40** are expected to yield sulfoxide **10.42**. Since the third

TABLE 10.2 NOE Enhancement Data for Olefins 10.35–10.37[27]

Compound	Solvent	$H_{vinyl}\{SMe\}$	$H_{vinyl}\{{}^tBu\}$
10.35	$CDCl_3$	−2%	24%
10.36	$CDCl_3$	12%	28%
10.37	CCl_4	21%	1%

product in fact yields sulfoxide **10.42** on degradation, this identifies it as structure **10.37**. It is worth noting that the assumption of 1,2 addition was *required* in this logic, since even the combination of NOE results and chemical degradation could not have distinguished structures **10.36** and **10.40**. Had experimental evidence been required for this distinction, it might have been provided by the expected difference in ^{13}C coupling partner over the three bond pathway ($^{13}C-S-C^1H_3$).

t-BuCH(Me)SO₂Me   t-BuCH₂CH₂SO₂Me

10.41   **10.42**

This example was included to show a "worst case" as far as the ambiguity involving *trans*-1,2 exchange of substituents is concerned, and the possible complications were described in their full detail. As previously mentioned, most cases would not be so complicated, since the chemical nature of the problem generally limits considerably the possible structures that need to be considered. The remaining examples in this section illustrate this.

In the next two cases, the only distinction sought was between *E* and *Z* isomers. The first involves another addition reaction to an acetylene. Reaction of bromine with *N*-propargylacetamide (**10.43**) in carbon tetrachloride produces both the *Z* and the *E* dibromides (**10.44** and **10.45**), which were distinguished by the observation, for the *Z* isomer (**10.44**) only, of a 15% enhancement of the vinylic proton on irradiation of the CH_2.[28] This experiment apparently employed a very simple 60 MHz CW spectrometer (Varian EM 360A), showing that in cases as favorable as this, NOE measurement need not necessarily be a specialist's technique!

MeCONHCH₂C≡CH

10.43

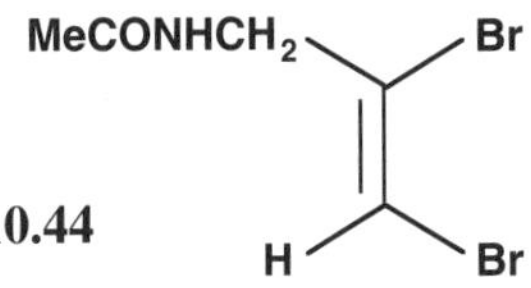

10.44 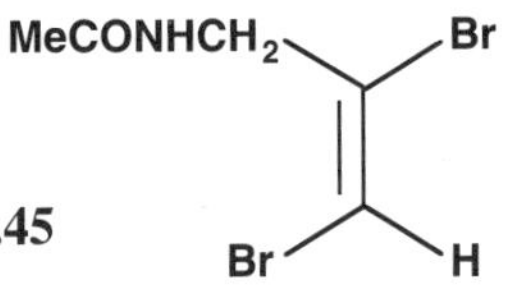   **10.45**

418 APPLICATIONS OF THE NOE TO STRUCTURE ELUCIDATION

A very similar approach was used to demonstrate that simple lithium ene-thiolates (e.g., **10.46**), formed from the corresponding thioamides, have the Z configuration, since the products of quenching with trimethylsilyl chloride are themselves Z. Thus, for the product, N,N-dimethyl-S-silylketene-S,N-acetal **10.47**, the Z configuration is established by the enhancements of the vinyl proton indicated on the structure.[29]

10.46

10.47

A somewhat more complicated example is provided by the structure elucidation of cyanobacterin, an antibiotic isolated from a species of freshwater

TABLE 10.3 NOE Enhancement Data for Cyanobacterin 10.51 and Anhydro Derivatives 10.53 and 10.54[30] [a]

Proton Irradiated	Protons Enhanced (%)		
	10.51	**10.53**	**10.54**
2	7(8), 11(8)		
5α	2(5), 7(4), 11(5), 14(2), 5β(25)		11(9), 7(7), 14 + 15(2)
5β	11(4), 13(7), 5α(23), 7(5)		
7	2(4), 5α(2), 5β(2)		
OH	2(10), 16(7)		
11	2(4), 5α(2), 5β(2)		
13	14(2), 15(2), 7(2), 5β(5), 11(3)	16(11), 14 + 15(2)	18 + 22(3), 14 + 15(2)
14	5α(3), 5β(4), 13(10)	7(3), 5(2), 11(13), 16(10), 13(15)	18 + 22(3), 5(5), 7(4), 11(6), 13(20)
15	13(12), 16(12)		
16	15(1), 18 + 22(9)	18 + 22(9), 13(12), 14 + 15(1)	Obscured
18 + 22	16(15), 19 + 21(14)	19 + 21(13), 16(12)	19 + 21 + 16(9), 13(6)
19 + 21	23(5), 18 + 22(13)		18 + 22(8), 23(10)
23	19 + 21(9)		

[a] Percentage NOE enhancements measured by difference at 361 MHz, quantitated by peak height (±10%). Compound **10.51** is in benzene-d_6, **10.53** and **10.54** are in $CDCl_3$. Protons H_5 and H_{19+21} were not irradiated in compound **10.53**.

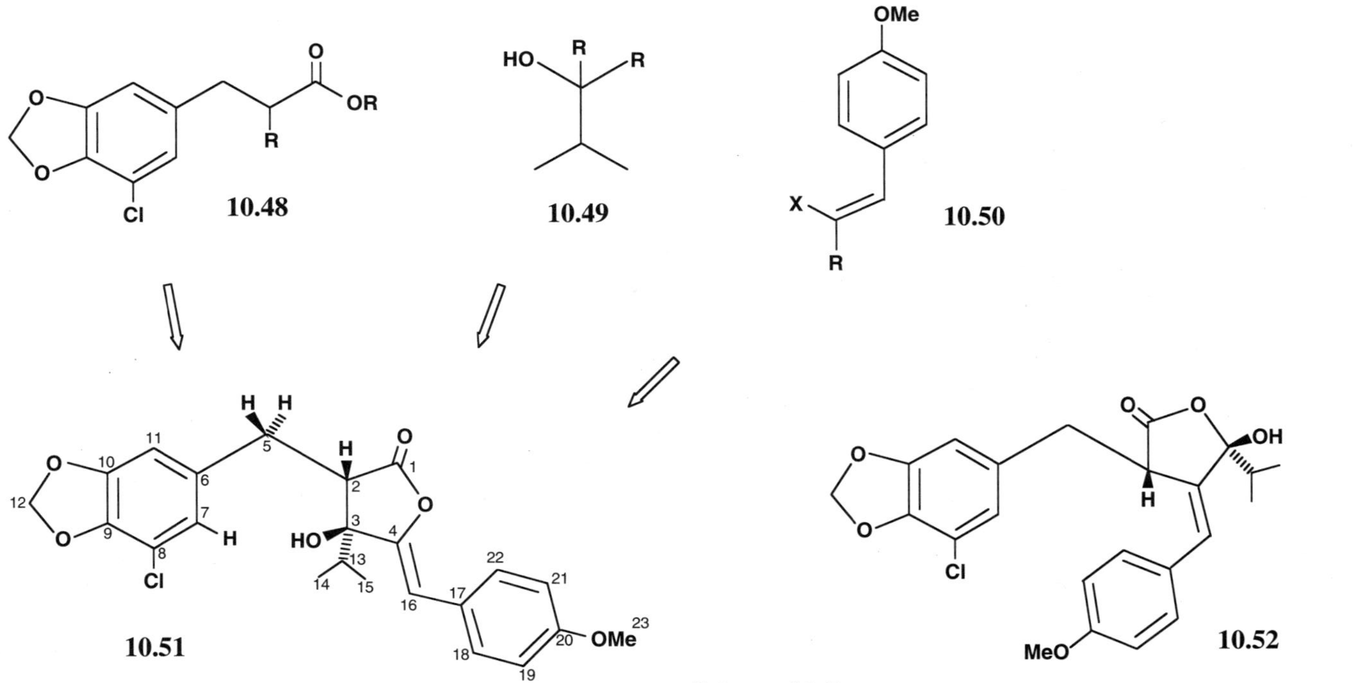

Scheme 10.5

blue-green alga.[30] As is often the case, NOE experiments were used at a fairly late stage in the study, to connect structural fragments already found by other means. Three partial structures, **10.48**, **10.49**, and **10.50** (Scheme 10.5), had been deduced using a combination of mass spectroscopy, IR, UV, and normal 1D NMR, and the substitution pattern of the aromatic ring in fragment **10.48** was established using NOE difference experiments in which the benzylic protons were irradiated. The correct assembly of these fragments, within the constraints of the molecular formula $C_{23}H_{23}O_6Cl$, was deduced largely using the NOE results summarized in Table 10.3. Enhancements linking the vinylic proton with both the *p*-methoxyphenyl group and the isopropyl group established the configuration of the double bond, while the various enhancements involving H_2 establish the relative stereochemistry of the five-membered ring. However, these results are consistent with two structures, **10.51** and the pseudoacid structure **10.52**. The distinction between them might have been possible by a very careful quantitation of all the enhancements and comparison with calculated enhancements for both structures, but, in spite of first appearances, the patterns of enhancements expected for structures **10.51** and **10.52** would probably not differ very much. The distinction was actually made on chemical grounds (failure of cyanobacterin to react with diazomethane, and dehydration to yield structure **10.53**). Similar studies were carried out for the two anhydro derivatives of cyanobacterin, **10.53** and **10.54**, isolated from the same source. Results for these compounds also appear in Table 10.3.

10.5. SATURATED RING SYSTEMS: SIMPLE CASES

Potentially, the title of this section probably includes within its scope a greater number of applications of the NOE than all other sections combined. Our introductory remarks as to the problem of reviewing all of chemistry are thus particularly applicable here. It would be extremely tedious, and (in the authors' view) also quite unhelpful, to attempt to cover this area according to chemical structure types. Instead, the examples have been chosen to try to illustrate as many aspects as possible of the *interpretation* of NOE results from simple saturated ring systems.

As before, one-dimensional difference experiments are generally the method of choice for these studies. The molecules almost invariably fall within the positive NOE regime, and their spectra are not so complicated that relevant and accessible preirradiation targets cannot be found; even one line of a multiplet will suffice as a target if necessary (Section 7.2.5). However, with such a variety of applications there must be exceptions to both points, even if none is included here, and in such cases 2D NOESY experiments may sometimes be preferable. The factors involved in choosing between the 1D and 2D methods are considered in Section 8.1.

In some molecules, there are simply not enough hydrogen atoms in the right places for a homonuclear NOE experiment to be informative. When this is the case, a heteronuclear NOE experiment may be useful. Such experiments are technically somewhat more demanding and, apart from enhancements involving ^{19}F, much less sensitive than interproton NOE measurements, but they are feasible on most modern FT NMR spectrometers. A few applications of selective ^{13}C{^{1}H} NOE experiments have appeared, and some are included in this section. In future the HETGOESY sequence (Section 7.5.3.3) may make ^{1}H{^{13}C} experiments attractive.

The interpretation in the examples below can sometimes be as simple as that for the largely "flat" problems considered in the preceding three sections. However, there is no doubt that the potential for complication is much greater in these cases because of the three-dimensional nature of the problems. While almost every issue in the preceding sections could be framed in terms of groups being either "adjacent" or "nonadjacent," the same is no longer necessarily true here. Geometrical relationships between groups may be quite subtle, and it is always wise to examine models of the structures under consideration, using either accurate physical models or a suitable molecular modelling computer program. It is particularly easy to be misled by conventions in structural drawings and formulas. Very often, the constraints of producing a clear two-dimensional picture, or of emphasizing some particular structural aspect, result in a grossly distorted impression of the true three-dimensional structure. This may seem an obvious point, but confusions of this sort can be surprisingly insidious.

However, the most significant reason for caution when dealing with NOE results from saturated ring systems is their greater scope for internal flexibility. The aromatic rings and double bonds of the preceding sections provided highly rigid frameworks, and little consideration had to be given to internal motions except of flexible side chains. The rigidity of acyclic systems cannot be taken for granted in this way. As we shall see, the greater the degree of internal

422 APPLICATIONS OF THE NOE TO STRUCTURE ELUCIDATION

flexibility in a molecule or fragment, the less use can be made of NOE enhancements to deduce its structure. Thus, while polycyclic caged structures are essentially rigid, and their NOE enhancements can be interpreted accordingly, monocyclic medium ring compounds often exist distributed among many different ring conformations in rapid mutual exchange, and their NOE enhancements are correspondingly ambiguous. One should therefore always consider the scope for internal flexibility when interpreting NOE results from saturated ring systems.

The examples that follow are divided rather arbitrarily according to the type of information gained from the NOE studies.

10.5.1. Substituent Stereochemistry

For monocyclic systems, applications of the NOE to problems of substituent stereochemistry are probably commonest for five-membered rings. Reasons for this are not hard to find. As commented earlier, seven-membered or larger monocyclic rings are generally too flexible for successful NOE studies, while three- and four-membered rings, although sufficiently rigid, are somewhat rarer than five-membered rings. For cyclohexanes, stereochemical analysis is uniquely facilitated by the characteristic vicinal coupling constants found for the chair conformation, so that very many stereochemical problems can be settled simply by analysis of couplings in the normal 1D NMR spectrum.

The first three examples are of applications to five-membered rings. In the first, the stereochemistry of the two diastereomeric thiazolidines **10.56** and **10.57** was established using the enhancements shown on the structures (the size of the enhancement for **10.56** was not reported).[31] These two molecules, prepared by condensation of β-mercapto-DL-isoleucine (**10.55**) with acetone, were then oxidatively cleaved to provide chromatographic standards (**10.58** and **10.59**) with which to correlate the stereochemistry of a penicillin degradation product (Scheme 10.6). The degradation product itself was not isolated preparatively. The assignment of the C_5 methyl signal irradiated in compound **10.57** was checked by repeating the condensation reaction with acetone-d_6.

The primary interest in this study was to establish the stereochemistry of the process by which the enzyme system converts the tripeptide **10.60** into the penicillin **10.61**. That this approach succeeded is a powerful demonstration of how chemical correlation can greatly extend the usefulness of the NOE. However, a more direct alternative approach is also possible in this case, and this will be considered at the end of this section.

Another, simpler, example of chemical correlation is provided by work on synthetic statine analogs.[32] The protected diamino acids **10.62** and **10.63** are synthetic precursors of the two diastereomers of aminostatine (3,4-diamino-6-methylheptanoic acid), and a means of distinguishing them was required. As in the previous example, it was necessary to eliminate the flexibility of these linear molecules before an NOE study was feasible; this was done by cyclization using phosgene, to give the diastereomeric cyclic ureas **10.64** and **10.65**

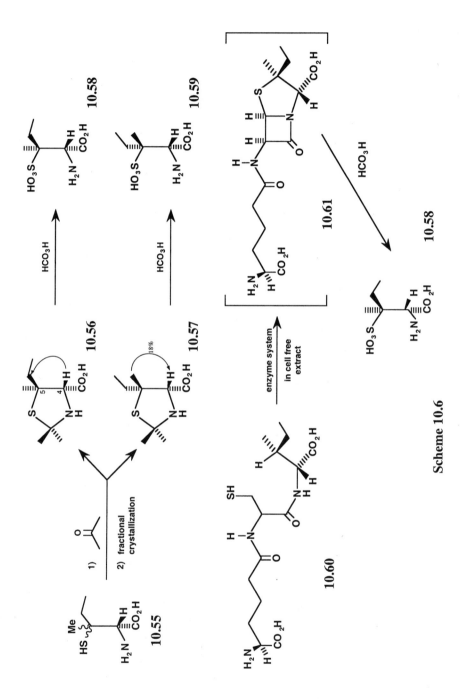

Scheme 10.6

(Scheme 10.7). The pattern of several enhancements shown around each structure quite conclusively proves the relative stereochemistry for these ureas, and hence also for the parent diamino acids. As commented in earlier examples, although some of these enhancements might at first appear redundant, measuring a large number of interlocking connectivities in this way leads to a much more secure result, and is certainly preferable whenever the structure and spectrum allow. In this particular case, additional confirmation of the result was available in the form of an X-ray crystal structure of the methylbenzylamide of compound **10.62**.

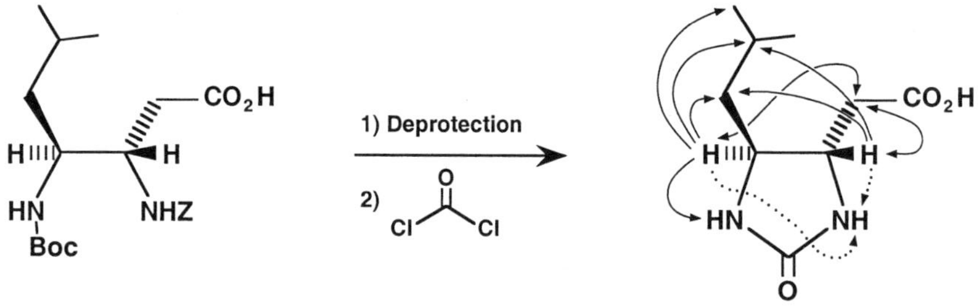

Scheme 10.7

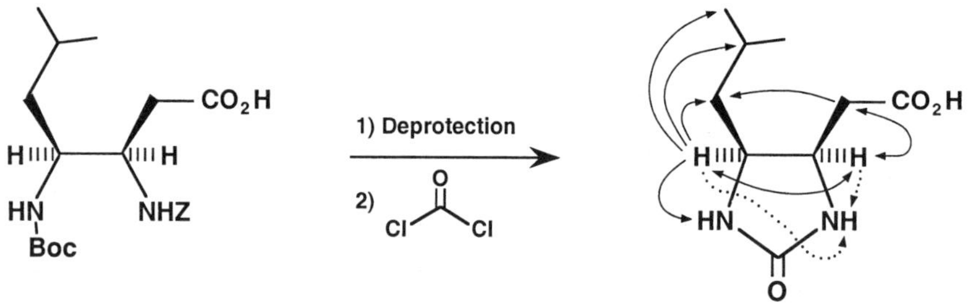

(dotted arrows indicate apparent enhancements actually caused by saturation transfer; see text)

Notice that no enhancements were observed into the NH portion adjacent to the CH_2CO_2H group in either urea; although it has no bearing on the structure determination, the cause of this is interesting more generally. It arises due to two effects of the carboxylate function. First, the rate of exchange between the CO_2H proton and the protons of the residual water in the DMSO is such that both resonances are extensively broadened, the water line being over 1 ppm wide (~$\delta 2.8-4.0$). Thus, irradiation of other resonances in this area (including the methine protons on the urea ring) also causes partial saturation of the water signal. Second, the NH in question must undergo much faster exchange with residual water than does the other NH. Presumably, the reason for this is specific catalysis of this exchange reaction by the carboxylate group, which is very accessible to the "right-hand" NH, but not to the other. Consequently, irradiation of either urea methine resonance causes significant saturation transfer from the underlying H_2O signal into the "right-hand" NH (marked with dotted lines on the structures), obscuring any possible NOE enhancement. Effects due to saturation transfer from even trace amounts of residual water are quite commonly encountered in NOE studies, and their elimination by rigorous drying of both solvent and solute is at best tedious and often impossible. In the present case, as mentioned earlier, the effects were irrelevant to the structure determination; nevertheless it is always important to be aware of possible complications from this source.

The next example shows another way in which conformational freedom has been reduced by chemical modification to aid an NOE study. Ephidrene (**10.66**) and pseudoephidrene (**10.67**) form oxazolidines such as **10.68** and **10.69** on condensation with aldehydes (Scheme 10.8.)[33] These are used as synthetic intermediates in various chiral transformations, so it is important to know the stereochemistry at C_2 in the adducts. Irradiation of the C_2 Me group gave a small enhancement of H_5 in compound **10.69**, but not **10.68**, consistent with the structures shown, but it was felt that more conclusive proof was desirable. This was provided by the NOE enhancements shown for the borane complexes **10.70**, **10.71**, and **10.72**. Complexation with BH_3 achieved two things. First, it fixed the configuration of the nitrogen atom, preventing inversion and making the enhancements involving the NMe group stereochemically specific. Second, it provided an additional convenient protonated group between C_2 and C_4, from which stereospecific enhancements could be measured.

For all of these examples, interpretation depends on the fact that, although the rings are not completely rigid, the various conformations available to them do not differ very much. Geometrical relationships between groups therefore do not differ *significantly* between conformations, and there is no need to consider conformations in detail. The next few examples show, to an extent, where a limit must be placed on such interpretation.

Compounds **10.73**–**10.76** are bicyclo[5.1.0]octane derivatives, prepared as part of a strategy directed toward sesquiterpene synthesis.[34] For the present purpose, they may be viewed as partially rigidified cycloheptane derivatives. The stereochemistry of the two methoxy compounds, **10.73** and **10.74**, was established by the observation, for **10.73** but not for **10.74**, of an enhancement to the C_2 methine proton on preirradiation of one of the methyl singlets. In compound **10.74**, neither methyl group gave an enhancement to the C_2 methine proton. The success of these NOE experiments probably depended largely on the rigidity imparted by the cyclopropane ring, which must limit considerably the possible relative movements of groups within this portion of the molecule. For compounds **10.75** and **10.76**, however, the stereochemical ambiguity lies within a more flexible region of the molecule, and the original authors state that "an unambiguous assignment of stereochemistry could not be made based solely on NMR." The distinction between them was in fact made using X-ray crystallography.

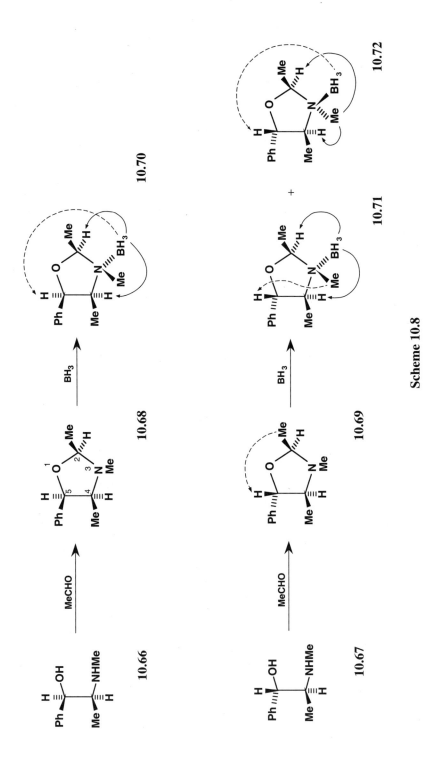

Scheme 10.8

(dashed arrows indicate enhancements of < 6%)

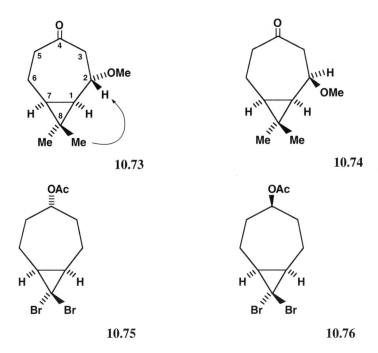

10.73 10.74

10.75 10.76

A somewhat similar example is provided by the hydroxyhomocepham **10.77**, isolated during work on the mechanism of penicillin biosynthesis.[35,56] In this case, a sufficiently clear set of enhancements (summarized on the structure) was available to establish the stereochemistry at C_3.

10.77

Quite possibly, the difficulties involved with acetates **10.75** and **10.76** had as much to do with spectroscopic features as with conformational ones, and it is very hard to assess the relative internal flexibilities of these compounds and the hydroxycepham system. Nonetheless, it is fairly clear that these examples are all quite close to a limit, set by internal flexibility, on what structural distinctions are feasible using the NOE.

The next example illustrates the determination of both stereochemistry and regiochemistry in a spiro-fused cyclohexane derivative (**10.78**).[37] As mentioned

428 APPLICATIONS OF THE NOE TO STRUCTURE ELUCIDATION

earlier, coupling constants are a particularly powerful tool for stereochemical analysis of cyclohexanes, but there are several types of problem to which they are unsuited, and for which NOE experiments succeed. The present case includes two such types. First, determination of stereochemistry at a disubstituted ring carbon, in this case a spiro ring junction, and second, identification of sites of attachment for proton-bearing substituents that do not show couplings to the cyclohexane ring protons.

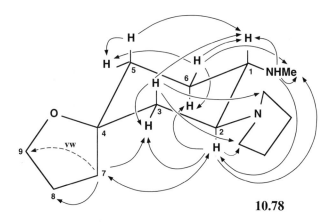

10.78

vw = very weak

This problem was attacked using a combination of 1D NOE difference and 2D phase-sensitive double quantum filtered (DQF)-COSY experiments. All the protons of spiro-diamine **10.78** were readily assigned using the DQF-COSY spectrum, and the pattern of NOE enhancements recorded in the 1D experiments is shown on the structure. The chair conformation is clearly established both by the characteristic 1,3 diaxial enhancements involving the axial protons at C_1, C_3, and C_5, and by the large diaxial couplings ($^3J \approx 11.5$ Hz) apparent in the DQF-COSY spectrum. The relative positions of the N-methyl and pyrrolidyl groups could not be determined from the spin-coupling network alone (the NH proton showed no detectable couplings), but were established from the NOE results, principally by the enhancement of the NMe singlet on preirradiation of the equatorial proton at C_6. Enhancements involving H_1 and H_2 would be ambiguous for this purpose, since either proton could be close to either substituent. The stereochemistry at the spiro-ring junction is shown by the enhancements between H_2 and the C_7 methylene multiplet. The latter is largely buried under nearby signals from the pyrrolidyl C^β protons, and the equatorial protons at C_3 and C_5 (Fig. 10.6). Nonetheless the DQF-COSY spectrum clearly shows that two exposed lines just to one side of this envelope are part of the C_7 methylene multiplet, so making a selective irradiation possible. Without this information from the 2D spectrum, it is unlikely that the problem could have been solved using 1D difference NOE experiments.

The next example also involves cyclohexane rings, but in this case two were fused together in a *trans*-decalin system. Compound **10.79** was synthesized as part of a general strategy directed toward diterpene insect antifeedants, particularly ajugarin and clerodin, and an NOE study was undertaken to check the stereochemical outcome of the synthesis.[38] The enhancements recorded are summarized on the structure. The most informative enhancements are those between the methine proton H_C and the epoxy proton H_B. These very strong enhancements clearly indicate that these protons are close together, an arrangement that is achieved only in **10.79** or in one of the other possible stereoisomers, **10.80**. However, structure **10.80** can be ruled out because it cannot account for the diaxial coupling between H_C and H_F ($^3J = 10$ Hz). The very short distances between H_A, H_B, and H_C imply that these spins form an almost self-contained system as far as the NOE is concerned; consequently, large three-spin effects are seen when either H_A or H_C is preirradiated. The strong enhancement between the axial methyl group and H_D further supports structure **10.79**, and also reveals a strong rotamer preference within the CH_2OAc group.

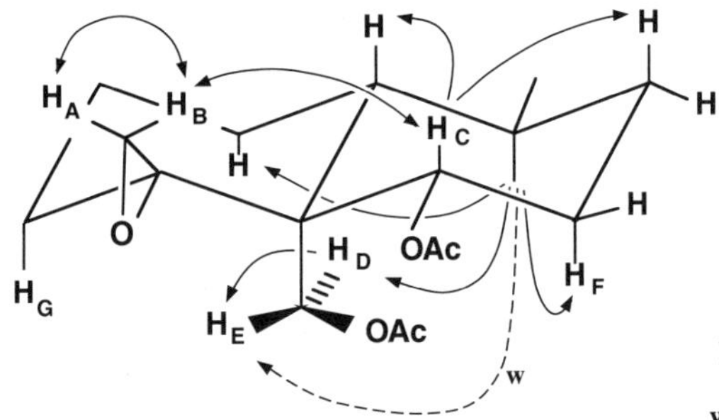

10.79

w = weak

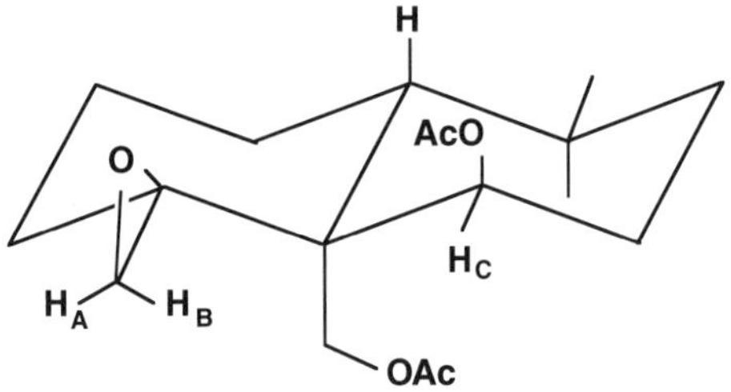

10.80

These conclusions are further evidenced by two long-range couplings that follow a favored "planar-W" pathway, that is $^4J(H_B, H_G) = 2$ Hz and $^4J(H_E, H_C)$ < 0.5 Hz (visible as a broadening of H_E that is removed on irradiation of H_C). Note that the latter coupling and the enhancement of H_D from the axial

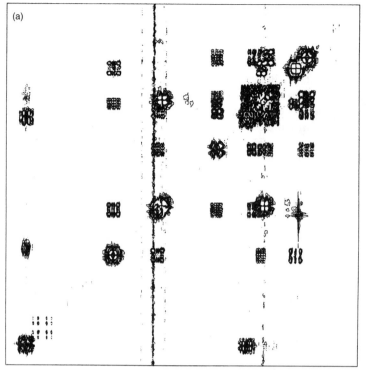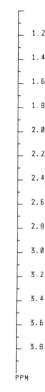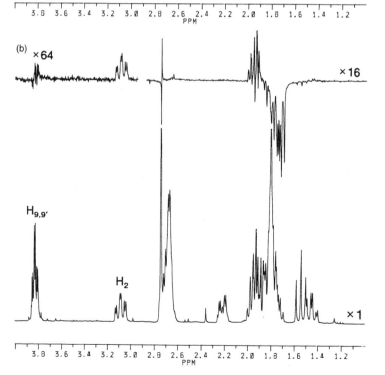

430

methyl group *together* constitute a stereospecific assignment of the prochiral CH$_2$OAc group resonances. This example well illustrates the combined use of enhancement and both vicinal and long-range coupling data, which approach is particularly well suited to the analysis of six-membered rings.

The power of the editing ability of NOE difference spectroscopy to produce useful results from complex mixtures was mentioned earlier (Section 10.2). The next example is a particularly striking demonstration of this ability, in that the stereochemistry of the oxidized penicillin derivative **10.81** was determined

10.81

directly within a highly complex biochemical mixture of cell constituents after only minimal sample preparation.[39] This experiment represents an alternative approach to exactly the same problem described in the first example of this section. The tripeptide **10.60** (Scheme 10.6) was incubated with a cell-free preparation from the microoganism *Cephalosporium acremonium*, then the mixture was deproteinated (acetone precipitation) and oxidized using sodium periodate. The motive behind the oxidation step was to alter the conformation of the thiazolidine ring; in the sulfone form, the C$_2$ α-substituent and H$_7$ are held more closely together than in the parent penicillin, so increasing their

---

Figure 10.6. (*a*) NOE difference spectrum (preirradiation 5 s), and (*b*) phase-sensitive double-quantum filtered COSY spectrum of the spiro-diamine **10.78** in CDCl$_3$, recorded at 300 MHz. Every proton in the spectrum can be assigned by analyzing the DQF-COSY spectrum. In particular, the multiplet due to the C$_7$ methylene group can be located using (among other things) the weak cross-peak due to long-range coupling with the trivially assignable H$_{9,9'}$ multiplet. This shows that the two exposed lines just to high field of the complicated envelope at δ1.7–δ2.0 form part of the H$_{7,7'}$ multiplet, and these lines can therefore be used as a selective preirradiation target to probe the stereochemistry of the spiro ring junction. The NOE difference spectrum in (a) shows the result of this preirradiation. Since a (composite) 90° observe pulse was used, the saturation *appears* to be spread into all lines of the H$_{7,7'}$ multiplet, but nonetheless the preirradiation was genuinely selective for the highest field line at δ1.70 ($\gamma B_2/2\pi = \sim 2$ Hz; cf. Sections 6.2 and 7.2.5). The enhancement of H$_2$ in this spectrum clearly demonstrates that the spiro ring junction has the stereochemistry shown, rather than that with the oxygen atom axial.

432 APPLICATIONS OF THE NOE TO STRUCTURE ELUCIDATION

mutual NOE enhancements. The NOE spectrum of the resulting mixture is shown in Figure 10.7, and the stereochemistry at C_2 is quite clearly established by the enhancement at H_7 on irradiating the methyl singlet. As may be seen, the subtraction process efficiently removes the myriad other signals present in the normal 1D spectrum, leaving only the information of interest.

10.5.2. Ring Fusion Stereochemistry

Unlike the examples of aromatic ring fusion in Section 10.2, saturated ring systems often include bridgehead protons or proton-bearing substituents. One may therefore tackle problems of ring fusion stereochemistry in two slightly different ways. In the first, ring fusion stereochemistry is deduced by determining the stereochemistry of bridgehead substituents or protons at the relevant ring branching points. In the second, protons further away from the ring junctions are used to relate those portions of two rings which come close to one another.

There is generally not much choice to be made between these approaches; they are dictated by the nature of the problem, and it is in any case wise to collect as many interlocking enhancements as possible. However, it is worth pointing out that bridgehead protons have a number of advantages in such NOE studies. They are necessarily methine protons, so that enhancements *into* them are quite likely to contain long-range information (Section 3.2.2.1). Second, they are likely to be at particularly rigid parts of the molecule, and are often held in close proximity to other groups.

In the first example, which was one of the first applications of NOE difference spectroscopy to a small molecule, the stereochemistry of ring fusion was deduced from the *cis* relationship between two bridgehead groups.[40] Figure 10.8 shows the normal and NOE difference spectra of flustramine A (**10.82**); both of the diastereotopic methyl groups of the inverse isoprene unit at C_{3a} were saturated. The strong enhancement of H_{8a} clearly proves the *cis* stereochemistry at the ring junction. Other enhancements are just those expected from the structure; note the three-spin effects at H_5 and H_{18} (*trans*), the latter being transmitted through *both* H_{17} and H_{18} (*cis*). Of course, this very direct approach of measuring enhancements between two bridgehead groups on the same ring junction will only give a positive result when the ring fusion is *cis*. The considerable dangers of interpreting negative results (i.e., absent enhancements) from NOE experiments were pointed out earlier.

Figure 10.7. ^{1}H NMR and NOE difference spectra of a cell-free incubation mixture (in D_2O) containing the oxidized penicillin derivative **10.81**, recorded at 300 MHz. Reproduced with permission from ref. 39.

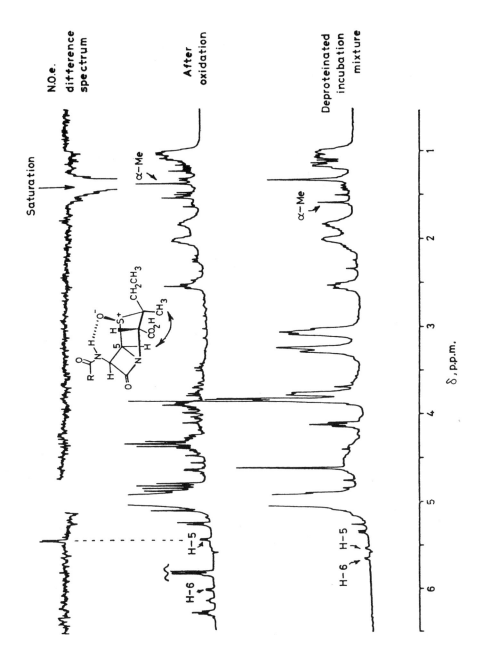

434   APPLICATIONS OF THE NOE TO STRUCTURE ELUCIDATION

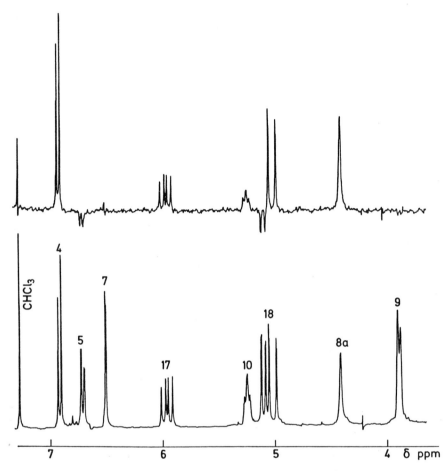

Figure 10.8. NOE difference spectra (low-field region only) of flustramine A **10.82** in CDCl$_3$, recorded at 270 MHz (preirradiation time not given). Simultaneous preirradiation of both C$_{14}$ methyl signals (off-scale to the right) causes a strong enhancement of H$_{8a}$, showing the ring fusion to be *cis*. Other enhancements are as expected from the structure. Reproduced with permission from ref. 40.

An example of the second approach is provided by some work on face selectivity in Diels–Alder reactions of caged dienes.[41] Reaction of tetracyanoethylene with the caged dichlorodiene epoxide **10.83** gave a mixture of the *endo* and *exo* products **10.84** and **10.85** (Scheme 10.9). These products were not separated, but in order to assess their relative proportions by NMR it was necessary to assign the resonances of each compound in the spectrum of the mixture. This was done using the NOE difference spectrum shown in Figure 10.9. The singlets near $\delta 6.3$ arise from the (equivalent) methine protons at C$_3$ and C$_6$ in the two compounds in the mixture, while the C$_{11}$ and C$_{12}$ methylene groups of both compounds contribute sets of signals near $\delta 2.1$ and $\delta 1.3$. The

SATURATED RING SYSTEMS: SIMPLE CASES   435

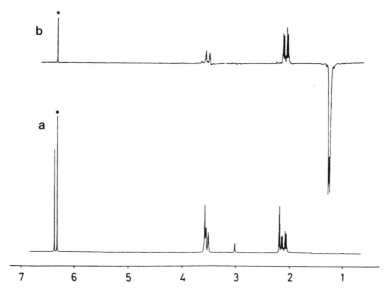

signals near δ1.3 correspond to those methylene protons further from the epoxide (in both compounds) and, when they are saturated, enhancements occur at the other methylene protons of both compounds and the bridgehead protons of both compounds. However, the key enhancement is that of the H_3/H_6 singlet, since this occurs only in the *endo* compound **10.84**. The relative assignment is thus established for the two H_3/H_6 singlets, and the face selectivity of the reaction can be assessed by comparing their integrals. Notice also the direction

Figure 10.9. NOE difference spectrum (*b*) of a mixture of the *endo* and *exo* products **10.84** and **10.85** in acetone-d_6, recorded at 360 MHz (preirradiation time not given). Only in the *endo* product **10.84** is the $H_{3/6}$ singlet (marked with an asterisk) enhanced on preirradiation of the methylene protons further from the epoxide. Trace (*a*) shows the preirradiated spectrum. Reproduced with permission from ref. 41.

of the enhancement; it is unlikely that the reverse enhancement (i.e., $H_{11}/H_{12}\{H_3/H_6\}$) would have been so easy to detect, since it would be *into* a methylene group (cf. Section 3.2.2.1). Several related reactions were investigated similarly.

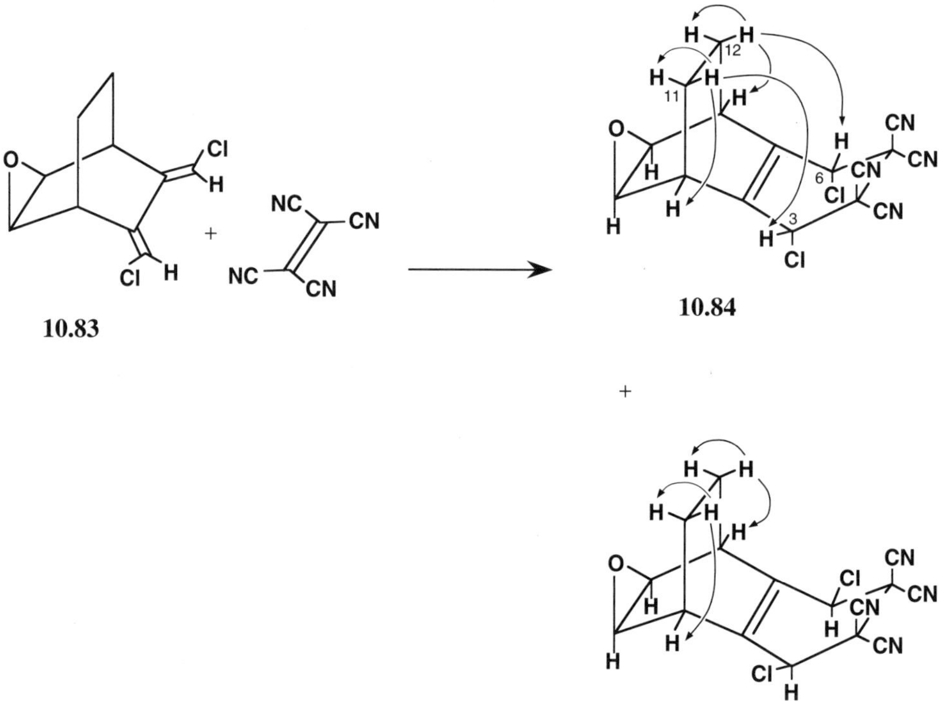

Scheme 10.9

The key enhancement in this example owes its existence to the particular arrangement of the rigid molecular skeleton in compound **10.84**. In effect, the bonding pathway between the interacting protons "turns a corner," bringing them close together in spite of the large number of intervening bonds. Quite clearly, this is a characteristic and necessary structural feature if inter-ring NOE enhancements are to be used to solve problems of ring fusion stereochemistry. The most striking example of such an inter-ring enhancement is probably that between H_A and H_B in the half-cage acetate **10.86**, featured in Anet and Bourn's original application of the NOE in 1965,[1] but some more recent work on spiro derivatives of penicillin also illustrates the point well.

Scheme 10.10 shows several trapping reactions of the penicillin-6-carbene species **10.87**, the products of which (**10.88–10.93**) obviously have considerable scope for stereochemical ambiguity. As in the previous examples, there are pairs of protons that are held close together in spite of being "distant" in

10.86

terms of the bonding pathway. These close contacts lead to particularly useful enhancements, summarized on the structures, which interrelate the two halves of the molecules and establish the ring fusion stereochemistry in each case. The several other enhancements observed in each molecule are here omitted for clarity.[4,42]

The structure determination of the proaporphine alkaloid misramine (**10.94**) provides a somewhat more complicated example of the determination of ring fusion stereochemistry, though the principles are just the same as in previous examples. Most of the structural features of the molecule were deduced from a combination of IR and UV spectroscopy, and a spin-decoupling study of the normal 1D NMR spectrum.[43] NOE enhancements around the isoquinoline portion (summarized on the structure) helped to piece together this part of the molecule, but the remaining partial spin systems could be fitted together within the constraints of the molecular formula $C_{19}H_{25}O_4N$ in either of two stereoisomeric structures, **10.94** and **10.95**. The interlocking set of enhancements summarized on the structure resolved this ambiguity. In particular, the enhancements involving $H_{6\alpha}$ establish that this proton is *cis* with respect to the C_8 methylene group across the five-membered ring, so ruling out structure **10.95**. This *cis* relationship is shown both by direct, cross-ring enhancements between $H_{6\alpha}$ and H_8 equatorial, and also by the enhancements that both these protons give to $H_{7\alpha}$. These and other enhancements also confirmed the connectivities between the three partial spin systems at $C_{6\alpha}$–C_7, C_8–C_{10}, and C_{12}. (For numbering scheme, see structure **10.95**. These numbers refer to the same absolute positions in both structures, so that for structure **10.94** the dioxygenated carbon is C_{11} and the CH_2CH_2CHOH fragment runs from C_8 to C_{10}.)

Another example in which the stereochemistry of several rings was simultaneously determined using the NOE is provided by the dimerization product, **10.97**, of 4a-methyl-4a*H*-fluorene (**10.18**).[44] Spin decoupling experiments established the coupling constants shown on the partial spin system **10.96**, which suggested that the product resulted from a Diels–Alder cyloaddition. However, the expected coupling between H_I and H_J was absent, and an NOE study was undertaken both to explain this, and to distinguish between the many plausible stereoisomers. The results appear in Table 10.4 and are summarized around the structure.

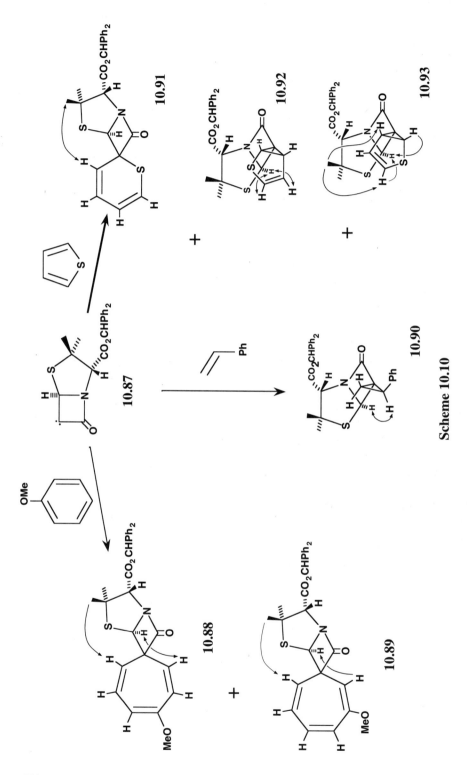

Scheme 10.10

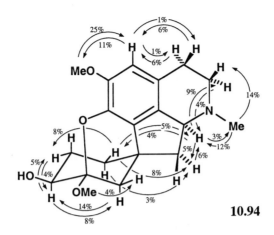

10.94

10.95

10.96

TABLE 10.4 NOE Enhancement Data for 4a-*H*-Fluorene Dimer 10.97[44]

Proton Preirradiated	Protons Enhanced (enhancements quoted to ±0.5%)
G	A(4.5%), D(8%), F(5%), H(4.5%)
H	A(3.5%), F(4.5%), G(4.5%), J(5%)
I	E(5.5%), J(3.5%)
J	H(5.5%), I(4%), Me$_A$(1%)
Me$_A$	B(2%), C(1.5%), H(6%), J(14.5%)
Me$_B$	E(5%), I(7.5%)

As H$_J$ (δ1.68) is clearly not olefinic, the strong enhancement H$_J${Me$_A$} shows that the 2π component in the Diels–Alder reaction was provided by the terminal double bond of the triene system. The remaining enhancements at H$_J$ show that this 2π component approached the other 4a*H*-fluorene in such a way that both methyl groups were facing away from the reaction center. Given the proximity of H$_J$ and Me$_A$, the spin system **10.96** establishes which way round the 2π component was during reaction, but this conclusion is extensively reinforced by all the other enhancements in this region of the molecule. This mutual orientation is in fact just what one might expect, since it minimizes both steric interactions and loss of conjugation on reaction. As for the missing coupling between H$_I$ and H$_J$, models of structure **10.97** show the torsional angle between these protons to be close to 90°, largely as a result of a severe twisting of the bicyclo[2.2.2]oct-2-ene moiety caused by fusion of the indene unit to one edge.

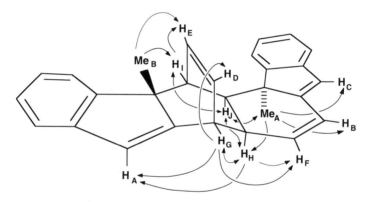

10.97

Finally, we consider a case in which selective ^{13}C{^{1}H} NOE enhancements were used to deduce the stereochemistry and ring fusion pattern of a naphthoquinone derivative.[45] Reaction of 2-methylnaphthoquinone with 2-diazopropane gives a bis-adduct. A preliminary inspection of the ^{1}H and ^{13}C spectra of the adduct showed that one diazopropane unit had, as expected, undergone a 1,3-cycloaddition reaction with the 2,3 double bond of the quinone, while the second had formed an epoxide by attacking one of the carbonyl groups. This still left a large number of possible regio- and stereoisomers to be distinguished. Given that, as usual, both the structure and assignments had simultaneously to be determined, and that the molecule is almost devoid of useful interproton couplings, it is not surprising that initial ^{1}H{^{1}H} NOE experiments gave ambiguous results. In order to resolve the regiochemical ambiguities, therefore, a selective ^{13}C{^{1}H} NOE spectrum was recorded in which H$_{3a}$ was irradiated (Fig. 10.10). Note that no enhancement is observed *or expected* at the directly bonded carbon C$_{3a}$, since in the isotopomers with ^{13}C at C$_{3a}$, proton H$_{3a}$ reso-

SATURATED RING SYSTEMS: SIMPLE CASES 441

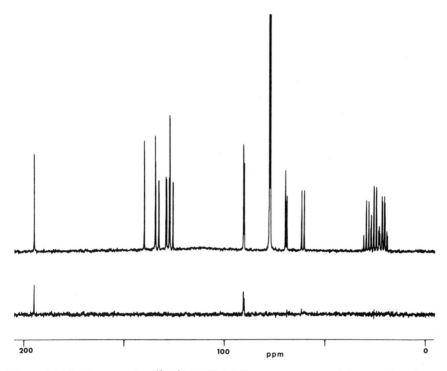

Figure 10.10. Heteronuclear $^{13}C\{^{1}H\}$ NOE difference spectrum of the naphthoquinone epoxide **10.98** in CDCl$_3$, recorded at 100.6 MHz (for ^{13}C; preirradiation period not given). The enhancements at C$_3$, C$_4$, and C$_{9a}$ on preirradiating H$_{3a}$ prove the regiochemistry of addition of the diazomethane fragment, and the position of the remaining carbonyl group. Reproduced with permission from ref. 45.

nates as a very widely spaced doublet, unaffected by the low-power irradiation at its center (cf. Section 9.1.2). Although (^{13}C, ^{1}H) couplings to H$_{3a}$ also occur in other isotopomers of interest (^{13}C$_4$, ^{13}C$_3$, and ^{13}C$_{9a}$), these are multibond (^{13}C, ^{1}H) couplings, which cause only narrow doublet splittings and do not interfere with saturation of H$_{3a}$.

The enhancements observed at C$_4$, C$_3$, and C$_{9a}$ prove two things: (i) the unreacted carbonyl was that adjacent to H$_{3a}$ rather than that adjacent to the C$_{9a}$ methyl group, and (ii) the quaternary carbon of the first diazopropane unit reacted at C$_{3a}$ rather than C$_{9a}$ (this numbering refers to the product in each case). With these facts established, it is possible to make sense of the interproton enhancements, which appear summarized on the structure (**10.98**) (the directions of these enhancements were not reported). The enhancements involving H$_{3a}$ clearly establish the ring fusion stereochemistry to be *cis*, and simultaneously allow stereospecific assignment of the two C$_3$ methyl groups.

10.98

The enhancement from the "lower" of the C_3 methyl groups to one of the (somewhat confusingly numbered) $C_{3'}$ methyl groups on the epoxide ring, together with the absence of any enhancement from either $C_{3'}$ methyl to the C_{9a} methyl, proves the stereochemistry of the epoxide.

10.6. SATURATED RING SYSTEMS: MORE COMPLEX CASES

The division between this section and the previous one is, of course, somewhat arbitrary, but there are common features to the examples that follow. They, and other studies like them, represent probably the most sophisticated applications of the NOE to structure determination (as opposed to conformational analysis). In some ways they are more complex even than the conformational studies of biopolymers described in Chapters 12 and 13, since there are no simple repeating residues or isolated units from which the structures are built up. Consequently, each structure demands its own unique strategy, without the benefit of general rules for possible relationships between repeating fragments.

The principles underlying interpretation in these cases are no different from those illustrated in previous sections. However, the overall interpretation of results from studies such as these is immensely more intricate and more involved than in simpler cases. This is partly because there is more chance of difficulties arising due to spectral overlap, but, just as importantly, it becomes much more complicated to assess all the *implications* of a given structural deduction as it is made. In other words, contradictions between a few of the very many interlocking pieces of evidence can easily escape notice.

This is made all the more serious since, as for the examples in previous sections, interpretation must in general establish both assignments and structure. Our earlier comments about making the maximum number of assignments prior to using the NOE data (Section 10.1.2) thus apply particularly here. In each of the cases described below, the first step in the NMR interpretation was to extract as much information as possible from coupling connectivities, so generating a set of partial spin systems. Newer methods would probably be used to accomplish these tasks today (e.g., phase-sensitive DQF-COSY to find ^{1}H spin-systems and HMBC to link at least some of them; cf. Section 10.1.2), but, overall, the strategy used now would not differ greatly from that in the studies discussed here.

Many of these molecules tumble at rates that put them near the zero-crossing region of the NOE ($\omega\tau_c \simeq 1.12$). Thus, some give positive enhancements, and others negative, depending on their molecular weight, and on conditions of solvent viscosity, solute aggregation, field strength, and temperature (Sections 2.2 and 9.4). On the whole, this is still the territory of 1D difference experiments, either TOE (for the negative NOE regime) or steady-state (for the positive NOE regime). However, ROESY experiments are clearly useful in these cases (Section 9.3), or NOESY experiments might be competitive for molecules far enough into the negative NOE regime. In the future, gradient-assisted experiments such as DPFGSE-NOE may also prove valuable in such studies.

Another earlier comment that applies particularly here concerns the way in which interpretation is reported (Section 10.1.3). Most of the molecules in this category are of natural origin, and their overall structure elucidation requires a combination of many techniques, such as normal NMR, IR, and UV spectroscopy, mass spectroscopy, and isotopic labeling studies. When these have to be reported in one paper together with details of isolation, the pressure to summarize results with minimal discussion of interpretative logic becomes very strong. Although perhaps unavoidable, this trend is unfortunate, since it is in just these cases that interpretation becomes most fallible. A structural diagram annotated with 50 or 100 enhancements may present a very precise summary of all the data, but it is also almost impossible to criticize, if one has no other insight into how it was deduced. As noted previously, such reporting fails to show how reasonable *alternative* structures were discounted.

This overall complexity also explains another feature of such studies; many of them are incomplete. Unlike an X-ray crystallographic structure determination, which generally gives an "all or nothing" result, an NOE-derived structure must be pieced together like a jig-saw puzzle, starting with the easiest fragments and working on to the more difficult. Sometimes the final pieces are too difficult for the NOE to establish. Conversely, if there *is* evidence concerning the whole structure, the ability to fit the last few pieces provides a very powerful indication that previous steps in the interpretation were correct. "Force-fitting" of the last pieces is usually a good indicator of earlier mistakes!

For all this caution, one should not take a negative view of these applications. They represent a remarkably complete characterization of complicated unknown materials. This achievement is the more significant because it is obtained from solution; many of these materials cannot be crystallized for X-ray studies.

A seminal paper in this area was that of Hall and Sanders on spectra of steroids.[3] Although, strictly, this was concerned only with spectroscopic assignment, as in the case of Anet and Bourn's work (Section 10.1), the potential

10.6.1. Pulvomycin

Pulvomycin is a complex macrocyclic antibiotic from *Streptoverticillium netropsis*, with a unique mode of action involving binding to prokaryotic elongation factor Tu. The original paper of Smith et al.[46] describing this work avoids all the pitfalls mentioned earlier, and presents an exceptionally clear and complete account of the way in which the structure of pulvomycin (**10.99**) was built up. The reasons for each step in the argument are elucidated, as are the limitations of interpretation at each stage. This was one of the main reasons for selecting this example, even though only a summary can be presented here.

Fast atom bombardment (FAB) mass spectroscopy showed that pulvomycin has a molecular mass of 838 Daltons, and decoupling difference experiments[3,4] established the partial spin systems **10.100–10.104**. On these diagrams, dashed lines represent allylic couplings (used, *inter alia*, to attach the vinylic methyl groups), and solid arrows represent NOE enhancements. Negative enhancements were observed in all cases, but, as would be expected on grounds of viscosity, these were more intense when DMSO-d_6 was used as solvent than when methanol-d_4 was used. For all seven disubstituted double bonds, the proton coupling data established an *E* configuration. The remaining three trisubstituted double bonds also had *E* configurations, as shown by the NOE data. All five hydroxy protons were identified using the difference spectrum in Figure 10.11, in which the residual water resonance was irradiated, leading to saturation transfer into all exchangeable resonances.

All but one of the interconnections between these partial spin systems were established using the NOE data summarized in partial structure **10.105**. Note that in all but one case, several enhancements establish each connection. The one remaining connectivity was made by deduction—since all but 44 mass

SATURATED RING SYSTEMS: MORE COMPLEX CASES   **445**

10.100   **10.101**

10.102

10.103   **10.104**

10.105

units are accounted for and only two bonding sites remain unspecified in partial structure **10.105**, the missing unit may be assumed to be a lactone unit ($-CO_2-$) bridging between C_2 and C_{21}. The sense of attachment is clear from the chemical shifts of the C_2 methylene protons ($\delta 2.89$ and $\delta 3.14$) and H_{21} ($\delta 4.99$). Arguments against alternative sites of lactone ring closure are given, both on chemical grounds and also using proton and carbon shift data.

CPK models of the macrocyclic ring suggest that it is moderately rigid, with a roughly triangular shape having vertices at C_{12-13}, C_5, and C_{20-21}. The coupling data for the first portion of the side chain (C_{21-24}; see partial structure **10.107**) also show considerably restricted mobility in this part of the molecule. Because of this relative rigidity, it was possible to establish the relative stereochemistries at all of the chiral centers within this part of the molecule, that is at C_5, C_{13}, and $C_{21}-C_{24}$. At C_5, mutual enhancements of H_5 and Me_{42} show the relative orientation of H_5 to the "triene slab" C_6-C_{11}, while the orientation of H_{13} relative to the other end of the same triene unit is shown by mutual enhancements of H_{13} and H_{10}. For the largely rigid portion running from C_{21} to C_{24}, the stereochemistry was established by the interlocking mesh of NOE and coupling data summarized in partial structures **10.106–10.108**; the rigidity of the macrocyclic ring allows no ambiguity about the relative stereochemistry of this portion and the C_5 and C_{13} centers. All of the large vicinal couplings (~ 10 Hz) correspond to antiperiplanar arrangements between the relevant protons,

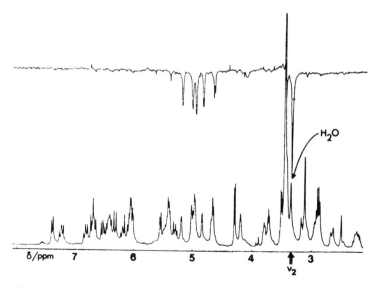

Figure 10.11. Saturation transfer difference spectrum of pulvomycin **10.99** (10–20 mg in 0.5 ml DMSO-d_6), recorded at 400 MHz (preirradition period not given). The water resonance (at $\sim \delta 3.3$) was preirradiated, resulting in negative peaks due to saturation transfer at the resonances of all five hydroxy protons of pulvomycin. Reproduced with permission from ref. 46.

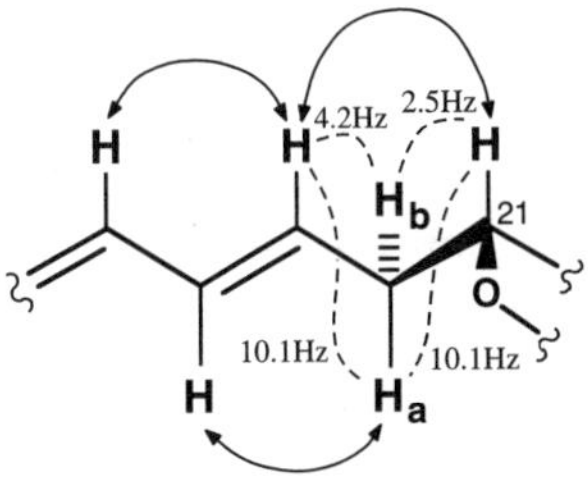

SATURATED RING SYSTEMS: MORE COMPLEX CASES   447

10.106

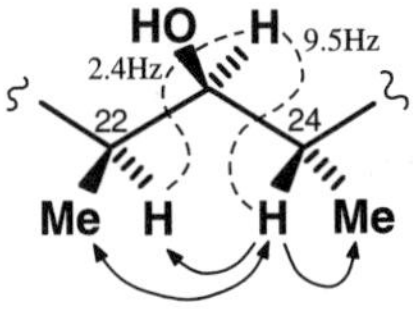

10.107

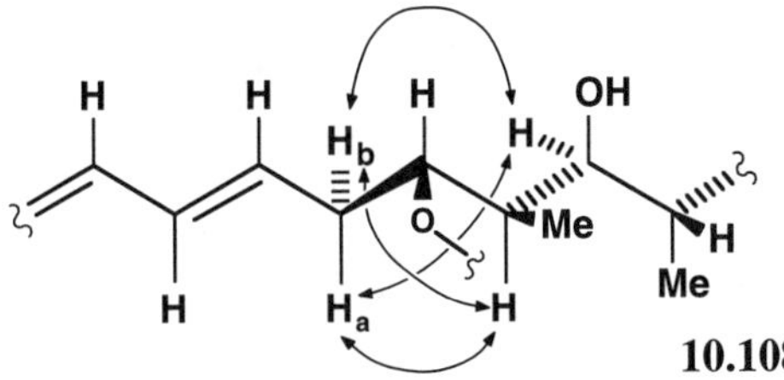

10.108

while gauche couplings (H_{20b}, H_{21}) and (H_{22}, H_{23}) are small (~2.5 Hz), in part due to the presence of an oxygen substituent antiperiplanar to one proton in each case. For the enhancements around the C_{21}–C_{24} region, approximate NOE kinetics were measured, in the form of half-times for the buildup to maximal enhancement. These were used to produce, for each proton, a rough rank ordering of its near neighbors according to distance. This allowed the close contacts to be more reliably assessed than could be done from the steady-state data, since the effects of spin diffusion were at least partially eliminated. Another useful ploy was the addition of small quantities of benzene-d_6 to manipulate chemical shifts, so allowing selective irradiation of the almost degenerate C_{22} and C_{24} methyl resonances.

At the end of this study, a very full picture of the pulvomycin molecule had emerged, comprising the complete covalent connectivity, the relative stereochemistry at 11 of the 13 chiral centers (the identity of the sugar unit was known previously), and a rough outline of the conformation of the rigid portion of the molecule. The only omissions were the relative stereochemistries of C_{32} and C_{33}, which from the discussion in Section 10.1.1 is not surprising, and the absolute chirality of the molecule, which NMR is, by its nature, unsuited to tackle. Most important of all, this achievement was made using directly isolated, unmodified material in solution, after only chromatographic preparation. Earlier structural studies of pulvomycin, employing chemical degradation, failed to detect any of the macrocyclic portion of the molecule.[47]

10.6.2. Penitrem A

Penitrem A (**10.109**) is one of a group of neurotoxic metabolites of *Penicillium crustosum* that causes tremor and convulsions in animals. In many respects, its structure determination followed a strategy similar to that for pulvomycin, the

448 APPLICATIONS OF THE NOE TO STRUCTURE ELUCIDATION

main difference being that long-range (^{13}C, ^{1}H) coupling connectivities, rather than NOE data, were used to interconnect partial spin systems. The full proton assignments were thus available *prior* to using the NOE data, allowing the interpretation of enhancements to concentrate entirely on stereochemical issues.[48,49]

10.109

The initial stages of the study did not use the NOE at all, and will therefore be outlined only briefly. The presence of the indole unit was inferred from the UV spectrum, and from feeding studies using labeled tryptophan. Signals due to an isolated aromatic CH proton, the indole NH proton, and five tertiary methyls were identified in the 1D NMR spectrum in acetone-d_6; three exchangeable peaks, assigned to hydroxy protons, were disclosed by saturation transfer from the residual water signal. Homonuclear decoupling experiments next showed that the remainder of the protons in penitrem A constituted the three partial spin systems shown in partial structures **10.110**–**10.112**. Furthermore, coupling constants within the two —CH$_2$—CH$_2$— fragments (in partial structures **10.110** and **10.112**, respectively) suggested that these each formed part of cyclohexane rings in chair conformations.

The major part of the study consisted of extending these partial spin systems to include the various quaternary centers and tertiary methyl groups, and then interconnecting them to build up the complete connectivity pattern of penitrem A. The detailed logic is too involved to present here, but the approach was to use two- and three-bond (^{13}C, ^{1}H) coupling connectivities (measured using ^{13}C{^{1}H} SPI experiments) and two- and three-bond isotope effects (measured after exchanging the OH and NH protons for deuterons) first to assign, and then to connect, the various quaternary carbon atoms. Directly bonded (^{13}C, ^{1}H) pairs were correlated using a series of single-frequency off-resonance spectra. The final few links, involving C_{19}, C_{30}, C_{31}, and C_{32}, had to be supplied by

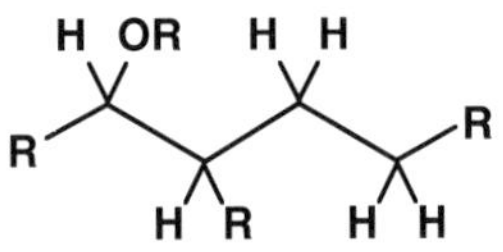

SATURATED RING SYSTEMS: MORE COMPLEX CASES   449

10.110

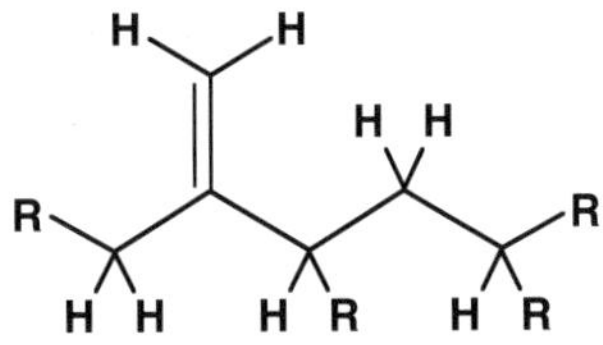

10.111

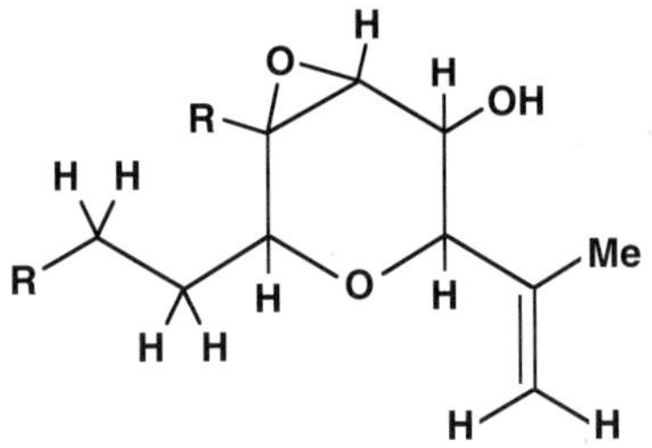

10.112

one-bond (^{13}C, ^{13}C) coupling connectivities, measured in multiply labelled material (available as a result of feeding experiments with [1,2–^{13}C$_2$] acetate). Nowadays, most or all of this information would probably be obtained using an HMBC experiment (cf. Section 10.1.2).

With the covalent connectivity fully established, the NOE was brought in to sort out the stereochemistry. Enhancements were measured using 1D steady-state experiments in acetone-d_6, and were presumably positive, although this is not reported. All the reported enhancement data are summarized on structure **10.109**. Starting from the cyclobutane ring, which cannot but be *cis* fused to the adjacent six-membered ring, the *trans* relationship between H$_{12}$ and H$_{14}$ was established by the enhancements connecting these protons to those of the C$_{13}$ methylene group. A similar sequence of enhancements from H$_{14}$ to H$_{17}$ via the C$_{16}$ *gem* dimethyl group established that these protons were also *trans* related. Proton H$_{17}$ was then further linked to the methyl at C$_{32}$, showing these groups to be *cis* related, while the absence of any enhancement from either of these groups to H$_{19}$ was taken to show a *trans* ring junction between the five- and six-membered rings. The ring junction at C$_{22}$–C$_{31}$ was shown also to be *trans* by the enhancements linking the C$_{32}$ methyl group to H$_{30\alpha}$, and the C$_{31}$ methyl group to H$_{19}$ (partial structure **10.113**). The mutual enhancements between H$_{24}$ and the C$_{31}$ methyl group showed the final ring junction to be *cis* (partial structure **10.113**), while the stereochemistry of the substituents at C$_{25}$ and C$_{26}$ was established using both NOE and coupling data for the 25-*O*-acetyl derivative. These conclusions have since been reinforced by a 2D NOESY experiment, which revealed further enhancements consistent with structure **10.109**.[50]

10.113

10.6.3. Other Examples

Many studies broadly similar to the two just described have appeared in the literature. Macrocyclic metabolites have proved a popular area, largely because the ring systems are often fairly rigid and have relatively amenable NMR spectra, so that structure and conformation can be established simultaneously. Examples include elaiophylin (Section 3.2.3) and hygrolidin (**10.114**).[51] A series of molecules related to penitrem A has been studied,[52,53] and many other papers concerned with terpenoid natural products have appeared. A series of norditer-

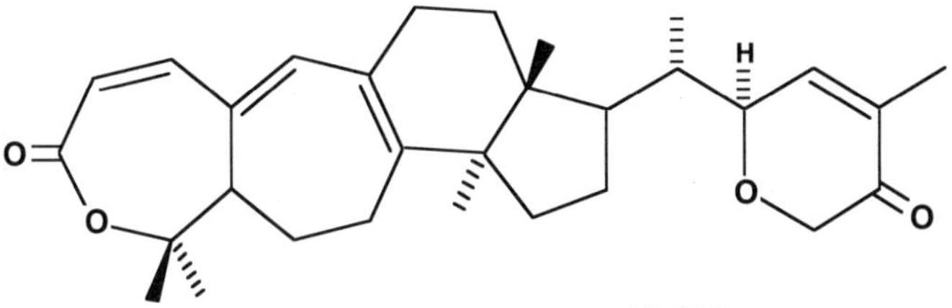

SATURATED RING SYSTEMS: MORE COMPLEX CASES   451

10.116

pene dilactones related to **10.115** was investigated by Matlin et al.,[54] while the stereochemistry of the triterpene Schisanlactone B (**10.116**) was determined using NOE data.[55] Another example emphasizes the importance (and difficulty) of saturating only what should be saturated: the originally proposed structure[56] (**10.117**) of the limonoid insect antifeedant azadirachtin was modified to the revised structure **10.118** following an NOE study.[57] Subsequent X-ray crystallographic analysis showed the true structure to be **10.119**, whereupon re-examination of the NOE difference spectra showed that a key enhancement, originally considered to be $Me_{18}\{H_{16a}\}$ in structure **10.118**, was in fact due to accidental partial saturation of Me_{30} or $Me_{4'}$, which are close in space to Me_{18} in structure **10.119**.[58] Meanwhile, another NMR study of azadirachtin, by Kraus et al.,[59] established structure **10.119** independently, using (^{1}H, ^{1}H) NOE data in conjunction with ^{13}C long-range coupling data and ^{13}C isotope shifts observed on partial deuteration of the OH groups (cf. structure determination of penitrem A, Section 10.6.2).

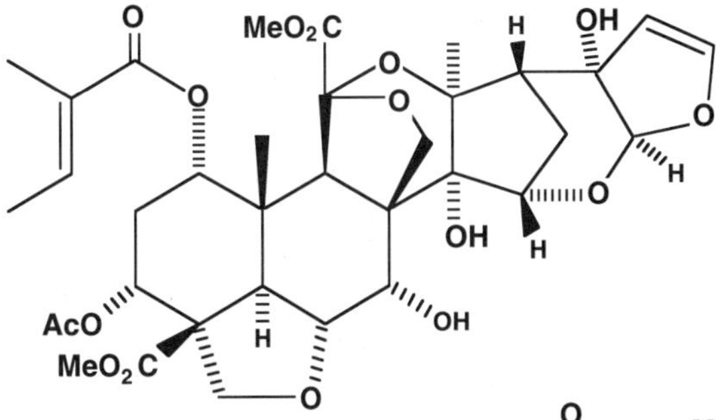

10.117

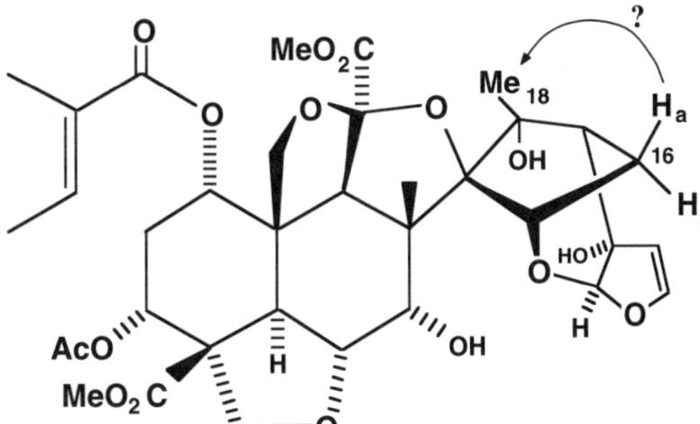

10.118

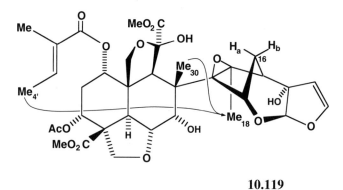

10.119

As a final example, Pfaltz et al.[60] elucidated most of the stereochemical features of the nickel-containing porphinoid factor F430 using the NOE data summarized around the structure **10.120**. (The few enhancements represented by dotted arrows are less certainly assigned because of signal-to-noise limitations or ambiguity as to the multiplet patterns of the enhanced signals.) This

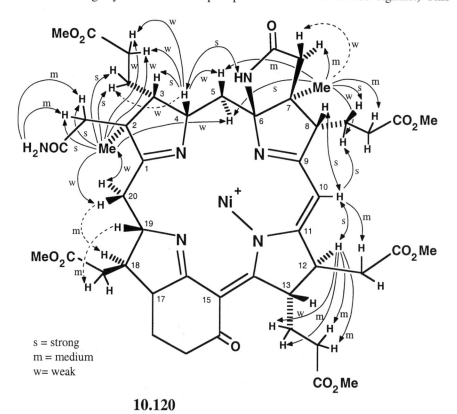

s = strong
m = medium
w = weak

10.120

example differs somewhat from most of the others mentioned here, in that the NOE experiments formed part of a wider study of the biogenesis of F430, so that much additional structural information and many assignments were available from the results of feeding experiments with labelled precursors.

REFERENCES

1. Anet, F. A. L.; Bourn, A. J. R. *J. Am. Chem. Soc.* 1965, *87*, 5250.
2. Noggle, J. H.; Schirmer, R. E. "The Nuclear Overhauser Effect; Chemical Applications," Academic Press, New York, 1971.
3. Hall, L. D.; Sanders, J. K. M. *J. Am. Chem. Soc.* 1980, *102*, 5703.
4. Sanders, J. K. M.; Mersh, J. D. *Prog. Nucl. Magn. Reson. Spectrosc.* 1982, *15*, 353.
5. Piantini, U.; Sørensen, O. W.; Ernst, R. R. *J. Am. Chem. Soc.* 1982, *104*, 6800.
6. Rance, M.; Sørensen, O. W.; Bodenhausen, G.; Wagner, G.; Ernst, R. R.; Wüthrich, K. *Biochem. Biophys. Res. Commun.* 1983, *117*, 479.
7. Neuhaus, D.; Wagner, G.; Vašák, M.; Kägi, J. H. R.; Wüthrich, K. *European J. Biochem.* 1985, *151*, 257.
8. Martin, G. E.; Crouch, R. C. *J. Natural Products* 1991, *54*, 1.
9. Rinaldi, P. L.; Keifer, P. A. *J. Magn. Reson. A* 1994, *108*, 259.
10. Colombo, R.; Colombo, F.; Derome, A. E.; Jones, J. H.; Rathbone, D. L.; Thomas, D. W. *J. Chem. Soc., Perkin Trans. 1*, 1985, 1811.
11. Hunter, B. K.; Russell, K. E.; Zaghloul, A. K. *Can. J. Chem.* 1983, *61*, 124.
12. Nechvatal, G.; Widdowson, D. A. *J. Chem. Soc., Chem. Commun.* 1982, 467.
13. Hutt, M. P.; MacKellar, F. A. *J. Heterocyclic Chem.* 1984, *21*, 349.
14. Moody, C. J.; Rees, C. W.; Tsoi, S. C.; Williams, D. J. *J. Chem. Soc., Chem. Commun.* 1981, 927.
15. Neuhaus, D.; Rees, C. W. *J. Chem. Soc., Chem. Commun.* 1983, 318.
16. Zoltewicz, J. A.; Baugh, T. D.; Paszyc, S.; Marciniak, B. *J. Org. Chem.* 1983, *48*, 2476.
17. Krane, J.; Skjetne, T.; Telnæs, N.; Bjorøy, M.; Solli, H. *Tetrahedron* 1983, *39*, 4109.
18. Chicarelli, M. I.; Wolff, G. A.; Maxwell, J. R. *J. Chem. Soc., Chem. Commun.* 1985, 723.
19. Ocampo, R.; Callot, H. J.; Albrecht, P. *J. Chem. Soc., Chem. Commun.* 1985, 200.
20. Ortiz de Montellano, P. R.; Kunze, K. L. *Biochemistry* 1981, *20*, 7266.
21. Guineadeau, H.; Freyer, A. J.; Shamma, M. *Nat. Prod. Rep.* 1986, 477.
22. Guineadeau, H.; Cassels, B. K.; Shamma, M. *Heterocycles* 1982, *19*, 1009.
23. Shamma, M. Personal communication.
24. Neuhaus, D.; Sheppard, R. N.; Bick, I. R. C. *J. Am. Chem. Soc.* 1983, *105*, 5996.
25. Shamma, M.; Lan, H.-Y.; Freyer, A. J.; Leet, J. E.; Urzúa, A.; Fajardo, V. *J. Chem. Soc., Chem. Commun.* 1983, 799.
26. Valencia, E.; Patra, A.; Freyer, A. J.; Shamma, M.; Fajardo, V. *Tetrahedron Lett.* 1984, *25*, 3163.

27. Capozzi, G.; Caristi, C.; Lucchini, V.; Modena, G. *J. Chem. Soc., Perkin Trans. 1*, 1982, 2197.
28. Capozzi, G.; Caristi, C.; Gattuso, M. *J. Chem. Soc., Perkin Trans. 1*, 1984, 255.
29. Goasdoue, C.; Goasdoue, N.; Gaudemar, M. *J. Organomet. Chem.* 1984, *263*, 273.
30. Pignatello, J. J.; Porwoll, J.; Carlson, R. E.; Xavier, A.; Gleason, F. K.; Wood, J. M. *J. Org. Chem.* 1983, *48*, 4035.
31. Bahadur, G. A.; Baldwin, J. E.; Usher, J. J.; Abraham, E. P.; Jayatilake, G. S.; White, R. L. *J. Am. Chem. Soc.* 1981, *103*, 7650.
32. Arrowsmith, R. J.; Carter, K.; Dann, J. G.; Davies, D. E.; Harris, C. J.; Morton, J. A.; Lister, P.; Robinson, J. A.; Williams, D. J. *J. Chem. Soc., Chem. Commun.* 1986, 755.
33. Santiesteban, F.; Grimaldo, C.; Contreras, R.; Wrackmeyer, B. *J. Chem. Soc., Chem. Commun.* 1983, 1486.
34. Taylor, M. D.; Minaskanian, G.; Winzenberg, K. N.; Santone, P.; Smith, A. B., III. *J. Org. Chem.* 1982, *47*, 3960.
35. Baldwin, J. E.; Adlington, R. M.; Derome, A. E.; Ting, H.-H.; Turner, N. J. *J. Chem. Soc., Chem. Commun.* 1984, 1211.
36. Derome, A. E. "Modern NMR Techniques for Chemistry Research," Pergamon Press, Oxford, 1987, pp. 124–127.
37. Neuhaus, D. Unpublished results.
38. Ley, S. V.; Neuhaus, D.; Simpkins, N. S.; Whittle, A. J. *J. Chem. Soc., Perkin Trans. 1*, 1982, 2157.
39. Bahadur, G.; Baldwin, J. E.; Field, L. D.; Lehtonen, E.-M. M.; Usher, J. J.; Vallejo, C. A.; Abraham, E. P.; White, R. L. *J. Chem. Soc., Chem. Commun.* 1981, 917.
40. Carlé, J. S.; Christophersen, C. *J. Org. Chem.* 1980, *45*, 1586.
41. Avenati, M.; Vogel, P. *Helv. Chim. Acta* 1982, *65*, 204.
42. Mersh, J. D. Ph.D. Thesis, University of Cambridge, 1983.
43. El-Masry, S.; Mahmoud, Z.; Amer, M.; Freyer, A. J.; Valencia, E.; Patra, A.; Shamma, M. *J. Org. Chem.* 1985, *50*, 729.
44. Neuhaus, D. Ph.D. Thesis, University of London, 1982.
45. Aldersley, M. F.; Dean, F. M.; Mann, B. E. *J. Chem. Soc., Chem. Commun.* 1983, 107.
46. Smith, R. J.; Williams, D. H.; Barna, J. C. J.; McDermott, I. R.; Haegele, K. D.; Piriou, F.; Wagner, J.; Higgins, W. *J. Am. Chem. Soc.* 1985, *107*, 2849.
47. Akita, E.; Maeda, K.; Umezawa, H. *J. Antibiot. Ser. A* 1964, *17*, 200.
48. de Jesus, A. E.; Steyn, P. S.; van Heerden, F. R.; Vleggaar, R.; Wessels, P. L.; Hull, W. E. *J. Chem. Soc., Chem. Commun.* 1981, 289.
49. de Jesus, A. E.; Steyn, P. S.; van Heerden, F. R.; Vleggaar, R.; Wessels, P. L.; Hull, W. E. *J. Chem. Soc., Perkin Trans. 1*, 1983, 1847.
50. Steyn, P. Personal communication.
51. Seto, H.; Akao, H.; Furihata, K.; Otake, N. *Tetrahedron Lett.* 1982, *23*, 2667.
52. de Jesus, A. E.; Steyn, P. S.; van Heerden, F. R.; Vleggaar, R.; Wessels, P. L.; Hull, W. E. *J. Chem. Soc., Perkin Trans. 1*, 1983, 1857.
53. Gallagher, R. T.; Hawkes, A. D.; Steyn, P. S.; Vleggaar, R. *J. Chem. Soc., Chem. Commun.* 1984, 614.

54. Matlin, S. A.; Prazeres, M. A.; Mersh, J. D.; Sanders, J. K. M.; Bittner, M.; Silva, M. *J. Chem. Soc., Perkin Trans. 1*, 1982, 2589.
55. Liu, J.-S.; Huang, M.-F.; Ayer, W. A.; Bigam, G. *Tetrahedron Lett.* 1983, *24*, 2355.
56. Zanno, P. R.; Miura, I.; Nakanishi, K.; Elder, D. L. *J. Am. Chem. Soc.* 1975, *97*, 1975.
57. Bilton, J. N.; Broughton, H. B.; Ley, S. V.; Lidert, Z.; Morgan, E. D.; Rzepa, H. S.; Sheppard, R. N. *J. Chem. Soc., Chem. Commun.* 1985, 968.
58. Broughton, H. B.; Ley, S. V.; Slawin, A. M. Z.; Williams, D. J.; Morgan, E. D. *J. Chem Soc., Chem. Commun.* 1986, 46.
59. Kraus, W.; Bokel, M.; Klenk, A.; Pöhnl, H. *Tetrahedron Lett.* 1985, *26*, 6435.
60. Pfaltz, A.; Juan, B.; Fässler, A.; Eschenmoser, A.; Jaenchen, R.; Gilles, H. H.; Diekert, G.; Thauer, R. K. *Helv. Chim. Acta* 1982, *65*, 828.
61. Nechvatal, G. Ph.D. Thesis, University of London, 1982.

CHAPTER 11

APPLICATIONS OF THE NOE TO CONFORMATIONAL ANALYSIS

11.1. GENERAL CONSIDERATIONS

11.1.1. Why Structural and Conformational Problems Are Different

Section 10.1.1 notwithstanding, there are some intrinsic differences between structural and conformational problems that are of importance when devising and interpreting NOE experiments.

The NOE is generally brought to bear at a fairly late stage in the overall process of structure elucidation. For synthetic unknowns, the number of possible product structures that is reasonable, given the structure of the starting material, is usually small, and the task of the NOE is to distinguish between these possibilities. Similarly, for natural unknowns, even though no guidance is available from a starting material structure, much information will have been gathered from other techniques before the NOE is brought in to sort out the remaining regiochemical and stereochemical ambiguities. It is thus fairly clear that the natural role of the NOE in structure elucidation is one of *selecting between predefined possibilities*. This is fortunate, since as previous chapters have emphasized, the NOE is fundamentally better suited to making such distinctions than it is to providing a quantitative measure of internuclear distance.

For conformational analysis the situation is less clear cut. Some conformational problems may be very easily framed in terms of predefined and clearly distinct possible conformations. An obvious example would be the *cis–trans* equilibrium of amides, while a more extreme example is provided by the two conformers of teucrin P1 (**11.23** and **11.24**; Section 11.3), which are sufficiently

GENERAL CONSIDERATIONS 457

stable to be separately isolated. In cases such as these, the NOE can be applied in very much the same way as for structural problems.

In many other cases, predefined possible conformations are not nearly so obvious. In principle, the calculated conformational energy minima of a molecule could always provide such a set of starting points for interpretation of the NOE data, but in practice it is clear that such an approach quickly becomes impractical for larger molecules. Thus, it might be easy to distinguish *cis* and *trans* amide conformers, or the various chair and boat conformers of a cyclohexane ring, and one might even imagine using the NOE to pick out the preferred conformer of a macrolide ring from among a small number of calculated low-energy conformers. However, it would plainly be impossible to tackle the conformational analysis of a protein in this way.

For such complicated molecules, with very many degrees of freedom, there is no option but to attempt to *build* a conformation based on the NOE data. In simple cases this may be done by hand (or rather, by eye), examining and altering models of various conformations to assess whether they are consistent with the NOE data. In more complex cases, computer-aided calculations fulfil the same role; these facilitate the "book-keeping" required to ensure that the various NOE-derived constraints are all satisfied simultaneously. In either case, there is the immediate problem that the translation of enhancement data to internuclear distances is a very uncertain business, for all of the reasons discussed in Part I. One must therefore accept that the distance constraints that NOE data generate are, by their nature, necessarily *approximate*.

This has some important consequences. It is easy to see that if a distance constraint is approximate, for example, that a particular internuclear distance must lie in the range 1.5–3.5 Å, then its significance depends heavily on how the two nuclei are connected through the covalent structure of the molecule. If they are separated by only a few bonds, then the constraint has little significance, as the two nuclei are incapable of getting far apart in any conformation. If, on the other hand, they are separated by many bonds, then the constraint has much more significance; since the relevant distance would be more than 3.5 Å in the great majority of possible conformations, these may all be excluded, leaving relatively only a few for further consideration.

We may see from this that the NOE is much better suited to determining the larger-scale features of conformation than it is to determining local detail. In principle, if the process of translating enhancement data into distance constraints were genuinely to become less approximate, then the degree of local detail revealed by the NOE would presumably increase. However, in practice this course is fraught with danger; one may simply assert that the level of approximation is reduced, so as to obtain a more "precisely" defined conformation, but the result may be that the conformation is no longer correct. A better approach is to try to use some other, more appropriate, means to establish the details of local conformation, the most obvious candidate being an analysis of scalar coupling data. Thus, NOE data and scalar coupling data are highly complementary in their information content, and their combined use represents a powerful strategy.

To a large extent, the foregoing discussion explains how the balance of examples in these final four chapters comes to be as it is. Since the NOE is more suited to determining large-scale conformational features than it is to determining local detail, it follows that the predominant field of application to conformational problems will be to larger molecules. This is the case. Although there are also successful applications to smaller molecules, such as macrolides and alkaloids, the number of applications to determining the local conformations of very small synthetic molecules remains relatively small. In contrast, since the role of the NOE in structure elucidation is mainly one of distinguishing between clear-cut, predefined possibilities, the predominant field of application is to small molecules. Far more efficient means are available for determining the *covalent* structures of biopolymers (with the possible exception of oligosaccharides), whereas the sorts of structural ambiguities that arise for small molecules, at least for those that are rigid, are just those that the NOE is well suited to solve.

11.1.2. Multiple Conformations

There is a further important difference between structural and conformational problems that should be mentioned. Structural problems, by their nature, have only one (correct!) solution. The same cannot be said for conformational problems, since most molecules exist in a dynamic equilibrium involving many conformers. The consequences of this depend to a large extent on the rate at which the conformers interconvert, and on the way in which the problems are posed. If the interconversion rate is slow enough, then separate NMR resonances will be observed for each conformer, and the relative population of each can be assessed by integration of the normal NMR spectrum. The role of the NOE is then to identify what each conformer is, either by assigning which signals arise from each of a predefined set of possibilities (as for instance with *cis* and *trans* amides), or by building up a conformation from distance constraints based on the NOE data for each set of resonances. In the limit of slow exchange, this process is entirely analogous to NOE experiments upon mixtures of different compounds. Note, however, that since spectral coalescence generally requires faster exchange rates than does efficient saturation transfer, in many cases conformers give separate resonances but are nonetheless *in fast exchange on the T_1 timescale* (cf. Section 5.2). In such cases, enhancements are completely averaged between conformers, even when specific resonances due to only one conformer are selectively irradiated.

If there is only a single set of resonances observed, one cannot tell from the normal spectrum alone whether this is because only a "single" conformer is present (within the limits of what can be determined using the NOE), or whether several conformers are present in fast exchange. This distinction emerges only later, from interpretation of the NOE data itself; the existence of multiple conformations is revealed by protons that appear, from their enhancements, to be in two places at once. In contrast to the slow exchange case, information as to the relative populations of the conformers is now present only in the quantitative magnitudes of the enhancements, and as will be seen, it is much harder to extract this information. Moreover, since all NMR properties are averaged by the fast exchange, NOE enhancements can no longer be used to *characterize* the individual conformers separately. If one is to derive conformer populations, one needs, at the very least, some independent knowledge (or assumptions) as to the nature of each conformer; as it was put earlier, the possible conformers must be predefined.

For a few simple cases, such as the methoxy group conformational preferences described in Section 11.2.3 and the statistical approach discussed in 5.3.2.3, such an analysis has been carried out, but more usually this level of detail is neither sought nor found. Highly mobile substituents are often not *expected* to have specifically preferred conformations in solution, and are therefore imagined as being rather indistinctly located when their enhancement data are interpreted. This assumption of conformational averaging is very often made almost unconsciously during interpretation.

In cases in which the possible conformers are not predefined, and the conformation must be built up from the NOE data, contributions from multiple conformers in rapid exchange are in general not treated explicitly. Rather they become one further reason why the finally derived "single" conformation must be regarded as approximate. If, as is often the case for biopolymers, the gross conformation is largely preserved while the local conformation may be in a state of flux, then the process of interpreting the NOE data may not be greatly affected. As we have seen, the approximate nature of the NOE distance constraints in any case imposes an uncertainty as to the local conformational detail. Alternatively, there may be regions of conserved conformation amid more flexible regions of a molecule. Interpretation of the NOE data would then be expected to succeed only for those conserved regions. However, in cases in which the rapidly exchanging conformers differ grossly throughout the molecule, there can be little hope that the NOE will be of much use. As with any useful technique, finding an answer depends on there being an answer to be found.

11.2. LOCAL CONFORMATIONAL DETAIL IN SMALL MOLECULES

11.2.1. Slowly Exchanging Equilibria

The simplest cases to deal with using the NOE are conformational equilibria in slow exchange. As discussed in Section 11.1.2, conformer populations may be assessed by integration of their separate, resolved resonances, and the NOE is required only to assign resonances to conformers, provided the possible conformers are known in advance. These features are well illustrated by the following simple example.

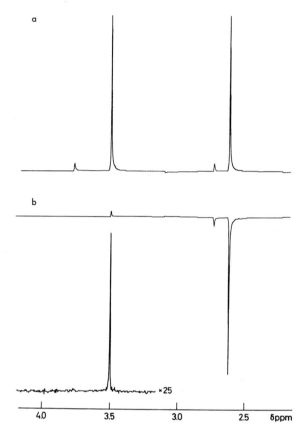

Figure 11.1. NOE difference spectra of *N*-methylthioacethydrazide (5% in CDCl$_3$), recorded at 270 MHz. Both thioacetyl singlets (δ2.6 and δ2.7) were irradiated by rapid cycling of the decoupler (combined preirradiation time 35 s per scan), but only the major *N*-methyl singlet (δ3.5) shows an enhancement. The major species is therefore the Z form (**11.2**). Reproduced with permission from ref. 1.

Figure 11.1 shows the normal and NOE difference spectra of *N*-methylthioacethydrazide, which exists as an equilibrium mixture of the *E* and *Z* forms (**11.1** and **11.2**, respectively).[1] Since these are in slow exchange, two sets of methyl resonances are observed, with an intensity ratio of about 6:94. For each conformer, the lower-field resonance (~δ3.5 major and ~δ3.8 minor) corresponds to the N-methyl group, while the higher-field resonance (~δ2.6 major and ~δ2.7 minor) corresponds to the thioacetyl methyl group. Presaturation of the major thioacetyl methyl peak causes a 4.4% enhancement of the major N-methyl peak, whereas saturation of the minor thioacetyl peak causes no enhancement (<1%) of the minor N-methyl peak. Quite clearly, the major conformer is therefore the Z form (**11.2**).

It might be thought from the appearance of Figure 11.1 that efficient saturation transfer was occurring between the two thioacetyl resonances, and indeed such saturation transfer would result in a similar appearance for the high-field part of this difference spectrum. In fact, presumably in order to save time, the experimenters chose to combine both experiments, cycling the irradiation between the two thioacetyl peaks so as to presaturate both simultaneously. Had efficient saturation transfer been occurring, the enhancement of the major N-methyl peak would itself have been carried across into the minor conformer, resulting in an enhancement of the minor N-methyl peak also, regardless of how the irradiation was carried out. That this did not happen at all shows that saturation transfer was negligible.

In general, saturation transfer progressively erodes the ability of the NOE to assign resonances to conformers in experiments such as these. The faster the rate of conformer interconversion (relative to T_1), the smaller the difference between enhancements of corresponding resonances in the different conformers (Sections 5.1–5.3). This can be tolerated to an extent, but for sufficiently fast interconversion rates, the usefulness of the NOE is destroyed.

Probably the most frequently encountered slowly exchanging conformational equilibria are those involving tertiary amides. The approach illustrated by the previous example (except perhaps the multiple simultaneous irradiation) is equally applicable to these, and a comparison of methods for assigning E/Z conformer ratios for tertiary amides concluded that the NOE was the only consistently reliable method.[2] The only point to add is that the improved sensitivity of NOE difference spectroscopy renders quite unnecessary the step of replacing CH_3 groups synthetically with CHD_2 groups.[1]

11.2.2. Rapidly Exchanging Equilibria: A Hypothetical Example, X—CH₂OH

As mentioned earlier, there have not been very many studies of the detailed nature of rapidly equilibrating *local* conformational preferences using the NOE. Nonetheless, many NOE spectra, possibly most, include features that are caused by such conformational preferences. Before dealing with specific studies, we should briefly consider why this should be so.

As a simple example, imagine a diastereotopic CH_2OH substituent attached to a relatively rigid molecular fragment X, where X might be, for instance, a steroid (Fig. 11.2). Assuming, as would usually be the case, that rotation about the X—CH_2OH bond is fast, then the various rotamers will contribute to an averaged NMR spectrum in which the two methylene protons, being diaste-

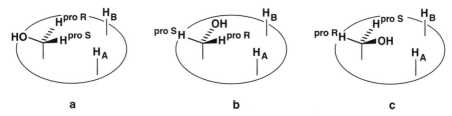

Figure 11.2. Rotamers for a diastereotopic methylene group. H_A and H_B are protons whose assignments and spatial positions within the rest of the molecule are known. Suppose (a) to be the preferred rotamer. If H_A is irradiated, $H^{pro\text{-}S}$ will be predominantly enhanced. However, if the methylene protons have not been stereospecifically assigned, this cannot be distinguished from an enhancement of $H^{pro\text{-}R}$ in rotamer (b). When H_B is irradiated in a further experiment, the distinction becomes possible, since only in rotamer (a) will $H^{pro\text{-}R}$ be enhanced by irradiation of H_B. Thus, two independent measurements are required to break the ambiguity.

reotopic, resonate separately. The enhancements of these two methylene resonances on irradiating various protons within X must be influenced by the conformational preference about the X—CH_2OH bond. If there is *no* rotamer preference, then for any proton irradiated within X, both methylene protons must be enhanced equally. However, if there is a rotamer preference, this symmetry will be lost, provided the proton irradiated within X spends on average more time close to one methylene proton than to the other.

In principle, there are three levels of information that may be extracted from NOE spectra in such circumstances:

1. Is there a conformational preference or not?
2. If there is, what is the preferred conformer?
3. Quantitatively, to what extent is it preferred?

Information of type 1 is frequently present in NOE spectra, and is often quite unrelated to the problem the NOE was intended to solve. In the present example, the existence of a rotamer preference would be shown by any asymmetry between the two methylene resonance enhancements when protons within X are irradiated. For less symmetrical structural fragments, equivalent evidence might be less clear-cut, but gross differences among the enhancements of various possible near neighbors of a mobile or flexible group usually give some indication that conformation preference is present.

To progress to information of type 2, assignments are needed. These may be relatively trivial to obtain, as for the vinyl and methoxy group conformations considered in Section 11.2.3 (there, it is the relative assignments of the protons on each side of the methoxy group that are required). For our hypothetical X-CH_2OH example, however, the relative assignments are the *stereospecific* assignments of the pro-*R* and pro-*S* methylene protons. These are much harder to obtain. In principle, if (separate) irradiation of *two* protons within *X*, whose assignments and spatial locations are known, gives *opposite* asymmetries between the enhancements of the two CH_2 protons, this information can be used to break the ambiguity and generate a stereospecific assignment (see Fig. 11.2). In practice, the structure of the molecule is rarely so obliging as to provide suitably placed potential irradiation targets within *X*. Other methods for making stereospecific assignments exist, but are usually still more demanding. For instance, combined interpretation of ($^1H,^1H$) and ($^{13}C,^1H$) three-bond coupling data is one useful method [especially for $NH(CO)CH-CH_2R$ groups in peptides[3]], but not only are the measurements difficult, they also generate the quantitative rotamer populations independently, rendering the NOE redundant.

The specific circumstance described here, that of partial rotamer preference in diastereotopic but flexible methylene groups, is probably one of the most commonly encountered conformational phenomena in NOE spectra (cf. also compound **10.79**). The difficulty of obtaining stereospecific assignments usually condemns this information in such spectra to remain uninterpretable.

If we assume assignments to be available, the next step is quantification. Clearly, for the $X\text{-}CH_2OH$ case this must depend on relating the measured magnitudes of the enhancements to the rotamer populations. In general, the translation of enhancement data into quantitative conformer populations is far from simple. The studies of methoxy group conformation referred to in Section 11.2.3 represent very favorable cases, and, even for these, significant difficulties were encountered in quantitation. If such favorable cases are difficult, there can be little prospect of success in the complicated situations encountered for more typical conformational equilibria.

11.2.3. Rapidly Exchanging Equilibria: Real Examples

Several features of the hypothetical example just discussed are illustrated by a study of the diene substituted glycopyranoside **11.3**.[4] In order to account for the diastereofacial selectivity that this diene shows in Diels–Alder reactions, its solution conformation (in $CDCl_3$) was probed by NOE difference spectroscopy, with particular reference to the $O-C_1$ bond rotamer preference. On the basis of the *exo*-anomeric effect,[5,6] the diene fragment would be expected to align itself in a plane at about 30° to the direction of the axial bonds of the sugar (i.e., with a $C_1-O_1-C_{1'}-O_{1'}$ torsion angle of about −90°), so that the lone pair in the *p*-orbital of the sp^2 hybridized O_1 atom is available for overlap with the $C_{1'}-O_{1'}$ σ^* orbital. If this lone pair is also to interact with the π^* orbitals of the diene, and if the diene has the s-*trans* geometry, this leads to two expected rotamers, **11.4** and **11.5**. These assumptions are supported by an X-ray crystal structure, which shows a solid-state conformation very similar to **11.4**.

11.4   **11.5**

On irradiation of the anomeric proton of the sugar ($H_{1'}$), a 13% enhancement of H_1 and a 4% enhancement of H_2 were observed (the assignments of these protons follow immediately from their shifts and multiplicities). The conclusion from this was that both rotamers **11.4** and **11.5** are present in solution, but that the major rotamer is **11.4**. In principle, it might be possible to quantify the rotamer populations on the basis of these enhancements, but, probably wisely, this was not attempted. The different overall relaxation of protons H_1 and H_2 would need to be explicitly included in such a treatment. For the present qualitative analysis, the conclusion rests on whether (assuming the s-*trans* geometry) the cross-relaxation that H_1 undergoes with the nine protons of the $OSiMe_3$ group *is not significantly less than* that which H_2 undergoes with H_{4a}. If, as seems likely, it is actually greater, then the rotamer ratio is, if anything, greater than the 13:4 enhancement ratio (this could probably be checked by comparing the enhancements $H_1\{OSiMe_3\}$ and $H_2\{H_{4a}\}$). Alternatively, the enhancements $H_{1'}\{H_1\}$ and $H_{1'}\{H_2\}$ could have been used (spectral overlap permitting), so as to avoid this particular difficulty. Of course, the NOE results in themselves do not confirm whether the details of conformers **11.4** and **11.5** are correct; these were predefined.

The next example is also concerned with conformational preferences related to the *exo*-anomeric effect.[7] Isoxazolidines **11.11**–**11.14** are derived from cyloaddition reactions of nitrones **11.6–11.8** with olefins **11.9** and **11.10** (Scheme 11.1). In each case, the regiochemistry of addition was apparent from the characteristically low chemical shift of H_5 (δH_5 = 5.15, 5.16, 5.20, and 6.57 ppm in compounds **11.11–11.14**, respectively). The *syn* stereochemistry of adducts **11.11** and **11.14** was established by the mutual enhancements of H_{4a} with *both* H_3 and H_5 in each case, while the stereochemistry of the remaining adducts **11.12** and **11.13** was shown to be *anti* by the mutual enhancements of H_{4a} with H_5 and of H_{4b} with H_3 in each case. For compound **11.11**, direct enhancements between H_3 and H_5 were also observed. These experiments parallel almost exactly those discussed for compounds **10.68** and **10.69** (Section 10.5.1).

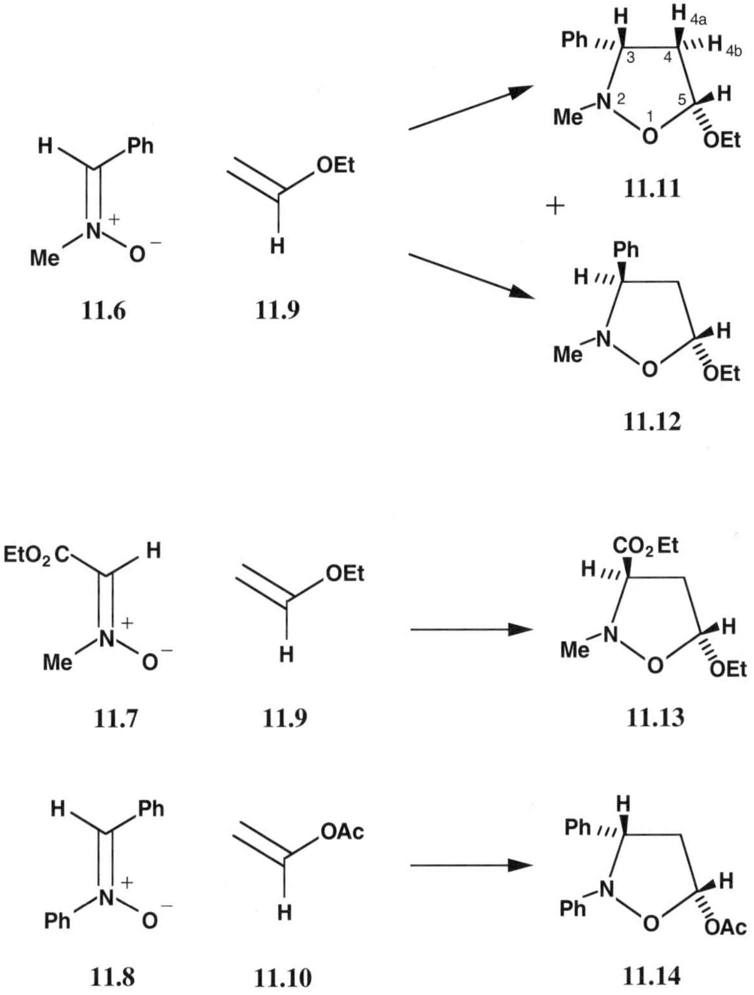

Scheme 11.1

For the two *anti* compounds, **11.12** and **11.13**, the H$_5$ proton was coupled detectably only to H$_{4a}$, implying that the torsion angle H$_5$—C$_5$—C$_4$—H$_{4b}$ must be close to ±90°. This limits the possible ring conformations to two, **11.15** and **11.16**, which are related by inversion at nitrogen. The enhancement of the (downfield portion of the) OC*H*$_2$Me multiplet on irradiating the *N*-methyl signal was taken as evidence that conformer **11.15** predominates in this conformational equilibrium. Further, it was proposed that the rotamers predominantly populated about the C—OEt bond are **11.17** and **11.18**, rather than **11.19**, based again on the enhancement OC*H*$_2$Me{NMe}. Arguments in favor of these conformers were given on the basis of the anomeric effect (**11.15** favored over **11.16**) and the *exo*-anomeric effect (**11.17** and **11.18** favored over **11.19**). For the two *syn* compounds, **11.11** and **11.14**, proton H$_5$ resonated as a double doublet, and conformational analysis was not reported.

As in the previous example, the difficult task of quantifying the conformer populations was not attempted. In fact, in the present case one might argue that the enhancement OC*H*$_2$Me{NMe} shows only that conformers **11.17** and **11.18** are *appreciably* populated. To make definite the proposal that they actually *predominate*, some comparison with the value expected for this enhancement if these conformers were populated exclusively would be needed.

The final part of this section concerns the quantitative analysis of conformer populations for an olefinic methoxy compound. Such analysis has proved to be something of a controversial area, not only in terms of the numerical results, but also in terms of the formulas used to derive them from the measured enhancements. For this reason, we will spend slightly longer on this topic and attempt to clarify some of the underlying theory (cf. also Section 5.3.2.1 for summary).

One motive for determining these conformer ratios is that they may reflect the bond order of the bonds immediately on each side of a methoxy group. If conformer fragments **11.20** and **11.21** represent the two possible "in-plane" conformers of an olefinic methyl ether, theoretical work suggests that conformer **11.20**, in which the methyl is s-*cis* to the double bond rather than the single, is preferred energetically because it allows more favorable interactions between the lone pair electrons on the oxygen and the π^* orbitals of the olefin.[5,6] In those aromatic systems where there is partial bond fixation, the same phenomenon is expected, but to a lesser degree. Measured methoxy conformer population ratios might therefore be used to report on the extent of bond fixation in such aromatic compounds, assuming any other factors affecting the conformer preferences to be negligible. Such a study has been reported by Kruse et al.[8]

11.20   **11.21**

As the methoxy conformers are in fast exchange, averaging occurs over the various rate terms σ and R, rather than over the enhancements themselves (Chapter 5). The relevant equations for the steady-state enhancements and for the buildup behavior of the direct term are therefore Eq. 5.26 and Eq. 5.27, respectively. Because the possible conformers have been predefined as being **11.20** and **11.21**, we may re-express all the averages in these equations explicitly. For example, the averaged cross-relaxation rate between H_A and the methyl protons of the methoxy group becomes

$$\langle \sigma_{AMe} \rangle = N^I \sigma^I_{AMe} + N^{II} \sigma^{II}_{AMe} \tag{11.1}$$

where conformer **11.20** is denoted I and conformer **11.21** is denoted II. Thus N^I and N^{II} denote the fractional populations of conformers **11.20** and **11.21**, respectively, and σ^I_{AMe} and σ^{II}_{AMe} are the cross-relaxation rates between H_A and the OMe protons *expected* for the isolated conformers **11.20** and **11.21**, respectively.

Using this notation we have also

$$\langle \sigma_{BMe} \rangle = N^I \sigma^I_{BMe} + N^{II} \sigma^{II}_{BMe}$$

$$\langle R_A \rangle = N^I \rho^I_{AMe} + N^{II} \rho^{II}_{AMe} + \sum_X \rho_{AX} + \rho_{AB} + \rho^*_A$$

$$\langle R_B \rangle = N^I \rho^I_{BMe} + N^{II} \rho^{II}_{BMe} + \sum_X \rho_{BX} + \rho_{AB} + \rho^*_B \tag{11.2}$$

where the terms ρ are defined by Eq. 2.23, and the summations over X include all spins other than A, B, and Me.

Note that since the conformers are in fast exchange, "separated" rate constants such as σ^I_{AMe} and σ^{II}_{AMe} *cannot be measured from the NMR data*, as they both contribute only to an averaged result during any NMR experiment. It may reasonably be assumed, however, that cross-relaxation of the methoxy group with H_B in conformer **11.20**, or with H_A in conformer **11.21**, is negligible, since the distances r^I_{AMe} and r^{II}_{BMe} are considerably shorter than the distances r^I_{BMe} and r^{II}_{AMe}. The terms σ^I_{BMe}, ρ^I_{BMe}, σ^{II}_{AMe}, and ρ^{II}_{AMe} may therefore be set to zero. Similarly, the distance r_{AB} is relatively long, so that the term ρ_{AB} may also be neglected, leaving

$$\langle \sigma_{AMe} \rangle \simeq N^I \sigma^I_{AMe}$$

$$\langle \sigma_{BMe} \rangle \simeq N^{II} \sigma^{II}_{BMe}$$

$$\langle R_A \rangle \simeq N^I \rho^I_{AMe} + \sum_X \rho_{AX} + \rho^*_A$$

$$\langle R_B \rangle \simeq N^{II} \rho^{II}_{BMe} + \sum_X \rho_{BX} + \rho^*_B \tag{11.3}$$

With these substitutions for the $\langle \sigma \rangle$ terms, Eq. 5.27 for the buildup of the direct enhancement yields

$$f_A\{Me\}(t) = N^I \frac{\sigma^I_{AMe}}{\langle R_A \rangle} [1 - e^{-\langle R_A \rangle t}]$$

$$f_B\{Me\}(t) = N^{II} \frac{\sigma^{II}_{BMe}}{\langle R_B \rangle} [1 - e^{-\langle R_B \rangle t}] \tag{11.4}$$

In the initial rate approximation (Section 4.4.1) this implies an initial slope for the buildup curves of these enhancements given by

$$\frac{d}{dt} f_A\{Me\}\Big|_{t \to 0} = N^I \sigma^I_{AMe}$$

$$\frac{d}{dt} f_B\{Me\}\Big|_{t \to 0} = N^{II} \sigma^{II}_{BMe} \tag{11.5}$$

For the steady-state enhancements, Eq. 5.26 yields

$$f^{ss}_A\{Me\} = \frac{N^I \sigma^I_{AMe} - \sum_X f^{ss}_X\{Me\} \sigma_{AX}}{\langle R_A \rangle}$$

$$f^{ss}_B\{Me\} = \frac{N^{II} \sigma^{II}_{BMe} - \sum_X f^{ss}_X\{Me\} \sigma_{BX}}{\langle R_B \rangle} \tag{11.6}$$

As noted in Sections 3.3.2 and 7.2.6, very long preirradiation periods are required to ensure that steady state is genuinely reached. The data in Figure 11.3 clearly show that for compound **11.22** (see below) the ratio $f_A\{Me\}/f_B\{Me\}$ is still changing appreciably after 5 s (which is over 2 times T_1 for the most slowly relaxing proton), and preirradiation periods of at least 5–10 times the longest T_1 are probably necessary for proper quantitation.

If, as is likely, no protons lie particularly near the region *between* the methyl group and either H_A (in **11.20**) or H_B (in **11.21**), then indirect effects will be negligible, leaving

$$f_A^{ss}\{Me\} \simeq \frac{N^I \sigma_{AMe}^I}{\langle R_A \rangle}$$

$$f_B^{ss}\{Me\} \simeq \frac{N^{II} \sigma_{BMe}^{II}}{\langle R_B \rangle} \qquad (11.7)$$

Armed with these expressions, we may now see what can be extracted from the NOE data. Figure 11.3 shows the buildup curves measured for the enhancements $H_A\{OMe\}$ (called $f_{12}\{OMe\}$ in the original paper) and $H_B\{OMe\}$ (called $f_{14}\{OMe\}$ in the original paper) in the methoxycycloheptatriene penicillin derivative **11.22**.[9] From Eqs. 11.5, it is clear that the *initial slopes* of these two

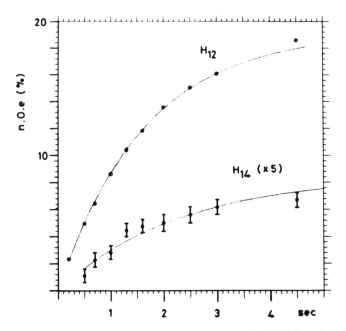

Figure 11.3. NOE buildup curves for the enhancements $H_A\{OMe\}$ and $H_B\{OMe\}$ in the methoxyheptatriene penicillin derivative **11.22** in $CDCl_3$ solution, recorded at 400 MHz. See text for discussion. Reproduced with permission from ref. 9.

curves reflect the conformer population ratio. Assuming for the moment that the terms σ^I_{AMe} and σ^{II}_{BMe} are equal, these initial slopes suggest a ratio of very roughly 1:15 in favor of the s-cis conformer **11.22**; however, the slopes are obviously difficult to gauge with any accuracy. Any difference between

11.22

Ar = 9-fluorenyl

the terms σ^I_{AMe} and σ^{II}_{BMe} would result in a small(ish) correction to this ratio, given by

$$\frac{\sigma^I_{AMe}}{\sigma^{II}_{BMe}} = \frac{\langle (r^I_{AMe})^{-6} \rangle (\tau_c)^I_{AMe}}{\langle (r^{II}_{BMe})^{-6} \rangle (\tau_c)^{II}_{BMe}} \qquad (11.8)$$

The τ_c terms account for possible differences in the detailed influence of anisotropic molecular tumbling on cross-relaxation in the two conformers. Details of how to estimate these τ_c ratios from ^{13}C T_1 values and the atomic coordinates of the molecular structure are given by Kruse et al.[8] Generally these ratios seem to be close to one for the cases they studied. The angular brackets in Eq. 11.8 refer to averaging over rotation of the methyl group itself, about the O—CH$_3$ bond in each of the conformers **11.20** and **11.21**. Clearly, the difference between these two averaged distances will be slight, but when raised to the sixth power, and in conjunction with the τ_c ratio term, this could nonetheless result in a correction of up to a factor of two or so [recall that $(0.891)^{-6} = 2$]. As with the τ_c ratio term, this distance correction term cannot be extracted from the NOE data, but must instead be supplied independently, for example, using estimates based on measurements of molecular models.

Measuring the initial slopes requires the least complicated formulas to translate NOE data into numerical conformer population ratios, but it must also involve a high degree of experimental uncertainty in the measurements, since the enhancements are inevitably very small in the initial rate region (Section 4.4.1). An alternative is to extract the conformer preferences from the steady-state enhancements, which may be much more accurately measured. Rearranging Eq. 11.7, we obtain

$$\frac{N^I}{N^{II}} = \frac{f^{ss}_A\{Me\}\langle R_A \rangle \sigma^{II}_{BMe}}{f^{ss}_B\{Me\}\langle R_B \rangle \sigma^I_{AMe}} = \frac{\langle \sigma_{AMe} \rangle \sigma^{II}_{BMe}}{\langle \sigma_{BMe} \rangle \sigma^I_{AMe}} \qquad (11.9)$$

Of the terms in this equation, the enhancements are known from measurements, and the ratio of the expected "separated" cross-relaxation terms (σ^{I}_{AMe} and σ^{II}_{BMe}) must be estimated from knowledge of the molecular geometry as discussed earlier. The $\langle R \rangle$ terms represent the overall, averaged, relaxation rates of H_A and H_B, respectively. Clearly, these will also depend on the conformer population ratio, since the proton that is closer to the methyl group for more of the time will relax faster. However, this dependence is not simple, since for both H_A and H_B the extent to which the methoxy group conformation influences their relaxation depends critically on the positions of their *other* near neighbors also. This is made explicitly clear if we write $\langle R_A \rangle$ and $\langle R_B \rangle$ in terms of distances. Thus, from Eqs. 11.3 and 3.55, we may write

$$\frac{\langle R_A \rangle}{\langle R_B \rangle} = \frac{N^{I} \langle (r^{I}_{AMe})^{-6} \rangle + \sum_{X} r^{-6}_{AX} + \mathscr{L}_A}{N^{II} \langle (r^{II}_{BMe})^{-6} \rangle + \sum_{X} r^{-6}_{BX} + \mathscr{L}_B} \qquad (11.10)$$

where the terms $\mathscr{L}_A$ and $\mathscr{L}_B$ represent the "extra" relaxation due to miscellaneous sources included in ρ^{*}_{A} and ρ^{*}_{B}, respectively (Sections 2.4.1 and 3.3.3).

Were it not for the unknown terms $\mathscr{L}_A$ and $\mathscr{L}_B$, substitution of Eq. 11.10 back into Eq. 11.9 would result in a closed, if clumsy, solution for the ratio N^{I}/N^{II}, since all the distances in Eq. 11.10 could, in principle, be measured using models. As things are, however, it is necessary to use some experimental measure for the relaxation behavior of H_A and H_B. It might seem that measured T_1 values would fulfill this role, but in fact there are difficulties with this. As discussed in Section 2.3.2, T_1 is not strictly defined for systems undergoing cross-relaxation. If, as is usual, approximate values are extracted from the results of inversion-recovery experiments in spite of this, then the results differ significantly according to whether the entire spectrum is inverted (nonselective T_1), or only one resonance is inverted (selective T_1). In fact, neither selective nor nonselective T_1 values (when measured outside the initial rate approximation) correspond exactly to the inverse of $\langle R \rangle$ as defined here, but selective T_1 values are probably a good approximation.

There is, however, an alternative measure for $\langle R \rangle$ available. As shown by Eq. 11.4, the growth curve for the NOE enhancements is exponential (in the absence of indirect effects), and the rate constant for this growth is $\langle R \rangle$. From the exponential fits to the data in Figure 11.3, values of $\langle R_A \rangle = 0.59$ s^{-1} and $\langle R_B \rangle = 0.43$ s^{-1} were obtained, giving a ratio $\langle R_A \rangle/\langle R_B \rangle$ of 1.37:1. When substituted into Eq. 11.9, and using the measured steady-state enhancement ratio of ~11:1, this yields a conformer ratio of about 15:1 (neglecting the terms in σ^{I}_{AMe} and σ^{II}_{BMe} as before). This agrees with our earlier, very approximate, estimate based on the initial slopes, but not with other literature interpretations of this or similar steady-state data. In one case, this may have been because the ratio $\langle R_A \rangle/\langle R_B \rangle$ was approximated using nonselective T_1 values,[8] while in the other case the nature of this correction was misinterpreted, with the result that it was applied in the reverse sense.[9]

Whatever the method used to translate enhancement data into conformer ratios, it is probably fair to say that the most severe limitation of this sort of analysis is the uncertainty in the correction terms in Eq. 11.8, particularly the distance terms. (Note that the population ratio of about 1:15 which was derived earlier *specifically ignored* these terms; to correct for them, the ratios $(r^I_{AMe}/r^{II}_{BMe})^{-6}$ and $(\tau_c)^I_{AMe}/(\tau_c)^{II}_{BMe}$ would have to be estimated, for example, by using measurements of models of conformations **11.20** and **11.21**). If comparisons within a series are required, errors of this sort may be a constant factor that merely bias the results without affecting internal trends. However, individual determinations of specific equilibria seem unlikely to be reliable within limits of less than a factor of roughly two.

As mentioned earlier, the present example represents a particularly favorable case for study, in that the equilibrium can be described using just two fairly well-defined conformations. The fact that, in spite of this, quantitation of conformer populations has proved so difficult does not bode well for applications to more complicated systems.

11.3. CONFORMATIONAL ANALYSIS OF MEDIUM-SIZED MOLECULES

This section covers what is really an area of common ground between conformational analysis and structure determination. The examples here are of applications to molecules whose size and complexity are broadly similar to those in Section 10.6. In fact, the divisions between these sections are largely arbitrary, and all the introductory comments made in Section 10.6 apply equally here. Just as there, the key issue is rigidity. For rigid systems, both structure and conformation often emerge from the NOE data together, whereas for highly mobile systems, neither is likely to be accessible (Section 10.1.1). This is well illustrated by the studies of pulvomycin (Section 10.6.1), penitrem A (Section 10.6.2), and elaiophylin (Section 3.2.3), for each of which conformational conclusions were closely tied to the structure determination.

The commonality between structure and conformation is particularly powerfully illustrated by the first example. Teucrin P1 is a caged polycyclic diterpene, which, as isolated, exists in the chair–boat–boat conformation **11.23**.[10] On treatment with a strong base (lithium diisopropylamide), and subsequent protonation on work-up, it is converted into the chair–boat–chair conformation **11.24**. Conformer **11.23** is the more stable, but conversion of **11.24** back to **11.23** in neutral chloroform solution is still incomplete after 2 weeks at 20°C, and to all intents and purposes the separated conformers behave as though they were different compounds. Enhancements $H_{18}\{H_{12}\}$ and $H_{20}\{Me_8\}$ showed characteristic differences in the two conformers, but in fact the NOE played only a limited role in their characterization, as both conformers were also analyzed by X-ray crystallography.

Another example of a boat–chair equilibrium is provided by 2,4-dioxabicyclo[3.3.1]nonane, for which the chair–chair (**11.25**), boat–chair (**11.26**), and chair–boat (**11.27**) conformations are possible.[11] Coupling data (particularly $^3J(H_1, H_{8\beta}) \leq 5$ Hz), ruled out the possibility that the cyclohexane ring was in

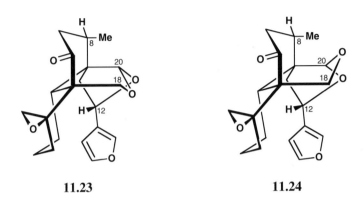

the boat form, while the strong mutual enhancements (~5%) between $H_{3\beta}$ and $H_{9\,syn}$, together with the absence of mutual enhancements between $H_{3\alpha}$ and $H_{7\alpha}$, established that it is the chair–boat conformer **11.26** that predominates in benzene-d_6 solution.†

For the triterpene friedelin (**11.28**), the chair–boat preferences of all five cyclohexane rings were found to be as shown using enhancements between the various angular methyl groups.[12] The resonances corresponding to Me_{23}, Me_{24}, and Me_{25} were assigned using enhancements from $H_{1\,axial}$ and $H_{6\,equatorial}$, while the remaining methyls were assigned using the intermethyl enhancements

---

†In a somewhat similar example, chair–boat equilibria in the 8-membered rings of substituted dibenzodioxaphosphocines were probed using variable temperature $^{31}P\{^1H\}$ NOE difference experiments.[20]

11.28

$Me_{24}\{Me_{25}\}$, $Me_{26}\{Me_{25}\}$, $Me_{26}\{Me_{28}\}$, and $Me_{28}\{Me_{30}\}$. These assignments were checked using T_1 measurements and additions of lanthanide shift reagent. The boat conformations of rings D and E follow particularly from the enhancements $Me_{26}\{Me_{28}\}$ and $Me_{28}\{Me_{30}\}$, together with the absence of any inter-methyl enhancements involving Me_{27}. All the relevant spectra are summarized in Figure 11.4.

The remaining examples in this section are all applications to the conformational analysis of larger rings. Unlike the chair–boat conformational problems for cyclohexane rings considered so far, these problems cannot easily be framed in terms of predefined possibilities, so that the alternative of building a conformation that is simultaneously consistent with all of the NOE data must be followed (Section 11.1.1). Macrocyclic fungal metabolites have proved a popular area, but the first two examples that follow are alkaloids.

Repanduline (**10.31**) is a bisbenzylisoquinoline alkaloid possessing a rather unusual macrocyclic ring, in which the two isoquinoline units are joined through a spiro-link. Experiments to determine its regioisomerism have already been discussed in Section 10.3.2. The full study also established the relative stereochemistry at all three chiral centers and the solution conformation of the macrocyclic ring, which are as shown in Figure 11.5. Only an outline of the interpretation can be presented here, and in particular the assignments, which were established using a series of many decoupling and NOE difference experiments, will all be assumed; more detailed arguments appear in the original paper.[13]

The relative stereochemistry of the spiro-link and C_1 was established mainly using the enhancements $H_1\{H_B\}$, $H_C\{H_{2''}\}$, and $H_D\{H_{6''}\}$, shown in Figure 11.6 (these experiments required the addition of benzene-d_6 to remove the chemical shift degeneracy of H_1, H_C, and H_D in $CDCl_3$ alone). Diastereomer **11.29** could not have produced this pattern of enhancements.

The distinction between the two remaining possible diastereomeric structures, **10.31** and **11.30** (differing in relative stereochemistry at $C_{1'}$), was a more difficult task, since these would show only relatively subtle differences in NMR properties arising from their different overall conformations. The approach was therefore to specify as many individual details of the actual conformation as possible, and then to examine Dreiding models of each to see which could best accommodate all these details simultaneously. In fact, sufficient detail was

CONFORMATIONAL ANALYSIS OF MEDIUM-SIZED MOLECULES 475

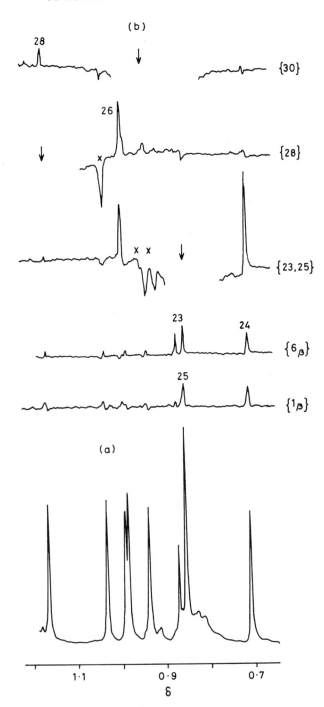

Figure 11.4. NOE difference spectra (methyl region only) of the triterpene friedelin (0.02 M in CDCl$_3$), recorded at 400 MHz (preirradiation period not given). Reproduced with permission from ref. 12.

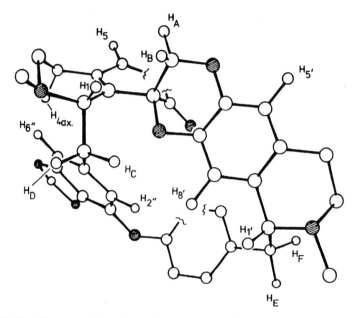

Figure 11.5. Solution conformation of repanduline **10.31**, as established by interproton NOE enhancements and coupling constants. Reproduced with permission from ref. 13.

available to specify the conformation of the macro-ring quite closely, and it was found that only diastereomer **10.31** could simultaneously fit all the NMR-based constraints.

The absence of detectable coupling between H_1 and H_D (Fig. 11.6) was particularly telling. In conjunction with the small but definite enhancement of H_5 on preirradiation of H_A (0.42%; Fig. 10.4), this specified the conformations of rings A, B, and E almost completely. Only in the conformation shown is the torsion angle (H_1, C_1, C_α, H_D) close to 90° while the distance H_5–H_A is simultaneously less than 4 Å. Other pieces of evidence that are consistent with the overall conformation illustrated in Figure 11.5 are (i) the features of ring G already discussed in Section 10.3.2, (ii) the large mutual enhancements between $H_{2''}$ and $H_{8'}$, and (iii) the pattern of enhancements and couplings for the

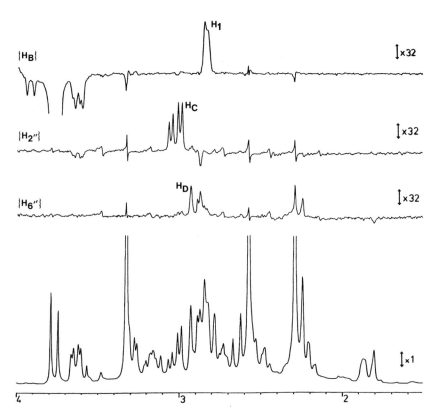

Figure 11.6. NOE difference spectra (high-field region only) of repanduline (0.26 M in CDCl$_3$ containing 20% benzene-d_6), recorded at 250 MHz (preirradiation period 5 s). Reproduced with permission from ref. 13.

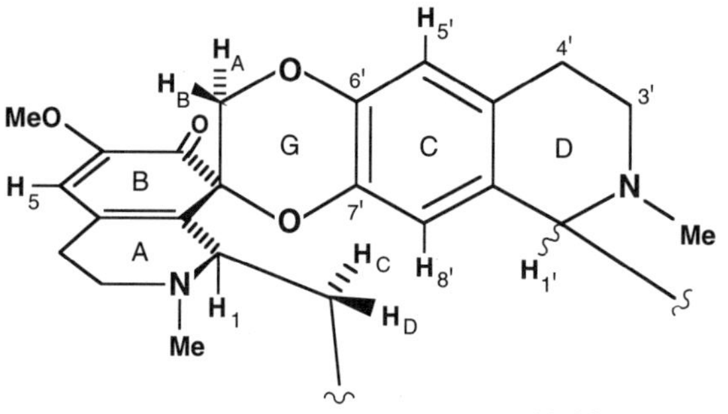

478 APPLICATIONS OF THE NOE TO CONFORMATIONAL ANALYSIS

11.29

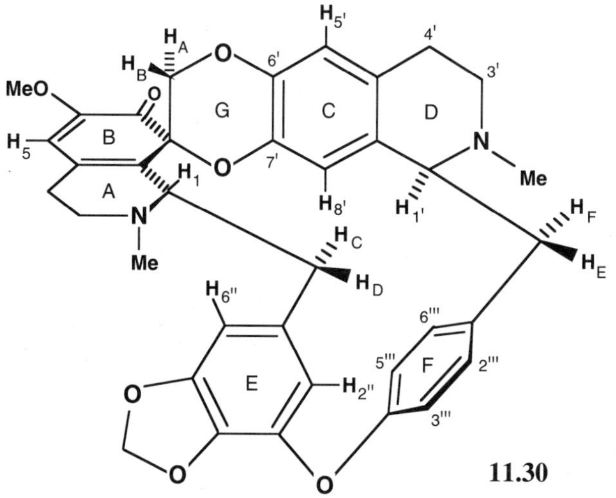

11.30

$H_{1'}$, H_E, H_F partial spin system, which show that $H_{1'}$ is pseudoequatorial and antiperiplanar to H_F. The spectra in Figure 11.6 also provide a striking illustration of the way in which the editing ability of the NOE difference experiment can reveal new and useful information (cf. Section 9.2.1); the resonances due to H_1, H_C, and H_D are all completely obscured by overlap in the normal 1D spectrum.

The conformation of the macro-ring in repanduline was specified using quite a small number of NMR-based constraints (although, of course, these represent the distilled result of interpreting a much larger body of data). The fact that so few constraints were sufficient is again due to rigidity. The 18-membered macro-ring actually has very limited flexibility, since it is composed of individually rigid aromatic units joined through a few relatively short flexible links.

A broadly similar approach was followed in an NOE study of the binary alkaloid vinblastine, **11.31**.[14] Using assignments based on a series of decoupling and NOE difference experiments, the conformation of the vindoline (i.e., bottom) half of the molecule was straightforwardly deduced from the enhancements summarized in Figure 11.7a. The more difficult task of specifying the conformations of the 9-membered ring in the catheranthine (i.e., top) half of the molecule, and the relative orientation of the vindoline and catheranthine fragments, was achieved using a combination of the enhancement data summarized in Figure 11.7b and c, and coupling data. In particular, the proximity of $C_{1'}$ and $C_{8'}$ largely defines the conformation of the 9-membered ring (Fig. 11.7b), while the flattening of the fused piperidine ring that this conformation necessitates was evident from the coupling constants in this part of the molecule.

11.31

The antibiotics vancomycin (**11.32**) and ristocetin (**11.33**) have been the object of much work using the NOE, not only to determine their structure and conformation, but also those of their complexes with a model cell wall analog, Ac-D-Ala-D-Ala.[15-17] This dipeptide mimics the site on growing bacterial cell walls where these antibiotics bind; binding to this site prevents cross-linking within the cell wall structure, so causing the death of the organism. The work is too involved to present in detail here, but some points of methodology are worth making, since the approach used differs substantially from previous examples in this section. When studied in DMSO-d_6 solution, these molecules and their complexes are some way, but not very far, into the negative NOE regime (the authors estimated that for vancomycin in dimethyl sulfoxide-d_6 at 35°C and 400 MHz, $\omega\tau_c \simeq 2.8$). This implies that spin diffusion, although a problem, will have a relatively limited effect on the NOE results (cf. Section 3.2.3). To accommodate this, enhancements were measured not only using steady-state experiments (5 s preirradiation) but also kinetic experiments (TOE). Spin diffusion was identified by the characteristic sigmoidal form that it causes in the buildup curves for the enhancements (Section 4.3.2). Since all

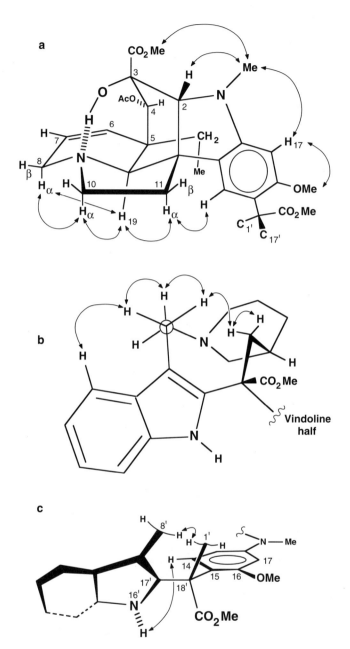

Figure 11.7. Solution conformation of the binary alkaloid vinblastine **11.31**, as established by interproton NOE enhancements and coupling constants. Selected direct enhancements are indicated around the structures. For the sequence of direct enhancements shown in (b), all but one of the corresponding three-spin effects were also observed. Fragment (*a*) shows the vindoline half of the molecule, fragment (*b*) the catheranthine half, and (*c*) shows the relative orientation of the two. Reproduced with permission from ref. 14.

11.32

11.33

enhancements were negative, saturation transfer was not distinguishable by its sign, and therefore complicated the interpretation considerably. Repeating key experiments after partial exchange with D_2O served to identify those effects that were transmitted via chemical exchange.

These studies established the key features of the bound conformation of Ac-D-Ala-D-Ala. Also, they showed that a vital feature of the complex, the carboxylate binding pocket, is absent in vancomycin or ristocetin alone, being *induced* only by the binding process itself. This work demonstrates that in favorable cases the NOE may be used to probe the conformations of drug-acceptor complexes. It must be said, however, that very few examples are as amenable to investigation using NMR as are these antibiotics; large enzymes or receptor proteins bound to cell walls would be quite impossible to study in this way.

The final two examples are of applications to macrocyclic fungal metabolites. Cytochalasin B, **11.34**, was analyzed using a combination of 1D and 2D

11.34

techniques.[18] A COSY spectrum was used to provide the bulk of the proton assignments, while a NOESY spectrum was used to provide a preliminary overview of the principal NOE interactions (Fig. 11.8). The molecule was studied in acetone-d_6 solution, and was in the positive NOE regime. Phase-sensitive NOESY (Section 8.3.4) would now be the preferred experiment here, because of its superior resolution and its ability to distinguish NOE cross peaks from exchange cross peaks (both of which are present in Fig. 11.8). Nonetheless, this absolute value NOESY spectrum simplified the later selection of 1D steady-state experiments to quantitate the important enhancements. Using a combination of these data and coupling constants, the conformation was deduced.

A very similar approach was followed in a study of erythromycin A.[19] The principal difference here was that extensive use was also made of molecular mechanics calculations when interpreting the NOE results, in order to check that proposed conformations were energetically viable. This combination of NOE experiments with conformational energy calculations undoubtedly represents a trend for the future in this area (cf. Section 5.3.2.3).

CONFORMATIONAL ANALYSIS OF MEDIUM-SIZED MOLECULES

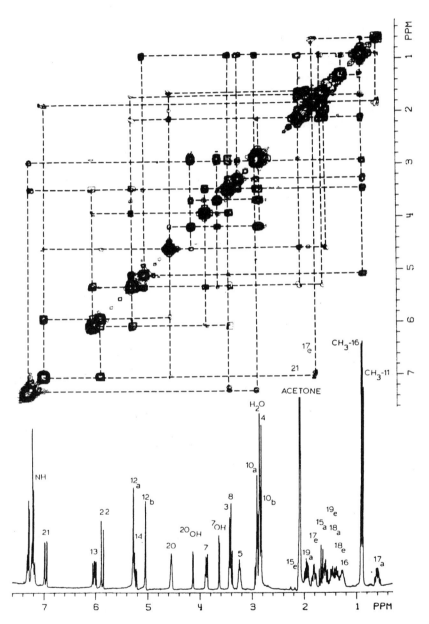

Figure 11.8. NOESY spectrum ($\tau_m = 0.45$ s) of cytochalasin B **11.34** (5 mg in 0.5 ml acetone-d_6), recorded at 360 MHz; 128 increments were used with 512 points in t_2. Reproduced with permission from ref. 18.

REFERENCES

1. Eggert, H.; Nielsen, P. H. *Tetrahedron Lett.* 1981, *22*, 4853.
2. Lewin, A. H.; Frucht, M. *Org. Magn. Reson.* 1975, *7*, 206.
3. Bystrov, B. F. *Prog. Nucl. Magn. Reson. Spectrosc.* 1976, *10*, 41.
4. Gupta, R. C.; Slawin, A. M. Z.; Stoodley, R. J.; Williams, D. J. *J. Chem. Soc., Chem. Commun.* 1986, 1116.
5. Kirby, A. J. "The Anomeric Effect and Related Stereoelectronic Effects at Oxygen," Springer-Verlag, New York, 1983.
6. Deslongchamps, P. "Stereoelectronic Effects in Organic Chemistry," Pergamon Press, Oxford, 1983.
7. DeShong, P.; Dicken, C. M.; Staib, R. R.; Freyer, A. J.; Weinreb, S. M. *J. Org. Chem.* 1982, *47*, 4397.
8. Kruse, L. I.; DeBrosse, C. W.; Kruse, C. H. *J. Am. Chem. Soc.* 1985, *107*, 5435.
9. Mersh, J. D.; Sanders, J. K. M. *Tetrahedron Lett.* 1981, *22*, 4029.
10. Morita, M.; Kato, N.; Iwashita, T.; Nakanishi, K.; Zeyong, Z.; Yamane, T.; Ashida, T.; Piozzi, F. *J. Chem. Soc., Chem. Commun.* 1986, 1087.
11. Peters, J. A.; Bovée, W. M. M. J.; Peters-van Cranenburg, P. E. J.; Van Bekkum, H. *Tetrahedron Lett.* 1979, 2553.
12. de Aquino Neto, F. R.; Sanders, J. K. M. *J. Chem. Soc., Perkin Trans. 1*, 1983, 181.
13. Neuhaus, D.; Sheppard, R. N.; Bick, I. R. C. *J. Am. Chem. Soc.* 1983, *105*, 5996.
14. Hunter, B. K.; Hall, L. D.; Sanders, J. K. M. *J. Chem. Soc., Perkin Trans. 1*, 1983, 657.
15. Williamson, M. P.; Williams, D. H. *J. Am. Chem. Soc.* 1981, *103*, 6580.
16. Williams, D. H.; Butcher, D. W. *J. Am. Chem. Soc.* 1981, *103*, 5697.
17. Williams, D. H.; Williamson, M. P.; Butcher, D. W.; Hammond, S. J. *J. Am. Chem. Soc.* 1983, *105*, 1332.
18. Graden, D. W.; Lynn, D. G. *J. Am. Chem. Soc.* 1984, *106*, 1119.
19. Everett, J. R.; Tyler, J. W. *J. Chem. Soc., Chem. Commun.* 1987, 815.
20. Rzepa, H. S.; Sheppard, R. N. *J. Chem. Res. (S)* 1988, 102.

CHAPTER 12

CALCULATING STRUCTURES OF BIOPOLYMERS

12.1. INTRODUCTION

For large biomolecules, the problem of finding a structure or conformation that is simultaneously consistent with as much as possible of the NMR data can, in practice, only be solved by computer calculations. Some of the methods used to calculate structures from NOE enhancements and other NMR data are discussed in this chapter, together with some important related topics, such as how the input to structure calculations is prepared, and how the quality of calculated structures is assessed.

Before going on there is a brief semantic point to make, concerning the distinction between "structure" and "conformation." In the preceding two chapters, it was helpful to maintain this distinction rather rigidly since, as discussed, there are important differences between the two where analysis of small molecules using NMR is concerned. However, for biological macromolecules the word "conformation" is used more rarely and less precisely, and the already somewhat fuzzy distinction between the two words seems largely to break down. For instance, the tertiary structure of a folded protein is maintained by a complex network of specific interactions that lead to a unique spatial arrangement; to maintain that this structure should be called a conformation just because most of the interactions that maintain it are noncovalent would be to fly against almost all of the literature. In keeping with this, within this chapter and the next we will refer to the three-dimensional arrangement of atoms in macromolecules as "structure."

To say that there is no clear consensus as to what is "the best" method for calculating structures from NMR data would be an understatement; indeed it can sometimes seem as if no two publications use the same method. However, behind this apparent complexity there are some unifying strands that make some approaches more similar than they first appear, and we shall try to bring these out in Section 12.3. Given the huge variety of methods in use and the rapid pace of change, it is only practical to present an overview of the area as it appears at present and without any pretence at being comprehensive. Further views can be found in a range of books and reviews, for instance references 1–6.

12.2. RESTRAINTS

The NMR restraint list, together with the covalent connectivity network, forms the only experimental input from which the NMR structure of a macromolecule is calculated. The eventual precision and accuracy of the calculated structure is therefore very largely determined by the quality of the NMR restraint list, and its importance cannot be overstressed. We shall return to the concepts of precision and accuracy in Section 12.4, but it is worth a brief reminder of their meanings here: The precision of a set of structures measures how similar they are to one another, whereas accuracy measures how similar they are to the "correct" answer. Accuracy can therefore only be assessed if a "correct" answer is known.

Given its fundamental role, it is worth summarizing briefly some of the attributes that make for a high-quality restraint list. As emphasized in other chapters, it is the number rather than the precision of the restraints that has the most impact on the outcome of the structure determination. However, the restraints also need to be nontrivial, in the sense that they should significantly reduce the conformational space available to the molecule. Thus, distance restraints that span a large number of covalent bonds in the covalent structure will be much more active in determining the calculated structure than will those that span only a few covalent bonds (cf. Section 11.1). The former are generally called "long-range" distance restraints, because they are long-range in terms of the bonding network between the constrained atoms (even though *through space* they may be no longer range than other distance restraints). As discussed in Section 11.1.1, because NOE restraints are inherently approximate, it is often difficult to define local details of a structure precisely using NOE data alone. Dihedral angle restraints derived from coupling constants are better suited to this problem, so it is desirable for a restraint list to contain a mixture of the two restraint types (see also Section 12.2.5). Further, the restraints should be distributed evenly throughout the structure, so as not to leave areas where the structure is poorly defined. Perhaps obviously, but certainly most important of all, the restraints all need to be correct! This must be true not only in the quantitative sense that a distance or angle should be constrained within a correctly specified numerical range, but also in the more fundamental sense that the restraint should be applied to the correct atoms. This will only happen if the cross peak(s) from which each restraint was derived are correctly assigned, emphasizing once again the imperative of making correct assignments.

In this section various issues concerning the restraint list will be discussed in turn.

12.2.1. Assigning NOE Restraints

Each NOE restraint in the list for a macromolecule is derived from one or more cross peaks in the corresponding NOESY spectra. Thus the most fundamental issue when using the restraint is that these NOESY cross peaks should be correctly assigned; if they are not, the restraint will be applied to the wrong atoms. Mistakes of this sort can have serious consequences, since such incorrectly assigned restraints are likely to be impossible to fulfil simultaneously with the correctly assigned ones. The structure-calculation algorithm must then attempt to find the least unacceptable compromise between mutually incompatible restraints, and the result may then owe as much to the properties of the particular algorithm being used as to the experimental data. Even a single misassigned restraint, if it attempts to bring together atoms that are held far apart in the correct structure, may grossly distort a set of calculated structures. However, gross errors of this sort are usually relative easy to detect and correct. More insidious are misassignments that correspond to only slightly erroneous distances, since they can be accommodated by relatively subtle distortions of the calculated structures.

Given this situation, how can one be sure that NOE restraints are assigned correctly? It is important to realize that there are two quite distinct stages in the assignment of NOESY cross peaks, *resonance* assignment and *cross-peak* assignment. Complete resonance assignment means that for every observed signal, a corresponding atom (or group of equivalent atoms) in the molecule has been allocated as the source of that signal. However, unless every resonance has a *unique* chemical shift, even this (rarely achieved) extent of assignment does not imply that all NOESY cross peaks can be assigned. Any cross peak for which one or both chemical shift ordinates corresponds to two or more accidentally degenerate resonances is ambiguous, since it could link any resonance from the set corresponding to one chemical shift ordinate to any resonance from the set corresponding to the other. Cross-peak assignment, at least as defined in this book, means the resolution of such ambiguities.

Resonance assignment is of course quite an involved process that may be carried out using one of several strategies, depending on the class of macromolecule involved. Some such strategies are mentioned in the appropriate sections of Chapter 13, but given the intricate and often lengthy nature of the process it would be impractical to give a detailed account. Although resonance assignment is obviously of crucial importance to assigning NOE restraints correctly, this section is more concerned with the (less frequently discussed) issue of cross-peak assignment. We will discuss several approaches that can be followed to resolve ambiguities of cross-peak assignment.

12.2.1.1. Using Only (^1H, ^1H) NOESY Data.
One key to minimizing the number of cross-peak ambiguities is to maximize spectroscopic resolution. For a conventionally acquired, 2D (^1H, ^1H) NOESY spectrum, t_{2max} will generally exceed t_{1max} by a factor of 4 or more, so that the resolution in F_2 will be substantially better than that in F_1 (cf. Section 8.3.2). Clearly, critical shift differences should therefore always be assessed in the F_2 dimension wherever possible. It may also be useful to examine spectra calculated using different processing parameters, since higher resolution obtained at the expense of sensitivity may allow critical shift differences to be determined, at least for the more intense cross peaks.

In relatively high resolution NOESY spectra of macromolecules with reasonably narrow lines, it is often possible to resolve partially the multiplet structures of resonances in F_2, or at least to observe differences in their apparent linewidths caused by unresolved multiplet structure. This can be a useful way to resolve cross-peak assignment ambiguities, since all the NOESY cross peaks originating from a particular resonance in F_2 must each show the same F_2 cross-section (excepting amplitude differences, noise, and any peak distortions arising through J-coupling effects such as zero quantum contributions). Thus, even when two cross peaks have identical F_2 chemical shifts, differences in F_2 multiplet structure or linewidth may allow them to be differentially assigned.

Another very important means of resolving cross-peak ambiguities is to compare NOESY spectra obtained under different conditions, usually of temperature, by searching for patterns of cross peaks that move together. Thus cross peaks that have the same F_2 chemical shift at one temperature may be resolved at another temperature, and even if there is a different ambiguity at the other temperature, the two spectra in combination may resolve both ambiguities. This approach works particularly well in resolving ambiguities between chemical shifts of exchangeable resonances, such as those due to amide protons in proteins, due to their large temperature dependencies. However, it is clearly limited by the stability range of the solute, and for non-exchangeable protons the temperature dependencies of chemical shifts often do not show sufficient differential variation to be useful. A further requirement for this approach is that the resonance assignments be checked under each set of conditions employed.

In addition to temperature, other parameters such as pH, concentration, and ionic strength may give useful differential shifts. A rather more drastic extension of this approach is to study spectra of a different molecule. This may not be so extreme a measure as at first appears, since in many cases there may be closely related molecules available. For proteins, slightly different constructs or mutants may exist, while for synthetic peptides, oligonucleotides, or carbohydrates, variant structures may have been synthesized for some other reason. Also, substances isolated from natural sources often contain closely related materials as unwanted impurities. Provided such molecules are sufficiently similar structurally (which should generally be clear from preliminary interpretation of the spectra), comparing their NOESY cross-peak patterns may help resolve cross-peak assignment ambiguities.

Finally, it is worth pointing out that homonuclear, 3D NOESY-NOESY spectra offer opportunities for avoiding ambiguities in assigning NOE interactions (cf. Section 9.2.3). In such spectra, a particular NOE interaction gives rise not to a single pair of cross peaks, but (at least potentially) to several cross-cross peaks, cross-diagonal and diagonal-cross peaks. Thus there is considerable redundant information in the spectrum, and a largely unambiguous set of assigned NOE interactions may emerge from a combined analysis of all the peaks.[7] Against this is the significant disadvantage that the method is rather insensitive, so that it is only applicable to molecules with fairly long relaxation times, and in practice not all possible peaks will be detectable.

12.2.1.2. Using Preliminary Structural Data. If preliminary structures exist, then one may use internuclear distances measured in such structures as a further means to resolve cross-peak ambiguities. If the internuclear distances corresponding to various possible assignments for a given cross peak are compared, then it will often be found that only one assignment corresponds to a sufficiently short internuclear distance for an NOE interaction to occur, and an assignment may be made on this basis. Thus a very commonly used strategy is to calculate a set of preliminary low-resolution structures using only those NOE cross peaks that can be assigned unambiguously, and then to use these structures to test possible assignments for the remaining NOE cross peaks. After assigning further cross peaks using the initial structures, higher resolution structures may be calculated using the improved restraint list. This cycle may be repeated, but it is to be expected that most of the benefit will be gained in the first one or two steps if the initial structures are at all reasonable. This approach lends itself to at least partial automation, and several programs now include tools to aid NOE cross-peak assignment by using preliminary structural data.

Obviously, this approach only works if the differences between internuclear distances corresponding to the various possible assignments are sufficiently clear-cut. Although one might perhaps risk making assignments based on smaller differences between internuclear distances when using a high-resolution structure, the process clearly becomes progressively more hazardous the smaller the difference in internuclear distance that is invoked. One improvement to the process might be to incorporate relaxation matrix calculations (Sections 4.3.1 and 12.5.2) to predict more closely the expected intensities of cross peaks corresponding to various possible assignments. This would presumably increase the confidence that could be placed in assignments based on somewhat smaller differences between internuclear distances, and it may be that this approach will develop in the future.

Preliminary structural information can also sometimes be used to check whether assignments already made are correct. Restraints that consistently give large violations in the calculated structures often turn out upon closer examination to have been misassigned, or even mistyped in the restraint file. Indeed, even restraints that regularly give small violations may repay more careful scrutiny, and, of course, this sort of checking may just as easily turn up resonance misassignments as cross-peak misassignments. However, one must always remember that consistent violation of a given restraint is not in itself sufficient cause simply to *discard* the restraint; such violations are generally a symptom of some other problem that should, if possible, be found and cured.

12.2.1.3. Using Heteronuclear Labeling. For biological macromolecules that can be uniformly labeled with ^{15}N and/or ^{13}C, use of heteronuclear experiments greatly reduces problems of cross-peak ambiguity (see Section 9.2.2). In these experiments, (^{1}H, ^{1}H) NOE interactions are characterized using not only the ^{1}H chemical shifts of the interacting protons as in normal (^{1}H, ^{1}H) NOESY, but also the chemical shifts of their corresponding directly bonded heteronuclei. This can be done either for just one of the two interacting protons, in which case the resulting spectrum has three frequency dimensions, or for both protons, in which case the spectrum is four-dimensional. Although 4D experiments are the least prone to overlap, such experiments are significantly less sensitive than 3D ones, and this disadvantage is at its most severe just when the experiment would be most useful, that is, for larger systems. Thus, 3D experiments have emerged as the "workhorses" amongst heteronuclear NOESY experiments. However, heteronuclear spectra can be extremely valuable even if only 2D spectra are acquired; for example, in proteins labeled with ^{15}N, NOE cross peaks involving NH protons may be characterized by either the ^{1}H or the ^{15}N shift of the NH group in F_1, and the two shifts in combination resolve very many ambiguities. Several variants of such experiments have been proposed, dealing with possible combinations of (^{1}H, ^{1}H) NOE interactions involving ^{1}H directly bonded to ^{13}C or ^{15}N, and some of these are discussed in Section 9.2.2.2.

Another important area of application for heteronuclear labeling is in the structural characterization of symmetrical dimers. The key difficulty in such cases always lies in making the distinction between intra- and interunit NOE interactions. Unless information from other sources is introduced (e.g., by using a plausible model for the structure, as was done during the structure determination of the Arc repressor protein,[8] or by allowing the structure determination protocol to assign the nature of each restraint dynamically[9]), this problem can only be overcome by breaking the symmetry of the dimer. Thus, if the dimer is prepared from a mixture of labeled and unlabeled monomer, then the symmetry is broken, and inter- and intraunit NOE interactions can be observed separately using the X-filtering techniques described in Section 9.2.2.1. If the mole fraction of monomer labeled with X nuclei is denoted a, and the status of the half-filter in each dimension of a 2D experiment is denoted either "*X*" (meaning that only signals coupled to the X nucleus are retained in that dimension) or "*N*" (meaning that only signals *not* coupled to the X nucleus are retained in that dimension), then intra- and interunit contributions may be separated by combining the differently half-filtered data sets as follows:

for interunit interactions,

$$[(F_1 = X, F_2 = N) + (F_1 = N, F_2 = X)] \qquad (12.1)$$

and for intraunit interactions,

$$[(F_1 = X, F_2 = X) + (F_1 = N, F_2 = N)] - \left(\frac{a^2 + (1-a)^2}{2a(1-a)}\right)$$
$$\times [(F_1 = X, F_2 = N) + (F_1 = N, F_2 = X)] \qquad (12.2)$$

The second term in the expression for the intraunit interactions (Eq. 12.2) is needed in order to remove interunit contributions from the $(F_1 = X, F_2 = X)$ and $(F_1 = N, F_2 = N)$ data sets. These contributions arise from dimers in which both monomer units are labeled or where both are unlabeled.

These expressions assume perfect selection by the pulse sequence, and ignore the natural abundance of X nuclei in the "unlabeled" molecules and the (presumably) small proportion of nonlabeled nuclei in the "labeled" molecules. However, the spread of different one-bond coupling values encountered in practice, particularly for (^{13}C, ^{1}H) couplings, means that suppression of the X-bound signals is quite difficult, compromising the separation of signals. Improved versions of half-filter pulse sequence elements have been developed recently that at least partially overcome these difficulties, including one that combines both the selections implied in Eq. 12.1 in a single difference experiment so as to increase sensitivity for symmetrical systems (cf. Section 9.2.2.1).[10]

Alternatively, mixing deuterated and nondeuterated monomers may be used to achieve a similar effect.[11] This simpler approach has been used for many years, but it cannot completely separate the contributions since it serves only to dilute the interunit interactions relative to the intraunit interactions. On the other hand, the dilution of the protons can have other beneficial effects in high-molecular-weight systems, such as reducing spin diffusion and enhancing the intensity of the remaining cross peaks in selectively deuterated proteins.[12] An interesting combination of these ideas has been published, in which a homotrimeric protein was prepared using a proteolytic protocol that results only in trimers having one chain ^{13}C labeled and the other two chains uniformly ^{2}H labeled; this allowed intramonomer NOE enhancements to be observed selectively.[13]

12.2.2. Measuring NOE Restraints

The simplest, and possibly still the most widely used, method for measuring the relative intensities of NOE cross peaks is just to measure the amplitude of the cross-peak maximum. This can be done by counting contour levels on a suitably contoured plot, or by using a suitable peak-picking algorithm to determine each maximum intensity. During contour counting, the contours are used essentially as though they were graduations on a ruler, so it is essential that they are spaced regularly, forming either a linear or a geometric series starting at zero (although the contour at zero intensity is not itself plotted). Cross peaks may then be classified just according to how many contour levels each of them crosses.

This approach has some severe shortcomings. It is not the peak height of a given NOE cross peak that is determined by the strength of the corresponding NOE interaction, but rather its *volume integral*. In its simplest form, the approach of measuring cross-peak amplitudes takes no account of this distinction, so that linewidth variations and even multiplicity differences between different signals are simply ignored. For instance, a cross peak connecting two singlets might have the same volume integral as one connecting two complicated multiplets, but it would certainly have a higher maximum than the latter, and would probably cross more contour levels; the former cross peak might therefore wrongly be classified as being stronger. To some extent, errors of this sort are reduced by low resolution. For many NOESY experiments t_{1max} and/or t_{2max} are sufficiently short that distinctions between different linewidths and multiplicities may be partly destroyed, while for 3D and 4D experiments the maximum acquisition time in each dimension is usually shorter still, making automatic intensity-picking relatively attractive as a method of quantitation. Even when intrinsic resolution is high, some attempt can be made to reduce such errors by deliberately introducing line-broadening, or by making empirical corrections based on visual or semiquantitative estimates of lineshapes. For example, the area enclosed by the contour nearest the cross-peak half-height may be used as an approximate linewidth correction factor, although this approach is not appropriate for multiplicity corrections.[14] However, even when such measures are used, problems of this sort still often represent a major source of uncertainty in the input to structure calculations when contour counting or amplitude measurement is used for quantification. Nonetheless, the relative simplicity of this method has ensured its continued popularity.

The alternative approach of using volume integration to measure NOE cross-peak intensities is undoubtedly more correct from a theoretical perspective, but has so far been very slow to displace amplitude estimation as the most popular method of quantitation. Partly this may be for reasons of perceived convenience, but another reason is that results achieved using much of the NMR processing software available until recently apparently offer little practical improvement over amplitude measurement. It is not entirely obvious why this should be, but at least in part it is because most volume integration routines work simply by adding all the intensity values within a defined area. This implies that the data must be extensively zero filled prior to Fourier transformation to achieve a reasonable representation of each cross peak, and accurate baseplane correction of the spectrum becomes of crucial importance. More sophisticated routines exist that can help to disentangle partially overlapped cross peaks, but these can be rather time-consuming in use and may require extensive manual interaction. Such routines generally work either by curve-fitting to cross-sections, or by using a library of "reference" cross-sections built up interactively from those cross peaks that are not overlapped.[14,15] It is to be expected that such software will become more widely used in the future.

12.2.3. Calibrating NOE Restraints

Before the measured NOE intensities can be used as input for a structure calculation program, they need to be interpreted to yield internuclear distances. However, because the NOE depends on molecular motions as well as internuclear distances, the relationship between NOE intensity and internuclear distance is unavoidably approximate. For this reason, an NOE-derived distance restraint can only usefully be expressed as a *range* of allowed distances, rather than one specific distance.

Usually, only the upper bound for a distance restraint range is set according to the NOE intensity, the lower bound being set either to zero or to the sum of the van der Waals radii for the interacting atoms (depending on the method of calculation employed, see Section 12.3).[16] When this is done, a strong NOE cross peak is taken to mean that the corresponding internuclear distance must be relatively short, but a weak NOE cross peak is not interpreted as *requiring* the distance to be long; this allows for the possibility that particular NOE interactions corresponding to short distances may be quenched selectively by internal motion (see Sections 5.5 and 5.6). Further approximation can be introduced by grouping restraints into categories such as "strong," "medium," "weak," and "very weak," rather than attempting to express each restraint as some individually calculated distance range. Thus the process of calibration usually becomes one of sorting the NOE cross peaks into groups according to intensity, and defining for each group the upper bound of a corresponding distance range. To some extent, the approximation deliberately introduced at this stage also allows for the uncertainties inherent in the actual measurement of NOE intensities (Section 12.2.2); in fact, this is one of the main motives for adopting this approach to calibration.

As discussed in Chapter 4, it is only during the period of validity of the initial rate approximation that the intensity of a particular NOE cross peak grows linearly and can be directly related to r_{ij}^{-6}; at longer mixing times, NOE intensities become a complicated function of many internuclear distances due to the onset of spin diffusion. On the other hand, if only short mixing-time spectra are used to obtain restraints, many potentially useful long-range restraints are excluded, since the corresponding cross peaks do not have detectable intensity in such spectra. A good compromise is therefore to measure NOE spectra at several mixing times (such a set of spectra is often called a "τ_m series" or a "mixing-time series"), and then to classify cross peaks according to the mixing time at which they first acquire a certain intensity. In this way each cross peak is classified using its intensity at or near the earliest practicable point during its development, and the validity of the initial rate approximation may be tested, at least roughly, by checking whether the continued growth of the cross peak is approximately linear.

Since structurally useful restraints generally correspond to internuclear distances in the range 2.2–5 Å, it is important that the spectra included in a τ_m series cover a range of mixing times within which NOE cross peaks corresponding to these distances reach appreciable intensity. The actual range of mixing times needed varies according to the solute tumbling rate, and the best guide is experiment. For instance, in many proteins the strongest $d_{\alpha N}(i, i + 1)$ cross peaks occur in regions of β-sheet and correspond to internuclear distances of about 2.2 Å, while $d_{\alpha N}(i, i + 3)$ connectivities in regular α-helices correspond to distances of about 3.4 Å (see Chapter 13, Table 13.1). The presence or absence of these cross peaks in NOE spectra of proteins that possess both regular β-sheets and α-helices gives a reasonable guide to the mixing times needed in a τ_m series, and can also conveniently be used subsequently to calibrate the restraint ranges. In cases where only one of these types of cross peak can be found, another calibration distance may be taken (see following), or else the mixing time at which longer-range interactions appear may be estimated from an assumed r^{-6} dependence (for example, if a particular intensity corresponds to an NOE interaction over 2.2 Å at one mixing time, then at four times that mixing time the same intensity would correspond to an NOE interaction over $[(2.2^{-6})/4]^{-1/6} \approx 2.77$ Å, provided that the initial rate approximation holds in both cases).

Unfortunately, many cross peaks corresponding to longer-range restraints only show up in spectra with significantly longer mixing times. The penalty for using any cross peaks from such spectra to generate restraints is that some may result from magnetization transmitted over one or more intermediate spins by spin diffusion, and thus correspond to significantly longer internuclear distances (Section 4.4.4). However, the genuine two-spin long-range restraints in such data are usually too valuable to ignore, as they often specify vital details of the structure. As a compromise, these various weak interactions are often collected together in the weakest restraint category, and the corresponding upper bound set rather long (e.g., to 5, 6, or even 7 Å) in an effort to ensure that, even if a cross peak does arise through spin diffusion, the corresponding restraint does not impose too short a distance. To an extent, one may further reduce the risk inherent in using such data by analyzing the spectra (perhaps in conjunction with preliminary structural data) to identify possible spin-diffusion pathways, so as to exclude any corresponding entries from the restraint list. For instance, a weak cross peak at long mixing time is almost certain to arise via spin diffusion if both partners show strong cross peaks to a common third signal; preliminary structures may then also confirm that the steps on such a pathway each correspond to a short distance.

Bearing all of the above in mind, typical choices for mixing times in a τ_m series for a protein having $\tau_c \approx 4$–8 ns might range from 20 ms to 100 ms, with "very weak" restraints included from 150 ms, 200 ms, or maybe even 250 ms data. For longer τ_c values, correspondingly shorter mixing times would be needed, although for very short mixing times the problem of zero quantum suppression may become severe (Section 8.4.2).

Many textbooks quote an equation for calibrating the distance dependence of NOE intensities with respect to a single reference cross peak (cf. Eq. 4.29):

$$r_{ij} = r_{\text{ref}}(I_{\text{ref}}/I_{ij})^{1/6} \qquad (12.3)$$

In practice, several problems can arise if this approach is used naïvely. Often, the reference distance recommended is one fixed by covalent geometry, such as that between two protons in a methylene group ($r \approx 1.8$ Å), or between two *ortho*-related protons on an aromatic ring of tyrosine ($r \approx 2.5$ Å). While it is clearly an advantage that these reference distances are known and fixed, their use can suffer some severe disadvantages. They necessarily involve protons that are scalar coupled to one another, which may cause systematic errors in measuring I_{ref}, for example, due to interference from zero quantum coherence (see Section 8.4.2) or from strong coupling (see Section 6.4.1). Also, methylene or aromatic protons may be located in relatively mobile side-chains, causing the cross peak between them to build up more slowly than would cross peaks originating in more rigid parts of the structure; again this can distort the simple relationship expressed in Eq. 12.3. Perhaps most significantly, these distances (particularly the CH_2 interproton distance) are shorter than most of the distances actually being determined, which can cause r_{ij} to be underestimated (Section 4.4.4).[17] When the initial rate approximation begins to fail, the initially linear growth of the NOE slows down at a rate dependent on the overall relaxation rates of the two interacting protons (see Eq. 4.3), so that the fall-off in NOE growth is necessarily faster if the protons are close together. Thus, if the reference distance is significantly shorter than r_{ij}, the intensity of the reference cross peak will be more severely diminished after a given time than will I_{ij}, leading to an underestimate of r_{ij} when Eq. 12.3 is used. Figure 12.1a–c illustrates this problem with some data simulated using an NMR structure of protein G.[14]

One way to limit this latter problem without abandoning the simplicity of the initial rate approximation is to use two reference distances (this is essentially a formalized version of the approach described at the beginning of this section). A modified form of Eq. 12.3 may then be used, where r_1 and r_2 are the two reference distances, and I_1 and I_2 the corresponding reference intensities:

$$r_{ij} = A + B(I_{ij})^{-1/6} \qquad (12.4)$$

with

$$A = \frac{r_1(I_1)^6 - r_2(I_2)^6}{(I_1)^6 - (I_2)^6} \quad \text{and} \quad B = \frac{r_1 - r_2}{(I_1)^{-1/6} - (I_2)^{-1/6}}$$

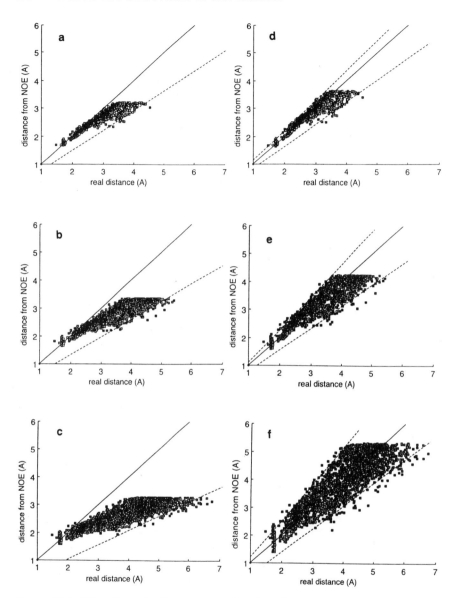

Figure 12.1. The effect of using a short calibration distance. Each point in these plots corresponds to an assigned NOE cross peak in the NOESY spectrum of a small protein (protein G) and correlates the true distance between the corresponding protons, as measured in the final set of NMR structures of protein G, with the distance obtained from the NOE intensity under various different assumptions. In plots (*a*), (*b*), and (*c*), the "distance from NOE" is calculated using a single calibration distance (methylene CH_2 = 1.7 Å), using Eq. 12.3. In plots (*d*), (*e*), and (*f*) the "distance from NOE" is calculated using two calibration distances (methylene CH_2 = 1.7 Å and adjacent aromatic ring protons = 2.4 Å), using Eq. 12.4. The NOE intensities used in these calculations were themselves calculated values, obtained using the full relaxation matrix method starting

Essentially, this equation corrects the "skew" that results when one short reference distance is used, as may be seen by comparing Figure 12.1a through 12.1c with figure 12.1d through 12.1f. Both approaches (i.e., using one reference distance or using two) attempt to describe the actual nonlinear NOE distance dependence as though it were linear, but when two reference distances are used, it is only the part of the curve corresponding to somewhat longer distances that is approximated, and the requirement that the fitted straight line pass through the origin is dropped. Clearly, either approximation will work best when the reference distance(s) are roughly equivalent in length to the unknown distances being determined.

Whatever approach is used, it is advantageous for the reference and unknown NOE interactions to be similar whenever possible. Thus, a widely used approach is to employ separate reference distances for different classes of interactions, dividing signals up into categories such as "rigid CH or CH_2," "CH_3," "aromatic," or "NH." Factors that differ between these categories, such as local mobility, multiplicity, and degree of partial isotopic substitution (if any), are then allowed for automatically, at least to a first approximation. For example, if a protein solution is made up in 15% D_2O/85%H_2O to provide a signal for the deuterium lock, cross peaks between NH and CH protons will be reduced by 15% relative to those between two CH protons (assuming that all exchange processes have reached equilibrium), and those between two NH protons by 27.8%. However, if separate reference distances are used for (CH, CH), (NH, CH) and (NH, NH) cross peaks, no correction need be calculated for this effect, as it is intrinsically included in the calibration. Table 12.1 shows some reference distances for different classes of interatomic distances in proteins and nucleic acids.

It would be unrealistic to expect to combine *all* of the various points just discussed. One would need, for each category, two fixed and known reference distances that neatly bracket the range of unknown distances in that category, all free from distortion caused by strong coupling or the various other artifacts considered above. In practice, different experimenters have adopted various different imperfect compromises, depending on the quality and type of data

---

from the coordinates of the protein G structure. In this way it was possible to use several different overall correlation times when calculating the theoretical NOE intensities: for plots (a) and (d) τ_c was set to 2 ns; for plots (b) and (e) τ_c was set to 4 ns; and for plots (c) and (f) τ_c was set to 8 ns. The dashed lines correspond to distance deviations of (a) 60%, (b) 100%, (c) 150%, (d) -10% and $+30\%$, (e) -20% and $+40\%$, and (f) -30% and $+40\%$.

As discussed in the text, when only one reference distance is used and this is shorter than most of the distances being determined, the actual distances are systematically underestimated. However, when two reference distances are used, this problem can largely be eliminated. Reproduced with permission from ref. 14.

TABLE 12.1 Reference Distances Used for Various Classes of NOE Interaction in Biological Macromolecules

Type of Interaction	Reference Interaction	Reference Distance (Å)
Proteins		
CH–CH	Gly $C^\alpha H$–$C^\alpha H$,	1.8
	AMX $C^\beta H$–$C^\beta H$	1.8
CH–methyl[a]	Ala $C^\alpha H$–$C^\beta H_3$	2.4
methyl–methyl[a]	Val $C^\gamma H_3$–$C^\gamma H_3$	3.1
	Leu $C^\delta H_3$–$C^\delta H_3$	3.1
NH–CH	$d_{\alpha N}(i, i+1)$ in regular β sheet	2.2
NH–methyl[a]	$d_{\beta N}(i, i+1)$ in regular α helix	3.3
NH–NH	$d_{NN}(i, i+1)$ in regular α helix	2.8
aromatic–aromatic[b]	Tyr $C^\delta H$–$C^\epsilon H$	2.8
Nucleic Acids		
sugar–sugar	H2′–H2″	1.8
sugar–base	conformation dependent	see Tables 13.2, 13.3, and 13.4
sugar–methyl	conformation dependent	see Tables 13.2, 13.3, and 13.4
base–base	conformation dependent	see Tables 13.2, 13.3, and 13.4
base–H1′	C H5–C H6	2.5
base–methyl	T H6–T Me5	2.7

[a]Distances involving one or two methyl groups are quoted to the corresponding pseudoatom(s).
[b]Used for degenerate signals averaged by ring flip motions fast on the chemical shift timescale.

available, and (probably) personal taste. Ultimately, the progress of the structure calculations themselves is one of the best guides to calibration. If the upper bounds for distance restraints have been set unrealistically short, this will lead to many restraint violations in the calculated structures, whereas if the bounds are too loose, the structures will be poorly defined. Also, errors in the classification of individual cross peaks often show up during refinement as they lead to consistent violations.

For proteins and other globular structures, approximate calibration is often acceptable because the quality of the final structure is determined principally by the number and distribution of a large and interlocking set of restraints, rather than by their individual precision. However, in the case of double-helical DNA, the molecule is essentially linear, the distances over which useful NOE interactions occur are generally quite long (>3 Å), and fewer restraints are available. Under these circumstances, defining distance ranges only in terms of an upper bound generally fails to constrain the molecule adequately, and a more restrictive calibration is often sought. Restraints are often expressed as a single distance or narrow range of distances, determined by intensity comparison with a reference distance. Of course, such an approach makes even the gross topology of the final structure rather heavily dependent on the accuracy of the restraints, so that this field has seen considerable effort expended on attempting to improve quantitation of the NOE, with mixed success; further discussion may be found in Section 13.2.3.1.

12.2.4. Averaging in Equivalent Groups

Certain protons that undergo rapid internal motions as a group give rise to a single averaged signal; obvious examples include methyl groups and symmetry-related pairs of aromatic protons on fast-flipping rings. When using NOE data involving such groups, two particular issues need to be addressed. First, one must decide how to test whether or not an NOE-derived distance restraint involving a group of equivalent protons is satisfied in the model structure. Second, a correction must be made for the fact that more than one proton contributes to one (or both) of the signals involved in the NOE interaction (unless, that is, the issue of multiplicity corrections has been dealt with by using separate calibration distances for different categories of signals such as CH's, methyls, and aromatics, as discussed in the previous section). These issues are considered in this section.

The issue of testing for restraint violations involving equivalent groups is closely related to that of how the NOE itself is averaged by internal motions. As discussed in Chapter 5, NOE enhancements that are averaged by an internal motion *slower* than overall tumbling of the molecule ($\tau_e \gg \tau_c$) correspond to an apparent average distance given by $\langle r^{-6}\rangle^{-1/6}$; this situation applies for most symmetric aromatic rings in proteins (i.e., rings of Phe and Tyr residues that flip fast enough to cause signal coalescence, $k_{flip} \gg \Delta\delta$, and which also flip fast on the T_1 timescale; see Sections 5.1 and 5.5.1). In contrast, the NOE enhancements that are averaged by an internal motion *faster* than overall tumbling of the molecule ($\tau_e \ll \tau_c$), such as rotation of a methyl group, correspond to an apparent averaged distance given by r_{Tropp} (Section 5.5.2). One particularly direct way to test for restraint violations in a model structure would thus be to determine whether or not the appropriate averaged distance in the model (calculated from the relevant individual distances either as $\langle r^{-6}\rangle^{-1/6}$ or as r_{Tropp}, depending on the type of internal motion involved) is within a particular range set according to the experimentally observed NOE intensity.

Several programs allow the option of r^{-6} averaging when testing the validity of restraints involving groups of equivalent protons, but the authors are not aware that any software packages at present include the more complicated option of calculating r_{Tropp} for the faster moving groups. However, on the assumption that methyl rotation is essentially the only case in which r_{Tropp} would be used, this has little practical consequence. It is true that r_{Tropp} can be significantly longer than the r^{-6} average distance for locations of an external spin at short distances close to the methyl axis, but elsewhere the difference between the functions is much smaller. Further, what error there is makes a restraint calculated using r^{-6} averaging over-conservative. An r^{-6} averaged distance is more strongly dominated by the shortest of the individual contributing distances and is thus somewhat shorter than the corresponding value of r_{Tropp}. Consequently, when the model structure is identical to the true structure, the model-derived r^{-6} averaged distance that is used to check restraint validity will still be shorter than the distance "sensed" by the NOE (r_{Tropp}) and used to set the restraint upper bound, implying that the methyl group could be further away from its NOE partner in the model structure, without the restraint being violated.

12.2.4.1. Pseudoatom Corrections.

Although r^{-6} averaging (or summation; cf. Section 12.2.4.3) is probably the most natural way in which to deal with internal motions of symmetrical groups, a computationally simpler approach is to define restraints involving such a group of protons in terms of the mean geometric position of the protons in the group. This approach, often called "center averaging," has been more widely used than r^{-6} averaging, at least until recently. Structure calculation programs that employ center averaging represent the centroid of a group of equivalent protons using a "pseudoatom" (often denoted Q); this is a fictional atom of zero size that exerts no forces on other atoms, and whose sole purpose is to define a given position relative to the molecular framework. However, it is a crucial feature of center averaging that the distance r_{SQ} from some other proton S to a pseudoatom Q is usually (though not always) *longer* than the distance corresponding to the NOE intensity between S and the group of protons I that the pseudoatom Q represents. This is because the NOE itself is averaged as either $\langle r^{-6} \rangle^{-1/6}$ or r_{Tropp} and these quantities are dominated by the shortest distance included in the average, whereas the distance r_{SQ} from S to the pseudoatom is not. For this reason, a restraint involving a group of equivalent protons requires the addition of a correction, called a "pseudoatom correction," to the upper bound if it is to be used in the context of center averaging, so as to prevent the restraint from being too restrictive.

Pseudoatom corrections were originally defined as the distance between any one of the equivalent atoms I_i and the pseudoatom Q.[18] Such a value for the correction must always be sufficiently long, since it represents the maximum distance that the pseudoatom could lie beyond the closest of the individual spins I_i in the equivalent group, as viewed from the external spin S. However, in effect this definition corresponds to the assumption that *only* the closest of the individual spins I_i contributes to the NOE interaction with S, whereas in reality the other spins in the group I also contribute something to the NOE interaction, with the result that the distance "sensed" by the NOE is made somewhat longer.

If, instead, one defines the pseudoatom correction as the maximum possible value of ($r_{QS} - r_{\text{effective}}$), where $r_{\text{effective}}$ is the appropriate averaged distance $\langle r^{-6} \rangle^{-1/6}$ or r_{Tropp}, this takes all members of the equivalent group into account. Under this definition, the pseudoatom correction is the maximum value by which the distance from S to the pseudoatom can exceed the averaged distance sensed by the NOE ($r_{\text{effective}}$) and used to set the restraint upper bound. This correction will always be smaller than that based on the original definition, to an extent that depends on how close together the atoms are within the equivalent group, and whether averaging occurs over r^{-6} values or r_{Tropp} values. However, the value of $r_{\text{effective}}$ also depends on the geometry of the interaction (i.e., the relative positions of all the interacting spins I_i and S), so defining a pseudoatom correction for general use implies finding the interaction geometry for which $r_{QS} - r_{\text{effective}}$ is a maximum. This geometry is necessarily that in which $r_{\text{effective}}$ is the most strongly dominated by the shortest of the individual distances r_{IiS}, and this property makes it trivial to find in several cases. For instance, for a methyl group the geometry corresponding to the maximum value of $(r_{QS} - r_{\text{effective}})$ has the external spin S in van der Waals contact with the nearest methyl proton, colinear with both it and the pseudoatom; this geometry corresponds to a pseudoatom correction of 0.4 Å.[19] A similar pseudoatom correction was obtained from molecular dynamics simulations.[20]

As we shall see shortly, pseudoatom corrections are also commonly applied to restraints involving diastereotopic groups for which no stereoassignments are available, so that restraints can be placed upon the overall position of a diastereotopic group without having to differentiate between the nonstereoassigned signals. For the case of a methylene group, the maximum value of $(r_{QS} - r_{\text{effective}})$ occurs when the external spin S is colinear with the two methylene protons and in van der Waals contact with the nearer one, leading to a pseudoatom correction of 0.7 Å. In the case of isopropyl groups of Val and Leu residues in proteins, the appropriate pseudoatom correction is more complicated to calculate, partly due to the more complicated geometry, and partly because the appropriate mode of averaging *within* each methyl group is given by r_{Tropp}, whereas *between* the two it is given by $\langle r^{-6} \rangle^{-1/6}$. Numerical simulations show that the maximum value of $(r_{QS} - r_{\text{effective}})$ corresponds to the geometry shown in Figure 12.2, and for this arrangement the pseudoatom correction is ~1.5 Å. A full set of pseudoatom corrections based on this approach has recently been published, and both this set and the original one appear together in Table 12.2.[19]

Although some of these newer pseudoatom corrections are shorter than the original values, it remains true that much of the time pseudoatom corrections act to weaken restraints *unnecessarily*. Pseudoatom corrections must always cater for the "worst case" interaction geometry, since the geometry of interactions cannot be known *a priori*, and this issue cannot, in general, be avoided in the context of center-averaged calculations. However, the whole issue is sidestepped when r^{-6} averaging is employed and pseudoatom corrections become unnecessary, allowing the true power of each restraint to be realized more fully. Figure 12.3 illustrates this point by showing the "validity boundary" for a restraint to a CH_2 group, and for a restraint to an aromatic ring, as evaluated using both center averaging and r^{-6} averaging.

Experience of comparing calculations using center averaging and r^{-6} averaging remains limited, but, as would be expected, the latter tends to produce

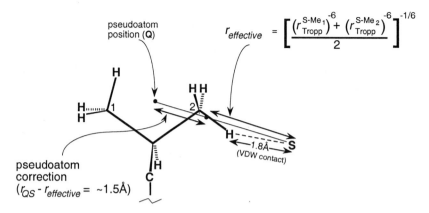

Figure 12.2. Pseudoatom correction for a nonstereoassigned isopropyl group of valine or leucine. The averaged distance sensed by the NOE is given by the expression for $r_{\text{effective}}$ in the figure, where $r_{\text{Tropp}}^{\text{S-Me}_1}$ and $r_{\text{Tropp}}^{\text{S-Me}_2}$ are the values of r_{Tropp} for the interaction of S with methyl groups 1 and 2 respectively (see text for discussion of r_{Tropp}). The geometry shown is that in which the distance from S to the pseudoatom Q exceeds the average sensed by the NOE by the greatest margin, namely ~1.5 Å. Adapted with permission from ref. 19.

TABLE 12.2 Pseudoatom Corrections for Groups of Equivalent or Non-stereoassigned Protons

Equivalent Group	Pseudoatom Correction	Conventional Value[a]
CH_3	0.4 Å	1.0 Å
NH_3^+ (e.g., of Lys)	0.4 Å[b]	1.0 Å
aromatic H_2/H_6	2.0 Å	2.0 Å
aromatic H_3/H_5	2.0 Å	2.0 Å
non-stereoassigned CH_2	0.7 Å	1.0 Å
non-stereoassigned $CH(Me)_2$	1.5 Å	2.4 Å
non-regioassigned $CONH_2$	0.7 Å	1.0 Å
non-regioassigned $N^\eta H_2$ of Arg[c]	1.8 Å	2.2 Å

In each case, the value shown is the maximum extent to which the distance between an external spin and the pseudoatom representing a group of equivalent spins exceeds the appropriately averaged distance between the external spin and the equivalent group as actually sensed by the NOE; each value has been rounded up rather than down.

[a] These values represent the maximum additional distance that the pseudoatom can lie beyond the nearest member of the equivalent group, as viewed from an external spin receiving the NOE from the equivalent group (see text).

[b] For an unprotonated NH_2 group this would become 0.7 Å, as for a CH_2 group.

[c] This value assumes the guanidinium group to be protonated at N^η, so forming a planar symmetrical group $C^\zeta(N^\eta H_2)_2^+$.

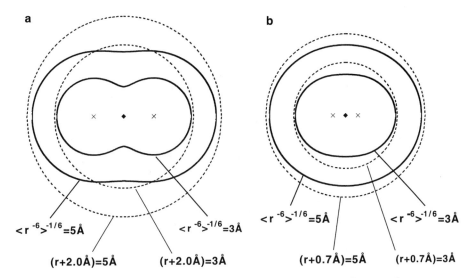

Figure 12.3. Validity boundaries for restraint upper bounds of 3 Å and 5 Å calculated using r^{-6} averaging (solid lines) and center averaging (dashed lines) for (*a*) a pair of symmetry-related aromatic protons, and (*b*) a CH$_2$ group. In each case the proton positions (I_1 and I_2) are indicated by crosses, and the pseudoatom position (Q) by a filled diamond. When the restraint upper bound is short relative to the interproton separation between I_1 and I_2, the volume bounded by the r^{-6} averaged restraint is significantly smaller than that bounded by the center-averaged restraint. The difference between the two restraints is at its minimum close to the line connecting the two protons, and at its largest close to the perpendicular plane through the pseudoatom. When the restraint upper bound is long relative to the interproton separation, the r^{-6} averaged restraint is shorter than the center averaged restraint, because no pseudoatom correction is included in the former. Reproduced and adapted with permission from ref. 19.

somewhat more tightly defined families of structures (starting from the same restraint list, but omitting pseudoatom corrections in the case of r^{-6} averaging). However, it appears that such differences are only appreciable in cases where the restraint list is relatively sparse and the structure relatively ill-defined.[19] This is not unreasonable, if one considers that the restriction on the position of a particular group is much more likely to arise from just one or two restraints if the restraint list is sparse. Removal of the pseudoatom corrections from the upper bounds of these few restraints is then more likely to restrict the spread of calculated structures than it would if the position of the group were determined by many interlocking restraints from a dense and mutually redundant set.

12.2.4.2. Multiplicity Corrections.
We turn next to the issue of correcting for multiplicity. In principle, this is very simply solved: The intensity of an NOE cross peak between signals I and S should simply be divided by $n_I n_S$

before it is converted into a distance restraint, where n_I and n_S are the numbers of equivalent spins in the groups corresponding to signals I and S respectively. Essentially, this correction is needed because absolute cross-peak intensities are being measured, rather than fractional enhancements.

If numerical values for NOE intensities are available, for instance as a result of volume integration of cross peaks, division by $n_I n_S$ is of course trivial. In still the majority of cases, however, NOE intensities are probably estimated semiquantitatively (e.g., by counting contour levels in an evenly contoured plot of the NOESY spectrum; see Section 12.2.2) and then classified into categories such as "strong," "medium," and "weak." In such cases, division of the intensity by $n_I n_S$ is not possible directly.

One way around this difficulty, often adopted, is to calibrate different classes of restraint separately; thus, all methyl–single proton restraints might be calibrated using $d_{\alpha\beta}(i, i)$ in alanine, methyl–methyl restraints using $d_{\delta1\delta2}(i, i)$ in leucine, and so on (see Section 12.2.3). However, if a common calibration is used for all interactions, it is necessary to correct individual restraints for different multiplicities. Quite commonly, 0.5 Å is added to the upper bound of all restraints involving a methyl group (and 1 Å to those of methyl–methyl restraints) to compensate for multiplicity (although the precise reason for introducing this correction originally is not completely clear).[21,22] No related corrections appear to have been adopted for restraints involving degenerate aromatic signals or methylene groups (although one can argue that it is less likely to lead to errors than would neglect of corrections for methyl groups[19]). However, it is clear that any such *addition* to the upper bound cannot in general correspond to a *division* of the NOE intensity, and that what is needed instead is a multiplicative correction to the upper bound. It follows directly from the r^{-6} dependence of NOE intensity that the appropriate multiplicity correction factors Z are given by

$$Z = (n_I n_S)^{1/6} \qquad (12.5)$$

For methyl groups, intensity correction by a factor of three then corresponds to lengthening the upper bound by 20%. For an intensity correction through a factor of 2, the upper bound would be lengthened by 12%, while for an intensity correction through a factor of 4 it would be lengthened by 26%, and for an intensity correction of 6 it would be lengthened by 35%.

As mentioned earlier, the concept of equivalent groups is also very widely used for a quite different purpose, namely allowing for missing stereoassignments. Stereoassignments, and methods for making them, are considered further in Section 12.2.5, but when no stereospecific assignment is available, restraints are usually referred to the whole set of protons (2 for a CH_2 group, 6 for an isopropyl group of valine or leucine), which is then treated *as though* it were an equivalent group. Thus, although there will generally be two cross peaks connecting X to the two signals from the diastereotopic group, in the absence of a stereoassignment only one restraint can be defined using these two cross peaks.

How should the upper bound for a restraint between such a diastereotopic group and another proton X be set? By analogy with the methyl and degenerate aromatic ring cases, the relevant intensity used for defining the upper bound should be the sum of the intensities of the two cross peaks between X and the two methylene protons, but divided by 2 to account for multiplicity; in other words it is the *average* of the two intensities. In the case of nonstereoassigned methyl groups of Val or Leu, the appropriate correction for multiplicity corresponds to dividing the summed intensity by 6, giving one-third of the averaged intensity. Of course, the structural significance of any intensity difference between the two cross peaks is thereby discarded, but this merely reflects the fact that such information cannot be interpreted when no stereospecific assignment has been made. The analogy to motional averaging is perhaps most clearly seen in the case where the diastereotopic signals are accidentally degenerate, since then the summed intensity is automatically what is measured directly, just as it would have been in the methyl or degenerate aromatic ring proton cases. As before, pseudoatom corrections will be required if center averaging is used (see Table 12.2 for values), but not if r^{-6} averaging is used.

12.2.4.3. r^{-6} Summation.
In this section we consider a subtle variation on r^{-6} averaging, called r^{-6} summation. In this approach, the distance measured in the structural model in order to test restraint validity is defined, not by the r^{-6} average, but instead by the (shorter) r^{-6} sum. This definition necessarily implies a change in the relationship between distance and NOE intensity. For r^{-6} averaging, the relationship between intensity and distance is given by

$$\frac{intensity}{n_I n_S} = k \frac{1}{n_I n_S} \sum_{i,j} r_{I_i S_j}^{-6} \qquad (12.6)$$

(where k is a constant of proportionality), whereas for r^{-6} summation, the division by $n_I n_S$ is simply removed from both side of Eq. (12.6), giving:

$$intensity = k \sum_{i,j} r_{I_i S_j}^{-6} \qquad (12.7)$$

Thus, when r^{-6} summation is used instead of r^{-6} averaging, the appropriate value to use for each NOE interaction intensity changes also from an average to a sum.

The implication of these equations is that no multiplicity corrections need to be made for NOE interactions involving genuinely equivalent spins when using r^{-6} summation. Uncorrected upper bounds should be used, just as though the corresponding cross peaks had connected two single-proton resonances; indeed, this simplicity is one of the attractions of the method. However, in the case of nonstereoassigned diastereotopic groups, the situation is more complex.

The intensity that should now be used is the sum over all the relevant NOE cross peaks, rather than their average intensity, as would be used in conjunction with r^{-6} averaging calculations. When numerical intensities or volume integrals are available, such sums may be readily calculated, but when intensities are only divided into categories, a procedure based on corrections to upper bounds is appropriate. The cases requiring a correction are those where two (or more) of the cross peaks corresponding to an interaction involving a nonstereoassigned diastereotopic group are in the same intensity category, since then one should *reduce* the upper bound to allow for the greater intensity that would have been found had the cross peaks been summed. Even though only a relatively small proportion of restraints are involved, it turns out that omitting these corrections can degrade the attainable precision of a structure when the NOE data are sparse; further details may be found in reference 19.

The r^{-6} summation approach was first applied to the treatment of symmetrical dimer structures (as already mentioned in Section 12.2.1.3).[9] More recently, it has been applied to deal with limited NOESY cross-peak assignment ambiguity.[23] Suppose two assignments are possible for a given cross peak, one corresponding to an NOE interaction between protons i and j, and the other to one between protons k and l. Despite the lack of this relative assignment, one could define a restraint that uses only the available information by stipulating that the *sum* of r_{ij}^{-6} and r_{kl}^{-6} should not exceed an upper distance bound corresponding to the measured NOE intensity of the ambiguous cross peak. Thus, depending on the structure, the restraint could be satisfied either by making r_{ij}^{-6} short, or by making r_{kl}^{-6} short, or both. This is clearly a less satisfactory situation than having the restraint fully assigned, but at the same time it is better than not using it at all (essentially it represents a generalization of the "floating" stereoassignment method to other cross-peak assignment ambiguities; see Section 12.2.5). This approach can be much more successful than one might at first expect, and its introduction may turn out to be a significant step forward in improving the efficiency and accuracy of structure calculations, particularly at the refinement stage.

Although these uses of r^{-6} summation may appear superficially to be very different from earlier applications of r^{-6} averaging to the treatment of equivalent groups, in fact, r^{-6} averaging could also be used in these contexts in an almost identical fashion to r^{-6} summation, provided only that NOE intensities are divided by the number of possible assignments involved for each cross peak before setting upper bounds. However, this point is probably only of theoretical interest, since it is very much more convenient to use r^{-6} summation precisely so as *not* to have to make these corrections.

12.2.5. Stereoassignments and Torsion Angle Restraints

Diastereotopic groups in a chiral molecule are (necessarily) chemically distinct, and in general will therefore give separate NMR signals. Thus the two protons of each CH_2 group in a protein will give separate signals unless the two are accidentally degenerate, as will the two methyls in the isopropyl group of each valine or leucine residue, and also the two protons of each methylene group in sugars of DNA, RNA, and oligosaccharide molecules. NOE enhancements into such diastereotopic groups from some other proton X may well differ between the two methylene protons (or methyl signals of valine or leucine), since X may be significantly closer to one than to the other. However, before such potentially valuable differential NOE enhancements can be used to constrain the correct proton (or methyl group), a stereospecific assignment of the two signals to the diastereotopic protons must be made. The following discussion will be limited to methods for stereoassignment applicable to proteins, since these involve the same general principles as for other cases. Figure 12.4 shows the IUPAC definitions and nomenclature used for diastereotopic groups in proteins.[24]

Stereoassignments for H^β protons are usually made as part of a more general process of analyzing local conformation, which can be carried out more or less independently of the rest of the structure determination. Generally, calculations involve some form of grid search over local conformational space at the dipeptide level, repeating the calculation for each dipeptide segment in turn along the whole sequence of the protein. In this way it is possible to analyze systematically all combinations of ϕ, ψ, and χ_1 for each residue (in discrete steps of, say, 10° or 30° for each angle) to see which conformations are compatible with the input data for that particular residue. Different programs use different input data (see the following), but typically they include various J-coupling values and NOE-derived distance restraints spanning only one or two residues [that is $d(i, i)$ and $d(i, i + 1)$], together with the covalent connectivity, geometry, and van der Waals radii. Since the stereoassignments are initially unknown, this analysis must be repeated for both possible stereoassignments for each residue. In cases where only one of the two stereoassignments leads to a reasonable fit to the input data, a stereoassignment is made on that basis.

In addition, the calculation reports the allowed ranges of ϕ, ψ, and χ_1 found for each residue, and these can be used to provide useful additional restraints when the overall structure calculation is carried out. Although such restraints will generally be tighter for residues where stereoassignments were made, some restriction may be possible even for nonstereoassigned residues. When two or more allowed ranges are found for a given torsion angle, some care is needed in specifying how penalties are imposed for violating the corresponding restraints; since no angle can be within more than one allowed range, all angles will violate one or more restraints in such cases, if the restraints are defined in the usual way. This problem may be avoided, either by defining one weakened restraint to include all the allowed ranges within one larger range, or by specifying restraints directly in terms of coupling constants and comparing these to predicted couplings for the evolving calculated structures, calculated using an appropriately parameterized Karplus relation.[25] This approach automatically handles any possible ambiguities in the angles, and also results in a better treatment of error ranges (since a given error in a coupling can imply very

508 CALCULATING STRUCTURES OF BIOPOLYMERS

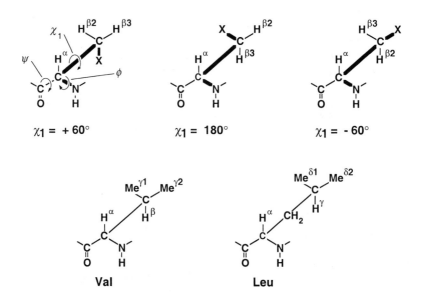

Figure 12.4. IUPAC scheme for stereospecific naming of diastereotopic protein side-chain groups. The top row shows the three regular staggered rotamers about the χ_1 torsion angle, which itself is defined by the bonds shown in bold (the ϕ, ψ, and χ_1 angles are also all indicated on the first rotamer). Looking away from the backbone along the C^α—C^β bond, the β substituent numbering always increases *clockwise* starting from the atom of the highest priority (highest atomic weight), which for C^β is always the γ heavy atom (shown as X). Exactly analogous rules apply to methylene groups located further away from the backbone, always viewing the prochiral center away from the backbone along the appropriate bond. For the isopropyl methyl groups of Val and Leu, these same rules result in the numbering shown, since the methyl groups take higher priority than H^γ.

[Note that the alternative pro-R and pro-S naming scheme based on the Cahn-Ingold-Prelog system, although it is more widely applicable than the IUPAC system, can be confusing when applied to proteins. Thus, although for most residues $H^{\beta 2}$ corresponds to $C^\beta H^{proR}$, this relationship is reversed for Cys, Ser, Met, Asp, and Asn residues, because in these cases (only) the γ atom takes priority over C^α in the Cahn-Ingold-Prelog system. Applying the same rules at C^γ, for Glu, Gln, Met, Arg, and Pro, H^2 is equivalent to H^{proS}, but for C^γ of Lys, H^2 is equivalent to H^{proR}. At C^δ of Arg, Pro, and Lys, and at C^ϵ of Lys, H^2 is equivalent to H^{proS}, but for $C^{\gamma 1}$ of Ile, H^2 is equivalent to H^{proR}. For methyl groups of Val and Leu, Me^2 is equivalent to Me^{proS}. In the authors' view, it is not impossible that some stereoassignments reported in the literature using the pro-R/pro-S nomenclature may not have taken account of these residue-specific differences, and we take this opportunity to discourage its use in this context.]

different uncertainties in the derived angle, depending on how steeply the Karplus curve varies near the particular coupling value in question).†

Generally, measured values of $^3J(\alpha, \beta)$ coupling constants are crucial to the process of making stereoassignments for H^β protons. They are related to the χ_1 torsion angle of the corresponding residue through a well parameterized Karplus relationship, and may be analyzed in various ways. If it is assumed that there is a single side-chain conformation present, then knowledge of both $^3J(\alpha, \beta)$ coupling constants may limit the possible values for χ_1 to only a few allowed ranges (taking into account residual uncertainty in the Karplus relation and in the measurement of the coupling constants). Typically, these allowed ranges form one or more pairs, such that, within each pair, either member fits the data equally well but with opposite stereoassignments. For example, the two regular staggered rotamers having $\chi_1 = 180°$ and $\chi_1 = -60°$ (see Fig. 12.4) form such a pair. Each would give rise to one large and one small value of $J_{\alpha\beta}$ ($\sim$10–12 Hz for the proton *trans* to H^α and $\sim$4–5 Hz for that *gauche* to H^α), but in one case ($\chi_1 = -60°$) the proton *trans* to H^α is $H^{\beta 2}$, while in the other case it is $H^{\beta 3}$. This ambiguity arises due to the symmetry of the Karplus relationship, which dictates that corresponding positive and negative internuclear torsion angles (in this case $+60°$ and $-60°$) give identical coupling constants.

In order to break this ambiguity, further information relevant to the χ_1 angle is required. Several programs, for instance HABAS[26,27] and STEREO-SEARCH,[28] use NOE data to achieve this (STEREOSEARCH also uses a database of known local conformations in high-resolution crystal structures of proteins). For instance, in the example just given, $d_{\beta N}(i, i)$ NOE cross-peak intensities may determine whether the backbone amide proton is close to both β protons (suggesting the $\chi_1 = -60°$ rotamer), or to only one (suggesting the $\chi_1 = 180°$ rotamer). However, $d_{\beta N}(i, i)$ distances also depend on the value of ϕ, which is part of the reason why a complete grid search using all the relevant NOE and coupling data is carried out. Other difficulties arise because of the nature of the NOE data. Efficient spin diffusion between the two β protons inevitably means that only NOE cross peaks measured at short mixing times will preserve the differential effects needed, although to an extent this difficulty can be reduced if the corresponding ROESY cross peaks are also measured. Also, effects of errors in estimating distances using NOE data are much more critical during a conformational grid search than they are during calculation of the full structure. This is because the number of conformations accepted during a grid search depends crucially on the exact definition of each distance restraint, conformations being rejected if they violate *any* restraint. Small changes in the definitions of restraint boundaries can therefore drastically affect the results of

---

†In principle, the problem could also be solved by changing the logic used by the program to test structures against constraints. One could specify that no penalty is imposed if *any* angular constraint relating to a particular torsion angle is satisfied, or else constraints could be listed as *excluded* ranges (which behave additively for a given torsion) rather than allowed ranges (which do not). To our knowledge, these approaches have not been implemented in practice.

grid search procedures, sometimes even reversing apparently secure stereoassignments. For this reason, it is wise to explore the stability of the results to small changes of this sort in the input data, and programs sometimes offer this facility built-in.

The most serious difficulty with this approach, however, is that it assumes that there is a single conformation to be found. For structured side-chains in a protein core, this is likely to be a valid assumption, but for many surface side-chains it certainly is not, and the grid search may then report that the data are "internally inconsistent" (i.e., not consistent with a single conformation). If the protein is not available in isotopically labelled form, then there is little alternative to the methods just described; however, several groups are developing alternative approaches to defining local conformation and stereoassignments in labelled proteins using a combination of homonuclear and heteronuclear coupling data. For instance, the $\chi_1 = 180°$ and $\chi_1 = -60°$ regular staggered rotamers considered earlier can be distinguished using either (^{13}C, ^{1}H) or (^{15}N, ^{1}H) coupling constants. Thus, for $\chi_1 = 180°$, $H^{\beta 2}$ is *trans* to the carbonyl carbon and would therefore show a relatively large value for $^3J(^1H^\beta, ^{13}C=O)$, whereas for $\chi_1 = -60°$, both $H^{\beta 2}$ and $H^{\beta 3}$ are *gauche* to the carbonyl carbon; similarly, for $\chi_1 = -60°$, $H^{\beta 3}$ is *trans* to the amide nitrogen, whereas for $\chi_1 = 180°$, both $H^{\beta 2}$ and $H^{\beta 3}$ are *gauche* to the amide nitrogen. This approach has the advantage that it can deal explicitly with motional averaging, by replacing the assumption that there is one fixed conformation with the assumption that the side-chain exists as a mixed population of the three regular staggered rotamers (Fig. 12.4).

As implied by the foregoing, the most important stereoassignments are usually those of $C^\beta H_2$ protons, partly because they are the most numerous methylene groups, and partly because they are the most likely to have a defined conformation involving separately resolved resonances and conveniently measurable ^3J-coupling constants. Related approaches exist for making stereoassignments of valine γ-methyl groups; while proline methylene protons can be analyzed in terms of likely ring conformations by using intraresidue NOE and coupling data.[29] There are also a number of methods based on specific labelling strategies. For instance, isopropyl groups of Val and Leu residues can be stereoassigned by growing protein samples using a mixture of $^{12}C_6$ and $^{13}C_6$ glucose, as this results predominantly in a one-bond $^{13}C-^{13}C$ coupling to the γ^1 (Val) or δ^1 (Leu) methyl but not to the γ^2 (Val) or δ^2 (Leu) methyl group in each such residue.[30] Similarly, methylene groups can often be stereoassigned by growing samples on sources containing appropriate stereospecifically deuterated amino acids (although this is a very labor-intensive and costly approach!). However, whatever combination of methods is used, it is usual to obtain only a partial set of stereoassignments.

In order to extend such a set, analysis of preliminary structures may be used. Stereoassignments of H^β protons can be made directly from the Karplus relationship for residues that have well-defined χ_1 angles in the preliminary structures and a clear-cut difference between the two $J(H^\alpha, H^\beta)$ coupling constants. Thus one may make further stereoassignments as part of an iterative process, checking each batch of successively refined calculated structures to see if further residues have acquired well-ordered χ_1 angles. Stereoassignments included in this way will improve subsequent rounds of calculation, particularly if there are differential NOE effects into the stereoassigned group from elsewhere in the protein. Occasionally, chemical shift data may replace NOE information to complete a stereoassignment once calculated structures are available. For instance, if one β proton is markedly upfield shifted, and the preliminary structures show clearly that the upfield shift must come from a particular aromatic ring of roughly known position, then the shift and $J(H^\alpha, H^\beta)$ coupling data together may settle the assignment, even though the χ_1 angle may not be well-ordered in the preliminary calculated structures.

For some residues that are well-ordered in preliminary structures, an analysis of all their NOE interactions may be sufficient to obtain a stereoassignment, without reference to coupling data. In addition to some remaining H^β proton and valine γ-methyl assignments, this approach can give stereoassignments for leucine δ-methyl groups and glycine H^α protons (the latter are virtually impossible to stereoassign by other means, at least in unlabeled proteins). However, such stereoassignments are likely to be less helpful than others in improving later rounds of structure calculations, since, by definition, the restraints that could be tightened following such a stereoassignment are mainly those that were already successful in defining the local conformation. Nonetheless, if center averaging is being used, such stereoassignments do help to minimize the number of pseudoatom corrections that are required.

Several programs employ the concept of "floating" stereoassignments, which allows individual stereoassignments to be reversed dynamically during the process of calculating the full structure, if doing so results in a better fit to the local restraints.[31,32]

12.2.6. Other Types of Restraints

In addition to restraints based on NOE enhancements and J-couplings, there may be other types of information available that can be used to supply restraints for NMR structural calculations. For proteins, these include particularly the locations of any disulfide bridges, bonds to metal ions, and hydrogen bonds. Other types of restraint that act mainly on local conformation and may be useful during refinement include lower-limit restraints, restraints based on chemical shift, and most recently, restraints based on residual dipolar couplings measured in weakly ordered media. We shall now briefly consider these various restraint types.

Disulfide bridges and connectivities through metals are essentially aspects of the covalent connectivity network, and can be included as part of the input to structure calculations, provided they have been firmly established beforehand. For certain types of structure calculations, there may be technical reasons why such restraints cannot be treated in quite the same way as other covalent bonds initially; for example, some programs that explore conformational space by varying backbone torsion angles in a macromolecular chain cannot easily treat cyclic systems directly. However, such problems may be overcome by treating disulfide or metal bonds simply as additional distance restraints until the global fold is largely established, and only then introducing the more stringent geometrical requirements that are applied to other covalent bonds.

Quite often, the locations of disulfide and metal linkages may be partially or completely unknown at the beginning of an NMR study, and in such cases preliminary structures may be used to resolve ambiguities. For instance, possible disulfide pairings may be assessed by measuring $C^\beta-C^\beta$ distances between cysteine residues in an ensemble of preliminary calculated structures, and rejecting pairings that consistently give implausibly long distances. The applicability of this approach of course depends on how many disulfides occur in close proximity in the 3D structure, and it is not common for the complete disulfide bonding pattern of a heavily disulfide-linked protein to be determined by NMR alone. However, NMR is often complementary to the conventional peptide mapping approach, for instance, where sequentially adjacent residues both form disulfide links in a region of β-sheet; such pairings usually cannot be resolved by peptide mapping, but may be trivial to distinguish using NMR, since the side-chains project on opposite sides of the sheet.

Hydrogen bonds represent a particularly important type of restraint, since of course it is hydrogen bonds that are largely responsible for stabilizing elements of secondary structure. They are often also among the tightest restraints, since they specify particularly short distances between atoms that are well separated in the sequence. For proteins, direct NMR evidence concerning hydrogen bond locations is generally limited to identification of some of those NH groups that act as donors; the corresponding signals are preferentially protected against exchange with water, so that spectra acquired from freshly prepared D_2O solutions can retain such signals for hours, days or even weeks, under favorable conditions. However, the feasibility of such studies depends strongly on the stability range of the protein; in some cases the conditions of pH and temperature required to maintain the folded state of the protein cause even hydrogen-bonded NH signals to exchange too quickly to be detected in this way.

Although amide groups that act as donors can often be identified through exchange experiments, there is generally no direct NMR evidence as to which carbonyl (or other) groups act as acceptors, let alone what the connectivity pattern is between donors and acceptors. In lieu of direct evidence, acceptors are therefore usually identified using preliminary calculated structures. For each slowly exchanging NH group, the preliminary structures are examined to determine which potential acceptor groups are positioned such that a hydrogen bond to that particular NH would be plausible. If only one such acceptor is found for a particular NH in most, or all, of the structures in the calculated ensemble, then a hydrogen bond is assigned on that basis, and corresponding distance restraints can be introduced in the succeeding round of structure calculations. In subsequent rounds, higher-resolution structures will be produced, perhaps allowing further ambiguities amongst the potential acceptor groups to be resolved, and so on. Normally, each hydrogen bond is specified by two distance restraints in order to favor roughly linear arrangements O $\cdots$ H—N; for hydrogen bonds between amide and carbonyl groups these restraints are typically set at or near 0 Å $< r_{H\text{-}O} <$ 2.3 Å and 2.5 Å $< r_{N\text{-}O} <$ 3.3 Å.

In some studies, hydrogen bond acceptors corresponding to particular slowly exchanging NH's are not identified experimentally, but rather are introduced into the restraint list based on an assumed regular secondary structure. Such an approach is clearly dangerous (one might, for instance, impose α-helical character on a region of 3_{10}-helix) and, once it has been used, one can never justifiably assert that the hydrogen bond network in the calculated structures was determined independently.

Lower-limit restraints can be useful under some circumstances, although in general it is dangerous to assume that the absence of an NOE interaction necessarily implies that the corresponding internuclear distance is long (cf. Sections 5.6, 10.1.3, and 12.2.3). For instance in proteins, if a particular sequential d_{NN} cross peak is demonstrably absent or very weak in NOESY spectra with long mixing times, and if both the relevant NH signals give normal NOE cross peaks to other signals, then a conservatively set lower-limit restraint (e.g., 3.0 Å $< d_{NN} <$ 100 Å) can be justified. Such a lower-limit restraint can be quite powerful in improving the definition of local conformation, since it limits considerably the accessible backbone angles for the corresponding dipeptide fragment; however, care is needed when using them (the criteria mentioned above should always be checked) and they should certainly not be applied indiscriminately.

Sometimes restraints are defined directly in terms of measurable NMR parameters, using some form of back-calculation to check agreement between the experimentally measured value and that predicted from the evolving model structure. For instance, in Section 12.2.5 it was mentioned that coupling-based restraints can be specified directly in terms of coupling constant values rather than by applying a derived torsion angle restraint; in this approach, J-couplings are predicted from the evolving model coordinates using a suitable Karplus equation and then compared directly with measured values from the spectra. Another example is the comparison of shifts predicted from coordinates of the evolving model structure (using semi-empirical calculations) against the experimentally measured shifts (Section 12.5.2). As with coupling-based restraints, such restraints act mainly to improve details of local geometry.

Recently, restraints based on measurements of residual dipolar couplings in weakly ordered media have been introduced, and these appear likely to become important new sources of information in future NMR structure determinations.[33] However, before we can discuss how such restraints are used, we must first say a little about the nature of dipolar couplings and how they are measured for biomacromolecules.

For two nuclei i and j in an ordered environment, the dipolar coupling constant D_{ij} depends on the internuclear distance r_{ij}, the angle of the ij vector relative to the molecular alignment tensor, and the extent of ordering. More formally, D_{ij} is given by

$$D_{ij} = (\xi_{ij}/r_{ij}^3) \left\langle A_a(3\cos^2\theta_{ij} - 1) + \frac{3}{2} A_r \sin^2\theta_{ij} \cos 2\phi_{ij} \right\rangle \quad (12.8)$$

where A_a and A_r are respectively the axial and rhombic components of the molecular alignment tensor **A**, θ_{ij} and ϕ_{ij} are the polar angles of the ij vector in the principal axis system of the alignment tensor, and ξ_{ij} is an interaction constant that reflects the extent of internal motion of the ij vector as well as the gyromagnetic ratios of spins i and j. Typically, for couplings between ^{1}H and ^{15}N in amide groups in proteins, ξ_{ij} has been approximated by $\gamma_i\gamma_j S_{ij}$ where S_{ij} is the square root of the order parameter S^2 determined by ^{15}N relaxation analysis (cf. Section 5.4). However, there may be some difficulties with generalizing this approximation, since S_{ij} relates only to internal motions faster than overall tumbling.

In normal (i.e., isotropic) solutions no ordering is present, so that both A_a and A_r are zero and no splittings due to dipolar couplings occur. In contrast, many liquid crystals can impose upon solutes almost complete order in one or two dimensions, resulting in dipolar coupling constants in the range of tens or even hundreds of kilohertz for short internuclear distances, and appreciable dipolar coupling constants for very many interactions over significantly longer distances. For a protein or other biomacromolecule, this makes the NMR signals very broad, difficult to detect, and generally uninterpretable. Thus, a key recent development was the introduction of *weakly* ordered media that scale down the dipolar interaction to a convenient size (e.g., to ca. 10–20 Hz for a directly bonded ^{15}N—^{1}H pair). Spectra in such weakly ordered systems are not very greatly perturbed from their normal appearance in isotropic media, and the number of interactions influencing the spectrum appreciably is kept down to a reasonable total. Measurement of dipolar couplings can then employ quite conventional experiments. If the sizes of dipolar splittings are required, these can be measured directly as splittings [e.g., for ^{15}N—^{1}H groups, the dipolar coupling appears as a perturbation to the value of $^1J(^{15}$N, ^{1}H)]; alternatively, if it is sufficient just to demonstrate the existence of particular dipolar couplings where it is known no J-coupling exists, then a TOCSY spectrum can be used.

Several systems that impose weak order on dissolved macromolecules have been suggested already, and doubtless more will follow. The original work used discoid lipid bicelles; these are essentially micelles constructed from two different lipids, one of which forms a highly curved surface, the other a largely flat surface.[33] Thus, instead of being approximately spherical as are micelles, the resulting bicelles are disc-like. Other systems proposed include solutions of rod-shaped viruses (e.g., tobacco mosaic virus),[34] filamentous phages (e.g., Pf1 and fd),[34,35] and also a lamellar liquid-crystal phase based on cetylpyridinium chloride and hexanol in brine.[36] In a strong magnetic field, any of these systems align and can then induce order on macromolecules present in the surrounding solution. Note that in all these cases, order is imposed on the macromolecular solute only by virtue of its collisions with the oriented particles (bicelles, viruses, or whatever), and that the resulting scaled-down dipolar couplings reflect an average over all positions that the macromolecules can occupy. Thus, if the concentration of oriented particles is increased, more macromolecules will be close to an oriented particle, the number of collisions between macromolecules and oriented particles will increase, and the average residual dipolar coupling will be increased.

Two types of application for residual dipolar coupling measurements have so far been proposed. In cases where the internuclear distance is fixed and known, for instance for backbone amide ^{15}N—^{1}H groups in proteins, then the variations in D_{ij} are determined largely by the angle of each ij vector relative to the molecular frame (the only other factor being any differences in the extent of the internal motions). Thus, for relatively rigid systems, residual dipolar coupling values effectively provide angular restraints, and, quite unlike other forms of NMR-based structural restraints, these are not just specified locally relative to other nearby nuclei, but rather in an absolute sense, relative to the molecular frame. Such restraints can be a powerful aid to determine the orientation of otherwise relatively unconstrained regions such as surface loops in proteins and double-helical stems in oligonucleotides; however, before they can be used, an analysis of the data must be carried out to determine the values of A_a and A_r.[37,38]

Another type of application is to use residual dipolar couplings to detect through-space interactions, just as with NOE enhancements. Because the dipolar coupling constant depends on r^{-3} rather than r^{-6}, effects can in principle be detected over longer distances than for NOE enhancements, and an early demonstration did indeed show a (^{1}H, ^{1}H) TOCSY cross peak resulting from an interproton interaction over about 7.5 Å in a small DNA oligonucleotide.[39] Such measurements are complicated by the fact that cross-peak intensities depend on internuclear distance, motion, and orientation, but this is little worse, in principle, than the dependence of the NOE on both internuclear distance and motion. Thus one could adopt a similar interpretative scheme, such that a relatively strong residual dipolar coupling interaction must imply a relatively short corresponding internuclear distance, but a weak or missing residual dipolar coupling need not mean a long internuclear distance. At the time of writing these experiments are still extremely new, and it is to be expected that there will be many developments in this area.

12.3. CALCULATING STRUCTURES

This section describes some of the approaches used to calculate macromolecular structures using NMR-derived restraints. In general, this process involves reaching some compromise between, on the one hand, satisfying simultaneously all of the NMR-derived restraints, and on the other, conserving reasonable covalent bonding geometries and nonbonded contacts. These two requirements are fundamentally quite distinct, and it is an unavoidable problem when designing methods for NMR-based structure calculation that one must somehow balance them against one another. For instance, if the covalent geometry and nonbonded contacts are assessed using a conformational energy calculated with a particular force field, then it becomes necessary to introduce new terms into the force field corresponding to energetic penalties for violating NMR-derived restraints. The weight of these terms relative to the others has no physical meaning, but represents the relative weight given to satisfying the NMR restraints as opposed to maintaining good covalent geometry and nonbonded contacts.

Many types of NMR structure calculation comprise two more or less distinct phases. During the initial "search" phase, the aim is to explore as much as possible of the conformational space accessible to the molecule, so as to ensure that all possible solutions are tested against the data. Subsequently, when a particular global fold has been adopted, the "optimization" or "refinement" phase seeks to define the local detail of the structure correctly. This second stage involves making relatively small-scale structural changes, rather than the much larger movements required to interconvert different global folding topologies. Of course, the transition between the two stages is handled differently in different methods, and in some there is a more gradual change in the character of the calculation.

Strategies of the type just discussed are necessarily "top-down" approaches, in that they attempt to find the global fold first, and then go on to establish smaller scale details of local conformation later. Although probably the majority of methods are of this type, there are also many proponents of "bottom-up" approaches, in which the local structure is determined *before* the global fold. Probably the best-known approach of this sort is torsion-space distance geometry (Section 12.3.1.3).

Generally, when determining an NMR structure a whole series of calculations is carried out, each employing randomly different starting conditions (the way in which the starting conditions differ between calculations depends on the method employed; see subsequent discussion). There are two main reasons for doing this. First, depending on the details of the method of calculation employed, the best-fit conformation may not be accessible from all sets of starting conditions. Thus, by using many different starting conditions, one increases the chance of finding the best-fit structure. Second, the variations between the members of such an ensemble of calculated structures gives an indication of the precision with which the data define the structure, and also of how this precision varies throughout the structure. However, this spread of structures also reflects how thoroughly the calculation searches conformational space. An ineffective search would leave calculations starting from randomly different conditions trapped in different local minima, rather than allowing them all to converge on similar structures. These two possible origins of variation among the calculated structures can usually be distinguished by analyzing how well the various structures fit the data (see Section 12.5).

Provided the structure calculation routine has searched conformational space adequately, low precision in part of a calculated structure generally reflects a lack of sufficiently specific NMR data for that region. However, this in turn may either reflect local flexibility in the molecule itself, or it may merely reflect some other purely spectroscopic difficulty in obtaining restraints. In the latter case, the real molecule might be just as rigid in the less precisely defined region as it is elsewhere. Of course, these two possible causes of poor local precision can often be connected, in that flexible regions often give rise to broadened NMR signals and limited chemical shift dispersion, but it is important to realize that there is no *necessary* connection. If one wishes to know how local flexibility varies within a molecule, some independent measure is needed; for proteins, the most commonly used measure at present is analysis of the ^{15}N NMR parameters for backbone NH groups (see Section 5.4).

Another important point is that specific details of individual structures within imprecisely defined regions have no real significance. It is a key assumption of most methods of calculating NMR structures that there is a single structure to be found (see text to follow). However, in cases where a lack of precision in the results of such calculations is due to flexibility in the real molecule, it may well be that the averaged NMR restraints employed in the calculations are simply not compatible with a single conformation. Even though an ensemble of structures is calculated, during each individual run the calculation must output just one set of coordinates, so that in each case it must make whatever compromises it can to resolve internal inconsistencies amongst the input restraints. Under such circumstances, differences between individual conformers in imprecisely defined regions of the structure are more likely to depend on details of the calculation protocol than they are upon the real behavior of the molecule, and it would be incorrect to regard them as having some experimentally determined significance.

12.3.1. Distance Geometry Calculations

As the name suggests, distance-geometry calculations aim to determine the 3D arrangement of a set of points starting only from a set of measured distances between the points. In the case where all of the distances are known exactly, there is an analytical solution to the problem that uniquely defines the corresponding geometry, and Section 12.3.1.1 will show how this calculation works. However, NMR-derived distance restraints are always approximate (Section 12.2.3), and are invariably very much fewer in number than the *total* number of distances within a macromolecule (given by $N(N-1)/2$, where N is the number of atoms in the macromolecule). In these circumstances, no analytical solution exists, and Section 12.3.1.2 will discuss what can be done in order to deal with this situation. A much more thorough treatment of the whole subject of distance-geometry calculations for macromolecular structure determination may be found in the book by Crippen and Havel.[40]

12.3.1.1. The Exact Case.

We deal next with the analytical solution for the case where all distances are known exactly. The essential translation between distances and coordinates that forms the core of the method is provided by the cosine rule. For two points (or atoms) i and j, whose coordinates are defined by the position vectors $\mathbf{r}_i$ and $\mathbf{r}_j$ respectively, the corresponding distance between them is given by

$$d_{ij}^2 = \mathbf{r}_i^2 + \mathbf{r}_j^2 - 2\mathbf{r}_i \cdot \mathbf{r}_j \tag{12.9}$$

The squares $\mathbf{r}_i^2$ and $\mathbf{r}_j^2$ may be replaced by the squared distances d_{i0}^2 and d_{j0}^2, where d_{i0} and d_{j0} are, respectively, the distances from atoms i and j to a point at the origin of the coordinate system. In general, if only d_{ij}^2 values are available, the origin is undefined unless one chooses to place an atom, say atom k, at the origin. Making this substitution and rearranging gives

$$\mathbf{r}_i \cdot \mathbf{r}_j = \frac{1}{2}(d_{ik}^2 + d_{jk}^2 - d_{ij}^2) \tag{12.10}$$

Thus, when all the distances are known, all of the scalar products $\mathbf{r}_i \cdot \mathbf{r}_j$ can be calculated from Eq. 12.10.

However, these scalar products are also defined in terms of the coordinates of the points. Let $\mathbf{R}$ be the $N \times 3$ matrix comprising the coordinates of all N points:

$$\mathbf{R} = \begin{pmatrix} x_0 & x_1 & \cdots & x_N \\ y_0 & y_1 & \cdots & y_N \\ z_0 & z_1 & \cdots & z_N \end{pmatrix} \tag{12.11}$$

By definition, the ($N \times N$) matrix $\mathbf{M} = \mathbf{R}^T\mathbf{R}$ (where the superscript T indicates the transposed matrix) consists of all of the possible scalar products:

$$\boxed{\mathbf{M}} = \boxed{\mathbf{R}^T} \boxed{\mathbf{R}} \tag{12.12}$$

The matrix $\mathbf{M}$ is called the metric matrix. Now, $\mathbf{M}$ can be calculated directly from the distances using Eq. 12.10, so the overall problem reduces to that of trying to find the coordinate matrix $\mathbf{R}$ by manipulating the matrix $\mathbf{M}$, which in turn reduces to the problem of diagonalizing $\mathbf{M}$. If the orthogonal matrix that diagonalizes $\mathbf{M}$ is $\mathbf{A}$, then

$$\mathbf{A}^T\mathbf{MA} = \begin{array}{|c|}\hline \mathbf{B}^T \\ \hline \mathbf{C}^T \\ \hline\end{array} \;\; \begin{array}{|c|}\hline \\ \mathbf{M} \\ \\ \hline\end{array} \;\; \begin{array}{|c|c|}\hline & \\ \mathbf{B} & \mathbf{C} \\ & \\ \hline\end{array} = \begin{array}{|c|c|}\hline \Lambda & 0 \\ \hline 0 & 0 \\ \hline\end{array}$$

(12.13)

Here Λ is a diagonal matrix containing the eigenvalues of **M**. Provided that the geometrical arrangement of points is contained within three dimensions, there are only three nonvanishing eigenvalues of **M**, implying that Λ is a 3 × 3 diagonal matrix, and that only the submatrix **B** (a 3 × N matrix) within **A** is needed to determine the coordinates of the points in three dimensions. Thus, using also Eq. 12.12, Eq. 12.13 may be rewritten as

$$\mathbf{B}^T\mathbf{R}^T\mathbf{RB} = \Lambda \tag{12.14}$$

Given that $\mathbf{B}^T\mathbf{B} = \mathbf{1}$ (the identity matrix), the matrix of coordinates **R** is given by

$$\mathbf{R} = \Lambda^{0.5}\mathbf{B}^T \tag{12.15}$$

The matrix **A** may be thought of as the rotation matrix that rotates the metric matrix **M** such that each atom of the structure lies along one of the N axes in N-dimensional space, with the origin of the coordinate system at atom k. Such a rotation will exist in N-dimensional space for any input set of distances between N atoms, but the resulting structure will only be contained in three dimensions if there are not more than three nonzero eigenvalues of **M**.

12.3.1.2. Distance Geometry Applied to NMR Structure Determination.
When applying the distance-geometry (DG) approach to NMR structure determination, the input data comprise three main categories of distance: the directly bonded and 1–3 distances, which are known precisely; the NOE-derived distance restraints, which are known only approximately; and all the remaining nonbonded distances, about which no information is available directly. It is worth bearing in mind through what follows that this last category generally constitutes the great majority of the total number of interatomic distances within a macromolecule.

Despite the unavoidable uncertainties in most of these distances, it is necessary to select *specific* values for *all* distances before a distance-geometry calculation is possible. For the NOE-derived restraints, this implies picking some specific distance, called a trial distance, from between the upper and lower bounds for each restraint. The same must be done for the other nonbonded distances, but in these cases the upper bound is unknown and must be set to some arbitrarily large value, larger than the maximum dimension of the molecule. In older implementations of the method, trial distances are chosen *at random* from between the upper and lower bounds, and the distance-geometry calculation is repeated many times using a different set of randomly chosen trial distances in each case. This generates an ensemble of structures that is supposed to represent the conformational space compatible with the NMR data. However, for those nonbonded distances where no NOE information exists, choosing trial distances *randomly* results in a heavy weighting towards unrealistically long distances, since the upper bounds are arbitrarily large in these cases. This causes a bias towards extended structures.[41] More recent implementations of the distance-geometry approach reduce this problem by weighting the choice of trial distances towards the lower bound.

Another problem is that the trial distances will not be geometrically self-consistent, meaning that they will not be mutually compatible with a three-dimensional object. Consider the simplest object involving more than one distance, namely a triangle. Of the three distances defined by a triangle, it is clearly impossible for any one distance to be longer than the sum of the other two. This simple fact is known as the triangle inequality, and it may be expressed more formally as

$$d_{ij} \leq d_{ik} + d_{jk} \tag{12.16}$$

However, if the three distances in a triangle are known only as allowed *ranges*, and if particular trial distances are picked arbitrarily from within these ranges, it is quite likely that such trial distances will break the triangle inequality, making them mutually incompatible with any three-dimensional object.

It is therefore usual to check for violations of the triangle inequality before carrying out distance-geometry calculations. For all possible sets of three atoms, the corresponding distances are individually checked; when a violation is found, the longest distance in that set is reduced accordingly, and the process is continued until no violations remain. This has the dual effect of partially eliminating internal inconsistencies in the distances, and of transferring the information contained in the constrained and bonded distances into the rest of the distance matrix, thereby limiting the tendency for unknown distances to be assigned unrealistically large values. At the simplest level, just the bounds are checked for violations, the upper bounds by using Eq. 12.16, and the lower bounds by using a slightly different inequality ($l_{ij} \geq l_{ik} - u_{jk}$, where l_{ij} and l_{ik} are lower bounds and u_{jk} is an upper bound). This process is called "smoothing" the bounds, and essentially all distance geometry calculations employ at least this level of checking. However, even when smoothed bounds are used, the specific trial distances chosen from between the upper and lower bounds may still break the triangle inequality, so checking of the trial distances is also desirable. This process, called "metrization,"[42] is much more time-consuming than smoothing the bounds, since it must be done independently for each structure calculation in an ensemble. Also, the order of checking is significant. If distances are checked in a particular order, then the local variation observed among structures is found to depend upon whether distances in that part of the molecule were checked early or late in the process. Thus, if the trial distances are checked in a random order, the sampling of conformational space is improved.[43]

Of course, checking just the triangle inequality does not guarantee geometrical self-consistency when more than three distances are involved. Analogous higher-order inequalities exist involving larger numbers of distances, but the computational burden of checking them rapidly becomes impossible. Some recent DG programs check distances within selected sets of four atoms for violations of the tetrangle inequality, but at present that represents the limit of checking that is practicable.[44]

Given that the input data may be geometrically inconsistent, it is not surprising that various problems can occur with the output of distance-geometry calculations. Typically, the matrix **M** has more than three nonzero eigenvalues, implying that the calculated structure is not fully contained within three dimensions. If these extra nonzero eigenvalues are small relative to the first three, then it is usual to ignore them and work with the projection of the structure into three dimensions that results when only matrix **B** is retained from the complete matrix **A** in Eq. 12.13. If the higher eigenvalues are comparable in size to the first three, this is a sign that the structure is not well defined by the data. Other signs of trouble include negative eigenvalues of **M**, which imply imaginary distances in the structure; such results are generally discarded.

The sequence of events in a distance geometry calculation is thus (i) smoothing the bounds, (ii) selecting trial distances, and (iii) solving Eq. 12.15. (This last step is called "embedding," since it results in a structure in which the input distances have been embedded.) However, the structures obtained directly from the embedding step always have grossly unreasonable covalent geometries. This is mainly due to one of the principal shortcomings of the distance-geometry calculations, namely, that it is not possible to give particular distances differential weights. Thus the bonded distances, which are known precisely, are not treated any differently from the NOE-constrained distances, or indeed from the very many other unknown nonbonded distances for which (probably grossly wrong) trial distances have been used in the embedding step. The resulting structure represents a compromise between all these input distances.

It is therefore essential to subject the structures to some form of optimization routine after the embedding step. Early versions employed energy minimization routines that simply found the nearest local minimum in the potential energy surface. Later versions often employ some form of dynamics or simulated annealing calculation, in an attempt to widen the conformational search to include other nearby potential energy minima; this approach is sometimes called hybrid distance-geometry/simulated annealing. Such calculations will be considered further below (see Section 12.3.3), since simulated annealing alone represents another approach to NMR structure calculation. However, an interesting twist specific to distance-geometry optimizations concerns the possibility of optimizing the structures in four- or five-dimensional space. If the fourth or fifth eigenvalues of **M** are comparable in size to the first three, but higher eigenvalues are smaller, then a reasonable structure may sometimes be obtained by retaining one or two higher dimensions in the initial part of the optimization calculation, provided the force field used to calculate energies is restricted only to distance terms. This may allow the structure to follow an easier pathway to a three-dimensional minimum, perhaps avoiding local minima that would exist if the structure were constrained to be within three dimensions throughout. Of course, for the final stage of optimization, the structure must be projected into three dimensions.

In summary, the main motive for using distance-geometry calculations is that they represent a relatively efficient and thorough method of searching conformational space, although in earlier implementations there were problems with their sampling properties as already discussed. Against this, distance-geometry calculations are rather inflexible in their input requirements, particularly in that they cannot handle angle-based restraints directly and that they cannot give differential weights to different categories of distance.

12.3.1.3. Distance Geometry in Torsion Angle Space. Programs based on this method are principally DISMAN[45] and its successors DIANA[27] and DYANA.[46] Structures are assessed using a "variable target function," which is essentially a summation of all distance violations of NOE-based restraints, together with violations of any torsion angle restraints and also van der Waals violations (suitably balancing the weights of the different types of violations). The program attempts to minimize the value of the target function by varying only the flexible torsion angles within the structure (i.e., for proteins, the backbone ϕ and ψ angles, and those side-chain angles not fixed by covalent geometry), while all other aspects of covalent geometry remain fixed. This process is repeated for many randomly different starting conformations and the results compared, thereby locating those conformations that best fit the NMR data. A version of this idea has been implemented in which the target function contains the NOE intensities themselves rather than the distance restraints derived from them.[47]

As with any method based on variation of the 3D structure, the principal difficulty is that of avoiding becoming trapped in local minima. However, several aspects of the programs are intended to limit this problem. First, the NOE restraints are not all included from the start, but are included gradually on the basis of increasing sequential "reach" as the calculation progresses (this is why the target function is called "variable"). Thus, initially only intraresidue restraints are considered, but, in later cycles of minimization, restraints involving sequential neighbors ($i, i + 1$) are included, then restraints involving residues separated by two in the sequence ($i, i + 2$), and later still ($i, i + 3$) restraints are included, and so on until eventually all the restraints become active. As we commented earlier, this results in the local conformation being defined first, with the global fold only emerging later in the procedure. Second, the van der Waals radii are kept small during the early part of the calculation, increasing to their final values only near the end. Note that, unlike the situation with a physical model, there is nothing to prevent the chain passing through itself during the early stages of these calculations, provided the van der Waals radii are small enough that the *atoms* do not collide.

As might be expected, results with this approach depend to an extent upon the type of secondary structure present in the molecule. For largely helical proteins, the secondary structure emerges early in the calculation, being largely defined by $(i, i + 3)$ restraints, so that later stages of the calculation mainly involve relative movements of preformed helical elements. However, in predominantly β-sheet proteins, secondary and tertiary structure are not necessarily separated in terms of the sequential "reach" of the restraints that define them, and it is more likely that such calculations for β proteins will become trapped in local minima.

This situation has been improved in later versions of the program DIANA, in which the convergence of the procedure is improved by progressively introducing redundant torsion angle restraints, which are themselves determined from the distance data during earlier stages of the calculation; this process is called the "REDAC" strategy.[27] In the most recent version, DYANA, the approach becomes much more similar to a simulated annealing calculation (see discussion to follow). The restraints are all introduced simultaneously, but with a gradually increasing weight, and the minimization procedure is replaced by a dynamics calculation in torsion angle space.[46] In this form the procedure is highly efficient, and problems previously associated with β proteins are essentially eliminated.

Although it is often referred to as a type of distance-geometry method, one should realize that the approach embodied by these programs is fundamentally and completely distinct from the distance-geometry calculations discussed in the previous section; indeed the concept of a torsion angle is not even defined within the context of a conventional distance-geometry calculation until the calculation is complete and the structure has been defined within three dimensions. Despite its name, at a methodological level this approach has far more in common with the simulated annealing methods considered next than it does with metric-matrix distance geometry; indeed, as we have seen, DYANA is now essentially a simulated annealing program.

12.3.2. Restrained Molecular Dynamics Calculations

True (i.e., unrestrained) molecular dynamics calculations are designed to simulate the dynamic behavior of a molecule using the equations of Newtonian mechanics, treating the molecule essentially as a sophisticated mechanical model. The initial atomic positions are defined by some chosen starting structure, and the atoms are each given some kinetic energy in the form of an initial velocity in a random direction; these initial velocities are calculated according to a Maxwellian distribution corresponding to a notional temperature for the molecule as a whole. Each atom is supposed to be subject to certain forces, so that, as the atoms move, their potential and kinetic energies change according to Newton's laws of motion. These forces, collectively called the "force field," are intended to represent the influence on atomic motions of the covalent bonding network and other properties of the molecule. Thus, as the atoms move during the simulation, some bonds stretch, others contract, bond angles are distorted, atoms collide, charged groups become closer or further apart, and so forth, and each of these changes causes the direction and magnitude of the total calculated force on each atom to change. Thus, by recalculating the forces and velocities for each atom repeatedly at very short time intervals (i.e., short relative to the highest frequency of motion in the system), one may simulate the movements of the atoms over time. The result is often called the trajectory of the system. In many implementations of molecular dynamics (MD), the kinetic energies of the atoms are also frequently checked and adjusted to maintain an approximately constant notional temperature for the molecule; this prevents atoms acquiring excessive kinetic energies that might lead to dissociation of the model molecule.

Of course, the extent to which such a simulation mimics the behavior of the real molecule depends crucially upon what the forces representing various molecular properties are set to, and upon what properties are represented within the force field. Different MD programs use different force fields, but most force fields break down into covalent and noncovalent contributions to V_{total} (the total potential energy) roughly as follows:

$$V_{\text{total}} = V_{\text{covalent}} + V_{\text{non-covalent}}$$

$$V_{\text{covalent}} = V_{\text{bond}} + V_{\text{angle}} + V_{\text{dihedral}} + V_{\text{improper}}$$

$$V_{\text{non-covalent}} = V_{\text{van der Waals}} + V_{\text{electrostatic}} + V_{\text{hydrogen bonding}} \qquad (12.17)$$

These terms are largely self-explanatory, except for the "improper" contributions, whose task is to impose those aspects of the covalent geometry that would otherwise not be represented; these include chirality of chiral centers and planarity of aromatic rings and peptide groups. Note also that the effect of a hydrogen bonding potential in the force field is nonspecific, in that it encourages any acceptors and donors to form hydrogen bonds whenever their trajectories bring them close together. In contrast, NMR-derived hydrogen bond restraints refer to specifically assigned donors and acceptors.

The force field specifies equilibrium values for properties such as bond lengths, bond angles, and van der Waals radii, and it must also include force constants that determine the energy penalty for deviations of the structure from these equilibrium values, as well as energy terms for electrostatic and van der Waals interactions. The origins of some of these numbers are quite clear, for instance, equilibrium values for bond lengths and angles are generally set using crystallographic or spectroscopic data, while the corresponding force constants are set so that the simulations mimic spectroscopically determined vibrational frequencies. However, others are sometimes more obscure, and the relative sizes of some force constants may often simply reflect the extent to which one wishes the final results to be influenced by one property rather than another. Bearing all this in mind, it is perhaps important not to be excessively awestruck by the results of dynamics calculations. They do not represent a "movie" of the behavior of the real molecule, but rather they involve a whole range of assumptions that conspire together to produce behavior that conforms with various expectations and prejudices contained within the force field. Nonetheless, dynamics calculations represent a very powerful tool.

So far this discussion has dealt only with unrestrained dynamics, where the aim is essentially to simulate molecular motions. In order to use dynamics calculations to determine or refine structures based on experiments, one must introduce restraints into the force field to represent the experimental data. Clearly, one problem is then how to balance the force constants specified for violating these restraints against the "real" force constants associated with distorting the molecular geometry. Indeed, it is sometimes argued that because the forces arising from restraint violations are so clearly fictitious, the whole method is somehow unsound. However, this view confuses the different purposes of restrained and unrestrained dynamics calculations. The trajectory that the atoms follow during a restrained dynamics calculation is indeed physically meaningless due to the action of the restraints, but the interest in such a calculation lies only in its final result. The trajectory itself is of interest only at an operational level, to ensure that local minima are avoided and that the calculation is efficient, and it is these criteria that are used to determine the appropriate balance between the new force constants representing the restraints and all others. One should also note that, if one were fortunate enough to find a structure that satisfied the experimental data completely, the final calculated forces on the atoms would no longer include any arising from restraint violations.

Before going further it is worth commenting briefly on the distinction between "restraints" and "constraints." In the context of crystallographic refinement, these two terms have acquired rather different meanings. Both refer to aspects of a calculation that restrict the freedom of an atom or group of atoms so as to conform with some imposed requirement, but a restraint achieves this by adding a term to the force field, whereas a constraint imposes the requirement directly and absolutely. To take an example, consider aromatic ring geometry. A constraint would cause the calculation to treat the whole group of atoms comprising the ring as a rigid body, so that the individual atomic positions are no longer independent variables and relative movements within the group are impossible. In contrast, a restraint would act more gently by imposing an energy penalty for deviations from ideal geometry, so that nonideal geometry is possible but discouraged. On this basis, there is little doubt that NMR data are formally represented by *restraints*, rather than constraints, in most, if not all, NMR structure calculations. We do refer to them as such in this book, but in practice the distinction has become all but lost in the wider NMR literature, and it is now common practice to speak of NMR constraints and restraints almost interchangeably (even in the context of *restrained* MD calculations).

Dynamics calculations can differ very substantially in their design, purpose, and outcome. For instance, if the notional temperature is maintained at a low level throughout the calculation, it is unlikely that atoms will have sufficient kinetic energy to surmount any but the lowest potential energy barriers. This would imply that the structure will not move far from its starting conformation, undergoing only limited excursions into neighboring conformational space. Such a calculation might be intended as a refinement protocol, changing a structure based purely on experimental data into a closely similar structure with properties "regularized" by the force field (one may of course debate whether such a refinement *necessarily* represents an improvement). If, on the other hand, the notional temperature is kept high for an extended period, atoms may cross many potential energy barriers, perhaps changing the structure utterly until, in the limit, it has lost all "memory" of the starting structure. Very slow cooling of the molecule might then allow the system to settle into one of its lowest potential energy minima, if not actually the lowest. Indeed, if the force field contained the whole truth about interatomic interactions in macromolecules, unrestrained dynamics calculations of this sort might be capable of predicting complete structures based only on sequence information. However, luckily for experimentalists, this is very far from being the actual situation at present.

Another way that dynamics calculations can differ is in which terms from the force field are active. An obvious case is the distinction between restrained dynamics, where experimentally derived restraints are included, and unrestrained dynamics, where they are not. However, one may make a wide range of other choices, such as whether to include electrostatic potentials, and, if so, how to define the dielectric constant; whether to include hydrogen bonding potentials; and whether to represent van der Waals terms as hard-sphere interactions or by including also a long-range attractive component, as in the Lennard–Jones "6-12" potential. One might imagine that the more information is included, the "better" the resulting structure is likely to be. On the other hand, the structure that is obtained always represents some compromise between the influence of the experimental data and other terms in the force field, so that by turning on these additional terms one may lose track of what has actually determined the final structure. A good compromise may be to carry out a calculation in stages, starting with a minimal force field so as to see how well the structure is defined by the experimental data alone, and only later using other terms in the force field. A minimal force field in this context means one restricted essentially to geometrical terms, that is $V_{\text{covalent}} + V_{\text{VDW(hard sphere)}}$. In general, the better a structure is defined by the experimental data, the less difference it will make to invoke other terms in the force field. At the opposite extreme, some calculations use the whole force field and some also simulate the effect of solvation by placing the model molecule in a "box" of several hundreds or thousands of (model) solvent molecules, the atoms of which are included when calculating the overall trajectory of the system. Inclusion of solvent can avoid problems such as the collapse of surface side-chains caused by electrostatic terms in the force field when only *in vacuo* calculations are used, and, more recently, calculations in biphasic solvent systems have been used to simulate conformations adopted by small molecules such as peptides at membrane interfaces.[48,49] As the power of computers grows ever larger, the number of such applications will doubtless increase.

One of the great advantages of dynamics calculations is their immense versatility. Thus, if a new criterion for refinement is introduced, it is usually a relatively simple matter to include it into dynamics calculations by adding an appropriate term into the force field. Another important recent development has been the introduction of restraint checking using an averaged dynamics trajectory.[50,51] Normally, if a restraint is violated, the corresponding atoms are subject to a corrective force calculated using the extent of the violation as it appears in the current instantaneous state of the evolving structure. When averaged dynamics restraint checking is employed, this force is calculated instead using an *averaged* value of the corresponding violation over some time interval along the trajectory, extending from the current moment into the past. In practice, the averaging is weighted exponentially away from the current moment, so that the extent to which the state of the system at one moment influences the forces thereafter decays as the simulation progresses; typical time constants for this exponential decay are around 20 ps, in the context of total simulation times of several nanoseconds. The reasoning behind using averaged dynamics calculations is that the NMR data are themselves inherently averaged measurements, so that if two or more conformations contribute, it might be impossible for a single structure to fulfil the NMR-derived restraints. In some such cases, the dynamics trajectory may interconvert between the different contributing conformations. If one then takes an average over a sufficiently long period of the trajectory and checks restraints against this average, the result may be similar to the real averaging that takes place in the sample. However, not all conformational exchange processes may be sampled in a single dynamics trajectory, given that, at present, it is only practicable to calculate trajectories lasting a few nanoseconds. An attractive alternative is to calculate an ensemble average over a number of parallel calculations, but of course this requires considerably more computational resources.[52]

Full restrained molecular dynamics (RMD) calculations are computationally rather demanding, so at present it is rare to use them to calculate structures from scratch. Instead, they are more commonly combined with some other calculation that handles the initial "search" phase of structure generation (e.g., distance geometry calculations), the RMD calculation being used as the "refinement" phase. Alternatively, RMD can be combined with model-building approaches, where the starting structure for RMD is obtained either by using an interactive graphics program to test hypothetical structures crudely against the data or by analogy with other known homologous structures.

12.3.3. Simulated Annealing Calculations

Simulated annealing calculations represent a particular type of restrained molecular dynamics calculation that has become very popular, particularly in protein and (more recently) RNA structure determination. It is hard to pin down exactly what distinguishes simulated annealing calculations from other forms of RMD, but three common factors are important:

1. The calculations use a force field that is restricted entirely or almost entirely to geometrical terms (i.e., those describing just the covalent bonding network, the van der Waals radii and the NMR restraints).
2. The dynamics protocol includes a high-temperature search phase followed by a slow cooling phase (the "annealing" step).
3. Various changes are made to the force field and/or the algorithm to make the calculation significantly faster than a full RMD calculation.

The technical changes that speed the calculation up need not concern us in detail, but they generally have the effect of limiting the number of calculations required for each step (particularly the time-consuming calculations of nonbonded interactions). This makes the method rather similar to distance-geometry in torsion space (Section 12.3.1.3), the principal difference being that in its original forms the latter method could not traverse maxima in its variable target function, being based on a minimization algorithm.

Early implementations of simulated annealing generally used it in combination with distance geometry (DG), relying on multiple DG calculations to search conformational space thoroughly. In the interests of speed, the DG step would often use a reduced atom representation, employing maybe one third or so of the atoms, and then fitting an all-atom representation over the resulting DG structure before proceeding with the simulated annealing stage. This approach is still quite widely used, but, more recently, improvements in the speed both of simulated annealing algorithms and of computers have made it feasible to carry out simulated annealing "from scratch." Such calculations start from random starting conformations, comprising either a template structure in which the backbone torsion angles have been randomized, or, more drastically, a set of completely random coordinates that take no account at all of covalent bonding (i.e., "atoms scattered in a box"). In either case, the initial search phase is characterized by a high temperature, by potential energy functions for NMR violations that increase only weakly with the extent of violation, and by much-reduced van der Waals interactions. The weak NMR restraint potentials ensure that large violations do not dominate the forces on the atoms early in the procedure, and the reduced van der Waals interactions allow the atoms to move past one another relatively freely. Slow cooling then follows with gradual increases in the penalties both for NMR restraint violations and van der Waals interactions, until, in the limit, the NMR restraint potential functions become essentially square wells, and the van der Waals interactions are fully restored. The key point is that, by decreasing the temperature only very slowly, the system is kept in a pseudoequilibrium throughout its cooling, thereby lessening the probability that it will become trapped in a local minimum. Probably the majority of such calculations to date have been carried out using the program X-PLOR.[53] The protocols supplied with the X-PLOR program have evolved over time; the version most commonly used at the time of writing is based on a protocol originally called YASAP (*Y*et *A*nother *S*imulated *A*nnealing *P*rotocol).[54–56] An interesting new protocol has recently been described, in which the torsion angles of the molecule (rather than the Cartesian coordinates of the atoms) become the variables.[57] In this form, the procedure has become very similar to the most recent methods described in Section 12.3.1.3.

It is worth being aware also that the term simulated annealing describes a general method of nonlinear optimization that is widely used in many areas unrelated to molecular structure determination. Such calculations differ from simple minimization of an error function in much the same way as molecular dynamics calculations differ from energy minimization. Small incremental changes are made to the system variables (whatever they may be) and an error function is computed at each step according to whatever optimization criteria are being used. For each step a decision is made whether or not to accept the changed system as the starting point for the next step. When the error function decreases the change is always accepted, but if the error increases there is an assigned probability that the calculation may still accept the changed system, making it possible for the system to traverse maxima on the error surface. This probability depends partly on how much the error function increased, but the intrinsic tendency of the calculation to accept states with increased error function is also varied through the calculation, being high during the first phase and then gradually decreased. Thus this intrinsic probability to accept states with increased error function corresponds to the temperature in a molecular dynamics calculation, and the sequence of steps made by changing the system variables corresponds to the trajectory. Calculations of this sort have been used, for instance, to optimize the performance of some NMR pulse sequences, including the well-known DIPSI isotropic mixing sequences.

12.3.4. Other Methods

There are various other methods for calculating macromolecular structures that have appeared from time to time, including one that uses a genetic algorithm as the tool for driving the optimization.[58] It would be inappropriate to attempt to review them all here, but one in particular stands out, the program PROTEAN from the group of Jardetzky.[59,60] This program differs fundamentally from others in this chapter because it involves a systematic search of conformational space, aiming to exclude conformers inconsistent with the data rather than to sample conformations consistent with the data. Such a systematic search is impractical at the all-atom level, and is therefore limited initially to the relative positions and orientations of secondary structural elements, which are themselves defined using a rule-based analysis of the NOE, coupling, and NH exchange-rate data. Once this phase of the search is complete, the variables are extended to include individual atomic coordinates in order to refine the details of the structure in a more limited conformational space.

At each stage in the calculation, the current best estimate of the structure is described by the "state vector," which contains the current coordinates of the objects being treated as search variables, and by the corresponding variance-covariance matrix, which expresses how precisely each of the coordinates is currently known both absolutely (the variance) and with respect to those of each other object (the covariances). Restraints act to limit the covariances, since they impose restrictions on the possible relative positions of different objects. Restraints are added sequentially, and the best estimate of the state vector and variance-covariance matrix is updated at each step using a probabilistic method called the extended double-iterated Kalman filter. As a consequence of this treatment, the result depicts not only the mean positions of the search objects (i.e., initially the secondary structure elements, later the atoms themselves), but also the spread of their absolute and relative positions that are compatible with the NMR data. So far, this approach has not been widely taken up outside the group that proposed it, perhaps because it requires significant human intervention and guidance when applied to larger systems.

12.4. ASSESSING AND DESCRIBING NMR STRUCTURES

Once a set of structures has been calculated, there are several criteria by which one may judge them. As always, it is crucial to be aware of the difference between precision (similarity of the structures to one another) and accuracy (similarity of the structures to the true structure). It is easy to establish the *precision* of a set of NMR structures, simply by assessing how similar the individual structures are to one another (although it is much less easy to know how to interpret this precision; see also Section 12.5). In contrast, genuinely establishing *accuracy* implies some knowledge of the "correct" structure, and herein lies an unavoidable paradox. Generally, there is no "correct" structure available to reassure us that the NMR structures are accurate; if there were, there would have been no need to do the NMR study in the first place. Usually, the best that can be done is to check that the calculated structures are consistent with the experimental data, and also that they are consistent with certain general expectations of macromolecular structures. We shall consider these various criteria in turn.

Overall, a balanced judgement of the quality of a set of NMR structures must be reached, using the restraint violations and geometric ideality on the one hand and the precision on the other. If the set of structures is precise, but shows many violations or bad geometry, it is clearly not properly consistent with the NMR data. On the other hand, if there are very few restraint violations or bad geometries, but the precision is poor, this shows that the NMR data are insufficient to define the structure.

12.4.1. Global Precision: Overall Root Mean Square Deviations

The commonest measure of overall precision among a set of NMR structures is the atomic root mean squared deviation (rmsd) after best-fit superposition. The rmsd between two structures a and b is defined as:

$$\text{rmsd} = \sqrt{\frac{1}{N} \sum_{i=1}^{N} (x_i^a - x_i^b)^2 + (y_i^a - y_i^b)^2 + (z_i^a - z_i^b)^2} \quad (12.18)$$

where x_i^a represents the x coordinate of atom i in structure a, y_i^b represents the y coordinate of the corresponding atom i in structure b, and so on, and the summation runs over the same corresponding set of N atoms in each of the two structures. Of course, the value of the rmsd depends on how the two structures are positioned relative to one another. Usually, the "best-fit" superposition of one structure on another is defined as that which corresponds to the smallest rmsd between them, and when one speaks of the rmsd between the structures one generally means the *minimum* rmsd obtained by least squares optimization of their superposition. However, it is important to realize that the set of atoms used to define a "best-fit" superposition need not include the entire molecule, nor need it necessarily be the same as the set used to calculate an rmsd. For example, one could optimize the superposition of two structures of a protein by minimizing the rmsd for the entire backbone (i.e., the N, C^α, and C' atoms), and then, maintaining the same relative positions for the two superposed molecules, calculate an rmsd for just the atoms of one residue; this will be dealt with further in Section 12.4.2.

When more than two structures are involved, several additional issues arise. First, defining the best-fit superposition among all of them is no longer trivial. The most widespread approach to superposing a set of NMR structures is to choose one of the set, often that with the lowest NOE energy, and superpose all the others on it in a series of pairwise superpositions. One may then calculate a corresponding average structure by averaging the coordinates over all the structures in their superposed positions, and repeat the superposition, this time using the averaged structure as the "template." This cycle may be repeated to obtain an "improved" average structure, but usually little change occurs after the first pass for a set of structures that are reasonably similar. However, this approach is obviously rather unsatisfactory in that the outcome may depend, if only slightly, upon the choice of the initial structure for the first round of superpositions. A more satisfying approach is to optimize the entire family of superpositions in one simultaneous operation, and methods for doing this have been published.[61]

Even when a global superposition has been achieved, there are (at least) three different methods of combining the individual pairwise rmsd's into one number characteristic of the whole ensemble of N structures, all of which give numerically different results. The resulting statistics may be named R^1, R^2, and R^3. R^1 is the average over all $N(N-1)/2$ pairwise rmsd's, R^2 is the overall rmsd of all structures to the average structure, and R^3 is the average over the N different individual rmsd's to the average structure. Notice that R^1 is systematically larger than either R^2 or R^3, since, *on average*, the distance between two members of the ensemble must be larger than the distance of either of them to the average structure; analytically, R^1 is larger than R^2 by factor of $\sqrt{2N/(N-1)}$ (so for a comparison of just two structures, the rmsd to the mean is half the value given by Eq. 12.18). The relationship between R^2 and R^3 is slightly more subtle, but mathematically straightforward; for R^2 the individual rmsd's (of each structure to the average structure) are squared before averaging and then the square root is taken of the resulting average, whereas for R^3 the individual rmsd's are averaged directly. In practice, R^2 and R^3 give quite similar outcomes numerically, but they differ in that only R^3 can be associated with a corresponding standard deviation. This standard deviation is useful since it expresses how similar the individual fits (to the average structure) are to one another, and hence gives some indication of how many "outliers" there may be among the structures. In addition, there is a further subtlety concerning R^1. As defined here, R^1 refers to the average pairwise rmsd when all the structures are simultaneously aligned for best overall fit. However, many programs only allow a pairwise rmsd to be determined by actually carrying out the relevant superposition, so that each pairwise fit would necessarily involve different orientations. This will give a result that is smaller than R^1, since each individual fit is separately optimized, but the fits cannot all be achieved simultaneously. The resulting statistic, which is probably what is usually meant in the literature by the phrase "the average over all pairwise rmsd's," may be called R^0.

A still more important factor that controls the outcome of superpositions is the set of atoms that define the superposition, which is generally only a subset of the total in the structure. For example, for proteins, it is common practice to quote two results, one for the backbone atoms (which may or may not include the carbonyl oxygens, according to the definition used) and the other for all heavy atoms. Any obviously disordered residues at the termini of the model are generally excluded, as may be some disordered loops (see text to follow). Note also that it is important to allow for the symmetry of some sidechain groups when calculating the all heavy-atom rmsd. For instance, even if the position of the carboxyl group of an Asp residue is perfectly defined in a set of structures, the $O^{\delta 1}$ and $O^{\delta 2}$ atom labels will be scrambled between the two possible locations in the results of any calculation that started from random initial conditions. This will lead to an unwanted contribution to the rmsd when all structures are compared. The problem can be avoided by swapping the two O^{δ} atom labels for some of the structures, the easiest criterion being to swap the labels for those structures having a negative χ_2 angle; similar issues arise for carboxyl groups of Glu, guanidinium groups of Arg, and the aromatic rings of Phe and Tyr residues in fast exchange on the chemical shift timescale.[62]

Needless to say, given this considerable complexity in the various ways that rmsd's may be defined, one should always make completely clear how a reported rmsd value was obtained. Nonetheless, in the literature it is very often completely unclear what has been done in a given case.

12.4.2. Local Precision: Local RMSD and Angular Order Parameters

Several approaches exist to reporting the local precision of a set of structures. The clearest and most direct is simply to present one or more figures showing a backbone superposition of the structures, since regions of poor local precision show up as areas of "fuzziness." The eye is very sensitive to such differences (particularly when it is the eye of the person responsible for the study).

The commonest method for reporting local precision numerically is to quote localized values for rmsd's, usually calculated on a per-residue basis. Here, the structures are still superimposed globally, but the rmsd is summed separately over the atoms of individual residues. This generates a different rmsd for each residue, characterizing how variable or disordered the position of that particular residue is relative to the structure as a whole. Various intermediate levels of "localization" for the rmsd may also be used. Some authors quote per-residue rmsd's that are actually calculated for a multi-residue "window" centered on the particular residue, thereby producing a smoothed version of the true curve of per-residue rmsd versus sequence. Others quote figures similar to overall rmsd values except that certain particularly disordered regions are excluded. However, it is important to be aware here of the possible distinction between the set of atoms used to superimpose the structures and the set used to calculate the rmsd (Section 12.4.1), since it is rarely stated which atoms define the superposition in such cases. Probably this is largely because many programs do not allow one the freedom to define a superposition over one set of atoms and report the rmsd over a different set, so that some authors may be unaware of the possibility of a distinction.

A quite different approach to defining local precision is to examine the conservation of corresponding torsion angles in the different structures. This may be measured either using the angular standard deviation $\sigma(\theta)$, or else the "angular order parameter," $S^{ang}(\theta)$, introduced by Hyberts.[63,64] In the latter case, each occurrence of a particular angle θ is represented by the phase of a vector of unit length from a common origin, and $S^{ang}(\theta)$ is defined as the normalized magnitude of the corresponding vector sum, given by

$$S^{ang}(\theta) = \frac{1}{N} \sqrt{(\Sigma \cos \theta)^2 + (\Sigma \sin \theta)^2} \qquad (12.19)$$

where N is the number of structures. $S^{ang}(\theta)$ reflects the degree to which θ is conserved across the N structures; if θ is identical in all N structures, $S^{ang}(\theta)$ has a value of 1, while if θ is completely random $S^{ang}(\theta)$ is expected to take the value $1/\sqrt{N}$. Values of less than $1/\sqrt{N}$ indicate anti-correlation, and $1/\sqrt{N}$ tends to zero for large N. The relationship between $S^{ang}(\theta)$ and the angular standard deviation has been determined numerically, and is given approximately by[63]

$$\sigma(\theta) = 2 \arccos(1 + 0.5 \log(S^{ang})) \qquad (12.20)$$

A shortcoming of angular order parameters is their relative insensitivity to small scale deviations in θ; S^{ang} remains close to 1 across quite a broad "plateau" as $\sigma(\theta)$ increases. Nonetheless, for the sorts of angular statistics encountered in all but the most precisely determined NMR structures, S^{ang} seems to be a useful parameter in practice.

Plots of S^{ang} and the per-residue rmsd versus sequence are shown for a small protein (the human complement regulatory protein CD59) in Figure 12.5.[65] The key attribute of these angle-based measures of local precision is that they allow one to distinguish whether a region of relatively high local rmsd is internally disordered throughout, or internally well-ordered and merely disordered relative to the rest of the structure. For instance, some regions of CD59 have high local rmsd values but do not show low values of S^{ang}; the disorder in these regions is presumably caused by concerted movements of parts of the structure that are, within themselves, well ordered. Similarly, if a protein were to show high values of S^{ang} except at two points on the backbone, this might suggest that the region between these dips in S^{ang} might be disordered as a result of "hinge" motions relative to the rest of the structure. More locally, if the ψ angle of one residue and the ϕ angle of the next show roughly equal values of S^{ang}, both significantly lower than other nearby values, this strongly suggests disorder arising through localized "crankshaft" motions of the intervening peptide bond as a more or less rigid unit; several examples of such crankshaft disorders are apparent in Figure 12.5b.[66] Such features can be valuable guides as to the nature of local disorder when one is refining or analyzing an NMR structure.

A further approach to reporting local precision has been suggested by Schwabe et al.,[67] based on C^α–C^α distance plots. Such a plot shows the C^α–C^α distance between each residue and every other residue in the structure in the form of a matrix. The C^α–C^α distance plot is a characteristic of the structure itself, but a similar matrix plot of the standard deviation of each C^α–C^α distance

---

Figure 12.5. Plots of (a) per-residue rmsd and (b) backbone angular order parameters for the 46 calculated structures of the small human complement regulatory protein CD59 shown in (c). In (a), the backbone rmsd (filled circles) is calculated for the N, C^α, and C′ atoms only, while the heavy atom rmsd (open circles) includes also the carbonyl oxygens and all side-chain heavy atoms. In (b), a value of $S^{ang} = 46^{-0.5} = 0.15$ indicates a random distribution. Note that ϕ (filled circles) is undefined for the N-terminal residue and ψ (open circles) is undefined for the C-terminal residue. Panel (c) shows the backbone traces of the 20 structures with lowest energy from among the 46 structures, superposed on the structure with the lowest restraint violations. Reproduced and adapted with permission from ref. 65.

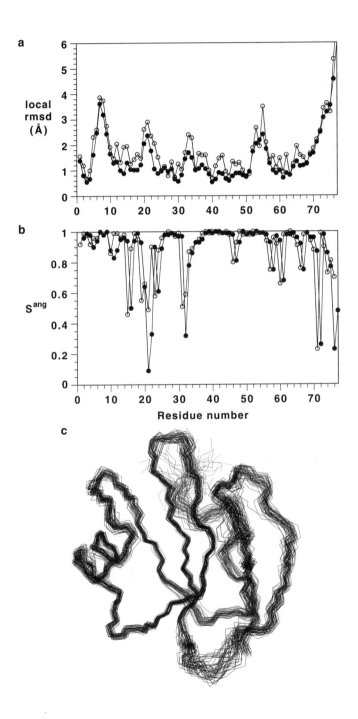

across the family of calculated structures reflects their local precision. However, because a larger distance would inherently display a larger absolute standard deviation, such a plot of absolute standard deviations would be biased by the magnitude of the $C^\alpha-C^\alpha$ distances themselves. It is therefore more helpful to normalize the standard deviations, so that they are each expressed as a percentage of the corresponding mean $C^\alpha-C^\alpha$ distance. The plot then reflects the order in different regions of the protein relative to one another. Also, as for the angle-based methods, this approach is completely independent of any superposition of the structures. An example of such a plot appears in Figure 12.6 for the structure of estrogen receptor DNA-binding domain.

12.4.3. Assessing the Quality of Structures

One of the most obvious and important questions to be asked about a set of calculated structures is, "Do they fit the experimental data?" This is usually assessed by examining violations of the NMR restraints. Typically, NMR-derived structures of globular proteins will shown many residual NOE restraint violations, but for a well-defined structure these should all be relatively small. A significant number of large violations (e.g., > 0.5–1 Å) suggests that the calculation has failed to converge on a valid solution, that some of the input data are incorrect, or that the molecule from which the data were obtained undergoes significant conformational averaging. One might suppose that, with all the various precautions used to interpret the NOE information conservatively (Sections 12.2.2–12.2.4), the "correct" structure for a relatively rigid molecule ought not to show any restraint violations at all, but in practice this is not so. Given the complexity of many structure calculations, it is often essentially impossible to trace the origins of specific violations, particularly if they are small; however, small residual violations of the order of a few tenths of an Ångström may be assumed to arise collectively from small errors in NOE calibration or classification, or from limited local motions in the actual molecule. In contrast, dihedral-angle restraints are often applied rather more loosely than the NOE restraints, their function then being to improve convergence rather than to influence the details of the final structure. For this reason, structures quite often have no dihedral violations.

Part of the explanation for these sorts of observations has been investigated by Kominos et al.,[68] who published an analysis of the effect of inaccuracies in distance restraints due to spin diffusion. They used a crystal structure to generate sets of simulated restraints, which were then used as the input for structure calculations. In some cases, a complete relaxation matrix calculation was used to generate simulated cross-peak intensities, which were then treated as though they were experimental measurements and used to generate restraints based on a two-spin approximation. Because they ignore spin diffusion, these restraints are too short; however, in common with the protocols followed in most macromolecule structure calculations from experimental NMR data, an additional upper range was added to the restraint upper bounds to compensate for the

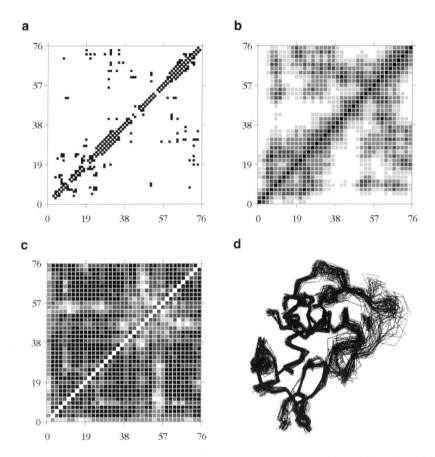

Figure 12.6. C^α–C^α distance plots for the estrogen receptor DNA binding domain. Panel (*a*) shows a diagonal plot representation of the NOE input data from which the structures were calculated (a filled square indicates that there is at least one NOE connectivity between protons in the connected residues). Panel (*b*) shows the mean C^α–C^α distances coded by gray-scale intensity; black, dark gray, medium gray, light gray, and white represent C^α–C^α distances of <5 Å, <10 Å, <15 Å, <20 Å, and >20 Å, respectively. Panel (*c*) shows the normalized standard deviations of these C^α–C^α distances, expressed as a percentage of the corresponding mean C^α–C^α distance; black, dark gray, medium gray, light gray, and white represent standard deviations of <4%, <8%, <12%, <16% and >16%, respectively. Data for plots (b) and (c) are plotted only for odd-numbered residues. Panel (*d*) shows a superposition of the structures for which these data were measured.

As discussed in the text, the C^α–C^α distance plot reflects the structure itself, and is therefore similar to the NOE diagonal plot in panel (a). The standard deviation plot, in contrast, reflects the order of different parts of the structure with respect to one another. Reproduced and adapted with permission from ref. 67.

"unknown" contribution of spin diffusion. The result was that the upper bounds were often quite close to the true distance. This has the consequence that internuclear distances in the calculated structures, which are a compromise between all the input restraints, tend to cluster around the upper bounds of the restraints. The authors conclude that this tendency can result in rather accurate structures, but only due to a near cancellation of errors within the restraints. In particular, it will imply that accurate structures may still be expected to contain residual restraint violations.

Although they are at present less widely used, there are several other methods of testing how well calculated structures agree with the NMR data. Brünger et al.[69] have proposed "complete cross-validation" as a statistical method of assessing how well the input data define the structure. Essentially, the procedure involves partitioning the restraints into a number of "test" sets, each typically comprising about 10% of the restraints chosen at random. For each test set, structures are calculated using only the remaining 90% or so of restraints (called the "working" set). Violations of the restraints in the *test* set are then measured in the resulting structures. This determines to what extent the distances corresponding to the test set can be predicted using only a knowledge of those in the working set. The procedure is repeated for each test set in turn, so that, by the end, every restraint will have been included in one of the test sets. Then, by summing the test-set restraint violations for each calculation, and summing the result for all calculations, one obtains a statistic including contributions from every restraint. This statistic is called the total cross-validated distance restraint violation. Brünger et al. suggest several applications for it, including its use a measure of structural "quality" since it is correlated to accuracy in test calculations (based on random deletion of correct restraints) where the accuracy can be quantified. However, essentially what the method does is to test how consistent the calculated structures are with the input data used to calculate the structures in the first place; in common with other tests based just on these input data, it is hard to see how in general it could address the problem of whether or not the input data are correct.

Another approach is to calculate the NOE intensities that would be expected theoretically for the calculated structures, using one of the relaxation matrix or iterative integration methods discussed in Section 4.3.1, and then to compare these to the experimental NOESY spectra directly. This method can highlight NOE interactions suggested by the structures but not observed experimentally. This could, in turn, suggest ways to improve the structures, but it may simply be that these particular NOE interactions are partially or completely quenched by internal motions, which are generally too difficult to take into account when calculating the theoretical NOE intensities. On the other hand, if the theoretical and experimental NOE data match closely, then this must increase confidence in the validity of the structures. More formally, by analogy with the R factor used in crystallography, the agreement between calculated and experimental NOE intensities may be expressed as an "R factor"

$$R_{\text{NOE}} = \frac{\sum_{ij} |A_{ij}^{\text{calc}} - A_{ij}^{\text{exp}}|}{\sum_{ij} |A_{ij}^{\text{exp}}|} \quad (12.21)$$

where A_{ij}^{calc} and A_{ij}^{exp} are respectively the calculated and experimental NOE intensities of the cross peak between spins i and j.[70] Refinement methods based on minimizing R_{NOE} are considered in Section 12.5.

A further test is to calculate theoretical chemical shifts for the structures and then to compare to those observed experimentally. Many chemical shifts in macromolecules are very difficult to calculate, but for certain favorable cases such as H^α protons or C^α carbons, quite reliable empirical relationships between shift and local structure have been developed. This approach has reached the stage that shift calculations based on NMR structures may, for instance, reveal occasional incorrect assignments.[71] Further developments in this area may be expected.

Another important consideration when assessing the quality of calculated structures is how well their covalent geometry conforms with general expectations. This may be judged in several ways. For a protein, one common approach is to plot a Ramachandran diagram for all the ϕ and ψ angles, in each residue, in each structure. For a well-determined structure the great majority of the angles should be in the allowed regions of (ϕ, ψ) space, but one must of course be prepared to recognize and accept the occasional untypical value that happens to be correct! Figure 12.7 shows such a Ramachandran diagram for the 46 calculated structures of the small complement protein CD59, in which two residues (Arg55 and Leu68) have well-ordered positive ϕ angles of 65° and 63° respectively.[65] For simulated annealing and RMD calculations, geometrical distortions most commonly appear in bond angles, so it may also be useful to analyze the number and distribution of bond angles that are distorted beyond some threshold value, say 5°. Alternatively, bond angle distortions can be assessed globally using the bond angle energy term.

A further criterion for a "good" structure is that it should have a low conformational energy. For structures calculated using RMD or simulated annealing, it is likely that the calculation can output the final energies directly, whereas for pure DG structures a separate calculation would be needed to assess conformational energy. The meaning of "low" conformational energy also depends on the nature of the calculation. Final energies from simulated annealing calculations are generally positive, since no attractive nonbonded terms are included in the force field, whereas, if a full force field is used, the conformational energy of a "good" structure should be negative. Generally, the dominant contribution to the conformational energy comes from the VDW interactions, and sometimes the VDW term is quoted alone. A somewhat more demanding way in which to test the conformational integrity of the final structures is to make them the starting point for a full force-field RMD run, or, even more demandingly, an *un*restrained dynamics run. Although most structures

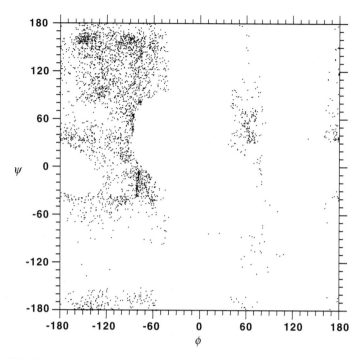

Figure 12.7. Ramachandran plot of the ϕ and ψ angles in the 46 final calculated structures of the small human complement protein CD59. Reproduced with permission from ref. 65.

will start to explore nearby conformational space during such a run, thereby becoming less precise, a "good" structure should not show a systematic tendency to turn into some other structure.

Clearly, the sizes of the restraint violations and geometrical distortions in a set of structures will depend, at least partly, on the way in which a structure calculation balances their relative significance. For instance, if one employs a high NOE force constant during the final stage of an RMD or simulated annealing calculation, the resulting structure may have relatively small violations but considerably distorted local geometries. On the other hand, with a low NOE force constant, the conflict is settled differently by allowing larger violations. This always needs to be borne in mind when analyzing the quality of structures.

12.5. REFINEMENT

In this section we consider the way in which initial low-resolution structures can be improved.

12.5.1. General

During the process of structure determination using crystallography, refinement usually represents a distinct and clearly identifiable stage, employing quite different procedures and programs from those used during initial analysis of the data. However, this is much less true for NMR structural refinement, which frequently consists simply of repeating the calculations already undertaken, but progressively incorporating successive improvements to the restraint list, each set of improvements being based on experience with the preceding round of calculations.

The procedures whereby initially ambiguous NOE cross peaks are assigned using preliminary structural information have already been discussed (Section 12.2.1.2), and this often represents by far the major part of the refinement process. It may be that in the future this process becomes more automated and integrated into the structure calculations, for instance through application of the r^{-6} summation protocols for using ambiguous NOE restraints (Section 12.2.4.3).[23] Additional stereoassignments can also be made based on intermediate structures (Section 12.2.5), as can additional hydrogen-bond restraints or even disulfide or metal-bonding restraints (Section 12.2.6). In this way the whole restraint list evolves through successive rounds of calculation, new restraints being added and perhaps a few errors being weeded out at each stage.

The biggest problem with this sort of refinement is that there is no clearly defined endpoint. There are a few purists who argue (generally in private) that any use of intermediate structures in assigning additional restraints cannot be justified. It is quite clear to any who have actually carried out a structure determination using NMR that, if no use whatsoever is made of intermediate structures in assigning initially ambiguous restraints, the information content of the spectra is greatly underutilized and the resulting calculated structures are grossly and unnecessarily imprecise. Nonetheless, the purists are right in at least one respect; once one starts to use intermediate structural information, one has crossed the only definable line, and the decision as to when to stop "refining" a structure becomes intensely subjective. There can all too easily come a stage at which further decisions on restraint inclusion are based on some earlier overinterpretation, and the process imperceptibly but increasingly begins to leave reality behind. The only things that stand in the way of such overinterpretation are sound judgement and competence, and probably the nature and extent of any overinterpretation is, in practice, almost completely undetectable by anyone not actually involved in the work (unless the work is repeated). However, this probably does not make NMR structural refinement particularly different from most other scientific activities.

This issue is complicated by and intimately bound up with the issue of how to interpret the precision of NMR structures. It is true that some proteins have intrinsic properties that prevent a structure being determined to high resolution. Sometimes this may be because the actual molecule has a high degree of internal motions, so that its true structure is imprecise. In other cases it may be for technical reasons, such as a paucity of NOE data due to lack of chemical shift dispersion, poor signal-to-noise ratios, or poor lineshapes. However, it is also likely that two groups working with identical NMR data might produce structures with different precisions, because of the myriad different decisions that they would each make during refinement, and particularly as to when they would each declare refinement to be over. It might even depend upon whether they knew of one another's activities.

Another subjective area is the choice of how many structures to use when reporting precision. Some methods of computing structures have intrinsically a much lower convergence rate than others, so that in such cases reported precision can only be based on a small subset of the total of calculations. However, it is also clear that one way in which to achieve instant refinement is to discard a larger number of "bad" structures from the calculated ensemble. Methods that achieve convergence in a high proportion of calculations have a clear advantage in this respect, since it is relatively easy to justify excluding a small proportion of conformers that are clear-cut statistical outliers in terms of NOE energy. One quite useful aid when deciding which structures to discard is a plot of rmsd versus ensemble size, where the rmsd is calculated independently for each ensemble size and the ensembles are built up by adding successive structures in order of increasing energy.[19] Such an "energy-ordered rmsd profile," in conjunction with a similar profile for the energies themselves, often shows a clear-cut plateau region that defines a statistically similar subset of the structures; the end of such a plateau region would then mark the division between the structures retained and those discarded (e.g., see Fig. 12.8).

For reasons such as those discussed in this section, the whole area of precision and accuracy of NMR structures, and their interpretation, is somewhat controversial at present.[69,72–74] Some take calculated precisions essentially at face value, while others feel that the rather high precision of some recent structures probably does not reflect the true character of the molecules in solution. Either way, it is important to remember that, for almost all of the methods discussed in this chapter (with the exception of time-averaged or ensemble-averaged dynamics), the results describe only the *mean structure*, and that precision in this context means the *precision with which the mean has been established*. This is not the same thing as knowing the true spread that the actual molecules adopt around their mean structure in solution due to their intrinsic conformational variability. In fact, the issue of how true conformational variability is related to the spread among calculated NMR structures is a complex one, which we have touched on earlier (Section 12.3). It is also important to remember that the mean structure is not necessarily *accurate*, even if it is precise, since there is no necessary connection between the accuracy and precision. Although there are some statistical relationships between accuracy and precision, the theory upon which these are based necessarily assumes that all errors are random, which is very far from the case in NMR structure determination. No statistical theory could be expected to compensate for the effects of unknown errors in restraint classification or assignment. It is thus

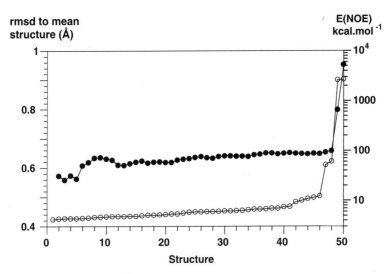

Figure 12.8. Energy-ordered rmsd profile (filled circles; scale on left-hand *y* axis) and the corresponding energy profile (open circles; scale on right-hand *y* axis) for ensembles of calculated structures of CD59 (corresponding to the data shown in Figs. 12.5 and 12.7). In the energy-ordered rmsd profile, the mean rmsd to the mean structure (R^3) for the ensemble was calculated for each ensemble size independently following a best-fit global superposition of the ensemble members. The plateau region in the rmsd profile corresponds to the first 48 structures, while that for the energies comprises the first 46 structures.

entirely possible (and possibly quite common) for precision to exceed accuracy, which of course would imply that the correct structure is not necessarily within the calculated ensemble at all.

12.5.2. Specific Protocols for Refinement

In Section 4.3.1 it was shown how relaxation matrix calculations can predict the intensities of NOESY cross peaks at any mixing time for a known or a model structure. Such calculations do not involve the commonly made "isolated two-spin approximation" (ISPA), but allow explicitly for the effects of spin diffusion; since all distances may be directly measured in the known or model structure, a complete relaxation matrix corresponding to that structure may be constructed. Usually, the only assumptions are that the molecule is rigid, that its tumbling in solution is isotropic and can be described by a single correlation time, and that effects due to the cross-correlation between motions of different spins may be neglected (Section 3.1.2).

The reverse of this calculation would be the prediction of distances directly from the NOE intensity matrix, so as to take spin diffusion explicitly into account. Such distances would be specified by a complete relaxation matrix, since of course the off-diagonal relaxation matrix elements correspond to the individual cross-relaxation rates σ_{ij}, themselves proportional to r_{ij}^{-6}. One might therefore imagine that a "back-calculation" of the relaxation matrix from the NOE intensity matrix might be the most direct way of obtaining optimal distance restraints for use in structure calculation. However, in practice this direct calculation is usually impossible since it would require knowledge of *all* the intensities in the NOESY spectrum, including all of the diagonal peaks. Such complete information is generally not available, except perhaps for small molecules under favorable conditions.

To tackle this problem, various iterative approaches have been developed. Probably the most widely used is that incorporated in the MARDIGRAS (*Ma*trix *A*nalysis of *R*elaxation for *DI*scerning *G*eomet*R*y of an *A*queous *S*tructure) program.[75,76] The input comprises an incomplete NOE intensity matrix, derived from the NOESY spectrum, and an initial model structure, which might itself typically result from a conventional structure calculation using distances estimated using the normal two-spin approximation. On the first pass through the MARDIGRAS program, theoretical NOE intensities are calculated for the model structure, and then all the intensities for which NOE cross peaks were measured are replaced with their experimental values. This results in a complete matrix of NOE intensities, some of which are experimental and some of which were calculated from the model structure. This intensity matrix then forms the input for a back-calculation step to generate a new relaxation-rate matrix. To complete the first pass, this new relaxation-rate matrix is modified by replacing those rates that correspond to spin pairs for which there is no corresponding experimental NOE intensity with rates calculated from the model structure. The sequence of forward and back-calculations is then repeated several times; after each forward step the experimental NOE intensities are always reinserted into the NOE intensity matrix, and after each back-calculation step the rates for which there is no experimental NOE intensity are always replaced with the corresponding model-derived rates, until no further significant changes occur.

If the two-spin approximation were valid, this iteration would have no effect on the distances for which NOE intensities are available, since the corresponding rates are never replaced during the above procedure. However, the whole basis of the relaxation matrix calculations is that the NOE intensity between spins i and j depends to some extent on all of the distances, not just upon r_{ij}. Thus, by the end of the iteration, the distances will in fact have altered so as to form a self-consistent set, at least within the context of the relaxation matrix equations (they may still not be consistent with a single structure, however). Although a complete set of distances is calculated, it would be imprudent to use those for which no corresponding cross peak was measured; rather the purpose of the calculation is to take the initial set of distances derived using the two-spin approximation and refine them so that their changed values are less affected by spin diffusion. The resulting improved distance restraints can then be used as the input to a DG or simulated annealing calculation. Similar approaches have also appeared in which no starting model structure is needed.[77,78]

An alternative and earlier approach is incorporated in the program IRMA (*I*terative *R*elaxation *M*atrix *A*pproach).[79,80] Here, the model structure itself is included in the iteration. As with MARDIGRAS, the initial intensity matrix is a composite of experimental values and theoretical intensities calculated from a model structure, which is then used to back-calculate a new relaxation rate matrix from which a set of distances may be extracted. In IRMA, these distances are then used as restraints to calculate a revised model structure, from which a new composite intensity matrix is derived and the cycle repeated until convergence is achieved. Because this method includes a full structure determination on every cycle, it is inevitably more computationally intensive than MARDIGRAS.

A quite different approach is to refine the structure directly against the NOE intensities. This can be done by defining an appropriate term in the force field:

$$E_{\text{relax}} = K \sum_{ij} (I_{ij}^{\text{calc}} - kI_{ij}^{\text{exptl}})^2 \qquad (12.22)$$

where K is a scaling factor controlling the weight of the relaxation term relative to the rest of the force field, k is a scaling factor between the calculated and experimental NOE intensities, and the sum over ij represents a sum over all measured NOE intensities. Refinement then consists of minimizing the term E_{relax}. The approach was first used to refine DNA oligonucleotide structures, where the nature of the problem allowed a rather specific implementation; in the program COMATOSE (constrained refinement of macromolecular structure based on two-dimensional nuclear Overhauser effect spectra), E_{relax} was minimized as a function of a minimal set of torsion angle variables chosen to represent the double-helical conformation.[81] For more general solutions, it is necessary to calculate the gradients of E_{relax} as a function of the atomic coordinates, armed with which it is possible to include E_{relax} as a term in molecular dynamics calculations. Yip and Case[82] have derived analytical expressions for these gradients, and show ways in which the potentially formidable size of the calculations can be limited (for a system of N atoms, there are in principle $3N$ gradients for every measured NOE intensity, but many of these need not be computed in practice). These expressions have also been extended to include information on internal dynamics.[83]

Along somewhat similar lines, refinement of structures based on comparison of experimental and calculated chemical shifts has been proposed. Although some chemical shifts, particularly of exchangeable spins, are rather resistant to prediction, there are favorable cases such as H^α and C^α shifts where quite reliable semi-empirical correlations are available. Again, a residual may be defined and introduced into the force field, and its value minimized during a molecular dynamics calculation. A recent and possibly very important innovation has been the introduction of restraints based on measurements of residual dipolar couplings in weakly ordered media. These can constrain the orientation of parts of a molecule relative to the overall molecular frame, and they may also be able to constrain atoms that interact over longer distances than can be "sensed" by the NOE (Section 12.2.6).

The whole area of refinement is a very active field of research, and one may expect to see rapid developments. However, in at least some cases there is something of a "take-up" problem, at least at present. Many of the methods are computationally highly demanding and require considerable additional analysis and treatment of the data beyond that required for a simple structure calculation. Also, while they improve the quality of the structure with respect to the particular refinement criterion being used, the atomic movements in the structure that achieve this improvement are often rather small. Probably for these reasons, few of the methods described in this section (with the possible future exception of using residual dipolar restraints) are yet in widespread use outside the groups that originated them.

REFERENCES

1. Hoch, J.; Poulsen, F. M.; Redfield, C., eds.; "Computational Aspects of the Study of Biological Macromolecules by NMR Spectroscopy," Plenum, New York, 1991.
2. Clore, G. M.; Gronenborn, A. M., eds. "NMR in Proteins," Macmillan, New York, 1993.
3. Wüthrich, K. "NMR of Proteins and Nucleic Acids," Wiley, New York, 1986.
4. Evans, J. N. S. "Biomolecular NMR Spectroscopy," Oxford University Press, Oxford, 1995.
5. Roberts, G. C. K., ed. "NMR of Macromolecules: A Practical Approach," IRL Press, Oxford, 1993.
6. Brüschweiler, R.; Case, D. A. *Prog. Nucl. Magn. Reson. Spectrosc.* 1994, *26*, 27.
7. Breg, J. N.; Boelens, R.; Vuister, G. W.; Kaptein, R. *J. Magn. Reson.* 1990, *87*, 646.
8. Breg, J. N.; van Opheusden, J. H. J.; Burgering, M. J. M.; Boelens, R.; Kaptein, R. *Nature* 1990, *346*, 586.
9. Nilges, M. *Proteins: Struct., Funct., and Genet.* 1993, *17*, 297.
10. Folmer, R. H. A.; Hilbers, C. W.; Konings, R. N. H.; Hallenga, K. *J. Biomolec. NMR* 1995, *5*, 427.
11. Arrowsmith, C. H.; Pachter, R.; Altman, R. B.; Iyer, S. B.; Jardetzky, O. *Biochemistry* 1990, *29*, 6332.
12. Pachter, R.; Arrowsmith, C. H.; Jardetzky, O. *J. Biomolec. NMR* 1992, *2*, 183.
13. Jasanoff, A. *J. Biomolec. NMR* 1998, *12*, 299.
14. Barsukov, I. L.; Lian, L.-Y. in "NMR of Macromolecules: A Practical Approach" (G. C. K. Roberts, ed.), IRL Press, Oxford, 1993, pp. 315–357.
15. Denk, W.; Baumann, R.; Wagner, G. *J. Magn. Reson.* 1986, *67*, 386.
16. Hommel, U.; Harvey, T. S.; Driscoll, P. C.; Campbell, I. D. *J. Mol. Biol.* 1992, *227*, 271.

17. Clore, G. M.; Gronenborn, A. M. *J. Magn. Reson.* 1985, *61*, 158.
18. Wüthrich, K.; Billeter, M.; Braun, W. *J. Mol. Biol.* 1983, *169*, 949.
19. Fletcher, C. M.; Jones, D. N. M.; Diamond, R.; Neuhaus, D. *J. Biomolec. NMR* 1996, *8*, 292.
20. Edmondson, S. P. *J. Magn. Reson. B* 1994, *103*, 222.
21. Wagner, G.; Braun, W.; Havel, T. F.; Schaumann, T.; Go, N.; Wüthrich, K. *J. Mol. Biol.* 1987, *196*, 611.
22. Clore, G. M.; Gronenborn, A. M.; Nilges, M.; Ryan, C. A. *Biochemistry* 1987, *26*, 8012.
23. Nilges, M. *J. Mol. Biol.* 1995, *245*, 645.
24. IUPAC-IUB Commission on Biochemical Nomenclature *Biochemistry* 1970, *9*, 3471.
25. Mierke, D. F.; Kessler, H. *Biopolymers* 1993, *33*, 1003.
26. Güntert, P.; Braun, W.; Billeter, M.; Wüthrich, K. *J. Am. Chem. Soc.* 1989, *111*, 3997.
27. Güntert, P.; Braun, W.; Wüthrich, K. *J. Mol. Biol.* 1991, *217*, 517.
28. Nilges, M.; Clore, G. M.; Gronenborn, A. M. *Biopolymers* 1990, *29*, 813.
29. Cai, M. G.; Huang, Y.; Liu, J. H.; Krishnamoorthi, R. *J. Biomolec. NMR* 1995, *6*, 123.
30. Senn, H.; Werner, B.; Messerle, B. A.; Weber, C.; Traber, R.; Wüthrich, K. *FEBS Lett.* 1989, *249*, 113.
31. Weber, P. L.; Morrison, R.; Hare, D. *J. Mol. Biol.* 1988, *204*, 483.
32. Folmer, R. H. A.; Hilbers, C. W.; Konings, R. N. H.; Nilges, M. *J. Biomolec. NMR* 1997, *9*, 245.
33. Tjandra, N.; Bax, A. *Science* 1997, *278*, 1111.
34. Clore, G. M.; Starich, M. R.; Gronenborn, A. M. *J. Am. Chem. Soc.* 1998, *120*, 10571.
35. Hansen, M. R.; Mueller, L.; Pardi, A. *Nature Struct. Biol.* 1998, *5*, 1065.
36. Prosser, R. S.; Losonczi, J. A.; Shiyanovskaya, I. V. *J. Am. Chem. Soc.* 1998, *120*, 11010.
37. Clore, G. M.; Gronenborn, A. M.; Tjandra, N. *J. Magn. Reson.* 1998, *131*, 159.
38. Clore, G. M.; Gronenborn, A. M.; Bax, A. *J. Magn. Reson.* 1998, *133*, 216.
39. Hansen, M. R.; Rance, M.; Pardi, A. *J. Am. Chem. Soc.* 1998, *120*, 11210.
40. Crippen, G. M.; Havel, T. F. "Distance Geometry and Conformational Calculations," Research Studies Press, Taunton, and Wiley, Chichester, 1988.
41. Metzler, W.; Hare, D. B.; Pardi, A. *Biochemistry* 1989, *28*, 7045.
42. Havel, T. F.; Wüthrich, K. *Bull. Math. Biol.* 1984, *46*, 673.
43. Havel, T. F. *Biopolymers* 1990, *29*, 1565.
44. Havel, T. F. *Prog. Biophys. Molec. Biol.* 1991, *56*, 43.
45. Braun, W.; Go, N. *J. Mol. Biol.* 1985, *186*, 611.
46. Güntert, P.; Mumenthaler, C.; Wüthrich, K. *J. Mol. Biol.* 1997, *273*, 283.
47. Xu, Y.; Sugár, I. P.; Krishna, N. R. *J. Biomolec. NMR* 1995, *5*, 37.
48. Guba, W.; Haessner, R.; Breiphol, G.; Henke, S.; Knolle, J.; Santagada, V.; Kessler, H. *J. Am. Chem. Soc.* 1994, *116*, 7532.

49. Guba, W.; Kessler, H. *J. Phys. Chem.* 1994, *98*, 23.
50. Torda, A. E.; Scheek, R. M.; van Gunsteren, W. F. *Chem. Phys. Lett.* 1989, *157*, 289.
51. Pearlman, D. A. *J. Biomolec. NMR* 1994, *4*, 1.
52. Kemmink, J.; van Mierlo, C. P. M.; Scheek, R. M.; Creighton, T. E. *J. Mol. Biol.* 1993, *230*, 312.
53. Brünger, A. T. "X-PLOR. Version 3.1. A System for X-ray Crystallography and NMR," Yale University Press, New Haven, CT, 1993.
54. Nilges, M. in "X-PLOR. Version 3.1. A System for X-ray Crystallography and NMR," (A. T. Brünger, ed.), Yale University Press, New Haven, CT, 1993.
55. Nilges, M.; Kuszewski, J.; Brünger, A. T. in "Computational Aspects of the Study of Biological Macromolecules by NMR" (J. C. Hoch, ed.), Plenum Press, New York, 1991.
56. Nilges, M.; Gronenborn, A. M.; Brünger, A. T.; Clore, G. M. *Protein Eng.* 1988, *2*, 27.
57. Stein, E. G.; Rice, L. M.; Brünger, A. T. *J. Magn. Reson.* 1997, *124*, 154.
58. Bayley, M. J.; Jones, G.; Willett, P.; Williamson, M. P. *Protein Sci.* 1998, *7*, 491.
59. Altman, R. B.; Jardetzky, O. in "Methods in Enzymology" (N. J. Openheimer and T. L. James, eds.), Vol. 177, Academic Press, San Diego, 1989, pp. 218–246.
60. Pachter, R.; Altman, R. B.; Jardetzky, O. *J. Magn. Reson.* 1990, *89*, 578.
61. Diamond, R. *Protein Sci.* 1992, *1*, 1279.
62. Neuhaus, D.; Nakaseko, Y.; Schwabe, J. W. R.; Klug, A. *J. Mol. Biol* 1992, *228*, 637.
63. Hyberts, S. G.; Goldberg, M. S.; Havel, T. F.; Wagner, G. *Protein Sci.* 1992, *1*, 736.
64. Detlefsen, D. J.; Thanabal, V.; Pecoraro, V. L.; Wagner, G. *Biochemistry* 1991, *30*, 9040.
65. Fletcher, C. M.; Harrison, R. A.; Lachmann, P. J.; Neuhaus, D. *Structure* 1994, *2*, 185.
66. Fadel, A. R.; Jin, D. Q.; Montelione, G. T.; Levy, R. M. *J. Biomolec. NMR* 1995, *6*, 221.
67. Schwabe, J. W. R.; Chapman, L.; Finch, J. T.; Rhodes, D.; Neuhaus, D. *Structure* 1993, *1*, 187.
68. Kominos, D.; Suri, A. K.; Kitchen, D. B.; Bassolino, D.; Levy, R. M. *J. Magn. Reson.* 1992, *97*, 398.
69. Brünger, A. T.; Clore, G. M.; Gronenborn, A. M.; Saffrich, R.; Nilges, M. *Science* 1993, *261*, 328.
70. Gonzales, C.; Rullmann, J. A. C.; Bonvin, A. M. J. J.; Boelens, R.; Kaptein, R. *J. Magn. Reson.* 1991, *91*, 659.
71. Williamson, M. P.; Kikuchi, J.; Asakura, T. *J. Mol. Biol.* 1995, *247*, 541.
72. Liu, Y.; Zhao, D.; Altman, R.; Jardetzky, O. *J. Biomolec. NMR* 1992, *2*, 373.
73. Clore, G. M.; Robien, M. A.; Gronenborn, A. M. *J. Mol. Biol.* 1993, *231*, 82.
74. Zhao, D.; Jardetzky, O. *J. Mol. Biol.* 1994, *239*, 601.
75. Borgias, B. A.; James, T. L. *J. Magn. Reson.* 1990, *87*, 475.

76. James, T. L. *Curr. Opin. Struct. Biol.* 1991, *1*, 1042.
77. Madrid, M.; Llinás, E.; Llinás, M. *J. Magn. Reson.* 1991, *93*, 329.
78. van de Ven, F. J. M.; Blommers, M. J. J.; Schouten, R. E.; Hilbers, C. W. *J. Magn. Reson.* 1991, *94*, 140.
79. Boelens, R.; Koning, T. M. G.; Kaptein, R. *J. Molec. Struct.* 1988, *173*, 299.
80. Boelens, R.; Koning, T. M. G.; van der Marel, G. A.; van Boom, J. H.; Kaptein, R. *J. Magn. Reson.* 1989, *82*, 290.
81. Borgias, B. A.; James, T. L. *J. Magn. Reson.* 1988, *79*, 493.
82. Yip, P.; Case, D. A. *J. Magn. Reson.* 1989, *83*, 643.
83. Dellwo, M. J.; Wand, J. *J. Biomolec. NMR* 1993, *3*, 205.

CHAPTER 13

BIOPOLYMERS

This chapter covers applications of the NOE to biopolymers, that is to say proteins, polynucleotides, and polysaccharides, and their complexes. There is an enormous literature in this area, mostly published within the past decade. To attempt anything approaching a comprehensive survey would demand (at least) another book, so we have tried instead to select a small number of examples that we feel are representative of current work and illustrate the ways in which the NOE is applied.

A few general remarks are applicable to all biopolymer work. All biopolymers are large enough that $\omega\tau_c$ >1.12 at commonly used magnetic field strengths. This means that NOE enhancements are negative and spin diffusion is extensive, so that steady-state enhancements contain no structural information (Section 3.2.3). Therefore, NOE measurements must be made using transient methods, usually NOESY or one of its analogs. NOESY is much more widely applicable to biopolymers than to small molecules, because it has higher resolution (essential for analyzing crowded spectra), higher sensitivity when applied to more slowly tumbling molecules, and it avoids the need for selective irradiation (Section 8.1).

For biomolecules, assignment of the spectrum is always a major hurdle. Reliance on the NOE as an assignment tool has been decreased by the recent explosive increase in the use of ^{13}C and ^{15}N isotopic labelling and the accompanying plethora of new experiments for making assignments based on heteronuclear J-coupling. Even so, the NOE is still used quite widely for assignment, particularly with smaller macromolecules at natural abundance.

In the remainder of the chapter, we discuss proteins, nucleotides, and oligosaccharides separately, and in each case we start by discussing typical structural parameters.

13.1. PEPTIDES AND PROTEINS

We begin by considering which of the interproton distances commonly found in proteins are likely to correspond to observable NOE enhancements. Table 13.1, which is based on reference 1, summarizes the short distances found in the major types of secondary structure in peptides and proteins, namely α-helix, 3_{10} helix, β-sheet, and β-turns type I and II (Fig. 13.1). The type III β-turn is essentially identical to a type I turn, or to part of a 3_{10} helix.[2] Peptide conformation is normally defined by the backbone dihedral angles ϕ and ψ (Fig. 13.2) and side-chain angles χ_1, χ_2, and so on, as needed. If the energy of a peptide residue is calculated as a function of ϕ and ψ, it is easily shown that

TABLE 13.1 Short Distances Found in Protein Structures[a]

Distance[b]	α-Helix[c]	3_{10} Helix	Antiparallel β-Sheet	Parallel β-Sheet	Type I Turn	Type II Turn
$d_{\alpha N}(i, i)$	2.7	2.7	2.8	2.8	2.7/2.8	2.7/2.2
$d_{\alpha\beta}(i, i)$[d]	2.2–2.9	2.2–2.9	2.2–2.9	2.2–2.9	2.2–2.9	2.2–2.9
$d_{\beta N}(i, i)$[d]	2.0–3.4	2.0–3.4	2.4–3.7	2.6–3.8	2.0–3.5	2.0–3.4/ 3.2–4.0
$d_{\alpha N}(i, i+1)$	3.5	3.4	2.2	2.2	3.4/3.2	2.2/3.2
$d_{NN}(i, i+1)$	2.8	2.6	4.3	4.2	2.6/2.4	4.5/2.4
$d_{\beta N}(i, i+1)$[d]	2.5–3.8	2.9–4.0	3.2–4.2	3.7–4.4	2.9–4.1/ 3.6–4.4	3.6–4.4
$d_{\alpha N}(i, i+2)$	4.4	3.8			3.6	3.3
$d_{NN}(i, i+2)$	4.2	4.1			3.8	4.3
$d_{\alpha N}(i, i+3)$	3.4	3.3			3.1–4.2	3.8–4.7
$d_{\alpha\beta}(i, i+3)$[d]	2.5–4.4	3.1–5.1				
$d_{\alpha N}(i, i+4)$	4.2					
$d_{\alpha N}(i, j)$[e]			3.2	3.0		
$d_{NN}(i, j)$[e]			3.3	4.0		
$d_{\alpha\alpha}(i, j)$[e]			2.3	4.8		
$^3J_{\alpha N}$	4	4	10	10	4/7	4/5

[a]Values based on ref. 1. Only distances up to 4.5 Å are shown. For convenience, values for $^3J_{\alpha N}$ corresponding to each specified backbone geometry are shown in the final row.
[b]$d_{AB}(i, j)$ means the distance between proton attached to heavy atom A on residue number i and proton attached to heavy atom B on residue number j.
[c]Secondary structure is constructed as follows: α-helix, $\phi = -57°$, $\psi = -47°$; 3_{10} helix, $\phi = -60°$, $\psi = -30°$; antiparallel β, $\phi = -139°$, $\psi = +135°$; parallel β, $\phi = -119°$, $\psi = +113°$; type I turn, $\phi_2 = -60°$, $\psi_2 = -30°$, $\phi_3 = -90°$, $\psi_3 = 0°$; type II turn, $\phi_2 = -60°$, $\psi_2 = 120°$, $\phi_3 = 90°$, $\psi_3 = 0°$. Further details are given in ref. 1. For turns, figures treat residue i as residue 2/residue 3, except for $d_{\alpha N}(i, i+3)$, where i is residue 1.
[d]Distances involving β protons are the lower and upper limits of the range of distances possible. A CH_2 fragment is assumed, and the nearest hydrogen atom of the two is always considered.
[e]Interstrand distance. The indices (i,j) indicate the residues that give the shortest distance of a given type (see Fig. 13.1).

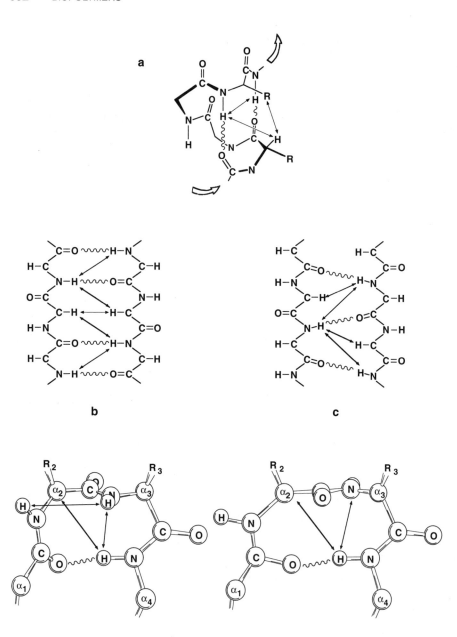

Figure 13.1. Regular secondary structures found in proteins. Some short interproton distances are indicated by double-headed arrows, and hydrogen bonds by wavy lines. (a) α-Helix. Direction of helix is indicated by large arrows. (b) Antiparallel β-sheet. (c) Parallel β-sheet. (d) β-turn type I. (e) β-turn type II. In (d) and (e), the double-headed arrows from NH(4) are to NH(3) and $H^\alpha(2)$ in each case.

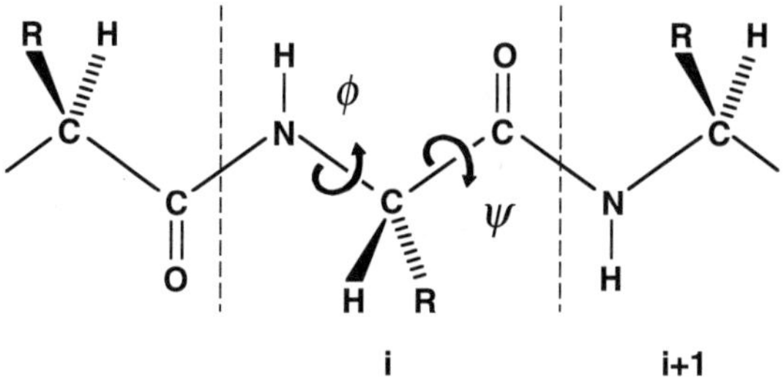

Figure 13.2. Definition of the backbone torsion angles ϕ and ψ.

only a few low-energy conformations are accessible, corresponding roughly to the α-helix and β-sheet geometries shown in Figure 13.3. A rather higher energy region is also accessible in the left-handed α region, while glycine, uniquely, can also adopt the mirror image positions with equal ease. In protein crystals, very few nonglycine residues are found outside the α-helix and β-sheet geometries,[2] and then only in particular local geometries, most significantly as residue 3 of a type II turn. The structural categories in Table 13.1 thus cover about 99% of all geometries found in crystalline proteins. Small cyclic peptides are somewhat more prone to contain unusual backbone geometries, which is a result of the steric strain induced by cyclization.

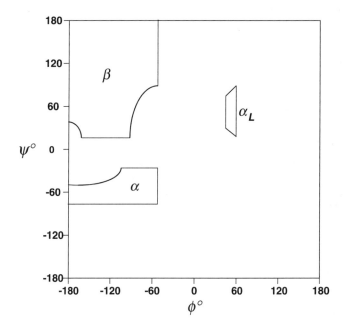

Figure 13.3. Low-energy regions of the peptide chain as a function of ϕ and ψ. Regions corresponding to α-helix and β-sheet, and left-handed α-helix are indicated.

A number of useful conclusions can be drawn from Table 13.1. First, the intraresidue vicinal distances $d_{\alpha N}(i, i)$, $d_{\alpha\beta}(i, i)$, $d_{\beta\gamma}(i, i)$ and so on, must always be short, and constitute a large proportion of the strong NOE enhancements seen in NOESY spectra of proteins. These enhancements can, in principle, be identified by comparing NOESY and COSY spectra. The intraresidue distance $d_{\beta N}(i, i)$ is also usually short, and a cross peak should appear at the same position in TOCSY spectra. Another common category of enhancement corresponds to the sequential distances $d_{\alpha N}(i, i + 1)$, $d_{NN}(i, i + 1)$, and $d_{\beta N}(i, i + 1)$, which are commonly abbreviated as $d_{\alpha N}$, d_{NN}, and $d_{\beta N}$, respectively. Note that these all involve the backbone amide proton, which is a crucial atom in the NOE-based sequential assignment procedure, discussed in Section 13.1.2. Finally, comparison of the distances for α-helix and β-sheet shows that regular secondary structure can be recognized and categorized fairly simply by their characteristically different NOE patterns.[1,3] α-Helices have particularly short sequential d_{NN}, and β-sheets have short sequential $d_{\alpha N}$. Even more characteristic are the short $d_{\alpha N}(i, i + 3)$ and $d_{\alpha\beta}(i, i + 3)$ found in helices, and the short interstrand $d_{\alpha N}(i, j)$, $d_{NN}(i, j)$, and $d_{\alpha\alpha}(i, j)$ found in β-sheets (Fig. 13.1). Indeed, even without any sequence-specific assignments, the presence of $d_{\alpha\alpha}$ enhancements can be used as a crude indicator for antiparallel β-sheets.[4]

13.1.1. Assignment: Heteronuclear Methods

Any detailed structural studies of large proteins requires uniform isotopic labeling with ^{15}N and ^{13}C. If a protein is uniformly double labeled, the most general method of assignment of the spectrum is to make use of the large one- (and sometimes two-) bond couplings along the backbone of the protein.[5] These couplings are large and almost independent of conformation, which means that the transfer of magnetization is rapid and efficient. A big advantage compared to NOE-based techniques is that the magnetization transfer is coherent; this means that most (sometimes almost all) of the net magnetization can be transferred, resulting in very sensitive experiments. Typically this approach involves running a series of 3D heteronuclear experiments. For example, the HNCA experiment relates the ^1H and ^{15}N signals of one amide group to the $^{13}C^\alpha$ signal of the same residue, while the HN(CO)CA experiment relates the same ^1H and ^{15}N signals to the $^{13}C^\alpha$ signal of the preceding residue. Thus, by combining the results of the two experiments and matching $^{13}C^\alpha$ frequencies, it is in principle possible to assign all H, N and $^{13}C^\alpha$ signals (with gaps at prolines). A large number of such triple-resonance experiments, sometimes collectively referred to as "backbone" experiments, is available, and the techniques have been reviewed.[5,6] The sensitivity and resolution of such methods means that in favorable cases they can be automated reasonably successfully, and it may be hoped that this trend will eventually make the laborious manual assignment of proteins a thing of the past.

For larger systems, slower tumbling leads to broader lines, and eventually even heteronuclear experiments based on one-bond couplings start to fail. Ironically, it could be that in this regime the NOE once again becomes a useful assignment tool. It has been suggested that one-bond (^{13}C,^{13}C) and (^{15}N,^{13}C) NOE enhancements might provide the basis for connecting resonances through the covalent framework of an isotopically labeled macromolecule.[7] At present such techniques do not compete favorably with triple-resonance experiments based on J-coupling, but the trend towards studying ever higher molecular weight species at ever higher field strengths may yet bring us to the point where this situation is reversed.

13.1.2. Assignment: (^{1}H,^{1}H) NOE-Based Methods

Use of the (^{1}H,^{1}H) NOE to carry out assignments is less desirable than the strategy just described, for two main reasons. First, experiments based on NOESY are less sensitive than "backbone" experiments based on scalar coupling, largely because cross-relaxation is usually slower than magnetization transfer through large one-bond couplings, and also because of the high efficiency of some individual through-coupling steps. Second, the magnitudes of sequential and intraresidue NOE enhancements depend on geometry, and so are much less uniform than one-bond coupling constants. It should be added that the resolution of 3D heteronuclear experiments is usually much better than that of 2D homonuclear experiments. However, it is often not possible to label proteins with ^{13}C and ^{15}N, either because of cost or because the source (animal/plant extract or synthesis) does not readily permit isotopic labeling. When only unlabeled material is available (or only ^{15}N labeled), either the classic sequential assignment strategy of Wüthrich[8] or the modified main-chain-directed strategy[9] may be used.

Both methods require a knowledge of the sequence. In the sequential assignment method,[8] through-bond experiments such as COSY and TOCSY are used to identify "spin systems." A spin system is a series of ^{1}H resonances connected by two- and three-bond (^{1}H,^{1}H) couplings. Because there are at least four bonds between protons on adjacent amino acid residues in a protein, a spin system corresponds essentially to an amino acid residue, with the proviso that some side-chain protons (namely the aromatic protons of Phe, Tyr, Trp, and His, the side-chain amide protons of Gln and Asn, and the C$^\varepsilon$H$_3$ protons of Met) form additional unconnected spin systems. The spin systems are then linked using the sequential NOE enhancements $d_{\alpha N}$, d_{NN}, and $d_{\beta N}$, and the partial sequences thus identified are matched to their correct locations on the known complete sequence. Of course, there is initially no way to distinguish sequential enhancements from the nonsequential enhancements that are also present, and it has often been objected that one must start by guessing which enhancements to take as sequential. This is true, but it is not the problem it might first appear, since the ability to fit the NOE data to the whole sequence ultimately provides a very high degree of confidence in the overall assignment. Moreover, most sequential links are supported by at least two of the three sequential NOE connectivities listed above. It is further worth noting that once a short fragment has been tentatively matched to its position on the protein sequence, the amino acid type of the next residue can be predicted, which strongly limits the choice of signals to use. From this point on, each predicted sequential connectivity that is actually found in the NOE data increases the confidence that a genuine stretch of sequential connectivities has been located, and once a stretch of more than five or six residues has been sequentially assigned, such confidence is very high. Indeed, one may even be able to locate a limited number of point errors in the sequence.

The method must be modified to deal with proline, which has no NH proton. However, the $C^\delta H$ protons of proline have a similar geometric dependence to those of normal NH protons, and can often be used in an analogous way:[1] this is one of the very few ways in which homonuclear assignment is simpler than heteronuclear methods.

The main-chain-directed strategy[9] places less reliance on initial recognition of amino acid types by their spin–spin coupling patterns, and more on recognition of cyclic networks of NOE connectivities characteristic of different types of secondary structure. Thus, in helices there is almost invariably found the cyclic NOE network NH_i–$C^\beta H_i$–NH_{i+1}–NH_i; the presence of such a cyclic network avoids ambiguous branch points and imposes an internal confirmatory check at each step. The sequential assignment and main-chain-directed approaches represent two extremes. In practice, one would generally identify first those amino acid types that are easy to find, then identify any obvious sequential assignments involving them and confirm these by looking for cyclic patterns. Only once the easier assignments were largely completed would one move on to the more ambiguous and difficult cases that might initially have been set aside. Such a "piecemeal" approach, although perhaps less satisfying to describe, comprises mixed elements of both the sequential and the main-chain-directed strategies.

13.1.3. Structure Determination of Protein Monomers

Methodology for structure calculations is covered in Chapter 12. Here, we aim to present typical examples of structure calculations applied to different types of problems. It should be stressed again that there is no single "best" method for structure calculation; indeed, there are probably almost as many different protocols for structure calculation as there are labs engaged in it.

13.1.3.1. Small Rigid Unlabeled Proteins. Of very many possible examples, we have chosen to discuss the structure calculation of bovine pancreatic trypsin inhibitor (BPTI).[10] Its NMR spectrum was assigned using the sequential assignment method, which results in not only a list of chemical shift assignments but also a list of sequential NOE connectivities. From this starting point, NOESY spectra (in H_2O and D_2O) were analyzed to provide a list of assigned NOE cross peaks, with volume integrals. The volumes were converted to distance ranges using an internal calibration based on NOE intensities in regular secondary structure (using different calibrations for methyl groups and NH protons, as discussed in Section 12.2.3). This resulted in a relatively small number of unambiguously assigned NOE restraints, which were used (together with angle restraints derived from spin–spin couplings) to calculate preliminary structures. Although the resulting initial structural ensemble had poor precision, it could be used to make further assignments of some of the ambiguous NOE enhancements, by eliminating clearly impossible assignments (cf. Section 12.2.1.2). Further, it became possible to make stereospecific assignments of some of the residues, based on spin–spin couplings and NOE enhancements (initially intraresidue and sequential, but later also long-range), and therefore assign some of the NOE enhancements involving diastereotopic protons (e.g., β-methylene protons and valine methyls) stereospecifically (cf. Section 12.2.5). These new NOE enhancements were then added to the existing list and a further set of structures was calculated. This cycle was repeated several times, until no new assignments could be made.

At the end of this process the restraint list comprised 916 upper-distance restraint limits, with lower limits set to the sum of van der Waals radii, plus 6 distance restraints for each of the three disulfide bonds, 41 restraints on ϕ, 41 on ψ, and 33 on χ_1. Distance restraints involving equivalent or nonstereoassigned diastereotopic protons employed pseudoatoms (Section 12.2.4.1). Of the 916 distance restraints, 274 were independent of the conformation (or there exists for them no conformation that would violate the restraint). Thus there was a total of 775 active experimental restraints. This represents a relatively high number of 13.6 restraints per residue, which were spread reasonably evenly over the sequence, and so might be expected to produce high-quality structures. In this case no hydrogen bond restraints were added, because it was felt that these might bias the resultant structures (cf. Section 12.2.6).

The structures were calculated using the DIANA variable target function distance-geometry program (cf. Section 12.3.1.3).[11,12] Of 50 structures, 35 had low violations of the input restraints (i.e., a low target function), of which the best 20 were selected (cf. Section 12.5.1 for a discussion of selecting representative structures). These 20 were energy minimized, to give structures with much lower energy, only slightly different geometry (0.25 Å rmsd) and slightly worse violations (although the sum of all violations increased from 3.0 to 4.6 Å on energy minimization, the maximum violation decreased from 0.22 to 0.09 Å and the number of violations greater than 0.2 Å decreased from 0.9 to zero per structure). The structures differ from the mean by 0.73 Å for backbone atoms (or 0.43 Å if the N- and C-terminal residues, which are poorly defined, are ignored), a value which is roughly typical for structures of this size calculated with this number of restraints. Many of the internal side-chains are as well defined as the backbone. A superposition of the 20 energy-minimized structures is shown in Figure 13.4.

The NMR structure was compared to three independent high-resolution crystal structures. Although the solution and crystal structures are very similar, there are differences between them in some regions. This is clear from a comparison

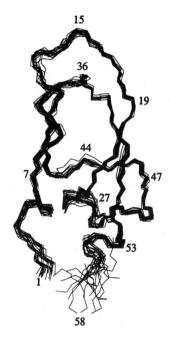

Figure 13.4. Twenty energy-minimized structures of bovine pancreatic trypsin inhibitor (BPTI). Only the backbone (N, C^α and C') atoms are shown. The structures were superimposed over residues 2 through 56. Reproduced with permission from ref. 10.

of NOE enhancements calculated from the crystal structures and the experimental enhancements, which revealed 28 violations greater than 0.5 Å distributed all over the protein. These are almost all due to displacements of surface loops. Residues that are well defined in the crystal are generally well defined in solution also. Thus, there are only small local differences between the solution and crystal structure. This is quite a typical result. Generally, there has been agreement, within their error limits, between NMR and X-ray structures of the same protein. Exceptions are two or three structures in which loops are genuinely in different locations (or, for example, where there is a helix in solution which is disordered in the crystal) and a small number of cases where there were errors, either in the NMR or the X-ray structure.[13] The NMR structures generally display much higher disorder for exposed surface residues than the corresponding crystal structures, as one might expect.

13.1.3.2. Larger Rigid Labeled Proteins. These proteins are harder to deal with by NMR because their spectra are more crowded. However, 3D and 4D heteronuclear experiments with labeled protein often relieve much of the overlap present in homonuclear 2D spectra. Indeed, the chemical shift dispersion in ^{15}N and to a large extent $^{13}C=O$ is almost uncorrelated with the ^1H chemical shift, implying a very useful degree of additional spectral dispersion in heteronuclear experiments. Although $^{13}C^\alpha$ and $^{13}C^\beta$ shifts are more strongly correlated with the shifts of their attached protons, they still provide enough resolution to resolve the majority of H^α and H^β protons even in reasonably large proteins. Thus in many cases the spectra actually used for larger labelled proteins, at least at the assignment stage, are significantly less overlapped than those used for smaller unlabelled proteins. In a 3D experiment, one of the heavy atoms attached to one of the two protons involved in an NOE enhancement is frequency-labeled, which not only allows the proton to be resolved from other ^1H signals at a similar chemical shift, but also allows it to be easily assigned. However, it provides no help with assignment of the other proton. Thus for example in a 3D ^{15}N NOESY-HSQC spectrum, which is most useful for providing NOE enhancements between NH and side-chain protons, the NH groups are characterized by both their ^{15}N and ^1H shifts, and so can often be readily assigned, whereas the side-chain protons are only characterized by a ^1H shift, and so are often ambiguous. The overlap can be lifted using a 4D $^{15}N/^{13}C$-resolved experiment, but at the expense of a long acquisition and poorer digital resolution in the indirect dimensions (and usually extensive folding in the ^{13}C dimension). These points are developed further in Section 9.2.2.

As an example of a larger, labeled protein studied by NMR, we have chosen the N-terminal fragment of urokinase-type plasminogen activator, a two-domain protein of 135 residues, ^{15}N- and ^{15}N, ^{13}C-labeled samples of which were obtained by growth in mammalian cell culture.[14] The spectrum was assigned using 3D heteronuclear techniques, and NOE enhancements were obtained from 3D ^{15}N- and ^{13}C-resolved NOESY spectra. Coupling constants were also obtained, using a variety of 2D and 3D heteronuclear experiments. Stereospecific assignments for glycine H^α and valine methyl groups were made using protein grown on selectively labeled feedstocks. No stereospecific assignments were made for β-methylene groups, because the corresponding linewidths were too large for the experiments used.

The investigators obtained 1299 nontrivial distance restraints from the 3D NOESY spectra, which were used with 27 ϕ angle restraints and restraints on 21 hydrogen bonds (only added later in the calculation), a total of 10.5 restraints per residue. The two domains of the protein are almost independent, with the result that they have different effective correlation times, and therefore the calibration of distances in the two domains was carried out independently. Structures were calculated using a combined metric matrix distance-geometry/simulated annealing protocol in XPLOR.[15] The most efficient method was found to be to calculate each of the domains separately and then link successful calculations together as starting positions for a final round of refinement. In the final structure, the two domains were well defined, but their relative positions were not, because of a lack of interdomain restraints.

With proteins larger than about 30 kD, there are problems of spectral crowding, even using 3D and 4D experiments, and spin–spin relaxation becomes very rapid, implying that the signal-to-noise ratio of multidimensional experiments is reduced. Currently the most useful technique to extend the size of proteins for which useful structural information can be obtained is partial deuteration of the protein. This can either mean that all carbon-bound protons are uniformly diluted by deuteration in the range of about 50–90%, or it can mean that selected sites (e.g., methyl groups and aromatic protons) are left fully protonated, while all other carbon-bound sites are 100% deuterated. Uniform partial deuteration can be achieved by growth in mixed H_2O/D_2O-based media (provided the microorganisms used are adapted to such conditions), while the site selective deuteration requires a specific biochemical strategy. Partial deuteration reduces spectral crowding and greatly reduces the dipolar spin–spin relaxation rate for the remaining protons.[16] Depending on the approach used, it may also reduce the number of potentially observable NOE enhancements, but calculations suggest that there will still be sufficient enhancements to permit reasonably accurate structure calculations, especially when assisted by backbone angle restraints based on ^{13}C chemical shifts. The recently proposed technique of measuring residual dipolar couplings in weakly ordered media may also make a useful contribution to NMR structure determinations of larger proteins in the future (cf. Section 12.2.6),[17] as may the TROSY technique, which markedly reduces relaxation rates for large proteins in many experiments and therefore potentially gives a large increase in sensitivity.[18]

Water is of course an essential element in protein structures, but it is not easy to locate bound water molecules using NMR because the signals from bound water molecules invariably resonate at the same frequency as bulk water (cf. Section 5.8). NOE enhancements to signals at the frequency of bulk water could arise from bound water, or from tyrosine, serine, or threonine hydroxyls, which might also be expected to resonate at the same shift as water. Therefore, bound water molecules can only be identified with any degree of confidence when the NOE partner is far enough away from potential tyrosine, serine, or threonine hydroxyls to exclude this possibility.

The sign of the enhancement to water will depend on the correlation time of the internuclear vector. Thus, very broadly, water that exchanges slowly will have a long residence time and therefore a negative NOE enhancement (i.e., a NOESY signal having the same sign as the diagonal), whereas water that exchanges rapidly will have a short residence time and a positive NOE enhancement. Both cases will have positive ROE enhancements. The exact details require a model for the motion, which can become very complex, but both positive and negative NOE enhancements are seen. These indicate a range of residence times that have been estimated at 10^{-2} to 10^{-8} s for buried internal water molecules, and 1×10^{-9} to 3×10^{-9} s for surface molecules.[19]

13.1.3.3. Conformationally Mobile Proteins.
This category includes linear peptides, which tend to be conformationally mobile in solution. As a result, although assignment of peptides and collection of restraints is usually simple, structure calculation is much more difficult than for well-structured proteins, and it must be said that at present there are no generally applicable methods for even confirming the presence of conformational heterogeneity, let alone

characterizing it.[20] The category also includes mobile parts of proteins, such as long N- or C-terminal extensions and long loops.

The way in which the NOE averages over internal motions (Sections 5.5 and 5.6) means that even small proportions of a conformer containing a short interproton distance can contribute significantly to the observed NOE enhancement. Other conformational parameters, such as coupling constants, are not averaged in such an extreme way, and therefore may be less biased by minor conformers. Short linear peptides often adopt a "random coil" conformation, which means that they populate ϕ-ψ space roughly according to the Boltzmann distribution; that is, each residue is most often in the β conformation though it spends a small proportion of its time in the α conformation.[20] In such circumstances sequential $d_{\alpha N}$ NOE enhancements are strong, and sequential d_{NN} enhancements are weak. A small increase in the proportion of helical conformation, therefore, gives rise to a significant increase in the intensity of d_{NN} enhancements, so it is not unreasonable to use NOE enhancements to identify residues where there is a high proportion of "helical" conformation. This is about as far as one can easily go, however; it should be clear that it would be meaningless to measure the intensity of enhancements in such a system and use them as restraints to calculate a single structure for the peptide (although, sadly, such attempts are not rare), because each enhancement represents a weighted average over unknown proportions of unknown conformers. Similarly, if large-scale conformational heterogeneity is suspected, NOE enhancements cannot safely be used to calculate structures directly; however, they can sometimes be used to show that a single structure cannot exist, when there is no single structure compatible with all the observed enhancements.

There have been several examples of cases where a single conformer does not satisfy all the NOE restraints. In a study of the acyl carrier protein, Kim and Prestegard[21] calculated conformations from restrained molecular dynamics calculations; individually these could not satisfy the experimental restraints, but they gave satisfactory fits when combined in pairs. In a similar study, Blackledge et al.[22] generated a set of structures, each of which was obtained by applying the NOE restraints in a random order until significant violations were encountered. Because each such structure satisfied only a subset of the restraints, the intention was to find pairs of such structures that *in combination* would satisfy the restraints. These structures were then combined together to produce satisfactory combinations. As these authors comment, any such fitting procedure will of necessity give better fits than using a single structure, but the mere fact that a better fit is achieved with two conformers does not prove that the conformers are real. Perhaps the most promising approach to this problem is complete cross-validation (cf. Section 12.4.3).[23] This can be used to test whether addition of a further structure is justified, by repeating the structure calculation with around 10% of the restraints (the test set) missing. If inclusion of the additional structure causes a reduction in violations of the test set of restraints, then this shows that adding the further structure is genuinely beneficial; if violations of the test set do not decrease, it is not.

13.1.4. Structure Determination of Symmetric Protein Oligomers

Symmetrical dimers present a particular problem for NMR structure determination. Since each resonance arises from two places in a symmetric dimer, every NOE enhancement is in principle ambiguous, there being *a priori* no way to distinguish intramonomer enhancements from intermonomer enhancements.

Several approaches exist to overcoming this problem. If homologous structures are known, one may use an assumed similarity to these to assign enhancements as intra- or intermonomer; this approach has been likened to that of molecular replacement in X-ray crystallography.[24] Alternatively, if the protein is available in labeled form and can be induced to exchange between dimeric and monomeric forms (if necessary by unfolding followed by refolding), one may prepare samples from a mixture of fully labeled and fully unlabeled molecules. Provided that complete exchange occurs, such a sample prepared from 50% labeled and 50% unlabeled material will contain dimer molecules of which 50% are mixed (i.e., dimers containing one labeled and one unlabeled monomer). One may then use half-filter techniques (Section 9.2.2.1) to distinguish between intramonomer and intermonomer enhancements (cf. Section 12.2.1.3).[25] However, although intellectually attractive, these techniques are inherently rather insensitive, because (i) they collect signal only from half the sample (the 50% that exists as mixed-labeled dimers) and (ii) the pulse sequences contain significant extra delays (relative to simple 3D HSQC-NOESY, for instance). The latter problem can be particularly serious because resonances at the interface are sometimes selectively broadened by intermediate-rate conformational exchange. Finally, one may leave the distinction between intra- and interunit enhancements to be handled during the structure calculation itself, allowing the protocol to switch the assignment of individual enhancements (those predefined to be "ambiguous") dynamically as the calculation proceeds, if by doing so a lower energy structure results.[26] This latter process has much in common with that of floating stereoassignment (Section 12.2.5), and experience so far suggests this is a very useful approach.

The actual consequences of these ambiguity problems depend strongly on the character of the dimer interface. If the interface is small and simple, for instance if it is formed just by an antiparallel interaction of two fairly short β-strands, one from each monomer, then assignment of NOE enhancements is likely to be fairly straightforward because geometric considerations within the sheet rule out some potential intramonomer enhancements. Enhancements that are clearly intermonomer can then be put into the restraint list, and the other initially ambiguous enhancements can be assigned using an iterative structure calculation procedure.[27] If the dimer interface is more complicated and extensive, a more sophisticated approach is likely to be needed. In trimers and tetramers the situation is more complicated still, and iterative methods are essential, perhaps used in combination with the "ambiguous restraint" protocol in molecular dynamics calculations (mentioned earlier).[28]

13.1.5. Through-Space Connections by Solid-State NMR Experiments with Proteins

In this section we give a few words about using NMR to measure dipolar interactions in proteins in the solid state. At first sight this represents quite a large deviation from the main topic of this book (and indeed we specifically excluded it in the preface to the first edition!). However, as studies of ever larger systems are undertaken using NMR, the "boundaries" between traditional solid phase and solution methodologies are becoming progressively more blurred, and techniques originating from both solid and solution phase NMR are increasingly being applied to samples that are strictly neither solid nor liquid in terms of their NMR properties. This is likely to be an area that develops rapidly in the next few years. All we can do here is give a cursory overview together with a brief look at some aspects that we have found interesting and which we feel have at least some connection with the NOE.

In principle, measurement of dipolar interactions in solids can yield information similar to that obtained from NOE measurements in liquids. However, dipolar couplings in solids can be very large, typically up to tens or even hundreds of kilohertz (depending on the nuclei involved and their separation). By contrast, although dipolar interactions are present in liquids, the couplings are averaged to zero by rapid molecular tumbling and their only remaining influence is to cause relaxation (which is essentially what most of the rest of this book is about). Thus, although solid-state experiments of the sort touched on in this section can be linked conceptually to liquid-state NOE experiments, at the NMR experimental level they differ fundamentally; the solid-state experiments are based on coherent transfer of magnetization through a (dipolar) coupling, whereas liquid-state NOE experiments are based on measurements of an incoherent effect (relaxation).

The presence of large dipolar couplings means that, without further manipulation, spectra of solids are generally extremely broad and consequently almost uninterpretable. Spectra with sharper lines can be obtained, either in the rather special cases of using single crystals or aligned samples, or else by using "magic angle spinning" (MAS), in which the sample is spun rapidly around an axis tilted by 54.7° relative to the applied magnetic field. This technique sharpens the NMR lines by helping to average away the dipolar interactions (and chemical shift anisotropy; see later discussion) in a similar way to that achieved by molecular tumbling in a liquid. It achieves this with a simple uniaxial rotation because dipolar coupling depends on the function $(3\cos^2\theta - 1)$, and this is zero when $\theta = 54.7°$. Very many solid-state NMR experiments make use of MAS, but it should be noted that the fastest attainable MAS spinning rates (at present around 35 kHz) are still not sufficient for the effective removal of (^{1}H-^{1}H) dipolar interactions in biological macromolecules. This is because of the large gyromagnetic ratio for ^{1}H, the usually high proton density, and the strong mutual dipolar couplings amongst them. Therefore very few solid-state NMR studies of macromolecules to date have used ^{1}H observation; the majority concentrate instead on observation of ^{13}C.

Although, in simple experiments, MAS is used to help *remove* dipolar interactions, over the last several years a flurry of methods has appeared that partly reintroduce some of the dipolar interactions in a controlled fashion. If the MAS rotation frequency is an exact submultiple of the difference in chemical shift between two nuclei, then residual dipolar interactions can be observed, a technique known as rotational resonance.[29,30] Although this technique is clearly useful, it is limited in application to macromolecules labeled in very specific patterns, usually having only two labeled sites per molecule. A more general method for reintroducing dipolar interactions is to apply sequences involving pulses synchronous with the rotor spinning period.[31–33] This can give high-resolution solid-state 2D spectra of low- or medium-γ nuclei that look very similar to NOESY spectra, and which can be interpreted in more or less the same way.[34] Internuclear distances of up to about 5 Å can be observed for ^{13}C-^{13}C interactions in labeled materials. However there is a problem when applying this method to macromolecules such as proteins, in that if the macromolecule is uniformly ^{13}C-labeled, the one-bond dipolar couplings between directly bonded nuclei swamp the weaker, and more interesting, long-range interactions. A number of possible solutions to this problem are currently being explored, but it is as yet too early to say how effective these will prove to be.

Turning to static samples, the most striking property of solid-state NMR spectra of static powder samples is that each nucleus gives a signal that is not a single sharp line as in a solution-state NMR spectrum, but rather a "powder pattern." For dilute ^{13}C spectra, each such pattern can be as much as 150 ppm wide. Powder patterns arise because chemical shift is in fact a tensor quantity; each nucleus has an associated chemical shift tensor, whose orientation is determined by the bonding arrangement around the nucleus. The observed chemical shift for each nucleus therefore depends strongly on the orientation of the chemical shift tensor with respect to the magnetic field, and in a powder sample all possible orientations will contribute. For instance, for polyalanine labeled with ^{13}C at all the carbonyl positions, the 1D ^{13}C spectrum of a powder sample looks similar to the skyline lineshape of Figure 13.5a, because each carbonyl carbon has a frequency somewhere between 90 and 240 ppm, depending on its orientation.

An interesting 2D experiment has been published that makes use of this orientation dependence to discover structural relationships in spider silk.[35] Spider silk is essentially a block copolymer comprising stretches of (Gly-Gly-X)$_n$ (n = 3–6, neglecting minor imperfections), interspersed with Ala$_m$ segments (m = 4–7), and samples can be prepared in which ^{13}C is introduced at *either* all

---

Figure 13.5. Spin-diffusion (NOESY) spectra of spider (*N. madagascariensis*) dragline silk ^{13}C-labelled at the carbonyl carbon of either alanine [panels (a) and (b)] or glycine [panels (c) and (d)]. Panels (*a*) and (*c*) show stack plots while (*b*) and (*d*) are the corresponding contour plots. Reproduced with permission from ref. 35.

PEPTIDES AND PROTEINS 565

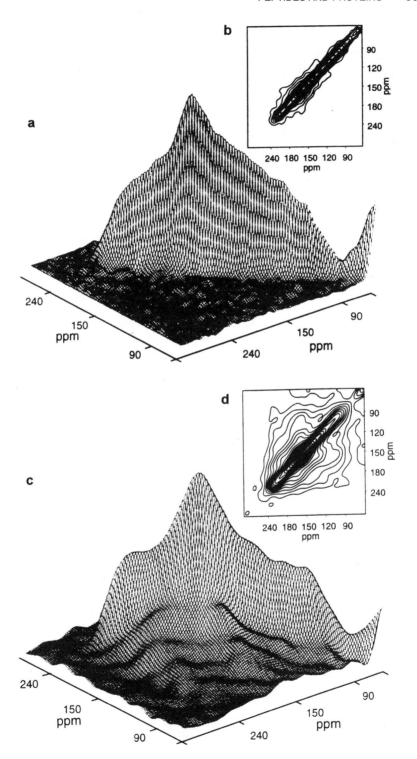

of the alanine carbonyl positions *or* all of the glycine carbonyl positions. The NMR experiment consists essentially of a homonuclear ^{13}C-^{13}C NOESY pulse sequence with a very long mixing time (10 s). Spin diffusion during this long mixing time causes magnetization to be exchanged efficiently between ^{13}C spins over distances of up to about 6 Å (or longer for relayed transfer). Thus, for a sample labeled at the alanyl carbonyls, magnetization will be exchanged via spin diffusion throughout each Ala$_m$ segment, and if multiple segments are brought close enough together (e.g., by formation of a β-sheet structure), it will also be exchanged between segments. The resulting spectrum, called a spin-diffusion spectrum, will thus show correlations between the chemical shift of a carbonyl in one alanine with all the other alanine carbonyls in the same Ala$_m$ segment and spatially close segments. Note also that dipolar interactions between labeled carbonyl carbons in these samples, although large enough to allow the spin-diffusion experiment to succeed, are relatively small because the ^{13}C atoms are quite far apart; the dipolar couplings therefore have little effect on the lineshapes in Figure 13.5.

As it turns out, the spin-diffusion spectrum of such a sample consists essentially only of the diagonal (Fig. 13.5a and b), implying that within each Ala$_m$ segment all alanine carbonyls have the same chemical shift and therefore the same orientation. Although a structure having a series of parallel carbonyl bond vectors is hard to envisage, one having a series of antiparallel carbonyls is simply achieved in a β-strand. Thus the spin-diffusion spectrum suggests that the alanines in each Ala$_m$ segment are arranged in β-strands, and would be consistent with adjacent strands forming a β-sheet. In contrast, the corresponding experiment for the sample labeled at all glycine carbonyl positions does show significant cross-peak intensity (Fig. 13.5c and d), implying that adjacent carbonyls have different orientations. Spectra were simulated for a range of possible simple geometries, and the best fit was for a 3_1 helix. An α-helical conformation could be definitely excluded. Therefore it was concluded that spider silk consists of stretches of 3_1-helical glycine-rich segments, interspersed with β-sheets made up of polyalanine stretches.

13.2. POLYNUCLEOTIDES

Early studies of nucleotide structure concentrated on DNA duplexes. In solution, these are almost invariably very close to the B-DNA structure (although sometimes there are larger distortions, and very occasionally they can be in the A- or Z-DNA forms, as discussed below); therefore the relevant question was usually "How is the canonical B-DNA structure distorted?" This was, and remains, a very hard question to answer, because in DNA duplexes the only interresidue contacts are sequential or cross-strand, and there is none of the long-range information that is so useful for limiting the conformational space of proteins. Calculations of distortions of duplex structure therefore rely on accurate quantitative interpretation of NOE enhancements. As should be evident from the rest of this book, such interpretation is difficult even in well-behaved cases, but in DNA duplexes, with problems of nonisotropic tumbling, conformational exchange, and solvent exchange (see following), it is even harder. Added to this, the structure of DNA duplexes means that there are rather few NOE enhancements along the DNA axis, and many of these correspond to long distances, which give rise to possible complications from spin diffusion. However, more recently the main interest in NMR studies of double-helical DNA has been in complexes with other macromolecules (cf. Section 13.4.2), where distortions of regular structure in the calculated conformers are often determined by the shape of the overall complex rather than by NOE enhancements within the DNA double helix itself.

In recent years there has been much interest in more complicated nucleic acid structures, including hairpins, triplexes, and quadruplexes; an enormous variety of RNA structures; and DNA–protein and RNA–protein complexes. Until recently, DNA has proved difficult to label with stable isotopes, so most studies on DNA complexes to date have used unlabeled DNA, often in combination with labeled protein or RNA. By contrast, RNA can be labeled, which has led to a big increase in NMR studies of RNA-containing complexes, which are discussed in Section 13.4.2. A recent review appears in reference 36. We begin, as with the proteins, with some definitions.

13.2.1. Structures and Conformations

DNA is made up of the four bases guanine, cytosine, adenine, and thymine (G, C, A, and T), while RNA contains uracil (U) in place of T. The structures of these bases are shown in Figure 13.6. G and A are purines, and C, T, and U are pyrimidines. Numbering of atoms is indicated in Figure 13.6, as is the amino-imino nomenclature. In almost all cases, bases are paired across the double helix either as A-T or as G-C, as shown in Figure 13.6. This is the standard Watson-Crick arrangement. Two other less common arrangements will also concern us: the Hoogsteen base pairing (Fig. 13.7a) and the G-U (wobble) pair, found in tRNA and characteristic of several unusual base pairings in RNA aptamers (Fig. 13.7b). As shown in Figure 13.8, deoxyribose and ribose protons are designated as 1′ through 5′ and 5″, and following the currently accepted definition[37] the chain direction is defined as going from the 3′ of one nucleotide through phosphorus to the 5′ of the next (i.e., considering the free hydroxyls at the chain termini, in the 5′ to 3′ direction). The five-membered ribose ring is puckered, and there are a number of ways of describing its conformation. The two sugar conformations most commonly found in oligonucleotides are known as ^3E and ^2E (or 3′-*endo* and 2′-*endo*, respectively), shown in Figure 13.9. The E/*endo* nomenclature denotes the atom that is out of plane, *endo* meaning that the out-of-plane atom is on the same side of the plane as C5 in D-ribofuranose. The glycosidic bond is the C1′—N bond, and its conformation is described by the angle χ, which is generally the angle O4′—C1′—N9—C4 for purines and O4′—C1′—N1—C2 for pyrimidines. Some earlier conven-

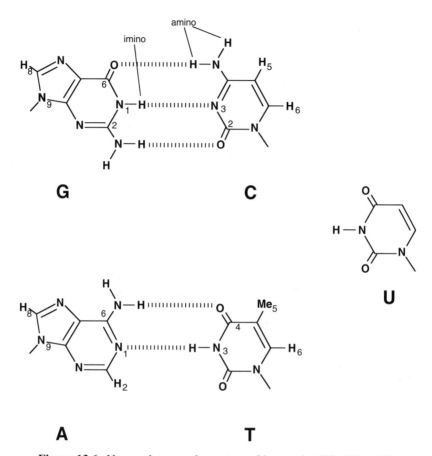

Figure 13.6. Nomenclature and structure of base pairs GC, AT and U.

tions have χ differing from this by 180°. χ angles of $0 \pm 90°$ are described as *syn*, and $180 \pm 90°$ as *anti* (Fig. 13.10). The region sometimes described as high-*anti* actually has $\chi = \pm -60°$, and is more properly described as $-syn$clinical.

Only three types of regular duplex are commonly found in solution: A, B, and Z. B is by far the most common for deoxyribonucleic acids and DNA/RNA hybrids, while A is commonly found in solution only for RNA double helices, though it is frequently seen in DNA crystals. The Z form is found occasionally for DNA, especially with substituents at the 5-position of cytidine and in high salt concentrations, usually in alternating purine-pyrimidine sequences.

B-DNA is a regular right-handed double helix, and some relevant short distances in B-DNA are indicated on Figure 13.11 and listed in Table 13.2. Distances vary somewhat according to the exact model used, and slightly different

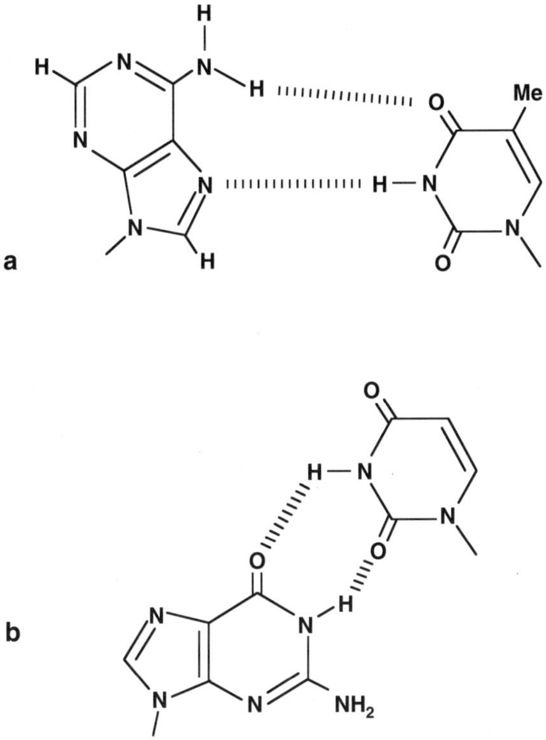

POLYNUCLEOTIDES 569

Figure 13.7. (*a*) AT Hoogsteen pair; (*b*) GU wobble pair.

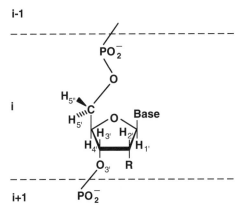

Figure 13.8. Nomenclature of sugar protons. For deoxyribose, R = H2″; for ribose, R = OH.

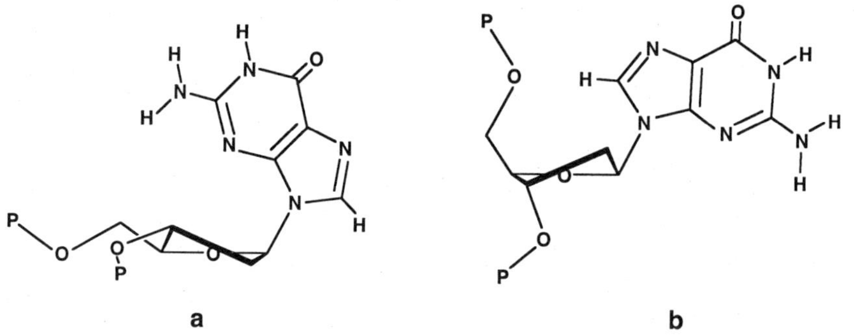

Figure 13.9. Glycosidic conformations and sugar puckers. The two structures show the conformations found most commonly in oligonucleotide crystal structures: (*a*) *syn* conformation about the glycosidic bond and a ^{3}E (also known as 3'-*endo*) sugar pucker; (*b*) *anti* glycosidic conformation and a ^{2}E (2'-*endo*) sugar pucker.

figures can be found elsewhere.[38–40] Both here and subsequently, the only distances reported are those involving exchangeable, base, and sugar 1', 2', and 2" protons, because signals from other protons (3', 4', 5', 5", and also 2' in ribonucleotides) are generally inaccessible because of overlap. One close contact that it was not possible to represent in Figure 13.11 is that between AH2 protons on opposite strands; this is discussed further in Section 13.2.2.1. As far as its surface is concerned, the key attribute of B-DNA is that there are two helical grooves running along its length. The "major groove" is about 12 Å wide and 8 Å deep and is in part formed by atoms of the bases, so that its nature depends on the local sequence. In contrast, the "minor groove" is about

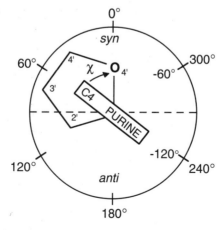

Figure 13.10. Definition of the glycosidic angle χ: purine in *syn* conformation.

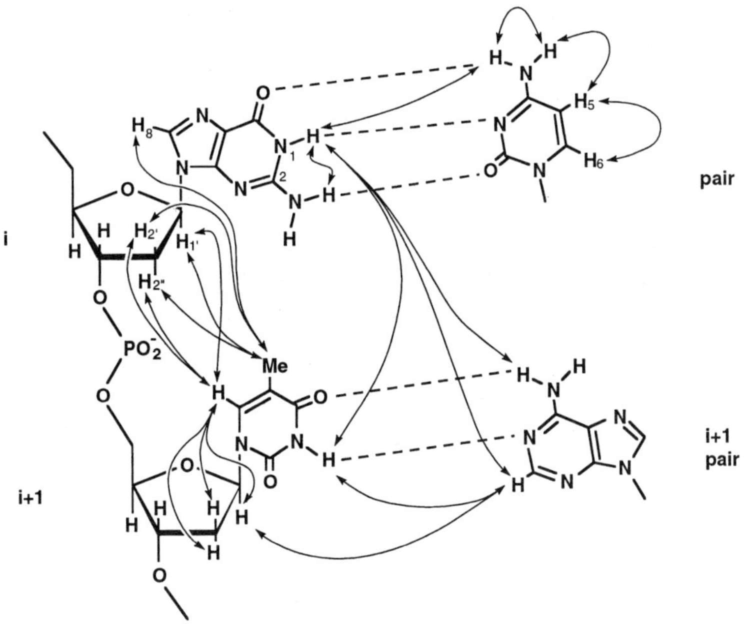

Figure 13.11. Interproton contacts in B-DNA of less than approximately 4 Å, involving base, 1′, 2′, and 2″ protons.

6 Å wide and 8 Å deep, and is formed only by backbone atoms that are independent of the local sequence.

A-DNA is also a regular right-handed double helix, and many of the short distances are similar. It differs from B-DNA mainly in that the centers of the base pairs are displaced away from the axis of the double helix (where they are in the B-form), resulting in the major groove becoming deeper and narrower

TABLE 13.2 Short Interproton Distances in Regular B-DNA[a]

		Residue i		Residue $i+1$			Pair		$i+1$ Pair		
		H2″	H6/H8	H6/H8	H5	Me5	Im	Am	Im	Am	AH2
i	H1′	2.6	3.8	3.1		3.8					
	H2′	1.75	2.1	3.9	3.2	2.8					
	H2″	–	3.5	2.3	2.8	2.4					
	H6/H8		–			3.2					
	Im						3.8	2.9	3.8	3.8	3.8

[a] Im, imino; Am, amino; AH2, adenosine H2. Covalently fixed distances are not included in this table or in Tables 13.3 and 13.4 (except H2′–H2″); they are H2′–H2″ 1.75Å, H5–H6 2.4Å, H5–Am 1.7Å, Im–Am 2.2Å. In this table and in Tables 13.3 and 13.4, distances longer than 4.0Å are omitted.

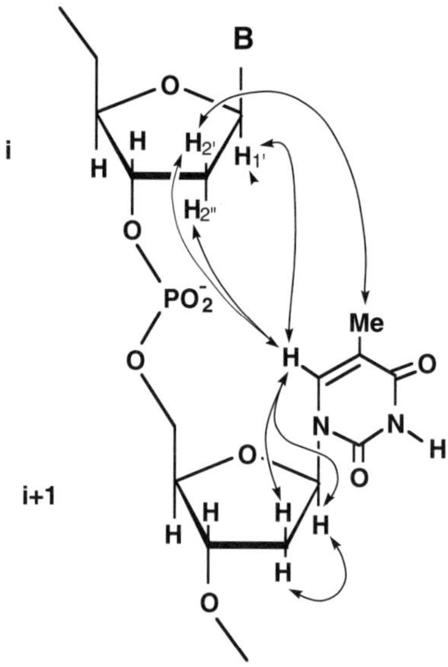

Figure 13.12. Interproton contacts in A-DNA of less than approximately 4 Å, involving base, 1′, 2′, and 2″ protons. Note that although almost the same contacts are marked here as in B-DNA (Fig. 13.11), the actual distances involved are very different (cf. Tables 13.2 and 13.3). Distances involving exchangeable protons are omitted.

while the minor groove is wider and shallower. Details of internuclear distances in A-form DNA are given in Figure 13.12 and Table 13.3. Z-DNA differs from A- and B-form DNA in two ways: (i) the repeat unit is a dinucleotide and (ii) the helix is left-handed. In A- and B-DNA, all bases are *anti*. In A-DNA the sugar is ^3E, while in B-DNA it is approximately ^2E. In the Z-DNA fragment d(GC)$_n$, the cytidines are all *anti* and ^2E as in B-DNA, while the guanosines are *syn* and ^3E. Short distances are shown in Figure 13.13 and Table 13.4. As is clear from the figure, there are fewer very short distances in Z-DNA than there are in A- or B-DNA.

TABLE 13.3 Short Interproton Distances in Regular A-DNA[a]

		Residue i		Residue $i + 1$		
		H2″	H6/H8	H6/H8	H5	Me5
i	H1′	2.3	3.8	4.0		
	H2′	1.75	3.8	1.6	3.0	2.8
	H2″	–		3.2		
	H6/H8		–		3.7	3.1

[a] Distances involving exchangeable protons are omitted, because of the paucity of data reported for A-DNA.

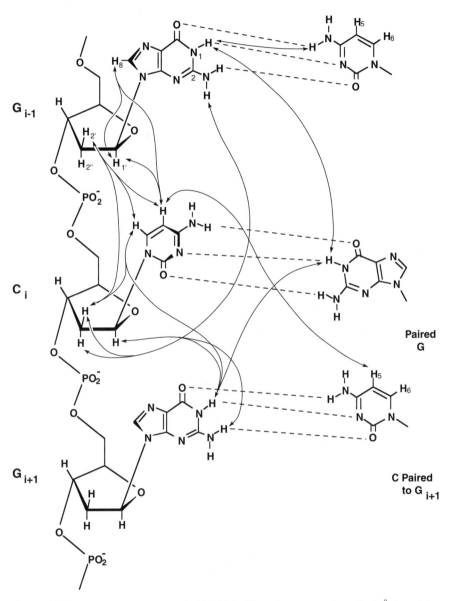

Figure 13.13. Interproton contacts in Z-DNA of less than approximately 4 Å, involving base, 1′, 2′, and 2″ protons.

TABLE 13.4 Short Interproton Distances in Regular Z-DNA

		Paired G			G_{i+1}		C paired to G_{i+1}	G_{i-1}				
		C_i										
		H6	Im	H8	Am	Im	H5	H1'	H2'	H8	Am	Im
C_i	H1'				3.2	3.6						
	H2'	3.1							4.0		2.0	
	H2''											2.9
	H5						4.5	4.8	2.8	4.9		
	H6	–				3.8			3.0			
	Am		2.8									
Paired G	Im	–										3.6
	H1'			2.2								

RNA exists in double helices in several globular structures, most significantly tRNA and ribosomal RNA. It is also capable of adopting a wide range of unusual structures whose structural rules are only now gradually being uncovered, including base triples and base stacks. In some cases, such as the RNA aptamers that bind to small ligands, the backbone follows an S-bend with two 180° reversals.

13.2.2. Assignment

Methods for producing labeled DNA have only recently become available,[41] so that assignment of DNA to date has generally followed the "classical" sequential NOE-based methods.[40] These methods generally work well, mainly because almost all DNA studied is B-form, so the sequential connectivities can be predicted well. By contrast, RNA can be made labeled with ^{13}C and ^{15}N isotopes, which allows more rigorous coupling-based assignment procedures to be used. This is particularly useful because RNA can adopt some very strange conformations, in which many NOE connectivities are nonsequential and (in the absence of a structure) unpredictable. The labeling procedure is more involved than the simple growth in labeled minimal media that is used for proteins, and hence has taken longer to develop and be adopted. It requires growth of bacterial cells on minimal medium, purification and digestion of the cellular RNA, enzymatic conversion of the 5' nucleotide monophosphates to triphosphates, and then polymerization, often done using *in vitro* translation from a DNA template.

13.2.2.1. Duplex DNA. Sequential assignment is much easier once it has been established whether the DNA is present as A-, B-, or Z-DNA. This is straightforward. The A- and B-forms of DNA can be distinguished by the NOE enhancements from purine H8/pyrimidine H6 to H2' and H2''.[38,39] In B-DNA, H8/H6 is very close to the intraresidue H2' and to the preceding H2'' (Table

13.2), whereas in A-DNA, H8/H6 is over 3.7 Å from both the H2′ and the H2″ on the same residue, but very close to the H2′ on the preceding residue (Table 13.3).[42] Z-DNA is very different, particularly because of the *syn* guanosines, which means that the intraresidue distance between GH8 and GH1′ is 2.2 Å, in contrast to 3.8 Å in right-handed DNA, and thus very strong enhancements are seen between these protons in Z-DNA.[43] On the other hand, the intraresidue GH8-GH2′ and GH8-GH2″ distances in Z-DNA are both around 4 Å, whereas in B-DNA the GH8-GH2′ distance is only 2.1 Å. The sequential connectivity between GH8 and the H2′, H2″, and H1′ of the preceding residue is also not present in Z-DNA.

Sequential assignment of duplex DNA relies on the short interproton distances shown in Figures 13.11 to 13.13.[13,26] and methods have been reviewed extensively.[44–48] In the past, observation of exchangeable protons was difficult and therefore many studies were conducted in D_2O and only used nonexchangeable protons. With the advent particularly of pulsed-field gradient methods for solvent suppression, it is now much simpler to observe spectra in H_2O. This has the big advantage that the signals from the exchangeable protons (particularly the base-paired imino protons) become simultaneously accessible; these come in an uncrowded region of the spectrum, are diagnostic of base pairing, and give easily observable sequential NOE enhancements. Thus, NOE enhancements across the base pairing can be observed, and the sequential NOE enhancements imino–imino, imino–AH2, imino–cytidine H-bonded amino, and AH2–AH2 can all be observed (Fig. 13.11).

These assignments are then confirmed and extended using nonexchangeable protons. Within each nucleotide, the H1′, H2′, and H2″ can be identified from COSY or TOCSY, as can CH5 and CH6, and usually TH6 and TMe5. The characteristic sequential NOE enhancements of B-DNA can then be used to make assignments, as indicated in Figure 13.14 for the d(CCGCTCA) strand of a non-self-complementary dimer d(TGAGCGG).d(CCGCTCA).[40] All intranucleotide H1′–H2′/H2″ NOESY cross peaks were observed. Each H8/H6 has connectivities to the H2′/H2″ on the same residue and on the preceding residue,

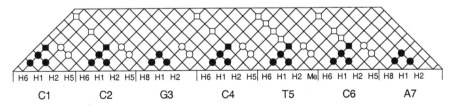

Figure 13.14. Summary of NOE contacts used for sequential assignment of the d(CCGCTCA) strand. Closed symbols denote intranucleotide contacts and open symbols denote internucleotide contacts. Cross-peaks indicated with squares involve the CH5 or TMe5 resonances of pyrimidine-containing nucleotides, and those with circles are found in both purine- and pyrimidine-containing nucleotides. Reproduced with permission from ref. 40.

so that there are four cross peaks for each H8/H6, and two to the 7–8.5 ppm region for each H2'/H2". As expected for B-DNA, the interresidue H8/H6–H2' cross peak is weaker than the corresponding cross peak to H2", and the intraresidue H8/H6–H2' cross peak is stronger than the corresponding cross peak to H2". This permits a distinction between H2' and H2", providing that short mixing times are used to avoid spin diffusion.

A further line of sequential assignments comes from H8/H6–H1' enhancements, which can be followed sequentially in the same way. However, because the intraresidue H8/H6–H1' distance is quite large, these cross peaks may often arise from spin diffusion rather than direct enhancements, and so may be regarded as not constituting an independent assignment route.

Sequential assignment of A-DNA and A-RNA proceeds in a similar manner, and uses the same NOE enhancements, although their relative intensities are different.[38,42] For Z-DNA, a sequential strategy has been proposed that relies on B-DNA-type NOE connectivities in the dG–dC step, but in the dC–dG step uses enhancements between H8 in residue i and H4', H5', and H5" on residue $i - 1$.[49]

Hairpin DNA structures can be assigned using similar methods. For a "simple" hairpin, the duplex region is assigned first and the remaining nucleotides can usually be assigned straightforwardly using sequential NOE enhancements.

If the DNA is not in a standard duplex or hairpin conformation, assignment is much more difficult, because the NOE enhancements are not easily predicted. It is noteworthy that the intranucleotide connection between base and deoxyribose must be made using NOE enhancements, and even this is difficult in nonstandard geometries, such as those found in quadruplexes. It is likely that methods based on isotopic labeling will have an impact on this area in the future.

13.2.2.2. RNA.
Duplex A-RNA can be assigned exactly as described for duplex DNA. However, very often RNA adopts other structures for which NOE enhancements are less predictable. If the RNA can be obtained doubly labeled, many more methods are available for assignment.[50]

For large RNAs, it is often necessary to compare unlabeled RNA, uniformly labeled RNA, and other more "specific" labeling patterns, such as base-specific uniform labeling, and single or double base mutations. Unlabeled samples are used to optimize conditions and can also be used to obtain assignments from the regions of regular secondary structure, as described for DNA. The next step is to assign signals to residue types, which can be done using base-specific labeling and also by heteronuclear correlation experiments. There is a growing number of experiments that are "tuned" to the specific structures of the different bases, and that can efficiently identify the signals from different bases. With very high molecular weight samples, rapid relaxation, particularly of transverse ^{13}C magnetization, reduces sensitivity substantially, and it may be necessary to deuterate samples to reduce ^{13}C relaxation rates. Although, in principle, each nucleotide can be assigned using through-bond correlations in labeled samples, in practice, rapid relaxation and small coupling constants may mean that the only practicable way of obtaining base-specific assignments is by using base-specific labeling. Similarly, it is sometimes possible to use through-bond (^{1}H,^{13}C,^{31}P) experiments to connect residues together, but often the small coupling constants and limited dispersion of ^{31}P mean that this approach is not successful. It is then necessary to resort to sequential NOE enhancements. As discussed above in Section 13.2.2.1, these can be hard to interpret if the structure is bent back on itself, because nonsequential NOE interactions can be observed. In such cases, sequential enhancements may be identifiable using a combination of base-specific labeling and heteronuclear filtered experiments. Alternatively, it may be necessary to use site-specific mutants.

13.2.2.3. Other Nucleotides. Most other nucleotides can be assigned using combinations of the above techniques. For example, a DNA–RNA triple helix was studied, which comprised a roughly standard Watson–Crick DNA hairpin, to which an RNA strand was hydrogen-bonded by Hoogsteen pairing.[51] Thus, only sequential NOE enhancements were observable, and the assignment is relatively straightforward once one knows which signals originate from the DNA and which from the RNA. This is most simply done using labeled RNA and unlabeled DNA. Using a variety of ^{13}C- or ^{15}N-filtered NOESY experiments, it was possible to identify enhancements as being DNA–DNA, DNA–RNA, or RNA–RNA. This gave the assignment and also the hydrogen-bonding pattern, from which the tertiary structure follows.

Even molecules as large as tRNA can be at least partially assigned. For tRNA, the overall structure and hydrogen-bonding pattern can be assumed, and hence one can make numerous assignments from observation of the imino protons. The base pair can often be identified from the characteristic cross-strand enhancements, as shown in Figure 13.15. Sequential imino–imino enhancements (seen at low intensity in Fig. 13.15) can then be used to make sequential assignments, which allows most of the base pairs in tRNA to be assigned.

13.2.3. Structure Calculation

13.2.3.1. Sequence-Dependent Conformation in Duplexes. High-resolution crystal structures of oligonucleotides have shown sequence-dependent conformational variation. For example, the structure of d(CGCGAATTCGCG)$_2$ exhibits propellor-twisting (Fig. 13.16),[53] the direction of which depends on the base pairs involved. These variations were rationalized by Calladine,[54] who showed that they maximized base stacking while keeping steric clashes to a minimum. Other X-ray studies of alternating purine-pyrimidine sequences have shown a "wrinkled" B-DNA, with a dinucleotide repeat unit.[55] These findings have prompted a number of detailed studies of oligonucleotides, which make extensive use of the NOE.

There are a number of problems to be tackled in attempting to derive detailed structures of oligonucleotides from NOE data. Because the double helix is

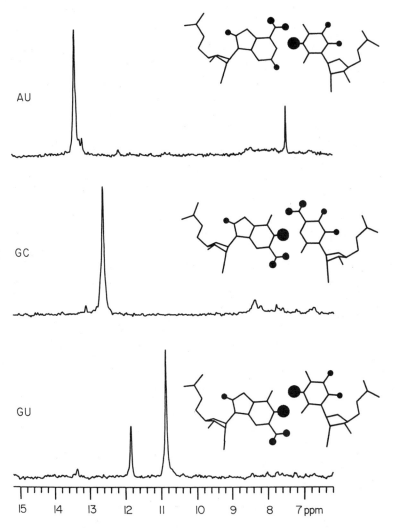

Figure 13.15. Characteristic intra-base NOE patterns of AU, GC, and GU pairs in tRNA. (*Top*) Saturation of AU imino proton gives an NOE enhancement to adenine AH2 as well as broad amino protons in the 6–9 ppm region. (*Middle*) Saturation of GC imino proton gives enhancements only to amino protons. (*Bottom*) Saturation of G imino proton in a GU wobble pair gives a strong enhancement to the U imino proton in the 10–15 ppm region. Imino protons are shown as large black circles, amino protons as medium-sized black circles, and aromatic protons as small black circles. Reproduced with permission from ref. 52.

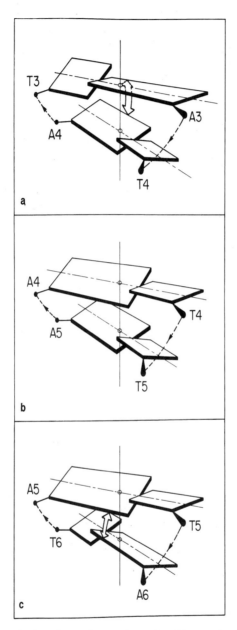

Figure 13.16. Schematic drawings of propellor-twisted base pairs, based on the crystal structure of d(CGCGAATTCGCG)$_2$. The propellor twists introduce a clash of purine rings on adjacent strands in the major groove in (*a*) and in the minor groove in (*c*), shown by curved double-headed arrows. Panel (b) shows a purine–purine step in which no clash occurs. Reproduced with permission from ref. 64.

essentially linear, distances derived from NOE enhancements must be fairly accurate to be of any value. Most proteins have much more globular folded structures, which is the reason why more qualitative NOE data can be used successfully to define their structures.

There are four major difficulties in obtaining useful distances for oligonucleotides. First, motional averaging means that average internuclear distances are not simply related to NOE intensities, even when the latter are known accurately. Second, different parts of the molecule may experience different correlation times, implying that the calibration between NOE enhancement intensity and distance may not be constant throughout the molecule. The third problem, much more obvious for oligonucleotide helices than for globular proteins, is that tumbling may not be isotropic, and so individual cross-relaxation rates may depend on the orientation of the internuclear vector relative to the helix axis. And fourth is the problem of spin diffusion, which is particularly severe for nucleotides because the distribution of protons in space is far from uniform; there are local "pockets" in which spin diffusion is rapid (such as the H1'/H2'/H2" cluster), with relatively long distances to other protons, in particular AH2 and CH5.

The problem of motional averaging is of course common to any structural study, and, as in most other cases, is normally largely ignored. It is however notable that interproton distances measured by NOE in oligonucleotides, particularly interresidue distances, tend to come out shorter than one would expect from static models, which could be caused by motional averaging.[56] There have been several observations of different correlation times in different parts of DNA helices, but there is no general agreement on how common this may be. The safest solution in the absence of better data is to apply generous distance ranges to each restraint.[57] As regards anisotropic tumbling, a B-DNA octamer has length 27 Å and diameter 26 Å, which means that isotropic spectral densities will produce distance errors of less than 10% for nucleotides of this length. However, a 20-mer has an axial ratio of 2.8, with calculated correlation times of 10.4 ns for end-to-end rotation and 4.2 ns for rotation around the helix axis. This can give distance errors of at least 10%.[58,59]

The question of spin diffusion in oligonucleotides is the one that has received most attention. A number of groups have come up with programs that calculate NOE enhancements from a structure using relaxation matrix calculations, then compare the results to the observed NOE enhancements, and in various ways iterate to a refined structure. Such approaches are clearly an improvement over the simplistic isolated spin pair approach (ISPA: Section 4.4.2), but have not won wide acceptance. There are several reasons for this, including the increased complexity of the method, and a general feeling that NOE enhancement data are too sparse and too imprecise (largely because of motional averaging) to justify sophisticated treatment.[60] Nevertheless, it is important to allow for spin diffusion. Van de Ven and Hilbers[61] have shown that spin diffusion in nucleotides can only safely be ignored if one uses mixing times so short that NOE enhancements over distances greater than 3 Å are not measurable. But, of course, many of the distances necessary for sequential assignment and structure calculation are longer than this. Hence, ignoring spin diffusion will necessarily result in errors. For example the only NOE enhancements that show significant variation with sugar pucker are (H2', H4') and (H2", H4'), and both of these are strongly affected by spin diffusion. Sugar pucker is in fact much better determined using J-values.

Despite these problems, it is undeniable that strong evidence has been obtained by a number of groups that the DNA helix is not regular in solution, and moreover adopts conformations roughly like those found in crystals and predicted by Calladine's principle (e.g., reference 62). The evidence for propelloring is particularly strong. Propelloring in an ATAT sequence causes the interstrand AH2–AH2 distance to be short in TA sequences and long in AT sequences (Fig. 13.16).[63] There is also considerable evidence for "wrinkling" of B-DNA in alternating purine-pyrimidine sequences. This is most apparent in the glycosidic torsion angle χ, which has been reported to be less negative for purines ($-90°$) than for pyrimidines ($-120°$), based on H8/H6–H2' intraresidue NOE enhancements.[65]

A large number of double-helical sequences have been studied by using NMR and are reviewed in reference 61. This includes references to structures with base-pair mismatches, bulges, hairpins, and chemical modifications, and should be required reading for anyone planning structure calculations on oligonucleotides. A great deal of information about oligonucleotide structure comes from measurement of J-couplings (ref. 66).

13.2.3.2. Nonhelical Conformations.

As nucleotide structures get less linear and more globular, it becomes possible to calculate more accurate structures.[67] Perhaps the most interesting oligonucleotide structures currently are those of RNA molecules that have been selected to bind selectively to small ligands, and are known as aptamers. They can adopt a wide range of structures, and in common with many RNAs, have different structures in the free and bound states. As an example, the structure of an ATP-binding aptamer bound to AMP has been determined, using an unlabeled RNA–AMP sample, a (^{13}C, ^{15}N)-labeled RNA complexed to unlabeled AMP, and an unlabeled RNA complexed to (^{13}C, ^{15}N)-labeled AMP.[68] The investigators observed 441 NOE enhancements within the RNA and 45 between RNA and AMP. The structure forms two A-type stems at an angle of about 106°, while the central region where the AMP is bound contains base-mismatch pairing (including an AG mismatch to the bound AMP) and stacking interactions.

There are currently many new types of nucleotide structure appearing, including triplexes and quadruplexes.[69] Triplexes involving a Watson–Crick DNA strand with a Hoogsteen RNA or DNA strand have possible roles in therapeutics and in chromosome mapping. Triplexes generally require the nucleotides to be in a roughly standard B-type conformation, and the hydrogen-bonding patterns can often be revealed straightforwardly by characterization of interbase NOE enhancements (Fig. 13.17). Quadruplexes are thought to occur in telomeres and

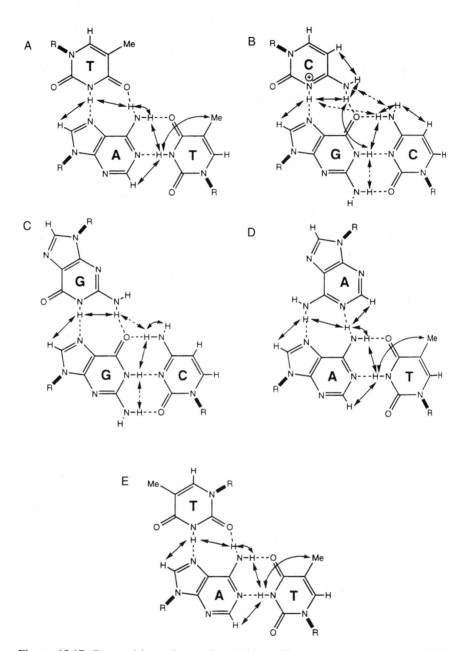

Figure 13.17. Base pairing schemes for triplexes. The arrows indicate typical NOE enhancements observable in triplexes; dashed arrows indicate enhancements that are not always observable because of exchange. Reproduced with permission from ref. 69.

possibly also in immunoglobulin switch regions, and are more complicated than triplexes because a strand reversal is required, implying that some bases must be *syn*. This creates problems with assignment, discussed in reference 69. Structure calculation, at least at an overall level, is based largely on the hydrogen-bonding pattern determined from NOE enhancements and is relatively straightforward once an assignment has been made.

13.3. OLIGOSACCHARIDES

The structure of oligosaccharides (and also of glycoproteins) is of great interest because of the role they play in cellular recognition, in antigenicity, and in formation of cellular structural components. However, oligosaccharides present peculiar problems in isolation, purification, and characterization, and their study has therefore lagged somewhat behind that of other biopolymers. This is true in NMR as much as anywhere else, especially as saccharides give rise to several particularly troublesome spectroscopic problems. The most obvious of these is spectral overlap, which is severe even at very high field. The problem is less severe if ^{13}C can be used to spread the spectrum into a second dimension, but isotope labeling of oligosaccharides is still difficult to achieve and many such studies are forced to use natural abundance ^{13}C. Assignment of the ^1H spectrum is difficult, particularly since strong coupling effects are often appreciable, making a first-order analysis invalid. Strong coupling also leads to a "spreading" of NOE enhancements between the protons involved, as described in Section 6.4.2 and noted by several authors.[70,71] Another major problem is the flexibility of saccharides at the glycosidic linkage, which means that averaging can occur over several conformations. In addition, the reliability of energy calculations is still rather uncertain, and so there is a lack of structural models for saccharides. A final problem is that, as for oligopeptides, $\omega\tau_c$ is close to unity for many small oligosaccharides, making the maximum observable enhancement small, regardless of geometry (Section 3.2.3). This problem can be alleviated by use of the rotating frame experiment (Section 9.3), which has been used widely in studies of oligosaccharides.

The conformations of saccharides are expressed by the angles ϕ, ψ, and ω (Fig. 13.18), where ϕ is the angle O5—C1—O'_n—C'_n (and O'_n is the oxygen

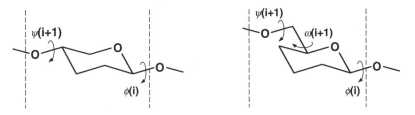

Figure 13.18. Definition of torsion angles in sugars.

of the adjacent saccharide attached to the anomeric carbon), ψ is the angle C1—O$'_n$—C$'_n$—C$'_{n-1}$, and ω, which is needed only for 1 → 6 linkages (in aldohexopyranoses), is the angle O6—C6—C5—C4.[72] Angles are positive when the bond to the front, viewed along the central bond, must be rotated clockwise to eclipse the bond at the rear; thus the angle ω in Figure 13.19 is −60°. The most commonly used conformational energy modelling method for oligosaccharides is known as the hard-sphere *exo*-anomeric approach (HSEA).[73] HSEA calculations generally predict that there will be one strongly preferred conformer about each glycosidic bond (except for linkages to the 6 position), with $\phi = 155 \pm 15°$ and $\psi = 0 \pm 25°$. In this conformation, the protons attached to the anomeric and attached aglyconic carbons are almost in van der Waals contact, in both α and β anomeric configurations. Thus, if oligosaccharides do adopt the conformation predicted by HSEA, then the interresidue enhancements from anomeric protons should indicate not only the sequence but also the linkage positions in the saccharide, with a high degree of certainty. As we shall see, this seems generally to be the case.

13.3.1. Sequence and Linkage Determination

Analysis of the constituent monomers in oligosaccharides is usually a fairly easy task. Analysis of the sequence and of the sites of linkage is much more difficult, because of the common occurrence of branching in saccharides. NMR is one of the most useful techniques for revealing these details.[74]

By far the most easily resolved signal in saccharides is from the anomeric proton (H1′), which is uncomfortably close in frequency to water, and therefore requires care in solvent suppression, or working in D_2O. Assignments start from the anomeric proton and work round the ring using COSY and TOCSY spectra. Oligosaccharides generally have long relaxation times (compared to protein spectra of similar complexity), and therefore multidimensional experiments can be useful; these have included a variety of (^1H,^{13}C) correlation experiments and homonuclear 3D experiments. NOE enhancements are not very useful in the assignment process, mainly because of the large number of interproton contacts possible.

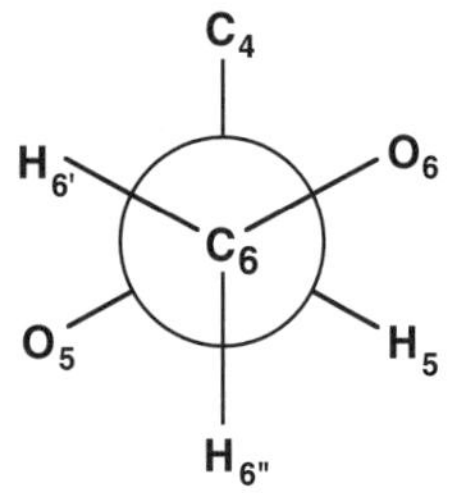

Figure 13.19. Nomenclature of H6′ and H6″. The structure is drawn looking down the C6—C5 bond, and the rest of the saccharide ring links O5 and C4.

Once the spectrum is assigned, the anomeric configuration can be determined most simply using chemical shifts (^{1}H and ^{13}C) and $^1J_{CH}$ values. In many cases, the primary sequence can be determined or guessed by comparison with a database of known structures. Otherwise, the sugar types can often be deduced from chemical shift and coupling patterns, and the linkage between saccharides can be determined using NOE enhancements. In β-linked sugars (with an axial anomeric proton) it is not surprising that strong enhancements can be seen between the anomeric proton and the proton attached to the aglyconic carbon, thus simultaneously providing sequence and linkage information.[75] In some cases enhancements of similar magnitude have been seen to other aglyconic protons as well, particularly in the case of 1 → 6 linkages (Fig. 13.20). Strong sequential enhancements from the equatorial anomeric protons in α-linked sugars are also found,[70] although, here again, enhancements are also seen to protons on neighboring positions. It is therefore wise to determine linkage positions using an alternative method such as gas chromatography–mass spectroscopy (GC-MS) methylation analysis. Linkage positions can also be checked from observation of coupling to OH protons. This requires the use of DMSO as a solvent (which can be messy), although OH protons have been observed by rapidly cooling saccharide solutions to below 0°.[77]

13.3.2. Conformation

Most studies indicate that conformational mobility is fairly limited in 1 → 2, 1 → 3, and 1→ 4-linked sugars, but more extreme in 1 → 6. Thus, in 1 → 2, 1 → 3, and 1 → 4 linkages, HSEA calculations predict a sharp single-energy

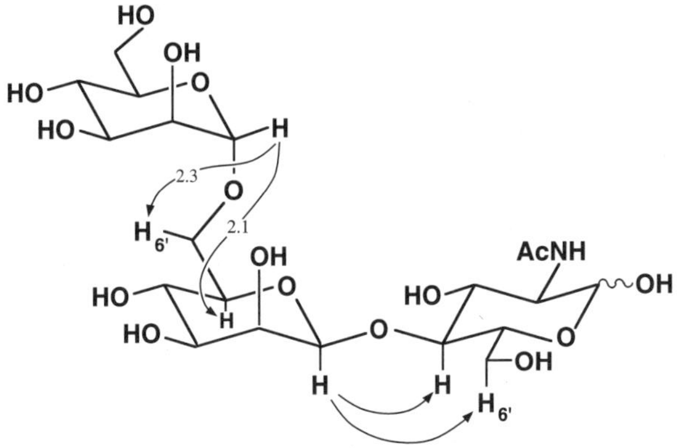

Figure 13.20. NOE enhancements observed in α-D-Man-(1 → 6)-β-D-Man-(1 → 4)-D-GlcNAc.[76] No figures are quoted for enhancements to the GlcNac H4 and H6′ because they are separated by only 0.01 ppm. The total enhancement seen of this combined signal was 6.8%.

minimum, and T_1 and NOE studies are consistent with a single conformation.[78,79] However, for more complicated structures, several minima may be predicted, and the conformation indicated by the NOE may not be the global minimum.[80] Conformational studies have tended to assume a single, fixed conformation, with a single correlation time. It may therefore be that the results from these studies are rather overprecise in their conclusions. NOE results for 1 → 6 linkages are incompatible with any single conformation, a classic sign of internal motion (Chapter 5).

Despite these potential problems with theoretical modelling, conformational studies of saccharides almost always rely heavily on modelling, because the NOE data are too sparse to define the conformation precisely. There are frequently only two enhancements observed between adjacent saccharides, which is not enough to give a precise structure, and (as in oligonucleotides) contacts between nonsequential units are rare. As an illustration of this, two groups have studied the conformation of the Man $\alpha(1 \rightarrow 3)$Man linkage in biantennary oligosaccharides and reported different conclusions, in spite of a thorough multispin analysis of the enhancements in both cases.[70,81] The tendency has therefore been to use energy calculations to derive structures, and then to test whether the NOE data fit the structures, which they almost always do. Clearly, the way to generate a more experimentally based structure will be to use more experimental restraints. There are currently efforts in this direction, such as the use of ^{1}H{^{13}C} NOE enhancements in ^{13}C,^{2}H doubly labeled saccharides,[82] or the observation of NOE enhancements to OH protons in supercooled solutions.[83]

Many eukaryotic proteins are glycosylated, a structural feature that plays a role in stabilization and recognition of the proteins. There has been a very limited number of studies on such proteins, partly because of the spectroscopic problems associated with oligosaccharides, described above, and partly because of difficulties of isotopic labeling of eukaryotic glycoproteins and purification of homogeneous samples. (Glycoproteins commonly have a range of different glycosylation states.) As an example, a study has been conducted on human CD2, which is a cell-surface glycoprotein that mediates cellular adhesion and signal transduction.[84] The protein was expressed in Chinese hamster ovary cells with (^{13}C,^{15}N)-labeling, and the resultant glycoprotein comprised a mixture of different glycoforms, as revealed by NMR and mass spectroscopy (see Fig. 13.21). From ^{13}C linewidths, it is apparent that the sugars get more mobile the further they are from the attachment site. Approximately 50 NOE enhancements were seen within the glycan, and 24 between protein and glycan, all of which involved the first two sugars. Only 2 of the enhancements within the glycan were nonsequential. These linked the acetyl group of GlcNAc2 and ManA, showing that for a substantial time the arm with the Man4' residue is folded towards the core. The NOE enhancements and linewidths indicate that the glycan core is ordered but that the termini are disordered, and that there are few specific interactions with the protein.

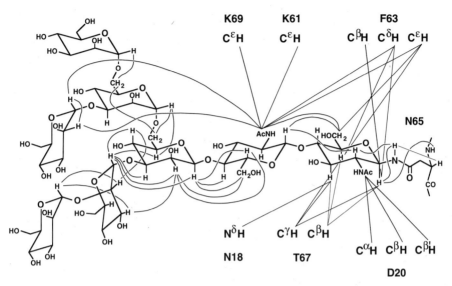

Figure 13.21. Summary of NOE enhancements observed for the high-mannose *N*-glycan of human CD2. For clarity, intrasaccharide enhancements are not shown. Amino acid residues are abbreviated using the single-letter code (D = Asp; F = Phe; K = Lys; N = Asn; T = Thr) and residue numbers. The glycoprotein sample was a mixture of several glycoforms; only those sugars present in all forms are indicated. Redrawn from ref. 84.

13.4. Complexes

Much of the interest in biological structures derives not from the structures themselves but from what they can reveal about function. Since an important part of biological function at the molecular level is the recognition of one molecule by another, this inevitably means that the study of complexes between biomolecules is of crucial importance. Some classes of biopolymer complexes, such as complexes of drugs with DNA, have been studied by NMR for many years. Over the past decade or so, however, techniques have improved to the point that more difficult categories of complexes, such as protein–DNA and protein–RNA complexes, can be successfully tackled by NMR.

The main factor that has allowed structures of complexes to be solved by NMR is the ability to obtain one or both components in isotopically labeled form. Given this ability, one may prepare samples of a complex containing only one of the components in labeled form, and use heteronuclear experiments to observe that labeled component selectively while it is complexed with an unlabeled partner. In principle, this ought to make assignment of the labeled component in the complex a task equally as easy (or difficult) as assigning the corresponding labeled free component, since the same number of resonances exist in each case. In practice, as discussed below, broader linewidths often make the NMR experiments less successful for the complex than they are when applied to the free components. Another crucial use for samples of complex having one labeled and one unlabeled component is in experiments to detect intermolecular NOE interactions specifically. This can be done using either suitable X-filtered sequences (Section 9.2.2.1) or X-separated sequences such as 3D-HSQC-NOESY (Section 9.2.2.2). In cases where one component cannot be labeled (as was until recently the case for DNA), one could also use X-filtered experiments to observe selectively only the unlabeled component, although such selection can be quite hard to achieve efficiently.

To date, the most successful approach to determining NMR structures of complexes of proteins with other biopolymers has been to solve the structures of the free components first (assuming they are both independently folded). This brings significant advantages. Most NMR experiments are likely to give better data for the free components than they do for a complex, because the complex generally has broader resonances. This in turn is largely because the complex has a higher molecular weight, although various types of exchange in the complex can occur at an intermediate rate on the chemical shift timescale (Section 5.1) and so cause further line-broadening. Such exchange processes could include ligand–complex exchange, solvent exchange of individual exchangeable protons, aromatic ring flipping, or other local conformational fluctuations, and sometimes these can even cause signals to disappear altogether from the spectrum of the complex. Data from the free components therefore provide a very important aid when analyzing data from the complex. Rather than facing the complete assignment of the complex from scratch, one can often use the assigned chemical shifts and, more importantly, the *patterns* of cross peaks already analyzed and assigned in the various spectra of the free components to help direct the analysis of the corresponding spectra from the complex. Such comparisons also make it much easier to proceed if data from the complex is incomplete. Still more importantly, comparison of the free and bound data helps show up any areas of genuine change that occur upon complexation.

In practice, there are two possible levels of characterization of complexes using NMR. The ultimate, of course, is a full structure determination, which has now been achieved in quite a number of cases. However, if all one has are *resonance assignments* for the free components and the complex (and structures for the free components), one may simply "map" the chemical shift changes seen upon complexation onto the free component structures to try to locate the interaction surfaces for each component. This approach is most commonly applied to ^{15}N-labeled protein samples, using shift changes in the ^{15}N and ^{1}H signals from amide groups to identify which residues are affected by interaction with some other species. In addition to shift changes, linewidth changes (including assumed linewidth changes observed only in the sense that certain signals disappear from the spectrum of the complex) are often also incorporated into such an analysis. However, it should be remembered that linewidth changes, even more so than chemical shift changes, have many possible origins and may be difficult to interpret. If one wants to look in more detail, (^{13}C,^{1}H) correlation spectra may be used to compare side-chain signals between free and bound states, provided the relevant assignments are available.

Many protein structure papers include such "mapping" experiments that allow a low-resolution footprint of some interaction with a partner molecule to be deduced. In some cases, such an interaction footprint may just represent an intermediate step on the route to a full structure of the complex, while in others it may represent the end of the study, if a complete structure determination is judged to be out of reach. Almost all NMR studies of protein–protein complexes to date fall into this category, which is why they have not been accorded a section in their own right in this book.

13.4.1. Drug–Protein Complexes

Perhaps the best known solution-state NMR studies in this area have been of complexes between the enzyme dihydrofolate reductase (DHFR) and various of its inhibitors (e.g., methotrexate and trimethoprim and many analogs), both in the presence and absence of the cofactor NADPH. These studies shed light on structure–activity relationships for DHFR itself (e.g., when apparently closely related drugs were found to bind to the same site in fundamentally different ways). They can also be viewed as providing a general model for drug–protein interactions, revealing some of the changes likely on complexation, for instance the way in which particular types of conformational exchange may be partially or completely frozen upon complex formation.[85,86] They also demonstrate the utility of isotopic labeling of the ligand, an approach also used in studies of other drug–protein complexes.[87,88]

13.4.2. Drug–DNA Complexes

Drug–DNA complexes have provided many interesting studies.[89] As always with complexes, exchange phenomena need to be carefully accounted for. In self-complementary nucleotides, each pair of equivalent bases gives only one signal. Addition of a ligand will in general make the two equivalent bases nonsymmetrical. This may or may not be apparent from the NMR spectrum, depending on the exchange rate of the ligand. If the ligand can bind in more than one site, still more possibilities arise, and the spectrum can become impossibly complicated. Bound signals may also be in exchange with free. Even if exchange is fast enough to give averaged signals, analysis of the structure is difficult, because the observed NOE enhancements will in effect be averages over all conformations present.

Many drugs bind to DNA in the minor groove, and their binding can be followed by monitoring changes in chemical shifts, particularly of the imino protons. Structures can be derived in the normal way using NOE enhancements. Other drugs intercalate. This can be detected by large chemical shift changes, by the loss of sequential NOE enhancements and the gain of enhancements of the type base–intercalator–sequential base, and usually by the slow exchange between free and bound forms. Some intercalators can force the base pairing into alternative arrangements such as Hoogsteen pairs.[89]

Closely related to this topic is that of RNA–aptamer studies, mentioned in Section 13.2.3.2.

13.4.3. Protein–Nucleic Acid Complexes

In protein–DNA complexes, the structure of DNA can range from undistorted B-DNA through to quite extreme bends, although it is more common to see largely undistorted DNA (indeed, some NMR structures of protein–DNA complexes have simply modelled the DNA as classical B-DNA, because it was judged that the NMR data did not permit detailed structural calculations; e.g., ref. 90). Similarly, the protein structure can be affected to various degrees by addition of DNA. Often, it has been observed that unstructured parts of the binding proteins become ordered when bound. NOE enhancements within the DNA helix provide only weak restraints on the structure; however, the structure of the protein in the contact region can provide strong restraints on the DNA conformation, as they can force it to follow the protein surface. Outside the contact region, the DNA structure is generally poorly defined, as discussed earlier for free DNA. The SRY–DNA complex provides an example of highly bent DNA (Fig. 13.22); the bend angle is 70–80°.[91] This structure was determined using both unlabeled samples and complexes in which the protein was labeled with ^{13}C and the DNA was unlabeled, so allowing the use of ^{13}C-filtered experiments that give a clear distinction between intraprotein, intraDNA and intermolecular NOE enhancements. The DNA has many A-like features, and is severely underwound. There is a partial intercalation of an Ile sidechain into the double helix (easily seen from NOE enhancements), which may assist in the bending.

As just mentioned, in protein–DNA complexes the DNA structure is frequently, though not always, largely unchanged. However, in protein–RNA complexes, the structures of both RNA and protein may undergo extensive structural rearrangements. An example is the complex between the U1A protein and a regulatory element present within the 3′-untranslated region of its own message RNA (U1A protein itself is a component of a complicated nuclear assembly called the spliceosome, involved in splicing together exons while removing introns during gene expression).[92,93] The assignment and structure calculation of this complex required unlabeled complex, plus complexes where one component was unlabeled and the other was uniformly (^{13}C,^{15}N)-labeled, and yielded 1710 intraprotein NOE enhancements, 591 intraRNA and 123 protein–RNA. Unlike protein–DNA structure determinations by NMR to date, in this case it was possible to use a calculation protocol that started from fully randomized conformations for the free components and proceeded to a fully folded complex structure in one step, without requiring any docking procedure. Of the final set of 50 calculations, 31 gave satisfactorily small violations of

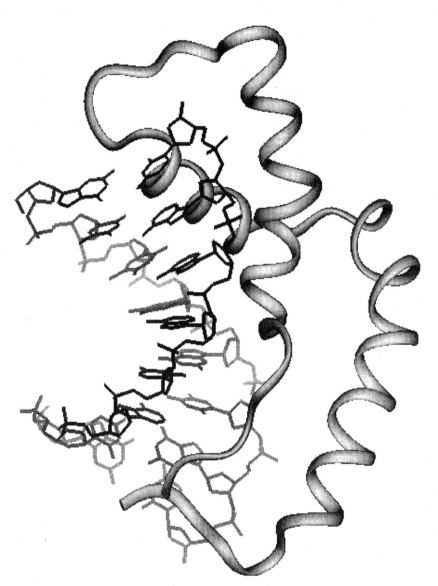

Figure 13.22. The structure of the complex between the human SRY protein (which is shown as a ribbon, and is largely helical) and the double-helical DNA octamer d(GCACAAAC.GTTTGTGC), on the left of the figure. The DNA double helix is very bent, and the DNA structure is mainly determined by the protein component. Drawn using coordinates of the complex submitted to the protein data bank.[91]

the NOE restraints. The protein contains a β-sheet, which forms a surface against which a single-stranded RNA segment folds, involving a range of stacking interactions of the bases with amino acid side-chains. The binding of the RNA also involves the rearrangement of a helix in the protein, which covers the RNA-binding site in the free protein but moves to a new location in the complex. The resulting interface is much more intimate than that seen for protein–DNA complexes, and, in fact, the overall architecture of the complex more closely resembles that of an enlarged protein structure, with the RNA phosphate groups exposed on the hydrophilic surface and the bases buried in the relatively hydrophobic interior.

REFERENCES

1. Wüthrich, K.; Billeter, M.; Braun, W. *J. Mol. Biol.* 1984, *180*, 715.
2. Richardson, J. S. *Adv. Protein Chem.* 1981, *34*, 167.
3. Williamson, M. P.; Marion, D.; Wüthrich, K. *J. Mol. Biol.* 1984, *173*, 341.
4. van de Ven, F. J. M.; de Bruin, S. H.; Hilbers, C. W. *FEBS Lett.* 1984, *169*, 107.
5. Whitehead, B.; Craven, C. J.; Waltho, J. P. in "Protein NMR Techniques" (D. G. Reid, ed.), Vol. 60, Humana Press, Totowa, NJ, 1997.
6. Clore, G. M.; Gronenborn, A. M. *Prog. Nucl. Magn. Reson. Spectrosc.* 1991, *23*, 43.
7. Fischer, M. W. F.; Zeng, L.; Zuiderweg, E. R. P. *J. Am. Chem. Soc.* 1996, *118*, 12457.
8. Billeter, M.; Braun, W.; Wüthrich, K. *J. Mol. Biol.* 1982, *155*, 321.
9. Englander, S. W.; Wand, A. J. *Biochemistry* 1987, *26*, 5953.
10. Berndt, K. D.; Güntert, P.; Orbons, L. P. M.; Wüthrich, K. *J. Mol. Biol.* 1992, *227*, 757.
11. Güntert, P.; Wüthrich, K. *J. Biomolec. NMR* 1991, *1*, 447.
12. Güntert, P.; Braun, W.; Wüthrich, K. *J. Mol. Biol.* 1991, *217*, 517.
13. Billeter, M. *Quart. Rev. Biophys.* 1992, *25*, 325.
14. Hansen, A. P.; Petros, A. M.; Meadows, R. P.; Nettesheim, D. G.; Mazar, A. P.; Olejniczak, E. T.; Xu, R. X.; Pederson, T. M.; Henkin, J.; Fesik, S. W. *Biochemistry* 1994, *33*, 4847.
15. Brünger, A. T. "X-PLOR 3.1 Manual," Yale University Press, Yale University, New Haven, 1992.
16. Smith, B. O.; Ito, Y.; Raine, A.; Teichmann, S.; Bentovim, L.; Nietlispach, D.; Broadhurst, R. W.; Terada, T.; Kelly, M.; Oschkinat, H.; Shibata, T.; Yokoyama, S.; Laue, E. D. *J. Biomolec. NMR* 1996, *8*, 360.
17. Clore, G. M.; Gronenborn, A. M. *Proc. Natl. Acad. Sci. U.S.A.* 1998, *95*, 5891.
18. Pervushin, K.; Riek, R.; Wider, G.; Wüthrich, K. *Proc. Natl. Acad. Sci. U.S.A.* 1997, *94*, 12366.
19. Otting, G.; Liepinsh, E.; Wüthrich, K. *Science* 1991, *254*, 974.
20. Williamson, M. P.; Waltho, J. P. *Chem. Soc. Rev.* 1992, *21*, 227.

21. Kim, Y.; Prestegard, J. H. *Biochemistry* 1989, *28*, 8792.
22. Blackledge, M. J.; Brüschweiler, R.; Griesinger, C.; Schmidt, J. M.; Xu, P.; Ernst, R. R. *Biochemistry* 1993, *32*, 10960.
23. Bonvin, A. M. J. J.; Brünger, A. T. *J. Mol. Biol.* 1995, *250*, 80.
24. Breg, J. N.; van Opheusden, J. H. J.; Burgering, M. J. M.; Boelens, R.; Kaptein, R. *Nature* 1990, *346*, 586.
25. Folmer, R. H. A.; Hilbers, C. W.; Konings, R. N. H.; Hallenga, K. *J. Biomolec. NMR* 1995, *5*, 427.
26. Nilges, M. *Proteins: Structure, Function, and Genetics* 1993, *17*, 297.
27. Clore, G. M.; Appella, E.; Yamada, M.; Matsushima, K.; Gronenborn, A. M. *Biochemistry* 1990, *29*, 1689.
28. Clore, G. M.; Omichinski, J. G.; Sakaguchi, K.; Zambrano, N.; Sakamoto, H.; Appella, E.; Gronenborn, A. M. *Science* 1994, *265*, 386.
29. Raleigh, D. P.; Levitt, M. H.; Griffin, R. G. *Chem. Phys. Lett.* 1988, *146*, 71.
30. Colombo, M. G.; Meier, B. H.; Ernst, R. R. *Chem. Phys. Lett.* 1988, *146*, 189.
31. Gullion, T.; Schaefer, J. *J. Magn. Reson.* 1991, *92*, 439.
32. Griffin, R. G. *Nature Struct. Biol.* 1998, *5*, 508.
33. Bennett, A. E.; Griffin, R. G.; Vega, S. in "NMR Basic Principles and Progress, Solid-State NMR IV," Vol. 33, Springer Verlag, Berlin, Heidelberg, 1994.
34. Fujiwara, T.; Sugase, K.; Kainosho, M.; Ono, A.; Ono, A.; Akutsu, H. *J. Am. Chem. Soc.* 1995, *117*, 11351.
35. Kümmerlen, J.; van Beek, J. D.; Vollrath, F.; Meier, B. H. *Macromolecules* 1996, *29*, 2920.
36. Wijmenga, S. S.; van Buuren, B. N. M. *Prog. Nucl. Magn. Reson. Spectrosc.* 1998, *32*, 287.
37. IUPAC-IUB Joint Commission on Biochemical Nomenclature *Eur. J. Biochem.* 1983, *131*, 9.
38. Haasnoot, C. A. G.; Westerink, H. P.; van der Marel, G. A.; van Boom, J. H. *J. Biomolec. Struct. Dynamics* 1983, *1*, 131.
39. Reid, D. G.; Salisbury, S. A.; Bellard, S.; Shakked, Z.; Williams, D. H. *Biochemistry* 1983, *22*, 2019.
40. Scheek, R. M.; Boelens, R.; Russo, N.; van Boom, J. H.; Kaptein, R. *Biochemistry* 1984, *23*, 1371.
41. Zimmer, D. P.; Crothers, D. M. *Proc. Natl. Acad. Sci. U.S.A.* 1995, *92*, 3091.
42. Uesugi, S.; Ohkubo, M.; Ohtsuka, E.; Ikehara, M.; Kobayashi, Y.; Kyogoku, Y.; Westerink, H. P.; van der Marel, G. A.; van Boom, J. H.; Haasnoot, C. A. G. *J. Biol. Chem.* 1984, *259*, 1390.
43. Feigon, J.; Wang, A. H.-J.; van der Marel, G. A.; van Boom, J. H.; Rich, A. *Science* 1985, *230*, 82.
44. Wüthrich, K. "NMR of Proteins and Nucleic Acids," John Wiley & Sons, New York, 1986.
45. Feigon, J.; Sklenar, V.; Wang, E.; Gilbert, D. E.; Macaya, R. F.; Schultze, P. *Methods Enzymol.* 1992, *211*, 235.
46. Reid, B. R. *Quart. Rev. Biophys.* 1987, *20*, 1.

47. Wemmer, D. E.; Reid, B. R. *Annu. Rev. Phys. Chem.* 1987, *36*, 105.
48. Patel, D. J.; Shapiro, L.; Hare, D. *Quart. Rev. Biophys.* 1987, *20*, 35.
49. Orbons, L. P. M.; van der Marel, G. A.; van Boom, J. H.; Altona, C. *Eur. J. Biochem.* 1986, *160*, 131.
50. Dieckmann, T.; Feigon, J. *J. Biomolec. NMR* 1997, *9*, 259.
51. van Dongen, M. J. P.; Heus, H. A.; Wymenga, S. S.; van der Marel, G. A.; van Boom, J. H.; Hilbers, C. W. *Biochemistry* 1996, *35*, 1733.
52. Hare, D. R.; Reid, B. R. *Biochemistry* 1982, *21*, 5129.
53. Dickerson, R. E.; Drew, H. R. *J. Mol. Biol.* 1981, *149*, 761.
54. Calladine, C. R. *J. Mol. Biol.* 1982, *161*, 343.
55. Arnott, S.; Chandresekaran, R.; Puigjaner, L. C.; Walker, J. K.; Hall, I. H.; Birdsall, D. L.; Ratliff, R. L. *Nucleic Acids Res.* 1983, *11*, 1457.
56. Lane, A. N. *Prog. Nucl. Magn. Reson. Spectrosc.* 1993, *25*, 481.
57. Mirau, P. A.; Behling, R. W.; Kearns, D. R. *Biochemistry* 1985, *24*, 6200.
58. Lane, A. N. *Biochim. Biophys. Acta* 1990, *1049*, 189.
59. Birchall, A. J.; Lane, A. N. *Eur. Biophys. J.* 1990, *19*, 73.
60. Pardi, A.; Hare, D. R.; Wang, C. *Proc. Natl. Acad. Sci. U.S.A.* 1988, *85*, 8785.
61. van de Ven, F. J. M.; Hilbers, C. W. *Eur. J. Biochem.* 1988, *178*, 1.
62. Lam, S. L.; Au-Yeung, S. C. F. *J. Mol. Biol.* 1997, *266*, 745.
63. Patel, D. J.; Kozlowski, S. A.; Bhatt, R. *Proc. Natl. Acad. Sci. U.S.A.* 1983, *80*, 3908.
64. Patel, D. J.; Kozlowski, S. A.; Nordheim, A.; Rich, A. *Proc. Natl. Acad. Sci. U.S.A.* 1982, *79*, 1413.
65. Broido, M. S.; James, T. L.; Zon, G.; Keepers, J. W. *Eur. J. Biochem.* 1985, *150*, 117.
66. Wijmenga, S. S.; Mooren, M. M. W.; Hilbers, C. W. in "NMR of Macromolecules: A Practical Approach" (G. C. K. Roberts, ed.), IRL Press, Oxford, 1993, p. 217.
67. Allain, F. H.-T.; Varani, G. *J. Mol. Biol.* 1997, *267*, 338.
68. Jiang, F.; Kumar, R. A.; Jones, R. A.; Patel, D. J. *Nature* 1996, *382*, 183.
69. Feigon, J.; Koshlap, K. M.; Smith, F. W. *Methods Enzymol.* 1995, *261*, 225.
70. Brisson, J.-R.; Carver, J. P. *Biochemistry* 1983, *22*, 1362.
71. Homans, S. W.; Dwek, R. A.; Fernandes, D. L.; Rademacher, T. W. *FEBS Lett.* 1983, *164*, 231.
72. IUPAC-IUB Joint Commission on Biochemical Nomenclature *Eur. J. Biochem.* 1983, *131*, 5.
73. Lemieux, R. U.; Bock, K. *Arch. Biochem. Biophys.* 1983, *221*, 125.
74. Homans, S. W. in "NMR of Macromolecules: A Practical Approach" (G. C. K. Roberts, ed.), IRL Press, Oxford, 1993, p. 289.
75. Koerner, T. A. W., Jr.; Prestegard, J. H.; Demou, P. C.; Yu, R. K. *Biochemistry* 1983, *22*, 2687.
76. Paulsen, H.; Peters, T.; Sinnwell, V.; Lebuhn, R.; Meyer, B. *Leibigs Ann. Chem.* 1984, 951.
77. Poppe, L.; van Halbeek, H. *Nature Struct. Biol.* 1994, *1*, 215.
78. Brisson, J.-R.; Carver, J. P. *Biochemistry* 1983, *22*, 3671.

79. Brisson, J.-R.; Carver, J. P. *Biochemistry* 1983, *22*, 3680.
80. Bush, C. A.; Yan, Z.-Y.; Rao, B. N. N. *J. Am. Chem. Soc.* 1986, *108*, 6168.
81. Homans, S. W.; Dwek, R. A.; Fernandes, D. L.; Rademacher, T. W. *FEBS Lett.* 1982, *150*, 503.
82. Kiddle, G. R.; Harris, R.; Homans, S. W. *J. Biomolec. NMR* 1998, *11*, 289.
83. Poppe, L.; van Halbeek, H. *J. Am. Chem. Soc.* 1991, *113*, 363.
84. Wyss, D. F.; Choi, J. S.; Li, J.; Knoppers, M. H.; Willis, K. J.; Arulanandam, A. R. N.; Smolyar, A.; Reinherz, E. L.; Wagner, G. *Science* 1995, *269*, 1273.
85. Birdsall, B.; Feeney, J.; Tendler, S. J. B.; Hammond, S. J.; Roberts, G. C. K. *Biochemistry* 1989, *28*, 2297.
86. Gargaro, A. R.; Soteriou, A.; Frenkiel, T. A.; Bauer, C. J.; Birdsall, B.; Polshakov, V. I.; Barsukov, I. L.; Roberts, G. C. K.; Feeney, J. *J. Mol. Biol.* 1998, *277*, 119.
87. Fesik, S. W.; Luly, J. R.; Erikson, J. W.; Abad-Zapatero, C. *Biochemistry* 1988, *27*, 8297.
88. Fesik, S. W.; Zuiderweg, E. R. P. *J. Am. Chem. Soc.* 1989, *111*, 5013.
89. Searle, M. S. *Prog. Nucl. Magn. Reson. Spectrosc.* 1993, *25*, 403.
90. Otting, G.; Qian, Y. Q.; Billeter, M.; Müller, M.; Affolter, M.; Gehring, W. J.; Wüthrich, K. *EMBO Journal* 1990, *9*, 3085.
91. Werner, M. H.; Huth, J. R.; Gronenborn, A. M.; Clore, G. M. *Cell* 1995, *81*, 705.
92. Allain, F. H.-T.; Gubser, C. C.; Howe, P. W. A.; Nagai, K.; Neuhaus, D.; Varani, G. *Nature* 1996, *380*, 646.
93. Howe, P. W. A.; Allain, F. H.-T.; Varani, G.; Neuhaus, D. *J. Biomolec. NMR* 1998, *11*, 59.

APPENDIX I

EQUATIONS FOR ENHANCEMENTS INVOLVING GROUPS OF EQUIVALENT SPINS

The equation of motion for I_z, where I is a resonance due to a group of N_I equivalent spins, is (Eq. 3.37)

$$\frac{dI_z}{dt} = -(I_z - I_z^0)[R_I^{DD} + (N_I - 1)\sigma_{II}] - N_I(S_z - S_z^0)\sigma_{IS} - N_I \sum_X (X_z - X_z^0)\sigma_{IX} \tag{A.1}$$

We need the following definitions, chosen for consistency with other sections:

$$R_I^{DD} = [2W_{1I} + N_S(W_{2IS} + W_{0IS}) + \sum_X N_X(W_{2IX} + W_{0IX}) + (N_I - 1)(W_{2II} + W_{0II})] \tag{A.2}$$

and

$$\sigma_{II} = W_{2II} - W_{0II} \tag{A.3}$$

where (by analogy with Eq. 3.21)

$$W_{1I} = \frac{3}{20} K_I^2 \frac{\tau_c}{1 + \omega_I^2 \tau_c^2}$$

$$W_{2II} = \frac{3}{5} K_{II}^2 \frac{\tau_c}{1 + 4\omega_I^2 \tau_c^2}$$

$$W_{0II} = \frac{1}{10} K_{II}^2 \tau_c \tag{A.4}$$

and (by analogy with Eqs. 3.22)

$$K_I^2 = \left(\frac{\mu_0}{4\pi}\right)^2 \left[N_S\hbar^2\gamma_I^2\gamma_S^2 r_{IS}^{-6} + \sum_X N_X\hbar^2\gamma_I^2\gamma_X^2 r_{IX}^{-6} + (N_I - 1)\hbar^2\gamma_I^4 r_{II}^{-6}\right]$$

$$K_{II}^2 = \left(\frac{\mu_0}{4\pi}\right)^2 \hbar^2\gamma_I^4 r_{II}^{-6} \quad (A.5)$$

N_S represents the number of equivalent spins contributing to resonance S, and similarly each N_X represents the number of spins in each X resonance.

By analogy with Eq. 3.11 we have

$$I_z^0 = \frac{\gamma_I N_I}{\gamma_S N_S} S_z^0 = \frac{\gamma_I N_I}{\gamma_X N_X} X_z^0 \quad (A.6)$$

and by analogy with Eq. 3.12

$$f_I\{S\} = \frac{I_z - I_z^0}{I_z^0} = \frac{\gamma_S N_S}{\gamma_I N_I} \frac{I_z - I_z^0}{S_z^0}$$

$$f_X\{S\} = \frac{X_z - X_z^0}{X_z^0} = \frac{\gamma_S N_S}{\gamma_X N_X} \frac{X_z - X_z^0}{S_z^0} \quad (A.7)$$

We now follow our earlier course, and use Eq. A.1 to generate an expression for the steady-state enhancement of I on saturating S (cf. derivation of Eq. 3.13). Setting dI_z/dt and S_z to zero, then using Eqs. A.7 and rearranging

$$f_I\{S\} = \frac{N_S\gamma_S}{\gamma_I} \frac{1}{R_I^{DD} + (N_I - 1)\sigma_{II}} \left[\sigma_{IS} - \sum_X \frac{N_X\gamma_X}{N_S\gamma_S} f_X\{S\}\sigma_{IX}\right] \quad (A.8)$$

Substituting back the transition probabilities (cf. Eq. 3.20)

$$f_I\{S\} = \frac{N_S\gamma_S}{\gamma_I} \left[\frac{(W_{2IS} - W_{0IS}) - \sum_X (\gamma_X N_X/\gamma_S N_S) f_X\{S\}(W_{2IX} - W_{0IX})}{2W_{1I} + N_S(W_{2IS} + W_{0IS}) + \sum_X N_X(W_{2IX} + W_{0IX}) + 2(N_I - 1)W_{2II}}\right] \quad (A.9)$$

At the extreme narrowing limit (cf. Eq. 3.23)

$$f_I\{S\} = \frac{N_S\gamma_S}{\gamma_I} \left[\frac{(\tfrac{3}{5} - \tfrac{1}{10})K_{IS}^2 - \sum_X (\gamma_X N_X/\gamma_S N_S) f_X\{S\}(\tfrac{3}{5} - \tfrac{1}{10})K_{IX}^2}{\tfrac{6}{20}K_I^2 + N_S(\tfrac{3}{5} + \tfrac{1}{10})K_{IS}^2 + \sum_X N_X(\tfrac{3}{5} + \tfrac{1}{10})K_{IX}^2 + (N_I - 1)\tfrac{6}{5}K_{II}^2}\right] \quad (A.10)$$

Assuming that rapid motional averaging makes all distances r_{IS} and r_{IX} effectively equivalent, and denoting the average distances so obtained $\langle r_{IS}\rangle$ and $\langle r_{IX}\rangle$, we have (by analogy with Eq. 3.24)

$$f_I\{S\} = \frac{N_S\gamma_S}{2\gamma_I} \left[\frac{\langle r_{IS}^{-6}\rangle - \sum\limits_{X}(\gamma_X N_X/\gamma_S N_S) f_X\{S\}\langle r_{IX}^{-6}\rangle}{N_S\langle r_{IS}^{-6}\rangle + \sum\limits_{X} N_X\langle r_{IX}^{-6}\rangle + \frac{3}{2}(N_I - 1)\langle r_{II}^{-6}\rangle} \right] \quad (A.11)$$

This equation reduces to Eq. 3.39 in the homonuclear case.

APPENDIX II

QUANTUM MECHANICS AND TRANSITION PROBABILITIES

The calculation of transition probabilities brings us into inevitable contact with quantum mechanics. A helpful introduction to quantum mechanics is given in Atkins,[1] and applications to NMR are described in Lynden-Bell and Harris.[2] A rather more detailed, but still very approachable, description of the derivation of transition probabilities can be found in Harris.[3]

In quantum mechanics, the system under study is described by a wavefunction, which is in general a complex function that varies with time. We are interested only in the nuclear spin system, and so our wavefunctions are confined to those needed to describe the spin states of this system. These are, for a one-spin system, the functions $|\alpha\rangle$ and $|\beta\rangle$; for a two-spin system, the four functions $|\alpha\alpha\rangle$, $|\alpha\beta\rangle$, $|\beta\alpha\rangle$, and $|\beta\beta\rangle$, and so on. We use the notation introduced by Dirac, in which a wavefunction is written as a *ket* $|\alpha\rangle$, and its complex conjugate $|\alpha\rangle^*$ as a *bra* $\langle\alpha|$: when a bracket is completed, integration over all variables is implied, so that (for example) $\langle\beta|\alpha\rangle$ is a shorthand notation for $\int \psi_\beta^* \psi_\alpha \, d\tau$, $\langle\beta|\hat{\mathcal{H}}|\alpha\rangle$ for $\int \psi_\beta^* \hat{\mathcal{H}} \psi_\alpha \, d\tau$, etc. Alternatively, the notation $\langle\psi_\beta|\hat{\mathcal{H}}|\psi_\alpha\rangle$ is sometimes used, which has exactly the same meaning as $\langle\beta|\hat{\mathcal{H}}|\alpha\rangle$. The wavefunctions are defined so as to have the convenient properties that they are orthogonal (e.g., $\langle\alpha|\beta\rangle = \langle\beta|\alpha\rangle = 0$) and normalized ($\langle\alpha|\alpha\rangle = \langle\beta|\beta\rangle = 1$). The labels α and β denote the value of the z component of spin angular momentum (m_s); α indicates $m_s = +1/2$ and β indicates $m_s = -1/2$.

The wavefunction contains all the information needed to describe the system. To obtain information about any particular *observable property* of the system (e.g., magnetization, energy, etc.), one must apply the corresponding *operator*, denoted in this Appendix by a hat (^), to the wavefunction, and then complete the integral with the corresponding bra. This calculation gives the *expectation value* of the operator, which is the numerical value expected for the corresponding observable property. For an operator $\hat{A}$, the expectation value is

$$\langle \hat{A} \rangle = \langle \psi_a | \hat{A} | \psi_a \rangle \tag{A.12}$$

Thus, by acting on a wavefunction with the operator $\hat{I}_z$ one obtains the expectation value of the longitudinal spin angular momentum (i.e., z magnetization); by operating on the wavefunction with another operator $\mathcal{H}$ (the Hamiltonian) one obtains the expectation value of the energy; and so on.

The above discussion refers to a single particle or spin system. In contrast, an experimentally measured value of an observable NMR property represents an average over all the spin systems in the sample, so that the relevant quantum mechanical expression becomes the *ensemble average* over the sample as a whole, denoted by a bar over the corresponding quantities:

$$\overline{\langle \hat{A} \rangle} = \overline{\langle \psi_a | \hat{A} | \psi_a \rangle} \tag{A.13}$$

It is very convenient if the wavefunctions are *eigenfunctions* of the operator being used, that is, if they obey the relationship

$$\hat{A} | \psi_a \rangle = a | \psi_a \rangle \tag{A.14}$$

where a is a number, the *eigenvalue*. For instance, the functions $|\alpha\rangle$ and $|\beta\rangle$ are eigenfunctions of $\hat{I}_z$, their corresponding eigenvalues being $+1/2$ and $-1/2$, respectively. Thus, for a system described by the wavefunction $|\alpha\rangle$, the expectation value of the longitudinal spin angular momentum is

$$\begin{aligned}\langle \hat{I}_z \rangle &= \langle \alpha | \hat{I}_z | \alpha \rangle \\ &= \langle \alpha | 1/2 | \alpha \rangle \\ &= (1/2)\langle \alpha | \alpha \rangle \\ &= 1/2 \end{aligned} \tag{A.15}$$

Likewise, $\langle \hat{I}_z \rangle$ for the wavefunction $|\beta\rangle$ is $-1/2$.

Besides the operator $\hat{I}_z$, there are also operators $\hat{I}_x$ and $\hat{I}_y$, whose expectation values are the x and y components of angular momentum. There are, however, two other operators $\hat{I}_+$ and $\hat{I}_-$, defined by

$$\begin{aligned}\hat{I}_+ &= \hat{I}_x + i\hat{I}_y \\ \hat{I}_- &= \hat{I}_x - i\hat{I}_y \end{aligned} \tag{A.16}$$

which are more convenient for our purposes than $\hat{I}_x$ and $\hat{I}_y$. This is because $\hat{I}_+$, known as the raising operator, acts on a wavefunction to raise its z angular momentum by one, while the lowering operator $\hat{I}_-$ lowers it by one; these two operators are therefore very useful in calculating the probabilities of transitions between states of the system. It also turns out (see, for instance, ref. 1) that $\hat{I}_+|\alpha\rangle = \hat{I}_-|\beta\rangle = 0$. To summarize:

$$\hat{I}_z|\alpha\rangle = (1/2)|\alpha\rangle$$

$$\hat{I}_+|\alpha\rangle = 0$$

$$\hat{I}_-|\alpha\rangle = |\beta\rangle$$

$$\hat{I}_z|\beta\rangle = -(1/2)|\beta\rangle$$

$$\hat{I}_+|\beta\rangle = |\alpha\rangle$$

$$\hat{I}_-|\beta\rangle = 0 \tag{A.17}$$

These equations can be extended to two-spin systems, where the operators act independently on the two spins, for example

$$\hat{I}_{1z}\hat{I}_{2+}|\alpha\beta\rangle = \hat{I}_{1z}|\alpha\rangle\hat{I}_{2+}|\beta\rangle$$

$$= (1/2)|\alpha\alpha\rangle \tag{A.18}$$

The other operator that we shall need is the operator that gives the energy of the system, namely the Hamiltonian $\hat{\mathcal{H}}$. This contains several terms, some of which are time-independent, while others (those that depend on molecular orientation) vary with time. This is expressed by separating the Hamiltonian into a static part $\hat{\mathcal{H}}_0$ and a time-dependent part $\hat{\mathcal{H}}_1(t)$:

$$\hat{\mathcal{H}}(t) = \hat{\mathcal{H}}_0 + \hat{\mathcal{H}}_1(t) \tag{A.19}$$

Generally, the static term $\hat{\mathcal{H}}_0$ dominates this expression because it includes the Zeeman term ($-\gamma\hbar B_0\hat{I}_z$; cf. Section 1.1). Our wavefunctions $|\alpha\rangle$ and $|\beta\rangle$ are eigenfunctions of the static Hamiltonian $\hat{\mathcal{H}}_0$; in fact, this is the reason for choosing these functions. Their corresponding eigenvalues are the *energies* of the $|\alpha\rangle$ and $|\beta\rangle$ states, respectively. For convenience, a factor of $1/\hbar$ is often included in the definition of $\hat{\mathcal{H}}$, so that these energies appear directly with units of angular frequency and are written ω_α and ω_β:

$$\hat{\mathcal{H}}_0|\alpha\rangle = \omega_\alpha|\alpha\rangle$$

and

$$\hat{\mathcal{H}}_0|\beta\rangle = \omega_\beta|\beta\rangle \tag{A.20}$$

Of interest here is the *dipolar Hamiltonian* $\hat{\mathcal{H}}^{DD}(t)$, which gives the dipolar interaction energy of the system. As the molecule containing the spins tumbles in solution, the dipolar interaction energy of the spins [and consequently $\hat{\mathcal{H}}^{DD}(t)$] changes with the varying relative orientation and separation of their nuclear dipoles. The action of $\hat{\mathcal{H}}^{DD}(t)$ can cause the $|\alpha\rangle$ and $|\beta\rangle$ states to interconvert, and of course this is precisely what happens during dipolar relaxation. In quantum mechanical terms, the rate of relaxation corresponds to the rate at which an initially pure state, for example, $|\alpha\rangle$, is converted into a mixture of $|\beta\rangle$ and $|\alpha\rangle$ states by the action of $\hat{\mathcal{H}}^{DD}(t)$. This rate can be found using perturbation theory, treating $\hat{\mathcal{H}}^{DD}(t)$ as a small perturbation of $\hat{\mathcal{H}}_0$. In essence, the derivation runs as follows.

The wavefunctions $|\alpha\rangle$ and $|\beta\rangle$ vary with time due to the action of $\hat{\mathcal{H}}^{DD}(t)$. Solution of the time-dependent Schrödinger equation (as demonstrated, for instance, in Harris[3]) gives

$$|\alpha\rangle(t) = |\alpha\rangle(0)\exp(-i\omega_\alpha t)$$

and

$$|\beta\rangle(t) = |\beta\rangle(0)\exp(-i\omega_\beta t) \quad (A.21)$$

The exponential terms represent a time-dependent *phase* that causes the wavefunctions to oscillate at ω_α or ω_β, the eigenvalues of $|\alpha\rangle$ and $|\beta\rangle$, respectively. In general, the wavefunction for the system, $|\psi\rangle(t)$, is a mixture of the $|\alpha\rangle$ and $|\beta\rangle$ states:

$$|\psi\rangle(t) = C_\alpha|\alpha\rangle\exp(-i\omega_\alpha t) + C_\beta|\beta\rangle\exp(-i\omega_\beta t) \quad (A.22)$$

where C_α and C_β are complex, time-dependent coefficients. The probability of finding the system in state $|\alpha\rangle$ at a given moment is given by

$$P(\alpha) = C_\alpha C_\alpha^* = |C_\alpha|^2 \quad (A.23)$$

(where all these quantities vary randomly with time). Thus, the rate at which a system initially in state $|\alpha\rangle$ is converted into a system in state $|\beta\rangle$ corresponds to the rate of change of $P(\alpha)$ with time, that is

$$W_{\alpha\beta} = \frac{d}{dt}P(\alpha) = C_\alpha \frac{dC_\alpha^*}{dt} + C_\alpha^* \frac{dC_\alpha}{dt} \quad (A.24)$$

$W_{\alpha\beta}$ is the transition probability per unit time for a single spin. Note that it too varies randomly, since its value depends on the state of the system at a given moment; the observable transition probability that we require is the ensemble average of $W_{\alpha\beta}$ taken over the sample as a whole, namely $\overline{W_{\alpha\beta}}$.

The result of this treatment, which is described in detail in references 3–5, is the equation

$$\overline{W_{\alpha\beta}} = \int_{-\infty}^{\infty} \overline{\langle\beta|\hat{\mathcal{H}}^{DD}(t)|\alpha\rangle\langle\alpha|\hat{\mathcal{H}}^{DD}(t+\tau)|\beta\rangle}\exp(-i\omega_{\alpha\beta}\tau)\,d\tau \quad (A.25)$$

where $\omega_{\alpha\beta}$ is the difference $\omega_\alpha - \omega_\beta$, and τ (the variable of integration) is a time interval.

It is convenient to decompose $\hat{\mathcal{H}}^{DD}(t)$ into two parts: a time-independent spin part $\hat{Z}$ that contains all the spin operator terms, and a time-dependent space part $F(t)$, which is a numerical factor describing the strength of the interaction at a given moment t, and containing distance terms such as r_{IS}^{-3}:[†]

$$\hat{\mathcal{H}}^{DD}(t) = \hat{Z}F(t) \quad (A.26)$$

With this substitution, Eq. A.25 becomes

$$\overline{W_{\alpha\beta}} = \langle\beta|\hat{Z}|\alpha\rangle\langle\alpha|\hat{Z}|\beta\rangle \int_{-\infty}^{\infty} \overline{F(t)F(t+\tau)}\exp(-i\omega_{\alpha\beta}\tau)\,d\tau \quad (A.27)$$

Since we know that $\overline{W_{\alpha\beta}}$ does not vary with time, it may reasonably be assumed that the values of the integrals in Eqs. A.25 and A.27 do not depend on the value of t, but only on the length of the interval τ: this corresponds to the statement that $\hat{\mathcal{H}}^{DD}(t)$ is a stationary random perturbation. Thus, we may replace $\overline{F(t)F(t+\tau)}$ by $\overline{F(0)F(\tau)}$. Using the definition of the correlation function $g(\tau)$ given in Section 2.2, we may simplify this term further:

$$\overline{F(0)F(\tau)} = \overline{|F(0)|^2}g(\tau) \quad (A.28)$$

We now have

$$\overline{W_{\alpha\beta}} = |Z_{\alpha\beta}|^2 \overline{|F_{\alpha\beta}(0)|^2} \int_{-\infty}^{\infty} g(\tau)\exp(-i\omega_{\alpha\beta}\tau)\,d\tau \quad (A.29)$$

where $|Z_{\alpha\beta}|^2$ is simply shorthand for $\langle\beta|\hat{Z}|\alpha\rangle\langle\alpha|\hat{Z}|\beta\rangle$. Completion of the integral [which here corresponds to Fourier transformation of $g(\tau)$] gives

$$\overline{W_{\alpha\beta}} = |Z_{\alpha\beta}|^2 \overline{|F(0)|^2} J(\omega_{\alpha\beta}) \quad (A.30)$$

If the correlation function $g(\tau)$ is assumed to have the exponential form

$$g(\tau) = \exp(-\tau/\tau_c) \quad (A.31)$$

[†] In many textbooks the symbol $\hat{A}$ is used in place of $\hat{Z}$; however, $\hat{Z}$ is preferred here to avoid confusion with the Hamiltonian component $\hat{A}$ in Eq. A.43.

then the spectral density function $J(\omega_{\alpha\beta})$ will be (Section 2.2)

$$J(\omega_{\alpha\beta}) = 2\tau_c/(1 + \omega_{\alpha\beta}^2\tau_c^2) \tag{A.32}$$

The important result expressed in Eq. (A.30) is that the transition rate depends on three terms:

1. The spin part, $|Z_{\alpha\beta}|^2$, which measures whether the Hamiltonian is capable of converting one wavefunction into the other, and can be regarded as a "selection rule." For example, $|\beta\rangle$ turns into $|\alpha\rangle$ at a rate determined by the matrix element $\langle\alpha|\hat{\mathcal{H}}^{DD}(t)|\beta\rangle$. This element is only nonzero if $\hat{\mathcal{H}}^{DD}(t)$ connects $|\alpha\rangle$ and $|\beta\rangle$, i.e., in this case if $\hat{\mathcal{H}}^{DD}(t)$ contains the operator $\hat{I}_+$ (cf. Eq. A.17).
2. The spatial part, F, which represents an ensemble averaged "snapshot" of the strength of the dipole–dipole interaction. This is the term that contains the distance information, since a shorter distance between the dipoles implies a stronger interaction.
3. The spectral density $J(\omega)$, which gives a measure of the power available from the lattice at the frequency ω.

As we shall see in Eqs. A.43 and A.44, in fact $\hat{\mathcal{H}}^{DD}$ is written as a sum of terms $\hat{Z}_i F_i$, rather than just one, but for any particular transition most of the terms do not contribute, because the $\hat{Z}_i$ term is zero. Thus, in the example worked through below, that of the single-flip transition rate W_{1X}, only one of the six terms needs to be considered.

It is appropriate here to insert a short digression concerning the effect of conformational change during the averaging represented by $\overline{F(0)F(\tau)}$, and in particular how to treat the effect of any alterations that occur in internuclear distance as a result of conformational changes. The function $F(t)$ is given explicitly in Eq. A.44, and in general consists of an angular part and a distance part:

$$F(t) = f_\theta(t) r^{-3}(t) \tag{A.33}$$

(the r^{-3} term is included in the factor K in Eq. A.44). In order to average $F(t)$, we shall therefore have to average over $r^{-3}(t)$. As usual in perturbation theory, we tackle this by writing

$$r^{-3}(t) = \overline{r^{-3}} + \delta r^{-3}(t) \tag{A.34a}$$

or, writing for simplicity $R \equiv r^{-3}$,

$$R(t) = \bar{R} + \delta R(t) \tag{A.34b}$$

where $\bar{R}$ is the time-averaged value of r^{-3}, and δR represents the instantaneous deviation of r^{-3} from its mean value. The conformational change occurs at a rate characterized by the correlation time τ_{conf}. We can then write

$$\overline{F(t)F(t+\tau)} \equiv \overline{F(0)F(\tau)}$$
$$= \overline{[\bar{R} + \delta R(0)]f_\theta(0)[\bar{R} + \delta R(\tau)]f_\theta(\tau)}$$
$$= \overline{(\bar{R})^2 f_\theta(0)f_\theta(\tau)} + \overline{\delta R(0)\delta R(\tau)f_\theta(0)f_\theta(\tau)} \quad \text{(A.35)}$$

(Note that there are no terms in $\overline{\bar{R}\delta R}$ since the average of δR is, by definition, zero.) These two terms are averaged as described above (Eqs. A.28–A.30); however, although the term in $(\bar{R})^2$ becomes uncorrelated [through its dependence on $\overline{f_\theta(0)f_\theta(\tau)}$] at the normal rate $\exp(-\tau/\tau_c)$, the term in $\delta R(0)\delta R(\tau)$ becomes uncorrelated faster, because of the additional conformational change occurring—in fact at a rate $\exp(-\tau/\tau_c)\exp(-\tau/\tau_{\text{conf}}) \equiv \exp(-\tau/\tau_{\text{av}})$, where

$$\tau_{\text{av}} = \tau_c \tau_{\text{conf}}/(\tau_c + \tau_{\text{conf}}) \quad \text{(A.36)}$$

Therefore in this case the overall averaging of W is given by

$$\overline{W} \propto \frac{(\bar{R})^2 2\tau_c}{1 + \omega^2 \tau_c^2} + \frac{\overline{(\delta R)^2} 2\tau_{\text{av}}}{1 + \omega^2 \tau_{\text{av}}^2} \quad \text{(A.37)}$$

(We have omitted the extra bar over $(\bar{R})^2$, since $\bar{R}$ is a constant.) If the conformational averaging is slow (namely, if $\tau_{\text{conf}} \gg \tau_c$), then $\tau_{\text{av}} \simeq \tau_c$, and W is proportional to $(\bar{R})^2 + \overline{(\delta R)^2}$, which is the same as $\overline{(\bar{R} + \delta R)^2} = \overline{R^2}$, because, as we have seen, $\overline{\delta R}$ is zero. This is the familiar result for most conformational averaging, that the conformationally averaged $\overline{W}$ is calculated from the $\overline{r^{-6}}$ average over all conformations (Section 5.5.1). [Elsewhere in this book, conformational averaging is expressed by angled brackets, e.g., as $\langle r^{-6} \rangle$; in this section the bar notation is preferred for clarity, and for consistency with Eqs. A.25–A.29.]

However, if the conformational averaging is fast ($\tau_{\text{conf}} \ll \tau_c$), then $\tau_{\text{av}} \approx \tau_{\text{conf}}$, and the second term in Eq. A.37 can be neglected. $\overline{W}$ is then proportional to $(\bar{R})^2$ only, or in other words to $\overline{(r^{-3})}^2$. In fact, as discussed in Section 5.5.2, this treatment gives only the upper limit for the conformationally averaged apparent distance, since for fast averaging the angular and radial parts of the space function must be treated together, and the angular part then gives rise to a geometry-dependent reduction relative to the result from just $\langle r^{-3} \rangle$ averaging.

To complete the calculation of transition probabilities, we need an expression for $\hat{\mathcal{H}}^{\text{DD}}$. Classically, the energy of interaction between two dipoles $\boldsymbol{\mu}_1$ and $\boldsymbol{\mu}_2$ is

$$U = \frac{\mu_0}{4\pi}\left[\frac{(\boldsymbol{\mu}_1 \cdot \boldsymbol{\mu}_2)}{r^3} - \frac{3(\boldsymbol{\mu}_1 \cdot \mathbf{r})(\boldsymbol{\mu}_2 \cdot \mathbf{r})}{r^5}\right] \quad (A.38)$$

where $\mathbf{r}$ is the vector connecting $\boldsymbol{\mu}_1$ and $\boldsymbol{\mu}_2$. By analogy, the quantum mechanical expression is

$$\hat{\mathcal{H}}^{DD} = K\hbar\left[\hat{\mathbf{I}}_1 \cdot \hat{\mathbf{I}}_2 - \frac{3}{r^2}(\hat{\mathbf{I}}_1 \cdot \mathbf{r})(\hat{\mathbf{I}}_2 \cdot \mathbf{r})\right] \quad (A.39)$$

with $K = (\mu_0/4\pi)\hbar\gamma_1\gamma_2 r^{-3}$. Here $\hat{\mathbf{I}}$ represents the column vector $(\hat{I}_x, \hat{I}_y, \hat{I}_z)$ and the dot product $\hat{\mathbf{I}}_1 \cdot \hat{\mathbf{I}}_2$ is (as usual) the scalar $\hat{I}_{1x}\hat{I}_{2x} + \hat{I}_{1y}\hat{I}_{2y} + \hat{I}_{1z}\hat{I}_{2z}$. We are interested in angular motion rather than linear motion, and so we transform the Hamiltonian into polar coordinates, substituting $r\sin\theta\cos\varphi$ for x, $r\sin\theta\sin\varphi$ for y, and $r\cos\theta$ for z. Moreover, we transform $\hat{I}_x$ and $\hat{I}_y$ to $\hat{I}_+$ and $\hat{I}_-$, as described earlier. As an example of these transformations, we consider the terms involving only one $\hat{I}_z$ operator. These turn out to be

$$-3K\sin\theta\cos\theta[(\hat{I}_{1y}\hat{I}_{2z} + \hat{I}_{1z}\hat{I}_{2y})\sin\varphi + (\hat{I}_{1x}\hat{I}_{2z} + \hat{I}_{1z}\hat{I}_{2x})\cos\varphi] \quad (A.40)$$

Using the definitions of $\hat{I}_+$ and $\hat{I}_-$, and noting that $\exp(\pm i\varphi) = \cos\varphi \pm i\sin\varphi$, it is easily shown that

$$2(\hat{I}_{2y}\sin\varphi + \hat{I}_{2x}\cos\varphi) = \hat{I}_{2-}\exp(i\varphi) + \hat{I}_{2+}\exp(-i\varphi) \quad (A.41)$$

Using this equality, the terms of Eq. (A.40) can be written as

$$-(3/2)K\sin\theta\cos\theta[\hat{I}_{1z}\hat{I}_{2+} + \hat{I}_{1+}\hat{I}_{2z})\exp(-i\varphi) + (\hat{I}_{1z}\hat{I}_{2-} + \hat{I}_{1-}\hat{I}_{2z})\exp(i\varphi)] \quad (A.42)$$

The rest of $\hat{\mathcal{H}}$ can be broken down in a similar (though more complicated!) manner, to give

$$\hat{\mathcal{H}}^{DD} = \hbar(\hat{A} + \hat{B} + \hat{C} + \hat{D} + \hat{E} + \hat{F}) \quad (A.43)$$

$$\hat{A} = -K\hat{I}_{1z}\hat{I}_{2z}(3\cos^2\theta - 1)$$

$$= -\hat{I}_{1z}\hat{I}_{2z}F_0 \quad (A.44a)$$

$$\hat{B} = (1/4)K(\hat{I}_{1+}\hat{I}_{2-} + \hat{I}_{1-}\hat{I}_{2+})(3\cos^2\theta - 1)$$

$$= (1/4)(\hat{I}_{1+}\hat{I}_{2-} + \hat{I}_{1-}\hat{I}_{2+})F_0 \quad (A.44b)$$

$$\hat{C} = -(3/2)K(\hat{I}_{1z}\hat{I}_{2+} + \hat{I}_{1+}\hat{I}_{2z})\sin\theta\cos\theta\exp(-i\varphi)$$
$$= -(\hat{I}_{1z}\hat{I}_{2+} + \hat{I}_{1+}\hat{I}_{2z})F_1 \quad (A.44c)$$

$$\hat{D} = -(3/2)K(\hat{I}_{1z}\hat{I}_{2-} + \hat{I}_{1-}\hat{I}_{2z})\sin\theta\cos\theta\exp(i\varphi)$$
$$= -(\hat{I}_{1z}\hat{I}_{2-} + \hat{I}_{1-}\hat{I}_{2z})F_1^* \quad (A.44d)$$

$$\hat{E} = -(3/4)K\hat{I}_{1+}\hat{I}_{2+}\sin^2\theta\exp(-2i\varphi)$$
$$= -(\hat{I}_{1+}\hat{I}_{2+})F_2 \quad (A.44e)$$

$$\hat{F} = -(3/4)K\hat{I}_{1-}\hat{I}_{2-}\sin^2\theta\exp(2i\varphi)$$
$$= -(\hat{I}_{1-}\hat{I}_{2-})F_2^* \quad (A.44f)$$

We have thus expressed $\hat{\mathcal{H}}^{DD}$ as a series of terms of the form $\hat{Z}F(t)$, as required. Moreover, these terms have a simple meaning: $\hat{A}$ causes no spin flips, $\hat{B}$ causes a flip-flop transition ($|\alpha\beta\rangle \leftrightarrow |\beta\alpha\rangle$), $\hat{C}$ causes a single spin flip $|\beta\rangle$ to $|\alpha\rangle$, $\hat{D}$ causes a flip $|\alpha\rangle$ to $|\beta\rangle$, $\hat{E}$ causes the double flip $|\beta\beta\rangle$ to $|\alpha\alpha\rangle$, and $\hat{F}$ causes the double flip $|\alpha\alpha\rangle$ to $|\beta\beta\rangle$. We can therefore use the simplification of Eq. A.30. The evaluation of $\overline{|F_n|^2}$ is quite simple. As shown in Harris,[6] the average of any angular function $f(\theta)$ over all volumes can be obtained by integrating over all angles θ, using the expression

$$\overline{f(\theta)} = \int_0^\pi (1/2)\sin\theta f(\theta)\,d\theta \quad (A.45)$$

Thus, for example, for F_1

$$\overline{|F_1|^2} = \overline{|F_1^*|^2}$$
$$= (9/8)K^2 \int_0^\pi \sin\theta\sin^2\theta\cos^2\theta\,d\theta$$
$$= (3/10)K^2 \quad (A.46)$$

Likewise, $\overline{|F_2|^2} = (3/10)K^2$, and $\overline{F_0^2} = (4/5)K^2$.

These results are made clearer when applied to a real example. The transition rate W_{12} in a simple AX spin system (Fig. A.1) is the rate for transitions between levels 1 and 2. This is a single flip, and therefore involves only term $\hat{C}$ or $\hat{D}$ in the Hamiltonian (depending on which way the flip goes). On application of Eq. A.30, the spin part gives a factor of $(1/2)^2$ (from the $\hat{I}_z$ operator in the spin part of $\hat{C}$ or $\hat{D}$), and therefore W_{12} is given by

$$W_{12} = \frac{1}{\hbar^2} \frac{1}{2^2} \overline{|F_1|^2} J(\omega_{12})$$

$$= \frac{3}{40} K^2 \frac{2\tau_c}{1 + \omega_X^2 \tau_c^2} \tag{A.47}$$

Similar calculations give

$$W_0 = W_{23} = \frac{1}{\hbar^2} \frac{1}{4^2} \overline{F_0^2} J(\omega_{23}) = \frac{K^2}{20} \frac{2\tau_c}{1 + (\omega_A - \omega_X)^2 \tau_c^2} \tag{A.48}$$

$$W_2 = W_{14} = \frac{1}{\hbar^2} \overline{|F_2|^2} J(\omega_{14}) = \frac{3K^2}{10} \frac{2\tau_c}{1 + (\omega_A + \omega_X)^2 \tau_c^2} \tag{A.49}$$

For strongly coupled systems, the only difference is that ψ_2 and ψ_3 now become

$$\psi_2 = \cos\theta |\alpha\beta\rangle - \sin\theta |\beta\alpha\rangle \tag{A.50}$$

$$\psi_3 = \sin\theta |\alpha\beta\rangle + \cos\theta |\beta\alpha\rangle \tag{A.51}$$

The spatial part is unchanged, but the spin part is different. For the 1,2 transition

$$\langle 2|\hat{\mathbf{Z}}|1\rangle = (\cos\theta\langle\alpha\beta| - \sin\theta\langle\beta\alpha|)\hat{\mathbf{Z}}|\alpha\alpha\rangle$$
$$= (\cos\theta\langle\alpha\beta| - \sin\theta\langle\beta\alpha|)(\hat{I}_{1z}\hat{I}_{2-} + \hat{I}_{1-}\hat{I}_{2z})|\alpha\alpha\rangle$$
$$= (1/2)(\cos\theta - \sin\theta) \tag{A.52}$$

and thus W_{12} is multiplied by a factor $(\cos\theta - \sin\theta)^2 = 1 - \sin 2\theta$, often written[7,8] $(1 - S)$ (cf. Section 6.4). Application to an *ABX* system is described in ref. 9.

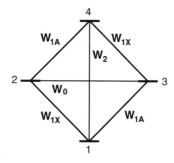

Figure A.1. Energy level diagram for an *AX* spin system.

For a three-spin system (spins I, S, and X, here labeled 1, 2, and 3) there are eight energy levels. The Hamiltonian is also more elaborate, as it involves a sum of three sets of operators: one series of operators $\hat{A}_{IS}$ to $\hat{F}_{IS}$ acting on spins 1 and 2, plus another series $\hat{A}_{IX}$ to $\hat{F}_{IX}$ acting on spins 1 and 3, plus a third series $\hat{A}_{SX}$ to $\hat{F}_{SX}$ acting on spins 2 and 3. Not only are the spin parts of the three sets different, but the space parts are also different, as they are governed by r_{IS}, r_{IX}, and r_{SX}, respectively. The effect of this can be appreciated by an explicit example.

The transition rate between levels 1 and 2, W_{12}, is governed by the single spin flip terms $\hat{C}$ and $\hat{D}$ in the Hamiltonian. The transition from $|\alpha\alpha\alpha\rangle$ to $|\beta\alpha\alpha\rangle$ can be caused by the operator $\hat{I}_{1-}\hat{I}_{2z}$ and also by the operator $\hat{I}_{1-}\hat{I}_{3z}$, or in other words by $\hat{D}_{IS}$ and $\hat{D}_{IX}$. When the Hamiltonian is split up into spin and space parts as given in Eq. A.26, and the subsequent steps from Eq. A.27 to A.30 are carried out, the spin parts of $\hat{D}_{IS}$ and $\hat{D}_{IX}$ both give the result 1/2, but the space part is now[9]

$$\overline{|F(0)|^2} = \overline{|(F_{1,IS} + F_{1,IX})|^2}$$
$$= \overline{|F_{1,IS}|^2} + \overline{|F_{1,IX}|^2} + 2\overline{F_{1,IX}F_{1,IS}} \quad \text{(A.53)}$$

The last of these terms is the cross-correlation term, the value of which depends on the ensemble average over all relative orientations of r_{IS} and r_{IX}. Inclusion of this term complicates the equations considerably, but as discussed in Sections 3.1.2 and 6.3 the problem is usually ignored by setting all such terms to zero. This corresponds to the assumption that the random fields at I arising from S and X are completely uncorrelated with each other. When this assumption fails, the various transition probabilities take different values. Notice that this results in several previously forbidden transitions taking finite probabilities in the presence of cross-correlation; in particular, terms involving different levels of spin order (e.g., $\hat{I}_z$ and $\hat{I}_z\hat{S}_z$; cf. Section 6.2) may become connected.

REFERENCES

1. Atkins, P. W. "Molecular Quantum Mechanics," 2nd ed., Oxford University Press, Oxford, 1983, Chapter 4.
2. Lynden-Bell, R. M.; Harris, R. K. "Nuclear Magnetic Resonance Spectroscopy," Nelson, London, 1969.
3. Harris, R. K. "Nuclear Magnetic Resonance Spectroscopy," Pitman, London, 1983, Appendix 3.
4. Ref. 1, Chapter 7.

5. Abragam, A. "The Principles of Nuclear Magnetism," Oxford University Press, Oxford, 1961, pp. 272–274.
6. Ref. 3, Chapter 2.
7. Freeman, R.; Wittekoek, S.; Ernst, R. R. *J. Chem. Phys.* 1970, *52*, 1529.
8. Noggle, J. H.; Schirmer, R. E. "The Nuclear Overhauser Effect," Academic Press, New York, 1971, Appendix I.
9. Keeler, J.; Neuhaus, D.; Williamson, M. P. *J. Magn. Reson.* 1987, *73*, 45.

INDEX

AB case, 206
Absence of NOE enhancements, 47, 80, 175, 397
 restraints based on, 513
Absolute value 2D, 304, 379
 inadvisability of using, in ROESY, 379
ABX case, 206
ACCORDION, 327
Accuracy
 of distances, 116, 542–543
 of structures, 486, 487, 530, 536ff
Acquisition time
 in 1D, 243
 in 2D, 296
Activation energy, for rotation, 147
Aggregation, 90, 224
Alkaloids, 90, 409–414, 432, 437, 474–479
Ambiguous NOE enhancements, restraints from, 489, 506, 562
Amplitude modulation, 286–293, 303–304
Angular order parameter, 533–534
Anisotropic chemical shift relaxation, 53, 163
Anisotropic tumbling, 38, 164, 580
Applicability of NOE experiments, 398
 see also Flexibility
Aqueous solutions, relaxation in, 222
Aromatic heterocycles, 90, 124, 238, 332, 375, 384, 399ff
Artifacts
 in 1D difference, 230–234
 in 2D, 309–313
Assignment
 of ^{13}C, 332

of ^1H, 394, 442, 462
of NOESY cross peaks, 487–491
of nucleic acids, 574–577
of proteins, 554–556
Attenuation of B_1 field, 380
Autocorrelation function, 31, 159
Averaging, effect on NOE, 140, 143, 167ff, 174ff, 499ff, 604
 see also Conformational averaging
Axes of tumbling, 38
Axial peaks, 266, 303, 313

Baseline correction, 321
Black body radiation, 12
Bloch–Siegert shift, 190, 228–229, 254
Boltzmann distributions, 5, 9
Brownian motion, 14
Buildup of NOE enhancement, 22, 101–102, 122, 469
 in heteronuclear NOE, 336
 in multispin systems, 108ff
 in NOESY, 107
 in ROESY, 120
 in TOE, 105–106, 111
 in transient experiments, 106, 120
 see also Mixing time

Calibration of decoupler strength, 237
Calibration of enhancement intensity to distance, 113–114, 116, 493–499, 556–557, 559
CAMELSPIN *see* ROESY, Rotating frame

611

Carbon-13
   applications of NOE, 332–333, 337–339, 343, 440
   C-C NOE for backbone assignment, 555
   maximum NOE enhancement, 54, 69
   proton couplings to, 399, 417, 448, 463
      see also Heteronuclear J-coupling
   relaxation mechanisms of, 53–56
   satellites and specific irradiation, 337–339, 440
   signal-to-noise improvement, 54
   three-spin NOE, 82, 339
Chemical equivalence see Equivalence, chemical
Chemical shift anisotropy, 53, 163, 167, 201–204, 564
Chemically induced dynamic nuclear polarization, 363
CIDNP see Chemically induced . . .
COCONOSY, 327
Complex Fourier transform, 300–303
Complexes, 587–592
   DNA, 589–592
   protein, 349, 562, 588–589
Composite pulse decoupling, 18, 55, 373, 323, 331
Composite pulse excitation, 242, 312
Concentration of solute, 224
Conformational averaging
   allowing for, 84, 148ff, 171–173, 174ff, 458, 604
   detection of, 148, 458, 510, 517, 560
   examples of, 150ff, 174, 463ff, 561, 580, 586
Conformational models, generation of, 149–154, 156–157, 456–459, 463
Conformational preference, 150, 466–472
Continuous wave (CW) NOE experiments, 144, 254ff, 417
Control spectrum in NOE difference, 93, 227–228, 231, 234–235, 239
Convection, compensation for in GOESY, 271–272
Correlation function, 31, 159, 603
Correlation time, 31, 159–162
   and anisotropic motion, 38, 470, 580
   definition of, 35
   definition of, for scalar relaxation, 213
   effect of, on NOE, 87ff, 224, 283–284, 343, 559
   effect of, on NOE calibration, 113ff, 494
   effect of, on rotating frame NOE, 368–369
   estimation of, 31, 113, 383, 470
   for internal motion, 159–162, 171, 175
   reduction of, by fast averaging rates, 158ff, 175–176
   variation of, 382ff
COSY
   DQF-COSY, 327, 395, 428

E-COSY, 199
Z-COSY, 199
Coupling
   effect on enhancements, 190–192
   restraints based on, 486, 507–511
   see also Decoupling, Selective population transfer, Strong coupling
Cross correlation, 66, 197, 201–204, 609
Cross-peak intensity, 293, 321–322, 491ff, 556–557
   see also Buildup of NOE enhancement
Cross peak, origin of in NOESY, 288–290
Cross-relaxation pathways, 25
Cross-relaxation rate σ
   definition, 29, 35, 39–40, 58
   definition of, in rotating frame (σ_2), 367
   measurement of, 113–114, 174
   relation to NOE enhancements, 25–26, 29
   see also Buildup . . . , Spin diffusion, Spin-lattice relaxation
Cross-validation, 538, 561
CSA see Chemical shift anisotropy
Cycling of experiments
   in 1D, 234
   in 2D, 296

DANTE pulses, 259
Data size in 2D, 297, 317
Decoupler, calibration of, 237
Decoupler power see Irradiation power
Decoupling
   difference spectroscopy, 190, 444
   noise, in ^{13}C spectra, 54–55, 69
   and proton spectra, 190, 343–345, 429–431, 437
Degassing, 46, 94, 226, 413
Deuteration of proteins, 167, 184, 362, 491, 560
Deuterium
   relaxation by, 52
   relaxation of, in protein methyl groups, 167
1,3 Diaxial interactions, xxvi, 429, 439, 450
Difference spectra, 227ff, 363
Diffusion, signal losses due to in GOESY, 271
Digital resolution, 317
Digitization
   of intensity by the ADC, 228
   of 2D spectrum, 317
Dimensionality
   1D vs 2D, 282–285, 443, 550
   3D vs 4D, 358
Dipolar coupling, 14, 513–515, 545–546, 560, 563ff
Dipole, definition, 3
Dipole-dipole relaxation, origin of, 14
   see also Longitudinal relaxation

Dipole moment, origin of, 3
Dirac notation, 599
Direct contribution to enhancements, 73, 76ff
Distance geometry calculations, 517–523, 557, 559
  in torsion angle space, 522–523
Distance measurement *see* Internuclear distances
Disulfide links in proteins, 165, 511
Double quantum coherence, 193ff, 262, 309
Double quantum term *see* W_2 term
Doubly selective NOE experiments, 276–279
DPFGSE *see* Gradient-selected NOE
Dummy scans, 232
Dwell time, 298

Echo selection, 304
Editing using the NOE, 341–343, 404, 431, 478
3/2 Effect, 84–85
Effective field, 15–17, 365–366, 373
Eigenstates, 4
Eigenvalues, 110, 519, 521, 600
Electric field gradient, 52
Electron spin, 50
Energy calculations, 153, 482, 551–553, 584
Energy exchange, 12–13
Energy level diagram
  for AB system, 205
  for ABX system, 207
  for AMX system, 63
  for AX system, 24, 608
  for 1 spin, 4, 9
Enhancements
  absent, 47, 80, 397
    lower limit restraints from, 513
  correlation with r^{-6}, 49, 169
  definition of, 25
  definition of, for heteronuclei, 58, 162–163
  origin of, 23–27
  quantitation of *see* Quantitation
  relative value of, 114, 332, 415
  size of, 29, 36, 239–240, 332, 395–396
  zero value of, due to $\omega\tau_c$, 36, 56, 87
  see also Small NOE enhancement
Equivalence
  chemical, 85
  magnetic, 82–84, 123, 499ff, 596–598
Evolution period, calculation of, 298
Exchange
  in ROESY, 185, 377
  in DMF, 143ff
  in 2D spectra, 293
  two-site
    in multispin systems, 150ff
    in a one-spin system, 131ff

  in a two-spin system, 136ff
  see also Conformational averaging, Fast exchange, Slow exchange, Transfer of saturation
Exchangeable protons, effect on NOE, 134–135, 185, 212–215, 222, 424, 444
Exponential decay, 11, 19, 21, 31, 42ff
External relaxation *see* ρ^*
Extreme narrowing condition, 35, 368
  deviation from, 86ff
  transient enhancements in, 106–108

Factors of two, 11, 37, 107, 112–113, 269, 315–316, 368
False transverse enhancements, 377
Fast exchange
  on the chemical shift time scale, 133, 139
  relative to molecular tumbling, 84, 172ff
  on the T_1 time scale, 133, 139–140, 148ff, 458, 461, 467, 605
Fictitious field, 15
FIDDLE, 249–254
Field-frequency lock, 222
Field-frequency lock phase, 233
Field strength (B_0), 5–6, 27, 52, 148, 383
Field strength (B_1) *see* Irradiation power
Filter function *see* Window function
Flexibility
  estimation of, by heteronuclear relaxation studies, 158ff
  reduction of, 422–425
  and structure solution, 392, 415, 421, 517, 561
  see also Rigidity
Flip-flop term *see* W_0 term
Floating stereoassignment, 506, 511
Fluorine-19
  NOE enhancements, 57, 214, 340
  relaxation of, 53, 203
Folding
  in 2D spectra, 298–299, 302
  in 3D spectra, 355–357
Forbidden transition, 25, 609
Force-field, in molecular dynamics calculations, 523–525, 528
Four-dimensional experiments, 358, 490, 559
Free induction decay, 8, 228
  truncation of, 318
Freeze–thaw degassing, 226
Frequency cycling *see* Irradiation cycling
Frequency labeling, 288, 559
Frequency of a transition, 27

Gated decoupling, 55, 190, 332
Gaussian pulses *see* Shaped pulses
Gaussian window, 319
Glycoproteins, 586

Gradient-selected NOE, 261ff, 310, 337
  DPFGSE-NOE, 100, 264–267, 274–275
  GOESY, 267–272
  HETGOESY, 272–273, 421
Gyromagnetic ratio, 4, 7
  definition of, 4
  negative, 57
  table of, 55

Half-filters, 347ff, 490–491, 562
Hamiltonian operator, 9, 170, 601
Heterocycles *see* Aromatic heterocycles, Isoquinoline alkaloids, Porphyrins, Saturated heterocycles
Heteronuclear J-coupling
  as alternative to NOEs for long-range connectivities, 395, 413, 443, 448, 451
  in stereoassignment, 510
Heteronuclear NOE, 331ff, 421, 440–441, 473
  with gradient selection, 272
  theory, 54–58
  three-spin, 82, 339
  use for characterization of mobility, 158ff, 334, 343
Heteronuclei, as editing tool, 325–327, 346–359
  X-filtered experiments, 347–350, 490–491, 562, 577, 581, 588, 590
  X-separated experiments, 350–359, 490, 588
HETGOESY, *see* Gradient-selected NOE
HOESY, *see* Heteronuclear NOE
HOHAHA transfer, 372
Homogeneous master equation, 22, 372
Homonuclear relaxation *see* Longitudinal relaxation, Transverse relaxation
Humor, importance of, 218
Hydrogen bonds, 338, 393, 512

I and S nomenclature, 24, 63
Incomplete saturation, 18–19, 92–93, 239
Increment of t_1 in 2D experiments, 298
Indirect contribution to NOE enhancements, 73–79ff
  effect on enhancement intensity, 102–105, 119
  see also Spin diffusion, Three-spin effect
Induction period, 102, 117
Inhomogeneity of B_0 field, 21
Inhomogeneity of B_1 field, 20–21
Initial rate approximation, 102, 112, 114, 118, 493
  validity of, 102, 115ff, 580
Integration of enhancements
  in difference decoupling, 190
  in 1D, 144, 248, 404–406, 417
  in 2D, 321, 492
  see also Cross-peak intensity, Quantitation
Interferogram, 287ff
Interleaving
  of COSY and NOESY acquisition, 328
  of frequencies in 1D, 231
Intermolecular NOE, xxv, 225, 275
  to water, 135, 185–187, 212, 424, 560
Intermolecular relaxation, 14, 41, 46–51, 225
Internal motion, 84, 92, 160, 167ff, 586
  if distance altered, 149
  if distance unaltered, 137
  effect on steady-state enhancements, 87–88, 91–92, 158ff
  faster than overall tumbling, 172
  of methyl groups, 172
  slower than overall tumbling, 171
  see also Longitudinal relaxation, spin-rotation
Internuclear distances
  from steady-state enhancements in 3-spin systems, 49, 71ff
  from steady-state enhancements in 2-spin systems, 46
  from kinetic NOE measurements, 111ff, 120
  in nucleic acids, 570ff
  in presence of motion, 116, 119, 167ff, 174ff
  in proteins, 494, 499, 551, 554
  skewed distance error function, 116, 120, 495–497
"Inverse" detection, 272, 336, 353
Inversion-recovery, 43ff, 471
Irradiation cycling, 242–243
  see also Interleaving
Irradiation power, 17, 235, 255–257
Irradiation selectivity, 17ff, 235ff, 394
Irradiation time in TOE, optimum, 245ff
Isochromats, 20, 263
Isolated spin pair approximation (ISPA) *see* Two-spin approximation, Initial rate approximation
Isoquinoline alkaloids, 90, 409ff, 437, 474ff
Isotopic editing, 346–359
Isotropic tumbling, 31, 158ff

J-coupling, effect on NOE
  see Coupling, Decoupling, Selective population transfer, Strong coupling
J-peaks
  in NOESY, 198–199, 309–313
  in ROESY, 372–378

Karplus curve, 507–510, 513

Lanthanides, effect on NOE enhancements, 51

Larmor frequency, 6–7
Lattice, definition of, 9, 12
Leakage term, see ρ^*
Line fitting, 322, 492
Linear prediction, 320
Lineshape
  and data processing in NOESY, 317–319
  in difference decoupling experiments, 191
  of subtraction artifacts in NOE difference experiments, 229
  in 2D experiments, 304–307
Linewidth see Transverse relaxation
Lipari-Szabo model-free approach, 158ff
Lithium-6, 57, 340–341
Local field, 13ff
"Local pulse," 13
Lock see Field-frequency lock
"Long-range" NOE enhancements, 93, 413, 432, 486, 515
Longitudinal magnetization, 5
Longitudinal 2-spin order, 194, 312
Longitudinal relaxation, 8ff
  application of, 124ff
  definition of, 43
  dependence on $\omega\tau_c$, 36, 38, 45, 382ff
  and equivalence, 85
  external see ρ^*
  incoherence of, 13
  intermolecular, 41, 222
  see also ρ^*
  mechanisms of:
    chemical shift anisotropy, 53, 56
    dipole-dipole, 50–52
    quadrupolar, 52
    scalar, 53, 213ff
    spin-rotation, 53, 56
  non-exponential, 43, 45
  non-selective, 43
  origin of, 12, 14
  requirements for, 12–13
  selective, 43, 471
Lorentzian function, 33, 229
Lower bounds for NOE-based restraints, 520
Lower limit restraints
  see Absence of NOE enhancements

Magnetic dipole, 3–5
Magnetic equivalence see Equivalence, magnetic
Magnetization, 8
  longitudinal, 5
  macroscopic, 6–8, 20
  pathway selection, 308
  transverse, 6, 18, 237
Magnetogyric ratio see Gyromagnetic ratio
Master equation for enhancements in multispin systems, 68–69
  see also Homogeneous master equation

Matrix equations for multispin NOE kinetics, 108–111, 545
  see also Relaxation matrix
Maximum NOE enhancement (η_{max})
  in rotating frame, 368–369
  in steady-state experiment, 29, 36, 42, 70
  in transient and NOESY experiment, 101, 107, 124
Methyl ethers, 150, 399, 467ff
Methyl groups
  effective distances to, 172–174, 499ff
  NOE enhancements to, 70, 78, 123–124, 173, 177–178, 497–498, 501–506
  spin-rotation relaxation in, 54
Methylene groups, enhancements to, 70, 78, 206ff, 461–463, 501ff, 506ff
Metric matrix, 518ff
Mixing time
  in heteronuclear experiments, 337
  in NOESY (τ_m), 118, 293ff, 493
  nomenclature, 100
  in rotating frame experiments, 368
  in transient experiments (τ), 117ff
  in variants of NOESY, 327
  variation of, in NOESY, 311, 327
  see also Buildup of NOE enhancements
Mobility
  see Flexibility, Internal motion
Model-free formalism for local flexibility, 158ff
Molecular charge, effect on tumbling rate, 90, 284
Molecular dynamics, 164, 176, 523ff
  see also Simulated annealing
Molecular mass, 31
Molecular models, importance of, 421, 586
Motions see Exchange, Flexibility, Internal motion, Rigidity
Multiple conformations see Conformational averaging
Multiple irradiation, 234ff
Multiplets, saturation of, 240ff, 394
  see also Selective population transfer
Multiplicity, correction for, 123, 503–506
Multispin systems
  and spin diffusion see Spin diffusion
  steady-state NOE enhancements in, 63ff
  time-dependent NOE enhancements, 108ff

Negative NOE enhancements, 26, 88
  in large molecules, xxv, 30, 36, 56, 90, 550
  with negative ratio of γ, 57
  scalar relaxation, 53, 213ff
  three-spin effects, 76, 79ff
  transfer of saturation, 135
Nitrogen-15, NOE enhancements, 57, 158, 334, 343
Nitrogen-14, relaxation by, 52

NOE *see* under separate headings, e.g.
  Absence of, Heteronuclear, Indirect,
  Integration, Long-range, Maximum,
  Negative, NOESY, Quantitation,
  ROESY, Steady state, Symmetry,
  Transferred, Transient, Truncated driven
NOESY, xvi, 282ff, 482, 550
  block decoupled, 323
  network edited, 322–327, 349
Noise decoupling, 54–55, 332
Non-exponential buildup, 102, 117
Normalization of spin functions, 599
Nuclear spin quantum number, 4, 58
Nucleotides, conformations of, 151ff, 567ff
Numerical values of enhancements, value of, 255
Nyquist equation, 298

Off-resonance effects *see* Resonance offset effects
Off-resonant irradiation, 16ff
Olefins, 415ff, 444ff, 463–464, 467
Operator, definition of, 599
Order parameter (S^2), 158ff, 175, 332
  angular (S^{ang}), 533–534
Oscillations, of irradiated signal during irradiation, 19–22, 94, 116, 240, 255ff
Overhauser effect, xxv
  generalized *see* Selective population transfer
Overlap, in spectra, 358, 393, 583
Oxygen, dissolved, 47, 50, 94, 226

Paramagnetics
  addition of, 50
  effect on NOE enhancement, 46, 50, 94
  paramagnetic proteins, 126, 283, 363
  removal of, 225–226
Partial saturation *see* Saturation, incomplete
Penicillins, 422, 427, 431, 436–437, 469
Phase
  in ROESY, 364, 377
  correction, 233, 304ff
  cycles, 232, 307ff
  gradient-induced, 262ff
  of lock, *see* Field-frequency lock
  random changes in, from spectrometer, 222, 307
  of 2D spectra, 290, 304ff
Phase modulation, 304
Phase sensitive 2D, 304ff, 379
Phase twist, 304, 307
Phosphorus-31, 53, 57, 473
Populations
  changes of, in 1-spin system, 9–11
    undergoing exchange, 132–133
  changes of, in 2-spin system, 27ff, 192ff
    undergoing exchange, 136
  changes of, in 3-spin system, 63ff
  and observed signal, 5, 7, 10, 25
Porphyrins, 38, 384, 407ff, 452
Pre-acquisition delay *see* Relaxation delay
Precision, of NMR structures, 116, 516, 530ff, 541ff, 557
Presaturation time (τ), 243ff
Product operators, 68, 193ff, 312
Pseudoatom corrections, 500–503, 551
Pulse angle, 7
  and artifacts in 2D, 198–199, 296, 312
  effect in coupled systems, 208ff, 240–242
  in NOE difference experiments, 245ff
Pulse sequence
  for editing using the NOE, 344
  for gradient-selected experiments, 265, 270
  for heteronuclear editing, 326, 347, 352
  for heteronuclear NOE experiments, 335
  for NOESY experiments, 100, 286ff, 294
  for 1D transient experiments, 100, 260, 277
  related to NOESY, 327
  for rotating frame NOE experiments, 370, 380–381
  for steady-state NOE difference experiments, 100, 227
  for 3-dimensional experiments, 353, 356, 360
  for TOE experiments, 100, 255
Pulsed field gradients *see* Gradient

Quadrature detection
  in 1D, 300
  in 2D, 302, 313
Quadrupolar relaxation, 52, 57, 216
Quality factor (Q), 232
Quantitation of NOE enhancements, 91, 94–97, 239, 248, 257, 260, 276, 395, 491ff
  and conformer ratios, 463
  see also Integration of enhancements
Quantization of angular momentum, 4
Quaternary carbons
  NOE enhancements to, 332–334, 337, 339
  relaxation of, 53
QUIET-NOESY, 278, 325

R_1, definition of, 11
  see also Longitudinal relaxation
R_I, definition of, 40–41
R_I^{DD}, definition of, 40–41, 66
R_{II}^{DD}, definition of, 85
R_2 *see* Transverse relaxation
R factor for NOE, 538–539
r^{-3} or r^{-6} averaging, 173, 499, 605

r^{-6} summation, 505, 541
RABBIT, 48, 384
Ratio of NOE enhancements, for distance determination, 114
Real and imaginary parts, in 2D, 305
Real Fourier transform, 301–303, 316
Receiver gain, 296
Reduced spectral density mapping, 165
Reference deconvolution, 249–254
Reference distance, 114–115, 116, 206, 495–498
Relaxation *see* Longitudinal relaxation, Rotating frame, Transverse relaxation
Relaxation agents, 50–51, 56
  see also RABBIT
Relaxation delay, choice of
  in kinetic experiments, 244, 258
  in steady-state experiments, 244
  in 2D experiments, 295
Relaxation matrix, 108–111, 118, 180, 183, 323, 489, 543, 580
Relayed NOE, 339
Reporting of results, 276, 395ff, 443, 533
Representative structures, selection of, 542, 557
Resolution, 243, 296–298, 320, 488
  and digital resolution, 296–298, 317, 322
Resonance offset effects, 15ff, 235ff
  in ROESY, 365, 371, 380
Response function, 17
Rigidity
  and local τ_c, 88, 116
  and structural determination, 392, 446, 472
Ring, 391–393
  4-Membered, 85, 126, 449
  7-Membered, 422, 425
  3-Membered, 425, 429
RMSD (root mean squared deviation), 531ff, 542, 557
ROESY, 364ff
  compared to NOESY, 185, 364, 375, 368–370, 377, 443
  JS-ROESY, 382
  one-dimensional, 272ff
  TOCSY transfer in, 372–378, 382
  T-ROESY, 382
  see also Rotating frame, Spin locking
Roof effect, from strong coupling, 205, 209–210
Root mean squared deviation (RMSD), 531ff, 542, 557
Rotating frame
  longitudinal cross-relaxation in, 371
  meaning of, 14
  relaxation in, 367–370
  see also ROESY

Saccharides *see* Sugars
Sample
  preparation of, 221ff
  spinning, 232
Saturated heterocycles, 341, 422ff, 437, 448, 465, 472–473
Saturation
  definition of, 20
  effect on saturated spin, 59–60
  incomplete, 17ff, 92, 235, 239
  of multiplets, 240ff, 394
    see also Selective population transfer
  off-resonant, 17, 235ff, 257
  onset of, 116, 255ff
  selective, 18, 235, 259, 394, 410, 428, 451
  of several lines, 235, 243, 461
  theory of, 19ff
  time, optimum in TOE, 245ff
  transfer *see* Transfer of saturation
Scalar coupling *see* Coupling
Scalar relaxation, 55, 213ff
Selection rules, 25, 197–198, 604
Selective irradiation field
  calibration of, 237
  effect of, 18
Selective population transfer, 191ff, 240ff, 404
Selective pulses, 17ff, 259ff, 322
Selective T_1, 43
Semiselective NOESY, 322ff
Sensitivity in 2D, 297
Shaped pulses, 259, 322
Shift reagents, 51
Shotgun approach, 234, 394
SI units, xxff
Sidebands, spinning, 238
Sigmoidal buildup curve, 117
Sign convention
  for definition of σ, 39
  for operator products, 193
  for plotting 1D NOE spectra, 248
  for plotting 2D NOE spectra, 107, 293
Signal-to-noise ratio, 78, 224, 228, 230ff, 248
Silicon-29, 54, 57
Simulated annealing, 521, 527–529, 559
Sinc wiggles, 318
Sine-bell window, 319
Single frequency X{^{1}H} irradiation, 337, 339, 440
SKEWSY, 327
Slow exchange
  on the chemical shift time scale, 131, 141
  on the T_1 time scale, 139–140, 171ff, 184
Small NOE enhancement, 46ff
  geometrical factors, 80–81
  interpretation of, 88ff, 397, 413
  $\omega\tau_c$ close to 1.12, 36, 224, 364, 383, 583

Solid state, dipolar interactions in, 563–566
Solomon equations
 for any number of spins, 65
 extension, to describe S spin behavior fully, 59–60, 277–278
 in presence of exchange, 137
 for three spins, 65
 for two spins, 29
Solvent
 choice of, 95, 221, 383
 transfer of saturation from, 134–135
Spectral density function $J(\omega)$, 9, 32, 160, 165, 169ff, 604
Spectral density mapping (reduced), 165
Spectral width in 2D, 297–299
Spherical harmonic functions, 37, 169ff, 177, 606–607
SPI-NOE experiment, 363
Spin diffusion, 45, 74, 88, 104, 117ff, 183, 343, 370, 447, 479, 494, 536, 550, 567, 580
 in extreme narrowing limit *see* Three-spin effect
 model for, in slow tumbling limit, 104
 reduction of, 276, 322
 in solids, 566
Spin-lattice relaxation *see* Longitudinal relaxation
Spin locking, 365ff
 practical requirements for, 379–382
 strength of field for, 366–367, 375
Spin polarization by ^{129}Xe, 363
Spin–spin coupling *see* Coupling
Spin–spin relaxation, 8
Spinning of sample, 232, 238
Spinning sidebands, 238
Spontaneous emission, 12
SPT *see* Selective population transfer
Spurious distance (a_i), 49, 95–96
Steady state
 failure to reach, 93
 reaching, 104, 334
Stereoassignment, 462–463, 506–511, 557, 559
Steroids, 210, 236, 392–393, 473–474
Stimulated emission, 12
Strong coupling, effect on NOE, 116, 204ff, 583, 608
Structure refinement, 540–546
 using relaxation matrix, 110
Subtraction
 of spectra, 229, 248
 of spectra *vs* FIDs, 22
Sugars, 224, 464, 583ff
Summation, *see* r^{-6} summation
Symmetric top, relaxation in, 38
Symmetrical structures, 85–86, 126, 358–359, 490–491, 562

Symmetrization of 2D spectra, 320
Symmetry of NOE enhancements, 78, 122ff, 320

t_1
 increment of, 296
 maximum value of, 297
 resolution and, 317
t_1 noise, 232, 314, 320
t_1 ridge, 316, 321
T_1
 applications of, 124
 meaning of, 12, 43, 85
 see also Longitudinal relaxation
$T_{1\rho}$, 366
T_2 *see* Transverse relaxation
$T_{2\rho}$, 19
Temperature
 effect on NOE enhancements, 88, 147–148, 383
 spin, 104
 variation, effect on NOE difference, 224, 233
Terpenes, 51, 200, 392, 429, 451, 473
Thiophenes, 124, 332, 403
Three-dimensional experiments, 354–362, 489–490, 554, 559
Three-halves (3/2) effect, 84
Three-spin effect, 74–76, 79ff, 104, 370, 429, 432
 heteronuclear, 82, 339
Tightly coupled spin system *see* Strong coupling
Time-proportional phase incrementation (TPPI), 301ff, 313
TOCSY, 274, 359, 514–515, 555, 575
 transfer during ROESY, 372–378, 382
"Top-up" irradiations, fallacy of, 247
Transfer of saturation, 131ff, 216, 402, 425, 444, 461, 482
Transferred NOE (TRNOE), 178ff
Transient NOE (1D), 43, 99, 101–102, 105, 258ff
Transition probability, 9, 41, 208
 calculation of, 33–34, 603ff
 relationship to longitudinal relaxation time, 11
 relationship to NOE enhancements, 29, 71
Transverse magnetization, 6, 18, 237
Transverse relaxation, 8–9
 spin-locked *see* ρ
Truncated driven NOE (TOE), 99, 105
 use of, 255ff, 479
Tumbling rate *see* Correlation time
Two cone picture, 5
Two-dimensional NOE *see* NOESY
Two-spin approximation, 98, 112, 543
 see also Initial rate approximation

Uniform averaging model, 175
Upper bounds for NOE-based restraints, 175ff, 493, 520

Viscosity
  of different solvents, 383
  effect on NOE enhancements, 31, 224

W see Transition probability
W_0 (flip-flop) term, 25–26, 29–30
  from scalar relaxation, 213
W_2 (double quantum) term, 24–26, 29–30
Water, enhancements to, 135, 185–187, 212, 424, 560
Window function
  in 1D, 23, 251
  in 2D, 318

Xenon-129, 363

Zero filling, 317
Zero quantum coherence, 310
Zero-quantum term see W_0 term
zz peaks, 194ff, 312

α-Helix, NOE enhancements in, 498, 551–554
β-Sheet, NOE enhancements in, 498, 551–554
η_{max} see Maximum NOE enhancement
ρ
  definition of, 29, 35, 40, 58
  in rotating frame (ρ_2), 367
ρ^*, 40, 42, 46ff, 95ff, 155, 206, 222–224, 471
σ see Cross-relaxation
τ_c see Correlation time
τ_e see Lipari-Szabo model-free approach
τ_m see Mixing time